Self-Organizing Systems
The Emergence of Order

LIFE SCIENCE MONOGRAPHS

SELF-ORGANIZING SYSTEMS
The Emergence of Order
Edited by F. Eugene Yates

TOXIC SUBSTANCES AND HUMAN RISK
Principles of Data Interpretation
Edited by Robert G. Tardiff and Joseph V. Rodricks

A Continuation Order Plan is available for this series. A continuation order will bring delivery of each new volume immediately upon publication. Volumes are billed only upon actual shipment. For further information please contact the publisher.

Self-Organizing Systems
The Emergence of Order

Edited by

F. Eugene Yates
Crump Institute for Medical Engineering
University of California, Los Angeles
Los Angeles, California

Associate Editors:

Alan Garfinkel
Department of Kinesiology
University of California, Los Angeles
Los Angeles, California

Donald O. Walter
Chemical Electrophysiology Laboratory
Neuropsychiatric Institute and Hospital
University of California, Los Angeles
Los Angeles, California

and

Gregory B. Yates
Crump Institute for Medical Engineering
University of California, Los Angeles
Los Angeles, California

Plenum Press • New York and London

Library of Congress Cataloging in Publication Data

Self-organizing systems.

(Life science monographs)
Includes bibliographies and index.
1. Self-organizing systems. I. Yates, F. Eugene. II. Title: Emergence of order. III. Series.
Q325.S48   1987   577   87-10151
ISBN 0-306-42145-3

© 1987 Plenum Press, New York
A Division of Plenum Publishing Corporation
233 Spring Street, New York, N.Y. 10013

All rights reserved

No part of this book may be reproduced, stored in a retrieval system, or transmitted in any form or by any means, electronic, mechanical, photocopying, microfilming, recording, or otherwise, without written permission from the Publisher

Printed in the United States of America

We gratefully dedicate this book to Mary Frances Thompson, whose encouragement and skill created the circumstances out of which this book could organize.

Contributors

Ralph H. Abraham   Division of Natural Sciences, University of California, Santa Cruz, California 95064

Philip W. Anderson   Department of Physics, Joseph Henry Laboratories, Princeton University, Princeton, New Jersey 08544

Michael A. Arbib   Department of Computer Science, University of Southern California, Los Angeles, California 90089

Francisco J. Ayala   Department of Genetics, University of California, Davis, California 95616

Nevenka Bajković-Moškov   Institute for Biological Research, Belgrade, Yugoslavia

Richard Bellman†   Department of Biomathematics, University of Southern California, Los Angeles, California 90089

Eduardo R. Caianiello   Faculty of Sciences, University of Salerno, 84100 Salerno, Italy

Radomir Crkvenjakov   Department of Biology, Faculty of Sciences, University of Belgrade, Belgrade, Yugoslavia

Caleb E. Finch   Ethel Percy Andrus Gerontology Center and Department of Biological Sciences, University of Southern California, Los Angeles, California 90007

†Deceased.

Alan Garfinkel Department of Kinesiology, University of California, Los Angeles, California 90024

Vladimir Glišin Genetic Engineering Center, Belgrade, Yugoslavia

Brian C. Goodwin Department of Biology, The Open University, Milton Keynes MK7 6AA, England

Stephen J. Gould Museum of Comparative Zoology, Harvard University, Cambridge, Massachusetts 02138

Hermann Haken Institute for Theoretical Physics and Synergetics, University of Stuttgart, 7000 Stuttgart 80, West Germany

Arthur S. Iberall Department of Oral Biology, University of California, Los Angeles, California 90024

Rolf Landauer IBM T. J. Watson Research Center, Yorktown Heights, New York 10598

Harold J. Morowitz Department of Molecular Biophysics and Biochemistry, Yale University, New Haven, Connecticut 06520

Charles Arthur Musès Editorial and Research Offices, Mathematics and Morphology Research Center, Albany, California 94706

Leslie E. Orgel Salk Institute for Biological Studies, San Diego, California 92138

Howard H. Pattee Department of Systems Science, State University of New York, Binghamton, New York 13850

Ramin Roosta Department of Electrical and Computer Engineering, California State University, Northridge, California 91330

Sabera Ruždijić Institute for Biological Research, Belgrade, Yugoslavia

Ana Savić Department of Biology, Faculty of Sciences, University of Belgrade, Belgrade, Yugoslavia

Peter Schuster Institute for Theoretical Chemistry and Radiochemistry, University of Vienna, A-1090 Vienna, Austria

Christopher D. Shaw Department of Mathematics, University of California, Santa Cruz, California 95064

Karl Sigmund Institute for Mathematics, University of Vienna, A-1090 Vienna, Austria

Harry Soodak Department of Physics, The City College of New York, New York, New York 10031

Edwin B. Stear Washington Technology Center, University of Washington, Seattle, Washington 98195

Daniel L. Stein Department of Physics, Joseph Henry Laboratories, Princeton University, Princeton, New Jersey 08544

Gunther S. Stent Department of Molecular Biology, University of California, Berkeley, California 94720

Rajko Tomović Department of Computer Science, University of Belgrade, EOX Belgrade, Yugoslavia

Christoph von der Malsburg Max Planck Institute for Biophysical Chemistry, D-3400 Göttingen, West Germany

Donald O. Walter Clinical Electrophysiology Laboratory, Neuropsychiatric Institute and Hospital, University of California, Los Angeles, California 90024

William B. Wood Department of Molecular, Cellular, and Developmental Biology, University of Colorado, Boulder, Colorado 80309

F. Eugene Yates Crump Institute for Medical Engineering, University of California, Los Angeles, California 90024

Gregory B. Yates Crump Institute for Medical Engineering, University of California, Los Angeles, California 90024

Preface

Technological systems become organized by commands from outside, as when human intentions lead to the building of structures or machines. But many natural systems become structured by their own internal processes: these are the self-organizing systems, and the emergence of order within them is a complex phenomenon that intrigues scientists from all disciplines. Unfortunately, complexity is ill-defined. Global explanatory constructs, such as cybernetics or general systems theory, which were intended to cope with complexity, produced instead a grandiosity that has now, mercifully, run its course and died.

Most of us have become wary of proposals for an "integrated, systems approach" to complex matters; yet we must come to grips with complexity somehow. Now is a good time to reexamine complex systems to determine whether or not various scientific specialties can discover common principles or properties in them. If they do, then a fresh, multidisciplinary attack on the difficulties would be a valid scientific task.

Believing that complexity is a proper scientific issue, and that self-organizing systems are the foremost example, R. Tomović, Z. Damjanović, and I arranged a conference (August 26–September 1, 1979) in Dubrovnik, Yugoslavia, to address self-organizing systems. We invited 30 participants from seven countries. Included were biologists, geologists, physicists, chemists, mathematicians, biophysicists, and control engineers. Participants were asked not to bring manuscripts, but, rather, to present positions on an assigned topic. Any writing would be done after the conference, when the writers could benefit from their experiences there.

At the end of the meeting the participants unanimously agreed that enough progress had been made in this area to justify a coordinated, multiauthor book. After agreeing to perform the task of securing manuscripts and arranging for their review, I invited Donald O. Walter, Gregory B. Yates, and Alan Garfinkel to assist as associate editors. Each chapter was thoroughly reviewed by at least two individuals, in addition to the editors. The result was a useful exchange of critical comments, all without the shield of anonymity.

To help understanding by the nonspecialist, we have written abstracts of each chapter. There are also introductions to the book and to each section, and an epilogue. This book can be read from start to finish as a coherent exposition on the emergence of order. Some sections, such as those on evolution and on physics, have an internal connectedness that permits them to stand well alone. Others do not. Although variety is inevitable in the presentation of a broad topic without clear-cut boundaries, we were pleasantly surprised to find that the book developed some conspicuous self-organizing tendencies of its own—the grouping into the nine sections seemed natural.

Some chapters restate previously published positions, but have been updated. We believe that such material takes on added significance from its juxtaposition with neighboring chapters. Much of the material is entirely new and cannot be found elsewhere. The topic of this book should appeal to any scientist possessing the qualities of technician and philosopher. We examine self-organizing systems to find unifying principles that might belong simultaneously both to biology and to physics. Such principles would enable these sciences to join forces in the next attack on the mystery of the spontaneous emergence of order that is the terrestrial biosphere.

The characteristics of self-organization invoke descriptions and images that depend upon words or phrases whose meanings are vague, multiple, or imprecise—"order," "level," "complexity," "organization," "system," "information," "optimization," "emergent property," "stability," "self," "evolution," and so on. The title of this book contains five words incapable of precise definition. These terms and the concepts they denote can be manipulated readily in a natural language, such as English, but they strain formal, mathematical languages.

Scientists attempt to achieve value-free relationships, statements, and concepts. Yet discussions of self-organizing or evolutionary systems often invoke a latent, if not blatant value calculus (as in "higher apes"), hinting at a belief in progress, increased complexity, or improvement with time. These faulty anthropomorphisms have damaged the scientific analysis of self-organizing systems. Although their incidence has been minimized in this volume, their intrusions can be so subtle that their absence cannot be guaranteed.

Complexity and self-organization can be found on technical agendas of fields as diverse as cosmology, geology, linguistics, artificial intelligence, communication networks, fluid mechanics, sociology, economics, embryology, paleontology, atmospheric science, and evolutionary biology. Because a single book cannot be that comprehensive, we have chosen to examine and display self-organizing systems from positions not frequently encountered together by the general scientific reader. Thus, presentations of modern biology were minimized because so much has recently been published elsewhere. The series *Biological Regulation and Development,* edited by Robert F. Goldberger (1980–1983, Plenum Press, New York), is an appropriate supplement to this book to introduce molecular, genetic, and cell biology. *Self-Organizing Systems* focuses on physics, epistemology, engineering, and mathematics as they might help us comprehend complexity, including the wonders of autonomous morphogenesis and descent with modification, which are unique, as far as we know, to terrestrial life.

F. Eugene Yates

Los Angeles, California

Acknowledgments

We owe much to the Research Council of Serbia; the U.S. National Institute for General Medical Sciences (Grant GM-23732); the Crump Institute for Medical Engineering of the University of California, Los Angeles; the Biomedical Engineering Center of the University of Southern California; and the Center for Multidisciplinary Studies of the University of Belgrade. All supported this effort in significant ways.

For special advice and help we want to thank Kirk Jensen and Eric R. Rippel.

Contents

General Introduction 1
F. Eugene Yates

I · EXAMPLES OF EVOLVING PHYSICAL SYSTEMS 15
F. Eugene Yates

1 · On Origins: Galaxies, Stars, Life 17
Harry Soodak

2 · On Rivers 33
Arthur S. Iberall

II · GENESIS AND EVOLUTION OF LIFE 49
Donald O. Walter and F. Eugene Yates

3 · A Hardware View of Biological Organization 53
Harold J. Morowitz

4 · The Origin of Self-Replicating Molecules 65
Leslie E. Orgel

5 · Self-Organization of Macromolecules 75
Peter Schuster and Karl Sigmund

6 · Is a New and General Theory of Evolution Emerging? 113
Stephen J. Gould

III · DIFFERENTIATION, MORPHOGENESIS, AND DEATH OF ORGANISMS 131
Gregory B. Yates

7 · Virus Assembly and Its Genetic Control 133
William B. Wood

8 · Molecular Biology in Embryology: The Sea Urchin Embryo 153
Vladimir Glišin (with Ana Savić, Radomir Crkvenjakov, Sabera Ruždijić and Nevenka Bajković-Moškov)

9 · Developing Organisms as Self-Organizing Fields 167
Brian C. Goodwin

10 · The Slime Mold *Dictyostelium* as a Model of Self-Organization in Social Systems 181
Alan Garfinkel

11 · The Orderly Decay of Order in the Regulation of Aging Processes . . 213
Caleb E. Finch

IV · NETWORKS, NEURAL ORGANIZATION, AND BEHAVIOR . . . 237
Donald O. Walter

12 · On a Class of Self-Organizing Communication Networks 241
Richard Bellman and Ramin Roosta

13 · Neural Circuits for Generating Rhythmic Movements 245
Gunther S. Stent

14 · Ordered Retinotectal Projections and Brain Organization 265
Christoph von der Malsburg

Contents xvii

15 · A View of Brain Theory 279
     Michael A. Arbib

V · EPISTEMOLOGY OF SELF-ORGANIZATION 313
     Gregory B. Yates

16 · Biological Reductionism: The Problems and Some Answers 315
     Francisco J. Ayala

17 · Instabilities and Information in Biological Self-Organization 325
     Howard H. Pattee

18 · Programmatic Phenomena, Hermeneutics, and Neurobiology 339
     Gunther S. Stent

VI · CONTROL THEORY VIEW OF SELF-ORGANIZATION 347
     F. Eugene Yates

19 · Control Paradigms and Self-Organization in Living Systems 351
     Edwin B. Stear

20 · Control Theory and Self-Reproduction 399
     Rajko Tomović

VII · PHYSICS OF SELF-ORGANIZATION 409
     F. Eugene Yates

21 · Synergetics: An Approach to Self-Organization 417
     Hermann Haken

22 · Role of Relative Stability in Self-Repair and Self-Maintenance 435
     Rolf Landauer

23 · Broken Symmetry, Emergent Properties, Dissipative Structures, Life: Are They Related? 445
Philip W. Anderson and Daniel L. Stein

24 · Thermodynamics and Complex Systems 459
Harry Soodak and Arthur S. Iberall

VIII · EXTENSIONS OF PHYSICAL VIEWS OF SELF-ORGANIZATION 471
F. Eugene Yates

25 · A Thermodynamic Approach to Self-Organizing Systems 475
Eduardo R. Caianiello

26 · Interfaces between Quantum Physics and Bioenergetics 489
Charles Arthur Musès

27 · A Physics for Complex Systems 499
Arthur S. Iberall and Harry Soodak

28 · A Physics for Studies of Civilization 521
Arthur S. Iberall

IX · TOPOLOGICAL REPRESENTATION OF SELF-ORGANIZATION 541
Gregory B. Yates

29 · Dynamics: A Visual Introduction 543
Ralph H. Abraham and Christopher D. Shaw

30 · Dynamics and Self-Organization 599
Ralph H. Abraham

EPILOGUE

Quantumstuff and Biostuff: A View of Patterns of Convergence in
Contemporary Science 617
F. Eugene Yates

Associative Index 645

Subject Index 649

Self-Organizing Systems
The Emergence of Order

General Introduction

F. Eugene Yates

> *Une montagne qui accouche d'une souris...*
> [A mountain labors and gives birth to a mouse...]
>
> In common language this expression has pejorative meaning—much ado about nothing. If we consider only the quantity of matter involved, we can understand the complaint; but if we are concerned instead with the richness of organization, then we take a different perspective. For all its millions of tons of rocks, a mountain is stolid, inert. It can do nothing. It stays there, impassively waiting for winds and rain to wear it away to nothing. The mouse, on the contrary, with its few wisps of matter, is a wonder of the universe. It lives, runs, eats, and reproduces. If some day a mountain could indeed give birth to a mouse, we would have to proclaim the greatest of miracles.
>
> Yet the actual history of the universe is indeed roughly that of a mountain that labors and gives birth to a mouse! This history, chapter by chapter, emerges from the many different scientific approaches to reality: physics, chemistry, biology, and astronomy.[1]

This book examines the resourcefulness of nature that can produce the infinite variety of all the beings, from mouse to man, quark to quasar. It explores the competence of science to comprehend variety, order, and their becomings. If the mountain (primitive terrestrial lithosphere, hydrosphere, atmosphere) did in fact parthenogenetically produce a mouse (the terrestrial biosphere), where did it get the means, the necessary tools to fabricate the parts it would need, when no causalities could be employed except those in and of the system itself and its

---

[1] From Reeves (1981), translated freely by the editor.

F. EUGENE YATES • Crump Institute for Medical Engineering, University of California, Los Angeles, California 90024.

dumb and noisy surroundings? The chapters that follow offer a new look at these old questions.

We contemplate some of the most remarkable processes of which we are aware: energy transforming into matter transforming into life; sperm and egg becoming a human being; rivers sculpting continents; molecules arranging for their own preservation; speciation of the biosphere; aggregation of the slime mold; self-assembly of a virus; carbon atoms reshuffling in a sea urchin embryo; rhythmic movements in a leech; the aging of mammals. In all these activities the processes of self-organization, self-maintenance, or self-repair appear to be fundamental. Order is emergent, created out of materials, potentials, fluxes, and principles at hand. To explain how this can be, we draw upon aspects of philosophy, mathematics, physics, and engineering.

Philosophy

It proves very difficult for scientists to write about philosophy or epistemology so that other scientists will not misunderstand or resent the attempt. Yet we have great need for metaphysics (and metabiology and metaphor) as pre-science to generate ontological theories and strategic views, out of which more specific theories and models can condense. Pepper's classification of all (Western) world hypotheses according to their dependence on one of four root metaphors (excluding absolute skepticism and absolute dogmatism, because they are destructive of intellectual activity) illustrates the method (Pepper, 1942). His irreducible set of root metaphors is shown in Table 1. Practicing scientists, being amateurs at philosophical discourse, are likely to drift unconsciously into one or the other of those metaphors—mechanism being, of course, very popular among physicists, chemists, and molecular biologists; organicism among physiologists; and contextualism among psychologists and linguists. These unexamined outlooks expose the culture- or metaphor-boundedness of our explanations.

Eclecticism, as an undisciplined mixing of metaphors, fails badly. No philo-

TABLE 1. Root Metaphors of Western Philosophy[a]

Names	Metaphor	Type	Some early exponents
Formism (Platonic idealism)	Similarity	Analytic (with inadequate precision)	Plato, Aristotle
Mechanism (naturalism, materialism, realism)	Machine	Analytic (with inadequate scope)	Democritus, Galileo, Descartes, Hume
Contextualism (pragmatism)	Acts in a context (the historical event)	Synthetic (with inadequate precision)	Peirce, James, Bergson, Dewey
Organicism[b] (absolute or objective idealism)	Integration as process; organic whole	Synthetic (with inadequate scope)	Schelling, Hegel, Royce

[a]From Pepper (1942).
[b]For a powerful attack on organicism, see Phillips (1976).

General Introduction

sophical position has ever achieved grand synthesis of the four root metaphors. Perhaps our minds cannot operate except with fragments of thought—thus, our pluralisms, complementarities, contradictions, cognitive dissonances, and perseverance of beliefs in the face of contrary facts. Yet I wonder if we can account for self-organization and the emergence of order without tapping the resources of all four metaphors. Section V of this book addresses some of those philosophical issues as raised by the processes of self-organization.

There is a trend toward a general philosophy of biology that hints at being a *philosophy of science*—previously a dignity reserved only to the physical viewpoint. Examples can be found in Smuts (1927), von Uexküll (1928), Simpson (1963), von Bertalanffy (1968), Jacob (1970), Monod (1971), Ruse (1973), Ayala (1975), Ayala and Dobzhansky (1975), Jantsch and Waddington (1976), Bunge (1977), Hofstadter (1979), Mercer (1981), and Allen and Starr (1982). [Some might include Koestler's *Janus* (1979) or his *The Ghost in the Machine* (1967). Koestler had a good intuition for some of the weaknesses of scientific explanation and for some of its important philosophical problems, but his handling of the material thereafter bears no resemblance to rigorous science or philosophy that I can see.]

Mathematics

The mathematics that currently seems most able to describe the emergence of order and the peculiar, marginal stability of living organisms is exemplified by Abraham in Chapters 29 and 30. This formalism is the topological, qualitative dynamics and general bifurcation theory. The reader interested in applications of differential topology and bifurcation theory to problems in physics and biology should also consult the collection edited by Enns *et al.* (1981). Its fourteen papers and seven notes, limited to mathematical topics, cover nonlinear phenomena in physics and biology. The two chapters by Stuart Kauffman (1981), entitled "Bifurcations in Insect Morphogenesis, I and II," have a strong bearing on the mathematical approaches to the topic of self-organization as presented in this book.

Davis and Hersh (1981), in their account of mathematical discovery, offer a short chapter entitled "Pattern, Order, and Chaos" that illustrates mathematical forms that generate order out of order, chaos out of order, chaos out of chaos, and order out of chaos. However we may define order, we can comfortably imagine or explain three of these processes: order can beget order; order can decay into chaos; chaos can beget chaos. Each of these three transformations withstands close examination, by science or even just by "common sense," without perturbing our satisfaction with our forms of explanation. But—order out of chaos?

As one example of the emergence of order out of chaos, Davis and Hersh point out that if a polygon with random vertices is transformed by replacing it with the polygon formed from the midpoints of its sides, then upon iteration of this transformation an ellipselike convex figure almost always emerges from the different initial polygons. (This example of the emergence of order is not isomorphic with any known physical process; mathematical possibility and physical possibility are not always the same.)

Physics

The physical analysis of self-organizing systems is not yet conventional. Sections VII and VIII present extensions or interpretations of physics designed to cope with complexity. The eight chapters in those sections, taken together, express the scientific ferment now gathering around the topics of stability and order, in which fluctuations play a constructive role in self-organization.

Engineering

The engineering sciences have enabled the designs of man, but not an understanding of man's design. The reason for this failure is that the engineering sciences work best when specifications, requirements, or goals can be stated in advance; but that is exactly what we do not have in the case of historical, evolutionary, chance-constructed, self-organizing natural systems. Our technological designs do not closely resemble those of life or nature, despite centuries of hope to the contrary. Section VI presents a detailed tutorial on engineering control theory, its beauty, and its inapplicability, in its present form, to biology. Forthright demonstrations of inapplicability are almost as valuable as those of applicability in a comparison of two different sciences. Thus, readers will find Section VI very instructive.

Order/Disorder

The brief accounts given above introduce the four perspectives from which we shall examine self-organization and the emergence of order. But great value should not always be put upon order. Too much order is as inimical to life as is too much disorder or uncertainty. Life is poised between absolute regularity and absolute chaos. If the heartbeat of a newborn child is extremely regular, like a metronome, the child is likely to die. If the pulse of a superannuated person is highly irregular, with rapid, ectopic, wandering ventricular pacemakers driving the erratic beat, sudden death is imminent. A normal heart beats quietly and regularly during deep sleep or relaxed sitting; it races during exercise, excitement, lovemaking, or dreaming; it skips occasionally, especially at high altitude or with a sneeze—and the record it leaves behind is of order-disorder, a signature of noisy order. That is the character of life, and the universal desire to optimize systems of governance of our societies somewhere between absolute, tyrannical order and absolute, chaotic anarchy is obvious to us all. In life, the order that emerges from chaos reminds me of Michelangelo's two overwhelming, unfinished statues "Prigione" (Prisoners)—powerful half-figures struggling to emerge from rough blocks of stone, but never becoming fully free of the unformed, crude matrix out of which they are composed.

Antecedents

A topic as rich as that of self-organizing systems deserves and receives recurring attention. Many of the themes treated in this book can be found also in Plato, Aristotle, Pythagoras, and Democritus (see Montalenti, 1975, for a reassessment of some of the Greek contributions to modern science). But for the modern reader, the first chapter in *On Growth and Form* (Thompson, 1917; see 1961 reprint) brilliantly sets a scene in which mathematics, physics, and biology act together to illustrate the emergence of form. By the middle of this century at the Hixon Symposium (Jeffress, 1951), several strong statements of the problem were made, particularly in von Neumann's now-famous paper, "The General and Logical Theory of Automata." In 1968 a "sourcebook" on systems research (Buckley, 1968) presented an impressive array of distinguished authors and ideas in 59 chapters, ranging from philosophy, mathematics, and engineering to neural science and psychology.

The immediate inspiration for the present book was the four-volume series arranged by C. H. Waddington, entitled *Towards a Theoretical Biology*. Again we hear a chorus of physics, philosophy, mathematics, and biology (Waddington, 1968, 1969, 1970, 1972). During the 1970s and early 1980s many books addressed biological complexity from philosophical, physical, mathematical, or engineering points of view (see, e.g., Prigogine and Stengers, 1984—but also the criticism of their book by Pagels, 1985). The present book benefits from these predecessors, which provide us with a mature scientific base. Our chapters present a foundation for a scientific analysis of self-organizing systems that I believe will be serviceable for several decades. Our exhibits—self-organizing processes of the cosmos, river systems, embryos, brains, viruses—do now, and will for the indefinite future, contain all the elements of the problem. But more than that, this book foreshadows a physical-mathematical reductionism that will be competent to encompass life without requiring that we reach outside of nature for an *élan vital* or a *deus ex machina*.

Reductionist Position

The reductionist stance throughout this book is not just the usual downward reduction of the ontological or methodological kinds (see Chapter 16). Rather, emphasis is placed on the lateral, side-to-side, epistemological reduction that seeks a principle to join many (not all) of the concepts, terms, and laws of biology to those of physics, under special boundary conditions—then to interpret the evolution of those special boundary conditions through physical accounts of fluctuations (random variations), natural selection (as tests of stability), and descent (copying by chemical reactivity) with modification.

In taking that reductionistic stance, however, we discover and expose the weakness of physics and mathematics with respect to historical, evolutionary, and informational processes—just those processes that dominate in biology! When a physicist or mathematician refers to the "evolution" of a system, he means very much less by the term than does the biologist. In physics or mathematics, evolution is usually the following of a given trajectory, often one determined in advance. In contrast, the usual meaning of evolution to biologists is that well expressed by Luria (1973) as follows:

> Many people, including some scientists, have refused to believe that a probabilistic process like natural selection could have worked with such precision to bring about the almost uncanny fitness of plants and animals to their natural environments, as well as the marvels of the human mind. They have suggested the possible existence of biological laws other than those of physics and chemistry in order to explain the direction, speed, and apparent purposefulness of evolution. But what seems to be purposefulness is only selection for superiority of performance, however slight, that leads to improved reproductive success. Invoking unknown biological laws to explain the efficiency of natural selection is a return to vitalism, the theory that tried to explain the uniqueness of living organisms by postulating a "vital force." Such explanations explain nothing, and, in the ultimate analysis, they can be traced to the metaphysical belief that each organism has a vital spirit or soul imposed upon it from outside.
>
> The modern theory of evolution, like all historical theories, is explanatory rather than predictive. To miss this point is a mistake that theoreticians of history have often made. Prediction in evolution would require not only a knowledge of the main force—natural selection—but also prescience of all future environmental conditions, as well as of future balances between the quasideterministic effects of the law of great numbers and the purely probabilistic role of genetic drift.
>
> For evolution, like history, is not like coin tossing or a game of cards. It has another essential characteristic: irreversibility. All that will be is the descendent of what is, just as what is comes from what has been, not from what might have been. Men are the children of reality, not of hypothetical situations, and the evolutionary reality—the range of organisms that actually exist—is but a small sample of all past opportunities. If a species dies out, evolution may, in an environment where the lost species would have been fitted, fashion a reasonable likeness of it. This process, named *convergence,* is the one that gave to dolphins and whales their almost fishlike shape and to bats their birdlike appearance. But evolution does not thereby retrace its steps. It makes the best of whatever genetic materials have managed to remain available to it at any one time.

The reader will, in the main, discover here the materialist outlook, updated and extended. This is not the place to find sympathy for such philosophical views as the (I think mystical) interactionist position regarding the mind–body problem, as expressed by Popper and Eccles (1977). To the small extent that the mind–body problem does appear (see Chapters 15, 18, and 26), the position adopted is (except for Chapter 26) much closer to that of Bunge (1977, 1980), i.e., (to overcompress the issues) the mind is a direct expression of brain function; it depends utterly on brain function, and stops when (or before) the electroencephalogram goes flat.

The materialist persuasion does not dispel, but indeed highlights the mystery of mind and the wonder of man. There are no claims here that all that is interesting about nature, including humans themselves, requires or is susceptible to scientific explanation. Science can be done only according to its own disciplined rules of evidence and testing; it is not an arbitrary game, but one whose moves are sufficiently open that discovery becomes autocatalytic. In this light, the question is not how to "reduce" man to physics, but to ask how physical principles can expose the intimate connectedness of human behavior (dynamics) to the rest of nature.

Entropy, Symmetry, and Symmetry-Breaking

It is remarkable, given our focus on self-organization and the emergence of order, that no equation explicitly addressing entropy production is presented in this book! Fifteen years ago it would hardly have been possible to treat these topics without saying more about entropy. Of course, changes in entropy are implicit in almost all of the processes discussed herein, but the authors have managed to keep entropy itself in the background. What has happened to the concept?

Entropy is usually introduced as a statistical concept applicable to an ensemble (classically, the isolated system provided the model) but not to the individual members or events comprising them. It is impossible to get the measure of macroscopic entropy by taking the ensemble average of some microscopic variable (in contrast to temperature, which can be specified in just that way). Entropy is an extensive variable that can be measured only very indirectly. It is related to probability, uncertainty (but this is not the quantum mechanical uncertainty), information, irreversibility and direction of time, dissipation (loss of free or useful energy to do work), and macroscopic symmetry. Entropy appears in two of the three fundamental laws of thermodynamics. The Second Law claims that all circumstances permitting natural systems to change states (evolve) are entropy-generating; the Third Law specifies that the entropy of a system is zero at the absolute zero of temperature.

In the statistical, probabilistic view of entropy, a bridge is made between microscopic and macroscopic aspects of a collection or ensemble:

$$S = - k \sum_{i=1}^{n} p_i \log p_i$$

where i designates a particular microscopic configuration, p is the probability of a microscopic configuration, and n is the number of possible states, under the constraints. S is the Boltzmann statistical mechanical, macroscopic entropy of the ensemble, and k is the Boltzmann natural constant. Conrad (1983) has very aptly discussed some of the many aspects of entropy (Table 2). A clear account can also be found in Mercer (1981).

Entropy cannot be directly transported, but undergoes virtual transport in the exchanges of matter or energy across the boundaries of systems. Tsuchida has discussed entropy "transports" for the earth in an interesting way (see the translation by Sibatani, 1982, with additional commentary). Tsuchida–Sibatani comment as follows:

> Entropy is a property of matter and energy, and an open system can exchange entropy in the form of matter and energy between itself and its surroundings. Hence it is a mistake to hold a *direct* analogy between chaos and high entropy. Thermal energy input to the Earth is made through the greenhouse effect of the atmosphere and sunlight in a ratio of 2:1, which would heat the earth to 31°C instead of the actual average of 15°C, were it not for air convection and water evaporation on the Earth's surface. The warm, light air and water vapor generated on the Earth's surface rise and undergo an adiabatic expansion because of the decrease in atmospheric pressure at the upper atmosphere with the result that the temperature falls to −23°C; the water vapor condenses into cloud, rain and snow. The entropy change of this process is equal to the difference between its *increase* at the earth surface ($\Delta S_b = \Delta Q/T_b$, $T_b = 288°K$; ΔQ is the latent heat of water evaporation) and its *decrease* at the upper atmosphere ($-\Delta S_a = -\Delta Q/T_a$, $T_a = 250°K$); $T_b > T_a$, hence $\Delta S_b - \Delta S_a < 0$. The heat released through water condensation ($-\Delta Q$) is dispersed outside of the Earth. Thus, the net entropy change of the earth is negative,

TABLE 2. Different Entropic Concepts and Their Characteristics[a]

Entropic concept	Characteristic	Chessboard analogy
Thermodynamic entropy	Measures ensemble of microscopic states	Ensemble of possible microstates *compatible with* macroscopic state of pieces and board
Macroscopic diversity (functionally distinct)	Measures ensemble of functionally distinct macroscopic states or (more generally) pathways of dissipation	Ensemble of possible board positions that are *distinct from* each other, from the standpoint of the rules of the game
Macroscopic diversity (functionally equivalent)	Measures ensemble of functionally equivalent macroscopic states or (more generally) pathways of dissipation	Ensemble of possible board configurations that are *equivalent to* each other, from the standpoint of the rules of the game
Behavioral uncertainties	Measures ensemble of behavioral sequences (i.e., measures of macroscopic diversity conditioned on the past and possible other, external factors)	Ensemble of possible board configurations, given previous position
Entropy of organization	Difference between size of initial set of microstates and size of final set, or, alternatively, thermodynamic entropy change concomitant with irreversible preparation of the system starting from standard state (in general an impractical and arbitrary concept)	Thermodynamic entropy change concomitant with formation of pieces and board in a reversible process
Structural diversity (diversity of components)	Measures number of types of components in the system and the evenness of their relative occurrence	Measures indicating number of different types of chess pieces (pawns, horses, kings) and their relative frequency (one king, two horses)

[a] From Conrad (1983).

the excess entropy being discarded into outer space, contributing to the total entropy increase of the universe. The low entropy in the form of condensed water is returned to the Earth's surface by precipitation. So the entropy of the Earth may continue to decrease by virtue of water circulation. This low entropy of water drives photosynthesis, using CO_2 and sunlight as auxiliary raw material, and is transformed into the low entropy of the organic world. Hence, the Earth has been able to support organic evolution throughout its history!

Sibatani then goes on to say:

> ... Earth is a "magnificent ice-producing" machine constantly being fed low entropy in the form of water precipitation. As long as the water cycle remains intact, Earth can be sustained in a low entropy state.... We should define the desired future as the Water Age, as it has always been on Earth, at least until the very recent past.... The Sun is primarily a source of thermal pollution, because most of its energy is not used by photosynthesis, while the low entropy of water is fully utilized by the Earth, which is nothing like a space ship because *it can discard entropy to the outside.*

This somewhat unconventional emphasis on entropic changes in the terrestrial hydrosphere and atmosphere as a basis for the order in the biosphere is refreshing.

In going from the powering of order in liquid water to entropic changes in the morphogenesis of living systems, we encounter systems that exhibit a spontaneous, strong tendency to break symmetry. The symmetrical state or pattern is unstable, and the asymmetrical patterns are more stable (see Chapter 23 by Anderson and Stein). This feature makes morphogenetic processes superficially seem to contradict the Second Law (recall that entropy "likes" symmetry). To give a proper, thermodynamic, physical account of morphogenesis requires that we follow *both* energy and entropy. It then becomes apparent that *morphogenesis simultaneously minimizes and dissipates energy,* under constraints. The cooperation of these two processes leads to the local creation of asymmetric "negentropic" forms (Conrad, 1983).

Descriptions of entropy and entropy production have now turned into discussions of symmetry and symmetry-breaking. In that way they have become more general and more profound, but the thermodynamic issues remain the same, and mountains can indeed still breed mice (with a little bit of luck and lots of time). A recent book by Brooks and Wiley (1986) keeps entropy in the forefront, but we prefer to focus on issues of symmetry, believing that they are more closely related to modern bifurcation theory, as elaborated in Chapter 30.

Determinacy and Indeterminacy: Chance in Physics and Biology

René Thom's work, discussed in Chapters 29 and 30, bears on the question of determinism versus indeterminism. As interpreted by Pattee (1972):

> A ... more mathematical view of self-simplifying processes is the topological dynamics of Thom. Here the simplification results from the assumption that it is the discontinuities in the vector field of trajectories which determine the simplified structure at a higher level. Furthermore, these discontinuities may exhibit *a kind of stability which is entirely deterministic in the evolutionary sense, but not predictable in the dynamical sense.* [Italics added.]

Of course, bifurcations or catastrophes are dynamically predictable in the sense that it is known that the dynamic regime will change at critical values of control parameters.

In a review of *The Foundations of Biological Theory* by Mercer (1981), Brian Goodwin remarked (1982):

> A perennial source of tension in science arises from the possibility of explaining phenomena in two distinct ways: either as a result of law, whereby they become intelligible in terms of principles at once simpler and more general than the phenomena themselves; or as a result of contingencies, of events that just happen to occur.

He concluded by saying:

> ... there will continue to be those who ... seeing clear evidence of systematic regularity as well as variety in the biological realm ... will continue to seek a rational foundation for biology whereby organisms and their evolution become intelligible in terms of universal laws of organization and transformation, not simply in terms of chance events and survival.

But this last hope does not seem likely to be fulfilled. The newer physical views of self-organization give chance a leading role. The overarching purpose of this book is to examine the proposition that self-organizing systems, particularly those of the terrestrial biosphere, accomplish their remarkable feats by incorporating chance into the substances of their forms, which then persist, maintained by their own functions. This is the modern synthesis of dynamics, not yet fully accomplished, that sees motion and change in both physical and biological systems as similar examples of a bounded indeterminism, a partial determinism—sometimes as a manifestation of "chaotic" dynamics, deterministic in principle, but not exactly knowable in practice.

Variety and chance have always seemed important in biology; what is new is that physics has relaxed its 18th and 19th century insistence on Descartes's absolute determinism. Quantum mechanical probabilities, fluctuations, and uncertainties, and their occasional manifestations as statistical mechanical fluctuations, are now believed to be essential ingredients of the physics of the emergence of order. Chapters 1, 5, 6, 10, 14, 17, 19, 21–28 all discuss chance, fluctuations, instabilities, and contingencies, and show that classical deterministic causalities fail to account for self-organization. They emphasize that macroscopic, regular phenomena arise out of the general principles of a probabilistic physics and the operations of chance. [The proper role of chaotic dynamics, if any, has not yet been decided. It can generate random–looking behavior out of fully noise–free, deterministic processes—see Abraham and Shaw (1983) and Chapters 29 and 30.]

The role of chance can be very subdued or even effectively suppressed in many of our models: $x = x_0 - \frac{1}{2} gt^2$ is still a useful model for the position of a dropped body at various times after its release, as it falls to earth in a vacuum. But the new discovery is that *self-organizing processes may have an absolute requirement for random fluctuations at some point in their histories*. To get the initial conditions and boundary conditions for a fully deterministic system that could imitate a self-organizing system, we would have to reach outside the system for the blueprint design and thereby violate the requirement that the system be self-organizing. (Again, although the new mathematical field of "chaotic dynamics" shows that seemingly random processes can be generated deterministically, as a practical matter we cannot prove the deterministic origins of any time history of a system and so we here assume as usual that the cosmos has independent sources of random noise that contaminate all measurements and observations, regardless of the underlying, perhaps unknowable, process.)

As long as physics insisted on full determinisms, biology had to stay away:

> To the biologist as to the philosopher the great inadequacy of determinism is its inability to account for the appearance of novelty. Biology above all seems rich in new and more complex structures and functions; the rigid determinist must either deny the reality of these novel appearances or claim that they are somehow derivable from the general laws.... The feeling grew that biology would have to choose either to follow physics and live with determinism or to strike out on its own. To many it seemed that biological phenomena were so different from physical that some element of indeterminism would have to be admitted. The schism thus established lasted well into this century and produced some notable debates. However, at the present time controversy has faded in perspective and a very different atmosphere prevails; physics and biology are together edging toward a unified theory. The change in atmosphere results from [three] circumstances: ... First, determinism has ceased to be dominant in physics, and even its necessity in classical physics has been reassessed. Second, the evolutionary outlook has spread from biology to the physical world. Finally, the extraordinary efflorescence of biochemistry in the midcentury has unraveled the intricate chemistry of energy transfer in biological systems and revealed the chemical basis of inheritance. While this last point is reductionist (or rather unificationist) in character, since it demonstrates that organisms are chemical systems necessarily operating in accordance with the principles of physics and chemistry, it no longer carries the same implication of ineluctable determinism as once thought. The fact that the generation of mutational novelty ... is by the random or chance chemical modification of the informational molecules, which are the genetic store of organisms ... underlines the evolutionists' stress on the creative contribution of pure chance and the impossibility of predicting the results of its intervention. [Mercer, 1981, pp. 16–18]

Today, neither physics nor biology has to be regarded as a fully deterministic science, yet they both address invariances, conservations, and regularities in the formation of variety. In previous discussions of causality and determinism in physics and biology (Yates, 1980, 1982), I have noted that causalities became less deterministic as physics progressed from Newtonian mechanics to statistical mechanics to quantum mechanics. In the latter theory, it is chiefly *probabilities* that are determined. In the biological domain we always make room for chance. As a causal, historical result of chance and arbitrary pruning in the form of blind selection, according to capricious circumstances:

> We expect a derepressed gene to direct synthesis of its enzyme. We expect a human ovum fertilized *in vivo* to produce a human being in due course. We expect the nucleus of an intestinal cell of a tadpole to produce an identical tadpole, if it is implanted into the experimentally enucleated, unfertilized ovum of a frog. We expect seeds and spores to germinate under simple conditions. We expect a frozen black mouse embryo to produce a black mouse when rethermalized and implanted in an experimentally prepared, virginal, albino (white) female 5 years later, under careful conditions of freezing and

> thawing. We expect insulin release when we eat a mixed diet. We expect our friends to know what we are like from day to day.
>
> All the above expectations are met with a high probability, and they all attest to short-term determinism in biological processes and structures. This is not to deny the generator of diversity in the immune system, nor the probabilistic aspects of the mixing of genes in sexual reproduction, but only to emphasize that random processes and uncertainty do not obscure the partial determinism of both microscopic and macroscopic biological processes.... Even human behavior, though amazingly varied in detail, globally wends its way through only about 20 different modes.... [Yates, 1982]

Biological systems, only loosely coupled in a causal sense, surprisingly act as if they were highly constrained and tightly coupled. Chance becomes necessity through selection and nearly invariant reproduction. Therefore, we seek physical explanations of biological invariance, regularity, and partial determinacy. Biology is ready for physics to contribute more strongly than ever before. We have outgrown the confusions of the past. As Mercer (1981, pp. 46–47) remarks:

> Both evolution and the change predicted by the Second Law of Thermodynamics are processes in time that are compatible with the basic laws [of physics] but cannot be deduced from them. The Second Law predicts a loss of order and an increase in entropy with time, whereas evolution has led to the appearance of more ordered, better organized systems of lowered entropy. When attention was first drawn to this difference, it seemed to many that biological systems had found a way to circumvent the Second Law and were creating order in defiance of its predictions.
>
> Such is the perennial appeal of vitalism that at one time this idea attracted much attention. It was, however, based on a misunderstanding of the entropy law, which is concerned with the trend in a[n] ... isolated system, whereas organisms exchange both mass and energy with their surroundings and are ... open systems. The entropy increase in an open system may be offset by the effects of an energy flow through it.... The conflict that arose between the interpretation of the Second Law and evolution is an example of the difficulties that have impeded efforts to develop a uniform science to include both the biological and physical sciences. Similarly the deterministic interpretation of Newtonian physics seemed to make the evolution of novelty impossible or illusionary. The history of vitalist explanations in embryology and metabolism has been equally unfortunate. These controversies are now history....

The Authors

The reader, by now, should be curious to see what our authors have to say. This is a full book, suitable for browsing, with no obligatory starting point. But we did not produce it for browsers; we imagined ourselves in rapport with curious readers whose powers of concentration are intact and strong. If we disappoint them, we have failed.

> We realized that a record ... would impose the restrictions ... inherent in written communication.... the content would express ideas which the form could not support....
> We thought of how to involve the reader so that his active mind would be working to engage the ideas.... Einstein ... then suggested that we were creating an experience about how we knew the world.... The open book becomes a new experiment with a new observer. [Schlossberg, 1973]

References

Abraham, R. H., and C. D. Shaw (1983) *Dynamics—The Geometry of Behavior* (Vol. 2): *Chaotic Behavior.* Aerial Press, Santa Cruz, California.
Allen, T. F. H., and T. B. Starr (1982) *Hierarchy: Perspectives for Ecological Complexity.* University of Chicago Press, Chicago.
Ayala, F. J. (1975) Introduction. In: *Studies in the Philosophy of Biology,* F. J. Ayala and T. Dobzhansky (eds.). University of California Press, Berkeley, pp. vii–xvi.
Ayala, F. J., and T. Dobzhansky (eds.) (1975) *Studies in the Philosophy of Biology.* University of California Press, Berkeley.
Brooks, D. R., and E. O. Wiley (1986) *Evolution as Entropy: Toward a Unified Theory of Biology.* University of Chicago Press, Chicago.
Buckley, W. (1968) *Modern Systems Research for the Behavioral Scientist.* Aldine, Chicago.
Bunge, M. (1977) Emergence and the mind. *J. Neurosci.* **2**:501–509.
Bunge, M. (1980) *The Mind–Body Problem: A Psychobiological Approach.* Pergamon Press, Elmsford, N.Y.
Conrad, M. (1983) *Adaptability: The Significance of Variability from Molecule to Ecosystem.* Plenum Press, New York.
Davis, P. J., and R. Hersh (1981) *The Mathematical Experience.* Birkhaüser, Basel.
Enns, R. H., B. L. Jones, R. M. Miura, and S. S. Rangnekar (eds.) (1981) *Nonlinear Phenomena in Physics and Biology.* Plenum Press, New York.
Goodwin, B. C. (1982) Life viewed in the particular. [Book review] *Nature* **296**:876.
Hofstadter, D. R. (1979) *Gödel, Escher, Bach: An Eternal Golden Braid.* Basic Books, New York.
Jacob, F. (1970) *La Logique du Vivant.* Editions Gallimard, Paris [English translation by B. E. Spillman, Pantheon Books, New York, 1973].
Jantsch, E., and C. H. Waddington (eds.) (1976) *Evolution and Consciousness: Human Systems in Transition.* Addison-Wesley, Reading, Mass.
Jeffress, L. A. (ed.) (1951) *Servomechanisms and Behavior: The Hixon Symposium.* Wiley, New York.
Kauffman, S. A. (1981) Bifurcations in insect morphogenesis, I, II. In: *Nonlinear Phenomena in Physics and Biology,* R. H. Enns, B. L. Jones, R. M. Miura, and S. S. Rangnekar (eds.). Plenum Press, New York, pp. 401–484.
Koestler, A. (1967) *The Ghost in the Machine.* Macmillan Co., New York.
Koestler, A. (1979) *Janus: A Summing Up.* Vintage, New York.
Luria, S. E. (1973) *Life: The Unfinished Experiment.* Scribner's, New York, pp. 22–23.
Mercer, E. H. (1981) *The Foundations of Biological Theory.* Wiley, New York.
Monod, J. (1971) *Chance and Necessity.* Knopf, New York.
Montalenti, G. (1975) From Aristotle to Democritus via Darwin: A short survey of a long historical and logical journey. In: *Studies in the Philosophy of Biology,* F. J. Ayala and T. Dobzhansky (eds.). University of California Press, Berkeley, pp. 3–19.
Pagels, H. R. (1985) Is the irreversibility we see a fundamental property of nature? [Review of *Order out of Chaos,* by I. Prigogine and I. Stengers, Bantam Books, New York, 1984] *Physics Today* **Jan**:97–99.
Pattee, H. H. (1972) The evolution of self–simplifying systems. In: *The Relevance of General Systems Theory,* E. Laszlo (ed.). Braziller, New York, pp. 31–42.
Pepper, S. C. (1942) *World Hypotheses.* University of California Press, Berkeley.
Phillips, D. C. (1976) *Holistic Thought in Social Science.* Stanford University Press, Stanford, Calif.
Popper, K. R., and J. C. Eccles (1977) *The Self and Its Brain: An Argument for Interactionism.* Springer-Verlag, Berlin.
Prigogine, I., and I. Stengers (1984) *Order out of Chaos—Man's New Dialog with Nature.* Bantam Books, New York.
Reeves, H. (1981) *Patience dans L'Azure: L'évolution Cosmique.* Éditions du Seuil, Paris, p. 15.
Ruse, M. (1973) *The Philosophy of Biology.* Hutchinson, London.
Schlossberg, E. (1973) *Einstein and Beckett: A Record of an Imaginary Discussion with Albert Einstein and Samuel Beckett.* Links Books, New York, pp. 3–4.

Sibatani, A. (1982) Review of "Entropy: A New World View" by J. Rifkin. *Q. Rev. Biol.* **57**:100.
Simpson, G. G. (1963) Biology and the nature of science. *Science* **139**:81–88.
Smuts, J. C. (1927) *Holism and Evolution*. Macmillan & Co., London [reprinted in 1973 by Greenwood Press, Westport, Conn.].
Thompson, D. W. (1917) *On Growth and Form*. Reprinted in 1961 by Cambridge University Press, abridged edition by J. T. Bonner, London.
von Bertalanffy, L. (1968) *General Systems Theory*. Braziller, New York.
von Uexküll, J. (1928) *Theoretische Biologie*. [English edition published by Kegan Paul, London.]
Waddington, C. H. (ed.) (1968, 1969, 1970, 1972) *Towards a Theoretical Biology*. (1) Prolegomena; (2) Sketches; (3) Drafts; (4) Essays. Aldine, Chicago.
Yates, F. E. (1980) Physical causality and brain theories. *Am. J. Physiol.* **7**:R277–R290.
Yates, F. E (1982) Systems analysis of hormone action: Principles and strategies. In: *Biological Regulation and Development*, Vol. 3A, R. F. Goldberger and K. R. Yamamoto (eds.). Plenum Press, New York, pp. 25–97.

I

Examples of Evolving Physical Systems

F. Eugene Yates

To illustrate the capability of a few physical forces to create variety of form and function, we present a synopsis of modern cosmological theory—how the universe came to have its present form—and an account of the physical issues involved in establishing the distribution, course, and volume flow of rivers. In both cases, competition among several forces or processes appears to be central to the "creative" abilities of physics. By selecting the cosmos and terrestrial rivers as exhibits of self-organizing systems, we have tried to capture the essence of the physical character of the creation of order, complexity, and variety.

Physical theory recognizes only three or four types of forces: gravitational, electromagnetic, nuclear strong interaction, and the weak interaction (now subsumed by electromagnetic). Furthermore, nature has provided only a few materials with which to work. Do a few forces, limited materials, and a small set of approximate physical laws become arranged so that their own interactions, without dependence on outside intelligence or design, imply the variety observable in nature? If so, under what boundary conditions? What are the origins of the constraints?

Later sections of this book focus on the origin of life. We ask whether or not it is understandable as an extension of the principles of self-organization illustrated here by the cosmos, and by rivers.

F. EUGENE YATES • Crump Institute for Medical Engineering, University of California, Los Angeles, California 90024.

1

On Origins
Galaxies, Stars, Life

Harry Soodak

ABSTRACT

This chapter shows application of a consistent set of clear physical principles to describe the beginnings of the universe by the (hot) Big Bang model. The ultimate beginning presents difficulties for the physicist, particularly when it comes to explaining why we find a very inhomogeneous cosmos, instead of a uniform, radiating gas. Stars, globular clusters, galaxies, and clusters of galaxies dot space, but we cannot at present explain their presence without invoking a "deus ex machina." We expect ultimately to rationalize the existence of these local inhomogeneities through gravitational contractions of regions of higher than average mass density that in turn arise out of fluctuations, but the case cannot yet be made completely. Meanwhile, the Big Bang (incomplete) model deals with cosmic evolution in the large, as a balance between gravitational attraction and cosmic expansion. Its relationships are described by two simple equations derived from Einstein's theory of general relativity via local conservations of mass–energy and of momentum. Unfortunately, this model does not by itself produce the observed inhomogeneities. Neither the assumption of thermodynamic fluctuations following a "smooth" beginning nor that of a chaotic beginning can itself account for stars and galaxies. A mystery remains. —THE EDITOR

This chapter describes the efforts of cosmologists (and nature?) to form galactic and stellar structures, comments on them from the viewpoint of the homeokinetic principles (Soodak and Iberall, 1978) discussed elsewhere in this volume, and pursues some parallels in the origin of life. It appears that there are no serious problems in understanding the origin of stars, given the previous formation of a galactic gas mass. A number of reasonable, even likely, paths lead to star formation. On the other hand, the origin of inhomogeneities required to initiate galaxy formation is not yet known. Appendix 1 remarks on some advances in observational and theoretical cosmology made since the original writing of this chapter

---

HARRY SOODAK • Department of Physics, The City College of New York, New York, New York 10031.

and on a number of new and speculative possible mechanisms for galaxy formation that have been opened up by these advances.

Expanding Universe and Standard Model

Observations beginning with Hubble's 1929 discovery of the proportional relation between the red shift of distant galaxies and their distance from us, and continuing to the more recent discoveries of the 2.75°K cosmic radiation and its high degree of isotropy, a uniform abundance ratio of hydrogen to helium in many galaxies, and measurements of the distribution in space of galaxies and clusters of galaxies, provide evidence for an expanding universe that is homogeneous and isotropic in the large and has been so for quite a long time. These observations led cosmologists to a standard cosmological model, called the (Hot) Big Bang, described in popular books by Weinberg (1977) and Silk (1980a). According to this model, the universe expanded and evolved from an initially singular infinite–density state at the beginning, cosmic time zero, some 15–20 billion years ago. The model satisfies the Cosmological Principle according to which the universe is and has been homogeneous and isotropic from the very beginning and appears the same from every location in space (these locations are assumed to be moving along with the cosmic expansion).

In the standard model, cosmic evolution is controlled by the balance between gravitational attraction and cosmic expansion. This is described by two equations derived from Einstein's general relativity theory, which express local conservation of mass–energy and of momentum (Appendix 2). These equations, which relate the rate and deceleration (due to gravity) of cosmic expansion to cosmic size, involve the mass–energy density and pressure of the cosmic fluid. Solving them for the cosmic size as a function of time therefore requires the thermodynamic equation of state relating the pressure and density of the cosmic fluid. The equation of state itself evolves because it depends on the composition of the cosmic fluid (the types and fractions of particles and field quanta), which in turn evolves through the action of all the forces of nature (gravity, strong nuclear, weak nuclear, and electromagnetic). Thus, for example, at the very earliest of times (by the "Planck" time of 10^{-43} after cosmic time zero), all possible forms of elementary particles and field quanta were present, presumably created by the intense gravity fields that existed at the near–infinite densities. As the universe expands and its density and temperature drop, the composition shifts toward those particles, particle combinations, and field quanta which are more stable in the new conditions (as in chemical reaction equilibrium). A partial chronological account is given in Table 1.

Apparent Failure to Produce Structure

The currently observed universe, although homogeneous in the large, contains structures such as stars, globular clusters of stars, galaxies of stars, clusters,

1. On Origins

and superclusters of galaxies. In units of solar mass, the mass range of these structures are several one-hundredths to several tens for stars, about 10^5 to 10^6 for globular clusters, 10^{11} to 10^{12} for galaxies, and up to 10^{14} to 10^{15} for galactic clusters. It is generally believed that the essential process in the formation of such structures is gravitational contraction of regions of higher than average density, i.e., a gravitationally driven instability transition initiated by density fluctuations or inhomogeneities. The rate at which such density contrasts grow or decay is determined by the competition between gravitational attraction on the one hand and cosmic expansion and internal pressure on the other, and also by dissipation (via viscosity and heat transport). Since all these factors vary with cosmic time, the outcome of the balance of forces also depends on cosmic time and on the spectrum of density contrast amplitudes in regions of various sizes.

At present it appears that the standard model, in which density inhomogeneities arise only as a result of normal thermodynamic fluctuations, does not lead

TABLE 1. A Partial Chronology

Cosmic time	Cosmic size scale	Occurrence
0	0	The beginning.
10^{-43} sec	10^{-32}	Planck time. All elementary particles and quanta created. Gravitons decouple.
10^{-6} sec	10^{-13}	Protons and neutrons (and their antiparticles) form by quark combination, and then annihilate with their antiparticles leaving behind the small excess of protons and neutrons. The universe now contains protons, neutrons, neutrinos and antineutrinos, electrons and positrons, and photons (in addition to the decoupled gravitons).
1 sec	10^{-10}	Neutrinos and antineutrinos decouple.
14 sec	3×10^{-10}	Electrons and positrons annihilate into photons, leaving behind the small excess of electrons (the same as the number of protons).
3 min	10^{-9}	Nucleosynthesis of protons and neutrons into helium nuclei takes place. The universe now consists of a plasma of protons, α particles, and electrons, along with about 10^8 photons/proton (and the decoupled gravitons and neutrinos).
10^4 years	10^{-4}	t_{eq}. The photon energy density which was larger than the particle rest mass density (in energy units) at earlier times is now equal to it, and falls below it at later times.
3×10^5 years	10^{-3}	t_{rec}. The recombination epoch. Electrons combine with protons and α particles to form neutral hydrogen and helium. Photons decouple.
$\sim 2 \times 10^9$ years to $\sim 5 \times 10^9$ years	$\frac{1}{20}-\frac{1}{4}$	Galaxies, clusters, and first-generation stars begin to form. Heavy elements are found in supernova explosions.
$\sim 10 \times 10^9$ years	$\sim \frac{1}{2}$	Second-generation stars (containing heavy elements) start to form.
$\sim 15 \times 10^9$ years	$\sim \frac{3}{4}$	Our sun forms.
$\sim 20 \times 10^9$ years	~ 1	Now.

to the formation of galaxies and stars! Thermodynamic fluctuations are too weak to initiate gravitational condensation into galaxies and stars at any stage of cosmic evolution. Thus, the simplest possible cosmological model that is consistent with many current observations seems to be too "smooth" or too close to macroscopic thermodynamic equilibrium to give rise to galaxy formation. As a result, the Aristotelian–like goal of Lemaître (1950), "to seek out ideally simple conditions which could have initiated the world and from which, by the play of recognized forces, that world, in all its complexity, may have resulted," is not yet realized.

The situation seems not to be saved by resort to so-called chaotic cosmological models in which inhomogeneity and anisotropy are present at cosmic time zero. Too many initial possibilities exist, and almost all, if not all, of them very likely lead to a current universe inconsistent with ours. Thus, the alternative Epicurean–like goal of deriving the present order and structure of the universe from initial chaos is also not realized.

Since the two ideal programs of generating existing structure—one from a smooth beginning and one from a chaotic beginning—seem to fail, cosmologists usually adopt what Silk calls a conservative approach, in which density inhomogeneities larger than the thermodynamic ones are introduced into the otherwise standard model at some early cosmic time or times by a process (or processes) presently unknown. They then show that gravitational instability of galaxy–sized structures is the evolutionary result of postulated inhomogeneities having appropriate amplitudes. This program also provides an observational consistency check because the postulated inhomogeneities should imprint the 3°K cosmic radiation with a slight amount of anisotropy; a prediction that is now being investigated.

A less conservative approach, called the cosmic turbulence theory (Jones, 1976), harks back to the vortices of Descartes and assumes that turbulence was somehow introduced into the otherwise homogeneous and isotropic universe at some early time. (A deus ex machina is again required because a transition to turbulence does not occur in the standard cosmological model.) At some later stage of cosmic expansion, the homogeneous isotropic turbulence is then shown to result in density inhomogeneities that are appropriately scaled so that they lead to gravitational instability of galaxy–sized structures. The postulated turbulence should also leave an anisotropic imprint on the 2.75°K cosmic radiation, but in this case it is very likely to be larger than what is currently observed (anisotropy of about one part in 10^4).

Recombination Epoch

In both approaches, the recombination epoch ($t_{rec} \simeq 3 \times 10^5$ years, $T \simeq 3000°K$, and the universe is approximately 1000 times smaller than at present) plays a central role. Before recombination the cosmic fluid consists of an ionized plasma of protons, α particles (helium nuclei), and electrons in interaction with

1. On Origins

a large number of photons. Because of the very large radiation pressure, the fluid is highly incompressible and the speed of pressure–density (sound) waves is an appreciable fraction of the speed of light. Consequently, the cosmic fluid resists gravitational contraction with a large and rapidly responsive force, and growth of density inhomogeneities is strongly inhibited.

When the temperature of the fluid drops to approximately 3000°K, the electrons recombine with the nuclei to form neutral hydrogen and helium atoms, a process that occurs "fairly rapidly," and is essentially completed during a cosmic expansion of about 20%. The neutral atoms interact only weakly with the photons; so weakly that at the low matter densities present, the photons are essentially decoupled from the matter. It is precisely these decoupled photons that are now observed as the 2.75°K background radiation. Although they were initially at $\simeq 3000°K$ at the recombination time, their wavelengths have expanded along with the cosmic expansion so that now when the universe is $\simeq 1000$ times larger than at t_{rec}, their observed temperature is 2.75°K. This radiation provides our earliest view of the universe, when it was $\simeq 1000$ times smaller than at present, and only about 3×10^5 years old.

Because the decoupled photons no longer exert pressure on the matter, the postrecombination matter is simply a low–pressure ideal–matter gas that is easily compressible, and the sonic speed is reduced to the low value of several kilometers per second. As a result, the pressure resistance to gravitational contraction is reduced and slowed by a very large factor, and regions of higher than average matter density can then collapse more rapidly.

In the conservative approach, therefore, the size and mass of the structures that begin collapsing by gravitational instability at later times (after t_{rec}) are determined by the scale of inhomogeneities that are present at t_{rec}. The main tasks of this approach are then to follow the evolution of postulated early density inhomogeneities up to the recombination time; to show that the surviving inhomogeneities present at t_{rec} include those having masses appropriate to galactic clusters, galaxies, and globular clusters; to show that gas clouds of such masses can actually condense into their observed sizes; and finally to show that they can fragment into smaller clouds that condense into stars. These ideas are discussed briefly in the next section. More extended discussions are those of Jones (1976), Reddish (1978), and Silk (1980b).

In the cosmic turbulence theory, the scale of postrecombination density inhomogeneities is determined by the scale of the turbulent eddies present at t_{rec}. Before recombination, the turbulent velocities are smaller than the sonic speed (an appreciable fraction of the speed of light) and consequently have little effect on density. After recombination, however, the same turbulent velocities are much higher than the reduced sonic speed (a few kilometers per second) and the turbulent motion is then highly hypersonic. This results in violent motions leading to density inhomogeneities at various scales up to the size of the largest turbulent eddies present at t_{rec}. One of the main tasks of the theory is then to show that this largest scale is on the order of a galactic or galactic cluster mass. Further discussion of this cosmic turbulence theory is given in a review article by Jones (1976).

Structure Formation (in the Conservative Theory)

Density inhomogeneities in the prerecombination era can be classed into two pure types, called adiabatic and isothermal. In an isothermal fluctuation, the radiation is not affected, but the matter component of the cosmic fluid is compressed. Thus, the temperature, which is dominated by the radiation, is unchanged. Because of the strong coupling between them, relative motion between the ionized matter and radiation is very strongly inhibited by successive rapid scatterings of the photons by the ions. As a result, the density contrast $\delta\rho_m/\rho_m$ of an isothermal inhomogeneity remains constant during the prerecombination era. Thus, the density contrast and mass contained within any isothermal inhomogeneity at t_{rec} are exactly those that were introduced into the universe at some early time by some deus ex machina.

Adiabatic fluctuations are those in which both matter and radiation are compressed together. Leaving aside the effects of cosmic expansion and dissipation, the evolution of such inhomogeneities is determined by the balance between gravity and pressure forces. If the size or mass of the density–enhanced region is large enough, gravity wins out and collapse is initiated. If the mass is too small, however, pressure wins out, and the result is simply an adiabatic vibration. The mass value at which gravity and pressure are in balance is called the Jeans mass, M_J (Appendix 3). It occurs when the size of the region is just large enough that the time it takes a sound wave to cross the region is about the same as the time it would take for the region to collapse by gravitational contraction in the absence of an opposing pressure force. In regions containing significantly more than the Jeans mass, the pressure response is too slow to counteract continued gravitational contraction.

In the prerecombination era, the Jeans mass increases with cosmic time, reaching a plateau value of about 10^{17} solar masses at cosmic time $t_{eq} \simeq 10^4$ years. (This is the time at which the temperature of the cosmic fluid has dropped to about 4000°K, and the mass equivalent of the radiation energy density is equal to the matter density.) The Jeans mass is then fairly constant until recombination. Thus, all adiabatic density inhomogeneities in the prerecombination era, up to scales including giant galaxy clusters, simply vibrate.

If it weren't for damping, the density contrast $\delta\rho/\rho$ of these vibrations would remain constant throughout the prerecombination era. The damping mechanism consists of diffusive transport of photons (which drag the matter along with them) from the region to the outside during each half cycle in which the region is compressed and into the region during the other half cycle. Because of the slow rate of diffusion, smaller–scale inhomogeneities are damped more rapidly. The net effect of damping is that the only inhomogeneity scales that survive dissipation by the time of recombination are those large enough to contain over 10^{12} solar masses (which is the order of a galactic mass).

After recombination and photon decoupling, the pressure and sonic speed are sharply reduced, and much smaller regions become gravitationally unstable. Stated quantitatively, the Jeans mass drops to about 10^5 to 10^6 solar masses (which is the scale of globular clusters). As a result, the inhomogeneities that can

1. On Origins

become gravitationally unstable after recombination consist of all isothermal ones of globular cluster mass or larger, and all adiabatic ones surviving at t_{rec} (which are those of galactic mass or larger). For these inhomogeneities to be unstable, the density contrast must be large enough to overcome the effect of cosmic expansion. (In the absence of cosmic expansion, even the smallest density contrast over a region larger than the Jeans mass eventually collapses.)

As indicated earlier, the density contrasts of normal thermodynamic fluctuations are too small to lead to gravitational instability on any scale. Thus, they do not result in formation of galaxies and stars, which requires much larger amplitudes. The task then becomes to postulate (or deduce) the spectrum of density contrast amplitudes present at t_{rec} that lead to present structure. This task is complicated by the fact that there is a multiplicity of possibilities. In one, called hierarchical clustering, smaller structures form first and larger structures are then formed by gravitational aggregation of the smaller ones. Another possibility, called gaseous fragmentation, is that large gas masses collapse first and then fragment into smaller ones. Finally, any combination of both processes is possible at various scales. Thus, the order of initial formation of stars, globular clusters, galaxies, and galaxy clusters is in doubt, both theoretically and observationally. (Note that a gas cloud in the stellar mass range is far below the Jeans mass of 10^5 to 10^6 solar masses. Therefore, stars can only form after the collapse of a larger gas cloud, as will be discussed later.)

One scenario is that globular cluster masses collapse first, initiated by gravitational instability of isothermal inhomogeneities. These collapsing gas clouds fragment into star–size masses that condense into stars and aggregate into large groups that might be the elliptical galaxies currently observed. Galaxies can then aggregate into clusters of galaxies, and clusters into clusters of clusters.

In another scenario, large gas clouds on the order of galactic cluster masses collapse first, then fragment into galaxy–size masses and then into globular cluster–size masses. The collapse of a galaxy–size mass can lead to the currently observed disk of spiral galaxies. The multiplicity of combinational possibilities is clearly large.

Although there is no single pathway from the onset of gravitational collapse to the formation of the final structures (and perhaps all possible routes were taken), there is a basic thermodynamic requirement for all such paths. Because stars and galaxies are gravitationally bound structures, an appreciable fraction of the kinetic energy of the infall motion must be dissipated (by viscous damping of the turbulent motion generated during collapse) and transported out of the gas cloud (by radiation) at various stages during the collapse. (This is true of all binding condensation processes.)

Thus, for example, it can be argued that efficient dissipation and radiation can take place during the collapse of a large gas cloud only from fragments that contain less than about 10^{12} solar masses and are less than about 200,000 light-years in diameter. This argument can explain why galaxies are not larger or more massive than they are. (Our galaxy contains about 10^{11} solar masses of visible mass and is 60,000 light-years in diameter—a typical spiral galaxy.)

The mass range of first-generation stars (consisting mainly of hydrogen) is similarly determined by dissipation and radiation of infall energy. Viscous damping of the turbulent motions of a collapsing gas cloud that contains many solar masses increases the temperature of the gas. As collapse continues, the temperature rises until it reaches $\simeq 10^4$ °K. At this stage, an appreciable fraction of the hydrogen becomes ionized, resulting in efficient radiation of energy out of the cloud. The temperature then holds at $\simeq 10^4$ °K. as gravitational contraction continues. This state of affairs persists until the density becomes high enough for the Jeans mass to be reduced to a fraction of the cloud size. (The Jeans mass decreases as density increases and the temperature remains constant, as seen in Appendix 3.) At such a stage, smaller pieces of the cloud become gravitationally unstable and the cloud fragments.

The process is repeated as gravitational compression of the smaller fragments continues, leading to fragmentation into still smaller and denser pieces. The process stops when the density of the fragments becomes high enough to prevent rapid escape of radiation. Thus, when the fragments become sufficiently opaque, cooling ceases to be efficient and no further fragmentation takes place. This point occurs at fragment masses on the order of tens of solar masses. What happens next is a slow gravitational contraction of these protostellar fragments until their cores become hot and dense enough to initiate nuclear fusion of hydrogen into helium. The energy so generated can then produce the pressure required to stop further contraction (and the star has entered the main sequence).

Thus, first-generation stars are large stars containing some tens of solar masses. Such large stars burn their core hydrogen into helium quite rapidly. After only some millions of years, their core hydrogen is exhausted and can no longer generate the pressure to balance gravity. Gravitational collapse of the core takes place until it is stopped by the fusion of helium into carbon. After several such stages, the core nuclei are fused into the most stable nuclear configuration, that of iron. At this stage, the core collapses violently, leading to a supernova explosion in which the elements heavier than iron are synthesized and most of the star mass is ejected into the surrounding space.

The heavy-element content (heavier than the primordial hydrogen and helium) of an evolving galaxy increases as more and more first-generation stars form and explode. These heavy elements lead to the formation of dust grains (typically 0.1 μm in diameter and thought to consist of refractory material like silicates on which is condensed a layer of ice made of water or other molecules). The presence of dust grains provides a new cooling mechanism by radiation at far-infrared frequencies. When their concentration in a collapsing gas cloud is high enough, radiative cooling is dominated by the grains rather than by the hydrogen, and the cloud can remain as cool as $\simeq 10$°K. As a result of the lower temperature, and therefore lower pressure, fragmentation can continue past masses of several tens of solar masses and extends down to about a tenth of a solar mass.

As a consequence, second-generation stars, such as those being formed now in our own galaxy, range in mass from about 0.1 to about 50 solar masses. The gas-dust clouds out of which they form contain a very large number of solar

1. On Origins

masses and may be set into gravitational collapse by compression due to a nearby supernova explosion or to a passing spiral density wave sweeping round our galaxy.

Comments

Two immediate historical–evolutionary morals are revealed by the discussion of cosmic structure formation. One is that from the galactic scale on down, matter condensation processes involve and in fact require an "up–down" or "in–out" (often radial) asymmetry of a flow process. Thus, gravitational infall energy is converted by internally delayed dissipation and leaves by radiation. This scheme is similar to the psychrometric process in liquid condensation (Iberall and Soodak, 1978). What results is a stably bound structure with a finite deformation range of resistance to stress (as in a galaxy, star, or liquid drop), rather than simply a flow pattern (such as a Bénard cell or turbulent eddy). The internal dissipative process responsible for the in–out asymmetry may be regarded as the bulk-viscosity gateway to the condensed stable form.

The second moral is that further internal cooperation emerges within the condensed forms. Thus, for example, birth, life, and death are not independent events from star to star within a galaxy but, rather, are strongly coupled by large-scale processes of a thermodynamic hydrodynamic nature within the galaxy.

It can also be seen that condensation in form seems to arise from a transformation of side–side (flow) processes into up–down binding processes. This picture is clear in the cosmic turbulence theory, in which density contrasts are the direct result of the flow patterns. It is equally clear in the process of gas fragmentation and in the formation of stars resulting from flow processes within the galaxy. However, the source of the primordial density inhomogeneities in the conservative approach is still relegated to some unknown initiation process or deus ex machina. In all cases, the existence of flow-initiated fluctuations is no guarantee for the formation of matter condensation, as witnessed by the failure of thermally initiated fluctuations to generate structures.

Life

Modern ideas regarding the origin of life involve a material field consisting minimally of condensed water, significantly below 100°C, prebiotic organic molecules, and various trace amounts of other elements required by present life forms. In some sense, this epoch in earth history (some 4 billion years ago) is analogous to the recombination epoch in cosmological history.

It may be surmised that condensation of some kind or kinds of biotic molecules is going to be the result of associational fluctuations coupled to flow processes. The biochemist has generally stressed the chemical associational process with insufficient emphasis on the flow aspects, especially those that might be involved in heterogeneity scaling.

The homeokinetic view is in agreement with that of the molecular biologist in emphasizing that life's scaling seems to originate at the size of molecular machinery (e.g., 10 to 40Å). But in contradistinction, because of the difficulty in achieving life's complex machinery in just a few steps, the homeokinetic view stresses that heterogeneous scaling of life's processes by externally scaled fields, as intermediaries, appears most useful, if not necessary. Heterogeneous catalysis seems more likely than homogeneous catalysis to lead to formation of life.

Appendix 1: More Recent Developments

Inflation

Many particle physicists believe in a Grand Unified Field Theory in which the basic laws of physics are totally symmetric, and in which gravity, strong, weak, and electromagnetic forces are all separate aspects of a single grand force. Theories have been constructed that unify the strong, weak, and electromagnetic forces. Experimental evidence, however, exists only for the unification of the weak and electromagnetic forces into the electroweak force. No theories exist that unify gravity with the other forces. A full-fledged quantum theory of gravity is still beyond reach.

According to these ideas, the decoupling of the gravitons at cosmic time 10^{-43} sec (see Table 1) accompanies the separation of gravity from the other three forces, which remain unified at the high energy levels existing at this epoch. The next separation occurs at 10^{-35} sec, when the energy level drops below those required to form the W and Z particles that mediate the weak interaction. At this time the strong force separates, leaving only the weak and electromagnetic forces unified. According to Guth and Steinhardt (1984), a phaselike transition in the properties of the quantum vacuum becomes possible at this time and, if it occurs, can in some possible way lead to a very rapid—inflationary—expansion of the universe by some 50 orders of magnitude in a period of some 10^{-32} sec. Expansion after this inflationary epoch is normal, controlled by the momentum of the expanding universe and by gravity. Inflation has provided a partial explanation of the large-scale isotropy of the universe.

Galactic Superclusters

Studies of the large-scale structure of the universe in recent years have demonstrated the presence of giant superclusters of galaxies that are made up of many clusters of galaxies (Burns, 1986). The clusters may contain hundreds or thousands of galaxies and are more or less spherically shaped. The superclusters, containing tens of clusters connected by low-density "bridges" consisting mainly of single galaxies, take the form of long filaments or shells, and are on the scale of a hundred million lightyears. Separating the superclusters are large spherical or elliptical void regions that are hundreds of times less dense than the supercluster regions and remarkably free of elliptical and spiral galaxies. One exceptionally large void, found in 1981, is about 120 million lightyears in radius and ringed by walls of galaxies.

Structure Formation

There is as yet no satisfactory scenario that explains the formation of galaxies, clusters, and superclusters. The origin and scale of the initiating density fluctuations leading to formation is not yet known.

It appears that there are serious problems with every suggested fluctuation whose consequences can be analyzed. In some cases, as described earlier and as discussed by Burns (1986) and by Wilkinson (1986), the density inhomogeneity would leave its imprint as an anistropy in the cosmic background radiation that is larger than the observed value of 1 part in 10^4. In other cases, analysis may indicate that the fluctuation does not grow large enough to initiate structure formation.

In the neutrino scenario, for example, neutrinos are assumed to have a non-zero rest mass. (It should be noted that the experimental value of the neutrino mass is still consistent with the value zero.) In this case, after their decoupling at cosmic time 1 sec (see Table 1), they can be slowed down from their initial near-light speeds, and can conceivably form local density inhomogeneities around which matter structure can condense by gravitational attraction. Calculations indicate, however, that their low mass prevents neutrons from slowing or cooling rapidly enough (as the universe expands) to allow for the formation of sufficiently strong nucleating centers at sufficiently early times to give rise to many existing galaxies.

The problem arising from the small mass of the neutrino is avoided by the possibility that more massive particles were decoupled at much earlier times in the universe, and that these particles could have produced the initiating density fluctuations. One such suggestion—speculation—involves particles such as gravitinos, photinos, and axions, particles that must exist in the so-called supersymmetric theories in which symmetry is postulated between fermions and bosons.

In another speculative suggestion, structure formation is initiated by so-called cosmic strings. These are one-dimensional faults, defects, or singularities in space that were conceivably formed during the inflationary epoch. If neighboring regions of space made transitions into different quantum vacuum states, they would then be separated by "domain boundaries," which could have taken the form of two-dimensional walls, monopoles, or cosmic strings. It is thought that walls are unstable, that the monopole scenario is inconsistent with present observation, and that the cosmic string scenario is not inconsistent. Cosmic strings are presumably formed with a very large mass-energy per unit length, allowing them to provide nucleation centers possibly by gravitational attraction or possibly by emission of electromagnetic radiation which compresses nearby regions to higher densities. A recent discussion is given by Gott (1986).

Appendix 2: The Expanding Universe

Consider any point in space moving along with the cosmic expansion. Select it as the reference frame and consider a sphere, of radius a, centered at that point. It is then true in general relativity, as it is in Newtonian physics, that the gravitational field at the surface of this sphere is due only to the gravitational mass within it. If the radius a, is small enough, Newtonian physics can be used.

The gravitational acceleration at the surface of this sphere is then GM/a^2, where G is the gravitational constant. In general relativity, M includes the mass equivalent of all forms of energy in the sphere in addition to rest mass and also the gravitational effect of stress, as measured by the pressure P of the cosmic fluid. The value of M can be expressed as

$$M = \frac{4}{3}\pi a^3 \left(\rho + \frac{3P}{c^2}\right) \tag{1}$$

where ρ is the total mass–energy density expressed in mass units and $3P/c^2$ is the additional gravitational effect due to stress. (Here, c is the speed of light.) Setting the acceleration $\ddot{a}$ (each dot denotes one differentiation with respect to time) of a mass element on the sphere surface equal to $-GM/a^2$ (radially inward) leads to

$$\frac{\ddot{a}}{a} = -\frac{4}{3}\pi G \left(\rho + \frac{3P}{c^2}\right) \tag{2}$$

This equation, which follows from Newton's second law of motion, expresses local conservation of momentum.

The equation of local energy conservation expresses the fact that the total mass–energy in the sphere decreases as the sphere expands by the amount of pressure work done. In energy units, the equation is

$$\frac{d}{dt}(\rho V c^2) = -P\frac{dV}{dt} \tag{3}$$

where $V = 4/3 \,\pi a^3$ is the sphere volume. Equation (3) can be written as follows:

$$\dot{\rho} + \frac{3\dot{a}}{a}\left(\rho + \frac{P}{c^2}\right) = 0$$

Using this result to eliminate P from equation (2) leads to

$$\ddot{a}\,\dot{a} = \frac{4}{3}\pi G(a^2 \dot{\rho} + 2\rho a \dot{a})$$

which can be integrated directly. The result can be expressed in the following form:

$$\frac{\dot{a}^2}{a^2} + \frac{k}{a^2} = \frac{8}{3}\pi G \rho \qquad (4)$$

where k is a constant of integration and is a measure of the curvature of the universe.

A zero k value denotes a flat universe of infinite extent in which expansion continues forever but the rate $\dot{a}$ goes asymptotically to zero at long times. A positive k describes a positively curved, closed, and finite universe in which the expansion is stopped in some finite time and the universe then collapses back upon itself. Finally, a negative k value describes an open, negatively curved universe in which the expansion rate $\dot{a}$ asymptotically approaches a fixed value $(-k)^{1/2}$.

These statements can be made evident by rewriting equation (4) as follows:

$$\frac{1}{2}\dot{a}^2 - \frac{4}{3}\pi G \rho a^2 = -\frac{k}{2}$$

where the left side can be regarded as the sum of kinetic and gravitational enery of a unit mass on the sphere surface. A positive value for the total energy (negative k) denotes an expansion at a rate greater than the escape speed. A negative value (positive k) describes a gravitationally bound situation in which expansion turns into collapse after some time. Finally, zero value describes the dividing case in which all mass elements are expanding at exactly the escape speed (from the sphere inside them).

Equations (2) and (4) together with a state equation relating P to ρ, and with the initial condition $a = 0$ at $t = 0$, determine the relative scale of cosmic expansion $a(t)$ at any time. The ratio $a(t_2)/a(t_1)$ then gives the factor by which the universe expanded in the interval from t_1 to t_2. As an example, if the equation of state at early times is $P = \rho c^2/3$, then $a(t)$ is proportional to $t^{1/2}$.

Consider any two galaxies (both moving along with the cosmic expansion) a distance R apart at some time t. Then, $\dot{R}/R$ is equal to $\dot{a}/a$. But $\dot{R}$ is the relative recessional velocity v of the galaxies, and thus Hubble's law is obtained:

$$v = \left(\frac{\dot{a}}{a}\right) R \qquad (5)$$

which states that the recessional velocities of galaxies are proportional to their separation. The proportional constant $\dot{a}/a$, which is a function of cosmic time, is called Hubble's constant. Its value at the present epoch of cosmic evolution is measured to be about 15 km/sec per million light–years.

The critical density ρ_c for the universe to be exactly flat—on the borderline of just being closed—can be obtained from equation (4) by setting k equal to zero. The result is

$$\rho_c = \frac{3}{8\pi G}\left(\frac{\dot{a}}{a}\right)^2$$

The present value of Hubble's constant $\dot{a}/a$, given above, leads to a present critical density of about three hydrogen atoms per cubic meter of space. The currently visible average density is perhaps one-tenth of critical, but the "missing mass" that would just about close the universe may yet be found.

Appendix 3: The Jeans Mass

The Jeans radius R_J and the mass contained within it,

$$M_J = \frac{4}{3}\pi R_J^3 \rho$$

can be estimated by equating the gravitational free-fall time t_f to the time t_s required for a sonic wave to travel the distance R_J. The sonic time is

$$t_s = R_J/v_s$$

where v_s is the sonic speed. Since the gravitational infall acceleration at the surface of a sphere with radius R_J is GM/R^2, the relation between R_J and t_f is approximately

$$R_J = \frac{1}{2}\left(\frac{GM}{R_J^2}\right) t_f^2$$

or

$$t_f = \left(\frac{3}{2\pi G \rho}\right)^{1/2}$$

Equating this to the sonic time gives

$$R_J = v_s \left(\frac{3}{2\pi G \rho}\right)^{1/2}$$

The Jeans mass is then proportional to

$$M \propto v_s^3 \rho^{-1/2}$$

Use of the formula $v_s = (P/\rho)^{1/2}$ for the sonic speed gives the Jeans mass in terms of pressure and density:

$$M_J \propto P^{3/2} \rho^{-2}$$

Finally, use of $P \propto \rho T$ for a matter gas gives

$$M_J \propto T^{3/2} \rho^{-1/2}$$

for the Jeans mass in terms of temperature and density.

References

Burns, J. O. (1986) Very large structures in the universe. *Scientific American* **July**:38–47.
Gott, J. R. III (1986) Is QSO 1146 + 11B,C due to lensing by a cosmic string? *Nature* **321**:420–421.
Guth, A. H. and P. J. Steinhardt (1984) The inflationary universe. *Scientific American* **May**:76–84.
Iberall, A., and H. Soodak (1978) Physical basis for complex systems—Some propositions relating levels of organization. *Collective Phenomena* **3**:9–24.
Jones, B. (1976) The origin of galaxies: A review of recent theoretical developments and their confrontation with observation. *Rev. Mod. Phys.* **48**:107–150.
Lemaître, G. (1950) *The Primeval Atom.* Van Nostrand, Princeton, N.J.
Reddish, V. (1978) *Stellar Formation.* Pergamon Press, Elmsford, N.Y.
Silk, J. (1980a) *The Big Bang: The Creation and Evolution of the Universe.* Freeman, San Francisco.
Silk, J. (1980b) In: *Star Formation,* I. Appenzeller, J. Lequeux, and J. Silk (eds.). Geneva Observatory, Sauverny, Switzerland.
Soodak, H., and A. Iberall (1978) Homeokinetics: A physical science for complex systems. *Science* **201**:579–582.
Weinberg, S. (1977) *The First Three Minutes: A Modern View of the Origin of the Universe.* Basic Books, New York.
Wilkinson, D. T. (1986) Anisotropy of the cosmic blackbody radiation. *Science* **231**:1517–1522.

2

On Rivers

Arthur S. Iberall

ABSTRACT
Among the most enchanting of the sculpting processes on earth (and apparently also at an earlier time on Mars) is water flow. The large-scale interactions among lithosphere, hydrosphere, and atmosphere are modeled in this chapter, on the basis of very few assumptions and only a few physical principles. Despite the mathematical detail, readers not mathematically inclined should be able to follow the thematic development that reveals rivers to be self-organizing, dynamic systems; complex, but analyzable. Rivers show the richness of near-equilibrium operations, the effects of scale, and the results of competition between coupled forces and processes. In this presentation about rivers is a hint that rivers are an appropriate metaphor for the stream of life.

The physical model of river systems has to take into account precipitation, evaporation, groundwater flow, river volume flow, and the changing course of rivers, from cascades to meanders. Particular emphasis is placed on the role of the bed load carrying capacity of the river. —THE EDITOR

Natural systems are not like man-made systems. They are not forced by hard-molded, hard-wired, hard-geared, man-made constraints. Instead, they are self-organizing as a field process and morphology within boundary constraints which generally are potentials and fluxes from outside. Within the field so loosely constrained, they sort themselves out in both form and function. A river system on a large land mass is an example. A river system is always part of a larger hydrological cycle that is strongly coupled to a comparably large meteorological cycle. This chapter focuses on the rivers, beginning with a rough sketch of the characteristics of such a system. Imagine a collection of piles of sand (as an analogue for continents), each of which has a sprinkler system mounted above its summit, and a fixed flow of water. When the water is turned on, it will fall upon each pile and carve a system of channels. Each evolving pattern will be different. The basic physical principle at work is simply that water runs downhill. But, in running

ARTHUR S. IBERALL • Department of Oral Biology, University of California, Los Angeles, California 90024. This chapter is dedicated to the memory of my dear friend, Jule Charney. Whatever the issue, we squabbled.

downhill, the water carries some of the sand downstream. The rivulets both carve channels and transport a silting bed load. These dual processes—running downhill and carving and carrying a bed load—constitute a river system. A river is not simply hydraulic flow in a fixed channel. A principal result to be inferred is that a river does not run down lines of steepest descent but meanders instead. Its characteristics are defined by the dual forces of gravity promoting water flow and of the bed load the river carries resisting the flow.

On a geological time scale, rivers behave in an almost lifelike way. One way to imitate these characteristics of a river system is to scribe a sand pile so that runoff begins in channels. But as channels are carved and silt up, they begin to meander, and the network that evolves adapts to the rain water supply and to the nature of the land. This adaptation by changing form becomes increasingly prominent if the supply flow is fluctuating, or impulsive, rather than constant.

The scale of real rivers is, of course, dependent on the characteristics of the earth upon which the water trickles and flows. Terrestrial hydrological cycles are tied to lithospheric processes of continent buildup (Wood, 1980; Iberall and Cardon, 1980–1981). The actual carving process of a river takes millions of years, through abrasion and erosion of the land followed by subsequent continental uplift (e.g., through isostasy). Yet both the faces of the earth and an ensemble of sand piles, despite different scales, are tied up with thermodynamic processes that are near equilibrium in local regions and that exhibit a fairly ergodic character. The distribution statistics of rivers, of rivulets on a collection of sand piles, or of an ensemble of progressive states of a large land mass appear to fill the available phase space and are essentially stationary. A large land mass is believed to exhibit a changing complexion over millions of years through the action of diastrophic, orogenic, tectonic, and volcanic forces, so that a sample of ergodiclike states is presented by the successive aspects of the evolution of a given land mass.

The Hydrological Cycle

The mass balance of the hydrological cycle is contained in the following statements:

At equilibrium:

Intensity of rainfall = evaporation + runoff

At disequilibrium:

Change in ground storage = intensity of rainfall − evaporation − runoff

This discussion does not cover instantaneous processes whereby short-term, nearly impulsive rainfall immediately increases the ground storage and delays the appearance of waters in the rivers. (The shortest time constant is on the order of a few days.) Instead, I shall consider performance averaged over a few years, i.e., at near-equilibrium conditions. The equilibrium equation requires that for any small area of the earth's surface, there is a given intensity of rainfall (inches of rainfall per year), some of which evaporates, while the remainder leaves the region in the form of runoff (ground and river flow). The assumption of near equilibrium over one or a few years implies that the vertical variation of groundwaters is negligible, i.e., that there is a near-stationary water table. If these considerations are applied to a large land mass, such as a continent or a river valley, the difference in volume between total rainfall per year and total evaporation will appear as river flow out of the boundary, usually to an ocean.

There are two conceptually relevant models of a land mass—one supposes a high, conical, glassy surface (imagine a newly formed volcanic atoll), in which case most of the rainfall quickly runs off; the other assumes a nearly flat absorptive surface, in which case most of the rainfall evaporates. For large continental expanses, with maximum elevations only a few miles above sea level and horizontal expanses thousands of miles wide, the second model seems more appropriate. For the United States, on average, 30 inches of rain falls each year, 22 inches evaporates, and 8 inches runs off in rivers. The question is, what is causal among these three factors, only two of which can be independent? That computation is one requirement for a theory of the hydrological cycle. This theory is presented in detail in Iberall *et al.* (1961-1962), but can be briefly described here by listing its basic components:

- A theory of rainfall
- A theory of average humidity
- Various other components of meteorological theory
- A theory of the water table
- A theory for groundwater flow
- A theory for bed load
- A theory of river flow

Essential Aspects of a Dynamic River System

The next sections contain discussions of rainfall, evaporation, the free convection process, forced convection, humidity, water table levels, and river runoff. By examining these essential aspects of a dynamic river system, and relating them to each other, the lifelike characteristics of the system emerge.

Rainfall

Rainfall is an independent variable of the meteorological cycle. If such a cycle could be predicted, then the prediction would have to include estimates of annual rainfall over any large land mass, the distribution of intensity of rainfall for each of the river valleys associated with large land masses, and the slow variation of annual rainfall in time. Meteorological theory is not able to do this yet. Instead, we have to depend on empirical records of rainfall. Still, it seems clear that the rainfall is related to many of the processes making up the average meteorological cycle—the vertical temperature distribution, the horizontal temperature distribution, the prevailing wind structure, the vertical pressure distribution, the average cloud cover, and the earth's heat budget.

Loosely speaking, the plot of the distribution of annual rainfall against latitude has an approximately flat top (40 inches per year average, with a small quadratic deviation) between 60° north and south latitudes, beyond which it drops quickly toward zero. (This statement neglects the fact that there is a high-intensity spike of rainfall at equatorial latitudes.) Thus, some sort of global model, particularly one related to nonlinear dynamic effects of the atmosphere, ought to be invoked. Because runoff is not a significant factor in oceans (although currents may be), and because two-thirds of the earth's surface is covered by water, the observed (approximate) overall balance between intensity of precipitation and evaporation indicates that a common mechanism dominates on land and sea.

For more detailed modeling of precipitation, continental effects over land masses have to be taken into account, particularly those of coastal mountain ranges in blocking or channeling rainfall. The production of deserts over an appreciable amount of the earth's surface by such mechanisms is of considerable concern.

Evaporation

Evaporation and its control over long periods is often erroneously presumed to be dependent upon plant life, as an evapotranspiration process. The following rudimentary argument shows that for regions with daily cloud cover, the annual evaporation is governed by a large-scale process in which ground cover offers negligible impedance to the evaporative flux. Basically, a psychrometric process governs: heat and mass (evaporative) transfers take place as a counterexchange, via the same hydrodynamic boundary layer,[1] between earth and cloud cover. A psychrometric process, as illustrated by a wetted bulb in a wind stream, is a hydrodynamic field process in which a fluid resistance element creates a thin boundary layer, across which heat and mass are adiabatically transferred. The two transfer processes are tied together by a heat-of-phase-transformation, e.g., heat of vaporization or sublimation. For the earth, the wetted surface is the water table at or near the ground surface, and the hydrodynamic boundary layer is a layer of convective air near the earth's surface. Plants do not have significant control of the impedance at that water surface. Instead, average atmospheric humidity is the governing variable. On the other hand, plant cover is an important process over arid regions without daily cloud cover (Charney *et al.*, 1975).

---

[1]The simple model presented here has as its basis the assumption of a thermal-mechanical zone of convective instability in the atmosphere. The complex meteorological coupling of vertical, free convective and horizontal, forced convective processes are modeled as independent heat transfer processes, and their relative contributions are judged. The results obtained do not differ much from those obtained by following the more specialized arguments used by meteorologists.

Free Convective Transfer

In a free convective field, heat and mass transfer depend on the product of the Grashof number, G, and the Prandtl number, Pr (Jakob, 1949; McAdams, 1954). I will analyze free convective field first, because, quite independent of any forced convection by winds, there will always be a free (gravitationally driven) convection at the surface.

Substituting values for the relevant parameters in the GPr product (fluid density, viscosity, gravity, coefficient of thermal expansion, specific heat, thermal conductivity, and an approximate mean surface temperature), one obtains (see, e.g., McAdams, 1954) the following:

$$GPr = 1.8 \times 10^6 \text{ (per °F, per ft}^3\text{)} \; L^3 \, \Delta t$$

where L is the characteristic dimension of the land mass, e.g., the breadth of the United States = 3000 miles = 15×10^6 ft, and Δt is the driving temperature potential between the surface and a plane remote from the surface. A characteristic difference of 30°F is assumed (the approximate difference between the 60°F mean surface temperature and the freezing temperature of water).

Thus:

$$GPr = 0.18 \times 10^{30}$$

This value is extremely large because of the very large size of the convective surface. It suggests that the heat transfer coefficient (given in dimensionless form by the Nusselt number N) is determined by a turbulent flow relation such as

$$N = hL/k = 0.13(GPr)^{1/3}$$

where h is the overall heat transfer coefficient and k is the thermal conductivity of the fluid.

For air at 60°F and earth atmospheric pressure, the free convective heat transfer coefficient becomes

$$h_c \cong 0.20(\Delta t)^{1/3}$$

where h_c is the convective heat flow in BTU/hr ft^2°F.

The total heat transfer power P is

$$P = h_c A \Delta t$$

where A is the transfer area.

Note that in turbulent flow, the power transfer is independent of scale dimension L and depends only on the heat flux area and Δt.

It is instructive to estimate the boundary layer thickness, s:

$$s = k/h = L/N$$

If N is $0.13(\text{GPr})^{1/3}$, GPr is 0.18×10^{30}, and L is 15×10^6 ft, then:

$$s = 0.02 \text{ ft}$$

which is a very thin boundary layer. This result is well within the scale of any possible ground inversion layer. Thus, we surmise that the actual governing boundary layer is the ground-to-low-cloud layer, i.e., that level at which the vertical temperature reaches freezing temperature. The rate of temperature drop (the so-called lapse rate) in the atmosphere is about 3°F per 1000 vertical feet. Thus, freezing temperature occurs at about 10,000 ft. Because the low cloud layer dominates both heat and mass transfer, as well as a significant portion of the radiation greenhouse effect, it creates conditions for the transfers to be almost perfectly psychrometric.[2] Furthermore, at these temperatures, heat transfer coefficients are not significantly affected by humidity (Iberall et al., 1961–1962).

Forced Convective Transfer

If now we add an estimate of any forced convective flow that may pertain, then that heat transfer coefficient can be expressed approximately (Jakob, 1949) by

$$N = 0.036 \text{Re}^{0.8} \text{Pr}^{1/3}$$

where Re is the Reynolds number.

For air, the Prandtl number is approximately 0.74, so that

$$N = 0.033 \text{Re}^{0.8}$$

The heat transfer coefficient, h_f, is given by

$$h_f = 0.033 \frac{k}{L} \left(\frac{LV\rho}{\mu} \right)^{0.8}$$

which is not completely independent of scale L. Substituting values, for atmospheric air, we have

$$h_f = 3.3 \times 10^{-5} V^{0.8}$$

where V is the horizontal wind velocity in miles per hour.

Although it is not absolutely rigorous to do so, we can add together the two estimates of the heat transfer coefficient ($h = h_c + h_f$). Thus, the combined heat

---

[2]The validity of that conclusion is attested to by the experimental near-constancy of the Bowen ratio which is a measure of the psychrometric constant found prevailing in atmospheric studies (Iberall et al., 1961–1962).

transfer coefficient, h, for free and forced convection, is approximately given by

$$h = 0.20(\Delta t)^{1/3} + 3.3 \times 10^{-5} V^{0.8}$$

Radiative Transfer

In a radiation field, the radiation heat transfer is given by

$$h_r = 0.171 \times 10^{-5} a(T_1^4 - T_2^4)F$$

where h_r is given in BTU/hr ft^2 and a is the absorptivity of the surface (here the ground). "All electrical nonconductors with a refractive index $m < 3$ emit 75% or more of the radiation of a black body. Because generally $m < 2$, the emissivity of many insulators exceed even 90% of the black radiation" (Jakob, 1949). Thus, a first approximation for the emissivity or absorptivity of the ground is unity. In this equation, F is a factor that takes into account absorption in the atmosphere. It may have the value $3.5/4 = 0.9$ (McAdams, 1954).

The smallness of the temperature difference permits replacing the fourth power difference required by radiation theory by a linear difference $(T_1 - T_2)$:

$$h_r = 0.171 \times 10^{-8} \times 3.5 a T_{\text{mean}}^3 (T_1 - T_2)$$

Assuming the radiation exchange takes place between ground at 60°F and a freezing 30°F cloud,

$$h_r = 0.77a$$

Combined Heat Transfer

The combined heat transfer is thus:

$$h = 0.77a + 0.20(\Delta t)^{1/3} + 3.3 \times 10^{-5} V^{0.8}$$

with contributions from radiation, free convection, and forced convection.

Because the forced convection contribution appears to be small (i.e., if average maximum wind velocities on the order of 10 mph are assumed), then

$$h \cong 0.77a + 0.20(\Delta t)^{1/3}$$

The temperature drop across the boundary layer is small ($< 30°F$) so that the one-third power of Δt is on the order of 1 to 3. Thus, the overall heat transfer coefficient approximates unity, e.g.,

$$h = 1.0\text{--}1.4 \text{ BTU/hr ft}^2 \text{ °F}$$

Similitude: Mass and Heat Flows

One of the great achievements of a theory of similitude is the transformability of a hydrodynamic heat transfer to an analogous mass transfer. If thermal diffusivity (α) is a governing factor in a thermal field, then in a mass diffusion field (e.g., water evaporating into air), mechanical diffusivity (δ) is the same governing factor.

Thus, if in a heat transfer field

$$h = (k/L)N$$

where N is a complicated function $\phi(\text{Re}, \text{Pr}, G, X)$, then the same function holds for mass transfer, with the Schmidt number substituted for the Prandtl number:

$$\phi = \phi(\text{Re}, S, G, X)$$

where the mass transfer is computed from

$$RT_m \frac{M_{wv}}{A\Delta p} = \frac{\delta}{L} \phi(\text{Re}, S, G, X)$$

M_{wv} is mass transfer (evaporation) per unit time, A is evaporating area, Δp is partial pressure difference, wetted ground to air, and RT_m is gas constant-temperature product, on the basis of the mean temperature ground-air.

The Prandtl number, $\text{Pr} = \nu/\alpha$, is the ratio of kinematic viscosity to thermal diffusivity. The Schmidt number, $S = \nu/\delta$, substitutes, instead, the mass diffusivity. In both cases the hydrodynamic field involves the kinematic viscosity, ν.

The mass transfer can be computed from

$$\begin{aligned} RT_m \frac{M_{wv}}{A\Delta p} &= \frac{\delta}{L} \phi \\ &= \frac{\delta}{L} N \\ &= \frac{\delta}{L} \frac{hL}{k} = \frac{\delta h}{k} \end{aligned}$$

where h is 1 BTU/hr ft^2 °F, δ is 0.77 ft^2/hr, k is 0.0145 BTU ft/hr ft^2 °F, and RT_m is 1.43×10^6 ft^2/sec^2 (from steam tables).

Converted to a rate of water height per unit time,

$$\left(\frac{dz}{dt}\right)_e = \frac{\delta h}{k} \frac{g}{\rho_w RT_m} \Delta p$$

where ρ_w (water density) is 62.4 lb/ft^3, g is 32.2 ft/sec^2, and $(dz/dt)_e$ is evaporation rate. Thus,

$$\left(\frac{dz}{dt}\right)_e = 2.02 \left(\Delta p \frac{\text{lb}}{\text{ft}^2}\right) \frac{\text{in.}}{\text{yr}}$$

The average surface temperature of the United States is approximately 58°F (an average of 67 cities). The mean relative humidity is estimated to be about 67%. The partial pressure of water vapor at 58°F is 0.238 lb/in.^{2} (= 34.3 lb/ft^2). If air and ground temperature are assumed equal, then the approximate partial pressure difference is 34.3 (1−0.67) = 11.3 lb/ft^2. The estimated evaporation rate then would be given by

$$\left(\frac{dz}{dt}\right)_e = 23 \text{ in. water/year}$$

This is a very good estimate of the actual evapotranspiration.

Humidity Disequilibria

The previous calculations assumed the persistence of a cloud cover, but as will appear later, this model depends not so much on the fractional cloud cover as on the average rainfall. It assumes that the dominant transport resistance is in the ground-to-cloud-cover air layer and that the diffusive resistance of the ground is negligible. Furthermore, it assumes that the average humidity in the boundary layer is high, but not at 100% saturation. The computation would be very similar over a large water-covered area in which the evaporation is equal to the rainfall. In both cases, the model is large scale and seems very little related to local meteorological conditions. However, there are circumstances under which this appearance is false.

Suppose the atmosphere *were* a thermodynamic equilibrium process at its large scale. Then, being largely exposed to a free water surface (25,000-mile circumference, spherical surface presenting to a 50-mile-high atmosphere), the atmosphere should be saturated (100% humidity). But the atmosphere is not saturated; its average humidity is about 60%. How can that be? Consider a closed pot of water as an analogue. Here the air space is saturated with water vapor. However, water condenses on the lid and occasionally drips off. Evaporation now takes place to make up for the "rainfall." Under these conditions, the humidity in the air space is no longer exactly 100%; yet the near equilibrium in a pot is good enough to serve almost as a standard atmosphere for 100% humidity. However, if the lid is cooled or shaken, an appreciable evaporation takes place to equilibrate the rainfall and the humidity becomes appreciably less than 100%, even though the hydrological equilibrium still holds.

Thus, the extent to which evaporation (balancing rainfall) appears is a measure of the gross thermodynamic disequilibrium in the system. In other words, either the intensity of rainfall or the average humidity are measures of disequilibrium. The horizontal temperature distribution of the earth at the top of the atmosphere is a function of the solar radiation, but then the temperature of the earth's surface, the vertical temperature distribution, the prevailing wind structure, the average cloud cover, and, as I have shown, the intensity of rainfall *or* the average humidity are all variables determined by the character of a nonlinear meteorological cycle, involving a complex hydrodynamic field that exhibits a prominent greenhouse effect. These variables become independent inputs to the hydrological cycle. Over an ocean (except for currents), then,

$$\text{Rainfall} = \text{evaporation}$$

with either term being regarded as "causal."

Although we have shown that the evaporation can be computed from the overall heat transfer coefficient as a global process, this computation is only true for cloud-covered regions. If we compute the overall heat transfer from actual temperature and humidity data for different regions, we find a transfer coefficient ranging from 0.95 to 1.95 BTU/hr ft^2 °F, with a mean value of 1.5 and a mean deviation of 0.2 *independent of rainfall*, wherever precipitation exceeds 20 in./year (data from 44 U.S. cities; Iberall *et al.*, 1961–1962). Regions with less rainfall are considered arid or semiarid, and may be presumed to lack persistent cloud cover. In those regions a different theory of mass transfer has to be developed—one that does depend on vegetation and ground cover as governing variables. Such a model has been developed by Jule Charney, in particular for the Sahel Desert. For that type of region, the calculated transfer coefficient increases linearly with rainfall. However, the scatter of data is considerable, indicating that we have only a poor correlative theory for arid regions; thus the need, instead, for a dynamic theory, such as Charney has developed (Charney *et al.*, 1975).

River Runoff and Water Table

Is there an independent theory of river runoff that relates only to the ground? The answer is yes.

Imagine a large land mass surrounded by a level ocean. Every point on that mass may be indexed by the earth's tangent plane coordinates (x, y at sea level). An associated z coordinate, the height above sea level, creates a topographic map of altitude contours associated with that plane surface, or alternatively, an orthogonal set of lines of steepest descent. The land mass contour and the runoff system change by erosion or uplift. However, we assume that the surface changes quite slowly and that the rivers—albeit only slowly changing themselves—adapt *relatively* rapidly to the slow upheavals of the land, with its nominal period on the order of approximately 100 million years (Iberall and Cardon, 1980–1981).

Each x, y coordinate on the map can be associated with a river. (It is either a point on a river or on a line segment that drains toward a river.) A river, in this modeling, is simply the smoothed line that a gross aerial survey sees scored upon the earth. Though rivers must course downhill, they do not flow in straight lines.

2. On Rivers

They meander at low inclines. Ridges divide the land mass into isolated river basins (water cannot flow over ridges without being pumped). These open up to the ocean. In the United States there are ten major, isolated river basins. Thus, a map of a land mass can be generated by an intrinsic coordinate system consisting of altitude contours and not-quite-orthogonal river lines. In addition, the map is divided into isolated river basins.

Various regions on the map can then be assigned their empirically known, average intensity of rainfall (e.g., U.S. regions can be associated with 10, 20, 30, 40, 50, and 60 in. of rainfall). The question then is, how does that rainfall distribute itself through the rivers? The basic assumption is that just as water runs downhill on the surface, so the groundwaters also must trickle downhill from the ridges toward the river bed along lines of steepest descent.

A new theory appropriate to groundwater flow has been developed (Iberall et al., 1961–1962; see Iberall and Cardon, 1980–1981, for an improved model) which is intermediate between the Dupuit–Forscheimer theory of ground flow and the Muskat theory (Muskat, 1937; Scheidegger, 1957). The former, a common but crude approximation for ground flow, assumes near-zero slope for the surface of the groundwater flow, whereas the latter "exact" theory, utilizing a method of contour integration, leaves the results in the form of elliptic integrals. Although the latter "exact" theory is necessary for seepage flow through a dam, or for waterfalls seeping out of a ground stratum where infinite tangent slopes can be encountered, seepage through long aquifer fields with moderate change in groundwater slope is more common for ground flow supplying rivers. An analysis can be conducted to provide the details of such a field flow. All three of these theories start with Darcy-type permeability laws of flow through porous media (linear laws):

$$u = -K \frac{\partial p}{\partial x}$$

$$v = -K \frac{\partial p}{\partial y}$$

$$w = -K \left(\frac{\partial p}{\partial z} + \rho g \right)$$

$$\frac{\partial u}{\partial x} + \frac{\partial v}{\partial y} + \frac{\partial w}{\partial z} = 0$$

where $u, v, w = x, y, z$ components of velocity, ρg is the gravity potential, and K is the Darcy coefficient, a measure of ground permeability.

The pressure field is Laplacian:

$$\frac{\partial^2 p}{\partial x^2} + \frac{\partial^2 p}{\partial y^2} + \frac{\partial^2 p}{\partial z^2} = 0$$

In the Dupuit–Forscheimer theory, the w velocity is neglected and the free surface is assumed to be parabolic. In our modified theory (Iberall et al., 1961–1962; Iberall and Cardon, 1980–1981), there is a small, higher-ordered vertical velocity,

$$w = -K\rho g \left(\frac{\Delta h}{L}\right)^2$$

where Δh is the differential height of the aquifer's surface from the crest of the ridge to the river and L is the horizontal path length.

Water percolates into the ground and joins the prevailing water table, nearly along a line of steepest descent from the ridge to the river segment. Thus, associated with each river segment there is a strip of drainage into the water table. Assuming that we know the ground permeability by empirical determination, or even as a crude average for the land mass, what can we say about the river flow, the runoff from the land?

Along a small strip of land within a river valley, there is a roughly constant intensity of rainfall. If none of it evaporated, the entire rainfall would have to appear in the river. But when the ground flow is computed from that total intensity, it is found that the water table would break out at a level much higher than the river. So we must decrement the net rainfall intensity until the water table breakout is found to occur at the marked position of the river. That *net* intensity that is available for river flow is given by rainfall less evaporation. The latter is that portion of the rainfall intensity that had to be decremented. Thus, we find that the river is defined as the contour at which the water table breaks out at the ground level. There are small adjustments for river bank properties and for the level at which the water table is found at ridges, but this model gives approximately valid answers (Iberall et al., 1961–1962; Iberall and Cardon, 1980–1981).

The theory was tested both for major river valleys and for individual portions of rivers. It would be expected to work down to fairly small scale, but only with increasing "noise." By use of this theory, I have been able to account for the fact that the incremental change of river flow from point to point is determined by the net rainfall integrated over its drainage basin. In effect, therefore, the existing average river flow is an independent variable. If rainfall is also independent, then either the evaporation or the average humidity of the region is a dependent variable.

If we ask what determines the existing or current river flow, the answer is yesteryear's rainfall (that of a previous epoch). The causation is both a long-range geological process that keeps changing the earth's form and meteorological processes of the earlier epoch that established the form of rivers and their conformity to the water table, which is seldom far from the earth's surface, unless the land becomes extremely arid. If we attend to erosion (Iberall and Cardon, 1980–1981), we find that earth and river systems remain conformal after continental uplift by dual erosion processes, above and below ground (at about a total erosion time scale to wear down a continent of about 10^8 years).

Some Additional Features of Rivers

There is much more to a theory of rivers and how they fit the topography and lithospheric geology. Just a few more details are added here with regard to

2. On Rivers

the time-averaged hydrological cycle. Bernoulli's theorem,

$$v^2 = 2gh$$

might leave an impression of water plunging precipitously, like a waterfall, from mountain ridges to oceans. This is not the case. An "effective" viscosity, dependent on the bed load, determines the river velocity. If the water velocity is high as a result of heavy, brief rainfalls or floods, its scouring action is high. The river quickly overflows its banks (rising water table) and begins to scour. In the near-equilibrium time scale of many years, the river scours out a river valley adapted to overflow, which produces only relatively low river velocities. (Young rivers, newly cut in rocky regions with little bed load, will exhibit much higher yearly variation in velocity.) Most rivers exhibit a low range of average velocities, e.g., 1–10 ft/sec, that correspond to their bed-load carrying competence. [In fact, it is that carrying competence, as developed first in a theory by DuBoys (1879) for a critical shear threshold, that has become a hallmark of non-Newtonian rheological flow.]

Using the settling velocity (hydrodynamic drag measure) as a criterion, one begins to find competence to move sand (particle size 0.1 to 2 mm), gravel (0.1 to 2.5 in.), or cobbles (2.5 to 10 in.) in the river velocity range of 1 to 10 ft/sec.

A variation in flow velocity over that full range may occur during any period on the order of 25 to 50 years. The details of cross-sectional form (variation of width and height with flow) of a river in a river valley depend very strongly on details of the scouring process. To some extent, the morphology can be accounted for (Iberall et al., 1961–1962). At low carrying competence (low velocity of 1 ft/sec), a river scours a series of meanders along its average smoothed pathline at the bottom of the river valley. The arc length of the river is thereby appreciably longer than the center pathline of the valley. These meanders are continuously being resculpted by the abrasion forces of the bed load. This action helps keep the flow velocity down.

An "effective viscosity," defined by Richardson (1950), depends upon the fact that a suspended bed load is carried by the motion of particle saltation whose density can be described by a concentration law of atmospheres. Richardson showed the following linear law, with concentration, c, for different-sized grains:

$$\ln c = a - (v/A)z$$

where v is particle velocity (i.e., free-fall velocity) and A is the coefficient of eddy viscosity ("Austausch coefficient"):

$$A = \tau \left/ \frac{dv}{dz} \right.$$

where τ is the shearing stress rate.

Richardson's data thus suggest an apparent viscosity, μ, of a mixed bed load that approximates an Arrhenius law

$$\mu/\mu_0 = e^{bc}$$

where b is a constant and μ_0 is the viscosity of particle-free medium concentration ($c = 0$).

Within the bed load boundary layer, the concentration may be expected to be fairly appreciable and nearly constant. A reasonable correlation with shear rate ($\tau = g$) is

$$\frac{\mu}{\mu_0} = B \left/ A \frac{dv}{dz} \right.$$

where B is a constant and

$$v = \frac{B}{A} \frac{\mu_0}{\mu} z$$

The local velocity gradient thus determines the anomalous (effective) viscosity. However, the concentration range, up to the largest particle, determines the boundary layer thickness (the thickness is the layer that carries a concentration of, say, 50% by volume). The free-fall velocity, v_0, for spheres in a turbulent region is approximately given by the following equation:

$$v_0 = \left[\left(1 - \frac{\rho_w}{\rho_p}\right) 0.03 \, g \, d \right]^{1/2}$$

where ρ_w is density of water, ρ_p is particle density, and d is particle diameter.

The anomalous viscosity represented by a bed load is a measure of the slope of the river and, thus, a measure of the amount of meander relative to the straighter river path that cuts through the meanders. Thus, the meander path arc-length, relative to the smoother river path arc-length, can be used to estimate the bed load, or conversely, the bed load can be used to estimate the meander length along the lines of steepest descent associated with the river.

Summary

Readers who wish to compare this discussion with a more standard geological presentation may examine Leopold et al. (1964), who refer to Einstein's early contribution on the meandering of rivers. This chapter has shown that a few fluid mechanical-thermodynamic laws and principles go a long way toward accounting for many of the interesting, average properties of a river system and of its coupling to other systems. The complex form of rivers arises because of the interactions of gravity force on the fluid flow and because of the drag force of a bed load. The overall, self-organization of the earth's surface must also include the following two processes: (1) "above water [sea level] the relief [of continental landscape] suffers continuous attack, mostly from the solar-powered water cycle" (Wood, 1980); (2) a *production* of continents and new mountain ranges results from orogenic, tectonic movement produced by thick sedimentary piles where the crust is thin and sinking on a boundary between continent and ocean (Wood, 1980; Iberall and Cardon, 1980–1981). In other words, the generation of rivers begets the generation of continents. And the generation of continents begins the generation of new river systems! Volcanoes help out, of course (Goss and Wright, 1980).

In terms of principles of homeokinetic physics (Chapter 27), the complex forms of rivers are born because of the entwinement of gravity force on the molecular constituents of the flow field and the mechanical drag and abrasion force developed by bed load constituents (Principle 4). The river constituents are tied up in time-delayed, internally complex diffusive modes (definition of a complex homeokinetic system). As a result of these forces and the emergent modes, the river system is born, it lives, and it dies (Principle 5).

References

Charney, J., P. Stone, and W. Quirk (1975) Drought in the Sahara: A biogeophysical feedback mechanism. *Science* **187**:434–435.

DuBoys, P. (1879) Le Rhone et les Rivières à Lit Affouillable. *Ann. Ponts et Chaussées* (Series 5). **18**:141–145.

Goss, I., and J. Wright (1980) Continents old and new. *Nature* **284**:217–218.

Iberall, A., and S. Cardon (1980–1981) Contributions to a Thermodynamic Model of Earth Systems, August 1980, November 1980, February 1981, Quarterly Report to NASA Headquarters, Washington, D.C., Contract NASW–3378.

Iberall, A., S. Cardon, and H. Schneider (1961–1962) A Study for the Physical Description of the Hydrology of a Large Land Mass Pertinent to Water Supply and Pollution Control, Four Quarterly Reports for DWSPC, Public Health Service, Department of Health, Education, and Welfare, Contract SAph 78640. [Division currently in the Department of the Interior.]

Jakob, M. (1949) *Heat Transfer*, Vol. 1. Wiley, New York.

Leopold, L., M. Wolman, and J. Miller (1964) *Fluvial Processes in Geomorphology*. Freeman, San Francisco.

McAdams, W. (1954) *Heat Transmission*. McGraw-Hill, New York.

Muskat, M. (1937) *The Flow of Homogeneous Fluids Through Porous Media*. McGraw-Hill, New York.

Richardson, E. (1950) *Dynamics of Real Fluids*. Arnold, London.

Scheidegger, A. (1957) *The Physics of Flow Through Porous Media*. Macmillan, New York.

Wood, R. (1980) The fight between land and sea. *New Sci.* **87**:512–515.

II

Genesis and Evolution of Life

Donald O. Walter and F. Eugene Yates

Biological evolution, whose progress from our present viewpoint seems so clearly oriented, must long ago have organized itself, oriented itself, and got under way, without the help of an organic chemist. Some steps in the evolutionary process are fairly well understood but much specific clarification is still being worked out. Morowitz and Orgel have here reconstructed different, previously unelaborated steps in the prebiotic scenario. Morowitz considers physical ("hardware") principles and constraints that might well have contributed to the natural selection of a restricted network of chemical reactions among early minerals, oceans, and atmosphere. Consideration of the interactive, cumulative effects of those principles and constraints reduces the list of reactions that might be plausible.

In discussion of a slightly later prebiotic step, Orgel proposes a solution chemistry in which enough self-reproducing (or approximate self-reproducing) chains of ribose nucleic acid could have grown long enough to become the material on which further, "higher-level" naturally-selective steps could begin to act. As he says, now that he has proposed one plausible mechanism, several others will probably be found.

Thus, some initial problems are brought closer to solution, and we can turn to later steps. The contribution of Schuster and Sigmund combines ratiocination with simulation to show ways in which Orgel's sufficiently long macromolecules could have interacted both geometrically and chemically (in ways like those suggested by Morowitz) to create vesicles that could have experienced increasingly effective kinds of natural selection, leading to their (still prebiotic) evolution into protocells. Terrestrial chemistry has now spontaneously transcended and brought its chemicals together into entities like cells.

DONALD O. WALTER • Clinical Electrophysiology Laboratory, Neuropsychiatric Institute and Hospital, University of California, Los Angeles, California 90024. F. EUGENE YATES • Crump Institute for Medical Engineering, University of California, Los Angeles, California 90024.

Subsequently, cells enrich their design, and multicellular plants and animals begin to arise. The rest of the story is usually told by the "modern synthesis" of Darwinism and Mendelism. Building on nongradual chromosomal processes often long known, but not considered central to the story, Gould presents the "punctuated evolution" view of speciation (the origin and fate of species). Step-like changes in the character and the efficiency of natural selection were noticed by Schuster and Sigmund, as emerging in their prebiotic simulations. Gould hypothesizes that analogous processes commonly occur in biotic evolution. He also calls our attention to a commonly overlooked constraint on evolutionary change, i.e., an animal is so complicated that many changes are lethal or disabling in one or more of its stages of development or life situations. Therefore, the range of changes *that can survive* is narrowed by the requirement that subsystems must remain adequately compatible with each other, so the whole animal can stay alive and survive competition. In Chapter 9, Goodwin exploits this same concept in a different application.

The authors do not address the current disagreement with the "modern synthesis" that claims that some microevolutionary process, itself not random, must have shaped macroevolution.

In a valuable article contrasting with the chapter by Gould, Stebbins and Ayala (1981) rebut challenges to the modern synthesis, pointing out that we do not know for sure whether the process of evolution is better described by a gradual or a punctualist model. Both modes of speciation may have occurred, and neither is inconsistent with the modern synthesis. The modern synthesis insists that microscopic, genetic changes underlie the evolution of organisms. These changes may affect genes or chromosomal arrangements of genes. Natural selection operates under a variety of constraints, including those arising from the preexisting structures and functions of an organism. Not all genetic changes need be adaptive. For example, protein polymorphisms are pervasive and according to some, they are largely neutral, at least initially. The "selectionist" versus "neutralist" views about molecular evolution are hypotheses that can compete within the framework of the modern synthesis. Speciation as a macroevolutionary phenomenon occurs on a time scale much greater (i.e., slower) than does molecular microevolution. A most difficult question concerns the dependence (or lack thereof) of macroevolution on microevolution. That is, can we extrapolate from microscopic allelic substitution in the genome to changes in populations that lead to reproductive isolation of a subgroup? What is the relative role of allelic substitution, occurring over short time periods, compared to the large-scale phenomena affecting different species, genera, and higher taxa, over long periods? Stebbins and Ayala answer this question by rephrasing it according to the views of various types of reductionism (see Chapter 16 and Ayala and Dobzhansky, 1975). They see three separate questions:

1. Have microevolutionary processes operated throughout the different taxa in which macroevolutionary phenomena are observed?
2. Are the microevolutionary processes such as mutation, chromosomal

change, random drift, and natural selection sufficient to account for morphological changes observed in higher taxa?
3. Can large-scale evolutionary trends and patterns be predicted from knowledge of microevolutionary processes?

They conclude that macroevolutionary patterns, whether gradual or punctuated, are compatible with microevolutionary processes as we currently understand them and, therefore, cannot be derived logically from them. No current microevolutionary hypothesis entails any particular macroevolutionary theory. Furthermore, when correctly stated, the modern synthesis tolerates all current versions of microevolution, as well as either mode of macroevolution. From the perspective of Stebbins and Ayala, macroevolutionary arguments and microevolutionary arguments would both be seen as special cases easily accommodated by the modern synthesis of Darwinism and, therefore, unnecessary additions.

The reader who examines all three discourses will come away with a strong sense of the useful ferment within modern biological science. The points of disagreement do not obscure the unanimous agreement on the fact of evolution, nor the unanimous regret that the general public is once again being assailed by the dangerous fantasies and misrepresentations of the "creationists." Creation did occur, but as a process of natural self-organization, the scientific understanding of which is the subject of this book.

We do not yet have an adequate scientific theory of evolution. But we do begin to see many of the requirements of such a theory. The theory must connect microevolutionary processes and macroevolutionary processes, and in that sense it would have to be a hierarchical theory in time and space. The theory will have to show in what way environmental changes become sources of "information" to the genome—a question raised formally 25 years ago by Kimura (1961). Both competition and cooperation among organisms inhabiting common environments will have to be accommodated in the theory. Even among relatively simple microorganisms occupying the same niche we find generalists and specialists; we find vicious attempts to poison neighbors; we find predation and parasitism; we find commensal and mutual interactions with the avoidance of competition (Frederickson and Stephanopoulos, 1981). The challenge is to provide a theory broad enough to cover the facts, yet formal enough to permit us to base models on theory that can be checked for correspondence with models of data, according to set theoretic rules (Suppes, 1969). We have a long way to go.

The glossaries added by the editors to Chapters 5 and 6, and an account of some simplified basic principles of molecular biology in Chapter 8 (Section III), will help nonspecialist readers of both sections.

References

Ayala, F. J., and T. Dobzhansky (Eds.) (1975) *Studies in the Philosophy of Biology*. University of California Press, Berkeley, pp. vii–xvi.

Frederickson, A. G., and G. Stephanopoulos (1981) Microbial competition. *Science* **213**:972–979.
Kimura, M. (1961) Natural selection as the process of accumulating genetic information in adaptive evolution. *Genet. Res.* **2**:127–140.
Stebbins, G. L., and F. J. Ayala (1981) Is a new evolutionary synthesis necessary? *Science* **213**:967–971.
Suppes, P. (1969) *Studies in the Methodology and Foundations of Science.* Humanities Press, New York.

3

A Hardware View of Biological Organization

Harold J. Morowitz

ABSTRACT

This chapter focuses on the nature of the physical ordering principles that presumably were operative in the prebiotic world and played a role in structuring the chemistry of the evolving planet. In looking at biological organization, we must first understand the hardware before we can move to the software. Examining the physics underlying prebiotic chemistry, we see a number of examples of high organization: vesicle formation, electron and proton coupling between chemical reactions, Pauli's exclusion principle applied to reacting electrons, and the principle of detailed material balances in a stable, recycling system.

Because of the specific strength of earth's gravity, hydrogen slowly escaped from our atmosphere, thus leading from a reducing to an oxidizing environment. Because of the spectrum of solar radiation, electronic transitions are readily excited. Without external planning, a particular set of reaction networks are naturally selected within such an environment: those involving primarily carbon, hydrogen, nitrogen, oxygen, phosphorus, and sulfur (CHNOPS). The initially reducing environment favors production of a "primordial oil slick" on the earth's oceans. This slick is likely to generate amphiphilic molecules that form into bilayer vesicles whose inner aqueous contents can easily differ from its outer aqueous surround. A plausible system for driving synthesis of pyrophosphate in such vesicles is illustrated. In conclusion, it seems that a simple hierarchy of physical organizing features underlies the spontaneous formation of the complex global biosphere, and these physical conditions are now reasonably well understood.   —THE EDITOR

The type of order to be discussed in this chapter has its genesis deep in the structure of matter. It relates to the distribution of electrical charge among electrons, protons, and heavier nuclei, and the quantum-mechanical rules of interaction among these particles. The motivation for considering biological order in this

---

HAROLD J. MOROWITZ • Department of Molecular Biophysics and Biochemistry, Yale University, New Haven, Connecticut 06520.

context comes from an overview of molecular biology and the dependence of function both on precise molecular structure and on much less highly specific phase behavior. We are inquiring about the nature of the physical ordering principles that were operative in the prebiotic world and played a role in structuring the chemistry of the evolving planet.

My viewpoint will be at variance, on philosophical grounds, with others in this volume; indeed, it departs somewhat from my own earlier views (Morowitz, 1968). The present approach stresses that in looking at biological organization, we must first understand the "hardware" before we can move to the "software." The study of hardware involves organic chemistry, colloid chemistry, and quantum chemistry. On first inspection, it is often messy and inelegant; however, appearances are sometimes deceiving. We shall gain much by concentration on molecular detail. This chapter reviews examples of rather high degrees of organization that emerge from relatively simple rules in the underlying physics. These examples and rules include vesicle formation, electrochemical and protochemical[1] coupling, reaction graphs, the Pauli exclusion rule, and the cycling theorem.

We start with the empirical generalization that all living cells possess a lipid bilayer plasma membrane that separates the aqueous interior of the cell from its aqueous environment. To search out the underlying organizational principle in membrane structure, we examine substances of biological interest and find that the dielectric constant ranges from 2 to 80. Two chief components enter into this parameter: the electron polarizability and the permanent dipole moment, which comes from the ionic contribution to covalent bonds, as well as the net charges arising from acid-base dissociations. The relationship between the relevant quantities is usually written as

$$\frac{\kappa - 1}{\kappa + 2} = \frac{4\pi n}{3}\left(\alpha + \frac{P_0^2}{3kT}\right) \tag{1}$$

where κ is the dielectric constant, n the number of molecules per unit volume, α the molecular polarizability, and P_0 the molecular dipole moment (Page, 1952).

The phase separations normally seen in biological materials are predominantly determined by the electrical asymmetry just noted. In general, miscibility will depend on substances having similar dielectric behavior. The structure of water is particularly interesting (Edsall and Wyman, 1958), because almost all biochemical activity takes place in aqueous surroundings. The water molecule in the gas phase consists of two hydrogen atoms covalently bonded to an oxygen with a 105° angle between the two O–H bonds. The structure and charge distribution are shown in Figure 1. The feature of water of particular importance in its solid and liquid states is the hydrogen bond that forms between neighboring water molecules (Figure 2). This bond results from the electrostatic attraction between the positively charged hydrogens and the neighboring negatively charged oxygens. The rather tight water structure resists disruption; consequently, materials are appreciably soluble in water only if they are sufficiently electrically asymmetric

---

[1]Using protons coupled to acid–base dissociations through solid-state proton conductors.

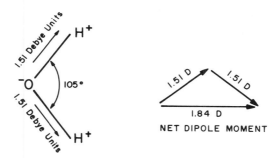

FIGURE 1. Structure and charge distribution of a water molecule in gas phase.

to have strong local electrostatic interactions. Electrically symmetric molecules, such as hydrocarbons, require energy to disrupt the water structure but do not yield much solvent–solute electrostatic interaction energy. Hence, such molecules are relatively insoluble in water. All of this has been known for some time and has been expressed in the old adage "oil and water don't mix." Most chemical phase separations, particularly in aqueous environments, are, as already noted, the result of electrical polarizations. Phase separation is the most elementary macroscopic spatial structure that follows from normal covalent chemistry. The structures are stable and follow, in every sense, the rules of equilibrium thermodynamics (Gibbs, 1906).

A class of molecules of special interest are the amphiphiles. These have separate regions that are polar and nonpolar. The existence of such structures leads to elaborate spatial ordering because of the formation of micelles, coacervates, vesicles, and a wide variety of other colloidal forms. Everything from bearnaise sauce to mitochondria depends on the properties of amphiphilic molecules. Note that these colloids are also equilibrium structures found at local free energy minima of these systems. In general, they arise from chemical reaction rules rather than from any special kinetics. They have been studied in considerable detail (Israelachivilli *et al.*, 1976; Gershfield and Tajima, 1977).

Of particular interest are molecules with a polar head and a long nonpolar tail. In aqueous solution these molecules may form monolayers at an air-water interface or polar lipid bilayers with the nonpolar regions in contact with each other and the polar regions facing the water structure. These are planar structures that attain a free energy minimum when they close into vesicles. Again note that this kind of structure follows directly from considering the minimum free energy

FIGURE 2. Hydrogen bond between two water molecules in liquid or solid phase.

states of systems of this type.

The role of lipid bilayer vesicles in biogenesis has been given some attention in recent years. Lasaga *et al.* (1971) have suggested that the primitive planetary oceans were covered with an alkane layer, "the primordial oil slick." Onsager (1974) has discussed how such a layer might lead to biologically interesting structure, and Folsome and Morowitz (1969) have demonstrated the synthesis of membranelike bilayers from the irradiation of such a slick with short-wavelength ultraviolet light. In any case, the origins of closed lipid bilayer vesicles follow directly from the physical chemistry of a class of amphiphilic molecules likely to have been present on the early planet. The same principles of membrane formation are likely to be operative on the present terrestrial surface (Stoeckenius and Engelman, 1968).

The presence of "bags" with a skin thickness in the range of molecular dimensions provides a structure whose functional potential can be utilized in subsequent biochemical processes. The importance of this particular hardware cannot be overstressed. Its existence follows in a highly probable way in an aqueous environment containing carbon compounds in a reduced, but not totally reduced, state. The membrane is a thermodynamically likely structure, and the thermodynamics of its formation is the same in highly developed eukaryotic cells as it was in early prebiotic sacs. Membranes are the least specific of biological entities because they embody this primitive yet highly efficient manner of generating a cell component. Even in mammals, the detailed fatty acid compositions of membranes are influenced by diet. These structures are thus seen to be much less specific than proteins and nucleic acids, which are under rigorous genetic control.

The functions that become available with the advent of closed membranes are the coupling of charge separation with chemical reactions. The lipid bilayer vesicle gives rise to a three-phase system: exterior aqueous, membrane, and interior aqueous phases. The configuration, consisting of an insulator between two conductors, is an electrical circuit element with a capacitance of about 1 $\mu F/cm^2$. A relatively small charge separation across a membrane leads to a substantial voltage and a very high field strength. The membrane can serve as a coupler between microscopic chemical events and a macroscopic potential difference between two phases. The existence of such a potential allows the following types of processes:

1. If electron conductors are available, the potential can be coupled to oxidation–reduction reactions.
2. If proton conductors are available, the potential can be coupled to acid–base dissociations.

In stressing the importance of transmembrane potentials in the origin of life, we are trying to incorporate fully the results of contemporary bioenergetics into our view of biogenesis. In 1941, Lipmann established that adenosine triphosphate (ATP) was a universal energy intermediate in cellular metabolism. Energy processing led to the synthesis of ATP. This mode of intermediate-stage energy storage is ubiquitous in the living world. A number of other widely occurring energy molecules were discovered, including reduced NAD (nicotinamide adenine dinucleotide), acetyl coenzyme A, and other nucleotide triphosphates; but the central role in energy processing almost always involved ATP. The synthesis of this molecule in aerobic systems was found to be associated with a membrane-bound sequence of oxidation–reduction enzymes. The chemiosmotic hypothesis set forth by Mitchell (1961) postulated an intermediate energy storage in terms of a transmembrane electrochemical potential of hydrogen ions. The role of this type of energy storage is now widely assumed in mitochondria, bacteria, and chloroplasts. Membrane-associated oxidative phosphorylation and reductive dephosphorylation connect the two methods of storage into one coherent scheme.

Recent advances in the study of photosynthesis (Witt, 1979) indicate that the initial event involves a charge separation and transmembrane transport of hydrogen ions. In *Halobacterium halobium* purple membranes, the photon absorption is directly related to transmembrane hydrogen ion pumping (Oesterhelt and Stoeckenius, 1973). Thus, the present-day energy inputs into the biosphere are all associated with charge separation across lipid bilayers.

Halobacterium provides a particularly illuminating example in that the cells normally respire oxidatively. Under low O_2 tension, the synthesis of photopigment is induced and purple membrane is formed. Absorption of photons leads to the transport of hydrogen ions from the interior of the cell to the exterior. The back flow of protons is associated with the synthesis of ATP, as in the case of mitochondria. What is in fact taking place is anaerobic photophosphorylation. It results from a phase separation that allows photochemical charge separation to give rise to thermodynamic potentials and the synthesis of ATP linked to the back flow of protons.

The possible flow of protons in biological material has recently been investigated in some detail and appears to be of general applicability in cellular systems (Morowitz, 1978a; Nagle and Morowitz, 1978). Photochemical processes in effect allow the reactions to take place at a low pH and the products to be released at a high pH. The thermodynamically favorable transport of protons is thus coupled to the thermodynamically unfavorable chemical reactions. Of particular importance in this case is that the reactants and products have a sufficient number of acid-base groups to allow this type of process to take place.

The polyphosphates are a particularly favorable set of compounds in that they present, in aqueous medium, a range of pK's spanning the entire region from strong acids to strong bases (Table 1). The detailed thermodynamics of the photochemical synthesis of phosphate bonds has been discussed in detail in an earlier publication (Morowitz, 1978a). The possibility thus presents itself that primitive systems stored energy by the photosynthetically driven, photochemical formation of pyrophosphate and higher phosphates from orthophosphate. Figure 3 diagrams such a system with the following three features:

1. There are membrane-bound chromophores that separate charge in the photoexcited state. Among the compounds formed in a CHNOPS (carbon, hydrogen, nitrogen, oxygen, phosphorus, sulfur) system, one would expect nonpolar molecules with π electrons. These dissolve in the lipid bilayer and become oriented because of the asymmetry caused by a difference in the radius of curvature

TABLE 1. Acid–Base Dissociation Constants of Polyphosphate Compounds[a]

pK >	Orthophosphate	Pyrophosphate	Teraphosphate
pK_1	2.14	1	Completely dissociated
pK_2	7.10	2	1
pK_3	12.32	6.57	2.30
pK_4		9.62	6.26
pK_5			8.90

[a]Data from Van Wazer (1958).

across the membrane. Under appropriate conditions the excited state will lead to charge separations. This result can be seen in the extreme case of the photoionization of an asymmetric molecule.

2. A transmembrane, proton-conducting channel, presumably made of a continuous chain of hydrogen bonds, acts to convert the charge separation into a protonmotive force that drives the synthesis of pyrophosphate within a vesicle (Figure 3). The energy of this phosphate ester bond is on the order of 3 kJ, which is the same order of magnitude as the dissociation free energies measured by the pK:

$$\Delta G = 2.3RT(pK) = 5.8(pK) \text{ kJ/mole} \qquad (2)$$

Thus, pyrophosphate has a labile ester bond, energetically commensurate with protomotive forces. The work of Baltscheffsky on *Rhodospirillum rubrum* suggests that the photosynthetic production of pyrophosphate may still be of significance in some contemporary bacteria (Baltscheffsky, 1977).

3. Vesicles with membranes, showing some degree of selectivity and containing high-energy phosphate bonds, will select from a very large chemical network a much smaller network, determined by the reactions involving phosphate-group transfer. In addition, the phosphorylated reaction products tend to stay within the lipid bilayer membrane because of their highly polar character. We thus envi-

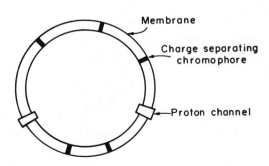

FIGURE 3. Diagram of a primitive system for photosynthetically driven formation of pyrophosphate.

sion an anaerobic photophosphorylator that will direct the intermediate chemistry. Now the problem of biogenesis can be viewed in a much more organized context: the operating principles of the molecular hardware are sufficient to direct the system along highly specific pathways toward biogenesis.

The creative potential of directed reaction networks can be seen in the analysis of the Urey–Miller class of experiments on random synthesis. These experiments, which are aimed at the problem of biogenesis, have many common features. They all start with a reducing atmosphere and low-molecular-weight compounds of C, H, N, and O; they all involve the inflow of electron-exciting energy; and they all result in a class of compounds that looks promising from a biological point of view. Indeed the reaction products are quite similar, somewhat independent of the details of the starting materials or the energy sources. This result has made these experiments of limited value in deducing the most probable prebiotic chemical pathways. It is perhaps surprising that such random experiments should lead to appreciable quantities of any compounds at all, because on a strictly Laplacian probability argument we would expect an entire Beilstein catalogue of reaction products, each in vanishingly small concentrations. The reason that this does not happen is instructive in showing the extreme ordering power of simple rules covering the behavior of system components. These experiments may be fundamental to the origin of life, not because of their specific outcomes but because of their general features!

Reaction graphs are convenient ways to analyze the pathways and to explore the system. Picking out sub-graphs of interest from a very large graph of possible reactions is one way of demonstrating organization (C. Pabo, M. Schmertzler, and H. J. Morowitz, personal communication, 1974). Attention is focused on gas-phase reactions in a reducing atmosphere. This condition prevails in most published experiments of this type. Such gas-phase systems, under input of electron-exciting energy, will be dominated by free radical reactions. Kochi (1973) has detailed the sense in which gas-phase chemistry is dominated by free radicals as reactive intermediates.

For definiteness we begin by postulating the simplest, fully-reduced starting system: CH_4, NH_3, and H_2O. As energy flow begins, the only accessible free radicals are those involving the release of hydrogen which is bonded to carbon (H–CH_3), nitrogen (H–NH_2), and oxygen (H–OH). The details will depend on the rates for those processes; however, lacking detailed data we can compare bond dissociation energies:

Bond	Energy (dissociation), kJ/mole
H–CH_3	435
H–NH_2	460
H–OH	498

Because of the similarities of these values, we can assume the same order of magnitude for concentrations of the free radicals · CH_3, · NH_2, and · OH. Because the concentration of starting materials would be much greater than that of free radicals, most reactive collisions would reasonably have to involve hydrogen exchanges with stable molecules such as:

$$\cdot C'H_3 + CH_4 \rightarrow C'H_4 + \cdot CH_3$$
$$\cdot CH_3 + H_2O \rightarrow CH_4 + \cdot OH$$

(the primes identify a particular C atom). Such reactions would at most entail a shift in the relative free-radical populations; they could not affect the number of different types of compounds involved.

In this system only the rarer radical–radical collisions can lead to new compounds. Following is a list of six possible compounds generated directly from the starting materials:

- C_2H_6
- CH_3OH
- CH_3NH_2
- NH_2OH
- N_2H_4
- H_2O_2

Note, however, that the bond energies of these molecules differ significantly:

Bond	Energy, kJ/mole
H_3C-CH_3	345
H_3C-OH	358
H_3C-NH_2	305
$HO-OH$	147
H_2N-OH	222
H_2N-NH_2	163

The lower dissociation energies of the N–N, N–O, and O–O bonds suggest that they readily undergo fission soon after formation. Hydrazine, hydrogen peroxide, and hydroxylamine are known to be relatively unstable compounds.

If the compounds we have discussed are treated as points (vertices) in a space and the reactions are represented by lines (edges), then the pattern of reactions can be formally represented as a graph. Figure 4 represents such a graph for a system that starts with CH_4, NH_3, and H_2O and synthesizes the proposed prebiotic starting materials: cyanide, formaldehyde, and aliphatic carbon chains. The graph is based on the following assumptions:

1. The dominant feature in gas-phase reaction is single-atom abstractions. None of the steps shown in Figure 4 involves more than a single-atom transfer per step. Reactions of the following type are not included:

$$\cdot CH_3 + H_2O \rightarrow H_3COH + \cdot H \qquad (3)$$

Benson (1973) offers support for this characterization of radical reactions.

3. A Hardware View of Biological Organization

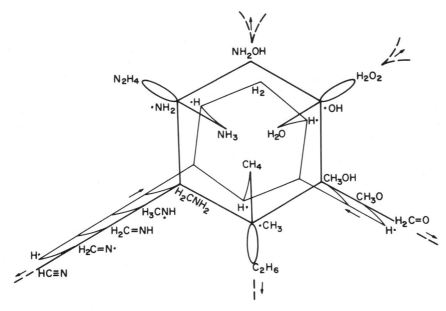

FIGURE 4. Reaction graph of a CHNO system under reducing conditions. The most probable free radical reactions structure the system and lead to the predominance of a small group of compounds.

   2. Evolution of hydrogen is a major process. Hydrogen atoms are abstracted as free radicals in first-order reactions. Subsequently, the remaining radicals stabilize themselves with the loss of another hydrogen or by bonding with other radicals. The former facilitates the formation of unsaturated bonds. In either case the evolution of relatively stable H_2 molecules constitutes a driving force for the system by keeping it from equilibrium and preventing the saturation of unsaturated bonds.

   If this scheme is used as a model for early planetary chemistry, then the loss of H_2 from the planetary atmosphere is a strong driving force leading to a more oxidized state of the system. For the earth, H_2 is the only gas-phase molecular species to have a mass small enough that appreciable quantities are lost from the planetary atmosphere. Thus, terrestrial chemistry is always slowly driving toward a less reducing state. This drive imposes a direction on the evolution of the planet's chemical features.

   3. The compounds hydrogen peroxide, hydroxylamine, and hydrazine are assumed to be much less stable than ethane, methanol, and methyl amine. They do not stay around long enough to be involved in subsequent free radical reactions. Hydrogen peroxide is a powerful oxidizing agent and hydrazine is a powerful reducing agent. The long-term behavior will be governed by the relative instability of the O–O, N–N, and O–N bonds.

   4. The graph shown in Figure 4 is terminated at chains of two atoms. The purpose of this analysis is to show the directedness imposed by hydrogen atom abstractions. In principle, the graph could be extended to a three-atom chain and higher, although the number of vertices rapidly becomes very large. Note also that

if the gas phase is in contact with an aqueous phase, hydrogen cyanide and formaldehyde will be partitioned into the liquid.

What emerges from this analysis is the fact that the dominance of free radical reactions selects a subgraph from the overall graph, and this subgraph is moving toward lipids through ethane and toward amino acids through the reactants for the *Strecker synthesis*: cyanide, ammonia, and an aldehyde. These results are independent of the energy source and relatively independent of the starting material, due to the reflexivity of the graph. Picking the appropriate subgraph depends on hardware assumptions about the class of allowable reactions.

Next we move to a somewhat more abstract view of how organization emerges as a result of operational rules. The primary example is the Pauli exclusion principle, which in effect shows how all chemical organization is generated from quantum mechanics. The discussion here is heavily influenced by the views of Margenau (1977), who points out that "the exclusion principle says that states which fail to have certain mathematical properties are not realized in nature." The principle has its roots in symmetry considerations and is extra-energetic in an important sense. It can be shown, for example, that atomic lithium would be in the lowest energy state if all three electrons were 1S rather than two 1S and one 2S. Exclusion gives a different answer than energy minimization. The exact requirement may be stated as follows: *all functions representing states of two electrons must be antisymmetric*. This behavior is responsible for chemical binding, the cohesion of solids, and the properties of crystals. The structure seen in the periodic table of the elements follows from this restriction.

In discussing these matters, Margenau writes, "Not too far ahead lies the field of biology with its problems of organization and function, and one is tempted to say that modern physics may hold the key to their solution. For it possesses in the Pauli principle a way of understanding why entities show in their togetherness laws of behavior different from the laws which govern them in isolation."

The final example is the cycling theorems that illustrate the power of network theorems (Morowitz, 1978b) in demonstrating a kinetic structure as distinguished from the kinds of spatial structures we have been discussing. Assume a system of CHNOPS in contact with an isothermal reservoir and otherwise adiabatically isolated from the rest of the universe (Figure 5). The equilibrium state of the system can be precisely determined and will represent some distribution among all pos-

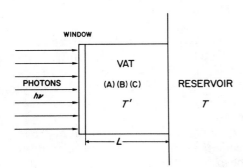

FIGURE 5. A system in contact with an isothermal reservoir. When the window is closed, the system is isolated from the rest of the universe. When the window is open to a photon flux, the system organizes under the flow of energy from the source to the reservoir.

sible compounds of CHNOPS. In some suitable descriptive graph, each point will correspond to a molecular species and each edge will represent a possible flow of material due to chemical reactions. At equilibrium, the flow along every edge is zero because of the requirement for detailed balance.

Next, open the system to a flow of electron-exciting energy and allow it to age until a steady state obtains. Each flow no longer has to be zero, but the sum of currents into each node must be zero. However, the inflow of electron-exciting energy assures us that some nodes must have some negative flows that must be balanced by positive flows. This requirement represents a kind of Kirchhoff's current law for the system. Because there are no sources or sinks for matter, the result above assures us that chemical cycles must be taking place in this system. (Compare these cycles with the great ecological cycles of carbon, nitrogen, and others.)

What emerges is that ecological-type material cycles are a network property of chemical systems pumped by electron-exciting energy. This result follows from the network constraints of conservation at the nodes in the steady state and the necessity of energy inflow being associated with electronic transitions, which lead, of course, to change of chemical species.

We can now begin to see a hierarchy of physical organizing features that underlie the development of active features on the surface of a planet. First, we have the astrophysical features, including solar flux of electron-exciting energy, and a planet of mass and temperature in which a steady loss of hydrogen acts as an evolutionary arrow moving from a reducing to an oxidizing system. Next is a reaction network and a high-probability subgraph that selects a limited chemical domain. The subgraphs are further constrained to those that possess material cycling. Inherent in a system of reduced carbon compounds in an aqueous environment is the tendency for phase separation, leading to lipid vesicles as a fourth state of matter. Phase separation opens up the possibilities of electrochemistry and protochemistry. This possibility permits the coupling of thermodynamic potentials with a specific class of chemical reactions, once again selecting a subgraph. When viewed in this light, the emergence of high degrees of order on an evolving planet looks quite causal and directed. The trajectory of the evolving system may have random features, but the trajectory occurs within a subvolume determined by the physical laws of interactions of the entities. Organization is clearly a property of nonequilibrium systems whose hardware elements interact by restrictive rules. We argue here for the importance of considering the detailed hardware rules as a necessary first step to a more global theory.

References

Baltscheffsky, M. (1977) Biological membranes as energy transducers. In: *Living Systems as Energy Converters*, R. Buvet (ed.). Elsevier, Amsterdam, pp. 199–207.
Benson, S. W. (1973) *Thermochemical Kinetics*. Wiley, New York.
Edsall, J. T., and J. Wyman (1958) *Biophysical Chemistry*. Academic Press, New York, pp. 27–46.
Folsome, C. E., and H. J. Morowitz (1969) Prebiological membranes, synthesis and properties. *Space Life Sci.*, **1**:538–544.

Gershfield, N. L., and K. Tajima (1977) Energetics of the transition between healthier monolayers and bilayers. *J. Colloid Interface Sci.* **59**:597–604.

Gibbs, J. W. (1906) The equilibria of heterogeneous substances. In: *The Scientific Papers of J. W. Gibbs*, Longmans, Green, New York.

Israelachivilli, J. N., D. J. Mitchell, and B. W. Ninham (1976) Theory of self assembly of hydrocarbon amphiphiles. *J. Chem. Soc. Faraday Trans. 2* **1976**:1525–1568.

Kochi, J. K. (1973) *Free Radicals*. Wiley, New York.

Lasaga, A. C., H. D. Holland, and M. O. Dwyer (1971) Primordial oil slick. *Science* **174**:53–55.

Lipmann, F. (1941) Metabolic generations and utilization of phosphate bond energy. *Adv. Enzymol.* **1**:99–162.

Margenau, H. (1977) *The Nature of Physical Reality*. Ox Bow Press, Woodbridge, Conn.

Mitchell, P. (1961) Coupling of phosphorylation to electron and hydrogen transfer by a chemi-osmotic type of mechanism. *Nature* **191**:144–148.

Morowitz, H. J. (1968) *Energy Flow in Biology*. Academic Press, New York.

Morowitz, H. J. (1978a) Proton semiconductors and energy transduction in biological systems. *Am. J. Physiol.* **235**:R99-R114.

Morowitz, H. J. (1978b) *Foundations of Bioenergetics*. Academic Press, New York.

Nagle, J. F., and H. J. Morowitz (1978) Molecular mechanism for proton transport in membranes. *Proc. Natl. Acad. Sci. USA* **75**:298–302.

Oesterhelt, D., and W. Stoeckenius (1973) Functions of a new photoreceptor membrane. *Proc. Natl. Acad. Sci. USA* **70**:2853–2857.

Onsager, L. (1974) Life in the early days. In: *Quantum Statistical Mechanics in the Natural Sciences*, S. Mintz and S. Wedmeyer (eds.). Plenum Press, New York, pp. 1–4.

Page, L. (1952) *Introduction to Theoretical Physics*. Van Nostrand, Princeton, N.J.

Stoeckenius, W., and D. M. Engelman (1968) Current models for the study of biological membranes. *J. Cell Biol.* **42**:613–646.

Van Wazer, J. R. (1958) *Phosphorus and Its Compounds*. Interscience, New York.

Witt, H. T. (1979) Energy conversion in the functional membrane of photosynthesis. *Biochim. Biophys. Acta* **505**:355–427.

4

The Origin of Self-Replicating Molecules

Leslie E. Orgel

ABSTRACT

How could self-replicating molecules have come into being—with sufficient length and concentration to get life started—without the interference of an organic chemist? Formation of the pieces—bases, sugars, and thence nucleotides—is fairly well understood. Hooking them together in strings of eight or ten nucleotides seems not too hard to imagine. The next steps require an understanding of the origins of a self-replicating molecule, and of the genetic code. Only the former is discussed in this chapter.

Strings of 20 to 100 nucleotides seem to have been necessary to get prebiotic chemistry going. The modern genetic code (thought to have been the original code) uses the four bases adenine (A), guanine (G), cytosine (C), and thymine (T) in their nucleotide forms (adenosine, guanosine, cytidine, and thymidine, respectively) polymerized as DNA, with the pairing rules between two DNA chains being: A-T, C-G. In RNA polymers, uracil (U) and uridine replace thymine and thymidine, and the RNA pairing rules are A-U, C-G. Experimentally, RNA-like nucleotide polymers containing only uracil [poly(U)] or only cytidine [poly(C)] as bases have been synthesized. The question was then asked: will poly(U) create strings of various nucleotide monomers containing only A's and do so without the presence of protein catalysts? Conditions were found in which poly(U) did indeed induce formation of strings of nucleotides containing A's, up to lengths of eight units. Poly(C) could not induce chains containing A's, but did induce chains containing G's. Addition of lead ions as inorganic catalysts, or zinc ions, favor longer chains—up to 30 units or more. All the nucleotide monomers being linked must have the same optical isomeric "handedness." Furthermore, the nucleotides must contain ribose, the pentose sugar; deoxyribose does not work. Finally, the modern RNA pairing rules hold: poly(U) induces A-containing chains, but not G-containing chains; poly(C) induces G-containing chains, but not A-containing chains.   —THE EDITOR

This chapter discusses how a self-replicating molecule like a nucleic acid first came to exist on the primitive earth. Although this is essentially a topic in organic

LESLIE E. ORGEL • Salk Institute for Biological Studies, San Diego, California 92138.

chemistry, I shall present it in a manner accessible to readers with no special training in that subject. More technical descriptions of this work, together with references to the literature, are found in an earlier review (Sulston *et al.*, 1968) and a later paper (Lohrmann *et al.*, 1980). (See Chapter 5 for a glossary that will help with some of the terminology in this chapter also.)

I assume that one of the important features of the origin of life on the earth was the appearance of a self-replicating molecule and, lacking any obvious alternative, I also suppose that the first self-replicating molecule was something like a contemporary nucleic acid. I have no very strong views as to whether peptides came into the picture early or not: it is obviously a very important historical question, but one that cannot be resolved on present evidence.

The formation of self-replicating molecules is supposed to involve three distinct problems. The first concerns the way that the pieces from which nucleic acids are composed came into existence on the primitive earth. Stanley Miller (see Miller and Orgel, 1974) and others have found that lightning, volcanoes, solar radiation, and other forces acting upon the primitive atmosphere very likely produced organic molecules. Organic molecules could also have been present in the dust from which the earth accreted, or they could have been formed in meteorites and brought to the earth in them. It is fair to say in review of this work that reasonable, though undetailed, models exist to explain how the five bases that occur in the polynucleotide nucleic acids DNA and RNA—adenine, guanine, uracil or thymine, and cytosine—came into existence. We have a rather good model of how the sugars formed and a modest, but not completely satisfactory understanding of how they may have joined together to form nucleotides, the basic components of nucleic acids. I do not want to underestimate the difficulties that remain: getting these molecules in large enough amounts, excluding other substances that interfere, and preventing dilution of reactants all remain severe problems. But in a very general way, we can see how the pieces might have formed.

Problem two is how to join nucleotides in a purely chemical, non-self-reproducing fashion. This process is not as well understood as nucleotide formation, but nonetheless we do see how strings of eight or ten nucleotides could have formed in relatively small yields and, by extrapolation, we can guess ways in which somewhat larger oligonucleotides could have come into existence. Although the understanding of the random joining of pieces to form the nucleic acids is not as satisfactory as is the understanding of the formation of the pieces, it does not present any very substantial problem, at least in principle.

The third problem, and the one that is the primary subject of this chapter, is the origin of the biological organization, that is, of molecules that not only can come together by chance, but also can go on to make more or less faithful copies of themselves and do other interesting things. The problem of biological organization divides intellectually into two events (chronologically, they may have occurred more or less simultaneously). One is the origin of a self-replicating molecule and the other is the origin of a genetic code, that is, of a mechanism by which nucleic acids came to control the sequences of proteins and thus to control their environment far more completely than they otherwise could have. This chapter is restricted to a discussion of the first of these two events—the origin of

4. The Origin of Self-Replicating Molecules

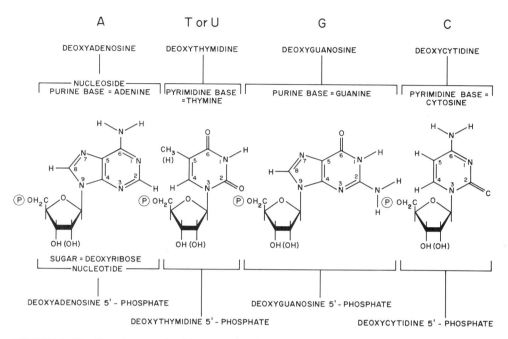

FIGURE 1. The four deoxynucleotides occurring in DNA. In RNA there is an extra OH group on each sugar, and the methyl group of the base thymine is replaced by hydrogen to give uracil. Both of these differences are indicated by parentheses. (Redrawn from Watson, 1970.)

a self-replicating molecule. I leave completely open the question of whether or not peptides had any part in the origin of replication as a historical process on the earth: they play no part in the process described herein.

Figure 1 depicts the actors in this particular play: the four nucleotides consisting of a phosphorylated sugar molecule and a purine or pyrimidine base. The nucleotides are adenosine, thymidine, guanosine, and cytidine. The two component bases, adenine (A) and guanine (G), have two rings and are purines, whereas the other two, thymine (T) and cytosine (C), have single rings and are pyrimidines. These are the nucleotide bases of DNA. In the nucleic acid RNA, which is very similar, each nucleotide contains an extra hydroxyl group in its sugar part, which is critically important to the detailed chemistry but which need not concern us here. Furthermore, in RNA thymine is replaced by a slightly different base, uracil (U).

In nature, these nucleotides are combined to form strings containing up to several million component bases. Figure 2 shows a piece of a typical string with the sequence TGCA (progressing toward the so-called 5′ end of the phosphorylated-sugar polymeric backbone). It is difficult to say just how many nucleotides must be put together to satisfy the needs of prebiotic chemistry. Clearly two or three is inadequate, and probably a thousand is unnecessary. It is usually guessed that an oligonucleotide containing somewhere between 20 and 100 bases is the smallest replicating molecule with much chance of becoming involved in a pro-

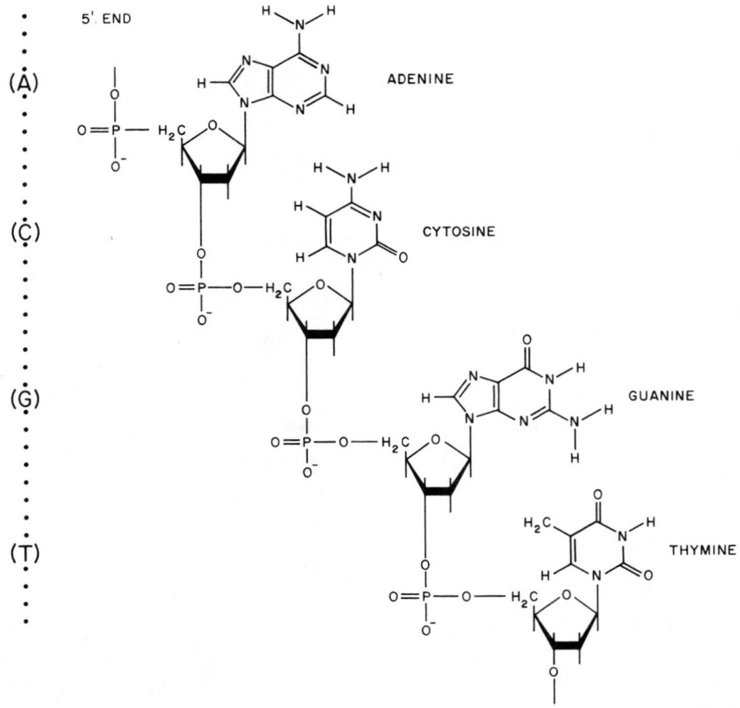

FIGURE 2. A section from a DNA chain showing the sequence ACGT. (Redrawn from Watson, 1970.)

cess of natural selection, but this conclusion does involve a fair amount of uncertainty.

Given nucleotide strings, we may then imagine their replication according to the classic Watson–Crick pairing rules, which require that T (or U) always pairs with A, and G always pairs with C. This principle, which is inferred from many experimental facts, is illustrated for a special case in Figure 3, showing something like an RNA double helix straightened out. It is a very special double helix owing to its sequence: one of the chains carries exclusively U, and the other, A. It has been shown experimentally that if molecules with these two sequences are dissolved in water and mixed together, they wrap around each other to form an RNA helix that has very much in common with DNA. Thus, homopolymers—ones that have only one base—will do exactly the same thing as real nucleic acids. When one of the strands is replaced with monomers, very similar chemistry is observed. This was first shown by two independent groups of workers who produced double and triple helices very much like DNA's, but in which one of the chains is not joined together (Howard et al., 1966; Ts'o, 1974). The chain of U's on the left side of Figure 3 is joined just like a piece of nucleic acid, but the A's need not be joined to form an organized structure. However, there is a price for not having the second chain joined, in that these structures are much less stable.

4. The Origin of Self-Replicating Molecules

FIGURE 3. Part of a poly(U), n times mono A chain. The A's are arranged head-to-tail, ready to be joined.

The structure of DNA characteristically melts at 60 to 100°C, depending on its composition. Organized helices with only a single joined chain melt at temperatures below 30°C. Almost all the experiments I describe here were carried out at 0°C. This is a very important point to note if one is interested in the origins of life, because chemistry of this type could not have proceeded at temperatures much above room temperature.

In Figure 3, all of the A's are organized on a chain of U's. If the vertical line represents the "tail" of a nucleotide, it is clear that each base is in constant orientation relative to the position of its neighbor. If that orientation happens to favor the joining of nucleotides, then the presence of poly(U)—a polynucleotide of uridine—should facilitate the joining of strings of A. Thus, A's, which when floating around in solution have little tendency to join, might be expected to join more readily when organized on a strand of poly(U). This is the principle behind our experiments.

Figure 4 shows the results of an experiment to test this principle. It is a chromatogram of radioactively labeled substances resulting when labelled A monomers are mixed either in the presence or in the absence of poly(U). The peak on the right (upper tracing) represents the proportion of A's that are joined to form strings two, three, or more in length. The important point to observe is that when there is no poly(U) present, only a very small number of A's join together. But

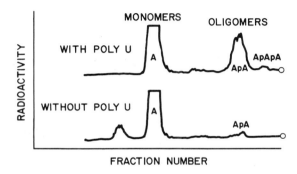

FIGURE 4. Condensation of ^{14}C-labeled A, with cold adenosine 5'-monophosphate (pA), in the presence and absence of poly(U). The labeled peaks represent oligonucleotide products. Clearly, the yields of ApA and ApApA are much larger in the presence of poly(U) than in its absence. (Redrawn from Sulston et al., 1968.)

with the addition of poly(U), about 10% of the A's join to form dimers, and about 1% form strings of three A's joined together. In control experiments we have found that if the poly(U) is replaced with poly(C), then the A's do not join. Poly(U) will do this trick with A's, but not with G's; poly(C) will do it with G's, but not with A's. Thus, we have established that the Watson–Crick pairing rules work in systems of this kind—a pyrimidine base will facilitate the condensation of the complementary purine but not of any other base. I must emphasize, however, that we cannot do this experiment in the opposite direction, using purines to link pyrimidines, because pyrimidines do not really form organized structures with polypurines. We are beginning to find ways around this difficulty by using short oligomers, but this is not discussed in this chapter.

The experiments described above made use of special chemical reagents that certainly could not have existed on the primitive earth. When we tried to replace them with the actual substrates of contemporary nucleic acid replication, e.g., ATP, our experiments did not give much product. This poor yield occurs partly because the reactions are very slow in the absence of enzymes. Extrapolation from experiments that lasted as long as a year suggests that in a sufficiently long time (perhaps 10 years), a reasonable yield of products could be obtained, but not as good as the yields obtained in experiments described below. It should be recalled that 10 years is a very short time on the geological scale, even if it is long enough to deter a graduate or postdoctoral student.

The compound we finally selected (adenosine 5'-phosphorimidazolide, ImpA) is shown in Figure 5. It is an analogue of ATP, with an imidazole group substituted for the pyrophosphate of ATP. There is no particular subtlety to the choice of this compound: it is simply a convenient compound that reacts at a rate that allows an experiment to be completed in a day or two instead of a year or two. We chose it from among many candidates because it resembles compounds found in modern biochemistry. Figure 6 illustrates the manner in which a chain is incremented by joining a pA (adenosine 5'-monophosphate) and one of these ImpA molecules.

When ImpA was substituted for ATP in our experiments with poly(U) templates, it produced almost an order-of-magnitude increase in the yield of resulting "A" chains: chains of up to eight units appeared in yields of more than 1%. With guanosine 5'-phosphorimidazolide (ImpG) condensation on poly(C), we did even better: 1% yields were obtained for chains up to ten units long.

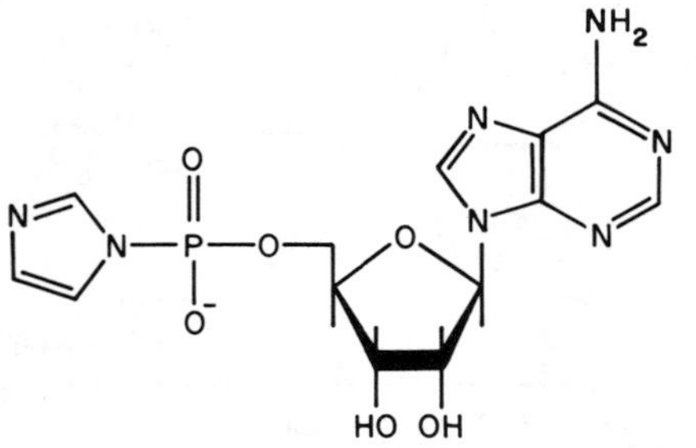

FIGURE 5. The structure of adenosine 5'-phosphorimidazolide, the activated derivative of A used in our work.

4. The Origin of Self-Replicating Molecules

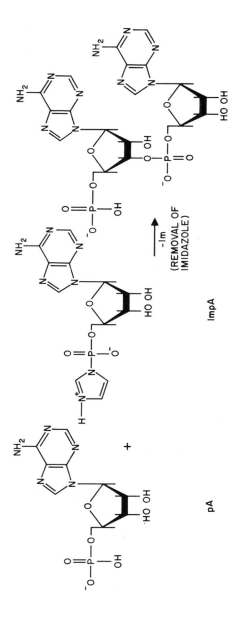

FIGURE 6. The condensation reaction of an activated A derivative with pA.

With experiments of the sort described, we have been able to resolve some interesting questions concerning the selection of sugars in nucleic acids. It is natural to ask what is special about ribose, the sugar that goes into natural nucleic acids. Why did nature use ribose rather than arabinose, for example? We have completed an extensive series of experiments using a variety of different five-carbon sugars, and the results are striking indeed: only ribose works. Of all the five-membered sugars, only ribose works in chain formation. Deoxyribose, the sugar of DNA, does not. This result is one of the hints that RNA may have preceded DNA in the evolution of nucleic acids.

Another result of interest concerns the optical activity of nucleic constituents: one wonders why no L sugars are found in natural nucleic acids. We can offer a partial answer: nucleotides of mixed optical specificity will not form chains in our system. One must be careful in interpreting this result. It does not imply that one optical isomer works better than the other, but only that if you wish to make a nucleic acid chain in this way, all of its constituents must be of the same handedness.

We have obtained very similar results in many experiments putting together, not monomers, but small nucleotide oligomers. In chemistry of this type a template organizes monomers or oligomers with a characteristic efficiency allowing five or six units to be joined, but not many more. Thus, if one starts with dimers instead of monomers, the resulting chains are 10 to 20 long rather than 5 to 10 long.

With respect to prebiotic chemistry, the major problem with the experiments described so far is their poor efficiency. We want to make chains about 20 to 100 long with reasonably high efficiency, whereas until recently we were only able to make much shorter chains with an efficiency of only 1%. This poor showing naturally makes a chemist think of turning to catalysis—of adding some small molecule that does not interfere with the course of the reactions, but makes them proceed faster. There had to be something on the primitive earth that increased these efficiencies at least 10-fold if we are to believe that this form of self-replication ever was a substrate for a process of natural selection. Obviously, there are enzymes that catalyze these reactions; but this fact begs the question, because to make enzymes one needs nucleic acids and ribosomes, which are organelles of proteins and special nucleic acids and could not have preceded the molecules we are trying to make.

I am not sure that we have found the substances in question, but we have some very interesting catalysts. The most efficient is the simple inorganic ion of lead, Pb^{2+}. A second ion that does almost as well in catalyzing the condensation, and is very interesting in other respects, is the zinc (Zn^{2+}) ion. With these ions we can make chains of G's more than 30 units long, and in appreciable yield (Figure 7). The details of this work are too complicated to describe here, so the reader is referred to the original publications (Lohrmann et al., 1980).

In conclusion, we have found that there are simple inorganic catalysts that greatly facilitate the joining of nucleotides on homopolymer templates. Thus, we are beginning to find systems that look like the prebiotic equivalents of modern

4. The Origin of Self-Replicating Molecules

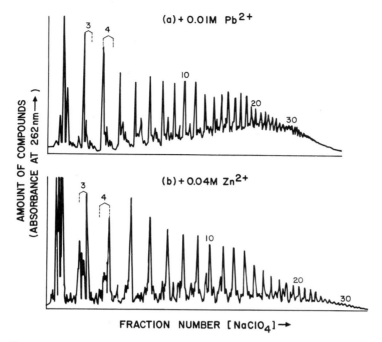

FIGURE 7. Elution profiles of products from the template-directed self-condensation of ImpG in the presence of 0.01 M Pb^{2+} (a) or 0.04 M Zn^{2+} (b). The positions of the major 10-, 20-, and 30-mer peaks are indicated. The positions of the peaks of all the [2′–5′]- and all the [3′–5′]-linked isomers of $(pG)_3$ and $(pG)_4$ are also indicated. Reaction conditions: 0.02 M ImpG* (0.25 mCi/mmole), 0.04 M poly(C), 0.4 M $NaNO_3$, 0.5 M $Mg(NO_3)_2$, 0.4 M 2,6-lutidine buffer, pH 7.0. After 12 days at 0°C, excess EDTA was added, and the pH adjusted to 7.9 with Tris. Pancreatic RNase [0.25 mg/μmole poly(C)] was added, and the solution was incubated at 37°C for 8 hr. Material on chromatogram (a) accounts for 86.2% of the products, and on (b), 74.6%. The remaining substance, monomeric pG, eluted with the void volume (not shown). (Redrawn from Lohrmann et al., 1980.)

polymerase enzymes. The amount of work required to get this far has been substantial—perhaps 30 postdoctoral years—but by the scale of many scientific efforts that is not very much. Perhaps another 100 postdoctoral years will enable us to produce efficient self-replicating systems.

More recently we have developed a new system for template-directed synthesis which has high efficiency and fidelity (Inoue and Orgel, 1982) and which, under optimal conditions, will incorporate more than one base (Inoue and Orgel, 1983). Using this system, we can, for example, show that a polymer composed of C and G residues will facilitate the incorporation of only G and C from a mixture of all four bases, while a polymer composed of C and U will lead to the incorporation of G and A. This same system enables us to copy some very simple short sequences. For example, CCGCC will facilitate the synthesis of its complement GGCGG (Inoue et al., 1984). However, much still remains to be done before self-replication can be achieved in the test tube.

This chapter endeavors to provide the background to the problem of RNA and the origins of life, but it does not offer the latest achievements. These have been reviewed recently (Orgel, 1986), where I argue that ribozymes—molecules in which RNA acts catalytically—are the key to theories that envision protein-free replication as preceding the appearance of the genetic code.

References

Howard, F. B., J. Frazier, M. F. Singer, and H. T. Miles (1966) *J. Mol. Biol.* Helix formation between polyribonucleotides and purines, purines, nucleosides and nucleotides. **16**:415–439.

Inoue, T., and L. E. Orgel (1982) Oligomerization of (guanosine 5′-phosphor)-2-methylimidazolide on poly(C). An RNA polymerase model. *J. Mol. Biol.* **162**:201–217.

Inoue, T., and L. E. Orgel (1983) A nonenzymatic RNA polymerase model. *Science* **219**:859–862.

Inoue, T., G. F. Joyce, K. Grzeskowiak, L. E. Orgel, J. M. Brown, and C. B. Reese (1984) Template-directed synthesis on the pentanucleotide CpCpGpCpC. *J. Mol. Biol.* **178**:669–676.

Lohrmann, R., P. K. Bridson, and L. E. Orgel (1980) Efficient metal-ion catalyzed template-directed oligonucleotide synthesis. *Science* **208**:1464–1465.

Miller, S. L., and L. E. Orgel (1974) *The Origins of Life on the Earth*. Prentice-Hall, Englewood Cliffs, N.J.

Orgel, L. E. (1986) Mini review: RNA catalysis and the origins of life. *J. Theor. Biol.* **123**:127–149.

Sulston, J., R. Lohrmann, L. E. Orgel, and H. T. Miles (1968) Specificity of oligonucleotide synthesis directed by polyuridylic acid. *Proc. Nat. Acad. Sci. USA* **60**:409–415.

Ts'o, P.O.P. (1974) *Basic Principles in Nucleic Acid Chemistry*. Academic Press, New York.

Watson, J. D. (1970) *The Molecular Biology of the Gene*, 2nd ed. Benjamin, New York.

5

Self-Organization of Macromolecules

Peter Schuster and Karl Sigmund

ABSTRACT

Evolution seems to occur in bursts, separated by relatively quiescent periods. From the simpler prokaryotes (organisms without a nuclear membrane around their genetic material) the fully nucleated eukaryote came—but how rapidly? Prokaryotic evolution was preceded by the chemical evolution leading to polynucleotide self-replication, although no fossils are recognized from that period. Autocatalysis must have been required at that stage. But autocatalytic chemical processes are not simple, *in vitro*. Today there is enormous structural variability of biopolymers and polynucleotide synthesis can occur on several different levels of increasing efficiency (faster rates). Systems for these syntheses are complex; models of them have ten-dimensional kinetic equation systems, even in fitting data from simple experiments. Under restricted conditions, these equations show stationarity of relative polynucleotide concentrations. However, a relatively simple equation system predicts an early phase of exponential growth followed by linear growth of polynucleotide concentrations. Very general equations governing the growth of polynucleotide concentrations can be analyzed, but that analysis is much simplified by studying systems having constant total nucleotide concentration. With increasing degrees of realism in the simulations, models predict that: (1) the polynucleotide having the largest excess of synthesis over degradation becomes dominant; (2) if only a few copying errors are made, a whole "quasispecies" of polynucleotides comes to dominate; (3) if many errors are made, the system merely drifts, thus limiting the length of duplicatable polymers; (4) other results will be obtained, such as symmetric self-replication, hypercyclic cooperation, "frozen accidents," Eigen and Schuster's elementary hypercycle, the gamelike dynamics of Maynard Smith, or Lotka-Volterra dynamics.

Specification of fidelity of copying, and of other relevant parameters, yields simulated genome lengths very close to those observed experimentally for enzyme-free oligonucleotide synthesis, for RNA synthesis in the Qβ bacteriophage, and for *in vitro* versions of bacterial replication. Polynucleotides having lengths of present-day tRNA's were presumably formed prebiotically in significant amounts. When templates arose, they were relatively so efficient that they came to dominate, and processes of microscopically random variation followed by natural selection of the more stably pro-

---

PETER SCHUSTER • Institute for Theoretical Chemistry and Radiochemistry, University of Vienna, A-1090 Vienna, Austria. **KARL SIGMUND** • Institute for Mathematics, University of Vienna, A-1090 Vienna, Austria.

ductive "quasispecies" naturally began. But enzyme-free replication has inherent upper limits to its effectiveness, leading to an evolutionary plateau with short RNA chains about the length of today's tRNA's. Emergence from this plateau had to await the creation of polypeptide catalysts—a development at present not well understood. A proposed partial model for this process contains constraints on polynucleotide composition that might reasonably have led to present-day codons for four simple amino acids. This model may guide the further formulation of research questions. In some unclear way, translation machinery arose, though no doubt in a form much less complex and less efficient than what we see today. Nevertheless, such a "factory" had a great relative advantage in self-replication rates, so that thereafter the story of only those factories need be followed. "Quality control" was essential to successful factory self-replication, allowing growth of longer polynucleotide sequences without destructive increase of error. A process description of quality control is given.

Eventually, dilution and export of self-replication products became a disadvantage to each factory, and encapsulation within lipid–polypeptide membranes emerged as an advantageous subprocess within the process of self-replication. These capsules are called protocells. Natural selection then begins to work on these units, and we can see the way cleared for the coming evolution of bacteria and blue-green algae. Other plateaus, followed by evolutionary bursts, might be explained in an analogous way, so that, for example, eukaryosis could have arisen through some process of symbiosis. But beyond this level, evolution, ever opportunistic, becomes more and more difficult for us to imagine according to principles like those used above. —THE EDITOR

(Editor's note: A glossary is provided at the end of this chapter.)

Self-Organization and the Evolution of the Biosphere

The evolution of the biosphere as it is revealed by present fossil records is often interpreted to be a kind of discontinuous process leading in steps usually (but not always) to more and more complex organisms, rather than a continuous or gradual development. Burstlike phases of rapid change seem to be interrupted by long, more quiescent periods; new capabilities were acquired in relatively short times. The changed design realized during rapid phases was usually fixed soon afterwards and not further subjected to large-scale variations. The enormous variety in the biosphere is a matter of details and differences in finish that seem to be established during the more quiescent periods.

The phases of rapid change are attractive to theorists who study the mechanisms of biological self-organization. For obvious reasons, the best documented epochs in paleontology are the more recent ones. They are characterized by fossils of "higher" organisms: plants, fungi, and animals. Compared to the bacteria and viruses, eukaryotic organisms are poorly understood as far as the molecular details of their multiplication are concerned. They are very complex. We need simpler objects of study. The oldest rocks carrying remnants of organic life are found in South Africa and Australia (Nisbet, 1980). They are dated to approximately 3.2 and 3.5×10^9 years ago. The first eukaryotic algae appeared about 2×10^9 years later (Schopf, 1978). Thus, there was a long quiescent period during which orga... ;ms similar to bacteria and blue-green algae were the only inhabitants of the primitive earth. Some biologists (Margulis *et al.*, 1976) suggest that it was the "invention" of the complex mechanisms of cell division, mitosis and meiosis, that gave the strong impetus toward the fast development of multicellular, differentiated, eukaryotic organisms. A biologically caused change in the environment may have contributed to the evolutionary burst in the late Precambrian period: through photosynthesis the blue-green algae created an oxidizing atmosphere containing free molecular oxygen.

It seems that the world of bacteria and blue-green algae was preceded by at least one earlier phase of fast development that led to these primitive organisms. There are no fossils known for this earliest epoch of biological evolution, but there are two reasons for concentrating on this phase. First, the machinery for polynucleotide replication and the language for their translation into proteins appear now to be universal in the biosphere. Some minor deviations from the genetic code have been found in mitochondria, but they represent marginal alterations rather than fundamental differences (Hall, 1979). Hence, the evolution of the replication–translation machinery must have been one of the burstlike developments that subsequently allowed little or no variation. Simulation experiments on possible prebiotic reactions have shown that molecules having the intrinsic capability to form biopolymers were probably abundant on the primordial earth. Numerous investigators are dealing with reactions that might have led to spontaneous formation of macromolecules, particularly those of polypeptides and polynucleotides. Others have looked at prebiotic self-organization from more formal points of view. For details of prebiotic chemistry and further references, see Orgel (Chapter 4, this volume), Miller and Orgel (1974), and our recent attempt to collect most of the available information on prebiotic evolution (Schuster, 1981b).

Second, although present knowledge in molecular biology is almost complete for the process of reproduction of some simple phage viruses, it is insufficient for an understanding of bacterial self-replication and rudimentary for that of eukaryotic organisms. Thus, the most primitive replicating systems are presumably the best objects to be studied now, on the molecular level, to gain understanding of self-organization in living systems.

This chapter discusses various features of molecular self-replication, particularly the properties of polynucleotides. Polynucleotide synthesis can occur with different degrees of sophistication. There is the elegant, but nonphysiological and "nonprebiotic," synthesis of oligonucleotides by the organic chemist (Köster, 1979), not to be considered here. The following processes are graded according to increasing complexity from (1) to (4):

1. Template-free RNA synthesis performed as a prebiotic simulation experiment
2. Template-induced RNA synthesis without specific catalysis by enzymes
3. Viral RNA replication and other closely related studies of enzyme-catalyzed RNA replication
4. Bacterial DNA replication

We shall not go into mechanistic details here, but we shall discuss some formal and general concepts under which these reactions can be subsumed.

Characteristic Features of Self-Replication: The Molecularity of the Process

The self-replication of molecules, in general, is a complicated process that commonly follows a multistep reaction mechanism. This generalization appears to be true for autocatalytic reactions in conventional chemistry as well as for the replication of polynucleotides. Moreover, autocatalytic reactions give rise to strange phenomena. Three examples of reaction mechanisms that have been studied in some detail are presented:

1. The oxidation of hydrogen in the vapor phase has been investigated extensively in the past (see, e.g., the review by Dixon-Lewis and Williams, 1977). The overall reaction $2H_2 + O_2 \rightarrow 2H_2O$ follows the scheme (after some simplification):

$$OH + H_2 \rightarrow H_2O + H \tag{1a}$$
$$H + O_2 \rightarrow OH + O \tag{1b}$$
$$O + H_2 \rightarrow OH + H \tag{1c}$$
$$H + O_2 + M \rightarrow HO_2 + M \tag{1d}$$
$$HO_2 \xrightarrow{\text{surface}} \text{destruction} \tag{1e}$$
$$HO_2 + H_2O_2 \rightarrow H_2O + O_2 + OH \tag{1f}$$
$$H_2O_2 + M' \rightarrow OH + OH + M' \tag{1g}$$
$$HO_2 + H_2 \rightarrow H_2O_2 + H \tag{1h}$$
$$HO_2 + H_2 \rightarrow H_2O + OH \tag{1i}$$

M and M' are some unspecific partners in inelastic collisions. Reactions (1a–i) represent the "autocatalytic core" of the mechanism:

$$OH + 2H_2 + O_2 \rightarrow 2OH + H + H_2O$$

In vapor-phase combustion, autocatalysis leads to chain branching and is responsible for the existence of low-pressure explosion limits.

2. Autocatalysis in homogeneous solution is not less sophisticated than are vapor-phase kinetics. The Belousov–Zhabotinsky reaction has been studied with great care. A detailed scheme of catalyzed oxidation and bromination of malonic acid has been proposed by Edelson et al. (1979). It consists of 22 individual steps. For many purposes the overall reaction

$$fA + 2B \rightarrow fP + Q \tag{2}$$

can be represented by a simplified five-step mechanism, called the Oregonator model (Field and Noyes, 1974):

$$A + Y \rightarrow X \tag{2a}$$
$$X + Y \rightarrow P \tag{2b}$$

$$B + X \rightarrow 2X + Z \qquad (2c)$$
$$2X \rightarrow Q \qquad (2d)$$
$$Z \rightarrow fY \qquad (2e)$$

where f is some stoichiometric factor. The compounds A and B react via three intermediates X ($HBrO_2$), Y (Br^-), and Z (one-electron, metal-ion redox couples of cerium, iron, manganese, and so on) to yield the products P and Q. The "hard core" of the reaction scheme is the autocatalytic step (2c). It is responsible for the unusual chemical phenomena observed with the Belousov–Zhabotinsky reaction in solutions far off equilibrium. These are oscillations in concentrations of intermediates and spontaneous spatial pattern formation (Tyson, 1976; Nicolis and Prigogine, 1977).

The two examples presented here demonstrate impressively the multistep nature of autocatalytic reactions in nonbiological systems. The new feature emerging in biology is not complexity of the basic mechanism of replication, but the enormous structural variety of biopolymers. The variety is a consequence of the universal building principle: a constant backbone and several, interchangeable side chains. Polynucleotides of only moderate lengths can exist in extremely large numbers of different sequences. As an example of self-replication, we consider these macromolecules.

3. Polynucleotide synthesis is the most relevant example of replication in biological systems. As expected, we find a highly complex reaction mechanism when we look at the molecular details. The order or molecularity of the reactions falls into three categories:

a. Template-free, spontaneous synthesis of polynucleotides is not actually a process of replication. It is mentioned here only for the sake of comparison and completeness. It represents the zero-order reaction, because no catalytically active polynucleotide is involved. Orgel and co-workers (see Chapter 4) have studied this process under possible prebiotic conditions and found it to be rather inefficient.

b. First-order polynucleotide synthesis is understood as template-induced replication. It can be studied by several *in vivo* and *in vitro* systems at very different levels of sophistication. We mention here only those approaches that allow a simple and straightforward analysis. Template-directed RNA synthesis without enzymatic catalysis has been studied extensively by Orgel and co-workers (see Chapter 4). They found effective metal ion catalysis: Pb^{2+} in the synthesis of oligoadenylates on a poly(U) template (Sleeper *et al.*, 1979) and Zn^{2+} in oligo(G) synthesis on poly(C) (Lohrmann *et al.*, 1980). The latter example led to polymers with chain lengths of up to 30 to 40 bases. The accuracy of this primitive replication process will be discussed later.

Polynucleotide replication in the presence of specific enzymes has been investigated extensively and under many experimental conditions. Examples are the studies on phage RNA replication summarized by Küppers (1979b). DNA replication is an exceedingly complicated process (Kornberg, 1974; Staudenbauer, 1978). Some *in vitro* investigations were performed by Loeb and co-workers (Kunkel and Loeb, 1979; Kunkel *et al.*, 1979), who attempted to determine the accuracy of this process.

In vitro studies on RNA replication have shed some light on the mechanistic details of this process. They will also be discussed below.

As first-order self-replication processes, *in vitro* investigations treat the enzyme as a kind of environmental factor capable of increasing the accuracy and rate of the processes. The enzyme is *added* to the solution of polynucleotides; it is *not* produced by translation of some gene that is part of the replicating system itself.

c. A new feature of self-replication becomes apparent when the replicating molecules succeed in acting both as templates and as specific catalysts. Polynucleotides are the only efficient templates in the biosphere. They may act as catalysts as well, but as such they are rather poor when compared to proteins. Only two classes of ribonucleotides act as catalysts: tRNA and rRNA. Interestingly, both are used exclusively in the formation of peptide bonds, especially in ribosomal protein synthesis.

The tRNA molecules act as transmitters of amino acids in ribosomal protein synthesis. Apart from their basic importance in translation, tRNA's play only a negligible role in the biochemistry of the cell. Some tRNA-like molecules specifically transport amino acids in glycopeptide synthesis. They are not recognized by codons.

The rRNA occurs as a structural factor in both subunits of the ribosome: one RNA molecule (16 S in bacteria, 18 S in eukaryotes) occurs in the small ribosomal subunit, and two (5 S and 23 S in bacteria) or three (5 S, 5.8 S, and 28 S in eukaryotes). RNA molecules are the building blocks of the large subunit. Both the tRNA molecules and the ribosomal ribonucleic acids are considered to be very old molecules in an evolutionary sense. Except for these two classes of catalytically active polynucleotides, cellular DNA and RNA are used exclusively for coding proteins or, in the intercistronic regions, for regulatory purposes. (The latter role is poorly understood.) This fact appears to be a legacy of their ancient history. After some chemical system had developed machinery to design catalysts by coding a message in a polynucleotide sequence and translating it into a polypeptide, there was no further advantage in exploring the use of other potential catalytic properties of polynucleotides. RNA synthesis is more involved than is protein synthesis; furthermore, polypeptides are much more efficient catalysts.

Proteins as excellent catalysts are used everywhere in the biosphere. Therefore, polynucleotides could exert their influence on their own replication most efficiently if they succeeded in achieving control over polypeptide synthesis. In this way, the rate of polynucleotide synthesis depends on polynucleotide concentration with an order higher than one. We shall use expressions like "second-order catalysis" or "second-order replication" for this kind of multiple-level influence on the replication process through direct action as template and, indirectly, via translation into catalysts.

Translation of polynucleotide sequences into polypeptides requires the existence of some kind of genetic code. Although some models of the origin of the genetic code have been proposed by several groups including ourselves (see, e.g., Crick *et al.*, 1976; Eigen and Schuster, 1979; Schuster, 1981; Eigen and Winkler-Oswatitsch, 1981a,b), this problem is far from being solved. The models presented so far may shed some light by suggesting new types of experiments.

5. Self-Organization of Macromolecules

Finally, it seems necessary to stress one important difference between first- and second-order replication, originating from the difference in complexity between these two processes. Suitable model systems for experimental *in vitro* studies of second-order catalysis in self-replication are inevitably very complicated. They must involve several genes and a complete translation machinery including ribosomes, tRNA molecules, and the corresponding aminoacyl-synthetase enzymes. It is not surprising, therefore, that no successful *in vitro* experiments have been reported so far for continuing replication, translation, and regulation. At present, we have to rely on the analysis of theoretical model systems similar to those found to be successful in first-order replication, or we have to refer to systems more complex than molecules. Such systems are outside the scope of this review.

Polynucleotide Synthesis *in Vitro*

RNA polymerization by suitable enzymes has been studied under conditions that allow us to draw some conclusions about the underlying mechanism. We refer to two investigations on "complementary" replication as representative examples. By "complementary" we mean a two-phase process: at first a negative copy (or minus strand, I^- is obtained by synthesis of RNA using as primer the single-strand template I^+. Then I^- is used as a template itself, and polmerization yields a new copy of I^+. In terms of overall reactions we may write:

$$I^+ + \sum_{\lambda=1}^{4} \nu_\lambda^{(-)} A_\lambda + E \rightarrow I^- + I^+ + E \quad (3a)$$

$$I^- + \sum_{\lambda=1}^{4} \nu_\lambda^{(+)} A_\lambda + E \rightarrow I^+ + I^- + E \quad (3b)$$

As indicated above, we denote plus and minus strands of some polynucleotide by I^+ and I^-; A_λ stands for one of four activated mononucleotides—namely, for one of the triphosphates ATP, UTP, GTP, or CTP—and $\nu_\lambda^{(+)}$ and $\nu_\lambda^{(-)}$ are the stoichiometric coefficients for plus and minus strands. E represents the enzyme used in RNA synthesis.

To obtain interpretable and computable experimental data, one has to apply suitable boundary conditions, such as constant triphosphate concentrations in a properly designed flow reactor. Then the concentrations of mononucleotides enter into the rate equations as constant factors, and the concentrations of polynucleotides are the only variables.

Schneider *et al.* (1979) have reported detailed kinetic studies on poly(A)-poly(U) synthesis in a stirred-flow reactor. The enzyme used is RNA polymerase from *Escherichia coli*. They propose a cyclic mechanism for complementary replication (Figure 1), which is able to explain all experimental findings observed so far. This multistep mechanism (Table 1) gives rise to a ten-dimensional system of ordinary differential equations, which is hard to analyze under general conditions. At low polynucleotide concentration, high and constant total enzyme con-

TABLE 1. Complementary Replication in the Poly(A)-Poly(U) System[a,b]

Reaction steps		Concentrations[c]
$I^+ + E$	$\underset{k_{-1}}{\overset{k_1}{\rightleftharpoons}} I^+ \cdot E$	$[E] = e;\ e_0 = e + y_1 + y_2 + z + w_1 + w_2$
$I^+ + E^-$	$\underset{k_{-2}}{\overset{k_2}{\rightleftharpoons}} E \cdot I^-$	$[I^+] = [\text{poly(A)}] = x_1;\ x_1^0 = x_1 + y_1 + z + w_1$
$I^+ + E \cdot I^-$	$\underset{h_{-1}}{\overset{h_1}{\rightleftharpoons}} I^+ \cdot E \cdot I^-$	$[I^-] = [\text{poly(U)}] = x_2;\ x_2^0 = x_2 + y_2 + z + w_2$
		$[I^+ \cdot E] = y_1,\ [E \cdot I^-] = y_2$
$I^+ \cdot E + I^-$	$\underset{h_{-2}}{\overset{h_2}{\rightleftharpoons}} I^+ \cdot E \cdot I^-$	$[I^+ \cdot E \cdot I^-] = z$
$I^+ \cdot E + U$	$\underset{g_{-2}}{\overset{g_2}{\rightleftharpoons}} I^+ \cdot E \cdot U$	$[I^+ \cdot E \cdot U] = w_1,\ [A \cdot E \cdot I^-] = w_2$
$I^+ \cdot E \cdot U + (n-1) U$	$\overset{f_2}{\rightarrow} I^+ \cdot E \cdot I^-$	$[A] = a,\ [U] = u$
$A + E \cdot I^-$	$\underset{g_{-1}}{\overset{g_1}{\rightleftharpoons}} A \cdot E \cdot I^-$	$A = \text{ATP},\ U = \text{UTP}$
$A \cdot E \cdot I^- + (n-1) A$	$\overset{f_1}{\rightarrow} I^+ \cdot E \cdot I^-$	

Differential equations

$$\dot{e} = \frac{de}{dt} = (k_{-1}y_1) + (k_{-2}y_2) - (k_1x_1 + k_2x_2)e$$

$$\dot{x}_1 = \frac{dx_1}{dt} = (k_{-1}y_1) + (h_{-1}z) - (k_1e + h_1y_2)x_1$$

$$\dot{x}_2 = \frac{dx_2}{dt} = (k_{-2}y_2) + (h_{-2}z) - (k_2e + h_2y_1)x_2$$

$$\dot{y}_1 = \frac{dy_1}{dt} = (k_1ex_1) + (h_{-2}z) + (g_{-2}w_1) - (k_{-1} + h_2x_2 + g_2u)y_1$$

$$\dot{y}_2 = \frac{dy_2}{dt} = (k_2ex_2) + (h_{-1}z) + (g_{-1}w_2) - (k_{-2} + h_1x_1 + g_1a)y_2$$

$$\dot{z} = \frac{dz}{dt} = (f_1a^{n-1}w_2) + (f_2u^{n-1}w_1) + (h_1x_1y_2) + h_2x_2y_1) - (h_{-1} + (h_{-2})z$$

$$\dot{w}_1 = \frac{dw_1}{dt} = (g_2uy_1) - (g_{-2} + f_2u^{n-1})w_1$$

$$\dot{w}_2 = \frac{dw_2}{dt} = (g_1ay_2) - (g_{-1} + f_1a^{n-1})w_2$$

$$\dot{a} = \frac{da}{dt} = (g_{-1}w_2) - (g_1y_2 + f_1a^{n-2}w_2)a$$

$$\dot{u} = \frac{du}{dt} = (g_{-2}w_1) - (g_2y_1 + f_2u^{n-2}w_1)u$$

Buffered triphosphate concentrations

$a = a_0 = \text{const.},\ u = u_0 = \text{const.}$
$\bar{g}_1 = g_1a_0,\ \bar{g}_2 = g_2u_0,\ \bar{f}_1 = f_1a_0^{n-1},\ \bar{f}_2 = f_2u_0^{n-1}$

[a] See mechanism shown in Figure 1. Rate constants are slightly modified.
[b] From Schneider et al., 1979.
[c] Total concentrations of polynucleotides and enzyme are denoted by x_1^0, x_2^0, and e_0 respectively. Const., Constant enzyme concentration, (e_0) = const.

5. Self-Organization of Macromolecules

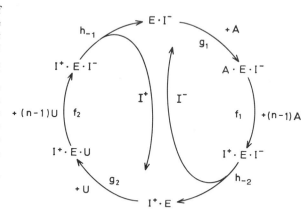

FIGURE 1. A cyclic mechanism of enzyme-catalyzed polynucleotide replication, according to Schneider et al. (1979). The symbols E, A, and U are used for RNA polymerase, ATP, and UTP, respectively. I^+ and I^- stand for poly(U) and poly(A). The mechanism distinguishes chain initiation and propagation as well as complex formation. The rate constants k_1, k_{-1}, k_2, and k_{-2} describe reversible complex formation between enzyme and template; f_1, g_1, f_2, and g_2 are the rate constants for template-induced RNA synthesis. (See Table 1.)

centration, and stationarity to trimolecular intermediates $I_1 \cdot E \cdot I_2$, $I_1 \cdot E \cdot U$, $A \cdot E \cdot I_2$, and so on, we obtain:

$$\dot{x}_1 = l_1 x_2; \quad l_1 = \frac{\bar{f}_1 \cdot \bar{g}_1}{\bar{f}_1 + \bar{g}_{-1}} K_2 e_0, \quad K_2 = \frac{k_2}{k_{-2}} \tag{4a}$$

and

$$\dot{x}_2 = l_2 x_1; \quad l_2 = \frac{\bar{f}_2 \cdot \bar{g}_2}{\bar{f}_2 + \bar{g}_{-2}} K_1 e_0, \quad K_1 = \frac{k_1}{k_{-1}} \tag{4b}$$

By x_1, x_2, and e_0 we denote the free polynucleotide and the total enzyme concentrations. Bars indicate mean values. For precise definitions and the meanings of the rate constants, see Table 1. The overall kinetics of RNA synthesis described by the differential equation (2) has been discussed previously (Eigen, 1971; Eigen and Schuster, 1979). The plus–minus ensemble of polynucleotides grows exponentially. Thereby it approaches a kind of internal stationary quality with respect to relative polynucleotide concentrations, $\bar{x}_1/\bar{x}_2 = \sqrt{l_1}/\sqrt{l_2}$:

$$X(t) = X(0) \cdot \exp(+\sqrt{l_1 l_2}\, t) \quad \text{with } X = \sqrt{l_2}\, x_1 + \sqrt{l_1}\, x_2 \tag{5a}$$

and

$$\xi(t) = \xi(0) \cdot \exp(-\sqrt{l_1 l_2}\, t) \quad \text{with } \xi = \sqrt{l_1 l_2}\, x_1 - \sqrt{l_1}\, x_2 \tag{5b}$$

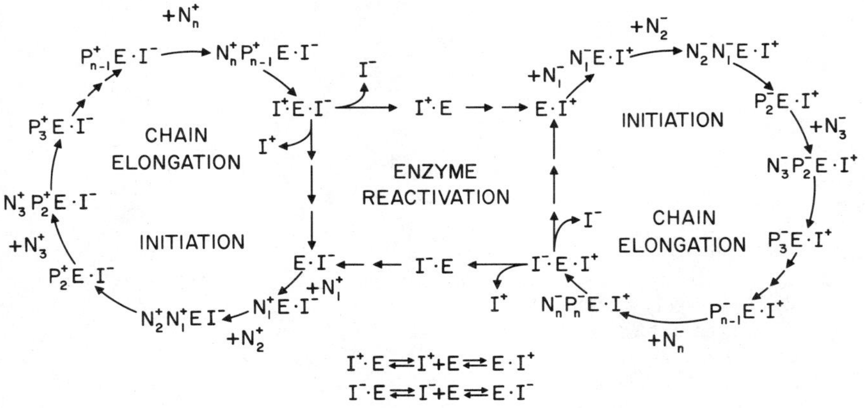

FIGURE 2. A cyclic mechanism for bacteriophage RNA replication *in vitro*, as used by Biebricher *et al.* (1980) to interpret the kinetic results obtained with the Qβ system. E, I$^+$, and I$^-$ represent free Qβ polymerase and plus and minus strands of the RNA to be replicated. $N_1^+, N_2^+, \ldots, N_n^+$ and $N_1^-, N_2^-, \ldots, N_n^-$ are the nucleotides (triphosphates) in the sequence as they appear in the plus and minus strands, respectively. The chain of the newly synthesized molecule grows only from the 5' to the 3' end of the phosphate connection to ribose. $P_2^+, P_3^+, \ldots, P_{n-1}^+$ and $P_2^-, P_3^-, \ldots, P_{n-1}^-$ are used as symbols for the growing chains. A nucleation length of two bases is assumed.

where the values of X and ξ at $t = 0$ are denoted by $X(0)$ and $\xi(0)$. Since $\lim_{t \to \infty} \xi(t) = 0$, the ratio of plus to minus strands approaches the given value $\bar{x}_1/\bar{x}_2 = \sqrt{l_1}/\sqrt{l_2}$.

Under stationary conditions, the plus–minus ensemble of complementary replication shows exponential growth just as we would expect in the case of direct replication. The relative concentrations are determined by the square roots of the individual replication rates.

The kinetics of phage RNA replication has been studied *in vitro* by Biebricher *et al.* (1980). Qβ-replicase was used as enzyme. The authors investigated RNA synthesis at constant total concentration of enzyme and buffered nucleotide triphosphate concentrations. The polynucleotides show three distinct phases of growth: the exponential increase in total RNA concentrations observed at low concentrations changes into linear growth at moderate concentrations and finally levels off to saturation or zero growth at large excess of template. The authors suggest the complicated multistep mechanism shown in Figure 2. This mechanism is somewhat more detailed than, but nevertheless closely related to, that mentioned before. It consists of three sets of reaction steps dealing with initiation of polymerization, chain elongation, and reactivation of the enzyme. The mechanism is able to explain all the kinetic details observed (Biebricher *et al.*, 1983).

It is sufficient for this discussion to consider a simplified, two-step version of the mechanism of polymerization that is consistent with both multistep reactions shown in Figures 1 and 2. It produces correctly the first two phases characterized by exponential and linear growth of RNA concentration:

5. Self-Organization of Macromolecules

$$I + E \underset{k^-}{\overset{k}{\rightleftharpoons}} I \cdot E \tag{6a}$$

$$I \cdot E + \sum_{\lambda=1}^{4} \nu_\lambda A_\lambda \overset{f}{\rightarrow} I \cdot E + I \tag{6b}$$

A typical solution curve of the corresponding differential equations is shown in Figure 3. These studies provide proof for the existence of a phase of exponential growth in complementary replication of polynucleotides. At high concentrations of template, the enzyme becomes saturated and the rate of synthesis changes to linear growth.

Some general results that apply to the phase of exponential growth are discussed later in this chapter.

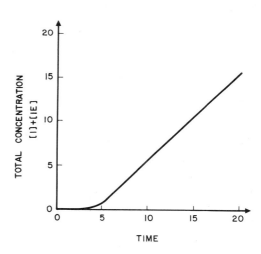

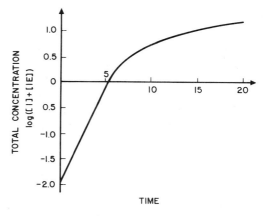

FIGURE 3. Solution curves for the differential equations corresponding to the simplified reaction mechanism for polynucleotide replication [Equations (4a) and (4b)]. The rate constants and initial conditions applied are: $f = 1$, $k = 1000$, $k^- = 10$, $[I]_0 = 0.01$, $[E]_0 = 1$, $[IE]_0 = 0$. Note that exponential growth changes into linear growth at the point of saturation: $[I] + [IE] \sim [E] + [IE]$. All quantities are given in arbitrary time (t) and concentration (c) units.

A Formal Theory of Self-Replication

To make a complicated problem such as molecular self-organization accessible to a formal mathematical treatment, we must minimize mechanistic diversity and environmental influence. The dialysis reactor shown schematically in Figure 4 is very well suited for this purpose. It allows control of the concentrations of low-molecular-weight material, especially those of the four nucleotide triphosphates ATP, UTP, GTP, and CTP. When the latter are constant, the concentrations of macromolecules are the only variables. Further control may be applied to the total concentration of polynucleotides $C = \Sigma x_i$, where the molar concentration of a particular polynucleotide is $[I_i]$, or x_i. The condition of constant C, which defines a stationary state, was denoted previously as "constant organization" (Eigen, 1971; Eigen and Schuster, 1979). It simplifies the mathematical analysis substantially.

For further analysis we split the time derivatives of the concentrations into two additive contributions:

$$\dot{x}_i = F_i(x_1, \ldots, x_n) = \Gamma_i(x_1, \ldots, x_n) - (x_i/C)\phi, \quad i = 1, \ldots, n \quad (7)$$

By Γ_i we denote the growth function that describes the evolution of the system in the absence of a dilution flux ϕ (typical units for ϕ would be concentration/time, or mass $\cdot l^{-3} \cdot t^{-1}$). Under constant organization, we simply obtain

$$\phi = \Sigma_i \Gamma_i \quad (8)$$

The restriction of constant organization does not imply a tremendous loss in generality, as might appear at first. In many cases the results obtained for the stationary state can be used to describe growing systems as well. The generalization is surprisingly simple when the growth functions Γ_i are homogeneous in the variables x_i.

According to mass-action kinetics, the growth functions Γ_i can be represented by a polynomial expansion:

$$\Gamma_i = k^{(i)} + \sum_{j=1}^{n} k_j^{(i)} x_j + \sum_{j=1}^{n} \sum_{l=1}^{n} k_{jl}^{(i)} x_j x_l + \cdots \quad (9)$$

The power of an individual term in this series refers to the molecularity of the polymerization process as far as macromolecules are involved.

$k^{(i)}$ is the rate of spontaneous, template-free formation of the polynucleotide I_i. The first-order diagonal term $k_i^{(i)} x_i$ represents the rate of template-instructed synthesis of I_i. Replication of molecules does not occur with ultimate accuracy. Hence, there will be error copies or mutations that give rise to off-diagonal terms of first order: $k_j^{(i)} x_j$, $i \neq j$ is the frequency at which I_i is synthesized as an error copy on the template I_j. The second-order terms require additional catalytic action of macromolecules: $k_{ij}^{(i)} x_i x_j$ represents the rate of template-instructed for-

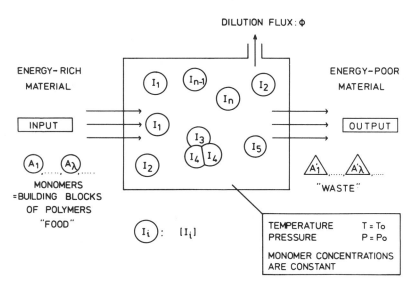

FIGURE 4. The dialysis reactor for evolution experiments. This kind of flow reactor consists of a reaction vessel that allows for temperature and pressure control. Its walls are impermeable to polynucleotides. Energy-rich material is poured from the environment into the reactor. The degradation products are removed steadily. Material transport is adjusted in such a way that the concentration of monomers is constant in the reactor. A dilution flux ϕ is introduced in order to remove the polynucleotides, I_i, produced by multiplication. Thus, the sum of the concentrations: $[I_1] + [I_2] + \cdots + [I_N] = \sum_{i=1}^{N} x_i = C$, may be controlled by the flux ϕ. Under "constant organization," ϕ is adjusted so that the total concentration C is constant.

mation of I_i with I_j acting as catalyst. Again we may distinguish diagonal terms $k_{ii}^{(i)} x_i^2$, which describe catalytic self-enhancement of second order, from off-diagonal terms with $i \neq j$. In principle, error copies can be produced in second-order replication as well: $k_{jl}^{(i)} x_j x_l$ is the rate of formation of I_i as an error copy of I_j with catalytic action of the molecules I_l. Overall reactions that give rise to the individual terms of the kinetic equations are summarized in Table 2.

Some concrete examples of the very general equation (9) have been analyzed in great detail. In the brief summary of the results obtained, we use an expression introduced by Dawkins (1976): a "replicator" is a molecule or system that is capable of replication under suitable environmental conditions. The simplest replicators in the biosphere today are the polynucleotides, with their participating enzymes.

The Replicator Equation with Constant Terms of Self-Replication

The equation

$$\dot{x}_i = x_i[k_i^{(i)} - (1/C)\phi] = x_i(E_i - \overline{E}); \quad i = 1, \ldots, n \quad (10)$$

TABLE 2. Overall Reactions and Rates of Polynucleotide Formation under Idealized Conditions as Described for the Dialysis Reactor[a]

Reaction[b]	Rate	Template	Order with respect to polynucleotides	Order with respect to autocatalysis
a $\quad \sum_{\lambda=1}^{4} \nu_\lambda^{(i)} A_\lambda \rightarrow I_i$	$k^{(i)}$	—	0	0
b $\quad I_i + \sum_{\lambda=1}^{4} \nu_\lambda^{(i)} A_\lambda \rightarrow 2I_i$	$k_i^{(i)} x_i$	I_i	1	1
c $\quad I + \sum_{\lambda=1}^{4} \nu_\lambda^{(i)} A_\lambda \rightarrow I_j + I_i$	$k_i^{(j)} x_i$	I_i	1	
d $\quad 2I_i + \sum_{\lambda=1}^{4} \nu_\lambda^{(i)} A_\lambda \rightarrow 3I_i$	$k_{ii}^{(i)} x_i^2$	I_i	2	2
e $\quad I_i + I_j + \sum_{\lambda=1}^{4} \nu_\lambda^{(i)} A_\lambda \rightarrow 2(I_i + I_j)$	$k_{ij}^{(i)} x_i x_j$	I_i	2	1
f $\quad I_i + I_l + \sum_{\lambda=1}^{4} \nu_\lambda^{(i)} \rightarrow I_j + I_i + I_l$	$k_{il}^{(j)} x_i x_l$	I_i	2	0

[a] See Figure 4.
[b] Reactions b, d and e involve template-induced synthesis of polynucleotide leading to an error-free copy; c and f describe mutations; and a represents the rate of spontaneous, template-free polynucleotide formation.

describes the action of selection within an ensemble of correctly replicating polynucleotides under constant environmental conditions. We use E_i instead of $k_i^{(i)}$ to eliminate the superscript. C is the total concentration of polynucleotides; ϕ/C is the first-order rate constant for mass removal of polynucleotides. The bar denotes mean value for all species i. In physical terms, E_i is the excess production rate of the polynucleotide I_i. By "excess" we mean the difference in rates between synthesis and degradation:

$$E_i = a_i - d_i \tag{11}$$

where a_i is the number of molecules I_i synthesized per unit of time, the rate constant of polynucleotide formation, and d_i is the rate constant of degradation. The expression "constant terms of self-replication" used in the title of this subsection indicates that neither a_i nor d_i nor E_i depend on the concentration of polynucleotides.

The mean excess production rate is denoted by

$$\overline{E} = \frac{1}{C} \sum_{i=1}^{n} E_i x_i = \frac{1}{C} \phi \tag{12}$$

It is adjusted by the flux ϕ as required by equation (8).

The analysis of equation (10) is straightforward. The results have been discussed elsewhere (Eigen and Schuster, 1979). Selection takes place in the reactor: the polynucleotide with the largest value of E_i survives the competition. A characteristic example is shown in Figure 5. The function $\overline{E}$ plays an important role. It increases steadily and approaches an optimum at the stationary state:

$$\lim_{t \to \infty} \overline{E}(t) = E_m, \quad E_m = \max(E_i; i = 1, \ldots, n) \tag{13}$$

The Replicator–Mutator Equation with Constant Terms of Self-Replication

The equation

$$\dot{x}_i = x_i(k_i^{(i)} - \frac{1}{C}\phi) + \sum_{j=1, j \neq i}^{n} k_j^{(i)} x_j = \sum_{j=1}^{n} w_{ij} x_j - \frac{x_i}{C}\phi, \tag{14}$$
$$i = 1, 2, \ldots, n, \; w_{ij} = k_j^{(i)}$$

is of particular interest in molecular self-organization. It is capable of describing the important features of Darwinian evolution. Originally proposed by Eigen (1971), equation (14) has been studied extensively (Thompson and McBride, 1974; Jones et al., 1976; Eigen and Schuster, 1979; Epstein and Eigen, 1979; Küppers, 1979a; Swetina and Schuster, 1982).

The physical meaning of the rate coefficients w_{ij} depends on whether it is a diagonal ($i = j$) or an off-diagonal term ($i \neq j$). In the former case we modify equation (11) by introducing a quality factor Q_i:

$$w_{ii} = a_i Q_i - d_i \tag{15}$$

Q_i is the probability that a polynucleotide produced on template I_i is a precise image: no error has occurred during the copying process. Somewhat as in the previous case, a_i is the number of copies, correct and erroneous, that are synthesized on the template I_i per unit of time, and d_i is the rate constant for the degradation. The coefficients w_{ij} with $i \neq j$ describe the rate at which I_i is obtained as an error copy of I_j. From equation (14) we derive

$$\frac{1}{C}\phi = \sum_{i=1}^{n} x_i(a_i - d_i) = \sum_{i=1}^{n} E_i x_i = \overline{E} \tag{16}$$

where E_i stands for the excess production as defined by equation (11) and $\overline{E}$ is the average value according to equation (12).

Equation (14) may be analyzed by a transformation of variables and second-order perturbation theory (Thompson and McBride, 1974; Jones et al., 1976). Apart from kinetic degeneracies caused by accidentally equal rate constants, we distinguish two situations:

1. *The accuracy of the replication exceeds a certain minimum value $Q_{\min}$.* In this case, selection takes place in the evolutionary reactor during the entire transient period. The mean excess production $\overline{E}(t)$ reaches a limiting value:

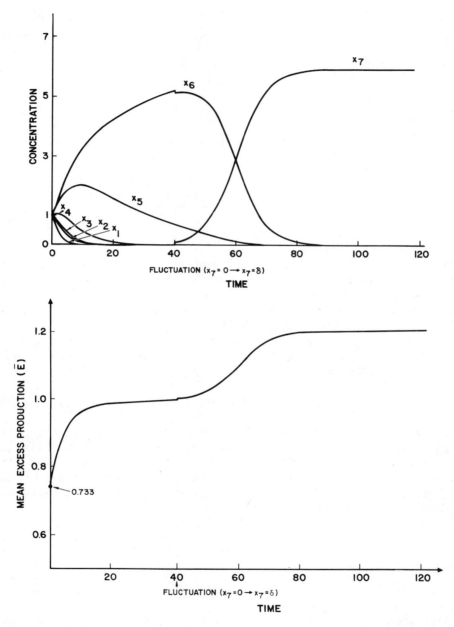

FIGURE 5. Solution curves for the differential equation $x_i = (E_i - \overline{E})x_i$ [see Equations (8)–(10)]; i = 1,2, ..., 7, with $E_1 = 0.4$, $E_2 = 0.6$, $E_3 = 0.65$, $E_4 = 0.8$, $E_5 = 0.95$, $E_6 = 1.0$, $E_7 = 1.2$, and $\overline{E} = \sum_i E_i x_i / \sum_i x_i$. Initial conditions: $x_1(0) = x_2(0) = \ldots = x_6(0) = 1$, $x_7(0) = 0$. The mean excess production $\overline{E}$ starts from an initial value of $\overline{E} = 0.733$ and increases steadily up to $\overline{E} \sim 1.0$ as the population becomes rather homogeneous: $x_1 = x_2 = \ldots = x_5 \sim 0$; $x_6 \sim 6$, $x_7 = 0$. One easily recognizes the cause for this increase in $\overline{E}$: the less efficiently growing polynucleotides are eliminated. They dis-

5. Self-Organization of Macromolecules

$$\lim_{t \to \infty} \bar{E}(t) = \lambda_{\max} = w_{mm} + \sum_{i \neq m} \frac{w_{mi} w_{im}}{w_{mm} - w_{ii}} \tag{17}$$

I_m denotes the species that has the largest value of the rate coefficient w, $w_{mm} = \max(w_{ii}, i = 1, \ldots, n)$. The results of second-order perturbation theory are good approximations to the exact solutions provided $w_{mi}, w_{im} \ll w_{mm} - w_{ii}$ for all $i \neq m$. The dynamical behavior observed shows close analogy to that of the error-free system, [cf. equation (13)]. There is one fundamental difference: instead of a single "best" sequence, a whole set of polynucleotides is selected. This set consists of a master sequence I_m and its most frequent mutants. We have called such an ensemble a "quasispecies," in analogy to a species in biology.

The stationary distribution of the "quasispecies" is given by

$$\bar{x}_m = C \frac{w_{mm} - \bar{E}_{-m}}{E_m - \bar{E}_{-m}} = \frac{C}{1 - \sigma_m^{-1}} (Q_m - \sigma_m^{-1}) \tag{18}$$

and

$$\bar{x}_i = \bar{x}_m \frac{w_{im}}{w_{mm} - w_{ii}}, \quad i = 1, \ldots, n, \, i \neq m \tag{19}$$

Stationary concentrations are indicated by bars ($\bar{x}$). For short notation, we use the quantities

$$\bar{E}_{-m} = \frac{1}{C - x_m} \sum_{i \neq m} E_i x_i \tag{20}$$

and the superiority factor

$$\sigma_m = a_m/(d_m - \bar{E}_{-m}) \tag{21}$$

From equations (16) and (17) we see that a stationary population can only exist if Q_m exceeds a certain minimum value:

$$Q_m > Q_{\min} = \sigma_m^{-1} \tag{22}$$

2. *The accuracy of replication is below the minimum value $Q_{\min}$.* In this case, no selection takes place in the evolution reactor. Because of mutations, new sequences are formed steadily, but they do not persist for longer than a few generations. There is no specific sequence. The system as a whole shows a kind of drifting behavior.

---

appear in the order of increasing excess production E_i. At $t = 40$ we introduce a perturbation: $x_7 = 0 \to x_7 = \delta$. The appearance of I_7 as a favorable mutant leads to further increase in $\bar{E}$, which finally approaches the value $\bar{E} = 1.2$ when the population again becomes homogeneous: $x_1 = x_2 = \ldots = x_6 \sim 0; x_7 \sim 6$.

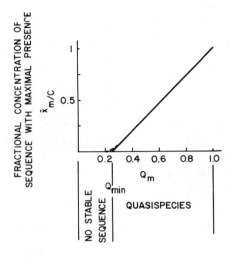

FIGURE 6. The dependence of the distribution within a quasispecies on the accuracy of replication. The percentage of the master sequence decreases linearly with decreasing quality factor, Q_m. We assume a superiority of $\sigma_m = 4$. The solid line gives the result of second-order perturbation theory according to equation (16). The curved broken line segment accounts for the influence of stochastic revertant mutations in a large-enough population. Q_{min} is the smallest value Q_m that will permit formation of stable sequences.

Equation (18) and its meaning for the existence of a quasispecies is illustrated in Figure 6.

Let us now consider template-induced replication in detail. The accuracy of replication of an entire sequence depends on the accuracy of the incorporation of single bases or "digits":

$$Q = q_1 \cdot q_2 \cdot q_3 \cdot \cdots \cdot q_\nu = \bar{q}^\nu \tag{23}$$

Here, q_1 is the accuracy of correct incorporation of the first base, q_2 that of the second, and so on. q_ν is the quality factor of the last base. By ν we denote the chain length of the polynucleotide or, as it is usually called in polymer chemistry, the degree of polymerization. The quality factors q_i are not the same. They will depend on the nature of the nucleotide (A, U, C, or G) and on its neighbors in the sequence. For the sake of convenience we define an average single-digit accuracy $\bar{q}$. For long enough sequences with similar nucleotide contents, in general, the values of $\bar{q}$ will be very similar and they will depend basically on the mechanism of replication.

The lower limit of the quality factor implies an upper limit for the chain length of polynucleotides that can be replicated with sufficient accuracy:

$$Q_m = \bar{q}^\nu_m\, m > Q_{min} = \bar{q}_m{}^{\nu_{max}} \tag{24}$$

or

$$\nu_m < \nu_{max} = -\frac{\ln \sigma_m}{\ln \bar{q}_m} \sim \frac{\ln \sigma_m}{1 - \bar{q}_m} \tag{25}$$

5. Self-Organization of Macromolecules

The approximation in the last part of equation (25) will be fulfilled in every case of interest, because $\bar{q}$ can be expected to be close to 1. Figure 7 shows the dependence of the concentration of the master sequence on the degree of polymerization.

The error threshold defined by equation (25) depends strongly on the average single-digit accuracy of the replication process. Because of the logarithmic function in this equation, the influence of the superiority factor is not so pronounced. Nevertheless, it seems worthwhile to illustrate the meaning of σ_m. According to equation (21), the master sequence I_m can be superior to the others either by replicating faster,

$$a_m > \bar{a}_{-m} = \frac{1}{C - x_m} \sum_{i \neq m} a_i x_i$$

or by being degraded more slowly,

$$d_m < \bar{d}_{-m} = \frac{1}{C - x_m} \sum_{i \neq m} d_i x_i$$

Let us assume identical rate constants of degradation $d_1 = d_2 = \cdots = d_n$. The superiority is determined by the replication rate constants exclusively: $\sigma_m = a_m/\bar{a}_{-m}$.

Finally, we shall try to visualize the development in a system whose members follow a growth law given by equation (14), assuming sufficiently high accuracy, $Q_i > Q_{min}$, of the replication process. At first, a quasispecies, say $X_1 = (\bar{x}_1, \bar{x}_2, \ldots, \bar{x}_n)$, is selected. Then, a favorable mutant not contained in X_1, e.g., I_{n+1}, appears (Figure 5). After a transient period, X_1 will be replaced by a new

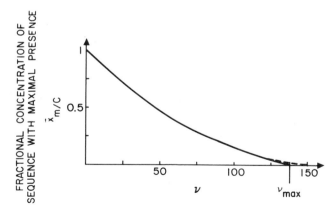

FIGURE 7. The dependence of the distribution within a quasispecies on the chain length ν of the polynucleotide. The average single-digit accuracy is assumed to be $\bar{q}_\nu = 0.99$, and the value $\sigma_m = 4$ is chosen for the superiority parameter. The solid line gives the results of second-order perturbation theory according to equations (16) and (22). The broken line accounts for the influence of stochastic revertant mutations in a large-enough population.

quasispecies X_2, which contains I_{n+1} as the master sequence. Unfavorable mutants outside the quasispecies, of course, are eliminated immediately after formation. This process of replacement may continue until we end up with a master sequence close to the optimum with the given environmental conditions. The stepwise approach toward this optimum is caused by a concerted mechanism of mutation and selection. It serves as a simple but instructive model for Darwinian evolution in systems of primitive organisms without genetic recombination. The scenario presented has been simplified to some extent because stochastic effects occurring at low population numbers have to be considered more carefully, but these effects are not likely to change the picture obtained.

A second difference between our model and the actual evolution of prokaryotes can be seen in the complexity of an entire bacterium or blue-green alga. The property to be optimized cannot be described by a few rate constants referring to fairly simple overall reactions. The quantities a_i, d_i, and Q_i are complicated functions of many variables. Furthermore, the "epigenetic landscape" is highly complex. Therefore, we may expect an enormous variety of different local maxima of the fitness function, which is optimized, whereas we have only a few closely related optima for polynucleotide replication in an evolution reactor. However, the basic mechanism—the interplay of mutation and evaluation of the mutant by selection—appears to be the same in both cases.

The Replicator Equation with Linear Terms of Self-Replication

Linear terms of self-replication are introduced by second-order catalysis:

$$\dot{x}_i = x_i \left[\sum_{j=1}^{n} k_{ij}^{(i)} x_j - \frac{1}{C}\phi \right] = x_i \left[\sum_{j=1}^{n} a_{ij} x_j - \frac{1}{C}\phi \right] \quad (26)$$

Because we neglect mutations in this case, we can eliminate the superscript: $k_{ij}^{(i)} = a_{ij}$.

Equation (26) expresses interaction between self-replicating elements in very general terms. Therefore, it is not surprising that special cases have been treated and are of importance in different fields. There are five relevant applications:

1. The symmetric case, $a_{ij} = a_{ji}$ is the basic equation of the continuous Fisher-Wright-Haldane model frequently used in population genetics (see, e.g., Hadeler, 1974, or Crow and Kimura, 1970).

2. A special case with cyclic symmetry (Figure 8), $a_{ij} = f_i \cdot \delta_{i,j+1}$ [i, j are counted mod n; δ represents the Kronecker symbol $\delta_{ik} = [{}^{1,i=k}_{0,i \neq k}]$], is the elementary hypercycle (Eigen, 1971; Eigen and Schuster, 1979). It represents the least complicated system that shows cooperative behavior of otherwise competitive ensembles of self-replicative elements. Hypercyclic coupling has been applied to study possible answers to certain questions of prebiotic evolution (Eigen and Schuster, 1979; Schuster, 1981a,b). Positive feedback loops of the same nature are used in models for biological pattern formation in morphogenesis and cell differentiation (Meinhardt and Gierer, 1980).

3. Equation (26) is obtained in game dynamics (Taylor and Jonker, 1978; Hofbauer et al., 1979; Zeeman, 1980; Schuster et al., 1981a,b,c) and can be used

5. Self-Organization of Macromolecules

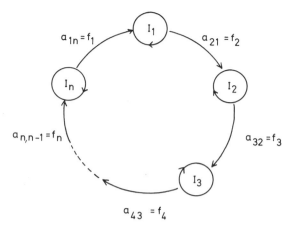

FIGURE 8. The catalytic hypercycle. The formation of a positive feedback loop of second-order catalytic interactions changes the properties of the system substantially and introduces cooperative behavior into an ensemble of otherwise competitive, replicating molecules.

to describe dynamical aspects in the social behavior of animals. The biological concepts underlying these studies are based on the game-theoretical model developed by Maynard-Smith and Price (1973).

4. Hofbauer (1980) presented rigorous proof for the equivalence of equation (26) and Lotka–Volterra systems (Kerner, 1961):

$$\dot{y}_i = y_i \left[\epsilon_i + \sum_{j=1}^{n-1} \alpha_{ij} y_j \right], \qquad i = 1, \ldots, n-1 \qquad (27)$$

where $y_i = x_i/x_n$, $\epsilon_i = a_{in} - a_{nn}$, and $\alpha_{ij} = a_{ij} - a_{nj}$. Lotka–Volterra equations have been used widely to analyze the dynamics of predator–prey and competition models and to discuss their ecological aspects. The analysis of equation (26) is often easier, despite the higher dimensionality. Therefore, equation (26) presents a certain advantage when compared to equation (27), particularly in dimensions $n > 3$.

5. Equation (26) is also useful for describing pattern formation in networks of the nervous system. Two examples are: (1) the ontogenesis of retinotopic mapping in the visual cortex may be modeled successfully by an equation of type (26) in high dimensions, where n is very large (von der Malsburg and Willshaw, 1981); (2) the self-organization of nervous networks has been investigated by Cowan (1970), using Lotka–Volterra-type equations closely related to equation (26).

Thus, it seems that many "nonlinearities" in self-replication can be deduced from a common origin that may be very well described by equation (26). Some of the properties of equation (26) have been studied in great detail (Jones, 1977; Schuster *et al.*, 1978, 1979; Epstein, 1979; Hofbauer *et al.*, 1980). There are two basic properties:

1. Hypercyclic coupling is a sufficient condition to guarantee cooperation between self-replicating elements. By cooperation we mean that none of the elements are extinguished (Eigen and Schuster, 1979; Schuster *et al.*, 1979). The elementary hypercycle

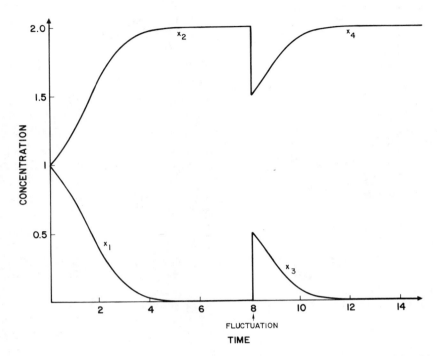

FIGURE 9. Solution curves for the differential equation $\dot{x}_i = [a_{ii}x_i - (1/C)\phi]x_i; i = 1,2,3$, with $a_{11} = 0.4$, $a_{22} = 0.8$, $a_{33} = 1.2$, and $\Phi = \Sigma_i a_{ii} x_i^2$. Initial conditions: $x_1(0) = x_2(0) = 1$, $x_3(0) = 0$. Very sharp selection leads to a homogeneous population $x_1 \sim 0$ and $x_2 = 2$. After the first transient period, there is almost no chance for a more efficient mutant to grow. We introduced a large fluctuation $x_3 = 0.5$ at $t = 8$. Even then the more efficient polynucleotide I_3 disappears very fast.

$$\dot{x}_i = x_i[f_i x_j - (1/C)\phi], \quad i = 1, \ldots, n, \quad j = i - 1 + n\delta_{i1} \quad (28)$$

is the simplest system having this property of cooperation.

2. A system of the type specified by equation (28), which essentially consists of cyclically coupled self-replicating elements, grows as an entity. It controls the relative concentrations of its constituents. Without restrictions, such a system follows a law of hyperbolic growth, i.e., its concentration goes to infinity faster than would an exponential process (Eigen and Schuster, 1979). A consequence of this growth behavior is intolerance of mutants, as illustrated in Figure 9.

Let us visualize how evolution might take place in a system in which the elements grow according to equation (26) or (28). In particular, we consider the development of a hypercyclic feedback loop. After such a system has been formed once, there is a negligibly small chance that favorable mutants will assert themselves against the established structure. A nonlinear, self-replicating system does not optimize its properties in the same sense that a Darwinian system like the one

5. Self-Organization of Macromolecules

discussed in the previous sections would. On the contrary, it tends to conserve the first successful attempts to form a complex dynamical organization and acts like a "frozen accident." However, there is an alternative for further development of hypercyclic systems, as indicated in Figure 10. Mutants belonging to one of the quasispecies within the hypercycle, and surviving because they are steadily produced, have a small but non-zero chance to be incorporated into the cyclic system if they possess the required kinetic properties. An analysis of such extensions of hypercycles by rigorous mathematical techniques would require investigations of the replicator–mutator equation with linear terms of self-replication of the type: $k_{ji}^{(i)} x_j x_i$. This has not been accomplished yet. The suggestions made here rely on numerical integration of the differential equations for various types of reaction networks (Eigen and Schuster, 1979) and on some general arguments described by Schuster *et al.* (1979). Recent summaries of the results obtained from the mathematical analysis of replicator equations can be found in Schuster (1981a) and Schuster and Sigmund (1983).

The simple monomial or polynomial expressions for the growth functions Γ_i as analyzed in this section are characteristic for single-step, overall reactions. They appear also in the asymptotic limits of more complicated multistep mechanisms. In particular, we are interested in the low- and high-concentration limits of replication kinetics. As we saw previously, RNA replication by a specific replicase can be described by mechanisms that lead to exponential growth in the presence of excess enzyme. Linear growth occurs after saturation of the enzyme at polynucleotide concentrations high enough to convert all the enzyme into the RNA–protein complex. An analogous behavior may be observed with multistep

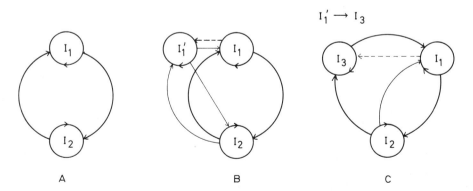

FIGURE 10. A mutation-incorporation mechanism for stepwise extension of hypercycles. (A) A two-membered hypercycle with translation. (B) a mutant of I_1, denoted by I_1', appears. It has a certain catalytic effect on the replication of I_1 and I_2. (C) Extension of the hypercycle by one member occurs if the mutant has the following two properties: (1) I_1' ($I_1'I_3$) is recognized better by I_2 than is I_1, and (2) the I_1' ($=I_3$) is a better catalyst for the replication of I_1 than is I_2. In other words, the new catalytic coupling terms have to fulfill a criterion as described by equation (20): mutual enhancement along the cycle $1 \to 2 \to 3 \to 1$ has to prevail over self-enhancement.

mechanisms of hypercyclic interaction: hyperbolic growth at low concentrations may change into exponential or parabolic growth in the concentration range above saturation (Eigen and Schuster, 1979; Eigen et al., 1980). Saturation, in a formal sense, reduces the apparent molecularity of the process. The evolutionary behavior and the properties discussed here are observed only in the low-concentration ranges.

Darwinian Selection and the Fidelity of Polynucleotide Replication

We shall now try to analyze the general results for the Darwinian systems derived above by means of some experiments on the fidelity of polynucleotide replication. The concrete examples include: template-induced, enzyme-free oligonucleotide synthesis (Sleeper et al., 1979; Lohrmann et al., 1980), RNA synthesis in the β-bacteriophage system (Domingo et al., 1976; Batschelet et al., 1976), and in vitro experiments on bacterial replication (Kunkel and Loeb, 1979; Kunkel et al., 1979). The data are compared with predictions from equation (25), and are shown in Table 3. Experimental evidence on superiority factors σ_m is almost nonexistent. We used three different values, $\sigma_m = 2$, 20, and 200, in our calculations (Eigen and Schuster, 1979). All realistic cases will fall between these limits.

Template-directed, enzyme-free synthesis of oligo(G) on poly(C) has been studied by Orgel and co-workers (see Chapter 4). The most impressive yields of oligomers were obtained in the presence of Zn^{2+} (Lohrmann et al., 1980): chain lengths of up to 30 to 40 were found. The fidelity of the replication process was determined to be 200:1 in a 50:50 mixture of ImpG (guanosine 5'-phosphorimidazolide) and ImpA (adenosine 5'-phosphorimidazolide) which are used as activated monomers (Orgel, personal communication, 1980). Under these conditions the average single-digit accuracy for the incorporation of G is $\bar{q}_G = 0.995$. The accuracy would be lower if all four bases were present as activated monomers. The nature of the metal ion has a dramatic influence. In the presence of Pb^{2+} instead of Zn^{2+}, but under otherwise identical conditions, the quality factor determined was much lower; $\bar{q}_G = 0.9$. As far as the order of magnitude is concerned, these experimental data agree well with previous theoretical estimates based on a knowledge of equilibrium constants for base pairing (Eigen and Schuster, 1979). No precise agreement can be expected because these estimates were made for the simultaneous presence of all four monomers in solution. Enzyme-free polynucleotide replication seems to be limited to molecules of about the size of present-day tRNA's (Table 3). Because of errors, longer polynucleotide sequences could not act as a storage for information.

One elegant series of in vitro and in vivo experiments provided a direct proof for the relevance of the model of equation (25) for RNA synthesis (Domingo et al., 1976). An error copy of the phage genome was produced in vitro by site-directed mutagenesis. The base change occurred extracistronically, and the mutant was infectious in in vivo experiments, i.e., in the host bacteria. Making use of a model from population genetics that gives the probability for revertant

5. Self-Organization of Macromolecules

TABLE 3. Single-Digit Accuracy, Superiority, and Error Threshold in Polynucleotide Replication[a]

Error rate per digit	Single digit accuracy ($\bar{q}_m$)	Superiority (σ_m)	Maximum digit content (ν_{max})	Biological examples
1 : 20	0.95	2	14	Enzyme-free RNA replication
		20	60	tRNA precursor, $\nu = 80$
		200	106	
1 : 3000	0.9997	2	2100	Single-stranded RNA
		20	9000	replication via specific
		200	16,000	replicases
				Phage Qβ, $\nu = 4500$
1 : 10^6	0.999999	2	0.7×10^6	DNA replication via
		20	3.0×10^6	polymerases and
		200	5.3×10^6	proofreading
				E. coli, $\nu = 4 \times 10^6$

[a]From Eigen and Schuster, 1979.

mutants in a growing phage population, Batschelet *et al.* (1976) were able to determine the accuracy of replication. An analysis of their data yields a selective advantage for the ratio of master copies, $w_{\text{wild-type}}/w_{\text{mutant}}$, of two to four (Eigen and Schuster, 1979) and a single digit accuracy of $\bar{q} = 0.9997$. Inserting these values into equation (25) we obtain a maximum chain length for Qβ RNA that is very close to the actual value of $\nu = 4500$ (Table 3). The bacteriophage Qβ thus approaches the upper limit of the information content as closely as possible. Domingo *et al.* (1976) provided a direct proof for the quasispecies-like nature of a Qβ bacteriophage population. The standard sequence master copy is present only as a relatively small fraction of the wild-type distribution. The majority of sequences thus are frequently occurring mutants. According to equation (18), such a situation is characteristic of an ensemble that grows close to the error threshold ($Q_m \sim \sigma_m^{-1}$). This finding agrees very well with the fact that ν is close to ν_{max}.

The successful analysis of Qβ replication by the model system described by equation (14) encourages extrapolation to prokaryotic DNA replication (Eigen and Schuster, 1979). This process is semiconservative (see, e.g., Kornberg, 1974), and at least two enzymes, roughly speaking a polymerase and a repair enzyme, are responsible for the average single-digit accuracy of the replication process as a whole. We may try an "off-the-cuff" estimate by assuming that both enzymes operate with similar accuracy in base recognition, as Qβ replicase does. Careful *in vitro* studies on different DNA polymerases have shown that they in fact synthesize with error rates of 1 in 1000 to 1 in 100,000 copies (Kunkel *et al.*, 1979). Further increase in fidelity depends on a single-strand binding protein. Our estimate yields a maximum chain length of $10^6 < \nu_{max} < 10^7$ base pairs (Table 3), a result in general agreement with the length of the bacterial genomes known. We have to be careful with this speculation, because the process of DNA replication is very complicated. A better guess would require more accurate quantitative data on the replication process than are available.

A Scenario for the Early Evolution of Life

This brief account of a model for prebiotic evolution is based on the general results of molecular self-replication, as well as on our knowledge of the conditions of the primordial earth (for further details, see Eigen, 1971; Eigen and Schuster, 1979; Schuster, 1981b; Eigen et al., 1981; Eigen and Schuster, 1982). This model tries to bridge the gap between prebiotic chemistry and the first primitive organisms. A diagram to make it easier to follow the text is given in Figure 11.

The general line followed here leads from the origin of molecular replication to individual "protocells." It is assumed that a process of gradual improvement is characterized by certain "landmarks." There are three important logical steps of molecular self-organization that, in principle, need not be separated by long intervals:

1. The origin of the first oligo- or polynucleotides introduces the capability of self-replication and changes the dynamics of the system by making Darwinian selection possible at the molecular level.

2. The origin of the genetic code and the molecular translation machinery allows the incorporation of instructed proteins into the self-replicating system. The system is able to design catalysts, particularly catalysts for its own replication.

3. The origin of self-instructed compartment formation creates the first individual organisms or protocells. This step enables the replicating unit to make optimal use of its own improvements. Straightforward evolution of the property obtained in step 2 by means of natural selection is possible from now on.

Polypeptides or protenoids presumably already played an important role in prebiotic chemistry before they were incorporated into the self-replicating machinery. At first they were present as environmental factors for polynucleotide self-replication, as were general catalysts like mineral surfaces or other inorganic materials. No direct modification of their catalytic properties by the replicating molecules was possible at this stage. After development of a translation machinery, we find a new class of polypeptides that are part of the replicating unit. The polynucleotides now instruct the synthesis of polypeptides, which act as enzymelike catalysts. Any modification of the primitive "gene" is translated into the polypeptide sequence—apart from the degeneracies of the genetic code. Mutants can be evaluated by selection according to their catalytic properties. Some problems that develop if diffusion is fast compared to polymer synthesis need to be examined in detail. Arguments similar to those applied to the prebiotic role of polypeptides may be used in the case of lipidlike compounds, bilayers, membrane fragments, or unicells. Presumably they were formed under prebiotic conditions and were present in appreciable amounts in the primordial soup. They exercised catalytic power on prebiotic chemistry by forming hydrophobic phases and interphases. Eventually, they formed primitive compartments. Spherules

5. Self-Organization of Macromolecules

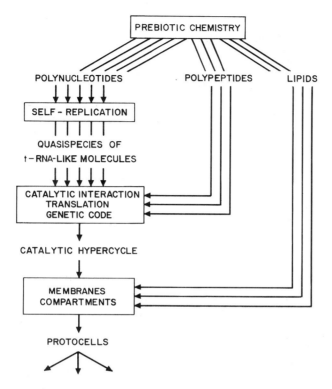

FIGURE 11. A scheme for the logical sequence of steps in prebiotic evolution, leading from unorganized macromolecules to protocells.

made from protenoids and hot water (Fox *et al.*, 1959) may have played an analogous role. These particles, from our point of view, are again important environmental factors, but not yet under the influence of the replicating system. The situation changes substantially when the replicating unit succeeds in achieving control of compartment or unicell formation. From now on the advantages of spatial isolation are subjected to selection. Any improvement of the catalytic efficiency of such a "protocell" is accessible only to the molecules inside the more prosperous compartments and to its progeny. These protocells represent the first individuals or spatially defined and bounded organisms. Here we concentrate on the properties controlled by the self-replicating unit. The role of proteins and spherules as environmental factors has been discussed extensively (see, e.g., Fox and Dose, 1972).

Prebiotic chemistry has been studied extensively in simulation experiments (Miller and Orgel, 1974). These investigations showed in a convincing manner that building blocks of biopolymers were probably abundant on the primordial earth. Polymer formation under prebiotic conditions is highly probable, but the most plausible sequences of reactions are not known yet. The results of some experiments, particularly those performed by Orgel and co-workers, are very promising and give strong support to the suggestion that polynucleotides of the size of present-day tRNA molecules were formed in significant amounts under prebiotic conditions. These molecules brought the capability of self-replication onto the primordial scene. Template-induced polynucleotide synthesis is so much more efficient than spontaneous polymerization that the latter played no role after the first templates had been formed. At the same time, a first process of optimization in the Darwinian sense starts. RNA molecules that replicate more efficiently are selected. All the properties relevant for replication are then steadily improved. These properties are determined by the sequences of bases, because these sequences determine stable secondary and tertiary structures. One of the most important features is the melting temperature T_M. This is the temperature at which double helices appear paired and dissociated in equal amounts (50%). The value of T_M is essentially determined by the relative GC content of the polynucleotide and the average lengths of double-helical regions (Pörschke, 1977). Optimal fitness requires an intermediate melting temperature: if T_M is high, only a very small portion of the polynucleotides is dissociated and thus ready for replication. The overall rate of synthesis will be low. If T_M is low, on the other hand, most polynucleotides are present as single strands and therefore exposed to hydrolysis. Then their rate of degradation is high. We do not yet know the temperature dependence of the rate constants for replication and degradation, so it is not possible to give a reliable estimate of the optimal value of T_M.

After enzyme-free replication has been optimized, the process of evolution temporarily comes to an end. That part of the prebiotic earth accessible to polynucleotides soon becomes populated as densely as the production of suitable energy-rich starting material and the rate of hydrolysis permit. Regional differences in the environment, such as temperature and solid-state catalysts, may introduce variations in which different polynucleotide sequences predominate locally.

The error threshold discussed in the previous section limits the lengths of polynucleotides to approximately the size of the present-day tRNA's. Now we can visualize the error threshold from a different point of view. Within the quasispecies the master copy is selected in preference to its own less efficient mutants. This selection breaks down if the maximum chain length is exceeded. Most biologists engaged in studies of molecular evolution share the opinion that tRNA molecules belong to the oldest class of biomolecules in living organisms. Attempts have been made to reconstruct phylogenetic trees from tRNA's in different organisms and to learn about the degree of relationship between different tRNA's in the same organism. A recent study (Eigen and Winkler-Oswatitsch, 1981a,b) was conducted to reconstruct a possible common ancestor of our tRNA's. This hypothetical polynucleotide has a number of relevant properties, including high GC content and a clear preference for a repetitive -RNY-,[1] in particular -GNC-, pattern. The extrapolated sequence allows the formation either of a cloverleaf or of a hairpin structure. The existence of complementary symmetry

---

[1] By R we denote one of the purine bases A or G; by Y, one of the two pyrimidines U or C; by N, *any of the four bases.*

at the 3'- and 5'-ends of the polynucleotide is of particular importance for replication. The minus strand then has an initial sequence identical with that of the plus strand—a condition that favors starting replication from one end.

How could the tRNA-like molecules overcome the dead end of their limited reactivity? Further development of the system would require an increase in the amount of information that could be stored in the sequence of bases. One way to achieve this would be to develop a more accurate mechanism of replication. Alternatively, cooperation of self-replicating molecules, each one present in a number of copies, would also allow an increase in the total length of the transferable genetic information. In the second case, competition among the cooperating elements has to be avoided; it remains necessary to preserve the property of selection against unfavorable mutants and other competitors. Earlier we discussed a formal way to fulfill these requirements by cyclic catalytic coupling. Any solution to the practical problem seems to combine both concepts for the increase of information: the tRNA-like molecules can improve the accuracy of replication only if they cooperate and start to design specific catalysts for their own needs. These specific catalysts are the proteins, as we know from the actual outcome of prebiotic evolution. Thus, cooperation of small polynucleotides, the origin of translation, and the structure of the genetic code are closely related problems. Various models for primitive translation have been conceived. One that was published more recently, and continues earlier ideas, was developed by Crick *et al.* (1976). We have modified this model slightly (Eigen and Schuster, 1979). Because an extensive discussion can be found in that paper, we omit details here. The essential results of our model are four predictions, given below, that are meaningful in the context of prebiotic chemistry and thus increase the plausibility of this approach.

1. In order to guarantee sufficient stability of mRNA–tRNA complexes, the primitive codons and anticodons would have to be rich in G and C.

2. In order to avoid sliding the tRNA along the messenger during ribosome-free translation, one needs a repetitive pattern of period three (Crick *et al.*, 1976). Actually, a pattern of the type -RNY- has been found in comparative studies of the known sequences of tRNA molecules (Eigen and Winkler-Oswatitsch, 1981a,b) as well as in the DNA of some viruses, of bacteria, and of eukaryotes (Shepherd, 1981). This pattern is presumed to be a remnant of regularities in primordial genes. It seems that mutations did not remove this pattern completely over billions of years, and it still exists in present-day polynucleotides.

3. Combining (1) and (2) in a straightforward way we obtain four codons that are now used for some amino acids:

glycine   = G G C   alanine = G C C
aspartic acid = G A C   valine = G U C

These four amino acids were the most abundant under prebiotic conditions. The corresponding adaptors, the tRNA's with complementary anticodons, are the four most closely related tRNA molecules. They may well be descendants of the same quasispecies.

4. Nucleotide binding probably was one of the most ancient properties of encoded polypeptides. Rossmann et al. (1974) made an alignment study of various dehydrogenases and other nucleotide-binding proteins in order to reconstruct a possible common ancestor protein. Walker (1977) analyzed the 21-site amino acid sequences of the nucleotide-binding surface and tried to identify the precursor amino acids coded into the ancestral surface, which is assumed to be 3×10^9 years old. He concluded that valine and one or both of the aspartic acid–glutamic acid group would be likely precursors, with valine predominating. Both constituents of the ancient recognition site are members of this group of the first four amino acids.

Our knowledge of the physics of polynucleotide–polypeptide interactions is fragmentary. Details of the processes and properties that might have played major roles in early translation are not known. The model considerations mentioned here may have their most important impact on experimental research through helping us to ask relevant questions about the origin of translation. For our further discussion, let us assume that the polynucleotides managed somehow to organize a primitive machinery for translation. Then what?

Using the primitive machinery for translation, polynucleotides were able to interfere in the replication of other polynucleotides by means of their translation products, the corresponding polypeptides. They might have exerted positive or negative influences, e.g., by coding proteins with specific replicase or nuclease activities. Indirect coupling via translation products is a realistic model for catalytic interaction in second-order self-replication. Cyclic coupling by means of proteins will yield the same dynamic properties of the system as those discussed for direct coupling. Kinetic models for hypercycles with translation have been proposed and analyzed before (Eigen and Schuster, 1979; Eigen et al., 1980). Moreover, we were able to prove that the property of cooperative interaction does not depend on a specific algebraic expression for the nonlinearity (Hofbauer et al., 1981). The only requirement is a plus sign on all the terms in the feedback loop.

Translation machinery of the complexity found today is an enormously rich chemical factory. In early evolution there was no need for such perfection. In principle, some of the molecules could play two roles at the beginning: the early transmitters or precursors of our tRNA's might have been early genes as well (Eigen and Winkler-Oswatitsch, 1981a,b). The first aminoacyl-synthetases—the enzymes that attach the correct amino acids to the corresponding tRNA's and consequently carry a specific recognition site for a single polynucleotide—might have acted as specific polymerases as well (C. Biebricher, personal communication, 1980). These suggestions indicate that an early translation mechanism using a primitive, highly redundant code for a handful of amino acids could arise from replicating polynucleotides. Admittedly, sufficient experimental background is not available to test this idea. We do not know precisely the probabilility that such an event occurred to bring the necessary parts together.

Let us assume that the primitive translation machinery was first formed by aggregation of information stored in a number of structurally independent genes that might well have originated from a single quasispecies. Their relative concentrations were controlled by hypercyclic organization. The first system that suc-

5. Self-Organization of Macromolecules

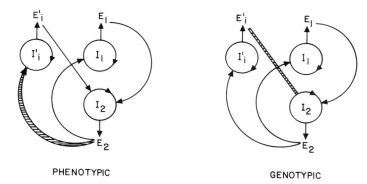

FIGURE 12. Two idealized classes of mutations in primitive replication-translation systems. The "phenotypic" mutations lead to mutants (I'_1) that are better targets for the specific replicase, whereas the properties of their translation products (E'_1) are about the same as in the wild type (E_1). The "genotypic" mutant, in contrast, is characterized by a better translation production but roughly unchanged recognition by the replicase.

ceeded in completing its design had an incredible advantage. It increased many times the rate of polynucleotide synthesis and the accuracy of the replication process. Longer genes became possible, and these in turn could code for better enzymes. The dynamical properties of such replication–translation machinery in aqueous solution follow a higher-order replication kinetics and do not allow mutants to grow, even if they are somewhat more efficient. Improvement was achieved by admission of new members to the feedback cycle. More and more amino acids were incorporated into the encoded ensemble, thereby improving the catalysts. Ultimately, replication became so accurate that gene multiplication was no longer necessary. The independent genes were eventually tied together to form a genome consisting of a single molecule.

Sooner or later, an inherent disadvantage started to outweigh the advantages of homogeneous solution. Any favorable "genotypic" mutation (Figure 12) is accessible to every beneficiary in the surrounding solution, whether or not it contributes an honest share to the common prosperity. Parasites may profit from the achievements of the hosts. A powerful counteractant to the dilution of favorable properties through diffusion is spatial isolation. This separation may range from very weak to complete, from hindered diffusion to packing into compartments. Compartments free of parasites are more prosperous and multiply faster than their companions that have to feed beneficiaries; such compartments will multiply and finally win the competition for resources. The positive and negative features of the isolation of genes in compartments have been summarized by Schuster (1981b). There one can also find a suggestion about how a set of genes may exert control on compartment formation by coding for and synthesizing a primitive membrane protein. Here, we mention the abstract, mechanistic aspect only. Compartmentalization brings the system back to first-order replication kinetics because the compartment multiplies as an entity. Darwinian selection is restored. During the intermediate period of nonlinear self-replication, the system was shifted from the level of individual molecules to the level of protocells, which can be understood also as the level of first, primitive—but already integrated and individualized—organisms. From now on a diversity of environments may act on the individuals and create the enormous variability in "finish," such as we observe in our studies on bacteria and blue-green algae.

Summary

Selection sets in at the level of polynucleotides as soon as template-induced replication becomes important. Integration of self-replicating elements into a functional unit requires higher-order kinetics in order to suppress competition between the individual members. This higher-order kinetics is not compatible with Darwinian evolution. During such a phase of "nonlinear" development we do not expect optimization of the newly achieved properties. Rather, we expect to find "frozen accidents" or "once-for-ever" decisions. Some primitive polynucleotides link their information content and form the coding capacity for a primitive translation machinery. Later in evolution, spatial isolation by compartment formation restores Darwinian behavior. From then on variation may start again. But the translation machinery as a whole is conserved as an entity, no longer subject to any serious change.

In principle, a formally similar mechanism may govern the other bursts in the evolution of the biosphere as well. The formation of eukaryotes could be a result of symbiosis. Some prokaryotes had to share their metabolic properties and their coding capacities in order to create a machinery for organizing the complex genome and the apparatus for mitosis and meiosis observed in all higher organisms. Again it seems very likely that the common features observed with eukaryotes are a kind of "frozen accident"—a remnant of a period lacking optimization.

We could push analogies further, but this chapter is directed only toward macromolecular self-organization. There is a good reason to stop the description at a molecular level, because the more highly organized the biological entities are, the more difficult it becomes to discover principles of development. Jacob (1977) has formulated this problem precisely in his article "Evolution and Tinkering." There is no reason for simplicity or intellectual elegance in nature. She does not design with the eyes of an engineer. The only thing that counts, and that is selected for, is functional efficiency. Moreover, selection can act only on the things that are present at a certain moment. The historical principles therefore are very hard to detect and are well hidden.

Glossary[2]

A Narrowly, the purine base adenine; broadly, a nucleotide unit containing that base, as in adenosine triphosphate (ATP).

Bacteriophage A virus that infects bacteria. Also called phage.

[2]Prepared by the editor.

5. Self-Organization of Macromolecules

C   Narrowly, the pyrimidine base cytidine; broadly, a nucleotide unit containing that base.

Cistron   The segment of a DNA molecule specifying one complete polypeptide chain. Synonymous with gene. (The gene may be interrupted by noncoding patches in eukaryotes.)

Codon   A string of three nucleotides in a polynucleotide that "represents" or "guides" attachment of a particular amino acid at a specific step in the assembly of a protein chain. The codon is recognized by the anticodon of a transfer RNA, which, by means of an enzyme (an aminoacyl-synthetase), is loaded specifically with the amino acid corresponding to the codon.

DNA   Deoxyribonucleic acids. These are polynucleotides with deoxyribose as the sugar and A, G, C, T as allowable bases.

Eukaryote   A cell containing a membrane-enclosed nucleus.

G   Narrowly, the purine base guanine; broadly, a nucleotide unit containing that base.

Genetic Code   The representation of each of the 20 natural amino acids by a codon. Additionally, there are three termination signals or "nonsense" codons that stop translation of the code into proteins.

Hypercycle, elementary   A "hypercycle" is a system of autocatalytic reactions arranged in a circle so that each reaction's product catalyzes production of its clockwise neighbor. The term "elementary" refers to the application of mass action kinetics, which means that the rate can be expressed by simple products of concentrations.

Molecularity   The kinetic order of a chemical reaction; the number of preexisting molecules that must come together to initiate a reaction.

Nucleotide   A three-component compound consisting of: (1) a pentose (a five-carbon sugar), either ribose or deoxyribose; (2) one to three phosphate groups; (3) a single purine or pyrimidine base (A, G, C, T, or U).

Oligo   A prefix meaning (in this chapter) "between few and approximately 40."

Phage   A bacteriophage (q.v.).

Poly   A prefix meaning (in this chapter) more than about 40.

Poly(A)   An adenine-containing polynucleotide. Adenine is the only base present.

Polynucleotide   High-molecular-weight polymers of nucleotides joined to form a sugar–phosphate repeating unit backbone, with bases projecting out at regular intervals.

Poly(U)   A uracil-containing polynucleotide. Uracil is the only base present.

Prokaryote   A cell containing no nucleus (e.g., a bacterial cell). Contrasts with eukaryote.

Quasispecies   A "master polynucleotide" sequence and its most frequent mutants.

Replication   Production of a copy of an original molecule, typified by DNA's replication during cell division. Contrasts with translation, which maps DNA (or RNA) into proteins.

Replicator   A molecule, or something more complex, that is capable of replication under suitable environmental conditions.

Ribosome   A cellular subsystem in which protein synthesis is carried out. The ribosome consists of many proteins and pieces of RNA. In protein synthesis, the ribosome guarantees correct reading of the codon by the anticodon of the RNA and catalyzes the formation of the peptide bounds.

RNA   Ribonucleic acids. These are polynucleotides with ribose as the sugar and A, G, C, U as allowable bases.

S   The Svedberg unit, a rough measure of molecular weight determined by ultracentrifugation (strictly, a sedimentation coefficient). The larger the number, the bigger the molecule (but the relationship is not linear).

T   Narrowly, the pyrimidine base thymine; broadly, a nucleotide unit containing that base.

Template   A molecular shape or form that is replicated.

Transfer RNA (tRNA)   A polyribonucleotide, about 70 to 80 nucleotides long, that acts as a specific transmitter between the codon and the amino acid. The correct amino acid is attached to the individual tRNA by a specific enzyme called aminoacyl-synthetase.

Translation   Production of a molecule, "directed" by the coded "instructions" contained in a molecule of different type. This is typified by translation of DNA's coded "instructions" into the series of amino acids constituting a protein. Contrasts with replication; see also Genetic Code, Transfer RNA.

U   Narrowly, the pyrimidine base uracil; broadly, a nucleotide unit containing that base, such as uridine triphosphate (UTP).

ACKNOWLEDGMENTS. Many stimulating discussions with Professor Manfred Eigen and William C. Gardiner are gratefully acknowledged. C. Biebricher, M. Eigen, J. Hofbauer, C. von der Malsburg, L. E. Orgel, J. C. W. Shepherd, and E. C. Zeeman made unpublished material available to us. Drs. R. Wolff and J. Swetina provided several computer plots. Financial support for these studies was provided by the Austrian Fonds zur Forderung der wissenschaftlichen Forschung, Project No. 3502. We also thank Dr. B. Schreiber and Mrs. J. Jakubetz for typing the manuscript and Mr. J. Konig for drawing the figures.

References

Batschelet, E., E. Domingo, and C. Weissmann (1976) The proportion of revertant and mutant phage in a growing population, as a function of mutation and growth rate. *Gene* **1**:27–32.

Biebricher, C., M. Eigen, and R. Luce (1980) Kinetics of template instructed and de novo RNA synthesis by Qβ-replicase. *J. Mol. Biol.* **148**:391–410.

5. Self-Organization of Macromolecules

Biebricher, C. K., M. Eigen and W. C. Gardiner, Jr. (1983) Kinetics of RNA replication. *Biochemistry* **22**:2544–2559.

Cowan, J. D. (1970) A statistical mechanics of nervous activity. In: *Lectures on Mathematics in the Life Sciences*, M. Gerstenhaber (ed.). American Mathematical Society, Providence, R.I., pp. 1–57.

Crick, F. H. C., S. Brenner, A. Klug, and G. Pieczenik (1976) A speculation on the origin of protein synthesis. *Origins Life* **7**:389–397.

Crow, J. F., and M. Kimura (1970) *An Introduction to Population Genetics Theory.* Harper and Row, New York.

Dawkins, R. (1976) *The Selfish Gene.* Oxford University Press, London, England, p. 16.

Dixon-Lewis, G., and D. J. Williams (1977) The oxidation of hydrogen and carbon monoxide. In: *Gas-Phase Combustion, Comprehensive Chemical Kinetics* Vol. 17, C. H. Bamford and C. F. H. Tipper (eds.). Elsevier, Amsterdam, pp. 1–248.

Domingo, E., R. A. Flavell, and C. Weissmann (1976) *In vitro* site directed mutagenesis: Generation and properties of an infectious extracistronic mutant of bacteriophage Qβ. *Gene* **1**:3–25.

Edelson, D., R. M. Noyes, and R. J. Field (1979) Mechanistic details of the Belousov–Zhabotinski oscillations: The organic reaction subset. *Int. J. Chem. Kinet.* **11**:155–164.

Eigen, M. (1971) Self-organization of matter and the evolution of biological macromolecules. *Naturwissenschaften* **58**:465–523.

Eigen, M., and P. Schuster (1979) *The Hypercycle—A Principle of Natural Self-Organization.* Springer, Berlin.

Eigen, M., and P. Schuster (1982) Stages of emerging life—five principles of early organization. *J. Mol. Evol.* **19**:47–61.

Eigen, M., and R. Winkler-Oswatitsch (1981a) Transfer-RNA: The early adaptor. *Naturwissenschaften* **68**:217–228.

Eigen, M., and R. Winkler-Oswatitsch (1981b) Transfer-RNA, an early gene? *Naturwissenschaften* **68**:282–292.

Eigen, M., P. Schuster, K. Sigmund, and R. Wolff (1980) Elementary step dynamics of catalytic hypercycles. *BioSystems* **13**:1–22.

Eigen, M., W. Gardiner, P. Schuster, and R. Winkler-Oswatitsch (1981) The origin of the genetic information. *Sci. Am.* **244** (4): 88–118.

Epstein, I. R. (1979) Coexistence, competition and hypercyclic interaction in some systems of biological interest. *Biophys. Chem.* **9**:245–250.

Epstein, I. R., and M. Eigen (1979) Selection and self-organization of self-reproducing macromolecules under constraint of constant flux. *Biophys. Chem.* **10**:153–160.

Field, R. J., and R. M. Noyes (1974) Oscillations in chemical systems. IV. Limit cycle behaviour in a model of a real chemical reaction. *J. Chem. Phys.* **60**:1877–1884.

Fox, S. W., and K. Dose (1972) *Molecular Evolution and the Origin of Life.* Freeman, San Francisco.

Fox, S. W., K. Haranda, and J. Kendrick (1959) Production of spherules from synthetic protenoid and hot water. *Science* **129**:1221–1223.

Hadeler, K. P. (1974) *Mathematik für biologen.* Springer, Berlin, pp. 96–97, 147–149.

Hall, B. D. (1979) Mitochondria spring surprises. *Nature* **282**:129–130.

Hofbauer, J. (1980) On the occurrence of limit cycles in the Volterra–Lotka differential equation. *Nonlinear Analysis, Theory, Methods & Applications* **5**:1003–1007.

Hofbauer, J., P. Schuster, and K. Sigmund (1979) A note on evolutionarily stable strategies and game dynamics. *J. Theor. Biol.* **81**:609–612.

Hofbauer, J., P. Schuster, K. Sigmund, and R. Wolff (1980) Dynamical systems under constant organization. II. Homogeneous growth functions of degree $p=2$. *SIAM J. Appl. Math. C* **38**:282–304.

Hofbauer, J., P. Schuster, and K. Sigmund (1981) Competition and cooperation in catalytic selfreplication. *J. Math. Biol.* **11**:155–168.

Jacob, B. (1977) Evolution and tinkering. *Science* **196**:1161–1166.

Jones, B. L. (1977) A solvable selfreproductive hypercycle model for the selection of biological molecules. *J. Math. Biol.* **4**:187–193.

Jones, B. L., R. H. Enns, and S. S. Ragnekar (1976) On the theory of selection of coupled macromolecular systems. *Bull. Math. Biol.* **38**:15–28.

Kerner, E. H. (1961) On the Volterra–Lotka principle. *Bull. Math. Biophys.* **23**:141–157.

Kornberg, A. (1974) *DNA Synthesis.* Freeman, San Francisco.
Köster, H. (1979) Chemical DNA-synthesis. *Nachr. Chem. Tech. Lab.* **27**:694–700.
Kunkel, T. A., and L. Loeb (1979) On the fidelity of DNA replication. *J. Biol. Chem.* **254**:5718–5725.
Kunkel, T. A., R. R. Meyer, and L. Loeb (1979) Single-strand binding protein enhances fidelity of DNA synthesis *in vitro. Proc. Natl. Acad. Sci. USA* **76**:6331–6335.
Küppers, B. O. (1979a) Some remarks on the dynamics of molecular self-organization. *Bull. Math. Biol.* **41**:803–812.
Küppers, B. O. (1979b) Towards an experimental analysis of molecular self-organization and precellular Darwinian evolution. *Naturwissenschaften* **66**:228–243.
Lohrmann, R., P. K. Bridson, and L. E. Orgel (1980) An efficient metal-ion catalyzed template-directed oligo-nucleotide synthesis. *Science* **208**:1464–1465.
Margulis, L., J. C. G. Walker, and M. Rambler (1976) Reassessment of roles of oxygen and ultraviolet light in Precambrian evolution. *Nature* **264**:620–624.
Maynard-Smith, J., and G. Price (1973) The logic of animal conflicts. *Nature* **246**:15–18.
Meinhardt, H., and A. Gierer (1980) Generation and regeneration of sequences of structures during morphogenesis. *J. Theor. Biol.* **85**:429–450.
Miller, S. L. and L. E. Orgel (1974) *The Origins of Life on the Earth.* Prentice–Hall, Englewood Cliffs, N.J.
Nicolis, G. and I. Prigogine (1977) *Self-Organization in Nonequilibrium Systems.* Wiley, New York.
Nisbet, E. G. (1980) Archaean stromatolites and the search for the earliest life. *Nature* **284**:395–396.
Pörschke, D. (1977) Elementary steps of base recognition and helix–coil transitions in nucleic acids. In: *Chemical Relaxation in Molecular Biology*, I. Pecht and R. Rigler (eds.). Springer-Verlag, Berlin, pp. 191–218.
Rossmann, M. G., D. Moras, and K. W. Olsen (1974) Chemical and biological evolution of a nucleotide-binding protein. *Nature* **250**:194–199.
Schneider, F. W., D. Neuser, and M. Heinrichs (1979) Hysteretic behaviour in poly(A)–poly(U) synthesis in a stirred flow reactor. In: *Molecular Mechanisms of Biological Recognition*, M. Balaban (ed.). Elsevier/North-Holland, Amsterdam, pp. 241–252.
Schopf, J. W. (1978) The evolution of the earliest cells. *Sci. Am.* **239**(3):84–102.
Schuster, P. (1981a) Selection and evolution in molecular systems. In: *Nonlinear Phenomena in Physics and Biology*, R. H. Enns, B. J. Jones, R. M. Miura, and S. S. Rangnekar (eds.). Plenum Press, New York, pp. 485–548.
Schuster, P. (1981b) Prebiotic evolution. In: *Biochemical Evolution*, H. Gutfreund (ed.). Cambridge University Press, London, England, pp. 15–87.
Schuster, P., and K. Sigmund (1983) From biological macromolecules to protocells—the principle of early evolution. In: *Biophysik*, W. Hoppe, W. Lohmann, H. Markl, and H. Ziegler (eds.). Springer-Verlag, Berlin, pp. 874–912.
Schuster, P., K. Sigmund, and R. Wolff (1978) Dynamical systems under constant organization. I. Topological analysis of a family of non-linear differential equations—a model for catalytic hypercycles. *Bull. Math. Biophys.* **40**:743–769.
Schuster, P., K. Sigmund, and R. Wolff (1979) Dynamical systems under constant organization III. Cooperative and competitive behaviour of hypercycles. *J. Diff. Eq.* **32**:357–368.
Schuster, P., K. Sigmund, J. Hofbauer, and R. Wolff (1981a) Selfregulation of behaviour in animal societies. I. Symmetric contests. *Biol. Cybern.* **40**:1–8.
Schuster, P., K. Sigmund, J. Hofbauer, and R. Wolff (1981b) Selfregulation of behaviour in animal societies. II. Games between two populations without selfinteraction. *Biol. Cybern.* **40**:9–15.
Schuster, P., K. Sigmund, J. Hofbauer, R. Gottlieb, and P. Merz (1981c) Selfregulation of behaviour in animal societies. III. Games between two populations without selfinteraction. *Biol. Cybern.* **40**:17–25.
Shepherd, J. C. W. (1981) Periodic correlations in DNA sequences and evidence for their evolutionary origin in a comma-less genetic code. *J. Mol. Evol.* **17**:94–102.
Sleeper, H. L., R. Lohrmann, and L. E. Orgel (1979) Template-directed synthesis of oligoadenylates catalyzed by Pb^{2+} ions. *J. Mol. Evol.* **13**:203–214.
Staudenbauer, W. (1978) Structure and replication of the colicin E1 plasmid. *Cur. Top. Microbiol. Immunol.* **83**:93–156.

Swetina, J., and P. Schuster (1982) Self-replication with errors: A model for polynucleotide replication. *Biophys. Chem.* **16**:329–345.

Taylor, P., and L. Jonker (1978) Evolutionarily stable strategies and game dynamics. *Math. Biosci.* **40**:145–156.

Thompson, C. J., and J. L. McBride (1974) On Eigen's theory of the self-organization of matter and the evolution of biological macromolecules. *Math. Biosci.* **21**:127–142.

Tyson, J. J. (1976) *The Belousov–Zhabotinsky Reaction. Lecture Notes in Biomathematics Vol. 10.* Springer-Verlag, Berlin.

von der Malsburg, C., and D. J. Willshaw (1981) Differential equations for the development of topological nerve fibre projections. *SIAM-AMS Proc.* **13**: 39–57.

Walker, G. W. R. (1977) Nucleotide-binding site data and the origin of the genetic code. *BioSystems* **9**: 139–150.

Zeeman, E. C. (1980) *Population Dynamics from Game Theory. Lecture Notes in Mathematics Vol. 819.* Springer-Verlag, Berlin.

6

Is a New and General Theory of Evolution Emerging?

Stephen J. Gould

ABSTRACT

The "modern synthetic" view of evolution has broken down, at least as an exclusive proposition, on both of its fundamental claims: (1) "extrapolationism" (gradual substitution of different alleles in many genes as the exclusive process underlying all evolutionary change) and (2) nearly exclusive reliance on selection leading to adaptation. Evolution is a hierarchical process with complementary, but different modes of change at its three large-scale levels: (a) variation within populations, (b) speciation, and (c) very long-term macroevolutionary trends. Speciation is not always an extension of gradual, adaptive allelic substitution, but may represent, as Goldschmidt argued, a different style of genetic change—rapid reorganization of the genome, perhaps nonadaptive. Macroevolutionary trends do not arise from the gradual, adaptive transformation of populations, but usually from a higher-order selection operating upon groups of species. Individual species generally do not change much after their "instantaneous" (in geological time) origin. These two discontinuities in the evolutionary hierarchy can be called the Goldschmidt break (change in populations is different from speciation) and the Wright break (speciation is different from macroevolutionary trending that translates differential success among different species).

A new and general evolutionary theory will embody this notion of hierarchy and stress a variety of themes either ignored or explicitly rejected by the modern synthesis: e.g., punctuational change at all levels, important nonadaptive change at all levels, control of evolution not only by selection, but equally by constraints of history, development, and architecture—thus restoring to evolutionary theory a concept of organism. —THE EDITOR

Previously published in slightly shorter form as Gould (1980a).

STEPHEN J. GOULD • Museum of Comparative Zoology, Harvard University, Cambridge, Massachusetts 02138.

The Modern Synthesis

In one of the last skeptical books written before the Darwinian tide of the modern synthesis asserted its hegemony, Robson and Richards (1936, pp. 370–371) characterized the expanding orthodoxy that they deplored:

> The theory of Natural Selection ... postulates that the evolutionary process is unitary, and that not only are groups formed by the multiplication of single variants having survival value, but also that such divergences are amplified to produce adaptation (both specializations and organization). It has been customary to admit that certain ancillary processes are operative (isolation, correlation), but the importance of these, as active principles, is subordinate to selection.

Darwinism, as a set of ideas, is sufficiently broad and variously defined to include a multitude of truths and sins. Darwin himself disavowed many interpretations made in his name (see, e.g., Darwin, 1880). The version known as the "modern synthesis" or "neo-Darwinism" (different from what the late 19th century called neo-Darwinism—see Romanes, 1900) is, I think, fairly characterized in its essentials by Robson and Richards. Its foundation rests on two major premises: (1) Point mutations (micromutations) are the ultimate source of variability. Evolutionary change is a process of gradual allelic substitution within a population. Events on a broader scale, from the origin of new species to long-ranging evolutionary trends, represent the same process, extended in time and effect—large numbers of allelic substitutions incorporated sequentially over long periods. In short this is gradualism, continuity, and evolutionary change by the transformation of populations. (2) Genetic variation is raw material only. Natural selection directs evolutionary change. Rates and directions of change are controlled by selection with little constraint exerted by raw material (slow rates are due to weak selection, not insufficient variation). All genetic change is adaptive (although some phenotypic effects due to, e.g., pleiotropy, etc., may not be). In short, this is selection leading to adaptation.

All these statements, as Robson and Richards also note, are subject to recognized exceptions—and this imposes a great frustration upon anyone who would characterize the modern synthesis in order to criticize it. All synthesists recognized exceptions and "ancillary processes," but they attempted both to prescribe a low relative frequency for them and to limit their application to domains of little evolutionary importance. Thus, genetic drift certainly occurred—but only in populations so small and so near the brink that their rapid extinction would almost certainly ensue. And phenotypes included many nonadaptive features by allometry and pleiotropy, but all were epiphenomena of primarily adaptive genetic changes and none could have any marked effect upon the organism (for, if nonadaptive, they would lead to negative selection and elimination and, if adaptive, would enter the model in their own right). Thus, a synthesist could always deny a charge of rigidity by invoking these official exceptions, even though their circumscription, both in frequency and effect, actually guaranteed the hegemony of the two cardinal principles. This frustrating situation had been noted by critics of an earlier Darwinian orthodoxy: for example, by Romanes (1900, p. 21) writing of Wallace:

6. A New and General Theory of Evolution? 115

> [For Wallace] the law of utility is, to all intents and purposes, universal, with the result that natural selection is virtually the only cause of organic evolution. I say 'to all intents and purposes,' or 'virtually,' because Mr. Wallace does not expressly maintain the abstract impossibility of laws and causes other than those of utility and natural selection; indeed, at the end of his treatise, he quotes with approval Darwin's judgment that 'natural selection has been the most important, but not the exclusive means of modification.' Nevertheless, as he nowhere recognizes any other law or cause of adaptive evolution, he practically concludes that, on inductive or empirical grounds, there *is* no such other law or cause to be entertained.

Lest anyone think that Robson and Richards, as doubters, had characterized the opposition unfairly, or that their two principles—extrapolation of small genetic changes to yield evolutionary events at all scales and control of direction by selection leading to adaptation—represent too simplistic or unsubtle a view of the synthetic theory, I cite the characterization by Mayr—one of the architects of the theory himself. In the first statement of his chapter on species and transspecific evolution, Mayr (1963, p. 586) wrote:

> The proponents of the synthetic theory maintain that all evolution is due to the accumulation of small genetic changes, guided by natural selection, and that transspecific evolution is nothing but an extrapolation and magnification of the events that take place within populations and species.

The early classics of the modern synthesis—particularly Dobzhansky's first edition (1937) and Simpson's first book (1944)—were quite expansive, generous, and pluralistic. But the synthesis hardened throughout the late 1940s and 1950s, and later editions of the same classics (Dobzhansky, 1951; Simpson, 1953) were more rigid in their insistence upon micromutation, gradual transformation, and adaptation guided by selection [see Gould (1980b) for an analysis of changes between Simpson's two books]. When Watson and Crick then determined the structure of DNA, and when the triplet code was cracked a few years later, everything seemed to fall even further into place. It had been known that chromosomes are long strings of triplets coding, in sequence, for the proteins that build organisms and that most point mutations are simple base substitutions. A physics and chemistry had been added, and it squared well with the prevailing orthodoxy.

The synthetic theory beguiled me with its unifying power when I was a graduate student in the mid-1960s. Since then I have been watching it slowly unravel as a universal description of evolution. The molecular assault came first, followed quickly by renewed attention to unorthodox theories of speciation and by challenges at the level of macroevolution itself. I have been reluctant to admit it—since beguiling is often forever—but if Mayr's characterization of the synthetic theory is accurate, then that theory, as a general proposition, is effectively dead, despite its persistence as textbook orthodoxy.

Reduction and Hierarchy

The modern synthetic theory embodies a strong faith in reductionism. It advocates a smooth extrapolation across all levels and scales—from the base substitution to the origin of higher taxa. The most sophisticated of the leading introductory textbooks in biology still proclaims:

> [Can] more extensive evolutionary change, macroevolution, be explained as an outcome of these microevolutionary shifts? Did birds really arise from reptiles by an accumulation of gene substitutions of the kind illustrated by the raspberry eye-color gene?
> The answer is that it is entirely plausible, and no one had come up with a better explanation.... The fossil record suggests that macroevolution is indeed gradual, paced at a rate that leads to the conclusion that it is based upon hundreds or thousands of gene substitutions no different in kind from the ones examined in our case histories [Wilson et al., 1973, pp. 793–794].

The general alternative to such reductionism is a concept of hierarchy—a world constructed not as a smooth and seamless continuum, permitting simple extrapolation from the lowest level to the highest, but as a series of ascending levels, each bound to the one below it in some ways and independent in others. Discontinuities and seams characterize the transitions; "emergent" features, not implicit in the operation of processes at lower levels, may control events at higher levels. The basic processes—mutation, selection, and so on—may enter into explanations on all levels (and in that sense we may still hope for a general theory of evolution), but they work in different ways on the characteristic material of diverse levels [see Bateson (1978) and Koestler (1978) for discussions on other inadequacies and on hierarchy and its antireductionistic implications].

The molecular level, which once seemed through its central dogma and triplet code to provide an excellent "atomic" basis for smooth extrapolation, now demands hierarchical interpretation itself. The triplet code is only machine language (I thank F. E. Yates for this appropriate metaphor). The program resides at a higher level of control and regulation—and we know virtually nothing about it. With its inserted sequences and jumping genes, the genome contains sets of scissors and pots of glue to snip and unite bits and pieces from various sources. Thirty to seventy percent of mammalian genes exist in multiple copies, some repeated hundreds of thousands of times. What are they for (if anything)? What role do they play in the regulation of development? Molecular biologists are groping to understand this higher control upon primary products of the triplet code. In that understanding, we will probably obtain a basis for styles of evolutionary change radically different from the sequential allelic substitutions, each of minute effect, that the modern synthesis so strongly advocated. The uncovering of hierarchy on the molecular level will probably exclude smooth continuity across other levels. (We may find, for example, that structural gene substitutions control most small-scale, adaptive variation within local populations, while disruption of regulation lies behind most key innovations in macroevolution.)

The modern synthesis drew most of its direct conclusions from studies of local populations and their immediate adaptations. It then extrapolated the postulated mechanism of these adaptations—gradual, allelic substitution—to encompass all larger-scale events. The synthesis is now breaking down on both sides of this argument. Many evolutionists now doubt exclusive control by selection upon genetic change within local populations. Moreover, even if local populations alter as the synthesis maintains, we now doubt that the same style of change controls events at the two major higher levels: speciation and evolutionary trends across species.

A Note on Local Populations and Neutrality

At the level of populations, the synthesis has broken on the issue of amounts of genetic variation. Selection, although it eliminates variation in both its classical modes (directional and, especially, stabilizing), can also act to preserve variation through such phenomena as overdominance, frequency dependence, and response to small-scale fluctuation of spatial and temporal environments. Nonetheless, the copiousness of genetic variation, as revealed first in the electrophoretic techniques that resolve only some of it (Lewontin and Hubby, 1966; Lewontin, 1974), cannot be encompassed by our models of selective control. (Of course, the models, rather than nature, may be wrong.) This fact has forced many evolutionists, once stout synthesists themselves, to embrace the idea that alleles often drift to high frequency or fixation and that many common variants are therefore neutral or just slightly deleterious. This admission lends support to a previous interpretation of the approximately even ticking of the molecular clock (Wilson *et al.*, 1977)—that it reflects the neutral status of most changes in structural genes, rather than a grand averaging of various types of selection over time.

None of this evidence, of course, negates the role of conventional selection and adaptation in molding parts of the phenotype having obvious importance for survival and reproduction. Still, it rather damps Mayr's enthusiastic claim for "all evolution . . . guided by natural selection." The question, as with so many issues in the complex sciences of natural history, becomes one of relative frequency. Are the Darwinian substitutions merely a surface skin on a sea of variation invisible to selection, or are the neutral substitutions merely a thin bottom layer underlying a Darwinian ocean above? Or are they somewhere in between?

In short, the specter of stochasticity has intruded upon explanations of evolutionary *change*. This represents a fundamental challenge to Darwinism, which holds, as its very basis, that random factors enter only in the production of raw material and that the deterministic process of selection produces change and direction (see Nei, 1975).

The Level of Speciation and the Goldschmidt Break

Ever since Darwin called his book *On the Origin of Species*, evolutionists have regarded the formation of reproductively isolated units by speciation as the fundamental process of large-scale change. Yet speciation occurs at too high a level to be observed directly in nature or produced by experiment in most cases. Therefore, theories of speciation have been based on analogy, extrapolation, and inference. Darwin himself focused on artificial selection and geographic variation. He regarded subspecies as incipient species and viewed their gradual, accumulating divergence as the primary mode of origin for new taxa. The modern synthesis continued this tradition of extrapolation from local populations and used the accepted model for adaptive geographic variation—gradual allelic substitution directed by natural selection—as a paradigm for the origin of the species. Mayr's (1942, 1963) model of allopatric speciation did challenge Darwin's implied notion of sympatric continuity. It emphasized the crucial role of isolation from gene flow and did promote the importance of small founding populations and relatively rapid rates of change. Thus, the small peripheral isolate, rather than the large local population in persistent contact with other conspecifics, became the incipient species. Nonetheless, despite this welcome departure from the purest form of Darwinian gradualism, the allopatric theory held firmly to the two major principles that permit smooth extrapolation from the *Biston betularia* model of adaptive, allelic substitution: (1) The accumulating changes that lead to speciation are adaptive. Reproductive isolation is a consequence of sufficient accumulation. (2) Although aided by founder effects and even (possibly) by drift, although dependent upon isolation from gene flow, and although proceeding more rapidly than local differentiation within large populations, successful speciation is still a cumulative and sequential process powered by selection through large numbers of generations. It is, if you will, Darwinism a little faster.

I have no doubt that many species originate in this way; but it now appears that many, perhaps most, do not. The new models, or the reassertion of old ones, stand at variance with the synthetic proposition that speciation is an extension of microevolution within local populations. Some of the new models call upon genetic variation of a different kind, and they regard reproductive isolation as potentially primary and nonadaptive rather than secondary and adaptive. Insofar as these new models are valid in theory and numerically important in application, speciation is not a simple "conversion" to a larger effect of processes occurring at the lower level of adaptive modeling within local populations. It represents a discontinuity in our hierarchy of explanations, as the much maligned Richard Goldschmidt argued explicitly in 1940.

There are many ways to synthesize the swirling set of apparently disparate challenges that have rocked the allopatric orthodoxy and established an alternative set of models for speciation. The following reconstruction is neither historically sequential nor the only logical pathway of linkage, but it does summarize the challenges—on population structure, place of origin, genetic style, rate, and relation to adaptation—in some reasonable order.

1. Under the allopatric orthodoxy, species are viewed as integrated units which, if not actually panmictic, are at least sufficiently homogenized by gene flow to be treated as entities. This belief in effective homogenization within central populations underlies the allopatric theory with its emphasis on *peripheral* isolation as a precondition for speciation. But many evolutionists now believe that gene flow is often too weak to overcome selection and other intrinsic processes within local demes (Ehrlich and Raven, 1969). Thus, the model of a large, homogenized central population preventing local differentiation and requiring allopatric "flight" of isolated demes for speciation may not be generally valid. Perhaps most local demes have the required independence for potential speciation.

2. The primary terms of reference for theories of speciation—allopatry and

sympatry—lose their meaning if we accept the first statement. Objections to sympatric speciation centered upon the homogenizing force of gene flow. But if demes may be independent in all geographic domains of a species, then sympatry loses its meaning and allopatry its necessity. Independent demes within the central range (sympatric by location) function, in their freedom from gene flow, like the peripheral isolates of allopatric theory. In other words, the terms make no sense outside a theory of population structure that contrasts central panmixia with marginal isolation. They should be abandoned.

3. In this context, "sympatric" speciation loses its status as an extremely improbable event. If demes are largely independent, new species may originate anywhere within the geographic range of an ancestral form. Moreover, many evolutionists now doubt that parapatric distributions (far more common than previously thought) must represent cases of secondary contact. White (1978, p. 342) believes that many, if not most, are primary and that speciation can also occur between populations continually in contact, if gene flow can be overcome either by strong selection or by the sheer rapidity of potential fixation for major chromosomal variants (see White, 1978, p. 17, on clinal speciation).

4. Most "sympatric" models of speciation are based on rates and extents of genetic change inconsistent with the reliance placed by the modern synthesis on slow or at least sequential change. Even Mayr (1963) acknowledged the essentially saltatory and sympatric formation of new insect species by switch in plant preferences for host-specific forms. (By granting this and no other exception to allopatry, he attempted to circumscribe departure into a limited domain.)

The most exciting entry among punctuational models for speciation in ecological time is the emphasis, now coming from several quarters, on chromosomal alterations as isolating mechanisms (Bush, 1975; Carson, 1975, 1978; Wilson *et al.*, 1975; Bush *et al.*, 1977; White, 1978)—sometimes called the theory of chromosomal speciation. In certain population structures, particularly in very small and circumscribed groups with high degrees of inbreeding, major chromosomal changes can rise to fixation in less than a handful of generations (mating of heterozygous F_1 sibs to produce F_2 homozygotes for a start).

Wilson, Bush, and their colleagues (Wilson *et al.*, 1975; Bush *et al.*, 1977) find a strong correlation between rates of karyotypical and anatomical change, but no relation between amounts of substitution in structural genes and any conventional assessment of phenotypic modification, either in speed or extent. They suggest that speciation may be more a matter of gene regulation and rearrangement than of changes in structural genes that adapt local populations in minor ways to fluctuating environments (see *Biston betularia* model in the glossary at the end of this chapter). Moreover, they have emphasized the importance of small demes and chromosomal speciation in maintaining that both speciation and karyotypical change are "fastest in those genera with species organized into clans or harems, or with limited adult vagility and juvenile dispersal, patchy distribution, and strong individual territoriality" (Bush *et al.*, 1977, p. 3942).

Carson (1975, 1978) has also stressed the importance of small demes, chromosomal change, and extremely rapid speciation in his founder–flush theory with its emphasis on extreme bottlenecking during crashes of the flush–crash cycle (see

Powell, 1978, for experimental support). Explicitly contrasting this view with extrapolationist models based on sequential substitution of structural genes, Carson (1975, p. 88) writes:

> Most theories of speciation are wedded to gradualism, using the mode of origin of intraspecific adaptations as a model.... I would nevertheless like to propose ... that speciational events may be set in motion and important genetic saltations towards species formation accomplished by a series of catastrophic, stochastic genetic events ... initiated when an unusual forced reorganization of the epistatic super-genes of the closed variability system occurs.... I propose that this cycle of disorganization and reorganization be viewed as the essence of the speciation process.

5. Another consequence of such essentially saltatorial origin is even more disturbing to conventional views than the rapidity of the process itself, as Carson has forcefully stated. The control of evolution by selection leading to adaptation is at the heart of the modern synthesis. Thus, reproductive isolation, the definition of speciation, is attained as a by-product of adaptation—that is, a population diverges by sequential adaptation and eventually becomes sufficiently different from its ancestor to foreclose interbreeding. (Selection for reproductive isolation may also be direct when two imperfectly separate forms come into contact.) But in saltational, chromosomal speciation, reproductive isolation comes first and cannot be considered as an adaptation at all. It is a stochastic event that establishes a species by the technical definition of reproductive isolation. To be sure, the later success of this species in competition may depend upon its subsequent acquisition of adaptations, but the origin itself may be nonadaptive. We can, in fact, reverse the conventional view and argue that speciation, by forming new entities stochastically, provides raw material for selection.

These challenges can be summarized in the claim that a discontinuity in explanation exists between allelic substitutions in local populations (sequential, slow, and adaptive) and the origin of new species (often discontinuous and nonadaptive). During the heyday of the modern synthesis, Richard Goldschmidt was castigated for his defense of punctuational speciation. I was told as a graduate student that this great geneticist had gone astray because he had been a lab man with no feel for nature—a person who had not studied the adaptation of local populations and could not appreciate its potential power, by extrapolation, to form new species. But I discovered, in writing *Ontogeny and Phylogeny*, that Goldschmidt had spent a good part of his career studying geographic variation, largely in the coloration of lepidopteran larvae. (During these studies, he developed the concept of rate genes to explain minor changes in pattern.) I then turned to his major book (Goldschmidt, 1940) and found that his defense of saltational speciation is not based on ignorance of geographic variation, but on an explicit study of it; half the book is devoted to this subject. Goldschmidt concludes that geographic variation is ubiquitous, adaptive, and essential for the persistence of established species. But it is simply not the stuff of speciation; it is a different process. Speciation, Goldschmidt argues, occurs at different rates and uses different kinds of genetic variation. We do not now accept all his arguments about the nature of variation, but his explicit antiextrapolationist statement is the epitome and foundation of emerging views on speciation discussed in this section. There

6. A New and General Theory of Evolution?

is a discontinuity in cause and explanation between adaptation in local populations and speciation; they represent two distinct, although interacting, levels of evolution. We might refer to this discontinuity as the *Goldschmidt break*, for he wrote:

> The characters of subspecies are of a gradient type, the species limit is characterized by a gap, an unbridged difference in many characters. This gap cannot be bridged by theoretically continuing the subspecific gradient or cline beyond its actually existing limits. The subspecies do not merge into the species either actually or ideally .... Microevolution by accumulation of micromutations—we may also say neo-Darwinian evolution—is a process which leads to diversification strictly within the species, usually, if not exclusively, for the sake of adaptation of the species to specific conditions within the area which it is able to occupy .... Subspecies are actually, therefore, neither incipient species nor models for the origin of species. They are more or less diversified blind alleys within the species. The decisive step in evolution, the first step towards macroevolution, the step from one species to another, requires another evolutionary method than that of sheer accumulation of micromutations. [Goldschmidt, 1940, p. 183]

Macroevolution and the Wright Break

The extrapolationist model of macroevolution views trends and major transitions as an extension of allelic substitution within populations—the march of frequency distributions through time. Gradual change becomes the normal state of species. The discontinuities of the fossil record are all attributed to its notorious imperfection; the remarkable stasis exhibited by most species during millions of years is ignored (as no data) or relegated to descriptive sections of taxonomic monographs. But gradualism is not the only important implication of the extrapolationist model. Two additional consequences have channeled our concept of macroevolution, both rather rigidly and with unfortunate effect. First, the trends and transitions of macroevolution are envisaged as events in the phyletic mode—populations transforming themselves steadily through time. Splitting and branching are acknowledged to be sure, lest life be terminated by its prevalent extinctions. But splitting becomes a device for the generation of diversity upon designs attained through "progressive" processes of transformation. Splitting, or cladogenesis, becomes subordinate in importance to transformation, or anagenesis (see Ayala, 1976, p. 141; but see also Mayr, 1963, p. 621, for a rather lonely voice in the defense of copious speciation as an input to "progressive" evolution). Second, the adaptationism that prevails in interpreting change in local populations gains greater confidence in extrapolation. For if allelic substitutions in ecological time have an adaptive basis, then surely a unidirectional trend that persists for millions of years within a single lineage cannot bear any other interpretation.

This extrapolationist model of adaptive, phyletic gradualism has been vigorously challenged by several paleobiologists—and again with a claim for discontinuity in explanation at different levels. The general challenge embodies three loosely united themes:

1. *Evolutionary trends as a higher-level process.* Eldredge and I have argued

(Eldredge and Gould, 1972; Gould and Eldredge, 1977) that imperfections of the record cannot explain all discontinuity (and certainly cannot encompass stasis). We regard stasis and discontinuity as an expression of how evolution works when translated into geological time. Gradual change is not the normal state of a species. Large, successful central populations undergo minor adaptive modifications of fluctuating effect through time (Goldschmidt's "diversified blind alleys"), but they will rarely transform *in toto* to something fundamentally new. Speciation, the basis of macroevolution, is a process of branching. Under any current model of speciation—conventional allopatry to chromosomal saltation—this branching is rapid in geological translation (thousands of years at most compared with millions for the duration of most fossil species). Thus, its results should generally lie on a bedding plane, not through the thick sedimentary sequence of a long hillslope. (The expectation of gradualism emerges as a kind of double illusion. It represents, first of all, an incorrect translation of conventional allopatry. Allopatric speciation seems so slow and gradual in ecological time that most paleontologists never recognized it as a challenge to the style of gradualism—steady change over millions of years—promulgated by custom as a model for the history of life. But it now appears that "slow" allopatry itself may be less important than a host of alternatives that yield new species rapidly even in ecological time.) Thus, our model of punctuated equilibria holds that evolution is concentrated in events of speciation and that successful speciation is an infrequent event punctuating the stasis of large populations that do not alter in fundamental ways during the millions of years that they endure.

But if species originate in geological instants and then do not alter in major ways, then evolutionary trends—the fundamental phenomenon of macroevolution—cannot represent a simple extrapolation of allelic substitution within a population. Trends must be the product of differential success among species (Eldredge and Gould, 1972; Stanley, 1975). In other words, species themselves must be inputs, and trends the result of their differential origin and survival. Speciation interposes itself as an irreducible level between change in local populations and trends in geological time. Macroevolution is, as Stanley (1975, p. 648) argues, decoupled from microevolution.

Sewall Wright recognized the hierarchical implication of viewing species as irreducible inputs to macroevolution when he argued that the relationship between change in local populations and evolutionary trends can only be analogical (Wright, 1967). Just as mutation is random with respect to the direction of change within a population, so too might speciation be random with respect to the direction of a macroevolutionary trend. A higher form of selection, acting directly upon species through differential rates of extinction, may then be the analogue of natural selection working within populations through differential mortality of individuals.

Evolutionary trends therefore represent a third level superposed upon speciation and change within demes. Intrademic events cannot encompass speciation because rates, genetic styles, and relation to adaptation differ for the two processes. Likewise, since trends "use" species as their raw material, they represent a process at a higher level than speciation itself. They reflect a sorting out of speciation events. With apologies for the pun, the hierarchical rupture between speciation and macroevolutionary trends might be called the Wright break.*

As a final point about the extrapolation of methods for the study of events within populations, the cladogenetic basis of macroevolution virtually precludes any direct application of the primary apparatus for microevolutionary theory: classical population genetics. I believe that essentially all macroevolution is cladogenesis and its concatenated effects. What we call "anagenesis," and often attempt to delineate as a separate phyletic process leading to "progress," is just accumulated cladogenesis filtered through the directing force of species selection (Stanley, 1975)—Wright's higher-level analogue of natural selection. Carson (1978, p. 925) makes the point forcefully, again recognizing Sewall Wright as its long and chief defender:

> Investigation of cladistic events as opposed to phyletic (anagenetic) ones requires a different perspective from that normally assumed in classical population genetics. The statistical and mathematical comfort of the Hardy-Weinberg equilibrium in large populations has to be abandoned in favor of the vague realization that nearly everywhere in nature we are faced with data suggesting the partial or indeed complete sundering of gene pools. If we are to deal realistically with cladogenesis we must seek to delineate each genetic and environmental factor which may promote isolation. The most important devices are clearly those which operate at the very lowest population level: sib from sib, family from family, deme from deme. Formal population genetics just cannot deal with such things, as Wright pointed out long ago.

Eldredge (1979) has traced many conceptual errors and prejudicial blockages to our tendency for conceiving evolution as the transformation of *characters* within phyletic lineages, rather than as the origin of the new *taxa* by cladogenesis (the transformational versus the taxic view in his terms). I believe that, in ways deeper than we realize, our preference for transformational thinking (with the relegation of splitting to the generation of variety upon established themes) represents a cultural tie to the controlling Western themes of progress and ranking by intrinsic merit—an attitude that can be traced in evolutionary thought to Lamarck's distinction between the march up life's ladder promoted by the *pouvoir de la vie* and the tangential deflections imposed by *l'influence des circonstances*, with the first process essential and the second deflective. Nonetheless, macroevolution is fundamentally about the origin of taxa by splitting.

2. *The saltational initiation of major transitions.* The absence of fossil evi-

---

*I had the honor—not a word I use frequently, but inescapable in this case—of spending a long evening with Dr. Wright. I discovered that his quip about macroevolution, just paraphrased, was no throwaway statement but an embodiment of his deep commitment to a hierarchical view of evolutionary causation. (The failure of many evolutionists to think hierarchically is responsible for the most frequent misinterpretation of Wright's views. He never believed that genetic drift—the Sewall Wright effect as it was once called—is an important agent of evolutionary *change*. He regards it as input to the directional process of interdemic selection for evolution within species. Drift can push a deme off an adaptive peak; selection can then draw it to another peak. Thus, many peaks can be occupied and evolution can occur by selection among demes. Yet if we fail to realize that Wright views allelic substitution within demes as a hierarchical level distinct from sorting out among demes—and if we regard all evolution as continuous with intrapopulational processes—then we must view drift as a directional agent and fail to understand Wright's position.)

dence for intermediary stages between major transitions in organic design, indeed our inability, even in our imagination, to construct functional intermediates in many cases, has been a persistent and nagging problem for gradualistic accounts of evolution. Mivart (1871), Darwin's most cogent critic, referred to it as the dilemma of "the incipient stages of useful structures"—of what possible benefit to a reptile is two percent of a wing? The dilemma has two potential solutions. The first, preferred by Darwinians because it preserves both gradualism and adaptationism, is the principle of preadaptation: the intermediary stages functioned in another way but were, by good fortune in retrospect, preadapted to a new role they could play only after greater elaboration. Thus, if feathers first functioned "for" insulation and later "for" the trapping of insect prey (Ostrom, 1979), a pro-towing might be built without any reference to flight.

I do not doubt the supreme importance of preadaptation, but the other alternative, treated with caution, reluctance, disdain, or even fear by the modern synthesis, now deserves a rehearing in light of the renewed interest in development: perhaps, in many cases, the intermediates never existed. I do not refer to the saltational origin of entire new designs, complete in all their complex and integrated features—a fantasy that would be truly anti-Darwinian in denying any creativity to selection and relegating it to the role of eliminating old models. Instead, I envisage a potential saltational origin for the essential features of key adaptations. Why may we not imagine that gill arch bones of an ancestral agnathan moved forward in one step to surround the mouth and form protojaws? Such a change would scarcely establish the *Bauplan* of the gnathostomes. So much more must be altered in the reconstruction of agnathan design—the building of a true shoulder girdle with bony, paired appendages, to say the least. But the discontinuous origin of a protojaw might set up new regimes of development and selection that would quickly lead to other, coordinated modifications. Yet Darwin, conflating gradualism with natural selection as he did so often, wrongly proclaimed that any such discontinuity, even for organs (much less taxa), would destroy his theory:

> If it could be demonstrated that any complex organ existed, which could not possibly have been formed by numerous, successive, slight modifications, my theory would absolutely break down. [Darwin, 1859, p. 189]

During the past 30 years, such proposals have generally been treated as a fantasy signifying surrender—an invocation of hopeful monsters rather than a square facing of a difficult issue (attempting to find or construct intermediates). But our renewed interest in development, the only discipline of biology that might unify molecular and evolutionary approaches into a coherent science, suggests that such ideas are not fantastic, utterly contrary to genetic principles, or untestable.

Goldschmidt conflated two proposals as causes for hopeful monsters—"systemic mutations" involving the entire genome (a spinoff from his fallacious belief that the genome acted entirely as an integrated unit) and small mutations with large impact upon adult phenotypes because they work upon early stages of ontogeny and lead to cascading effects throughout embryology. We reject his first proposal, but the second, eminently plausible, theme might unite a Darwinian insistence upon continuity of genetic change with a macroevolutionary suspicion of phenetic discontinuity. It is, after all, a major focus in the study of heterochrony (effects, often profound, of small changes in developmental rate upon adult phenotypes); it is also implied in the emphasis now being placed upon regulatory genes in the genesis of macroevolutionary change (King and Wilson, 1975)—for regulation is fundamentally about timing in the complex orchestration of development. Moreover, although we cannot readily build "hopeful monsters," the subject of major change through alteration of development rate can be treated, perhaps more than analogically, both by experiment and comparative biology. The study of spontaneous anomalies of development (teratology) and experimental perturbations of embryogenic rates explores the tendencies and boundaries of developmental systems and allows us to specify potential pathways of macroevolutionary change. See, for example, the stunning experiment of Hampé (1959) on re-creation of reptilian patterns in birds, after 200 million years of their phenotypic absence, by experimental manipulations that amount to alterations in rate of development for the fibula. At the very least, these approaches work with real information and seem so much more fruitful than the construction of adaptive stories or the invention of hypothetical intermediates.

3. *The importance of nonadaptation.* The emphasis on natural selection as the only directing force of any importance in evolution led inevitably to the analysis of all attributes of organisms as adaptations. Indeed, the tendency has infected our language, for, without thinking what it implies, we use "adaptation" as our favored, *descriptive* term for designating any recognizable bit of changed morphology in evolution. I believe that this "adaptationist program" has had decidedly unfortunate effects in biology (Gould and Lewontin, 1979). It has led to a reliance on speculative storytelling in preference to the analysis of form and its constraints, and, if wrong in any case, it is virtually impossible to break through because the failure of one story leads to invention of another rather than the abandonment of the enterprise—and the inventiveness of the human mind virtually precludes the exhaustion of adaptive possibilities.

Yet, as I argued earlier, the hegemony of adaptation has been broken at the two lower levels of our evolutionary hierarchy: variation within populations and speciation. Most populations may contain too much variation for selection to maintain; moreover, if the neutralists are even partly right, much allelic substitution occurs without controlling influence from selection and with no direct relationship to adaptation. If species often form as a result of major chromosomal alterations, then their origin—the establishment of reproductive isolation—may require no reference to adaptation. Similarly, at this third level of macroevolution, both arguments previously cited against the conventional extrapolationist view require that we abandon strict adaptationism.

- If trends are produced by the unidirectional transformation of populations (orthoselection), then they can scarcely receive other than a conventional adaptive explanation. After all, if adaptation lies behind single allelic substitution in the *Biston betularia* model for change in local populations, what else but even stronger, more persistent selection and adaptive orientation can render a trend that persists for millions of years? But if trends represent a higher-level process of differential mortality among species, then a suite of potentially nonadaptive explanations must be considered. Trends, for example, may occur because some kinds of species tend to speciate more often than others. This tendency may reside in the character of environments or in attributes of behavior and population structure bearing no relationship to morphologies that spread through lineages as a result of higher speciation rates among some of their members. Or trends may arise from the greater longevity of certain kinds of species. Again, this greater persistence may have little more to do with the morphologies that come to prevail as a result. I suspect that many higher-order trends in paleontology—a bugbear of the profession because we have been unable to explain them in ordinary adaptive terms—are nonadaptive sequelae of differential species success based on environments and population structures.

- If transitions represent the continuous and gradual transformation of populations, then they must be regulated by adaptation throughout (even though adaptive orientation may alter according to the principle of preadaptation). But if discontinuity arises through shifts in development, then directions of potential change may be limited and strongly constrained by the inherited program and developmental mechanics of an organism. Adaptation may determine whether or not a hopeful monster survives, but primary constraint upon its genesis and direction resides with inherited ontogeny, not with selective modeling.

Quo Vadis?

My crystal ball is clouded both by the dust of these growing controversies and by the mists of ignorance emanating from molecular biology, where even the basis of regulation in eukaryotes remains shrouded in mystery. I think I can see what is breaking down in evolutionary theory—the strict construction of the modern synthesis with its belief in pervasive adaptation, gradualism, and extrapolation by smooth continuity from causes of change in local populations to major trends and transitions in the history of life. I do not know what will take its place as a unified theory, but I would venture to predict some themes and outlines.

The new theory will be rooted in a hierarchical view of nature. It will not embody the depressing notion that levels are fundamentally distinct and necessarily opposed to each other in their identification of causes, as the older paleontologists held in maintaining that macroevolution could not, in principle, be referred to the same causes that regulate microevolution (e.g., Osborn, 1922). It will possess a common body of causes and constraints, but will recognize that they work in characteristically different ways upon the material of different levels—intrademic change, speciation, and macroevolutionary trends.

As its second major departure from current orthodoxy, the new theory will restore to biology a concept of organism. In a most curious and unconscious bit of irony, strict selectionism (which was not, please remember, Darwin's own view) debased what had been a mainstay of biology—the organism as an integrated entity exerting constraint over its history. Mivart expressed the subtle point well in borrowing a metaphor from Galton. I shall call it Galton's polyhedron. Mivart (1871, pp. 228-229) writes:

> This conception of such internal and latent capabilities is somewhat like that of Mr. Galton . . . according to which the organic world consists of entities, each of which is, as it were, a spheroid with many facets on its surface, upon one of which it reposes in stable equilibrium. When by the accumulated action of incident forces this equilibrium is disturbed, the spheroid is supposed to turn over until it settles on an adjacent facet once more in stable equilibrium. The internal tendency of an organism to certain considerable and definite changes would correspond to the facets on the surface of the spheroid.

Under strict selectionism, the organism is a sphere. It exerts little constraint upon the character of its potential change; it can roll along all paths. Genetic variation is copious, small in its increments, and available in all directions—the essence of the term "random" as used to guarantee that variation serves as raw material only and that selection controls the direction of evolution.

By invoking Galton's polyhedron, I recommend no return to the antiquated and anti-Darwinian view that mysterious "internal" factors provide direction inherently and that selection only eliminates the unfit (orthogenesis, various forms of vitalism and finalism). Instead, the facets are constraints exerted by the developmental integration of organisms themselves. Change cannot occur in all directions, or with any increment; the organism is not a metaphorical sphere. But if adjacent facets are few in number and wide in spacing, then we cannot identify selection as the only or even the primary control upon evolution. For selection is channeled by the form of the polyhedron it pushes, and these constraints may exert a more powerful influence upon evolutionary direction than the external push itself. This is the legitimate sense of a much-maligned claim that "internal factors" are important in evolution. They channel and constrain Darwinian forces; they do not stand in opposition to them. Most of the other changes in evolutionary viewpoint that I have advocated throughout this chapter fall out of Galton's metaphor: punctuational change at all levels (the flip from facet to facet, since homeostatic systems change by abrupt shifting to new equilibria); essential nonadaptation, even in major parts of phenotype (change in an integrated organism often has effects that reverberate throughout the system); channeling of direction by constraints of history and developmental architecture. Organisms are not billiard balls, struck in deterministic fashion by the cue of natural selection and rolling to optimal positions on life's table. They influence their own destiny in interesting, complex, and comprehensible ways. We must put this concept of organism back into evolutionary biology.

> Immediate use and old inheritance are blended in Nature's handiwork as in our own. In the marble columns and architraves of a Greek temple we still trace the timbers of its wooden prototype, and see beyond these tree-trunks of a primeval sacred grove; roof and eaves of a pagoda recall the sagging mats which roofed an earlier edifice; Anglo-Saxon land-tenure influences the planning of our streets, and the cliff-dwelling and cave-dwelling linger on in the construction of our homes! So we see enduring traces of the past in the living organism—landmarks which have lasted on through altered function and altered needs. [Thompson, 1942]

Glossary[1]

Allele   One of a pair of genes (or of multiple forms of a gene) located at the same locus of homologous chromosomes.

Allometry   The quantitative, usually nonlinear, relationship between a part and the whole (or another part) as the organism increases in size.

Allopatric   Organisms that inhabit separate and mutually exclusive geographic regions.

Anagenesis   Evolutionary change by gradual transformation, usually construed as "improvement" in design.

***Biston betularia* Model**   *Biston betularia* is the famous English moth that was whitish in color before the industrial revolution, but evolved, by a single gene change, to a black color when industrial soot blackened the trees on which it rests, thereby giving it protection from bird predators that hunt by sight. This model therefore refers to evolutionary change by gradual, gene-by-gene substitution within a single population, continually guided by natural selection. This is the conventional view of microevolution—that all evolution can be explained by extrapolating what happened to *Biston*. That is, evolutionary events of larger magnitude merely involve more genes and more time. But the process is the same as for *Biston*—gene-by-gene change, all mediated by natural selection.

Cladistic (Cladogenesis)   Evolution by splitting of a descendant from the ancestral line.

Deme   A local population in which individuals freely breed among themselves but not with those of other demes.

F_1   Notation for the first filial generation resulting from a cross breeding. Sibling.

F_2   Notation for the progeny produced by intercrossing (inbreeding) members of the F_1 generation.

Gnathostomes   Vertebrates with a developed jaw apparatus. Includes all modern vertebrates, except the jawless lampreys and hagfishes.

Heterochrony   Evolution by a change in the timing of development of one organ or of the whole body between ancestors and descendants.

Heterozygous   Having different alleles at one (or more) gene locus, therefore producing germ cells of more than one kind.

Homozygous   Having identical alleles at one (or more) gene locus, therefore producing germ cells that are identical (with respect to those genes).

Karyotype   The complement of chromosomes characteristic in form, size, and number of an individual, a species, or other grouping.

Mictic   Produced by sexual reproduction.

[1]Prepared by the editors.

Parapatric   Organisms living in adjacent regions.

Phenetic   Pertaining to the physical characteristics of an individual (regardless of genetic makeup).

Pleiotropy   The effect of a single gene on more than one part of the body.

Saltation   Evolution by abrupt change in a single generation (saltus is Latin for "jump").

Sibs   Short for "siblings."

Sympatric   Occupying the same geographic area as another species.

References

Ayala, F. J. (1976) Molecular genetics and evolution. In: *Molecular Evolution*, F. J. Ayala (ed.). Sinauer Associates, Sunderland, Mass., pp. 1–20.
Bateson, G. (1978) *Mind and Nature*. Dutton, New York.
Bush, G. L. (1975) Modes of animal speciation. *Annu. Rev. Evol. Syst.* **6**:339–364.
Bush, G. L., S. M. Case, A. C. Wilson, and J. L. Patton (1977) Rapid speciation and chromosomal evolution in mammals. *Proc. Natl. Acad. Sci. USA* **74**:3942–3946.
Carson, H. L. (1975) The genetics of speciation at the diploid level. *Am. Nat.* **109**:83–92.
Carson, H. L. (1978) Chromosomes and species formation. *Evolution* **32**:925–927.
Darwin, C. (1859) *On the Origin of Species*. Murray, London.
Darwin, C. (1880) Sir Wyville Thomson and natural selection. *Nature* **23**:32.
Dobzhansky, T. (1937) *Genetics and the Origin of Species*. Columbia University Press, New York.
Dobzhansky, T. (1951) *Genetics and the Origin of Species*, 3rd ed. Columbia University Press, New York.
Ehrlich, P. R. and P. H. Raven (1969) Differentiation of populations. *Science* **165**:1228–1232.
Eldredge, N. (1979) *Alternative Approaches to Evolutionary Theory*. Carnegie Museum of Pittsburgh, Pittsburgh.
Eldredge, N., and S. J. Gould (1972) Punctuated equilibria: An alternative to phyletic gradualism. In: *Models in Paleobiology*, T. J. M. Schopf (ed). Freeman, Cooper and Co., San Francisco, pp. 82–115.
Goldschmidt, R. (1940) *The Material Basis of Evolution*. Yale University Press, New Haven, Conn.
Gould, S. J. (1980a) Is a new and general theory of evolution emerging? *Paleobiology* **6**:119–130.
Gould, S. J. (1980b) G. G. Simpson, paleontology and the modern synthesis. In: *The Evolutionary Synthesis*, E. Mayr and W. B. Privine (eds.). Harvard University Press, Cambridge, Mass.
Gould, S. J., and E. Eldredge (1977) Punctuated equilibria: The tempo and mode of evolution reconsidered. *Paleobiology* **3**:115–151.
Gould, S. J., and R. C. Lewontin (1979) The spandrels of San Marco and the Panglossian paradigm: A critique of the adaptationist program. *Proc. R. Soc. London Ser. B* **205**:581–598.
Hampé, A. (1959) Contribution à l'étude du développement et de la regulation des déficiences et des excédents dans la patte de l'embryon de poulet. *Arch. Anat. Microsc. Morphol. Exp.* **48**:345–378.
King, M. C., and A. C. Wilson (1975) Evolution at two levels in humans and chimpanzees. *Science* **188**:107–116.
Koestler, A. (1978) *Janus: A Summing Up*. Random House, New York.
Lewontin, R. C. (1974) *The Genetic Basis of Evolutionary Change*. Columbia University Press, New York.
Lewontin, R. C., and J. L. Hubby (1966) A molecular approach to the study of genic heterozygosity in natural populations. II. Amount of variation and degree of heterozygosity in natural populations of *Drosophila pseudoobscura*. *Genetics* **54**:595–609.
Mayr, E. (1942) *Systematics and the Origin of Species*. Columbia University Press, New York.

Mayr, E. (1963) *Animal Species and Evolution.* Harvard University Press (Belknap), Cambridge, Mass.
Mivart, St. G. (1871) *On the Genesis of Species.* Macmillan & Co., London.
Nei, M. (1975) *Molecular Population Genetics and Evolution.* American Elsevier, New York.
Osborn, H. F. (1922) Orthogenesis as observed from paleontological evidence beginning in the year 1889. *Am. Nat.* **56**:134–143.
Ostrom, J. H. (1979) Bird flight: How did it begin? *Am. Sci.* **67**(1):46–56.
Powell, J. R. (1978) The founder-flush speciation theory: An experimental approach. *Evolution* **32**:465–474.
Robson, G. C., and O. W. Richards (1936) *The Variation of Animals in Nature.* Longmans, Green, New York.
Romanes, G. J. (1900) *Darwin and After Darwin,* Vol. 2, *Post-Darwinian Questions. Heredity and Utility.* Longmans, Green, New York.
Simpson, G. G. (1944) *Tempo and Mode in Evolution.* Columbia University Press, New York.
Simpson, G. G. (1953) *The Major Features of Evolution.* Columbia University Press, New York.
Stanley, S. M. (1975) A theory of evolution above the species level. *Proc. Natl. Acad. Sci USA* **72**:646–650.
Thompson, D. W. (1942) *Growth and Form.* Macmillan & Co., New York.
White, M. J. D. (1978) *Modes of Speciation.* Freeman, San Francisco.
Wilson, A. C., G. L. Bush, S. M. Case, and M. C. King (1975) Social structuring of mammalian populations and rate of chromosomal evolution. *Proc. Natl. Acad. Sci. USA* **72**:5061–5065.
Wilson, A. C., S. S. Carlson, and T. J. White (1977) Biochemical evolution. *Annu. Rev. Biochem.* **46**:573–639.
Wilson, E. O., T. Eisner, W. R. Briggs, R. E. Dickerson, R. L. Metzenberg, R. O'Brien, M. Susman and W. E. Boggs (1973) *Life on Earth.* Sinauer Associates, Sunderland, Mass.
Wright, S. (1967) Comments on the preliminary working papers of Eden and Waddington. *Wistar Inst. Symp.* **5**:117–120.

III

Differentiation, Morphogenesis, and Death of Organisms

Gregory B. Yates

The development of form by a sexual, multicellular organism, between fertilization and death, is a major example of self-organization. With only the encouragement of the elements a seed becomes a tree and an egg a tadpole. If either happened only once it would be regarded as an outright miracle, but the phenomena of differentiation and morphogenesis are commonplace and reproducible. This reproducibility makes the mechanism of self-organization far more susceptible to study in the case of the development of individuals of a species than in that of the once-only evolution of a species itself.

The central question of development is whether or not the later form, and perhaps also the death, of an adult organism is implied by a zygote and its environment. There are two general mechanisms by which this implication is commonly supposed to be realized, and these differ in the presumed relationship of the whole organism to its parts. In the first type of mechanism (often called "atomistic") the parts of the organism are viewed as building blocks that aggregate in the manner of a raindrop condensing. The parts—the molecules—with only simple rules governing their interaction, imply the whole. The second type of mechanism embodies the "field view" of self-organization, and in it the whole is defined by invariant relations; the parts emerge as the result of changes that conserve these relations. Thunder clouds form in the stratified atmosphere in this way. Here the whole, with simple boundary conditions, implies the parts. These two views of self-organization draw much of their respective inspirations from the particle and field (e.g., wave) metaphors of physics.

In the first chapter of this section, Wood works within the "particle" paradigm to account for the formation of viruses from their parts. A simple, equilibrium thermodynamic model adequately describes the self-assembly of small

---

GREGORY B. YATES • Crump Institute for Medical Engineering, University of California, Los Angeles, California 90024.

viruses like the tobacco mosaic virus, but this model must be augmented with kinetic controls to account for the assembly of the larger T4 bacteriophage. Glišin considers the problem of self-organization in the larger and more complex sea urchin embryo. In embryogenesis the atomistic view is encouraged by the observation that, at certain points in development, artificially dissociated cells can partially reassemble. But the field view is supported by the observation that in early development, cells can be experimentally interchanged and a healthy organism nevertheless often emerges. Goodwin finds atomistic theories of development inadequate in the face of repeated demonstrations that many constraints determining form do not arise from the genome. Following an approach sympathetic to modern structuralism, he defends the primacy of morphogenetic principles at the level of the organism and proposes a Laplacian field model for the events of early embryogenesis. Goodwin concludes that the field view must receive greater emphasis in theories of self-organization at all levels. This opinion is also reflected by Garfinkel in his description of slime mold aggregation behavior. The slime mold *Dictyostelium* is a creature composed of amoebae that spend much of their lives as independent multicellular organisms but sometimes aggregate to form a highly differentiated multicellular organism. Forms occurring during slime mold aggregation give clues to the dynamics of the aggregation process and reveal the importance of symmetry-breaking and entrainment in self-organization. Garfinkel concludes that slime mold aggregation is a good analogy of self-organization at much higher levels of complexity. (Additional discussion of differentiation, in neural networks, will be found in Chapter 14 by von der Malsburg).

In the final chapter of the section, Finch examines the tendency of organisms to age and die and compares atomistic (genetic error) theories to more global, systemic views.

The contributors to this section agree that the significant phenomena of differentiation, morphogenesis, and death are scientifically explicable in principle: all find natural mechanisms behind these processes. Although the two major views, the atomistic and the field view, are in superficial conflict, they actually answer somewhat different questions and therefore may be complementary in the same sense as are the metaphors that inspired them.

7

Virus Assembly and Its Genetic Control

William B. Wood

ABSTRACT

The genetic specification of virus parts is relatively well understood, but mechanisms of assembly to form complete viruses remain somewhat obscure. A simple virus like the tobacco mosaic virus can form by the process of self-assembly, in which all participating proteins are included in the final structure. Such viruses can spontaneously reassemble following dissociation of their parts. T4, a more complex virus, cannot do so. T4 has three major body parts: tail fibers, a tail, and a head composed of a DNA-containing capsid (head). These parts are assembled independently. Several T4 genes code for nonstructural proteins that are essential to the assembly of one of these parts but are not present in the finished virus. This dependency represents a clear departure from simple self-assembly. At least two features of T4 assembly cannot be explained in terms of thermodynamic equilibria alone: many assembly steps can occur in only one particular order, and only one capsid structure generally forms despite the possibility of alternative structures with roughly equal thermodynamic stability. Both features can be explained by rate controls on assembly steps. In T4, different kinds of protein subunits may affect each other's association by means of induced conformational changes, so that each assembly step increases the rate of the next step. This hypothetical process is called heterocooperativity. (It has features reminiscent of the kinetics discussed by Schuster and Sigmund in Chapter 5.) Maintenance of protein ratios may confer another kind of kinetic control on capsid formation. At least three kinds of nonstructural accessory proteins are found in T4: scaffolding proteins, proteolytic enzymes, and promoters of noncovalent bonding. Self-assembly, kinetic controls, and accessory protein functions provide significant insight into the mechanism of self-organization in viruses. (The glossary in Chapter 5 may be helpful for this chapter also.) —THE EDITOR

How structure arises in living systems is a fundamental biological problem that is still poorly understood. At the subcellular level, progress is being made in elucidating the manner in which supramolecular protein and nucleoprotein struc-

WILLIAM B. WOOD • Department of Molecular, Cellular, and Developmental Biology, University of Colorado, Boulder, Colorado 80309.

tures are assembled under genetic control. This chapter reviews recent results from studies on the assembly of complex bacterial viruses. These findings are providing general insights into the morphogenesis of subcellular structure. I shall focus on the bacteriophage T4, which has been one of the most informative model systems and is an object of my own research.

T4 has played a major role in the progress of molecular biology over the past 40 years (Wood, 1980), since the physicist Max Delbruck began his pioneering work on bacterial viruses in the late 1930s. During the 1950s and early 1960s, T4 provided, among other things, strong corroborative evidence for DNA as the genetic material (Hershey and Chase, 1952); initial proof for the existence of RNA (Brenner *et al.*, 1961); and the first clear understanding of functional relationships among the internal elements of a gene (Benzer, 1959). By the early 1960s, the major goals of Delbruck and his early co-workers had been realized with the general solution of what we can call the genetic problem: how is the phage DNA replicated and how does it direct the synthesis of viral proteins? Of course, important details of DNA replication and protein synthesis remained to be learned and still are being investigated. But in 1960, experiments by Edgar and Epstein, then associates of Delbruck at Caltech, started bacteriophage research in a new direction (e.g., Epstein *et al.*, 1963).

With the general solution to the genetic problem at hand, there remained, and still does remain, what we can call the epigenetic problem: how is the temporal and spatial organization of gene product functions controlled to form the phage particle or, in the general case, to form any living organism? This question, as basic to biology as the genetic problem, is only now becoming the focus of research as the genetic problem became in the 1950s.

Edgar and Epstein pointed phage research toward the epigenetic problem with their discovery and exploitation of conditionally lethal mutants. Under permissive conditions, for example, low growth temperature or presence of a genetic suppressor in the bacterial host, these mutants can be propagated normally. Under nonpermissive conditions—high temperature or absence of a host suppressor—the mutationally altered gene product fails to function, and the infection aborts at the point where that gene product ordinarily is required. Analysis of such mutants using biochemical, electron microscopic, and other approaches allows a systematic investigation of essential viral gene functions, such as those required for assembly of the phage particle.

Before proceeding to that topic, I shall briefly describe T4 and its genome as they are currently understood. Figure 1 is a high-resolution electron micrograph of T4. The phage is a baroque contraption with a complex contractile tail that allows transfer of its DNA from the elongated phage head into the bacterial cell. Operation of the tail components during infection is shown schematically in Figure 2.

Like the morphology of the virus, the genetic repertoire of T4 and its strategy for host cell takeover are also complex. Figure 3 shows a taxonomy of the known T4 gene functions (in 1980). Approximately 150 phage genes have been genetically identified (reviewed in Wood and Revel, 1976; see also Mathews *et al.*, 1983, for more recent information). The functions are known, or can be surmised, for

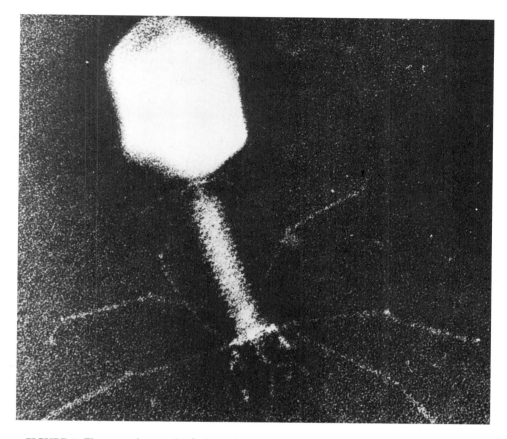

FIGURE 1. Electron micrograph of a bacteriophage T4 particle. (Courtesy of Robley Williams.)

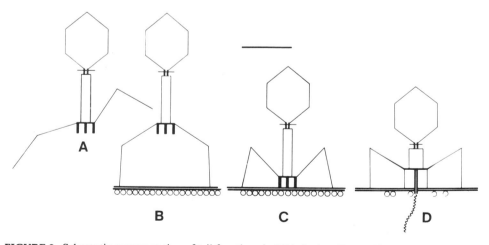

FIGURE 2. Schematic representation of tail functions in T4 infection. Bar = 10 nm. A: free virus; B: contact with bacterial cell surface; C: preparation for penetration; D: insertion of DNA. (From Simon and Anderson, 1967.)

III. Differentiation, Morphogenesis, and Death

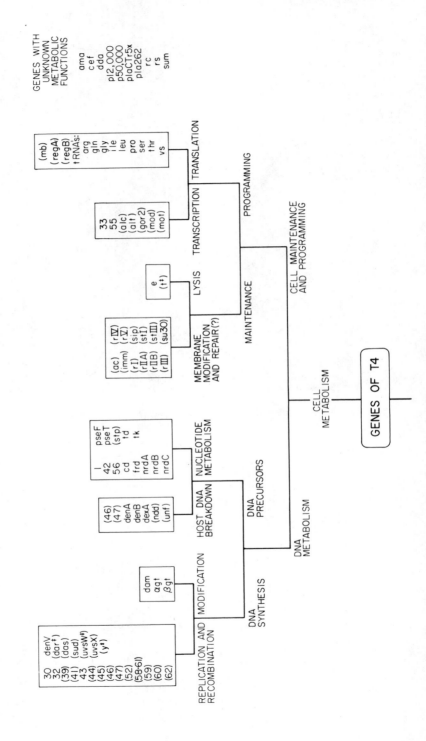

7. Virus Assembly and Its Genetic Control

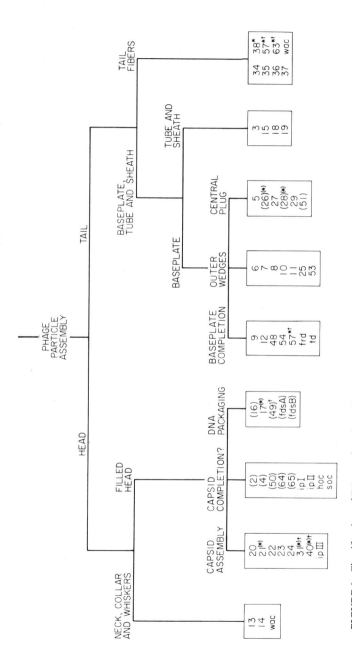

FIGURE 3. Classification of T4 gene functions. Genes are designated by numbers or letters. (From Wood and Revel, 1976.) Genes with unknown metabolic functions: *ama, cef, dda, p12,000, p50,000, plaCTr5x, pla 262,rc, rs,* and *sum*.

about 130 of the identified genes. These functions can be divided into two major categories: those at the top of the chart function in cell metabolism and those at the bottom control the assembly of phage particles.

Many of the metabolic genes function in the synthesis of the phage DNA. Remarkably, 60 of the 80 metabolic genes shown are nonessential under standard laboratory conditions. In any particular bacterial host, most of these genes are not required. They allow the phage to have a more extended host range and to increase the production of progeny by supplementing the host metabolic machinery.

The bottom half of Figure 3 shows the genes involved in particle assembly. The morphological complexity of the virion and the availability of conditionally lethal mutations turned out to be a fortunate combination of circumstances for investigating morphogenesis of the phage particle. The general problem of morphogenesis can be stated simply as follows: how does the linearly arranged information in a DNA sequence specify the three-dimensional structure of a biological object like T4? Such understanding as there was about this process until recently was confined to multimeric proteins and simple viruses such as the rod-shaped tobacco mosaic virus. These structures can form by *self-assembly*, which can be defined as a form-determining process in which all the necessary energy and information are contributed by the assembling components that comprise the final product. In other words, no tools are required; the parts can assemble spontaneously by themselves.

The approach to studying assembly of these simple structures had been to dissociate them by mild chemical denaturing conditions (see, e.g., Fraenkel-Conrat and Williams, 1955; Klug, 1972). There are two disadvantages to this approach. First, the re-formation may occur by a pathway different from that of the original assembly process in the cell. Second, this approach will not work with anything much more complex than tobacco mosaic virus. T4 cannot reassemble following dissociation. Fortunately, the conditionally lethal mutants of T4 presented an opportunity to dissect genetically the original process of assembly, from virgin proteins, of a considerably more complex structure.

Genetics and Physiology of T4 Assembly Functions

Early insight into the nature of T4 assembly came from the initial genetic studies of Edgar and Epstein (see, e.g., Epstein *et al.*, 1963). Analysis of defective phenotypes for a large number of essential gene defects provided clues to corresponding gene product functions. Figure 4 shows a simplified genetic map of T4, emphasizing the genes essential for particle assembly. The boxes on the outer circle contain schematic diagrams of the structures seen in electron micrographs of cells infected with mutants defective in a particular gene. Many gene product functions can be inferred from these pictures. For example, mutation in gene 23 or one of its neighboring genes leads to accumulation of tails and tail fibers but no heads; therefore, these genes are assumed to be essential for head formation.

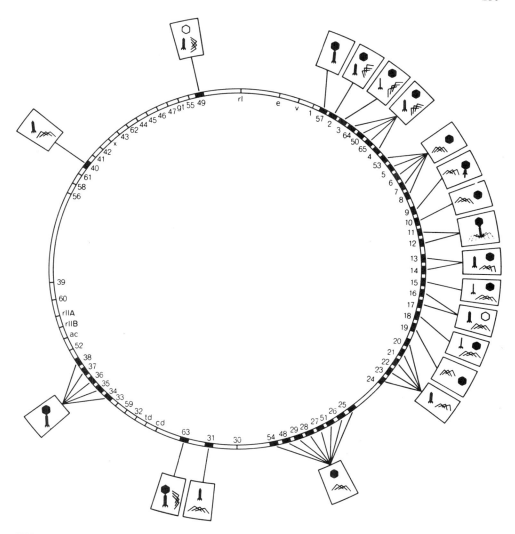

FIGURE 4. A simplified genetic map of T4 emphasizing the genes essential for particle assembly. (From Wood, 1979.)

By similar reasoning, other groups of genes appear to control tail fiber and tail assembly, respectively.

These results showed that assembly of each of the three major viral subcomponents (head, tail, and tail fibers) is independent; a defect that eliminates one does not affect the other two. In addition, these studies showed that a large number of genes are involved in the assembly process and suggested that this number exceeds the number of proteins in the completed phage particle, thus raising the possibility of the use of nonstructural accessory proteins during assembly.

In Vitro Complementation and the Pathway of T4 Assembly

Further progress required more detailed understanding of the functional defects in assembly mutants. The opportunity for acquiring it was provided by the discovery that many of the steps in assembly could proceed *in vitro* with production of infectious virus when extracts of mutant-infected cells lacking different phage components were incubated together (Edgar and Wood, 1966). Moreover, phage components isolated from one extract could be assayed for the ability to complement another, i.e., to correct its defective function by substituting for a defective part. This technique made possible the assignment of gene functions that had been unclear from electron microscopy alone. For example, several gene defects lead to accumulation of unassembled heads, tails, and tail fibers, all of which appear morphologically normal. However, analysis of the corresponding defective extracts by *in vitro* complementation showed that the heads produced by these mutants are defective, whereas the tail and tail fibers are functional. Many of the assembly-defective mutants accumulate incomplete precursor structures that had been previously identified from electron micrographs as probable intermediates in assembly and could be tentatively ordered on the basis of their appearance. *In vitro* complementation made it possible to demonstrate that most of these structures could in fact be converted to infectious virus. In some cases it also allowed ordering of assembly steps in which changes in the structure of a component had not been detected by electron microscopy (reviewed in Wood *et al.*, 1968).

From these results, it was possible in 1968 to formulate the pathway of T4 assembly shown in Figure 5. At that time only the terminal steps in assembly had been demonstrated in extracts (solid arrows); early steps in head, tail baseplate, and tail fiber assembly were inaccessible to *in vitro* complementation analysis (dashed arrows). Nevertheless, several important characteristics were evident. (1) The branches of the pathway are indeed functionally independent, as inferred from earlier findings. (2) Most of the steps are constrained to a unique sequential order. Because all the structural proteins of T4 are synthesized simultaneously during the latter half of the infectious cycle, this order cannot be due to temporal regulation of gene product synthesis; it must be imposed at the level of gene product interaction. (3) The levels of finished components do not appear to exert feedback controls on their own production or on the production of other components. Instead, the system appears set to produce components in fixed relative amounts with tails and tail fibers in excess over heads, presumably as a means of maximizing the probability that finished heads containing the viral genes will become converted to infectious virus.

Analysis of Gene Functions in Assembly

The pathway as formulated in 1968 represented only a partial sequence of assembly steps, and little was known about the nature of these reactions. During the subsequent decade, efforts in a number of laboratories were directed toward

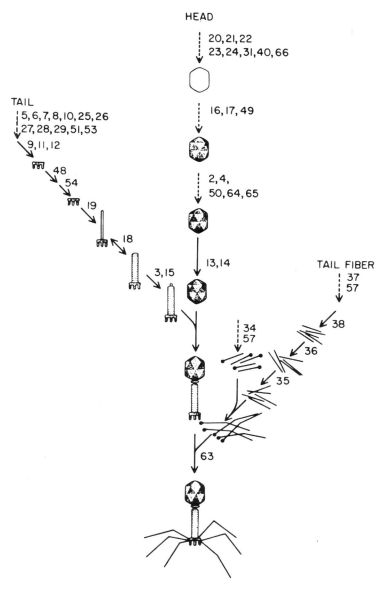

FIGURE 5. An early representation of the pathways for T4 assembly. Gene functions required for each step are indicated by numbers corresponding to those in Figures 3 and 4. (From Wood, 1979.)

pushing the analysis backward to earlier steps in the assembly process and extending it to explore the mechanisms and the functional roles of individual assembly gene products. In addition, similar analyses of assembly in other phages provided useful comparative information.

A powerful tool in these investigations was the technique of denaturing poly-

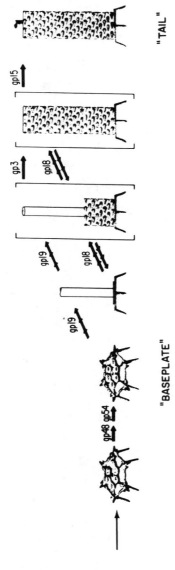

FIGURE 6. The sequence of steps in T4 tail assembly. The 70 S baseplate precursor is formed from a central "plug" (upper pathway) and six "wedges" (lower pathway). In this and in subsequent figures, gp stands for gene product (i.e., a protein). (Adapted from Wood et al., 1968.)

acrylamide gel electrophoresis of radioactively labeled phage proteins, followed by autoradiography of the dried gel to determine the positions of polypeptide bands. This technique exploits two useful characteristics of the T4 system. First, T4 infection shuts off host protein synthesis so that subsequent administration of a radioactive precursor labels only phage-induced proteins. Second, amber mutations, the most widely used class of conditional lethal mutants, cause premature polypeptide chain termination under nonpermissive conditions and thus lead to disappearance of mutant gene products from the normal positions in the gel pattern. Disappearance of a band resulting from amber mutations in a particular gene usually identifies the band as the polypeptide coded by that gene. Once particular gene products have been identified in this manner, their presence or absence in isolated phage components can be assayed electrophoretically and the structural contributions of assembly gene products can be ascertained. For example, the gene 19-dependent conversion of free baseplates to the tube-baseplate structure (see Figure 5) was shown by electrophoresis to involve addition of many copies of gene product 19 (*gp*19) to the baseplate structure, indicating that *gp*19 is the structural subunit of the tube. By contrast, an early step in tail fiber assembly involves the gene 38-dependent dimerization of *gp*37 to form a rod-shaped precursor that does not contain *gp*38.

Results of research in this field have given us a richly detailed, although still incomplete, picture of how several bacterial viruses are assembled. The particulars have been reviewed by Wood and King (1980). Following is a brief summary of the major findings.

Tail Assembly

Current knowledge of the T4 tail assembly sequence is due almost entirely to King and his colleagues (Kikuchi and King, 1975). These investigators were able to demonstrate the entire sequence *in vitro*—from individual gene products to the completed tail. They isolated almost all the assembly intermediates and characterized their sedimentation properties and gene product compositions. The resulting detailed pathway is shown in Figure 6.

The pathway is branched, with independent formation of the central plug and the peripheral wedges, which combine to form the hexagonal baseplate structure. Three of the gene products required for plug formation are proteins of unknown function that do not appear to become part of the structure, but all the remaining steps are self-assembly reactions in which a gene product is added to the precursor structure. Remarkably, almost all these steps are constrained to occur in a strict sequence. Completed baseplates can be dissociated only by stringent denaturing conditions, such as treatment at 100°C in the presence of 1% sodium dodecyl sulfate. Therefore, associations between the baseplate proteins are strongly favored thermodynamically. Nevertheless, with only three exceptions (additions of *gp*9, *gp*11, and *gp*12), none of the associations in the pathway will occur unless the preceding one has taken place. If *gp*10 is absent, blocking the initial assembly step, all the remaining gene products remain unassociated as soluble monomers. Only the sheath subunit, *gp*18, eventually polymerizes late in the infectious cycle into long tubes of "polysheath" in the contracted configuration (see Berget and King, 1983, for recent review).

Head Assembly

The assembly of the phage head (capsid) and the packaging of nucleic acid within it are more general features of viral maturation than is assembly of tails which are found only on the complex bacteriophages. Head assembly also has been more difficult to understand than tail assembly. Several capsid proteins are capable of forming other stable aggregates besides the correct one, so that the structures that accumulate in mutant-infected cells often are aberrant by-products, rather than normal assembly intermediates. Moreover, in T4 the earlier steps in head assembly have remained largely intractable to *in vitro* analysis. Nevertheless, considerable understanding of this process has come from the early studies of E. Kellenberger and his colleagues and from the more recent work of Laemmli, Showe, Black, and others (reviewed by Black and Showe, 1983). In addition, valuable information resulted from research on head assembly in somewhat simpler phages, in particular the *Salmonella* phage P22 and coliphages lambda, T7, and P2, in which *in vitro* packaging of DNA was first achieved and exploited (see Wood and King, 1980, for references).

In considering head assembly, it is instructive to compare the process in several phages. All raise two major questions in common: First, what is the mechanism for construction of a capsid shell, which is usually, but not always, a regular icosahedron? Second, how does nucleic acid become encapsulated within the shell in such a way that it can spontaneously exit during infection? The head assembly pathways of four phages are sketched in Figure 7. All the phages appear to assemble the capsid shell first and to incorporate the nucleic acid into it afterward. The capsid assembly process involves prior or concomitant formation of an inner protein scaffold or core in addition to the outer protein shell. At about the same time that nucleic acid is being packaged, the core protein either exits as intact monomers, as in phage P22, or is destroyed by proteolysis and partially lost, as in T4 assembly. In addition, the outer capsid shell undergoes a cooperative lattice transition that expands its volume. In T4 and lambda, but not in P22 and T7, several capsid structural proteins undergo proteolytic cleavages, which represent clear departures from self-assembly.

The relative timing and the mechanistic relationships of these events are not yet clear. Neither are the energetics of the packaging process. If condensation of the DNA occurs by association with basic proteins in the head, then some subsequent reaction must convert the complex to a metastable one that can spontaneously release the DNA upon infection. Alternatively, packaging may be somehow coupled to an energy-yielding reaction, such as hydrolysis of nucleoside triphosphates. The latter alternative is probably correct (Earnshaw and Casjens,

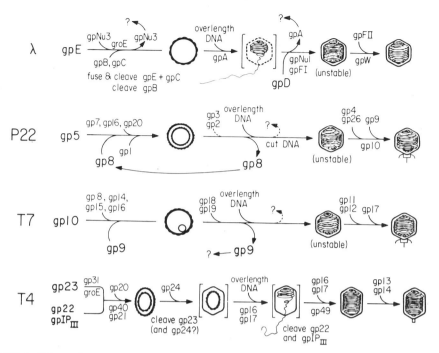

FIGURE 7. Schematic representation of head assembly pathways of four bacteriophages. (From Wood, 1979.)

1980; Black and Showe, 1983), at least for most phages, although details of the packaging mechanism remain poorly understood.

Tail Fiber Assembly

Assembly and attachment of the long fibers are processes found in only a few types of complex bacteriophages. Nevertheless, they have revealed unexpected features that may be of general interest for other macromolecular assembly processes as well (reviewed in Wood and Crowther, 1983). The tail fiber has the shape of a bent rod, which attaches to the baseplate at its proximal end and carries at its distal tip the primary host-cell recognition site, specific for a bacterial cell-surface lipopolysaccharide.

Tail fiber assembly occurs by a branched sequence of six reactions, as shown in Figure 8. This sequence represents the simplest portion of the T4 morphogenetic pathway. However, it presents a new challenge: although the finished product contains only four structural proteins (products of the adjacent genes 34 through 37) and no covalent bonds appear to be formed or broken during their assembly, eight phage genes are essential for building the structure and attaching it to the phage.

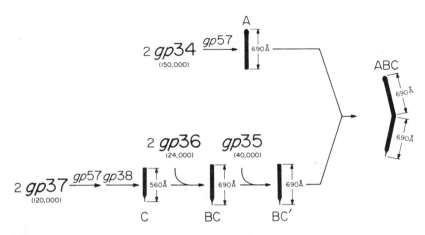

FIGURE 8. The sequence of steps in T4 tail fiber assembly. A, B, and C are characteristic antigens present on the intermediates shown. Molecular weights of the four structural proteins are indicated in parentheses. (From Wood, 1979.)

The products of genes 38 and 57 are required as nonstructural accessory proteins in the two earliest assembly steps, each of which involves association of two copies of a large polypeptide to form a rod-shaped dimer. Both steps require *gp57*, a small protein produced throughout infection and not found in completed tail fibers or in the completed phage. The role of *gp57* remains unknown; it may involve interactions with a host component based on the finding that it is possible

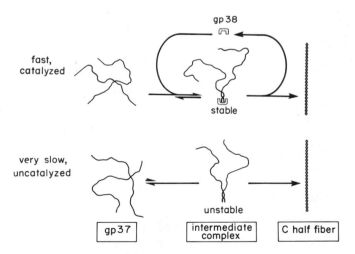

FIGURE 9. A possible function for the accessory protein *gp38* in dimerization of *gp37* to form the precursor of the distal half of the tail fiber.

to isolate bacterial mutants in which *gp57* is not required (see Wood and King, 1980, for other references).

Formation of the distal half fiber precursor also requires the function of *gp38*, which is not found in tail fibers or phage. There is indirect evidence that *gp38* may act by transient association with the C-terminal end of *gp37*. The two molecules of *gp37* in the distal half fiber associate in parallel to form what is probably a helix with extensive internal cross β structure (Earnshaw et al., 1979). The *gp38* protein could act to promote the formation of this structure by transiently stabilizing a complex between the C-terminal ends of two *gp37* molecules, thereby allowing the remainder of the structure to "zipper up" (Figure 9).

After the remaining structural proteins are added to the distal half fiber and the two halves are joined, the whole fiber is attached to the bacteriophage particle, as shown schematically in Figure 10. This process involves two more accessory proteins. One, *gp63*, is a soluble protein synthesized throughout phage infection and not found in completed phage particles. The other, *gpwac*, is the protein of the phage whiskers, slender filaments that can be seen in high-resolution electron micrographs to extend outward from the neck region where the tail joins the head (see Figure 1). There are six whiskers. The available evidence suggests that these two accessory proteins, *gp63* and *gpwac*, convert what otherwise would be a very slow bimolecular reaction to a fast bimolecular reaction-interaction of a tail fiber with the cloud of whiskers around the phage, followed by a fast unimolecular reaction promoted by *gp63*. Because no covalent bonds appear to be broken or formed during the attachment process, *gp63*, like *gp38*, appears to serve the novel function of promoting the noncovalent association of two structural proteins. How it does so is not yet clear. One possibility is that during the association reaction, one of the structural components may have to assume an energetically unfavorable conformation, which could be stabilized by transient binding to *gp63*, as schematized in Figure 11.

Summary, Conclusions, and Remaining Questions

Two of the most important insights to come from research on morphogenesis of complex bacteriophages have been realization of the importance of kinetic con-

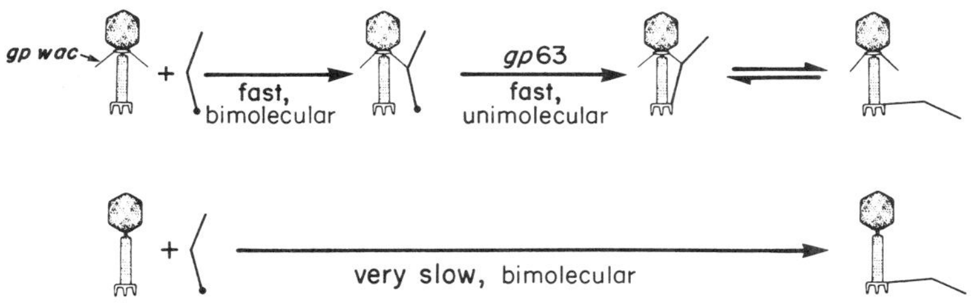

FIGURE 10. The proposed roles of whiskers *(gpwac)* and *gp63* in tail fiber attachment.

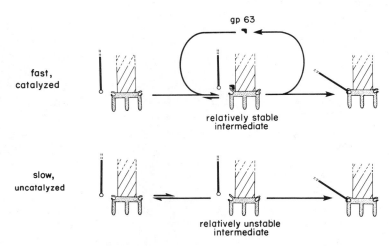

FIGURE 11. A fanciful model for the catalytic action of *gp*63 tail fiber attachment.

trols in assembly and recognition of accessory protein functions. The early attempts to understand the structure and assembly of small icosahedral viruses by Crick and Watson in the 1950s (1957) and by Caspar and Klug in the early 1960s (1962) assumed that in such simple structures the free energy of subunit association favored one form, the correct one, over all others, so that spontaneous assembly could be explained in terms of thermodynamic equilibria alone. The resulting concept of self-assembly largely ignored pathways of subunit association.

The recent work on more complex viruses reviewed here has corroborated the importance of self-assembly. Most of the T4 gene products produced late in infection, for example, are self-assembling structural proteins. However, this work also has shown us at least two features that cannot be explained on the basis of thermodynamic equilibria only. The first is the rigidly controlled temporal sequence of assembly steps, most of which can occur only in one particular order. In many cases maintenance of this order probably is important for the assembly process. (For example, sheath proteins must not be allowed to assemble with each other until the baseplate structure is ready to receive them.) The second feature, which is difficult to explain by thermodynamic equilibria alone, concerns the association of proteins that are capable of forming polymorphic assemblies, i.e., alternative structures of roughly equal thermodynamic stability. For example, the ability of capsid subunits to undergo quasiequivalent bonding, which makes possible the assembly of a large icosahedral shell (Caspar and Klug, 1962), also means that these subunits must be capable of associating into alternative stable structures. In T4 the capsid proteins can form aberrant "polyheads" as well as true heads. However, the phage is somehow able to direct subunits into correct structures rather than aberrant ones, at least most of the time.

These two features can be explained by controls on relative rates of individual assembly steps. Intriguingly, many of these controls in T4 seem to be built into the self-assembling structural proteins themselves. Each of these proteins forms highly stable associations with others, yet the order of associations is fixed. No assembly step will occur *at a significant rate*, until the preceding one is completed. We can still only speculate about the control mechanisms, but the following hypothesis seems likely. Studies with model systems indicate that the rate-limiting step in many protein-protein association reactions is subunit conformational change. In an assembly sequence, we can imagine that each association requires conformational changes in the interacting proteins and that these changes represent increases in free energy. These increases constitute energies of activation, which act as kinetic barriers to association. However, the proteins have evolved in such a way that, as assembly proceeds, some of the binding energy at each step is used to alter the subunit conformation, thereby lowering the activation energy of the next step. This kind of explanation accounts for the familiar phenomenon of cooperativity encountered in the formation of repeating structures made of a single kind of subunit, such as the tobacco mosaic virus protein helix. However, T4 illustrates a novel form of this phenomenon that might be called *heterocooperativity*: different kinds of protein subunits affect each other's association in such a way as to create an assembly pathway, in which each step increases the rate of the next one, relative to all other possible associations.

Heterocooperativity also can explain how formation of a correct repeating structure (e.g., the tail sheath) can be favored over formation of other polymorphic structures (polysheath) of equivalent stability. Different polymorphic associations require different subunit conformations. One such conformation may be favored by binding of the first subunit of the repeating structure to a different assembly component, thereby lowering the kinetic barrier to the correct, but not to the incorrect, polymerization. In the example mentioned, binding of the first sheath subunit to the core-baseplate could induce cooperative polymerization of the remaining subunits into the correct sheath structure, thereby kinetically favoring this process over formation of free polysheath.

The demonstration of such mechanisms could have important implications for organellogenesis and morphogenesis of other subcellular structures. Experiments with mutants blocked early in T4 tail assembly indicate that if the first protein in the pathway is missing, none of the subsequent associations will occur. Thus, all the components but one of a subcellular structure could be synthesized and available as soluble proteins; synthesis of the missing protein at an appropriate time could trigger the entire assembly sequence. Moreover, if the initiating protein were located on a membrane surface, then location and direction of growth of the structure could be determined as well as time of assembly. It seems quite likely that such mechanisms operate, for example, in the control of microtubule formation in the cytoarchitecture of cells. So far, however, the notion of heterocooperativity is hypothetical. More work is needed to establish the mechanism of sequential ordering in assembly and to test the hypothesis that it is achieved by specific conformational changes that remove kinetic barriers to association.

Another kind of kinetic control may be involved in the finer discrimination process that selects between alternative icosahedral forms in head assembly. Evidence has been presented that in T4, formation of the normal prolate icosahedral head rather than a smaller regular icosahedral head may depend on maintaining a certain ratio between the rate of shell protein synthesis and the rates of vertex protein and core protein synthesis. Other experiments indicate that the satellite phage P4 induces formation of a smaller icosahedral shell from the head proteins of its helper phage P2 by altering the transcriptional regulation of P2 head proteins in such a way as to change their relative rates of synthesis (see Wood and King, 1980, for other references). These findings emphasize the point that even at the level of rather simple supramolecular protein structures, biological form may be kinetically as well as thermodynamically determined.

In addition to kinetic controls in self-assembly, another significant general insight to come from studies of phage morphogenesis has been recognition of nonstructural accessory proteins, which represent departures from the classical self-assembly concept. These proteins so far appear to be of at least three kinds: scaffolding proteins, true enzymes, and promoters of noncovalent associations. The best-characterized scaffolding protein is the gene 8 product in phage P22 capsid assembly. One of its roles may be to increase precision of bonding to the degree necessary for assembling a closed icosahedral shell. The most evident enzymes in assembly are those responsible for the proteolytic cleavages of structural proteins that occur, for example, in T4 head assembly. The physiological importance of these cleavages is unclear, but conceivably they could serve, as in the case of zymogen-to-enzyme conversions, to convert stable structures to metastable ones, thereby allowing subsequent lattice transitions or new associations. The third class of accessory proteins, exemplified by *gp*38 and *gp*63 in the T4 tail fiber assembly pathway, remains the most difficult to understand. These proteins could represent a new kind of biological catalyst that helps to overcome kinetic barriers to noncovalent association. It is tempting to speculate, by analogy to enzymes, that such proteins might function by transiently binding to, and thereby stabilizing, otherwise unstable or improbable conformations that represent transition states of protein–protein association reactions (Figures 9–11). Nonstructural accessory proteins could be essential elements in many biological assembly processes. If so, identifying them will require either genetic analysis or assays for *in vitro* assembly, such as have been provided by the bacteriophages.

References

Benzer, S. (1959) On the topology of the genetic fine structure. *Proc. Natl. Acad. Sci. USA* **45**:1607.

Berget, P. B., and J. King (1983) Tail morphogenesis. In: *Bacteriophage T4*, C. K. Mathews, E. M. Kutter, G. Mosig, and P. B. Berget (eds.). American Society for Microbiology, Washington, D.C., p. 246.

Black, L. W., and M. K. Showe (1983) Morphogenesis of the T4 head. In: *Bacteriophage T4*, C. K. Mathews, E. M. Kutter, G. Mosig, and P. B. Berget (eds.). American Society for Microbiology, Washington, D.C., p. 219.

Brenner, S., F. Jacob, and M. Meselson (1961) An unstable intermediate carrying information from genes to ribosomes for protein synthesis. *Nature* **190**:576.

Caspar, D. L. D. and A. Klug (1962) Physical principles in the construction of regular viruses. *Cold Spring Harbor Symp. Quant. Biol.*. **27**:1.

Crick, F. H. C., and J. D. Watson (1957) Virus structure: general principles. In: *The Nature of Viruses*. G. E. W. Wolstenholme and E. C. P. Millar (eds.). Churchill, London, p. 5.

Earnshaw, W. C., and S. R. Casjens (1980) DNA packaging by double-stranded DNA bacteriophages. *Cell* **21**:319.

Earnshaw, W., E. B. Goldberg, and R. A. Crowther (1979) The distal half of the tail fibre of bacteriophage T4: Rigidly linked domains and cross-β structure. *J. Mol. Biol.* **132**:101.

Edgar, R. S., and W. B. Wood (1966) Morphogenesis of bacteriophage T4 in extracts of mutant-infected cells. *Proc. Natl. Acad. Sci. USA* **55**:498.

Epstein, R. H., A. Bolle, C. M. Steinberg, G. Kellenberger, E. Boy de la Tour, R. Chevalley, R. S. Edgar, M. Susman, G. H. Denhardt, and I. Lielausis (1963) Physiological studies of conditional lethal mutants of bacteriophage T4D. *Cold Spring Harbor Symp. Quant. Biol.*. **28**:375.

Fraenkel-Conrat, H., and R. C. Williams (1955) Reconstitution of active tobacco mosaic virus from its inactive protein and nucleic acid components. *Proc. Natl. Acad. Sci. USA* **41**:690.

Hershey, A. D., and M. Chase (1952) Independent functions of viral protein and nucleic acid in growth of bacteriophage. *J. Gen. Physiol.* **36**:39.

Kikuchi, Y., and J. King (1975) Genetic control of bacteriophage T4 baseplate morphogenesis. III. Formation of the central plug and overall assembly pathway. *J. Mol. Biol.* **99**:695.

Klug, A. (1972) Assembly of tobacco mosaic virus. *Fed. Proc.* **31**:30.

Mathews, C. K., E. M. Kutter, G. Mosig, and P. B. Berget (eds.) (1983) *Bacteriophage T4*. American Society for Microbiology, Washington, D.C.

Simon, L. D., and T. F. Anderson (1967) The infection of Escherichia coli by T2 and T4 bacteriophages as seen in the electron microscope. I. Attachment and penetration. *Virology* **32**:279.

Wood, W. B. (1979) Bacteriophage T4 assembly and the morphogenesis of subcellular structure. *Harvey Lec.* **73**:203.

Wood, W. B. (1980) The rise and decline of T4 phage biology at Caltech. In: *Genes, Cells, and Behavior, Proceedings of the 50th Anniversary Symposium of the Biology Division, California Institute of Technology*, N. H. Horowitz and E. Hutchings (eds.). Freeman, San Francisco.

Wood, W. B., and R. A. Crowther (1983) Long tail fibers: Genes, proteins, assembly, and structure. In: *Bacteriophage T4*, C. K. Mathews, E. M. Kutter, G. Mosig, and P. B. Berget (eds.). American Society for Microbiology, Washington, D.C., pp. 259.

Wood, W. B., and J. King (1979) Genetic control of complex bacteriophage assembly. In: *Comprehensive Virology*, Vol. 13, H. Fraenkel-Conrat and R. R. Wagner (eds.). Plenum Press, New York, pp. 581–633.

Wood, W. B., and H. R. Revel (1976) The genome of bacteriophage T4. *Bacteriol. Rev.* **40**:847.

Wood, W. B., R. S. Edgar, J. King, I. Lielausis, and M. Henninger (1968) Bacteriophage assembly. *Fed. Proc.* **27**:1160.

8

Molecular Biology in Embryology
The Sea Urchin Embryo

Vladimir Glišin
with
**Ana Savić, Radomir Crkvenjakov,
Sabera Ruždijić, and Nevenka Bajković-Moškov**

ABSTRACT

Modern work in embryology puts the old controversy between preformation and epigenesis in a new light. The rapidly dividing cells of embryos contain many maternally preformed, high-molecular-weight components, including both enzymes and molecular building blocks to be used long after fertilization. All presently investigated cellular macromolecules, with the exception of DNA, are found in excess in fertilized egg cells, compared to levels in mature somatic cells. Many of these macromolecules, including an interesting population of stable RNA's, remain constant in total amount through many successive cell divisions. Identical sets of genes in the cells of the early embryo later give rise to different cell types and lineages. Although some intermediate steps in this differentiation process have been identified, the ultimate cause of differential gene expression remains obscure. It is clear, however, that most differentiation is *not* the result of loss of nuclear genetic information. Artificially induced changes in the pattern of cell division in some cases create a disorganized embryo, but in others development proceeds normally. The polarity of molecular organization in an egg's cytoplasm appears to be critical to normal development. Cytoplasmic polarity may in turn result from the orientation of material in the cell nucleus. There seems to be an intrinsic geometry of chromosomes that permits a spatial display of information there to create significant, stable, and essential inhomogeneities and regionalizations of the cytoplasm. These, in turn, determine subsequent cleavage planes as cells divide, and the partitioning of information between daughter cells. The glossary in Chapter 5 will be helpful here also. In addition, for the nonbiological specialist, we offer the following, brief account of the main processes that relate DNA to proteins through RNA and that permit copying of genes during cell division.

VLADIMIR GLIŠIN • Genetic Engineering Center, Belgrade, Yugoslavia. ANA SAVIĆ and RADOMIR CRKVENJAKOV • Department of Biology, Faculty of Sciences, University of Belgrade, Belgrade, Yugoslavia. SABERA RUŽDIJIĆ and NEVENKA BAJKOVIĆ-MOŠKOV • Institute for Biological Research, Belgrade, Yugoslavia.

Replication: DNA is copied, and the copy checked for accuracy, by enzymes.

Transcription: a particular, structural gene segment of DNA, usually including any noncoding, intervening sections, is copied by enzymes into an RNA molecule, called a primary (gene) transcript.

Posttranscriptional processing: processing that includes cutting out noncoding sections of the primary transcript to produce a shorter RNA molecule (mRNA) that can leave the nucleus and enter the cytoplasm, where it is attached to the protein-synthesizing factory (the ribosome or, when very active, the polysome). This factory has intrinsic RNA of its own (rRNA) and accepts the string of codons from mRNA, as well as the anticodon tagged amino acids brought to it by RNA's. The amino acids are then joined on the mRNA template into a linear string that constitutes a polypeptide (protein). This completes "translation."

Posttranslational processing: the reader should know that much more can sometimes happen after translation, but the details need not be specified here. Anyone wanting to pursue a technical account of all these processes should consult Volumes 1 and 2 of *Biological Regulation and Development*, edited by R. F. Goldberger, Plenum Press, New York, 1979, 1980. —THE EDITOR

Embryology deals with the events associated with the formation of an adult individual from a fertilized ovum. Traditionally, the emphasis has been placed on succession of structures. The fertilized egg consists of a single cell, whereas an adult organism like the sea urchin consists of a few billion cells (the adult human has about 10^{14} cells). Embryology begins with the fusion of the egg and sperm to form a zygote. Division of the zygote and its daughter cells leads to the aggregation of cells into tissues, tissues into organs, and organs into organisms, all in a predetermined form and size.

The sea urchin zygote quickly develops into a hollow ball of cells. The cells multiply and migrate, and the embryo attains its final stage of development as a result of a complicated series of cell movements (morphogenetic movements), including the in-and out-pocketings of layers of cells. In these complex processes of embryonic development, five main features may be identified: pattern formation, differentiation, polarity, induction, and determination (Moore, 1972). The process that orders the distribution of cells in space is called pattern formation. Differentiation involves the construction of specialized cells with particular biochemical capabilities. Polarity refers to initial regional differences in molecular organization of the unfertilized egg that lead to regionalization in later embryological stages. When a cell or a group of cells act on other cells to alter their courses of development, this action is spoken of as induction. By altering the course of development, the inducer changes a cell from an undetermined to a determined path of differentiation.

We consider these five processes independently, for convenience. In the embryo, these processes are overlapping aspects of one continuous process. In this chapter, they are examined on the molecular level.

Historical Background

Developmental biology is a subject rich in philosophical difficulties. Many of the confusions of 18th and 19th century speculations, with their two rival views of development, i.e., preformation and epigenesis, are still with us today. The theory of preformation states that adult structures, including the most minute ones, are already differentiated at the very beginning of embryonic life; development then consists only of growth. Epigenesis indicates that the adult parts are not present initially but, rather, are developed during embryonic life (Browder, 1980).

Among epigenetic theories there have been two underlying concepts. The "mosaic" concept regards the embryo as an association of independent cells, each developing along a predestined path to form a specific part of the adult. The "regulative" concept supposes that the individual egg and its progeny can adjust to new situations at each embryonic stage and sculpt a whole embryo and, later, an adult. These two concepts were combined by Spemann (1938) in a single, organizer concept. According to Spemann, the change from an undetermined state to a determined one is in most cases the result of an influence external to the cells being determined. The main difference between the mosaic and the regulative concepts is the time at which external influences cause determination. In the mosaic model, this occurs when the egg is being formed in the ovary, and the developing embryo is a mosaic from the outset. In regulative models, on the other hand, determination occurs during development.

Modern work in genetics and embryology has put the old controversy of preformation and epigenesis in a new light. The preformed entities that are transmitted are the genes and the cytoplasm in the egg (Gurdon and Woodland, 1962). A fertilized sea urchin egg gives rise to a sea urchin embryo because it possesses particular genes and the specific cytoplasm that controls development. The hereditary basis of development is preformed in the structure of the gametes; the phenotype of adult parts is epigenetic. The genetic system specifies what a cell may do; nongenetic phenomena influence what it actually does. It would then follow that further specialization of cells through development, i.e., differentiation, is not necessarily a consequence of loss of genetic information. Such a notion is supported by two independent observations: (1) regeneration of whole organisms from small parts or individual cells is commonly encountered in plants and in lower animals (carrot plant, hydra) (Steward, 1970) and (2) generation of an adult toad can be accomplished by transplanting a somatic nucleus into an unfertilized ovum (Gurdon, 1962). Specifically, a haploid nucleus (containing the genes of only the female) of a mature but unfertilized ovum is replaced by the nucleus obtained from a specialized diploid cell, e.g., a skin or intestinal tract cell that contains the full number of genes for an individual of the species, constituted of the haploid contributions of both male and female. The ovum with the transplanted, diploid nucleus develops as if it were fertilized, and a mature organism is formed. In view of the fact that almost all the hereditary material of DNA is found in the cell nucleus (except for mitochondrial DNA, which is found in the cytoplasm and is transmitted exclusively through the mother), the renucleated ovum and the individual into which it develops are genetically identical to the mature organism that was the original donor of the nucleus. This experiment has been further generalized by saying that all the nuclei of an adult organism contain the full genetic potential of the original zygote (Gurdon, 1962). By generating whole organisms from individual somatic cells, or by transplanting somatic cell nuclei into an unfertilized egg, the roulette of sexual recombination is circumvented and asexual reproduction copies already-developed units.

In an oversimplified analogy, the developmental process and differentiation can be visualized with the aid of the "piano" model, as we call it (Glišin and Savić, 1971). An identical keyboard of a piano is present in every cell. The keys of these keyboards represent either individual genes or gene parts[1] and are played in chords and harmonies as development advances. These keys, i.e., the genes, are played/expressed through the synthesis of corresponding proteins at different times and in different cells according to some programmed schedule, which is still mysterious. The players, i.e., gene activators or deactivators (one can call them repressors or depressors), are the cytoplasmic factors whose involvement or noninvolvement at a given time and in a given cell depends on preceding processes (the part of the melody played before). Although each cell plays its own tune, there must be a basic harmony similar to the harmony in an orchestra.

This analogy is generally apt, but recent discoveries of the structure of mammalian immunoglobulin genes necessitate a modification of the piano analogy. During differentiation of undifferentiated stem cells into an antigen-specific, differentiated line of lymphocytes, a translocation of segments of DNA occurs by excision of the intervening segments (Maki et al., 1980). The selection of nucleotide sequences to be biochemically expressed occurs epigenetically. As a consequence of these rearrangements, after a nucleus from a highly differentiated, antibody-producing cell is transplanted into an enucleated egg, the new developing organism would not have the capacity to produce all antibodies. The result is a restriction in cell potential. The genetic organization of the immune system resembles more a violin than a piano. We do not yet know whether this manner of cell differentiation is confined to the immune system. The nuclear transplantation experiments and regeneration of whole organisms from somatic cells suggest that the behavior of the immune system might be the exception.

During the first few hours following fertilization, the sea urchin embryo divides rapidly and repeatedly, so that the number of cells increases from 1 to about 2^{10}. The biomass of the developing embryo remains constant. During these early stages, the molecular events within a cell cycle are characterized by an almost continuous synthesis of DNA (S phase of the cell cycle). The cells of the developing embryo have a particularly short G_1 phase of the cell cycle, although this is the phase in which major anabolic processes take place (Gross, 1967). This

---

[1]Many structural genes in eukaryotic cells are interrupted by very long, noninformational, intervening segments (except for the genes of histones and interferons). The role of the "noncoding" segments is unclear. However, what is already known for certain is that some of these intervening sequences are transcribed into RNA and that, in the course of the processing of these primary RNA transcripts, the intervening sequences are subsequently cut out. Thus, for example, the gene of β-globin of the mouse contains a segment of about 1500 nucleotides, which is not present in the cytoplasmic β-globin RNA (Konkel et al., 1978). Similarly, in the case of the ovalbumin gene of chickens, one finds an insert about 5000 nucleotides long (O'Malley et al., 1976). Accordingly, a eukaryote gene does not necessarily represent a continuous series of nucleotides (as is the case in prokaryotes) but, rather, consists, as Gilbert (1978) says, of "exonic" regions corresponding to the structural part of the genes and of "introns" that represent inserted, noninformational parts of DNA.

peculiarity makes the cell cycle of the cleaving embryo much shorter than any cycle of dividing cells in the adult organism.

The extraordinarily high division rate of embryonic cells immediately raises two questions: What enables the cells of the embryo to divide so rapidly without relying on the usual long synthetic period of a normal cell cycle? Should we not expect an exponentially increasing rate of total protein synthesis, because major subcellular components (like chromatin with its histones and nonhistones, plasma membranes, microtubule proteins, ribosomes, and so on. or enzymes needed to synthesize DNA or RNA precursors) should follow the increase of the cell number per embryo? In fact, the rate of protein synthesis, after an initial increase in the immediate postfertilization period, remains practically constant to the end of the cleaving period and is not affected by inhibition of transcription (Gross and Cousineau, 1963, 1964)! These results provide a strong indication that the rapidly dividing cells of these embryos contain many maternally preformed, high-molecular-weight components serving either as building blocks for higher structures or as ready-made, preformed enzymes to be utilized in the postfertilization period. Therefore, the ultimate step of gene expression, the posttranslational regulation of assembly of maternally synthesized macromolecule precursors into biologically meaningful structures, should be dramatically emphasized in the sea urchin embryo. These predictions proved to be correct. Furthermore, the same situation has been encountered in other oviparous embryos.

The Preformed Macromolecules

Many examples of macromolecules are present in large amounts in the unfertilized eggs of sea urchins. Actually, all the macromolecules investigated so far, except the chromosomal DNA, are found in excess, compared to the amount typical of a somatic cell of an adult.

One of the first results contributing to the notion of preformed enzymes was the measurement of the activity of deoxycytidylate aminohydroxylase (dCMP deaminase), an enzyme that regulates the concentration of pyrimidine nucleotide precursors for DNA replication. This enzyme is present in the same total quantity in the unfertilized single-celled egg as it is in the rapidly proliferating, cleaving, multicellular embryo (Scarano and Maggio, 1959). Similarly, the total activity of each of the four different deoxynucleotide kinase activities (enzymes that phosphorylate deoxynucleotide monophosphates into corresponding di-and triphosphates) is the same in the unfertilized egg, in the 2-cell embryo, and in the 100-cell embryo (Scarano *et al.*, 1968).

DNA polymerase has also been found in large quantities in unfertilized eggs, despite the fact that such eggs do not synthesize DNA. The unfertilized egg of *Strongylocentrotus purpuratus* contains about 5×10^8 molecules of the enzyme. This is enough for one polymerase molecule to bind at every 1600 nucleotide pairs of DNA in the hatching blastula, which contains about 300 cells, i.e., 300 genomes (Loeb, 1970). In the sea urchin, the amount of activity of each of the three classes of RNA polymerases remains constant from fertilization through gastrulation (Roeder and Rutter, 1970).

Another interesting case in which amounts of an enzyme could be followed during development is that of the plasma membrane enzyme, adenylate cyclase (Trams et al., 1974). We first calculated the surface area of the egg of *Arbacia lixula* and the sum of the surfaces of the cells at different stages of development. We assumed that all cells have a spherical shape. The total surface area of cells of an embryo increases about tenfold from the unfertilized egg up to the stage of 2^{10} cells. If the plasma membrane in any stage of development of the sea urchin necessarily had the same content of adenylate cyclase per unit area, then an increase in the activity of adenylate cyclase in the total embryo would have to parallel the calculated increase in the plasma membrane surface area. However, our results indicated that, except for a slight increase of activity in the immediate postfertilization period, the adenylate cyclase activity remained constant as cell number and membrane area increased. This result suggests that the enzyme is of maternal origin, existing in the unfertilized egg and gradually incorporated into the newly formed plasma membranes (Trams et al., 1974).

Perhaps an even clearer example of previously formed molecules of high molecular weight are the histones. Early embryogenesis requires continual synthesis of chromosomal DNA. Because in all eukaryotes chromatin consists of roughly equal quantities of DNA, nonhistone proteins, and histones, one would expect that the rate of synthesis of the protein part of chromatin should parallel the rate of DNA synthesis and correlate with the number of new cells. If this is not the case, then one of the conclusions might be that in the unfertilized egg both histone and nonhistone proteins must exist in a form that is not an integral part of the chromatin. In other words, they would have to be stored somewhere, most probably in the large cytoplasm of the egg—an unusual location for nuclear materials. We found that, indeed, cytoplasm of the unfertilized egg contains a quantity of nucleosome histones sufficient for the formation of approximately 50 diploid, complete chromatins (Savić, 1972; Solomun et al., 1978; Salik et al., 1981). In other words, histones preexist as a store (formed during oogenesis) later used in the immediate postfertilization period. Other examples of previously formed histones are known (Adamson and Woodland, 1974).

Finally, all types of cytoplasmic RNAs (tRNA, rRNA, mRNA) are present in surplus amounts in the unfertilized egg (Glišin and Glišin, 1964; Slater and Spiegelman, 1966; Giudice and Mutolo, 1967; Guidice et al., 1968). Ribosomes are not synthesized before the gastrula stage of embryogenesis, but the ribosome-associated RNAs are present from the beginning.

Thus, the large cytoplasm of egg cells is not only a food storehouse but also is the storehouse of components that enter the subcellular structures later: the main theme of gene expression during very early embryogenesis is the control at the posttranslational level.

Preformed mRNA—the Maternal mRNA

One of the most interesting results in the molecular biology of embryogenesis stems from the investigation of the metabolism, function, and fate of the maternal mRNA in the early development of the sea urchin egg.

Fertilization of the sea urchin egg is not obligatorily accompanied by *de novo* synthesis of mRNA (Glišin and Glišin, 1964). Furthermore, the sea urchin embryo, when cultured in the presence of an RNA synthesis inhibitor (actinomycin D), may develop to the blastula stage without a decrease in the rate of protein synthesis (Gross and Cousineau, 1963) that is dependent on mRNA's. These two results together led to the view that there are stable, long-lived mRNA's in eukaryotic cells. (Rapidly dividing prokaryotic cells have only short-lived mRNA's.) In eukaryotic cells, the mRNA's could therefore be synthesized in one parental cell, stored, transferred, and ultimately biochemically expressed in some of the daughter cells a few generations later. Because the egg's mRNA must have originated during oogenesis, these RNAs are called maternal (oogenic) mRNA's. Oogenic mRNA constitutes approximately 1.5 to 3% of the total RNA of the sea urchin egg (Jenkins *et al.*, 1978).

Because maternal mRNA's do exist in the unfertilized, metabolically dormant egg, new questions arise: how are the maternal mRNA's stored in the nonfunctional state, and what is the underlying molecular mechanism that allows a sudden (at fertilization) or gradual (during embryogenesis) release from an inactive to a biochemically active state? No definitive answers to these questions have been obtained. However, some experiments have indicated that attention should be focused on the protein moiety of the mRNA-protein complex, mRNP (Spirin, 1966), the complex in which all free or bound mRNA's exist in eukaryotic cells. It has been well documented that the oogenic RNA is present in the form of mRNP particles in the egg's cytoplasm (Spirin and Nemer, 1965; Ruždijić and Glišin, 1972). When these particles are isolated intact, the mRNA remains untranslatable in an *in vitro* translational system (Jenkins *et al.*, 1978; Ruždijić, 1980). In contrast, the same RNA separated from its protein moiety could be readily translated *in vitro*. This result shows that *in vivo*, at fertilization and during embryogenesis, an unmasking of oogenic mRNA's takes place.

Embryonic mRNA

The analysis of the organization of the sea urchin genome (and of other animal genomes) has revealed that DNA sequences can be grouped into two general classes, based on the frequency of their occurrence in the genome: (1) those sequences that are present in the genome only once and are called unique sequences (most mRNA's are transcripts of unique sequences) and (2) those sequences that are represented many times. These can be further subdivided into those moderately represented (the middle-repetitive) and those repeated many more times (the highly repetitive sequences). Middle-repetitive sequences are transcribed, but the highly repetitive sequences in sea urchins are never transcribed (Davidson and Britten, 1979; Davidson *et al.*, 1982).

The abundance of different mRNA's is not strictly correlated with the abundance of corresponding DNA segments from which they are transcribed. This fact has led Davidson and Britten to define three classes of RNA's: a limited-copy class of mRNA's that appear in the cell at levels of one to several copies; moderately prevalent mRNA's that are present up to a few hundred copies per cell; and superprevalent mRNA's of more than 10^4 copies per cell (Davidson and Britten, 1979).

Measurement of classes of sea urchin embryonic mRNA's has shown that about 20% of these DNA transcripts are present in 1 to 15 copies per cell; the remaining 80% are moderately prevalent, ranging between 15 and 300 copies per cell (Davidson and Britten, 1979). A great similarity exists between the oocyte structural gene sequence sets and the corresponding gastrula RNAs as well as the RNA of the developmental stages in between. In nuclei of embryos of all these stages, the RNA transcripts (in varying numbers of copies) of about 14,000 unique DNA gene-equivalents are present.

How does the stable set of unique DNA and RNA sequences in the nuclei of the early embryo give rise to different cell types and cell lineages and the appearance of certain protein species while others disappear during development (Brandhorst, 1976)? Specifically, the H1, H2A, and H2B histones of early developmental stages are replaced by the synthesis of new subsets of these proteins (Newrock *et al.*, 1978). As expected, these changes of patterns of histones are accompanied by changes in mRNA templates on polyribosomes (Hieter *et al.*, 1979).

Only a small fraction, about 10 to 20%, of single-copy sequences in nuclear RNA (nRNA) are also represented in cytoplasmic mRNA at a given time, even though single-copy DNA structural gene sequences are completely represented in an embryo's nRNA's. All the nRNA molecules bearing these structural gene sequences are potential mRNA precursors (Davidson and Britten, 1979).

All these results led Davidson and Britten to postulate a new model of gene expression of eukaryotic cells (Davidson and Britten, 1979). The model supposes that the selection of a given mRNA and transport to polysomes in the cytoplasm is regulated by families of middle-repetitive nRNA sequence transcripts. These are complementary to middle-repetitive DNA sequences flanking the unique structural gene sequence. The model further postulates that the genome is composed of constitutive transcriptional units (CTUs) and integrating, regulatory transcriptional units (IRTUs). CTUs are transcribed in all cell types and include structural gene segments, intervening sequences, and flanking sequences. IRTUs provide middle-repetitive sequences that combine intranuclearly to yield IRT–CT complexes. This duplex formation is a prerequisite for further processing of nRNA's to real template mRNA's appearing on polyribosomes. The main role of middle-repetitive sequences is then quantitatively and qualitatively to regulate gene expression. The sequence specificity of IRTUs regulates the ultimate sequence specificities of mRNA's (cell differentiation), while their abundance in the nucleus regulates the quantity of given mRNA sequences to be exported into the cytoplasm.

Applying this scheme directly to sea urchin development, the specific topological (spatial) distribution of middle-repetitive maternal RNA transcripts in the unfertilized egg substantiates the regulatory program of development. Consequently, different programs of gene expression in different cells of the embryo originate from various parts of the egg (polarity). Not all the necessary information is in DNA.

The fate of the maternal mRNA's during early embryogenesis also raises interesting questions. Are all RNA's present in the zygote identical to those of all cells during early development? Is the distribution of maternal mRNA's among progeny cells of the embryo nonuniform? The detailed molecular mechanism of embryonic differentiation based on polar distribution of specific maternal messages to specific progenic blastomeres is not yet resolved. Some experimental evidence presented indicates that such a sequestration of mRNA's in the embryo is not taking place (Angerer and Davidson, 1984). It is of importance for further understanding of the molecular basis of the polarity to resolve this problem unequivocally. However, the Davidson-Britten model provides an intellectual framework for understanding the biological role of the interspersion of unique and repetitive sequences in the genome. Simultaneously, this model of gene expression gives us some ideas of possible molecular mechanisms underlying the five main features of embryonic development: polarity, induction, cell differentiation, determination, and pattern formation. The attractiveness of the model stems also from the fact that it can be further tested experimentally, but the model does not tell us how the spatial information (polarity) comes about. It merely rephrases the fundamental question.

Cell Lineage

If we accept as a working hypothesis that the unfertilized egg of a sea urchin represents a spatial gradient (a field) of preformed informational macromolecules, then the orientation of spindles within the embryo becomes of utmost importance for determining the cleaving patterns of normal development. The developmental fates of various cells have been mapped by the vital dye marking method (Vogt, 1929) and, more recently, by the aid of the enzyme horseradish peroxidase (Weisblat *et al.*, 1978). Unfortunately, the results of fate-mapping experiments seem contradictory. We believe, however, that the contradiction is not real. The major experimental results that have to be reconciled are outlined below.

Hörstradius (1939) performed a simple experiment in which, by shaking fertilized eggs, he was able to delay the first cleavage and also to change the orientation of the first furrow formation. In cases in which the first cleavage was delayed (until the time that unshaken controls reached the third furrow formation), the spindles were oriented in a position perpendicular to that normal for the first cleavage. The next division was equatorial, as expected for later divisions. The result was a T-shaped configuration, with mesomeres at the animal pole, one macromere in the center, and one micromere at the vegetal pole. A disorganized embryo developed. If the same sea urchin eggs were mounted between glass plates, animal poles upward, and gently compressed, the direction of the third cleavage plane was altered; it was now vertical. When the pressure was released at the eight-cell stage, the fourth cleavage plane was horizontal. In this case a normal embryo developed (Driesch, 1892). Both these results can be interpreted to mean that polarity does exist in the unfertilized egg. In the first case the polarity was disturbed and an abnormal embryo developed; in the latter case an undisturbed polarity led to normal embryonic development.

In contrast, the results of Rappaport suggest that cytoplasmic organization is *not* required for induction of the cleaving furrows. He found that stirring the cytoplasm of an egg with a fine needle did not disrupt the orderly cleavage and a normal embryo developed (Rappaport, 1966). But, if one assumes that artificially disturbed cytoplasmic organization can be quickly and spontaneously restored—after the abuse—to the spatial arrangements in the untreated egg, the difficulty is resolved. Experiments by Capco and Jeffrey (1981) showed that the vegetal pole mRNA when injected into the fertilized *Xenopus* egg finds its way back to the vegetal pole. This then strongly indicates that the most obvious cytoplasmic determinant of differentiation spontaneously restores its place in cytoplasmic organization after disturbance.

The basic question still remains open: What is the nature of the effectors of polarity in the egg? We offer an idea, based on two testable assumptions: (1) A spatial orientation, an anisotropy, a polarity of chromatin itself within the nucleus, must exist. Then the chromatin can serve as a structural primer that directs the recruitment of preexisting microtubule proteins and, in turn, the spatial distribution of spindles and centrioles and the formation of regular furrows. (2) The polar organization of the cytoplasm is directed by the nucleus/chromatin (Glišin, 1976). The genome has an intrinsic geometry!

Results of some experiments seem to favor such an assumption. When the nucleus (chromatin) is removed, an artificially activated, enucleated half-egg undergoes cleavage, but the cleavages are abnormal without centriole and spindle formation (Harvey, 1940). In contrast, the regular cleavage pattern is retained under other conditions in which all transcriptions are prevented in a full egg. Because enucleated half-eggs contain the same maternal mRNA's as the whole embryo, they should not lack any newly synthesized proteins needed to maintain the regular patterns of cleavage. The only difference between the normal and enucleate "embryos" is the presence or absence of chromatin.

The role of chromatin as a coordinator for regular furrow formation was tested in an additional experiment in which the embryo was chemically enucleated. If sea urchin eggs are incubated in the presence of a synthetic analogue of cytidine (5-azacytidine, a drug that inhibits uridine kinase and orotidilic acid decarboxylase), replication of chromosomal DNA as well as transcriptions are prevented. Morphological development is not hampered for the first two cleavages, but a completely malformed embryo develops later on (Crkvenjakov *et al.*, 1970).

Conclusion

It is clear that the understanding of complex developmental processes must depend on an overall understanding of the organization and function of eukaryotic genomes themselves. At present, however, the endpoints of genes in eukaryotic cells are not well defined. Not enough is known about the higher orders of chromatin structure. Little is known about the control of transcription in eukaryotic cells; more is known about translation. The importance of posttranslational control has been appreciated only recently. Work with the sea urchin embryo has brought posttranslational control into focus (Glišin, 1976).

Self-assembly during embryonic development is not mediated by direct gene intervention. When all the transcriptions have been prevented (e.g., by the action of actinomycin D or α-amanitin), the regular cleavage patterns are retained. However, the polarity of molecular organization of both the egg's cytoplasm and its nucleus (chromatin) are essential for normal development. Hence, the main features of embryogenesis—cell differentiation, induction, determination, and pattern formation—all stem from the oogenetically originated, spatial distribution of preformed informational macromolecules. *The initial condition of embryogenesis is oogenesis. The epigenetics of embryonic development is built on the topological self-organization and orientation of macromolecules of the total egg.*

So, despite new insights about the molecular basis of embryonic development, the results accumulated are still fragmentary and insufficient to permit an unequivocal explanation of the mechanisms of embryogenesis. After a century or more of intensive work and ideological debates, we have built only a realistic framework within which to further our efforts. Therefore, embryology is still the science of tomorrow.

References

Angerer, R. C. and E. H. Davidson (1984) Molecular indices of cell lineage specification in sea urchin embryos. *Science* **226**:1153–1160.

Brandhorst, B. P. (1976) Two dimensional cell patterns of protein synthesis before and after fertilization of sea urchin eggs. *Dev. Biol.* **48**(2):458–460.

Browder, W. L. (1980) *Developmental Biology*. Saunders, Philadelphia, pp. 2–32.

Capco, D. G., and W. R. Jeffrey (1981) Regional accumulation of vegetal poly(A)$^+$RNA injected into fertilized *Xenopus* eggs. *Nature* **294**:255–256.

Crkvenjakov, R., N. Bajković, and V. Glišin (1970) The effect of 5-azacytidine on development, nucleic acid and protein metabolism in sea urchin embryos. *Biochem. Biophys. Res. Commun.* **39**:655–660.

Davidson, E. H., and R. J. Britten (1979) Regulation of gene expression: Possible role of repetitive sequences. *Science* **204**:1052–1059.

Davidson, E. H., B. R. Hough-Evans, and R. J. Britten (1982) Molecular biology of the sea urchin embryo. *Science* **217**:17–26.

Driesch, H. (1892) Entwickelungsmechanisches. *Anat. Anz.* **7**:584.

Gilbert, W. (1978) Why genes in pieces? *Nature* **271**:501.

Giudice, G., and V. Mutolo (1967) Synthesis of ribosomal RNA during sea urchin development. *Biochim. Biophys. Acta* **138**:276–285.

Giudice, G., V. Mutolo, and G. Donatuti (1968) Gene expression in sea urchin development. *Wilhelm Roux Arch. Entwicklungsmech. Org.* **161**:118–128.

Glišin, V. (1976) The cleaving sea urchin embryo: A biological model system for analysis of posttranslational gene expression. *Bull. Mol. Biol. Med.* **1**:1–11.

Glišin, V. R., and M. Glišin (1964) Ribonucleic acid metabolism following fertilization in sea urchin eggs. *Proc. Natl. Acad. Sci. USA* **52**:1548–1553.

Glišin, V., and A. Savić (1971) Informational macromolecules during the early development of sea urchins. *Prog. Biophys. Mol. Biol.* **23**:193–201.

Gross, P. R. (1967) The control of protein synthesis in embryonic development and differentiation. *Curr. Top. Dev. Biol.* **2**:1.

Gross, P. R., and G. H. Cousineau (1963) Effect of actinomycin D on macromolecule synthesis and early development in sea urchin eggs. *Biochem. Biophys. Res. Commun.* **10**:321–326.

Gross, P. R., and G. H. Cousineau (1964) Macromolecule synthesis and the influence of actinomycin on early development. *Exp. Cell Res.* **33**:368–395.

Gurdon, J. B. (1962) Adult frogs derived from the nuclei of single somatic cells. *Dev. Biol.* **4**:256.

Gurdon, J. B., and H. R. Woodland (1962) The cytoplasmic control of nuclear activity in animal development. *Biol. Rev.* **43**:233–267.

Harvey, E. B. (1940) A comparison of the development of nucleate and nonnucleate eggs of Arbacia punctulata. *Biol. Bull.* **71**:101–121.

Hieter, P. A., M. Hendricks, K. Hemminki, and E. Weinberg (1979) Histone gene switch in the sea urchin embryo: Identification of late embryonic histone messenger ribonucleic acids and the control of their synthesis. *Biochemistry* **18**:2707–2716.

Hörstradius, S. (1939) The mechanisms of sea urchin development studied by operative methods. *Biol. Rev.* **14**:132–179.

Jenkins, N. A., J. F. Kaumeyer, E. M. Young, and R. A. Raff (1978) A test for masked message: The template activity of messenger ribonucleoprotein particles isolated from sea urchin eggs. *Dev. Biol.* **63**:279–298.

Konkel, D. A., S. M. Tighman, and P. Leder (1978) The sequence of the chromosomal mouse β-globin major gene: Homologies in copying, splicing and poly(A) sites. *Cell* **15**:1125–1132.

Loeb, L. A. (1970) Molecular association of DNA polymerase with chromatin in sea urchin embryos. *Nature* **226**:448.

Maki, R., J. Kearney, C. Paige, and S. Tonegawa (1980) Immunoglobin gene. Rearrangement in immature B cells. *Science* **209**:1366–1369.

Moore, A. J. (1972) *Heredity and Development.* Oxford University Press, London, pp. 256–288.

Newrock, K. M., C. R. Alfageme, R. V. Nardi, and L. H. Cohen (1978) Histone changes during chromatin remodeling in embryogenesis. *Cold Spring Harbor Symp. Quant. Biol.* **42**:421–431.

O'Malley, B. W., S. L. C. Woo, J. J. Monahan, L. McReynolds, S. R. Harris, M. J. Tsai, S. Y. Tsai, and A. R. Means (1976) The synthesis, isolation, amplification and transcription of the ovalbumin gene. In: *Molecular Mechanisms in the Control of Gene Expression,* D. P. Nierlich, W. J. Rutter, and C. F. Fox (eds.). Academic Press, New York, pp. 309–329.

Rappaport, R. (1966) Experiments concerning the cleavage furrow in invertebrate eggs. *J. Exp. Zool.* **161**:1–8.

Roeder, R. G., and W. J. Rutter (1970) Multiple RNA polymerases and RNA synthesis during sea urchin development. *Biochemistry* **9**:2543–2553.

Ruždijić, S. (1980) Metabolism and structure of cytoplasmic particles in sea urchin embryos. Doctoral thesis, Faculty of Sciences, University of Belgrade.

Ruždijić, S., and V. Glišin (1972) Towards a total analysis of polyribosome-associated ribonucleoprotein particles of sea urchin embryos. *Biochim. Biophys. Acta* **269**:441–449.

Salik, J., L. Herlands, H. Hoffman, and D. Poccia (1981) Electrophoretic analysis of the stored histone pool in unfertilized sea urchin eggs: Quantification and identification by antibody binding. *J. Cell Biol.* **90**:385–395.

Savić, A. (1972) The role of cytoplasmic pool of nuclear proteins in early embryonic development of sea urchins. Doctoral thesis, Faculty of Sciences, University of Belgrade.

Scarano, E., and R. Maggio (1959) The enzymatic deamination of 5'-deoxycytidilic acid and of 5-methyl–5'-deoxycytidilic acid in developing sea urchin embryos. *Exp. Cell Res.* **18**:333.

Scarano, E., M. Rossi, and G. Geraci (1968) The regulation of the activity of deoxycytidylate aminohydrolase. In: *FEBS Symposium on Regulation of Enzyme Activity and Allosteric Interactions,* E. Kramme and A. Pihl (eds.). Academic Press, New York, p. 145.

Slater, D. W., and S. Spiegelman (1966) An estimation of genetic messages in the unfertilized echinoid egg. *Proc. Natl. Acad. Sci. USA* **56**:164–170.

Solomun, M., R. Radojcić, A. Savić, and V. Glišin (1978) Determination of a histone pool in sea urchin eggs—A three-dimensional electrophoretic procedure for histone purification. 12th FEBS Meeting, Dresden, Abstr. 2145.

Spemann, H. (1938) *Embryonic Development and Induction.* Yale University Press, New Haven, Conn. Republished by Hafner, New York.

Spirin, A. S. (1966) On "masked" forms of messenger RNA in early embryogenesis and in other differentiating systems. In: *Current Topics in Developmental Biology 2,* Vol. 1, A. Monroy and A. A. Moscona (eds.). Academic Press, New York.

Spirin, A. S., and M. Nemer (1965) Messenger RNA in early sea urchin embryos: Cytoplasmic particles. *Science* **150**:214–217.

Steward, F. C. (1970) From cultured cells to whole plants: The induction and control of their growth and morphogenesis. *Proc. R. Soc. London Ser. B* **175**:1–30.

Trams, E. G., C. J. Lauter, G. J. Koval, S. Ruždijić, and V. Glišin (1974) Plasma membrane marker enzymes in developing sea urchin embryos. *Proc. Soc. Exp. Biol. Med.* **147**:171.

Vogt, W. (1929) Gestaltungsanalyse am Amphibienkeim mit ortlicher Vitalfarbung. II Gastrulation und Mesodermbildung bei Urodelen und Amuren. *Wilhelm Roux Arch. Entwicklungsmech. Org.* **120**:385–406.

Weisblat, D. A., B. T. Sowyer, and G. S. Stent (1978) Cell lineages. Analysis by intracellular injection of a tracer enzyme. *Science* **202**:1295.

9

Developing Organisms as Self-Organizing Fields

Brian C. Goodwin

ABSTRACT

The way in which parts are related to wholes is crucial to any theory of self-organization, since it is necessary to know what is being organized and what collective properties they reveal. The analytic tradition of science uses an atomistic description of this relationship: parts ("atoms") are assumed to be primary and the whole is generated by their interaction. This approach dominates modern embryology and evolutionary biology. However, there is an alternative view that is very useful in understanding complex order, which is the structuralist tradition: a whole entity is defined by invariant relations within which parts emerge and behave in accordance with transformational principles. Evidence that the constraints determining biological form arise from organizational levels other than the genome and gene products emphasizes the primacy of morphogenetic principles at the level of the organism, and the inadequacy of atomistic theories on which the concept of the genetic program is based.

Many field theories in physics illustrate structuralist principles in that descriptions of global order are primary to the analysis (e.g., the Pauli exclusion principle, or Bell's theorem and nonlocal connectedness), so that the behavior of the system is not reducible to interactions of "atomic" constituents such as electrons or photons. In biology, field concepts have been used for many years to describe the behavior of developing embryos, which reveal high-level order and systematic spatial transformations suggestive of principles of global organization. An application of a particularly simple type of field theory to the earliest stage of embryonic development in many species, involving the use of harmonic functions to describe the systematic transformations of the fertilized egg through cleavage stages, illustrates the general principles as they apply to embryogenesis. The specific set of harmonic functions describing the cleavage planes are specified by selection rules which are related to particular aspects of embryonic organization, some arising from properties of the whole and some from properties of parts. This type of field analysis is then extended to a qualitative description of how the next stage of embryogenesis, gastrulation, could emerge from the conditions resulting from cleavage. Such field descriptions allow for comparisons of morphogenetic principles between unicellular and multicellular organisms, despite the fact that at one level of analysis their "parts" appear to be different. This illustrates the principles of a structuralist analysis, which leads one to seek organismic homologies at the

BRIAN C. GOODWIN • Department of Biology, The Open University, Milton Keynes MK7 6AA, England.

level of primary generative principles (deep structure) such as those defining field properties, and not in terms of genetic composition or historical relatedness. —The Editor

Self-organization is a property of a structurally and functionally integrated entity considered to be made up of, or to have, "parts." There are essentially two ways of relating the whole and the parts: either the parts are regarded as given (with or without some temporal sequence of presentation or production) and the entity is generated by their interaction; or the entity is regarded as a whole defined by certain invariant relations, the "parts" coming into being as a result of systematic transformations that preserve invariance while generating heterogeneity ("parts") within a functional and structural unity. The most extreme form of the first description is "atomism," a view assuming that all the information necessary for generating an entity is resident in the parts, so that spontaneous assembly occurs simply as a result of their interaction. This hypothesis, in various forms and with various modifications, is a dominant theme in contemporary models of evolutionary, developmental, behavioral, and even cognitive processes. This conceptualization is characterized by the assumption that there are no laws of self-organization other than those governing the interaction of the parts or constituents, so that the whole is reducible to these parts and their interactions.

In developmental biology, the constituents are usually considered to be, ultimately, molecules, although there are theories in which cells are used as intermediate "atoms" in the analysis. Because genes are generally considered to be the determinants of which (macro)molecules are present in organisms, it follows that organisms are reducible to genes. The self-organizing process of embryogenesis is then regarded as a consequence of two activities: (1) the operation of the "genetic program" that determines the types of molecular constituents in the organism, as well as the sequence and spatial location of their appearance, and (2) the interaction of these constituents in accordance with the short-range forces of crystallization and self-assembly.

Such an approach to embryogenesis means that the specific forms generated by morphogenetic processes, defining different species of organism, are irreducibly complex, because there are no higher-level laws or principles that constrain the forms resulting from the interactions of the molecules produced by the "genetic program." This program is the result of random permutation and natural selection—purely contingent processes as far as organisms are concerned. The only constraint is that the organism specified by the program must be able to survive and leave offspring in some environment. In such a view, there can be no general laws of biological form. Each species is, speaking more than metaphorically, a law unto itself. This conclusion is reflected in the primacy of the species concept in neo-Darwinism and the emphasis placed upon competition and survival as the expression of the individual species' success in establishing a unique relationship of order and stability within itself and with its environments. As one might expect, because cognitive constructs are themselves self-organizing, there is a clearly defined continuity between the way organisms are conceptualized in neo-Darwinism as "survival machines" and the way in which they are considered to be generated from their molecular parts.

There is no *a priori* reason why such a description of self-organization should not be valid, and indeed it is clear that there are special cases in both the animate and the inanimate realms in which atomistic explanations appear to be appropriate. These are the instances of a crystallization or self-assembly type of process leading to a unique structure. However, even in inorganic chemistry one encounters many instances of polymorphism in which the same substance can crystallize into different forms—familiar examples being graphite and diamond, or rhombic and monoclinic sulfur. Thus, in general, composition alone does not determine form. A similar polymorphism is observed in biological structures at different levels of organization: at the molecular level (Oosawa *et al.*, 1966), the cellular level (Sonneborn, 1970), the tissue level (Saunders *et al.*, 1957), and so on. Much of embryology consists in generating experimentally a variety of abnormal forms out of cells of identical genotype, e.g., the induction of supernumerary limbs in an amphibian by simple manipulations of tissue in the embryonic limb bud—manipulations involving no addition or deletion of cells but simply a change of relative position and no change in the external environment. Furthermore, many categories of morphological abnormality in organisms carrying mutant genes, called homeotic mutants, can also be produced in genetically normal individuals by nonspecific environmental stimuli such as heat or ether shock delivered at specific stages in embryonic development, a phenomenon called phenocopying (Goldschmidt, 1938). All these instances of organic polymorphism at different levels of structure in the organism can be classified into many fewer categories of form than the number of distinguishable perturbations, whether genetic, environmental, or spatial, so that the causal relations are not one-to-one but many-to-one. This suggests that there are organizational principles in developing organisms that cannot be reduced simply to molecular composition, so that self-assembly theories together with a genetic program are an inadequate basis for a generative theory of biological self-organization (see also Webster and Goodwin, 1982).

We must now consider whether or not there is empirical evidence relating to general organizational principles, or laws of form in biology, manifesting as regularities of morphology in large taxonomic groups. We have seen above that neo-Darwinism provides no basis for understanding any such regularities, because biological form in this theory is determined by contingency—not by law. Neo-Darwinism takes the view that "the chief part of the organization of every being is simply due to its inheritance" (Darwin, 1859). In contemporary usage, "inheritance" is the genetic program. However, if ordering constraints do exist in the biological realm, then this fact must be taken into account in any theory of biological self-organization.

We are thus led to the work of the pre-Darwinian morphologists, who were animated by a belief in the possibility of a rational, intelligible ordering or classification of organisms that would provide an insight into the laws of organic creation (i.e., generative rules). This tradition reached its peak during the late 18th and early 19th centuries in the work and insights of the great comparative morphologists such as St. Hilaire, Cuvier, Reichert, and Owen, who searched for and discovered empirical regularities of organismic structure (Russell, 1930). These regularities appeared as invariant structural relations or "typical forms" defining that which is common to a variety of particular realizations of the same type. Owen's demonstration of the structural homologies that exist between the great variety of vertebrate limbs, leading to the concept of the pentadactyl limb as the typical form, is characteristic of this work. Each specific member of the invariant set, such as the limb of the horse, of the bat, of the frog, and so on, can be seen as equivalent to every other member under a transformation, revealing a common plan that unifies the diversity of manifest forms.

This discovery is analogous to the realization that the different forms of motion shown by bodies under the action of a central attracting force, obeying Newton's laws, all belong to the same invariant set known as the conic sections. Indeed, the rational morphologists were inspired by the same vision as Newton, which was the Enlightenment Ideal of a mathematical natural science. Their conviction was, and they provided good evidence for the belief, that the morphological complexity of organisms is not irreducibly complex, but that rational principles or laws of form render the diversity intelligible.

Despite the fact that this tradition was largely eclipsed by Darwinism, which adopted the diametrically opposite view that organismic form is determined not by rational law but by historical accident, by contingency, a few rather isolated and sometimes misunderstood biologists have pursued this approach further. Among these, the embryologist Driesch stands out. His work is very relevant to the view of self-organization that will be developed below. It was Driesch who introduced into embryology the *field* concept by his demonstration that relative position in the whole embryo is an important determinant of cell fate. He used the concepts of wholeness, self-regulation, and transformation to define the properties of tissues that respond to a variety of disturbances (e.g., removal or addition of cells or spatial reordering of parts) by a reorganization resulting in the generation of the normal form. Examples of such fields are the amphibian embryo from fertilization up to about the gastrula stage, the limb and eye primordia, and a variety of other tissue domains that define secondary fields. Within such domains relative position is a primary determinant of cell fate and the parts that emerge during individuation and differentiation come into being as a result of local and global ordering principles, generating a structural and functional unity. Driesch (1929) assumed, as did the rational morphologists, that there are organizing rules operating within organisms to constrain or limit the forms that can be generated, but, again like his predecessors, he failed to give them any mathematical formulation. The question to be addressed now is: what type of mathematical description may be appropriate for these organizing principles for which there is clear biological evidence?

Organisms as Fields

The proposition that emerges from the above discussion is that living entities are wholes or *structures* defined by internal relations, which remain invariant under certain categories of transformation. These categories limit the possible generative processes that can result in organisms of specific form (species). Organisms are not, in this view, generated as a result of the interaction of "atomic" constituents, whatever these may be construed to be. Heterogeneity ("parts") arises as a result of systematic transformations of the organized whole, which may be described as the manifestation of states selected from a potential set that satisfies a primary property of invariance characteristic of organisms. Thus, the organism is not so much a self-organizing system that generates an ordered state from disordered or less ordered parts; it is more a self-organized entity that can undergo transformations preserving this state. The problems faced in this conceptualization are those of making explicit the nature of the invariant internal relationships that define the whole; the type of transformation the entity can undergo; and the relationships between whole and parts that confer upon it the properties of generation (reproduction) and regeneration.

Following Driesch's insight that developing organisms have field properties, we may proceed to the questions: what type of field is it, and how may it be characterized mathematically? A very extensive body of experimental work since Driesch has led recently to the observation that an appropriate arithmetic description of the spatial smoothness characteristic of developmental fields is a spatial averaging or intercalation rule applied to field values (French et al., 1976). This rule states simply that the field value at any point within the boundaries of a developmental field is the arithmetic mean of the values at equidistant neighboring points. Mathematically, the rule leads to the most general field equation used in physics, namely Laplace's equation. The question then naturally arises whether or not one can use solutions of this and related equations, known as harmonic functions, to describe developmental fields and hence biological form. Preliminary essays in this direction have been published (Goodwin, 1980; Goodwin and Trainor, 1980, 1985). This approach will now be illustrated by an analysis of the earliest stage of amphibian embryogenesis, following the treatment of Goodwin and Trainor (1980), and then certain conclusions regarding self-organization in biology will be drawn.

A Field Description of the Typical Cleavage Process

The first five stages of the typical cleavage pattern as described by classical investigation are shown in Figure 1, starting from the 2-cell stage after the first division of the egg. From the 32-cell stage, cell divisions continue to show an

FIGURE 1. The holoblastic radial cleavage pattern of the amphibian embryo. (A) 2 cells, (B) 4 cells, (C) 8 cells, (D) 16 cells, (E) 32 cells.

alternation between vertical and horizontal cleavage planes, but there is ultimately a loss of spatial and temporal order (synchrony) in a manner that differs among various species. Because our interest is in the geometry of the cleavage planes, we project the typical pattern shown in Figure 1 onto the initial, spherical egg to get the schematic sequence in Figure 2, which illustrates the first seven cleavages up to the 128-cell morula. This new pattern is suggestive of a sequence of harmonic functions on the sphere, the cleavage lines corresponding to nodal lines of spherical harmonics. Accordingly, we develop an "eigenfunction" description of cleavage on the basis of a minimization principle, wherein the eigenstates of a morphogenetic surface field describe the successive stages of the cleavage process in the early embryo. The cleavage process is then seen as a series of transformations to successively higher characteristic states of the morphogenetic field as metabolism proceeds. A crude analogy to this scheme may be found

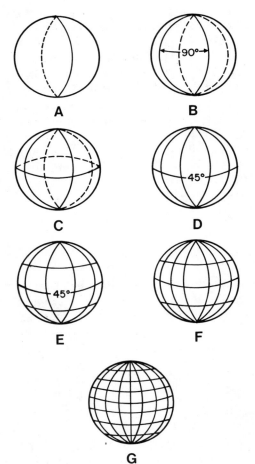

FIGURE 2. The geometry of typical cleavage planes up to the 128-cell stage. In (F) and (G) only, the silhouette forms one of the eight longitudinal sections. (A) 2 cells, (b) 4 cells, (C) 8 cells, (D) 16 cells, (E) 32 cells, (F) 64 cells, (G) 128 cells.

9. Developing Organisms as Self-Organizing Fields

in the electron density distributions of the hydrogen atom, in which the transitions to successively higher energy states result from the action of some external optical pumping field. A characteristic biological feature is that the cleavage transformations are "pumped" or induced internally, the system being self-generating.

A Variational Principle for Cleavage Planes

Proceeding with the analysis at a fairly abstract level, let us now introduce a field function $u(\theta, \phi)$ over the surface of the sphere (where θ and ϕ are spherical coordinates defining longitude and latitude) and adopt the convention that its nodal lines represent lines of least resistance to a furrowing process, preliminary to the development of cleavage planes. This function may be taken to be some kind of order–disorder parameter relating to the organization of microfilaments at the cell surface. The basic stability of the typical cleavage pattern suggests the use of a minimization principle on this field function. An appropriate surface density function, which in a physical problem would be the energy density, is

$$E(\theta,\phi) = A\left[\left(\frac{\partial u}{\partial \theta}\right)^2 + \frac{1}{\sin^2\theta}\left(\frac{\partial u}{\partial \phi}\right)^2 + \beta u^2\right] \tag{1}$$

where the constants A and β incorporate relevant physiological units. Then suppose that the characteristic cleavage planes correspond to a minimum of the integral of this density function over the surface energy $E(\theta, \phi)$:

$$\delta \int_0^{2\pi}\int_0^{\pi} E(\theta,\phi) \sin\theta \, d\theta \, d\phi = 0 \tag{2}$$

subject to a conservation law on u^2:

$$\int_0^{2\pi}\int_0^{\pi} u^2(\theta,\phi) \sin\theta \, d\theta \, d\phi = 1 \tag{3}$$

which amounts to a normalization condition on the field variable u. Equations (2) and (3) require satisfaction of the Euler–Lagrange equation (see Trainor and Wise, 1979):

$$\frac{1}{\sin\theta}\frac{\partial}{\partial \theta}\left(\sin\theta \frac{\partial u}{\partial \theta}\right) + \frac{1}{\sin^2\theta}\frac{\partial^2 u}{\partial \phi^2} - \alpha u = 0 \tag{4}$$

where α incorporates the parameter β and an undetermined multiplier of equation (3).

The usual conditions on u, i.e., that it be finite, single-valued, and continuous over the sphere, restrict the possible solutions, u, to a characteristic (eigenfunction) set, viz. the spherical harmonics (real part taken):

$$u(\theta,\phi) \to Y_{jm}(\theta,\phi) = \sqrt{2}\, N_{jm} P_j^m (\cos \theta) \cos m\phi \tag{5}$$

where j takes on the integral values 0, 1, 2, ..., and for given j, the m values are integers ranging from $-j$ to $+j$. The parameter α is restricted to the corresponding characteristic values (eigenvalues) $j(j + 1)$. In equation (5), the P_j^m values are associated Legendre polynomials (Hobson, 1955), and the N_{jm} values are the normalization constants given by

$$N_{jm} = \left[\frac{(2j + 1)(j - |m|)!}{4\pi(j + |m|)!} \right]^{1/2} \tag{6}$$

It is easy to calculate the surface "energy" corresponding to each characteristic cleavage state as follows:

$$E_j = A\{j(j + 1) + \beta\} \tag{7}$$

However, not every characteristic state (5) is realized in the cleavage process, since the mitotic apparatus imposes a biological constraint (somewhat analogous to superselection rules in physics; Wick et al., 1952) so that the number of cells is doubled in each cleavage stage.

According to the ideas set out above, the nodal lines of the characteristic function $Y_{jm}(\theta, \phi)$ on the sphere are in correspondence with the furrow lines of a characteristic cleavage state, except that the set of characteristic states is limited by the requirement that the number of cells is given by 2^p where p is the number of cell divisions. It is easily shown that the number of cells in a state characterized by Y_{jm} is $2m(j-m+1)$ unless $m = 0$, in which case the number is $j + 1$. (It is sufficient to choose the real part of Y_{jm}, i.e., the $\cos m\phi$ solutions, so that we need consider only $m \geq 0$.) Hence, the biological constraint requires that

$$\begin{aligned} 2^p &= 2m(j - m + 1) \quad &\text{if } m \neq 0 \\ &= j + 1 \quad &\text{if } m = 0 \end{aligned} \tag{8}$$

In general this equation, for a given characteristic cleavage state corresponding to p divisions, is satisfied by more than one set of (j,m) values. It is natural to suppose that the choice is made primarily on the basis of lowest j value, since according to equation (7) this value minimizes the "energy." The choice is then nearly unique, except for a twofold degeneracy every second division. It is assumed that the animal–vegetal polarity of the embryo defines a secondary polar field weaker than the primary field, which removes this degeneracy in favor of the highest m value for a given j in much the same way that a magnetic field removes the $(2j + 1)$-fold degeneracy of magnetic states in the hydrogen atom.

Table 1 shows the correspondence between cleavage states and characteristic functions $Y_{jm} = 2N_{jm} P_j^m \cos m\phi$ by listing the number of cells to be expected

TABLE 1. Correspondence between $Y_{jm}(0,\phi)$ and Cell Number [a]

j value	m value	Cell number	(j,m) pair selected
1	0	2	
	1	2	(1,1)
2	0	3	
	1	4	
	2	4	(2,2)
3	0	4	
	1	6	
	2	8	(3,2)
	3	6	
4	0	5	
	1	8	
	2	12	
	3	12	
	4	8	
5	0	6	
	1	10	
	2	16	
	3	18	
	4	16	(5,4)
	5	10	
6	0	7	
	1	12	
	2	20	
	3	24	
	4	24	
	5	20	
	6	12	
7	0	8	
	1	14	
	2	24	
	3	30	
	4	32	(7,4)
	5	30	
	6	24	
	7	14	
11	8	64	(11,8)
15	8	128	(15,8)

[a] Defined by $j + 1$, if $m = 0$, and $2m(j - m + 1)$, if $m \neq 0$.

from each set of (j,m) values up to $j = 7$. The appropriate (j,m) pair is then selected uniquely by the conditions expressed in equations (8), together with the minimization condition (7), except for the twofold degeneracies at the first, second, fourth, and so on, cell divisions. As remarked above, a unique correspondence is achieved by assuming that a weaker polar field selects (1,1) over (1,0), (2,2) over (2,1), and (5,4) over (5,2); that is, it favors the highest m value for a given j.

In Table 1 the selected states for the sixth and seventh cleavages corresponding to the 64- and 128-cell stages have been included without listing all of the rejected (j,m) values. Figure 2 illustrates the nodal lines for the successive stages in a typical cleavage process. A further study of this model is in Goodwin and LaCroix (1984).

Fields and Self-Organization

The foregoing example clarifies some of the abstract concepts introduced in the discussion of organisms as fields, and it is now of interest to elaborate on these and their relationship to the concept of self-organization in biology. The first point to emerge is the primacy of the *organism as the fundamental biological entity*, replacing the usual definition of the cell as the unit of life. This conclusion follows from the fact that the *field* is the self-organizing entity, and this field is coextensive, according to the above description, with the whole organism. The orderly geometry of the cleavage planes is a reflection of this organization at the level of the developing embryo. This fact does not mean that the parts, which in this case are cells, have no properties of their own. On the contrary, the constraint of binary division, a cellular property, is a major source of the distinctive geometry of the cleavage process, because this constraint defines one of the three selection rules that determine the harmonic functions allowable as descriptions of this process. It thus becomes apparent that the spatial organization of the whole derives from principles relating to global field behavior, together with constraints coming from the properties of the entities generated as parts. Such a description avoids the limitations of atomistic reduction and also of holistic description that identifies the whole with some (conceptually or materially) isolable essence. The genome is such an isolable essence, traceable historically to its roots in idealistic holism via Weismann's conceptualization of the organism as being separable into germplasm (essence) and somatoplasm (expression of the essence; see Webster and Goodwin, 1982). A field description of developing organisms depicts spatial organization as the expression of distributed influences, global order being constrained to give specific morphology as a result of autonomous properties of parts. This description defines a "decentered structure."

The ambiguity between the concepts of "cell" and "organism" can be resolved in terms of the above description. The fertilized egg is both a cell and a developing organism. It is an organism insofar as it is totality describable by a field; it is a cell insofar as it embodies the specific constraints (e.g., binary division) characteristic of such an entity. As cleavage proceeds, the organism continues to be identified with the whole field (the embryo), while cells are identified as parts that play specific roles within a transforming context. After gastrulation, more complex parts such as neural plate, limb fields, and eye fields, come into existence. These consist of aggregates of cells, so an integrated hierarchy of parts emerges. The emergence occurs within the context of the organism as the global field that continues to impose organizational constraints upon the whole, while the parts impose reciprocal constraints so that a specific form arises.

This type of description allows one to make comparisons between developmental processes in acellular (or unicellular) and multicellular organisms, and to understand them in the same terms. That is the classical view. If one adopts a position such as that of Wolpert (1971), that development is to be understood in terms of the responses of cells (really, of their genomes; Wolpert and Lewis, 1975) to "positional information" established over multicellular domains, then such comparisons become problematical, as discussed by Frankel (1974). The view of organisms as fields overcomes this difficulty and leads to the suggestion of some unexpected homologies between the morphology of ciliate protozoans and of amphibian gastrulae (Goodwin, 1980). This suggestion is very much in the tradition of rational morphology, because it makes clear that one must seek homologies at the level of "deep structure," i.e., at the level of generative principles (such as those defining field properties)—not in terms of "surface structure" (such as whether or not an organism is partitioned into cells). This claim does not deny that surface structure imposes constraints that affect manifest form. It simply states that in comparing field effects at the level of the whole organism, surface structure constraints are secondary.

A very important component of the study of embryogenesis is the search for the means whereby new or emergent aspects of morphogenesis are initiated at specific times in the process. For example, the relatively simple, recursive process of cleavage in amphibian embryos is followed by the dramatic phenomenon of gastrulation that transforms the hollow blastula into the three-layered, late gastrula. The transition from cleavage to gastrulation has been described as the expression of a cortical or surface field, which is radially symmetric in the unfertilized egg but which then develops bilateral symmetry as a result of sperm entry (Goodwin, 1980). The description of such a surface field in terms of harmonic functions reveals that with bilateral symmetry there appears a special point on the surface—a saddle point identified with the future dorsal lip of the blastopore. However, it is assumed that the influence of this singularity on the morphogenetic process cannot become significant and be expressed until cleavage transforms the initially solid sphere of the egg into a hollow spherical shell—the blastula. Then the saddle point can make its presence felt as a locus on the blastula, where surface polarity is absent, by acting as a radial force causing bottle cell formation and the initiation of invagination. Thus, cleavage is seen as a process that establishes a necessary condition for the emergence of a new phase of morphogenesis. The new phase results from the interaction between a singularity in the surface or cortical field and the residual radial component of the cleavage field, described by solid harmonics.

Generation and Regeneration

There is another basic property of organisms relating the whole and the part: the capacity of a part to generate or to regenerate the whole. So far, in discussing embryogenesis, the view has been developed that the fertilized egg is a whole that undergoes transformations resulting in the appearance of parts that have distinctive properties but that are not, within the context of the whole organism, autonomous in the sense that atomistic theories would have them be. However, there was a time when the egg was a cell within the ovary, itself a part of the parent organism. The oocyte during its maturation develops the capacity to develop into a new whole. Such a transformation can be achieved also by parts (multicellular fragments) of hydroids, such as *Hydra*, or parts (noncellular fragments) of ciliate protozoans, such as *Stentor*, or a cell of a carrot, any of which can regenerate the whole organism. The capacity of plants to propagate from leaves and stems, of insects and urodeles to regenerate limbs from stumps, and of higher organisms to regenerate skin and liver are other manifestations of this same property of parts to transform into wholes. It is evident that different organisms vary widely in their regenerative capacities, but it is true of all organisms that from particular parts, wholes can be produced. This property defines reproduction or generation, as well as regeneration, and it is one of the fundamental self-organizing properties that living creatures display. What does a field description of organisms have to say about such behavior?

An interesting property of harmonic functions suggests precisely this capacity: if such a function is described over any part of its domain of definition (e.g., the surface of a sphere), then the function can be recovered uniquely over the whole domain by analytic continuation. Thus, in a particular sense, the part contains the whole. This gives us a kind of existence theorem for the generative and regenerative properties of organisms, defined as fields describable by harmonic functions. It is just this type of property that needs to be embodied in a description of biological self-organization, although the specific property of harmonic functions described here is neither necessary nor sufficient to account for the actualities of generation and regeneration in the living realm.

Structuralist Biology

It may have become apparent to the reader before this point that the general context within which this chapter has been constructed is that of contemporary structuralism, as defined by Levi-Strauss (1968) and Piaget (1968), and developed in another, more extensive analysis (Webster and Goodwin, 1982). A field as described above is an example of a *structure* in that it belongs to an invariant set (the harmonic functions) defined by internal relations (those defining the field equation over a domain), each member of which is a transformation of the other (by a change of boundary values). The particular field functions that have been used in this treatment are not as important as the more general structuralist principles that inform the analysis. These principles emphasize the necessity to identify what is specific to a particular area of study, such as biology, before attempting to develop a theory to "explain" the phenomena. This necessity requires that in approaching the problem of self-organization in embryogenesis, we first identify any empirical regularities emerging from the study of biological form. Such regularities can suggest the existence of principles of organization or invariant relationships in organisms. The evidence points clearly in this direction: organisms, and hence embryos, are indeed structures in the technical sense. They are entities with the defining characteristics of wholeness and self-organization that are capable of undergoing transformations that preserve these deep properties while changing manifest form. This view is an elaboration of an earlier view clearly articulated by those who insisted that the primary problem of biology is that of organization and form—not of composition or heredity, the latter finding their place within the context of the former (cf. Russell, 1930; Needham, 1936). The use of the field concept and the more specific description of biological form in terms of harmonic functions makes more explicit the implications of this earlier view. Gene products can in certain instances stabilize specific form, such as left- or right-handed spiraling in the third cleavage planes and, consequently, in the shells of the snail, *Limnea*. However, in general we have seen from the field description of cleavage that the *constraints determining specific form arise from organizational levels other than simple molecular composition*, these constraints being such processes as binary cleavage, animal–vegetal polarity, and a minimum "energy" condition. This description emphasizes once again the primacy of organizational principles in biological processes and the inadequacy of any theory based solely upon genes and molecules.

In this view, the role played by the genes is to specify the potential molecular composition of an organism and to define some temporal sequences in which molecular components are made. These actions constitute constraints that impose some limitations on the forms organisms can assume and in certain instances may actually determine higher-order form. But in general the relationship between "genotype" and "phenotype" is one of causal necessity, not sufficiency, since gene products act within an organized context whose properties must be defined before the generated morphology can be described. A linguistic analogy would be that genes essentially determine the set of words out of which a text can be constructed. Words alone are clearly insufficient to define a text, which embodies higher-order syntactical, semantic, and contextual constraints or rules. These are rules of organization that limit the set of allowed arrangements of the words within the text. Such organizational principles are described herein as field constraints, which limit the allowed range of biological forms. The generative principles of organismic morphogenesis then consist of organizational constraints or rules (laws of form) common to all organisms in the form of fields (thus defining biological universals), together with specific constraints characteristic of individual species (which then define particulars). Genes specify many of the latter, but none of the former.

One of the most significant aspects of this structuralist approach is the deliberate avoidance of any *a priori* material reduction of the organism to parts such as cells, molecules, or genes. Once the problem of biological self-organization is clearly and explicitly described, and an appropriate description is available, then it *may* be possible to carry out a relevant material reduction. Certainly it will be possible to achieve an *abstract* reduction to laws and rules, such as those that have emerged from the analysis and description of cleavage, given above. This description was itself inspired by a paper that was, in my opinion, a landmark in the discovery of general rules of morphogenesis—expressed in terms that make no reference to composition or inheritance, but are of a purely relational nature (French *et al.*, 1976). To start with the assumption that one knows what the basic parts of the organism are is to make a strategic error at the outset. This assumption can lead one badly astray in seeking at once the most economical and the most rigorous analytical treatment of the problem. The description of organisms as fields that embody self-organizing properties and that can undergo transformations preserving invariant relationships is, despite its limitations, a step toward a biological science of form and self-organization that relates part to whole in a manner preserving organismic unity throughout the diversity of manifest morphology.

ACKNOWLEDGMENT. I am indebted to my colleague, G. C. Webster, for the intellectual stimulus which has led to many of the ideas developed in this chapter.

References

Darwin, C. (1859) *On the Origin of Species*. Reprint of the first edition, 1950, Watts, London.
Driesch, H. (1929) *The Science and Philosophy of the Organism*. Black, London.
Frankel, J. (1974) Positional information in unicellular organisms. *J. Theoret. Biol.* **47**:439–481.
French, V., P. J. Bryant, and S. V. Bryant (1976) Pattern regulation in epimorphic fields. *Science* **193**:969–981.
Goldschmidt, R. B. (1938) *Physiological Genetics*. McGraw-Hill, New York.
Goodwin, B. C. (1980) Pattern formation and its regeneration in the Protozoa. *Symp. Soc. Gen. Microbiol.* **30**:377–404.
Goodwin, B. C., and N. H. J. LaCroix (1984) A further study of the holoblastic cleavage field. *J. Theoret. Biol.* **109**:41–58.
Goodwin, B. C., and L. E. H. Trainor (1980) A field description of the cleavage process in embryogenesis. *J. Theor. Biol.* **86**:757–770.
Goodwin, B. C., and L. E. H. Trainor (1983) The ontogeny and phylogeny of the pentadactyl limb. In *Development and Evolution*, B. C. Goodwin, N. J. Holder, and C. C. Wylie (eds.). Cambridge University Press, Cambridge, England, pp. 75–98.
Goodwin, B. C., and L. E. H. Trainor (1985) Tip and whorl morphogenesis in *Acetabularia* by calcium-regulated strain fields. *J. Theoret. Biol.* **117**:79–106.
Hobson, E. W. (1955) *The Theory of Spherical and Ellipsoidal Harmonics*. Chelsea, New York.
Levi-Strauss, C. (1968) *Structural Anthropology*. Basic Books, New York.
Needham, J. (1936) *Order and Life*. Yale University Press, New Haven, Conn.
Oosawa, F., H. Kasai, S. Hatano, and S. Asakura (1966) Polymerisation of actin and flagellin. In *Principles of Biomolecular Organisation*, G. E. W. Wolstenholme and M. O'Connor (eds.). Little, Brown, Boston, p. 273.
Piaget, J. (1968) *Structuralism*. Routledge & Kegan Paul, London.
Russell, E. S. (1930) *The Interpretation of Development and Heredity*. Oxford University Press (Clarendon), London.
Saunders, J. W., J. M. Cairns, and M. T. Gasseling (1957) The role of the apical ridge of ectoderm in the differentiation of the morphological structure and inductive specificity of limb parts in the chick. *J. Morphol.* **101**:57–87.
Sonneborn, T. M. (1970) Gene action in development. *Proc. Roy. Soc. London Ser. B* **176**:347–366.
Trainor, L. E. H., and M. B. Wise (1979) *From Physical Concept to Mathematical Structure*. Toronto University Press, Toronto.
Webster, G. C., and B. C. Goodwin (1982) The origin of species: A structuralist view. *J. Social Biol. Struct.* **5**:15–47.
Wick, G., A. Wightman, and E. Wigner (1952) The intrinsic parity of elementary particles. *Physiol. Rev.* **88**:101.
Wolpert, L. (1971) Positional information and pattern formation. *Curr. Top. Dev. Biol.* **6**:183–224.
Wolpert, L., and J. Lewis (1975) Towards a theory of development. *Fed. Proc.* **34**:14–20.

10

The Slime Mold *Dictyostelium* as a Model of Self-Organization in Social Systems

Alan Garfinkel

> ... what wise hand teacheth them to doe what reason cannot teach us? ruder heads stand amazed at those prodigious pieces of nature, Whales, Elephants, Dromidaries and Camels; these I confesse, are the Colossus and Majestick pieces of her hand; but in these narrow Engines there is more curious Mathematicks, and the civilitie of these little Citizens more neatly sets forth the wisedome of their Maker.
> Sir Thomas Browne (c. 1663), partially quoting from Henry Power, *Experimental Philosophy*, 1663 (Winfree, 1980)

ABSTRACT

The phenomenon of aggregation in the cellular slime mold *Dictyostelium* has been studied intensely, both as a model for the self-organization of single cells into multicellular organisms and as a source of information about the extracellular function of the chemical messenger, cyclic AMP. There are two basic classes of models: models of individual cells in discrete sets and field theoretic (continuum) models of aggregation. These two kinds of model are designed to answer somewhat different questions, but any explanation of the nature and form of aggregation requires the field approach. An examination of the forms occurring during slime mold aggregation provides important clues to the dynamics of the process and indicates the importance of symmetry breaking and entrainment in self-organization. Analogies are proposed between slime mold aggregation and human social processes.
—THE EDITOR

ALAN GARFINKEL • Department of Kinesiology, University of California, Los Angeles, California 90024.

Associated with the term *entropy* is a tendency to disorder: structure tends to degenerate into uniformity. In the thermodynamic world, governed by the law of increasing entropy, the Parthenon crumbles into dust and order dissipates. But the opposite tendency is also regularly observed: processes in which structure is built out of the less structured. Such processes built the Parthenon, and, in a similar local defiance of entropy increase, biological evolution produces organisms of increasing complexity. Examples of emergence of complex structure can be found throughout the physical world: if a homogeneous fluid is subjected to sufficient heat or motion, forms develop. The forms are patterns of vortices in the fluid flow, of regularly spaced cells of convection currents, or of waves. In the social world, too, we find examples of emergent forms.

We need concepts with which to understand and explain these kinds of structure-building processes. In the case of the Parthenon, we recognize a guiding intelligence, the unfolding of whose conscious intention explains the development of the structure. But in other cases, such as biological evolution, we want to be able to say what Laplace said to Napoleon: "Sir, I have no need for that hypothesis." What we need are models or theories of how systems can develop or evolve into higher levels of structure or organization.

The term *self-organization* has recently come to be used as the general term for the processes by which order and structure emerge. One of its most important sources is the work of the mathematician Turing (1952), who considered some simple examples in which chemicals, reacting with each other and diffusing through space, produced regular patterns. The point was to show how it is possible for a uniform distribution of substances to display what he called *morphogenesis*: the qualitative leap to a regular pattern or order. His paper has been the inspiration for much recent work in self-organization, and has provided some of its basic philosophical themes.[1] In Turing's scheme, a system of equations has, for one range of its parameters, a homogeneous (i.e., uniform or constant) solution. But as these parameters pass critical points, the homogeneous solution becomes unstable and undergoes *bifurcation*: a splitting-off of a new solution. In these cases there is less homogeneity and more structure than in the original solution. The new solution violates the symmetry of the original configuration, and displays regular patterns in space and/or time.[2]

This approach gives us a general model for explaining pattern formation: find an equation (typically a partial differential equation) that describes the system's interactions, and study the bifurcations of solutions to this equation, especially those in which the prebifurcation state has greater symmetry than the postbifurcation state. These equations may be chemical reaction–diffusion equations, as in Turing's model, the Navier–Stokes equations of fluid dynamics, or more gen-

---

[1] Turing is better known as the first theoretician of the digital computer: he proposed the basic definitions of machine computability and proved theorems on its limitations. He also wrote about the "thinking" ability of such machines. His areas of interest are linked by the concept of emergence: how can systems display characteristics that do not seem to be inherent in their components?

[2] The idea of explaining morphogenesis as pattern formation in reaction-diffusion equations was also developed by Rashevsky (1940) and by Kolmogorov *et al.* (1937). See also Kametaka (1975).

eralized equations of probability density distributions, such as the various Fokker–Planck equations. The picture that emerges is something like this: several forces interpenetrate a medium. At one extreme (of parameter values), one force completely dominates; at another extreme, another force dominates. But in the twilight regions in which no single force completely dominates, patterns and morphologies form, a moiré of the interactions between forces. For example, in fluids it may be the interaction of viscosity (momentum diffusivity) versus convection, or viscosity versus heat transport. At one extreme, viscosity dominates, and the fluid is (macroscopically) quiescent. But as the parameters pass into critical regions, patterns of vorticity or heat transport are produced.

One kind of pattern formation is the phenomenon of *aggregation*, in which a uniformly distributed substance aggregates into clumps. Aggregation phenomena are found throughout nature. On the cosmic scale, the formation of galaxies is the result of aggregation of stellar material, producing *mesogranularity*: middle-range structures between the scale of individual particles, at one extreme, and the whole space, on the other. On a smaller scale, aggregation phenomena are found in the condensation of vapors into drops and, generally, in processes such as nucleation and granulation. In this chapter we shall be concerned with models of aggregation in biology (and to a lesser degree, in sociology). One example in particular has captured the imagination of many theoreticians, and will be the focus of this chapter: aggregation in the cellular slime mold *Dictyostelium discoideum* (Dd).

The slime mold is a remarkable creature. It spends one phase of its life as individual amoebae—single-celled animals that move around, eat bacteria, and in other ways lead their own lives. When deprived of food, however, the amoebae aggregate: they come together and fuse into colonies of thousands of cells. Once the merger is complete, the fused colonies migrate over macroscopic distances, and in other ways act as a single organism. This organism, initially homogeneous, later undergoes differentiation: part becomes a foot or base rich in cellulose, and part becomes a fruiting body rich in polysaccharides. Later still, the fruiting body bursts, releasing individual amoebae and completing the life cycle. (A good background reference is Loomis, 1975).

The process of aggregation has been intensively studied. Intrinsically fascinating, it also suggests the possibility of learning some basic truths about higher forms of life. First, the biochemicals involved (ATP and cyclic AMP) form the building blocks for metabolism and communication in all multicellular life forms. Second, the aggregation process seems to offer a model for the evolution of complexity. Cohen (1977) remarks: "In the development of *D. discoideum*, one may really be watching a replay of the basic kinds of events responsible for the appearance of the first multicellular organisms." Third, there is an obvious potential for social analogy. People have already suggested that aggregation in *Dictyostelium* can serve to model the formation of insect colonies, and the analogy to human aggregation, in towns or other collectives, has been hinted at.

Here, I shall consider various models of aggregation in the slime mold. Several philosophical themes will emerge which I think characterize self-organization processes. One is the need for an antireductionist or holistic approach; another is the role of symmetry-breaking in self-organizing systems. The possibility of analogy to human social processes is also explored.

How can we explain aggregation in *Dictyostelium*? Everyone seems to agree that the amoebae move *chemotactically*, i.e., their motion is in response to a chemical attractant. In this case the attractant, dubbed acrasin,[3] is secreted by the individual cells. In the individual-cell phase, this attractant is ineffective in producing aggregation, and the cells move randomly. Then something happens, and the cells aggregate in response to the attractant. The problem is to show how this occurs. The relevant literature reports empirical research as well as mathematical models. The mathematical models fall into two classes: global models of the aggregation field and models of the individual amoeba cell. Their relationships to each other and to the data are problematic.

The Field Model

The global or field model was developed by Keller and Segel (1970) (see also Segel, 1972) to study the global behavior of the whole system of amoebae, rather than that of the individual cells. They postulate several quantities, which are assumed to be continuously distributed over space. Then they form a set of partial differential equations for the system as a whole. By studying these equations they attempt to explain the formation of patterns of aggregation as a bifurcation in the solution to the field equations. It is therefore an example of the approach of Turing (1952): pattern formation and morphogenesis is explained as the instability of the homogeneous solution to a set of partial differential equations, and the symmetry-breaking bifurcation to a more structured stable configuration.

In this case, the parameters studied are the density of amoebae, $\alpha(x,t)$ at a point x and time t, and the density of acrasin, $\rho(x,t)$. The full model also requires incorporating an acrasinase (destroyer of acrasin) and the reaction product of acrasin and acrasinase, but these are ignored in the simplified model. The model assumes the following: (1) a tendency for amoebae to diffuse, (2) for amoebae to move chemotactically, (3) for acrasin to diffuse, and (4) for amoebae to produce acrasin.

If all the tendencies were connected, then, for certain parameter values, aggregation would not occur, because

$$(2) + (4) < (1) + (3)$$

On the other hand, if the reverse were true, then aggregation *would* occur, as acrasin → aggregation → more acrasin → more aggregation, and so on.

Mathematically, the Keller–Segel model takes the form of a partial differential equation in the two variables $\alpha(x,t)$ and $\rho(x,t)$. The equation governing $\alpha(x,t)$

---

[3]For *dictyostelium discoideum*, acrasin has been identified as adenosine 3':5'-cyclic monophosphate, or cAMP for short. See Bonner (1959, 1969) for a good introduction.

10. The Slime Mold *Dictyostelium*

is the sum of the two terms representing (1) and (2) above. The diffusion is assumed to follow Fick's law: diffusion follows the gradient $\nabla \alpha$ of the density α from more dense to less dense areas. This term is therefore

$$\nabla \cdot (D_2 \nabla \alpha)$$

where D_2 is the diffusion coefficient for amoebae.[4] The chemotactic aspect is modeled by assuming that the amoebae follow and move along the negative gradient of acrasin density. Hence, this term is

$$-\nabla \cdot (D_1 \nabla \rho)$$

where D_1 measures the sensitivity of amoebae to a given gradient of acrasin. Combining these two, we get the equation for amoeba movement:

$$\frac{\partial \alpha}{\partial t} = -\nabla \cdot (D_1 \nabla \rho) + \nabla \cdot (D_2 \nabla \alpha)$$

Turning to the acrasin density, ρ, they combine the various terms representing the tendencies for acrasin to be used up and to decay, the tendency for acrasin to be produced by the amoebae, and the tendency for acrasin to diffuse, yielding

$$\frac{\partial \rho}{\partial t} = -K(\rho)\rho + \alpha f(\rho) + \nabla \cdot (D_\rho \nabla \rho)$$

where D_ρ is the diffusion coefficient for acrasin.

Keller and Segel assume a homogeneous (i.e., uniform) equilibrium solution α_0, ρ_0, and study by a linear stability analysis its stability with respect to infinitesimal perturbation. This analysis reveals a condition for the instability of the homogeneous solution: the solution instability will increase if

$$\frac{D_1 f(\rho_0)}{D_2 \overline{k}} + \frac{\alpha_0 f'(\rho_0)}{\overline{k}} > 1$$

where $\overline{k}$ is a constant.

Examining this result qualitatively, they observe that several different mechanisms can produce instability leading to aggregation:

- An increase in D_1 (the chemotactic sensitivity of amoebae)
- An increase in $f(\rho_0)$ (the rate of acrasin production)
- An increase in $f'(\rho_0)$ (the rate at which acrasin stimulates further acrasin production)

---

[4] If D_2, is constant, this becomes the familiar $D_2 \nabla^2 \alpha$ of Fick's law. The authors retain the form above, presumably to allow for a variable D_2, although they do not actually consider such a case.

They remark that the first two factors have been observed at the beginning of aggregation. This fact is taken to provide "qualitative evidence for the applicability of our theory."

Critique of the Field Model

The most important thing about the Keller–Segel model is its general philosophical outlook. It models aggregation as morphogenesis in the sense of Turing: first a set of equations is postulated, similar to Turing's reaction–diffusion model, but is complicated in this case by the presence of chemotaxis. Aggregation is then explained as the result of the instability of the homogeneous solution and the bifurcation of a new solution. Their analysis provides a nontrivial criterion for the instability and gives an account of the relative effects of the tendencies that cause it.

Is it a good model? What do we want from a model of aggregation? Any model will explain some phenomena and not others; the problem is to determine which elements of the aggregation process are essential and which can be ignored. Several features of aggregation that do seem important are not explained in the Keller–Segel model. One key feature is the fact that aggregation territories are finite—the amoebae in the medium do not all collect into one single mass but, rather, form a number of territories of roughly equal size spaced throughout the field. This is an important part of the morphology—a stable mesogranularity that is larger than the cell and smaller than the whole field. We ought to be able to give an account of it, but in the simplest versions of the Keller–Segel model, the whole field aggregates into a single mass.

The question then becomes: how can the model be amended to account for this fact? Once aggregation begins, there is nothing in the simplified model to act as a counterforce or countertendency to the collection process. The simplified model ignores the activity of the acrasinase—the enzyme that destroys the attractant. Perhaps the finiteness of territory size is due to the action of this acrasinase. There is some empirical support for this conjecture: it has been observed that high acrasinase levels reduce territory size and that mutants with low acrasinase activity have large territories (Robertson and Cohen, 1974). When we try to include the acrasinase reaction in the full Keller–Segel model, we get a better model, but one that is not mathematically tractable. So far, it has not been demonstrated that the relevant phenomena can be derived from the full Keller–Segel model (Segel, 1972).

Another strategy for explaining the finiteness of territory size, suggested by Keller and Segel, is to assume that there is a threshold in the chemotactic response of the cells to the attractant. These investigators have explored the possibility of adding this assumption to their model and found that it does reproduce some of the desired features but that there are some problems with it. Most importantly, it creates difficulties for their stability analysis. In constructing their basic argument, Keller and Segel used the widely accepted technique of linear stability analysis. Adding a threshold makes instability depend on the amplitude of the perturbation and requires that it not be small. This need for a good-sized perturbation vitiates the earlier reasoning and invalidates the linear stability analysis they employ.

The entire question of the use of linear stability analysis needs reexamination, for despite its near-universal use, there are serious problems with it. Keller and Segel use it in the usual, accepted fashion: once they have obtained their homogeneous equilibrium solution, linear analysis gives the criterion for instability, which amounts to testing the system for destabilization by infinitely small perturbations. This approach has two drawbacks. First, it provides only local information about the qualitative direction of movement *at* the very point of instability. What happens after that is beyond the scope of this kind of analysis. But questions about what happens after the point of instability (especially: does it oscillate?) are crucial for the success of the model. Second, linear stability analysis is limited in that it tests only the sensitivity of the system to destabilization by *infinitely* small perturbations. It might well be that a given system is stable under vanishingly small perturbation, but unstable under finite perturbation greater than some ϵ, say on the order of the Brownian movement! This criticism illustrates an important fact about continuum models: in order for the techniques of infinitesimal calculus to be applied to any real situation and to produce meaningful results, some kind of scale or gauge must be imposed on the model so that the notion of "small" in the model is not absurdly small from a practical point of view. Any physical continuum disappears on a fine-enough scale, and it is important that the mathematical model of the continuum not commit us to things going on in such a small space as to be physically impossible.

These criticisms should not detract from the basic interest of the Keller–Segel model. In fact, it received very little attention, and no critique ever appeared. Instead, attention shifted away from field-theoretic models toward models of the individual cell. In part, this shift of interest may have been due to the failure of the Keller–Segel model to predict wavelike global oscillation of the aggregation field—a striking and well-confirmed observation. But even more, the shift was probably due to a philosophical antipathy toward holistic models. The essence of the field model lies in viewing the process of aggregation globally rather than focusing on activity inside the individual cell. But most workers in the field seemed convinced that in order to understand the mechanism of aggregation, it is necessary to study in detail exactly what is going on locally. The main argument for this perspective depends on pulsatory oscillation in the individual cell.

During aggregation, the individual amoebae change their characteristic emission of cAMP from a steady low-level output to an output consisting of periodic *pulses*. The duration of these pulses is quite short (< 2 sec) and the interval between them is quite regular (~ 300 sec). A second important feature of the individual cell dynamics is *excitability*: the individual cell during aggregation can emit pulses of cAMP in response to receiving a pulse from another cell above a certain magnitude threshold. There is a time delay of approximately 15 sec in this response (Robertson, 1972).

These phenomena cannot be easily represented in the field model. At the very least, they would require major modifications in the type of equation representing the overall aggregation to account for threshold, delay, and oscillation.

The existence of the threshold, together with the fact that achieving the threshold results in a relatively large output, implies that the acrasin production function has discontinuous slope at the threshold value. This implication means that we are dealing with a partial differential equation with discontinuities of slope sprinkled liberally through space and time. The existence of the time delay for relay creates a further problem: in a field model, a time delay appears as a discontinuity at the individual cell; the cause propagates continuously to the point of the cell, then nothing happens for 15 sec, then a pulse is emitted. The alternative to this discontinuity is to model the process by time-delay equations in which the output at a point x is a function, not just of the instantaneous state at x, but also of the state at x some t seconds past. This alternative enormously complicates the equation. Consider also the function $f(\rho)$, which describes how the local level of cAMP affects its production. If production is oscillatory, then the shapes of f and its derivative are going to be very complicated. These factors make it difficult to represent local oscillation on a field-theoretic model. Because local oscillation seems important in chemotaxis, we see another reason for the focus on the individual cell. Recent work on slime mold aggregation has centered on just the question avoided by the field model: why do the individual cells oscillate?

The Individual Cell Model

Oscillation is often regarded as undesirable, a dysfunction of a system. Examples like business cycles, pipe flutter, bridge and airfoil flutter, wheel shimmy, muscle tremor, as well as periodic diseases like anorexia nervosa and manic-depressive disease made oscillations seem to be unwanted aberrations, rather than a basic function. Recently, there has been recognition of the importance of oscillatory processes in life. For an introduction, see Chance *et al.* (1973) and the excellent survey by Winfree (1980).

In the slime mold, the role of oscillation in the production of cAMP by the cell has several features that are important functionally. Oscillatory production is a much more efficient process from the point of view of resource utilization and of signaling. Given that cells respond to rates of change in cAMP levels, it is more effective to produce cAMP in pulses (Nanjundiah, 1973).

Gerisch and Hess (1974), Goldbeter (1975), Goldbeter and Segel (1977), and Cohen (1977) modeled the workings of the individual cell, taking as the fundamental data to be explained the capacities for pulsing and for relay. They developed models of internal metabolic mechanisms to show how cells can emit periodic cAMP pulses and also how small pulses of *extra*cellular cAMP can elicit large pulses of *intra*cellular cAMP, thereby establishing a "relay" mechanism.

In the model of Goldbeter and Segel (1977), these two phenomena are modeled by an ordinary differential equation in α = ATP concentration, β = intra-

10. The Slime Mold *Dictyostelium*

cellular cAMP, and γ = extracellular cAMP, and the reaction is occurring inside a single cell. The equations are

$$\frac{d\alpha}{dt} = v - \sigma\phi$$
$$\frac{d\beta}{dt} = q\sigma\phi - k_t\beta$$
$$\frac{d\gamma}{dt} = (k_t\beta/h) - k\gamma$$

where $\phi = \alpha(1 + \alpha)(1 + \gamma)^2/[L(1 + \alpha)^2(1 + \gamma)^2]$, and everything else is constant.

By numerical integration of the equations, they showed that their model yielded two important behaviors: a small pulse of γ caused a large pulse in β for an appropriate domain of the parameter space, and then, as the parameters changed critically, the same system underwent a transition to regular autonomous pulsing.

There were some problems in reconciling this model with other data, however. In particular, the model called for cyclic variations in ATP level. These are not found in practice, according to Cohen (1977).

An alternative model was suggested by Cohen (1977), who employed qualitative dynamics to develop a model in which the individual cell's cAMP concentration is subject to several interrelated controls, producing a qualitatively described system.[5] As one of its control parameters of this model is slowly increased, the system exhibits successive bifurcations, so that its behavior progresses from

1. Steady-state low-level leakage of cAMP, to
2. Excitability, i.e., the ability to produce the pulses of cAMP in response to external pulses, to
3. Autonomous oscillation of pulses of cAMP, and then to
4. Rapid continuous release of cAMP.

There are some problems with this model. The control parameter that increases slowly with time, causing the bifurcations, is the equilibrium value of cAMP. This assignment yields several false predictions, e.g., that supplying large amounts of cAMP to a cell should induce the sequence of bifurcations. This is not so (Gerisch, 1978). None of the individual cell models takes into account the fact that the cell seems to respond not so much to the cAMP level as to rapid changes in it. It is not clear how to amend such models to represent the dynamics of a pulse input.

---

[5]The qualitative picture has a certain advantage over writing a single differential equation and then exploring parameter space by computer. The qualitative model is more robust, because it focuses on the general *forms* of the curves involved rather than on the specific equations. Consequently, this approach displays the important causal relations.

My purpose here is not to criticize these models of the individual cell, but rather to ask how, given any model of the cell, we can account for the collective phenomena of aggregation. Let us assume that we have a model of the oscillation and relay properties of a single cell. Explaining aggregation then takes the form of considering various aspects of global behavior, and showing how a model derived by connecting N copies of a single cell can display aggregation morphologies plausibly related to those observed.

The simplest possible model for collective behavior is that of Cohen and Robertson (1971a,b). A single autonomous oscillator has its pulses relayed sequentially by a field of relay-competent cells. This model faces a number of difficulties. To begin with, we have the problem of explaining the levels of cAMP produced during aggregation. The models of the individual cell and the empirical data agree that the relative increase in cAMP level that is brought about by a single cell's pulses is about 20:1 (Goldbeter and Segel, 1977). On the other hand, observations of cAMP levels in the aggregation field indicate that during aggregation overall cAMP levels increase 100:1 (Bonner et al., 1969). Therefore, something that is going on is not simply the sum of independent individual contributions. The basic phenomena of aggregation are essentially *collective* phenomena. Serious difficulties stand in the way of deducing these phenomena from models of the individual cell.

Waves

The most important macrophenomenon is the occurrence of *waves* of aggregation. All experimentalists use this term in describing their observations, and it should be possible to explain how it happens. Two attempts to derive wavelike phenomena from detailed models of the individual cell are found in Cohen and Robertson (1971b) and Durston (1973, 1974a,b).

The scheme of Cohen and Robertson is that of a discrete set of relaying cells surrounding a periodic pacemaker. The pacemaker fires, the next cell relays the pulse, and so on, and a wavelike disturbance passes over the field. In the simplest case, we imagine the autonomous pacemaker a_0 signaling to a_1 which signals a_2, and so on. The relay is not immediate, however: it is well established (Cohen and Robertson, 1971b; Gerisch and Hess, 1974) that there is a 15-sec delay between the receipt of a pulse by a cell and its subsequent relay. Compared to this, the cell-to-cell diffusion times are negligible (Cohen and Robertson, 1971b). But this fact has a surprising consequence: the velocity of propagation of the signal ought to be inversely proportional to the density, because higher density implies smaller interamoeba spacing, which implies more relay delays per unit distance. Robertson t al. (1972) report a signal propagation velocity of 42 μ/min. In their experiments, they used a preparation with a density of 6×10^4 cells/cm^2. (The laboratory vessel in which the experiments are done is two-dimensional—a plate. Therefore, densities are defined per unit area of field, not per unit volume.) Their velocity observations agree well with those of Gerisch (1968). But in Gerisch's experiments, the densities in question were on the order of 5×10^6 cells/cm^2, or

10. The Slime Mold *Dictyostelium*

two orders of magnitude larger. We would have to conclude that propagation velocity is independent of density, which cannot be reconciled with the cell-to-cell relay model.

Several other reports of wave propagation velocity give different results. Gross *et al.* (1976), using densities of 2.5×10^5 cells/cm^2, report propagation velocities ranging from 240 to 467 μ/min, or 6 to 10 times the velocities reported above. It is hard to see how this discrepancy can be explained. Moreover, the authors report a striking fact: the velocity of successive waves decreases linearly with time, a phenomenon for which no explanation exists. Alcantara and Monk (1974) also report velocities on the order of 350 μ/min, which they are unable to reconcile with the data of Gerisch.

The more we examine the cell-to-cell model, the less attractive it becomes. Consider a fairly dense packing of cells. Let a_0 be the autonomous pacemaker, and assume that two cells, a_1 and a_2, are both near a_0, with a_1 slightly closer. The pulse from a_0 reaches both a_1 and a_2, but a_1 slightly ahead of a_2. Fifteen seconds later a_1 fires, just ahead of a_2. What happens to a_2? If a_2 does fire, then by extension to a_3, and so on, there will be huge amounts of cAMP produced immediately. But experiments suggest that being hit by a pulse just as the cell is ready to pulse itself results in the suppression of the pulse (Gerisch and Hess, 1974). But then, if a_2 does not fire (and hence neither do the other cells in the immediate area), the signal will either die out or leapfrog over the suppressed cells until it is too late to suppress the next pulse. In order to predict which of these will happen, we would have to know the exact rate of diffusion, the precise time delay, and exact times of pulse suppression. Moreover, we would need detailed information about the effects on a relay resulting from the receipt of multiple pulses. Consider the following array:

If a_1 and a_2 are equidistant from the pacemaker a_0, they will relay its pulses simultaneously, resulting in *two* pulses hitting a_3. We do not have any information on this, nor any model, to predict what will happen.

Cohen and Robertson (1971a,b) are aware of the leapfrog effect. Using a slightly more complex model of rows of amoebae signaling simultaneously to the next row, they estimate that the pulse will activate the second row away at densities as little as twice the minimum density necessary for signaling. They also estimate that at higher densities there will be multiple row effects. Indeed, Alcantara and Monk (1974) have shown experimentally that in dense populations, "the range of the relayed signal encompasses 4 to 6 cells."

The leapfrog effect obviously matters, and even more important is the multiple-pulse effect, since it is a question not only of the magnitude of the received impulse but also of its direction. What happens when pulses from two different directions hit a cell within a short period? We do not know, either theoretically or empirically.

Some particularly thorny obstacles lie the way of a successful model of wave propagation from the level of the individual cell. If N individual cells are spread across a plane in a discrete model of the aggregation field, the formation of waves with the right properties will be extremely dependent on the position of the cells and diffusion times. There is a sense in which even very good data cannot help us, because the sensitivity of the model calculations will typically be greater than standard experimental error. The calculations necessary to derive the wave phenomena are therefore of such delicacy and complexity that there is little reason to have any faith in them. We must go beyond such local/mechanical theories, at the very least, to a statistical mechanics of ensembles.

Autonomy: The Emergence of "Centers"

Another aspect of global behavior is the small number of aggregation territories. The standard model accounts for this by postulating that only a small number of cells achieve autonomy. These become the center of aggregation. All the other cells have only the capacity to relay and the chemotactic ability to move in response to cAMP. The autonomous center cells give the order to aggregate by pulsing cAMP, and the masses of undistinguished cells obey by moving and relaying the order to others more distant. This is the essence of the best known models of aggregation (Shaffer, 1962, 1964; Cohen and Robertson, 1971a,b; Robertson and Cohen, 1974).

Why do only a small number of cells achieve autonomy, whereas all cells achieve relay-competence? In other words:

- What causes the differentiation into master cells (centers) and slave cells (relays)?
- Why do these two types of cells exist in the proportion that they do?

Perhaps autonomy requires extreme values of some cellular parameters—a property possessed only by a few cells. This conjecture explains the differentiation in the field by appealing to a prior differentiation of the individuals. But any conjecture that locates the field difference in a postulated underlying difference among the individual cells faces a series of objections. First, the emergence of autonomous cells depends both on the density of cells and on their absolute number. Raman *et al.* (1976) report that larger population size at a given density results in relatively *fewer* cells achieving autonomy. For example, at a density of 6×10^5 cells/cm^2, the percentage achieving autonomy was inversely proportional to the population size! On the other hand, for the same densities, the time from first signaling to early aggregation *decreased* with increasing size. So a larger population implies fewer cells achieving autonomy but faster aggregation. This result is obviously inconsistent with a model in which a number of predestined centers emerge and control the others through direct relay. Second, when the center is removed from an aggregation field, reaggregation occurs (Cohen and Robertson, 1971b). Moreover, suspension cultures of nonautonomous but relay-capable cells agglomerate when gently agitated and then develop into potentially active centers (Gerisch, 1968). A very significant observation is that of Gerisch and Hess (1974), who reported that applications of a few pulses of cAMP to suspension cultures of preaggregation, relay-competent-but-not-oscillating cells produced "precocious onset of oscillations."

It is obvious that these phenomena cannot be reconciled with the idea of an intrinsic difference as the explanation for the emergence of autonomy. The proponents of the individual cell models sometimes acknowledge this but do not account for it. Thus, Raman *et al.* (1976) acknowledge that the emergence of autonomy is "a cooperative phenomenon" but do not explain how this might come about. Cohen (1977) admits that "there seems to be some control mechanism on the emergence of autonomy operating at the population level" but can only appeal to "an as yet unspecified anabolic pathway." If we are to understand the relationships responsible for the emergence of autonomy, we clearly must consider models of the overall field.

The Aggregation Field

Let us consider the differences between a relay-capable device and an autonomous oscillator. Relay-capable devices are triggered responders. Given a small but finite perturbation above a certain threshold, they produce a quick burst of output and then slowly return to equilibrium. Relay behavior can be produced by a cusp catastrophe[6] in the dynamics governing the equilibrium level of the cell, with a stable equilibrium (point attractor) on the upper sheet (Figure 1). Pushing the state sufficiently off the equilibrium causes the jump response (here represented by the state falling off the upper sheet), and then the slow dynamic carries the system back to equilibrium—the relaxation phase.

There are two possible models for explaining the onset of oscillation in a population of coupled relays. The first model assumes that the individual cells, or some of them, as a result of internal processes develop a Hopf bifurcation (point attractor → closed cycle) in their internal dynamics, so that their equilibrium state becomes oscillatory. (See Abraham and Shaw, this volume, for an account of the Hopf bifurcation.) If the amplitude of the oscillator is sufficient to kick the state over the edge, regular spontaneous oscillation will result. Evidence suggests that the oscillating internal variable is adenylate cyclase concentration (Gerisch, 1978). The second model assumes that we are dealing with relay-capable devices coupled so that the state of each can be perturbed by the outputs of others. Then, under the right circumstances, a single random pulse can set up a self-sustaining oscillation in the population. These are inequivalent representations and imply different mechanisms, but it may be very difficult in a given case to tell which one is operating.

---

[6]Readers unfamiliar with elementary catastrophe theory should consult Thom (1972) and Poston and Stewart (1978), as well as Abraham and Shaw (Chapter 29, this volume).

We now have three possibilities for modeling the aggregation field: (1) as a field of relay devices, (2) as a field of relays driven by a number of autonomous oscillators, or (3) as a field of autonomous oscillators.

Oscillation in the overall field, or even oscillation in individual cells in the field, does not necessarily imply that the individual cells are autonomous oscillators. In fact, fields of relay devices, (1), above, are capable of overall oscillatory behavior of a kind that could be thought to imply case (2) or (3). Small fluctuations in fields of relay devices can set them into an oscillation that mimics the behavior of fields of oscillators. Winfree (1980) has provided an extensive discussion of the relationship between these two different models and offers some criteria for distinguishing between them, but his criteria do not clearly indicate one or the other in the particular case of *Dd* aggregation. Nevertheless, certain clues can be gleaned from the collective behavior of the aggregation field.

It seems likely that the early aggregation field is of type (1) above—a field of relays. In early aggregation, single pulses often propagate through the field (Gerisch, 1968). Kopell and Howard (1974) studied pattern formation in a model chemical reaction and reported that "the excitable (nonoscillatory chemical) reagent is known to produce both periodic wave trains and single pulses, whereas the oscillatory reagent produces only periodic wave trains." Furthermore, the early aggregation field can produce *spiral* waves (Gerisch, 1968; Durston, 1973, 1974a,b). Kopell and Howard (1974), in their study of chemical reaction–diffusion processes, remark that "spirals, as opposed to concentric rings, seem to be easier to achieve in the excitable reagent than in the oscillatory one." But if the early aggregation field is a field of relays, it does not long remain that way. The evidence is strong that the impact of external pulses quickly converts a relay into an autonomous oscillator. For example, Gerisch and Hess (1974) studied the

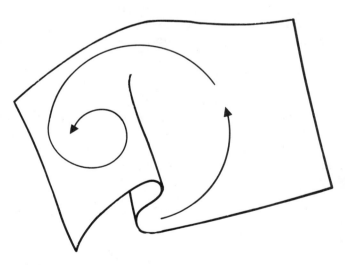

FIGURE 1. Cusp catastrophe representation of relay behavior.

results of applying pulses of cAMP to suspensions of cells at various states of development. They found that application of just two pulses of cAMP to cells at an early stage of development produced "precocious onset of oscillations." Even though none of the cells should have been capable of autonomous oscillation, soon all began oscillating. It seems as if the presence of an oscillator quickly turns the cells around it into oscillators, too. So it seems reasonable to conjecture that the evolution of the aggregation field consists of development from a field of excitable relay elements to a field of oscillators.

It can be objected that the results of Gerisch and Hess imply only that the *field* has become an oscillator, not that the individual cells have done so. Oscillating fields may be composed of nonoscillating relays. But features peculiar to the slime mold *do* indicate that the individual cells become oscillators. The 15-sec delay in the relay response of the individual cell implies that a field of such relays could never achieve global synchrony, because there would always be that 15-sec phase lag between stimulus and response, between signal and relay. A pair of *autonomous oscillators can become entrained or synchronized*—but *not* a pair consisting of *an oscillator and a delayed relayer*. Thus, the most important piece of evidence that we are dealing with a field of autonomous oscillators is the phenomenon of *coherence* (or *entrainment*), in which two or more oscillators bring each other into the same phase, or into an orderly succession of phases, as in the case of traveling waves.

It seems clear that some kind of entrainment does occur in the slime mold aggregation field. Among other things, it is necessary to produce the 100-fold increase in cAMP production that is observed during aggregation, whereas there is only a local 20-fold increase due to the activity of any individual cell. If there were no entrainment, there would be no constructive addition of pulses in the field. In addition, there is direct evidence that phase entrainment takes place among signaling cells. Gerisch and Hess (1974) and Malchow *et al.* (1978) have shown that stimulation of an oscillator by an external pulse tends to advance or retard the phase of the oscillator in just such a way as to produce entrainment: pulses arriving in advance of a scheduled pulse tend to elicit a "premature" pulse from the recipient and to shift the phase of the recipient oscillator forward from that point on. Conversely, pulses received after a regular pulse has been emitted tend to retard the phase of the recipient oscillator.

These considerations suggest that some kind of entrainment is taking place in the developing aggregation field, but they do not tell us what kind or what its role is. Further clues can be gained from a study of the morphologies of aggregation, in keeping with the general approach of inferring dynamics from the shapes and forms of the overall process.

Macromorphologies

Surprisingly little attention has been paid to the macromorphologies found in aggregation fields. Two exceptions are Durston (1973, 1974a,b) and Winfree

(1980). Macromorphologies are important for understanding the mechanisms of aggregation, and significant inferences can be made from their forms. Here is a taxonomy of forms as observed:

- *Target patterns*: concentric rings of aggregation radiating outward from the center.
- *Blobs*: clumps of aggregated cells.
- *Point spirals*: spiral waves of aggregation, emanating from apparently undistinguished points.
- *Whirlpools*: rings of cells (doughnuts) surrounded by spiral spokes of cells, curling into them. These are spiral waves of aggregation surrounding an *empty* core—one with no cells at all! Durston (1974a) describes their development: "The aggregating cells progressively collect into spirally oriented streams and aggregate into a 'doughnut' surrounding the central space. The space diminishes." After a few hours, the doughnut of cells implodes, and the cells form a mound-shaped aggregate "sited at the location of the original open space." The overall morphology is quite striking (Durston, 1973; Figure 2).
- *Spoke patterns*: blobs with radial spokes.
- *Strings*: linear aggregates of cells. These become the spokes of whirlpools and spoke patterns but they are also found free-floating.

There are a number of systematic relationships among these forms at all stages. In the early aggregation field, there are two morphologies: target patterns and point spirals. Both are found in the typical field: target patterns predominate over spirals by about 5:1. Both are stable, although in the late aggregation field, the spirals change to concentric propagation (Durston, 1974a). Both Durston and Winfree suggest that the target patterns are associated with autonomous pacemakers[7] at their center, but spiral waves lack such an autonomous pacemaker. One observation supports this view: in a mutant strain (91A), which makes few pacemakers, spiral waves predominate over target patterns (Durston, 1974b). This observation by itself is not decisive, but both Winfree and Durston seem to feel that there are theoretical reasons for asserting that autonomous pacemakers give target patterns, while spirals arise when there is no autonomous pacemaker. It is not clear what these theoretical reasons are. Why should this be true? The only relevant arguments I can find in Durston and Winfree do not seem to be valid. Durston (1974a) says: "A point source pacemaker region (for example, an autonomously firing cell) which propagates waves through a uniform isotropic medium should propagate wavefronts which are concentric rings." This is apparently an argument from symmetry. But this argument, which is valid for a point source in a *perfectly* uniform, *perfectly* isotropic medium, may fail in a *slightly* nonuniform or nonisotropic medium, which, of course, is what any actual medium is. One must be careful about this, or else we could imagine arguing that whirlpools cannot form in cylindrical tubs of water with a hole at the center of

---

[7]*Pacemaker* is the generic term for an entraining oscillator.

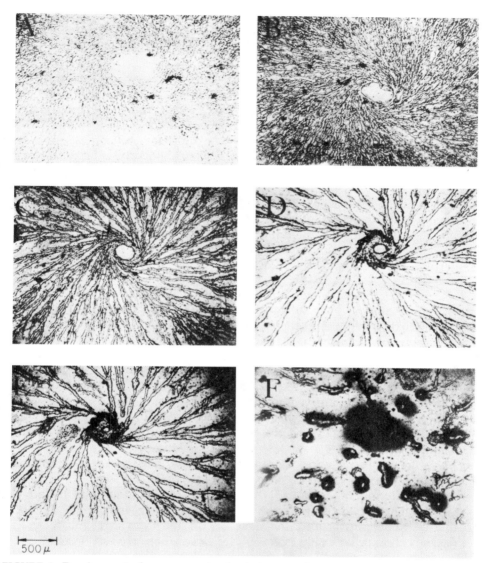

FIGURE 2. Development of an open-centered spiral aggregation center. The figure shows single frames from a low-magnification time-lapse film ($\times$ 2.5 to the film frame) of development of the center. Individual cells can only just be seen; the figure shows some details of the coordinated cell movements. (A) Early in development of the center, a cell-free space in an otherwise homogeneous lawn of Dd cells on an agar surface has become the focus for a spiral aggregation wave. The wave itself is not visible in the figure, but it induces a visible spiral orientation of the surrounding cells. The space will become the center of an aggregate. (B-E) After 23, 60, 79, and 86 min, respectively, the aggregating cells progressively collect into spirally oriented streams and aggregate into a "doughnut" surrounding the central space. The space diminishes. (F) After 150 min, most of the aggregating cells have entered a hemispherical mound sited at the location of the original open space. (From Durston, 1974a, with permission.)

the bottom. Small perturbations break the symmetry of the initial configuration and produce spiral forms in that case and could here, too.

Winfree (1980, p. 66) says, "It would seem that each wave must first appear somewhere, and that point is its source. But what is manifestly true of concentric ring waves is not necessarily true of rotating spiral waves." This is confusing. Of course, a set of concentric rings has a *geometric* center. But is the geometric center the *cause* of the propagation? Not necessarily. For example, in the case of chemical reaction–diffusion equations with local oscillators, Kopell and Howard (1973) challenge the analogous claim, widely made, that target patterns must have some special cause at their center. Their "working hypothesis" is "that the target patterns exist as a consequence of the interaction of chemical kinetics and diffusion, i.e., that no other physical mechanism plays a significant role. (Most other investigators appear to believe that there is a catalyst in the center of each pattern which affects the frequency of the oscillation . . . . We tentatively believe that the target patterns may form without the intervention of catalysts. . . .)" They do not claim to have proven this, because their derivation of target patterns from overall properties of the field is valid for the whole space except for the point at the center; their space has a hole in it.

Keener (1980) studied reaction–diffusion equations assuming excitable elements, not oscillatory ones, and found solutions by singular perturbation techniques applied to the small parameter ϵ—the ratio of the growth rates of the two state variables in the medium. He found that these methods do not yield any self-sustaining solutions unless the medium has holes (i.e., is not simply connected). In a medium with holes, he shows how self-sustaining spiral waves can arise and then combine to give target patterns. Once again, however, the question of pattern formation in a medium without holes is left open.

What is needed is a study of pattern formation in the slime mold along the lines of the work of Kopell and Howard on chemical reaction patterns. The key questions are:

1. Is the conjecture correct that target patterns have "centers" (autonomous pacemakers) whereas spirals do not?

2. How are spirals generated? [Durston (1973, 1974a,b) suggests that they are created by signals circulating around closed loops of cells, but there is a problem in that all the spirals acquire the same period and form early in aggregation, before the cell-to-cell contacts necessary for closed loops are established.]

3. What are "centers"? Are they special cells distinguished by some preexisting developmental difference? Are they merely fluctuationally different from their neighbors? Or do they have "centerness" thrust upon them by the developing field? Do pacemakers make aggregation fields, or do aggregation fields make pacemakers?

In any event, it seems as if the early aggregation field is a field composed largely of relay elements, with the outward gradients of phase (radial or spiral) characteristic of such elements. In the late aggregation field, actual aggregates of cells form. This involves direct cell-to-cell contact. These forms include blobs, doughnuts, strings, and whirlpools. It is in the late field that we see the evidence of widespread autonomy and entrainment of autonomous oscillators. Gerisch

(1968) took suspension cultures of late, contact-capable cells and transferred them to a plate. They did not form the spiral wave or target pattern configurations characteristic of the early field but, rather, formed into *strings*. This result supports the idea that signal relaying is the key to the radial patterns of the early field and that the late field does not depend on such relays. Instead, these cells, being themselves oscillators, quickly entrained each other and achieved contact. The strings in Gerisch's experiment rapidly proceeded to aggregate by forming a whirlpool with a doughnut at the center. The speed of aggregation is explained by the prevalence of autonomy (and hence the capacity for entrainment). The form of the aggregation into a whirlpool is explained by the absence of strong pacemaker aggregates (blobs). In the absence of a center, the aggregation field creates one: "In this process, aggregation centers usually form through a curling motion of streams."

Interestingly, the one case I have seen of radial (not curled) spokes arose when there was a powerful blob. An experiment of Robertson *et al.* (1972) consisted of placing a microelectrode in a plate of amoebae that pulsed cAMP at a rate simulating a blob. As soon as contact capability was achieved (about 8 hr), the amoebae formed well-defined strings that pointed directly at the center (no curling). Consequently, we can infer that blobs at the center are antagonistic to spiral formation. (This is a feature shared by whirlpools in water and by tornadoes, which also have "empty centers.") On the other hand, where there is high potential for entrainment, but aggregations (blobs) are prevented from forming, the system undergoes oscillation as a whole. This claim is demonstrated by an experiment of Gerisch and Hess (1974) on dense cultures in suspension: a few pulses set the suspension into coherent (simultaneous) global oscillation.

Entrainment helps to explain these morphologies. Were it not for entrainment, there would be a time lag between every pulse and its relay. This fact implies that there would be a constant gradient of phase radiating outward from any signal source. Therefore, the existence of actual aggregates along radial components implies that entrainment must be taking place.

Nanjundiah (1973) developed a mathematical model in the spirit of the Keller–Segel analysis in order to study the phenomenon of streaming. Streaming is the formation of strings of aggregating cells pointing to the center. This breaking of radial symmetry is explained in the Turing manner as an instability to azimuthal fluctuations. Most interestingly, he demonstrates that such streaming depends essentially on "cooperative signaling": "Whenever signalling is *not* restricted to one or a few of the cells, the radially symmetric mode of aggregation is itself unstable against azimuthal fluctuations, and the periphery of an aggregate breaks up to form *streams*." Although the model used is one in which the cells signal steadily, rather than by oscillation, it seems as if this phenomenon also supports the idea that there is widespread autonomy and cooperativity in the late aggregation field.

The common thread uniting these observations is that regions of aggregation are regions of high entrainment of signal, and, conversely, aggregation takes place wherever such regions of high entrainment form. This view helps to explain the patterns observed: the regions of entrainment become the *attractors* of the aggregation dynamic. Therefore, aggregation morphologies can be classified by the nature of the attractors involved. In particular, the spiral patterns and spoke patterns can be identified as marking out *isochrons* (in Winfree's terminology), i.e., regions in which phase is constant. When we look at an aggregation pattern, we are seeing a physical realization of entrainment: the foliation of the basin of attraction into lines of equal phase.

Entrainment

If what I have been saying is correct, then the notion of entrainment is crucial for understanding slime mold aggregation. Entrainment is essential for dynamics of many oscillatory processes, from the circadian rhythms of insects to ovulatory cycles in humans. The problem of modeling entrainment is far from solved and requires much more attention. In particular, the word *entrainment* is used in too many senses. In its most abstract sense it is used to refer to any situation in which small interactions among the individuals of a system have the effect of confining the total state of the system to some limited region of the global state space. In this most general sense, it is simply motion that is not ergodic—that does not wander all over the state space.

Even if we restrict our attention to populations of coupled oscillators, there are still many senses of the word. Most often *entrainment* means *frequency* entrainment, i.e., two or more oscillators interacting in such a way as to equalize their frequencies or bring them to a whole number ratio. Frequency entrainment therefore includes the notion of musical harmony. The idea that harmony consists of a whole number ratio, a rational relationship of frequencies, was known to the Pythagoreans. Abraham and Shaw (Chapter 29, this volume) present a model of frequency entrainment among simple harmonic oscillators.

Although frequency entrainment does occur in slime mold aggregation (and might explain the steady increase in the frequency of oscillation), the notion we have been using is that of *phase* entrainment. Sometimes, this is called "phase-locking" or "synchronization," although both these terms are also used, confusingly, to refer to frequency entrainment. True phase entrainment has not been discussed in any generality, although there are a number of examples of it in various sciences.[8] Winfree (1980) discusses these:

- Populations of crickets entrain each other to chirp coherently
- Populations of fireflies come to coherence in flashing
- Yeast cells display coherence in glycolytic oscillation
- Populations of insects show coherence in their cycles of eclosion (emergence from the pupal to adult form)

---

[8]See Dewan (1972) for a discussion of types of phase-locking or entrainment for the special case of van der Pol oscillators. A simplified model of phase-locking in biological oscillators can be found in Glass and Mackey (1979) for entrainment by a sinusoidal stimulus.

In addition, we can add:

- Populations of women living together may show phase-entrainment of their ovulation cycles
- Populations of secretory cells, such as in the pituitary, pancreas, and other organs, release their hormones in coherent pulses.

In a laser, there are many atomic oscillators, each emitting light. The transition from ordinary light output to laser light is caused by phase entrainment of the oscillators, so that they pulse light coherently (Haken, 1978).

For slime mold aggregation, a number of simplifying assumptions can be made. First, we are dealing with *pulse* oscillators (so-called "relaxation oscillators"), which are all of roughly the same period and communicate by diffusion. We can therefore give a general form to the problem of slime mold aggregation: we have a large population of pulse oscillators $0_1, \ldots, 0_N$, initially at locations $x_1, \ldots, x_N$ and emitting pulses at phases $w_1, \ldots, w_N$. The pulses then diffuse through the medium by Fick's law, with an added decay term due to the enzyme action of acrasinase. The effect on each phase w_i is then given by the sum of the diffused pulses from the other 0_j, which is then inserted for each 0_i into the known phase response relation. At the same time, the spatial locations of the oscillators are changing due to chemotaxis, so we must add an equation describing the dynamics of the chemotactic movement of each oscillator in response to the sum total of pulses received from neighboring oscillators. In addition, a condition must be added to represent the refractory periods during which the oscillators, after emitting a pulse, are immune to chemotaxis and to being signaled.

These features would comprise the most complete model of slime mold aggregation. The model would be, mathematically, completely intractable, and absolutely nothing can be said about it. The complexity involved is at least that of the general problem of the entrainment of N oscillators, on top of the problem of pattern formation in reaction–diffusion systems added to the N-body motion problem. Any *one* of these problems is insuperable in general form; simultaneously they render the analytic situation hopeless. Our hope thus consists of making numerous simplifying assumptions to bring the problem at least within range of known or imaginable mathematical techniques. Several obstacles stand in the way of a useful representation for this problem. One simplification that is common in oscillator models is to study only the limit behavior as it approaches infinity ("final motions") and to ignore transients. But in this case, it is the transients that we care about, because we are interested in representing the process by which small pockets of entrainment appear and grow. For the same reason, we are prohibited from making another of the standard assumptions used to simplify entrainment problems: throwing out precise spatial information and assuming that our oscillators are coupled by a generic, small, nonlinear perturbation. It is exactly the spatial development of entrainment that we want to study.

But the most serious problem is one that besets any discrete model, i.e., any model in which the aggregation field is represented as a set of N oscillators located at points spread over a plane. Whether or not entrainment takes place in such a population is extremely dependent on the exact locations, phases, diffusivities, and the like. If we attempt to follow the precise pattern of phase interaction, we can see that the trajectories in question are extremely unstable: if one phase or one location is changed by a small amount, the resulting pattern becomes destabilized. Consequently, we face a problem similar to the one encountered in deriving the phenomenon of wave propagation: the model that reduces the collective phenomenon to the interaction of N discrete individuals has the defect that the (global) overall effect becomes unstable with respect to initial conditions. Moreover, the sensitivity of the overall effect is greater than the degree of error built into the individualistic model by its simplifying assumptions. Consequently, the individualistic model of the collective process becomes useless.[9]

The only hope for a useful model of the phase-entrainment process therefore lies in rising above the level of the individual cells and finding some global model in which the local information is thrown out in favor of some order parameters that describe the average conditions obtaining in spatial neighborhoods. In this respect the situation resembles coherence in the laser or magnetization, in which there is the coherence of orientation of many magnetic dipoles interacting. In cases like these, the strategy that has proved effective is to find global parameters describing the overall state of the system, e.g., the local mean field. These *order parameters*, as Haken (1978) calls them, reflect the degree to which coherence has been achieved in some region. The analogous move in this case would be to consider as our order parameter something like the *distribution of phase* over the aggregation plane. The concentric rings and spirals of the aggregation field correspond to isophase lines (Winfree's "isochrons"). (Alternatively, we could look at the parameter *local degree of coherence of phase*.) It might then be possible to write dynamical equations representing the various influences on local phase coherence, in a manner similar to the laser equations (see, e.g., Haken, 1978).

Some Philosophical Principles

Several methodological or philosophical precepts emerging from this discussion seem to be typical of self-organization phenomena. The first is *antireductionism*. We have seen that modeling aggregation requires us to transcend the level of the individual cells to describe the system by holistic parameters. But in classical reductionism, the behavior of holistic entities must ultimately be explained by reference to the nature of their constituents, because those entities "are just" collections of the lower-level objects with their interactions. Although it may be true in some sense that systems "are just" collections of their elements, it does not follow that we can *explain* the system's behavior by reference to its parts, together with a theory of their connections. In particular, in dealing with systems of large numbers of similar components, we must make recourse to holistic concepts that refer to the behavior of the system as a whole. We have seen

---

[9]This argument is made in a more general context by Garfinkel (1981).

here, for example, concepts such as entrainment, global attractors, waves of aggregation, and so on. Although these system properties must ultimately be definable in terms of the states of individuals, this fact does not make them "fictions"; they are causally efficacious (hence, *real*) and have definite causal relationships with other system variables and even to the states of the individuals.

Self-organization is therefore inherently antireductionist in that global system parameters are essential for its explanation. The general model of self-organization proceeds by analysis of the qualitative dynamical properties of the overall system variables: the forms of organization are characterized as attractors of the global dynamic, and structural changes from one form of collective organization to another are characterized by bifurcations of the governing equations, as one global attractor gives way to another.

In particular, we have found that we are forced to adopt a field-theoretic, rather than a particle-theoretic, model of the self-organization process. As soon as we begin talking in terms of *waves* and their various properties, we are necessarily talking about a continuous medium and have transcended the level of the individual cell. To return to the subject with which we began, the relationship between the global (Keller–Segel) model and the individual cell model, we see that the essential approach of the global model has been vindicated: there is a need to consider aggregation as a phenomenon occurring in a continuous medium—not one occurring among a discrete set of individual cells. The first version of the Keller–Segel model may have been oversimplified, but some more sophisticated version of it, capable of modeling various kinds of wave propagation, would seem to be necessary for a satisfactory explanation of the patterns of slime mold aggregation.

There may be objections that amoebae are, obviously, discrete entities and that continuum-flavored analyses are mere metaphors, which can in principle be discarded in favor of an underlying model of discrete cells communicating by pulses, in the manner of the Cohen–Robertson (1971a,b) articles. I do not believe that this is right. The methods of analysis now employed depend essentially on the continuum assumption, since they use differential and topological arguments that only make sense in continua. If we increase the magnification so that the continuum disappears, so does the explanation for the phenomena of aggregation. As Winfree observes (1980, p. 270): "In media composed of discrete cells communicating by pulses, the arguments of continuum mechanics and topology have no clear application."

In a certain sense, this commitment to the reality of the continuum is paradoxical. The mathematics used to generate our explanation absolutely requires that the key functions are smooth or continuous in *arbitrarily* small neighborhoods. And yet we know this is false: at sufficiently small distances, the continuum breaks up. The general procedure for resolving such paradoxes is something like this: there is a gauge δ such that for $\delta < \delta_0$ the equations "break down" (lose their validity). In that scale we have to interpolate a finer-grained theory which is valid in the "boundary layer" $\delta < \delta_0$ and reduces to the old theory for $\delta > \delta_0$.

The second major principle operating in the self-organization paradigm is the notion of *symmetry-breaking*. This principle is also a form of anti-reductionism, because the explanation of global structure proceeds without appealing to underlying differences among the individuals. For example, consider the onset of ferromagnetism in a solid. For temperatures above the critical temperature, there is a uniform distribution of orientations. But as the temperature drops below the critical point, the uniform distribution bifurcates into a sharply bimodal distribution, corresponding to the two possible orientations of the elementary dipoles. How a molecule happens to assume one orientation rather than the other (opposite) orientation is unexplained. Again, if we consider a fluid heated from below, for temperature differences above a critical point, the uniform fluid changes into one with a regular pattern of convection cells, so that some molecules are at the center of convection cells. But there is no intrinsic difference between those molecules and others. The relevant self-organization model here takes the form of equations governing heat transport in the fluid. It shows that for values of the temperature gradient R (called the Rayleigh number) less than the critical R_c, the equations have a solution in which the fluid is at rest, while for R greater than R_c, heat transport overcomes viscous influences, and the steady-state solution bifurcates into a solution with regularly spaced cells of convection currents.

These principles become especially important and take on added significance when the structures in question are social structures, and the individuals are individual people.

Explaining Social Structure: The Emergence of Cooperation

We now turn to social explanation in order to assess the prospects of the self-organization paradigm. First, a word of caution about the dangers of analogy: the history of biological analogy in social theory is not a happy one. From Plato's organic model of the state to the periodic outbreaks of social Darwinism (most recently in discussions of alleged genetic bases for class stratification), biological theories of society have ranged from crackpot to worse. I do not wish to join this tradition of biological reductionism. Rather, I want to suggest that the general lessons and principles of self-organization in the biological sphere are applicable as well in social systems. Therefore, I am investigating a loose analogy, not a reductive explanation.

Even so, this investigation has some interest, for in each historical epoch there has been a rough correspondence between the concepts of biology and physical science, on the one hand, and social theory, on the other. Consider the theory of evolution. The traditional account depicts evolution as gradualist (a continuous gradation of small changes) and individualistic (evolution selects the fittest individuals to survive). With the development of thermodynamics in the late 19th century, the metaphor of entropy percolated quickly into both biology and social thinking as an all-purpose explanation for biological and social degeneracy (see Brush, 1978). Contemporary evolutionary theory has changed radically (see Gould, Chapter 6, this volume). In particular, recent discussions of evolution have stressed its holistic, rather than individualistic, nature, and treat it as a saltatory, rather than gradualist, process.

A similar claim can be made about the other areas of theoretical biology. The self-organization paradigm, with its concepts of antireductive holism, symmetry-breaking, bifurcation, and the like, offers a number of suggestive models for explaining the evolution of social structure. What follows is a somewhat haphazard survey of some topics of current interest in social theory, presented with a view toward exploring the possibilities of analogy.

The fundamental problem of social theorists is to explain the emergence of social structures. As Hobbes saw it, this meant explaining how cooperation occurs, given the fact that it is in each person's individual interest to betray the social contract while everyone else obeys it. For Hobbes, the only solution was the establishment of a *sovereign* to enforce the contract on each individual, but more recent attempts have tried to explain the emergence of a social contract without the authoritarian presence of a sovereign.

In its modern formulation, the problem of explaining how cooperation occurs is expressed as the *Prisoners' Dilemma*[10]: the maximum gain for each individual is to betray the social contract, yet if we all do that, we all lose. How can cooperative behavior possibly arise? The game-theoretic answer to this problem is to define some version of the Prisoner's Dilemma and to study its *Nash equilibria*, i.e., points at which players cannot improve their payoff by making changes just in their own strategies. The question then becomes: is cooperation a Nash equilibrium? If not, is there some weaker form of cooperation that is? How stable is it? What initial states will be drawn to it? Is it subject to bifurcation?

Massive literature on the subject formulates various interpretations and versions of the game and finds various equilibria. Smale (1980) gives a precise formulation of a two-person Prisoner's Dilemma with discrete time and describes a family of Nash solutions that converge over time to cooperation, and that possess global stability (all initial conditions flow to it). The solutions are, roughly, to cooperate as long as our cooperation has not been exploited by the other. Smale exhibits a subfamily of such solutions, S_μ, depending on a parameter μ, so that for $\mu > 0$, S_μ is one of these cooperative Nash solutions, but for $\mu < 0$, the solution bifurcates to noncooperative behavior. This description gives us a simple example of an important phenomenon: a single game, a single set of rules, may have one kind of behavior (competition) for one range of conditions and another (cooperation) for other conditions. That result explains how both responses are possible (without attributing either to "human nature") and how one can change or bifurcate into the other.

It would be useful to have a generalization to continuous choices: suppose each player chooses an amount of money to bet, a time of day to arrive, or a direction to follow (note that the last two are continuous, angular variables). Smale says that his framework can accommodate these conditions, but he does not carry out the proof.

---

[10] So called because in its standard formulation it concerns two prisoners each separately offered the following deal: Confess and testify against the other, and you will go free and the other will get the maximum sentence. If both confess, they both get fairly stiff sentences; if neither does, they both get a light sentence.

More importantly, it is absolutely necessary to have a generalization of this analysis to an *N*-person game. *N*-person games are as different from two-person games as the *N*-body problem is different from the two-body problem. In particular, coalitions arise in which some people unite to "exploit" others. These coalitions do not have to stem from agreements or even to be known to their members; tacit coalitions can form in a stable way.

A preliminary sketch of some aspects of an *N*-person Prisoner's Dilemma, from an informal point of view, can be found in Schelling (1978). Applications to some problems in social theory of another version of an *N*-person Prisoner's Dilemma can be found in Taylor (1976). Neither treatment uses dynamical methods, or establishes stability, bifurcation, and size of attracting basin. As a consequence, it is difficult to extract from their analyses any qualitatively useful conclusions or to tell which effects are stable under which perturbations of the model assumptions.

The key to modeling cooperation is that cooperative solutions in games are *entrainments*, i.e., mutually coherent behaviors. In a model of cooperation as entrainment, global attractors model overall equilibria. Because social conventions (like the working day, driving on the right, feudalism, or speaking English) are essentially coordination equilibria (see Lewis, 1974), the topology of the various attractors in a given system provides the kinematic foundation for theories of social change. We might say, with apologies to Marx, that all history is the history of phase transitions of coordination equilibria.

The global attractor model of the emergence of cooperation displays the characteristic features of the self-organization paradigm. First of all, it is antireductionistic in character. It establishes that the total state of the system will move to a certain attractor despite the absence of individual intentions to achieve that attractor. The global dynamics of the process, especially the existence of a constructive, nonlinear payoff to regions of entrainment, explains the development toward the attractor. This model does not attempt to explain the development of those regions of entrainment by assuming special intentions of "organizers." This removal from intentions enables us to apply game theoretic models to areas in which there are no real intentions or strategies in the conscious sense. Game theory can be introduced into biological evolutionary theory, following Lewontin (1961). We imagine a species playing a game with the environment: one or another genetic strategy is "chosen" (e.g., adaptability to wet environments) and then nature makes its move (lots of rain, or very little rain, and so on). The "payoff" to the species is the reproduction rate achieved. Lewontin (1970) argues that the units of selection in such games are not genes, but whole chromosomes.

Generalizing to the case of several species, we can imagine them playing such a game with each other. This abstraction leads to the notion of an *evolutionarily stable strategy*. The payoffs are again reproductive rates, and an evolutionarily stable strategy is one that evolves toward some stable configuration. Evolutionary stable strategies are Nash equilibria. Taylor and Jonker (1978) present rigorous definitions and prove basic theorems. Auslander *et al.* (1978) examine evolutionarily stable strategies for predator-prey populations and find various situations in which there is no stable strategy. In one case, the prey population displayed chaotic behavior, which they suggest may be an adaptive evolutionary strategy, preventing the coevolving predator population from "tracking" it. Chaotic behavior cannot entrain.[11] The concept of an evolutionarily stable strategy, such as a Nash equilibrium in the evolutionary game, allows us to examine the patterns of coevolution of various kinds of species or even of kinds of traits.

One problem that has attracted attention is explaining the evolution of *altruism*. How is it that altruistic tendencies can evolve, if they do not also confer advantage on the individual? For example, in many species of prey (birds, gazelles), the first individual to see an approaching predator will go into a ritualized warning behavior that is very useful to the flock as a whole but that has negative survival value to the individual. Any explanation of this fact will have to go beyond the Darwinian conception of evolution as selecting the fittest *individuals*. The evolution of altruism can be represented as a game in which genetic tendencies play the Prisoner's Dilemma on an evolutionary scale. We imagine a set of genes for cooperation competing with a set for selfish behavior. Can cooperation arise as an evolutionarily stable strategy? The answer seems to be that it can, although controversy still exists and no definitive model has yet emerged (Hamilton, 1964; Trivers, 1971; Axelrod and Hamilton, 1981).

A similar approach has been successful in explaining early stages of molecular evolution. The origin of life in the self-organization of nucleic acids was presumably an evolutionarily stable "strategy" on the part of prebiotic chemicals (Eigen and Schuster, 1977; Schuster *et al.*, 1978; Schuster and Sigmund, Chapter 5, this volume).

Social Structure: Emergence in Economics

In the prototypical market described by Adam Smith, a large number of traders, each having a negligible effect on the market, interact independently to produce a market and a price system. Can we describe the evolution of coalitions, such as cartels and trade unions, in the economic market? There are several discussions of the formation of unions (in the general sense), but they apply only to games in which there are agreements among the players enforced by the analog of Hobbes' sovereign. This is a drawback, since in typical cases there are no agreements to cooperate at all, or there are agreements, but no enforcer. In the case of holders of capital, however, entrained coalitions can form in the absence of any intentions to conspire. Such tacit coalitions can be modeled as stable attractors of the global dynamics, furnishing a basis for a theory of oligopoly. In the case of

---

[11]For an introduction to chaos, see Abraham and Shaw (Chapter 29, this volume). Chaotic attractors have been found in population dynamics (May, 1976), in a model for the weather (Lorenz, 1963), and have been proposed as models of turbulence in fluids (Ruelle and Takens, 1971). There have as yet been few applications to social theory, despite the fact that the first known instance of chaotic behavior was in a model of the price system, due to Cournot, in 1843—but see Benhabib and Day (1981) for a model in which the effects of actions on preferences, and vice versa, produce chaotic behavior.

trade unions, there is typically an agreement to cooperate, but since the agreement is not enforceable, the situation is really more similar to the self-organizing coalition than it is to the Hobbesian case. Here the self-organization approach can be used to model the development of small regions of entrainment in various areas or industries as well as their subsequent development.

The problem of stable entrainment is relevant also to the "problem of the second-best." A "second-best" problem arises when a system has an equilibrium solution that is optimal with respect to some desirable quantity, but finite perturbations applied to this optimal configuration may produce results much worse than those of other, "second-best" equilibria (Lipsey and Lancaster, 1956; Allingham and Archibald, 1975). Here the problem is a bifurcation of an equilibrium solution to another qualitatively different attractor.

In the market as described by Adam Smith, the situation of the traders is essentially homogeneous. As is well known, however, the homogeneous market, if it ever existed, surely no longer does. Instead, we represent the actual situation, to a first approximation, as a stratified market in which there are two types of traders, owners of capital (buyers of labor power) and nonowners of capital (sellers of labor power). How can this stratification be explained? One approach, based in classical determinism, asserts that there must be some systematic difference between the individuals who end up as capitalists and those who end up as labor. Social Darwinists assert that such differences exist and explain stratification. In the words of William Graham Sumner, for example, the individual differences are that the capitalists possess greater "industry, prudence, continence, or temperance"; for the neosocial Darwinists, it is "intelligence." Despite the difficulty of locating these traits in actual individuals, it is asserted nevertheless that there *must* be some such difference, for how else could stratification be explained? How can there be differences without a difference?

The alternative approach to explain the evident stratification is to assume that there are no significant individual differences. We explain stratification as the increasing instability of a homogeneous solution under increasing competitive pressure, leading to a symmetry-breaking bifurcation to a more structured global state—in this case a system of two strata in stable interaction (Garfinkel, 1981).[12] Social Darwinism can therefore be seen as denying the existence of symmetry-breaking bifurcation in social systems. But what if everyone possessed the desirable individual traits? If the structural features making the homogeneous situation unstable are still present, stratification would take place anyway. In contrast, the broken-symmetry approach locates the explanation of the stratification in structural properties of the system itself. Consequently, it offers an explanation with some theoretical stability as well as greater predictive power.

The general strategy of self-organization as treated here is to explain pattern formation in a macrosystem as resulting from small nonlinearities in the coupling of the individuals. These nonlinearities are negligible for the homogeneous state, whose behavior therefore resembles a (linear) system of independent individuals.

---

[12] In the language of dynamics, increasing competitive pressure introduces a cusp catastrophe into the distribution of income, thereby transforming it from a unimodal to a bimodal distribution. (See Garfinkel, 1981.)

But as some parameter, such as an external driving force or a density, passes into a critical region, the homogeneous state becomes unstable and the nonlinearities become significant.[13] In this image, the aggregation of capital into a small number of holdings, and the resulting differentiation that takes place, is explained as the development of an entrainment pattern among holders of capital, which in turn is explained as the result of nonlinear payoffs to collections of cooperating (i.e., entrained) holders of capital.

Orthodox economists have trouble explaining this aggregation because of the linear models to which it is attracted. In part, this is due to the ease of solution of linear models, but the attraction is also ideological: linear models obey *superposition*, guaranteeing the independence of individual solutions. This is the algebraic expression of the laissez-faire principle. As can be expected, pattern formation does not arise in such systems. For example, the orthodox economic approach to the two-sector character of the economic market is embodied in the neoclassical production function

$$O = f(K,L)$$

which expresses, for a given production process, the amount O of output that can be produced by combining K units of capital with L units of labor. In the standard treatments of this subject, a theory of distribution (payoff) is derived, according to which the day's output of O is paid back to the contributors of K and L at a rate that is proportional to their marginal contributions to the production process, $\partial f/\partial K$ and $\partial f/\partial L$. In the standard treatments, this theory of distribution according to marginal contribution is obtained from an application of Euler's theorem to the production function, yielding the desired payoff relation:

$$O = \frac{\partial f}{\partial L} L + \frac{\partial f}{\partial K} K$$

From this it follows that the payoffs to contributors of L and K are determined by purely technical factors (and therefore *not* by political factors, combinations such as cartels or unions, or other cooperative effects).

The trouble with this derivation lies in the assumption that is necessary to apply Euler's theorem: that the production function is *homogeneous of degree one*, i.e.,

$$f(aK,aL) = af(K,L)$$

This restriction says that multiplying the inputs by some arbitrary factor results in multiplying the output by the same factor. This implies complete linear *scaling*: doubling inputs produces a doubled output, and so on. A model that makes

---

[13] For convection cells in fluids, it is the driving force of heat transport (Rayleigh number); in fluid flow models of vortex formation, it is the driving velocity (Reynolds number); in the transition from an ideal gas, to a van der Waal gas, it is the density; in population dynamics, it is also density, and so on.

this assumption will be unable to explain pattern formation, because payoffs to combinations are exactly equal to what the individual combinants could earn separately. In contrast, the theme of self-organization is that macroscopic pattern formation is the result of small *nonlinear* coupling terms.

Complete (infinitely extendable) scaling never exists in reality. Sometimes we can invoke scaling in regions where the nonlinearities can be neglected. Economic theory has largely confined its attention to those areas, and whatever success it has had is due to the fact that economic reality displayed, until recently, at least some linear growth patterns. Further progress will necessarily involve coming to terms with economic and social nonlinearities.

Conclusion

A number of philosophical themes have emerged from the self-organization paradigm presented herein. All of them represent heresies from the point of view of traditional philosophical concepts. In my opinion, the correct themes are:

1. *Holism.* To explain self-organization one must study the dynamics of global system variables and attractors.

2. *Catastrophic change.* Although Leibniz claimed for all natural processes that *Natura non facit saltum* (nature does not make jumps), the concepts of catastrophe and bifurcation give a richer scientific content to the notion of emergence and make possible models of physical, biological, and social processes in which periods of continuous change are punctuated by episodes of emergence.

3. *Symmetry-breaking.* This principle requires us to explain stratification structurally, without appeal to underlying differences. It denies classical determinism.

In addition to the above themes, I would add the concepts that seem essential in discussing aggregation in the slime mold, especially the notions of *oscillation* and *entrainment*. In the future, the N-torus may be as familiar a model for biological or social space as Euclidean space is now, with pattern formation among coupled oscillators expressed in terms of various forms of entrainment or coherence. Entrainment seems to be the crucial concept in understanding the evolution of social, as well as biological, aggregation and cooperation. This seems to be the real lesson of the slime mold.

ACKNOWLEDGMENTS. I would like to thank the National Endowment for the Humanities for their support, and Gene Yates, Don Walter, and Tim Poston for many helpful suggestions.

References

Alcantara, J., and M. Monk (1974) Signal propagation during aggregation in the slime mould *Dictyostelium discoideum*. *J. Gen. Microbiol.* **85**:321.
Allingham, M., and G. Archibald (1975) Second best and decentralization. *J. Econ. Theory* **10**:157.

Auslander, D., J. Guckenheimer, and G. Oster (1978) Random evolutionarily stable strategies. *Theor. Pop. Biol.* **13**:276.

Axelrod, R., and W. D. Hamilton (1981) The evolution of cooperation. *Science* **211**:1390.

Benhabib, J., and R. Day (1981) Rational choice and erratic behavior. *Rev. Econ. Stud.* **48**:459.

Bonner, J. T. (1959) Differentiation of social amoebae. *Sci. Am.* **201**:152.

Bonner, J. T. (1969) Hormones in social amoebae and mammals. *Sci. Am.* **220**:78.

Bonner, J. T., D. S. Barkley, E. M. Hall, T. M. Konijn, J. W. Mason, G. O'Keefe, III, and P. B. Wolfe (1969) Acrasin, acrasinase, and the sensitivity to acrasin in Dictyostelium discoideum. *Dev. Biol.* **20**:72.

Brush, S. (1978) *The Temperature of History.* Franklin, New York.

Chance, B., E. K. Pye, A. K. Ghosh, and B. Hess (1973) *Biological and Biochemical Oscillators.* Academic Press, New York.

Cohen, Ma. (1977) The cyclic AMP control system in the development of Dictyostelium discoideum. *J. Theor. Biol.* **69**:57.

Cohen, Mo., and A. Robertson (1971a) Chemotaxis and the early stages of aggregation in cellular slime molds. *J. Theor. Biol.* **31**:119.

Cohen, Mo., and A. Robertson (1971b) Wave propagation in the early stages of aggregation of cellular slime molds. *J. Theor. Biol.* **31**:101.

Dewan, E. M. (1972) Harmonic entrainment of van der Pol oscillations. *IEEE Trans. Autom. Control* **17**:655.

Durston, A. J. (1973) Dictyostelium discoideum aggregation fields as excitable media. *J. Theor. Biol.* **42**:483.

Durston, A. J. (1974a) Pacemaker activity during aggregation in Dictyostelium discoideum. *Dev. Biol.* **37**:225.

Durston, A. J. (1974b) Pacemaker mutants of Dictyostelium discoideum. *Dev. Biol.* **38**:308.

Eigen, M., and P. Schuster (1977) The hypercycle, a principle of natural self-organization. *Naturwissenschaften* **64**:541.

Garfinkel, A. (1981) *Forms of Explanation.* Yale University Press, New Haven, Conn.

Gerisch, G. (1968) *Dictyostelium*: Aggregation and differentiation. *Curr. Top. Dev. Biol.* **3**:157.

Gerisch, G. (1978) Cell interactions by cyclic AMP in Dictyostelium. *Biol. Cell.* **32**:61.

Gerisch, G., and B. Hess (1974) Cyclic-AMP-controlled oscillations in suspended Dictyostelium cells. *Proc. Natl. Acad. Sci. USA* **71**:2118.

Glass, L., and M. C. Mackey (1979) A simple model for phase-locking of biological oscillators. *J. Math. Biol.* **7**:339-352.

Goldbeter, A. (1975) Mechanism for oscillatory synthesis of cyclic AMP in Dictyostelium discoideum. *Nature* **253**:540.

Goldbeter, A., and L. Segel (1977) Unified mechanism for relay and oscillation of cyclic AMP in Dictyostelium discoideum. *Proc. Nat. Acad. Sci. USA* **74**:1543.

Gross, J. D., M. J. Peacey, and D. J. Trevan (1976) Signal emission and signal propagation during early aggregation in Dictyostelium discoideum. *J. Cell Sci.* **22**:645.

Haken, H. (1978) *Synergetics.* Springer-Verlag, Berlin.

Hamilton, W. D. (1964) The general evolution of social behavior. *J. Theor. Biol.* **7**:1.

Kametaka, Y. (1975) *On the Nonlinear Diffusion Equation of Kolmogorov-Petrovskii-Piskunov Type.* Lecture Notes in Physics Vol. 39. Springer-Verlag, Berlin.

Keener, J. (1980) Waves in excitable media. *SIAM J. Appl. Math.* 39.

Keller, E. F., and L. Segel (1970) Initiation of slime mold aggregation viewed as an instability. *J. Theor. Biol.* **26**:399.

Kolmogorov, A., I. Petrovski, and N. Piskunov (1937) Étude de l'équation de la diffusion avec croissance de la quantité de matière et son application à une problème biologique. Bulletin de l'Université d'État a Moscou, Série International. Vol. I, Section A, pp. 1-25.

Kopell, N., and L. Howard (1973) Plane wave solutions to reaction-diffusion equations. *Stud. Appl. Math.* **12(4)**:291-328.

Kopell, N., and L. Howard (1974) *Pattern Formation in the Belousov Reaction.* Lectures on Mathematics in the Life Sciences Vol. 7. American Mathematical Society, Providence, R.I.

Lewis, D. (1974) *Convention: A Philosophical Study.* Harvard University Press, Cambridge, Mass.

Lewontin, R. (1961) Evolution and the theory of games. *J. Theor. Biol.* **1**:382.
Lewontin, R. (1970) The units of selection. *Annu. Rev. Ecol. Syst.* **1**:1.
Lipsey, R. G., and K. Lancaster (1956) The general theory of second best. *Rev. Econ. Stud.* **24**:11.
Loomis, W. (1975) *Dictyostelium discoideum: A Developmental System.* Academic Press, New York.
Lorenz, E. (1963) Deterministic nonperiodic flow. *J. Atmos. Sci.* **20**:167.
Malchow, D., V. Nanjundiah, and G. Gerisch (1978) pH oscillations in cell suspensions of *Dictyostelium discoideum*. *J. Cell Sci.* **30**:319.
May, R. M. (1976) Simple mathematical models with very complicated dynamics. *Nature* **261**:459.
Nanjundiah, V. (1973) Chemotaxis, signal relaying and aggregation morphology. *J. Theor. Biol.* **42**:63.
Poston, T., and I. Stewart (1978) *Catastrophe Theory and Its Applications.* Pitman, London.
Raman, R. K., Y. Hashimoto, M. H. Cohen, and A. Robertson (1976) Differentiation for aggregation in the cellular slime molds. *J. Cell. Sci.* **21**:243.
Rashevsky, N. (1940) An approach to the mathematical biophysics of biological self-organization and of cell polarity. *Bull. Math. Biophys.* **2**:15, 65, 109.
Robertson, A. (1972) *Quantitative Analysis of the Development of Cellular Slime Molds. Lectures on Mathematics in the Life Sciences* Vol. 4. American Mathematical Society, Providence, R.I.
Robertson, A., and M. Cohen (1974) *Quantitative Analysis of the Development of Cellular Slime Molds II. Lectures on Mathematics in the Life Sciences* Vol. 6. American Mathematical Society, Providence, R.I.
Robertson, A., D. Drage, and M. Cohen (1972) Control of aggregation in *Dictyostelium discoideum* by an external periodic pulse of cyclic adenosine monophosphate. *Science* **175**:333.
Ruelle, D., and F. Takens (1971) On the nature of turbulence. *Commun. Math. Phys.* **20**:167.
Schelling, T. (1978) *Micromotives and Macrobehavior.* Norton, New York.
Schuster, P., K. Sigmund, and R. Wolff (1978) Dynamical systems under constant organization I. *Bull. Math. Biol.* **40**:743.
Segel, L. (1972) *On Collective Motions of Chemotactic Cells. Lectures on Mathematics in the Life Sciences* Vol. 4. American Mathematical Society, Providence, R.I.
Shaffer, B. M. (1962) The Acrasina. *Adv. Morphog.* **2**:109–183.
Shaffer, B. M. (1964) The Acrasina. *Adv. Morphog.* **3**:301–322.
Smale, S. (1980) The prisoner's dilemma and dynamical systems associated to non-cooperative games. *Econometrica* **48**:1617–1634.
Taylor, M. (1976) *Anarchy and Cooperation.* Wiley, New York.
Taylor, P., and L. Jonker (1978) Evolutionarily stable strategies and game dynamics. *Math. Biosci.* **40**:145.
Thom, R. (1972) *Structural Stability and Morphogenesis.* Addison-Wesley, Reading, Mass.
Trivers, R. L. (1971) The evolution of reciprocal altruism. *Q. Rev. Biol.* **46**:35.
Turing, A. M. (1952) The chemical basis of morphogenesis. *Philos. Trans. R. Soc. Ser. B* **237**:37. London.
Winfree, A. (1980) *The Geometry of Biological Time.* Springer-Verlag, Berlin.

11

The Orderly Decay of Order in the Regulation of Aging Processes

Caleb E. Finch

ABSTRACT

Processes of aging are discussed in a comparative context, with emphasis on reproductive neuroendocrine functions of mammals. Two general hypotheses of aging are discussed. The first proposes that aging results from processes intrinsic to each cell, which may involve mutations and other types of random macromolecular damage. The second proposes that aging in the organism is regulated by factors extrinsic to most cells, such as hormones. The present evidence is interpreted to argue against random damage as a major factor in mammalian aging, since some aging changes can be manipulated by hormones or diet. The dissociability of chronological and physiological age by experimental manipulations suggests that biological senescence in some organisms mainly involves changes that are "event-dependent" rather that "age-dependent." Then, the temporal plasticity of some developmental and aging processes is compared to heterochronic changes during evolution, in which changes in the timing of developmental events are viewed as major substrates for evolutionary change. Knowledge of the genetic loci that control developmental timing may give insights into controls on the timing of age changes. —THE EDITOR

The ability of multicellular organisms to maintain most genotypically regulated characteristics throughout their lives is a cause for wonder, given the onslaught of external perturbations and diverse influences of radiation, diet, infections, climate, reproductive experience, and other social interactions. This remarkable stability of most cellular functions during the life span of higher organisms indicates the power of the mechanisms of "self-organization" and maintenance possessed by living systems. However, aging changes do occur, and these changes are generally characteristic for each species in a variety of environments. The highly pre-

---

CALEB E. FINCH • Ethel Percy Andrus Gerontology Center and Department of Biological Sciences, University of Southern California, Los Angeles, California 90007.

dictable expressions of aging in mammals and most complex organisms strongly suggest that most phenomena of aging are under a high degree of genetic control. This leads me to consider that relatively few body elements express a genetic program of "primary" aging. In this chapter, I will develop three views: (1) that at least some aging processes are dissociable from chronological time *per se*, (2) that many aspects of aging involve highly selective alterations in cells and their functions, and (3) that the brain plays an important role in regulating some mammalian aging process.

The reader should be aware that my views on aging have substantially different biases than those of some of my colleagues, especially with regard to the relatively *little* importance I ascribe to age-related damage at the genomic level (accumulation of mutations and other disorders in genomic information processing), whose occurrence during aging remains poorly documented.

Causes of the Finite Life Span

Mortality Risk Factors

The finite life span of multicellular organisms derives from two phenomenologically different factors of mortality risk: (1) a relatively fixed (intrinsic?), mortality risk factor that may represent the probability of "accidental" death at any time in the life span and (2) an age-related risk factor, which usually increases exponentially with adult age, the "Gompertz function" (Sacher, 1977). Given a limited cohort, either risk factor leads to a limited life span: if the "fixed risk" factor of human preadolescents were maintained indefinitely as the only risk, then the average human life span would be about 800 years, and the maximum life span would be about 20,000 years (Simms, 1946). The exponentially increasing mortality risk factors are similar in a given genotype or species across a variety of conditions, and in humans they lead to similar maximum life spans throughout the world—110 to 120 years. The causes of increasing mortality are not just the result of microbial infections, because "germfree" conditions (i.e., living without intestinal bacteria) do not increase the maximum longevity of inbred mice; rather, germfree life introduces unexpected dangers, because the absence of the intestinal flora increases the risk of constrictive intestinal looping—a common cause of death in germfree rodents (Gordon *et al.*, 1966).

Although finite life spans associated with exponentially increasing mortality rates are widespread in eukaryotes, there are apparent exceptions. For example, indefinite vegetative (nonsexual) proliferation occurs in some protistans (e.g., *Tetrahymena* strains lacking a micronucleus required for sexual reproduction) and in some plants (e.g., buffalo grass, grape grafts) (Finch, 1976; Thimann and Giese, 1981). These cases suggest that aging processes leading to senescence are *not universal or inevitable* features of eukaryotic cells or organisms. Thus, cellular senescence may have been acquired during evolution. It may be a characteristic that permits some developmental or regulatory processes to be introduced, e.g., the cellular death that occurs in the morphogenesis of the avian wing (Saunders, 1966). It may underlie sex differences of neuronal number in the peripheral nervous system or in the sex-controlling centers of the rodent brain (Arnold and Gorski, 1984). The much-discussed, limited *in vitro* proliferative capacity of diploid human fibroblasts, in contrast to the infinite replicative powers of "established" heteroploid cell lines (Hayflick, 1977), needs critical reevaluation as a model for cell senescence in the organism. Diploid human fibroblasts that have reached their proliferative limit show extensive further viability: some senescent cultures have survived for more than 1 year in a stationary state (Bell *et al.*, 1978). Moreover, proliferatively exhausted fibroblast cultures can support the metabolic demands of virus infections with the same effectiveness as "young fibroblasts" (Holland *et al.*, 1973; Tomkins *et al.*, 1974). Furthermore, many types of nondividing neurons in the human brain show no cell death even in the tenth decade (Brody, 1976). Thus, the finite proliferative ability of some diploid cells cannot be simply equated with cellular senescence. Finite proliferative capacities may be viewed as an aspect of differentiation. Other examples of cells programmed for finite proliferation during differentiation include precursors of red blood cells and antibody-producing plasma cells.

Issues in Molecular Aging

Many researchers have sought molecular correlates of the exponentially increasing mortality rate with age. The life-shortening consequences of ionizing radiation (reviewed in Strehler, 1977) gave major support during the past 20 years to theories that aging was *caused* by an accumulation of random, deleterious mutations (Szilard, 1959) and by other genomic impairments that killed cells and limited the potential duration of cell function. The positive correlation between the capacity for DNA repair (after ultraviolet-induced damage) in primary fibroblast cultures and species longevity (Hart and Setlow, 1974; Tice and Setlow, 1985) is intriguing, but does not prove that mutations actually accumulate during aging. At present, there is little direct supporting evidence that random impairments of genomic function and of macromolecular biosynthesis do occur and accumulate.

At the level of the cell nucleus, there is no evidence for advancing impairments of gene transcription. For example, we characterized rat brain RNA [poly (A) mRNA] at various stages in the animal's life span—2 to 32 months (Colman *et al.*, 1980). No age differences were detected within an experimental error of $\pm 5\%$ (by the technique of RNA-driven hybridization with unique DNA sequence) in the numbers of different types of mRNA (the RNA "complexity"). The amount of mRNA per brain also was not altered with age. Such measurements in whole tissue, of course, cannot identify select cell populations that may have specific patterns of aging. There is evidence for such selective targets of aging in some brain regions, e.g., in the striatum of rodents the total RNA is decreased 15% by the average life span. This region contains 2% of the neurons of the brain (Chaconas and Finch, 1973; Shaskan, 1977). Moreover, some neurons of the human brain show decreasing cell body RNA and nucleolar volume (Mann *et al.*, 1978; Mann and Yates, 1979) with age. The growing evidence for selectivity of age changes in cells of the brain and immune system argues strongly against the view that aging results from global, generalized defects of nuclear genomic function (Finch, 1976).

Some changes of enzyme levels and biophysical properties of enzymes have been observed in aging tissues (Gershon and Gershon, 1973; Rothstein, 1983). For example, liver aldolase from old rats has a decreased thermostability and an increasing (coisolated) pool of immunologically similar (cross-reactive) but enzymatically inactive molecules. The catalytic properties of the active enzymes remain unchanged (Gershon and Gershon, 1973). The age changes in the levels of specific proteins so far described are selective (Finch, 1971, 1976), and this is difficult to ascribe to the result of random genomic damage.

Attempts to detect increasing mistakes or errors in protein synthesis with age have yielded only negative results (Hirsch et al., 1980). Even investigators searching for mutations involving amino acid substitutions in hemoglobin of Marshall Islanders, possibly associated with exposure to radiation 20 years before, have not found such effects (Popp et al., 1976). Furthermore, the accuracy of protein synthesis in reticulocytes does not vary significantly between short- and long-lived species (Hirsch et al., 1980). Similar "negative" outcomes have been obtained from studies of protein charge heterogeneity, as detected on two-dimensional gel electropherograms (Wilson et al., 1978; Harley et al., 1980; Parker and Friesen, 1980). This technique is sensitive enough to detect errors at rates of 1% per codon (Parker and Friesen, 1980). Such results indicate the limits on the extent to which unrepaired errors may accumulate during aging processes.

Nevertheless, final conclusions cannot yet be drawn. It is noteworthy that streptomycin-induced increases in amino acid misincorporation in the bacterium *Escherichia coli* reached a maximum stable value of 50 times greater than normal after six generations, without loss of cell viability or the appearance of obvious abnormalities. However, the treated cells grew more slowly (Edelmann and Gallant, 1977). The discovery of a limiting value of misincorporation was used to argue against the indefinite expansion of an error cascade, leading to an error "catastrophe" according to Orgel's hypothesis (Orgel, 1963). This study suggests that cells have a great capacity to tolerate errors without loss of viability and that there may be yet undefined surveillance mechanisms limiting the growth of error cascades. Of course, extrapolation of these results to characteristics of eukaryotic cells must be done cautiously.

The correlations between UV-induced DNA repair capacity or rates in primary fibroblast cultures and species longevity (Hart and Setlow, 1974; Tice and Setlow, 1985), described above, may be interpreted as an evolutionary adjustment to provide protection over the time scale of the life span. Such correlations cannot prove that errors *do* accumulate during aging. As noted above, error rates of protein synthesis have not yet been correlated with species longevity (Hirsch et al., 1980). Examples of biological reserve or factors of safety have long been known (Meltzer, 1906), and the capacity for repair may be considerably in excess of the average endogenous requirement for maintaining genomic stability. Consequences of errors in amino acid sequence could potentially "cascade" throughout the cell and organism. Excess repair capacity may be an essential feature of self-organization. A variety of repair mechanisms are now being discovered; some of them may involve particular enzymes, but others may involve more basic macromolecular processes, such as rejection of less than "best of fit" interactions at the level of chemical bonding (Hendry *et al.*, 1977; Witham *et al.*, 1978; Hendry and Witham, 1979).

The susceptibility of protein macromolecules to posttranslational changes merits serious attention. Several types of spontaneous change are documented (reviewed by McKerrow, 1979). The *in vivo* "life span" of a specific protein is inversely correlated with the percentage of glutamine and asparagine over a time range of hours to years (Robinson *et al.*, 1970; Robinson, 1974). It appears that spontaneous deamidation occurs in proteins, thereby leading to increased proteolytic degradation. The rate of deamidation may be a molecular clock. In the case of cytochrome-*c*, successive deamidation of asparagine and glutamine residues leads to conformational changes and loss of a major function—the oxidation-reduction activity (Flatmark, 1967). Other studies show that altered proteins are preferentially degraded in bacterial and mammalian cells (Goldberg and St. John, 1976). It is argued that the small extent of asparagine and glutamine in "long-lived" proteins such as collagen and lens crystallins may result from evolutionary selection. In rapidly metabolized proteins, there may be greater choice. However, it cannot be concluded that rapid protein turnover is exclusively a consequence of glutamine and asparagine deamidation.

Another posttranslational change in proteins is progressive racemization, observed in the increase of D-aspartic acid in human tooth enamel (Masters-Helfman and Bada, 1975) and lens crystallins (Masters *et al.*, 1975). The rate of racemization is slow compared to deamidation, leading to accumulation of 10% of total aspartate as D-aspartic residues during the average life span. These considerations may apply to other macromolecular loci with chiral carbon atoms (DeLong and Poplin, 1977). The deamidation and racemization phenomena suggest that posttranslational error correction is accomplished mainly by turnover (new synthesis) in most circumstances. Ongoing biosynthesis may thus be a major strategy in the prevention of senescence in living systems. Molecules that are not renewed regularly, such as the DNA of nondividing cells, may have other, special mechanisms to remove or protect against errors.

Comparative Biological Changes in Aging

A different point of view results from considering comparative aspects of aging. The characteristic increase of mortality with aging in widely different organisms (annual plants, flatworms, mammals) is not, however, accompanied by the same types of physical or cellular changes in each case. It may not be possible to deduce the real causes of death in most cases beyond the correlation with gross pathological lesions. Apparently there is a variety of *different* mechanisms leading to increasing mortality, e.g., some insects lack complete digestive functions as adults and rely on larval foodstores; Pacific salmon at spawning have huge elevations of adrenal steroids, producing effects that are strikingly similar to the adrenal steroid toxicosis of Cushing's disease. These and other cases are reviewed in Finch (1971, 1977) and Van Heukelem (1978).

Because of the phylogenetic diversity in the pathophysiological events of aging, it seems most rewarding to focus on related organisms that share major

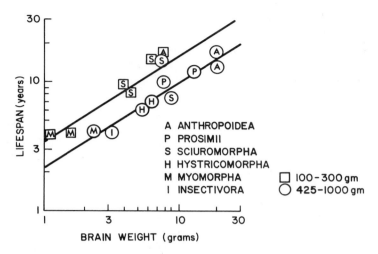

FIGURE 1. Correlation of longevity and brain weight for two size classes. (Redrawn from Sacher, 1975.)

features of development and adult organization. Among mammals, aging changes appear to be broadly similar. It is remarkable that despite nearly 50-fold differences in maximum life span, the major phases and phenomena of age changes throughout life occur in corresponding fractions of the life span, including changes in reproduction, immune functions, thermoregulation, and psychomotor reaction time (Finch, 1976). There are also species differences in aging patterns among mammals, e.g., in hair loss and graying (Finch, 1973a), and in extent of sex hormone production after female reproductive cessation (Finch and Flurkey, 1977; Finch *et al.*, 1984). But species differences in physiological function are found among young adults too. The *scaling* of overall patterns of functions in relation to the life span implies that the timing-of-aging mechanisms are under a high degree of genetic control. Thus, a key issue is to determine the levels of biological organization in which the genotypic control over the pattern of aging resides. There is much discussion of the possibility that the brain may be a major locus of control in aging, as in puberty (Reiter and Grumbach, 1982) and many other physiological processes of adults.

Allometric relationships between longevity and brain and body weight were clearly established by George Sacher. (The ratio of brain weight to body weight is also known as the "index of cephalization".) In a multivariate analysis of 85 species (from all orders except Chiroptera and other taxa that undergo torpor or hibernation), the variance of maximum longevity was most completely accounted for by brain weight (70% of longevity variance), whereas body weight accounted for less than 55%. Remarkably similar correlations are obtained even when various orders (e.g., Rodentia, Artiodactyla, and Anthropoidea) are analyzed separately (Sacher, 1975, 1976) over huge ranges of longevity (40×), brain weight (10,000×), and body weight (1,000,000×). Further analysis of variables with metabolic rate and body temperature yielded a multivariate regression accounting for slightly more (88%) of the longevity variance (Sacher, 1976), illustrated in

Figure 1 for two samples of small mammals with distinctly different body weights.[1] Other allometric relations with longevity are also known, however (Lindstedt and Calder, 1981). These striking correlations give rise to difficult questions about the mechanisms by which the brain could influence or control the maximum longevity or the rate at which the sequence of age changes occurs. Sacher notes that the onset of menarche in the chimpanzee and human follows the same relationship to brain growth as does gestation time. The brain is the most slowly growing organ, and gestation time in a wide range of species increases as the cube root of brain weight at birth (Sacher and Staffeldt, 1974). Thus, it may be considered that the brain either itself controls or is closely linked to the control over the rate of development and aging throughout the mammalian life span. This view is supported by evidence for involvement of the brain in regulation of hormonal changes of age, an example of which will be described later.

Neural and Hormonal Factors in the Regulation of Aging

A fundamental role of neural and humoral (endocrine) mechanisms in regulating aging processes and diseases has emerged as a major premise. The regulation of cell activities through neural and endocrine controls is a cardinal characteristic of mammals and other higher animals, whereas the basic machinery of the nucleated cell itself has changed relatively little during metazoan evolution. Because neural and humoral controls are deeply embedded within nearly all organismic functions, perturbations of any element (molecular, cellular, organ) will have an impact on *some* aspect of the neural–humoral regulatory network. No body element can be considered completely independent of the internal milieu.

Two types of mechanisms of aging have been postulated: (1) those in which an *intrinsic*, autonomous cellular or molecular aging change occurs and (2) those in which changes are regulated by interactions with *extrinsic* neural, humoral, or other physiological factors. In mammals, both types of mechanism could involve cascading effects or causal chains (Finch, 1976; Finch *et al.*, 1984). Thus, changes in a hypothalamic locus might alter pituitary functions and thereby influence various endocrine target cell functions. These changes would be expected to involve further responses from feedback or homeostatic mechanisms. At present it is not known whether intrinsic cellular or molecular aging events underlie *any* neural or endocrine extrinsic changes. In some cases, cellular functions allegedly showing "age" changes respond normally when directly acting stimuli are applied. For example, during adaptation to various stresses, the induction of liver tyrosine aminotransferase and glucokinase is delayed in aging rodents. But these enzymes can be rapidly induced in old animals by hormones, such as insulin or corticosterone, that act directly on the liver cells (Finch *et al.*, 1969; Adelman, 1970;

---

[1]Although these relationships suggest that longevity is a continuously varying function, analysis of brain–body size relationships showed that no mammal had a brain–body weight ratio greater than 4% (Sacher, 1975). This result suggests that there are upper limits to the index of cephalization, above which survival as an embryo or adult is unlikely.

Adelman *et al.*, 1978; Finch, 1979). The delayed induction during adaptation to stress appears to reflect altered hormonal controls, rather than an intrinsic defect in the old liver. More direct evidence for altered neural regulation of hepatic enzymes in old rodents is the much reduced increase of liver phosphorylase α after electrical stimulation of the ventromedial hypothalamic nucleus of 24-versus 2-month-old rats (Shimazu *et al.*, 1978; Shimazu, 1980). Identification of neural and humoral involvement in even a wide variety of aging changes probably will not completely explain or describe mammalian aging processes. In some cases, we may still need to identify the extent of the "residual" process not accounted for by neural and humoral influences (whether immediate or delayed). Moreover, internal cellular compensation may correct for some defects and cause them to be hidden in experiments of this kind.

There are few age changes of hormones as dramatic as the great decrease (more than 90%) of estradiol and progesterone production after menopause in humans; male mammals do not usually have a major drop in testosterone production (Finch and Flurkey, 1977). Another example is the major decrease in dehydroepiandrosterone (DHEA) observed in men and women (Orentreich *et al.*, 1984). Of equal importance is that many neural and endocrine factors show shifts in the patterns of secretion, involving changes in timing of secretion, rate of clearance, and amplitude of discharge or secretion—e.g., growth hormone (Sonntag *et al.*, 1980). These shifts are seen in diurnal patterns and in secretion thresholds as observed for insulin, growth hormone, gonadotropins and other pituitary hormones, glucocorticoids, and mineralocorticoids (Finch, 1976; Andres and Tobin, 1977; Adelman *et al.*, 1978; Finch and Landfield, 1985; Finch, 1987).

In some cases, changes in the regulation of neurohumoral factors occur progressively throughout the adult life span: the thyroid-dependent component of minimal oxygen consumption decreases gradually in adult rats (Denckla, 1974), as do the levels of thymic hormones (Goldstein *et al.*, 1974; Lewis *et al.*, 1978). In other cases, changes are relatively abrupt. The major decrease at puberty of the thyroid-dependent component of minimal oxygen consumption in rats (Denckla, 1974), the large decrease of DHEA (mentioned above) which often occurs between 20 and 40 years (Samuels, 1956; Yamaji and Ibayashi, 1969; Orentreich *et al.*, 1984), and the large decreases of estradiol and progesterone after menopause (Finch and Flurkey, 1977) are all examples of midlife changes. It should be stressed that changes in neural and humoral regulation are not restricted to the end of life and that an understanding of neurohumoral causality cannot be obtained only from study of the terminal stages of aging.

Taken together, the presence of alterations in neural and humoral mechanisms suggests that there may be no fixed set of homeostatic characteristics that is stable throughout the adult life span. This view does not necessarily challenge the fundamental homeostatic axioms of physiology; however, it does raise the possibility that the homeostatic "envelope" varies considerably during adult life. That is, there may be a set of metastable optima within which homeostatic tendencies operate, rather than a unique homeostatic configuration kept throughout adult life.

A substantial variety of age changes in cell functions are identified with changes in hormones or neural factors (Finch, 1976; Adelman, 1977). A key issue is the extent to which these changes form *causal chains* with ramifying and cascading consequences in other systems. As examples of cascades, note that the loss of estradiol after menopause in women causes an increase of gonadotropin-releasing factor production by the hypothalamus and a subsequent increase in pituitary output of gonadotropin hormones. The loss of estradiol also causes a reduction in the morning output of growth hormone (Frantz and Rabkin, 1965)—a diminution that could influence diverse metabolically sensitive loci. It is possible that a limited number of neural and endocrine loci have a major role in establishing the preconditions for a broad spectrum of age-related diseases.

Another example of a hormonal effect cascade leading to disease is associated with reductions of DHEA. DHEA, a major plasma steroid, is also a precursor of sex steroids and urinary 17-ketosteroids. Low production of DHEA during middle age was prospectively identified as a risk factor in human mammary cancer (Bulbrook *et al.*, 1971). Administration of DHEA to the C3H mouse strain reduced its high spontaneous mammary-tumor incidence from more than 50% to negligible levels (Schwartz, 1979). The mechanisms are obscure but could involve any of several different stages in tumorigenesis. DHEA directly and noncompetitively inhibits glucose-6-phosphate dehydrogenase (Oertel and Benes, 1972) and, hence, reduces the availability of the coenzyme NADPH. Less NADPH, in turn, lowers the metabolic activation of carcinogens (e.g., DMBA and aflatoxin B1) by reducing the activity of the mixed-function oxidases that require NADPH (Schwartz and Perantoni, 1975). These effects of DHEA on mixed-function oxidases lead also to the possibility that some aspects of mutagenesis may be subject to extrinsic, humoral influences. Another consequence of lowered NADPH is reduced lipogenesis, which also prevents obesity in mice of two strains without affecting food consumption (Yen *et al.*, 1977; Schwartz, 1979).

Hormonal influences on thymic changes during aging could also involve a cascade leading to preconditions for pathogenesis. The causes of the major reductions of thymic function and thymic hormones during aging are obscure. Because the hypothalamus and pituitary influence immunological functions at many levels (Piantanelli *et al.*, 1980; Fabris, 1982; Marx, 1985) a neuroendocrine basis for immunologic changes with age also seems possible. The networks of interactions that may regulate immune responses (Jerne, 1976) could interface with neural and humoral controls.

Not all neural and humoral changes can have ramifying consequences, because the degree of coupling between regulatory networks must vary and may be relatively slight in many cases. The analysis of physiological network interactions poses major theoretical problems for which there are few penetrating models. If many neurohumoral systems have shifting regulatory characteristics throughout life, it must be asked how the shifts are related and if there are major common denominators of change in physiological regulators that could be viewed as pacemakers of aging. Few such regulators are well understood in mammalian development. Although specific hormonal cues or requirements for development are known for thymus-dependent immunological development and sexual maturation, the triggers for their onset remain unclear. The major transitions of aging are even less well described than are those of postnatal development.

Circadian organization is another problem with some aspects similar to those of rate-controlling events in development and in aging. Is there only one pacemaker, or are there nested (hierarchical) pacemakers? If there are multiple pacemakers, are they in mosaic (noninteractive) or heterarchical (competing) pacemaker systems? It now appears that most mammalian 24-hr rhythms are entrained through the suprachiasmatic nucleus of the hypothalamus, which hierarchically controls diverse other systems through neural and humoral effectors and is entrained by either light–dark cycles or food availability cycles (Moore, 1978).

Temporal Plasticity in Cellular and Physiological Aging in Mammals

The premise that some cellular aging changes result from extrinsic factors rather than intrinsic aging processes leads to a major prediction: *if cellular aging changes are caused by extrinsic (e.g., neurohumoral) factors, it should be possible to experimentally manipulate the timing of age changes in either direction, to slow or accelerate aging.* The variations of timing in age changes observed in comparing long- and short-lived mammals also suggest an extensive *temporal plasticity* or *dissociability*, whose scaling is subject to modification during evolution by relatively few factors or genes. In some cases, as described below, experimental control has been achieved over the timing of cellular and physiological age changes; these studies confirm that some cellular aging processes may result from changes of extrinsic factors and are not strictly time-locked because of intrinsic cellular or molecular aging processes.

Pituitary Thyroid Manipulation

Maturation in rats has been associated with a new pituitary activity, according to Denckla and colleagues. The decreasing calorigenic response to thyroxine (Denckla, 1974) and the decreased β-adrenergic relaxation of the aorta (Parker *et al.*, 1978), among other age changes, can be arrested (apparently indefinitely) by prepubertal hypophysectomy. They are clearly not strictly time-dependent. In adults, these changes are reversed some long time after hypophysectomy. Denckla hypothesizes that the pituitary secretes a hormone (so far unidentified) that impairs diverse cell responses to thyroxine and other hormones. The secretion of this putative hormone appears to be triggered prepubertally by the thyroid (Denckla, 1974).

Dietary Manipulation

Moderate dietary restriction in rodents not only increases longevity and postpones the onset of kidney diseases and other major lesions but also modifies specific cellular responses. A striking example is seen in cellular immune responses of aging mice, which can be modified even if diet restriction is begun at midlife (Weindruch *et al.*, 1982). Similarly, the loss of hormonally stimulated lipolysis in fat cells can be postponed 12 months or more by restricting food intake (Bertrand *et al.*, 1980). Age-related changes in tissue responses to thyroxine are also postponed by dietary restriction (Denckla, 1974). The mechanisms involved are obscure, but nutritional influences on the pituitary are a possibility. Other changes in hormonal responsiveness of target cells, such as the reduction in number of corticosteroid receptors and of glucose oxidation by rat adipocytes (Roth and Livingston, 1976), may also prove to be experimentally manipulable through diet. The age-related loss of dopaminergic receptors in the neostriatum of the rodent brain, another change that can be detected by midlife (Severson and Finch, 1980; Severson *et al.*, 1982; Morgan *et al.*, 1987), was reduced by alternate day feeding (Levin *et al.*, 1981).

Steroid-Dependent Hypothalamic Aging

In most laboratory rodent strains, ovulatory (estrous) cycles become less frequent with aging and eventually cease altogether, as shown in Figure 2 (top) for the C57BL/6J mice studied in my laboratory. Individual profiles from a longitudinal study of estrous cycles are shown in Figure 3. Although the ovary irreversibly loses oocytes, a substantial number, at least several hundred, of potentially functional primordial follicles and oocytes are still present after cycles lengthen and cease in rodents (Jones and Krohn, 1961a; Gosden *et al.*, 1983). Postmenopausal women may also retain substantial numbers of potentially functional oocytes and follicles, although the status of this claim is controversial (reviewed in Finch, 1976). The most common, initial noncycling state of aging rodents (persistent vaginal cornification) differs significantly from human menopause (Finch and Flurkey, 1977), because in mice, substantial amounts of ovarian estrogens are still produced (Nelson *et al.*, 1981). Subsequently, aging rodents reach a status similar to menopause ("persistent anestrus") in which blood levels of ovarian steroids are very low and in which levels of gonadotropins are elevated (Gee *et al.*, 1983).

A major cause of the loss of cycles in aging rodents is an impaired ability of the hypothalamus to produce the preovulatory surge of gonadotropins in response to the rising levels of estradiol produced by the ovarian follicles (Mobbs *et al.*, 1984a,b). Despite the loss of most (greater than 75%) ovarian oocytes by 12 months, normal numbers of eggs can still be shed at ovulation (Harman and Talbert, 1970; Gosden *et al.*, 1983) and blood levels of estradiol at proestrus are normal (Nelson *et al.*, 1980). This loss of hypothalamic responsiveness to estradiol can also be shown in mice whose ovaries were removed, followed by injection of estradiol and other steroids to simulate the conditions of the intact mouse just before ovulation (Mobbs *et al.*, 1984a; Finch *et al.*, 1984). As illustrated in Figure 2 (bottom), there is a progressive reduction of estradiol-induced surges of luteinizing hormone (LH) that generally parallels the reduced frequency of cycles as a function of age. The impairment appears to reside at the hypothalamic level, because in aging female rodents, at the time cycles cease, the pituitaries are still responsive to the gonadotropin-releasing hormone LH-RH (reviewed in Finch *et al.*, 1984).

The cause of the lengthening of ovulatory cycles as rodents approach midlife

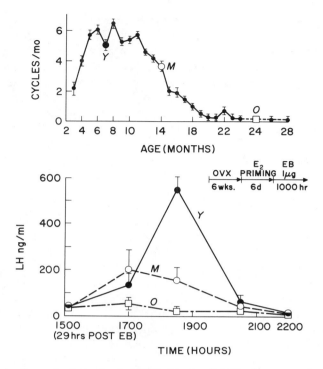

FIGURE 2. Events of reproductive aging in C57BL/6J mice. (A): The frequency of estrous (ovulatory) cycles as determined from longitudinal studies of the vaginal cell types (vaginal smears) (Nelson et al., 1980). See also Figure 3. (B): The estradiol (E_2)-induced surge of luteinizing hormone (LH). Mice were ovariectomized and injected with a pattern of E_2, which induces a surge of gonadotropins very much like that of the preovulatory surge at proestrus. The progressive impairments of the LH output generally paralleled the reduced frequency of cycling (see A) (Finch et al., 1980; Mobbs et al., 1984a).

is unknown. It appears that old mice with longer cycles require commensurately longer times to reach the same elevation of plasma estradiol that triggers ovulation in young mice (Nelson et al., 1980). The delay could derive from ovarian aging changes (e.g., fewer growing follicles) or from hypothalamic-pituitary aging (e.g., reduced output of gonadotropins from the preceding proestrous stage). Recent evidence for prenatal influences on cycle length in rodents (vom Saal and Bronson, 1980b) suggests a hypothalamic locus, as discussed below.

Further support for a role of hypothalamic loci in the loss of cycles is derived from studies showing that ovulatory cycles can be transiently reactivated in intact but aging rodents by a variety of maneuvers, including systemic injections of progesterone (Everett, 1939), L-dopa, and other catecholaminergic agonists or electric stimulation of the hypothalamus (Meites et al., 1978). Implantation of the catecholamine precursor L-dopa into the brain in the preoptic region of the hypothalamus also reactivates cycles (Cooper et al., 1979). (The preoptic area is important in the regulation of rodent estrous cycles.) This result suggests that local, age-related alterations of the hypothalamic loci contribute to the loss of cyclicity. The ability of adrenergic (catecholaminergic) drugs to reactivate cycles has led to the hypothesis that hypothalamic neurotransmission in catecholaminergic neural pathways is altered during aging (Finch, 1972, 1976; Quadri et al., 1973). A critical role for hypothalamic catecholamine neurotransmission in the control of the preovulatory surge of gonadotropins is indicated by much evidence, particularly that for a relationship between norepinephrine and LH (Weiner and Ganong, 1978; Barraclough and Wise, 1982). Although reductions in catecholamine levels, turnover, and receptors do occur in aging rodents (Finch, 1973b, 1979; Simpkins et al., 1977; Makman et al., 1978; Severson and Finch, 1980; Wise, 1982), it is very difficult to prove that these changes are sufficient causes of cycle loss, because of the multiplicity of neurotransmitter and hormonal influences on hypothalamic functions. Additional evidence for an extraovarian locus in reproductive aging comes from ovarian transplantations between hosts and donors of different ages: in general, young ovaries do not cycle when grafted to old, previously noncycling hosts (Aschheim, 1965, 1976; Peng and Huang, 1972; Felicio et al., 1983; Mobbs et al., 1984a).

The hypothalamic age changes that cause the loss of estrous cycles appear to be *themselves* regulated by exposure to ovarian steroids. This portentous possibility derives from studies showing that the loss of cycles is not strictly age-related, but may be slowed or accelerated in relation to steroid exposure. Aschheim was the first to report the intriguing finding that long-term ovariectomized rodents (animals whose ovaries were removed at 2 to 4 months of age), when given young ovaries even late in life (24 to 30 months), can sustain normal estrous cycles (Aschheim, 1965, 1976). We have recently verified these results for C57BL/6J mice (Nelson et al., 1980; Felicio et al., 1986; Mobbs et al., 1984a). The lengths of cycles of old, long-term ovariectomized mice with young ovarian grafts are very similar to those in "young to young" control transplants and, in most cases, continued until just before death. These results imply ovary-depen-

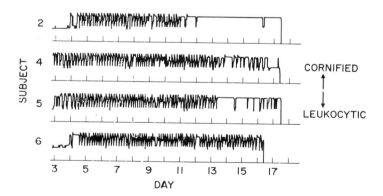

FIGURE 3. Individual patterns of estrous cycles during aging in C57BL/6J mice. In a longitudinal study of mice throughout their reproductive life spans, the cell types present in daily vaginal smears indicated the hormonal state. In the printout of data, the various cell type distributions (ranging from cornified to leukocytic) are represented by varying heights. (Redrawn from Finch et al., 1980.)

dent aging processes in the hypothalamus that are interruptible by removal of ovarian steroids.

Conversely, reproductive senescence can be accelerated by sustained exposure to estradiol. Estradiol implants that elevate blood levels for > 2 months in young rodents cause the rapid loss of cycles (Brawer et al., 1978; Mobbs et al., 1984b) and permanent impairments in LH regulation (Brawer et al., 1980; Mobbs et al., 1984b; Finch et al., 1984). The results of the long-term ovariectomy studies when considered with these studies then imply a critical limit or threshold for exposure to steroids which, when reached, prevents normal female hypothalamic responses to estradiol (Figure 4). About the same cumulative exposure to estradiol in adults may result in loss of estrous cycles, whether the exposure is experienced through the 50 or so estrous cycles of normal rodent reproductive life or if the dose is accumulated over a shorter term (Finch et al., 1984). Recent data indicate that ovary-induced neuroendocrine aging can be separated into effects on cycle lengthening and cycle cessation (Felicio et al., 1986; Finch et al., 1984). The particular effects of ovarian steroids associated with cycle cessation do not appear to be cumulative over all cycles.

It may be possible to determine estradiol "strength duration" functions (e.g., plasma estradiol integrated over time) that yield the hypothesized critical threshold for damaging effects of estradiol. Preliminary data suggest that the estradiol dose for irreversible hypothalamic damage in mice is in the range 1000–3000 pg-days estradiol/ml blood. Such a putative "strength duration" function for estradiol would obviously not be commutative for all values, e.g., a physiological lifetime of estradiol exposure could not likely have the same effects if experienced in a millisecond. These properties of reproductive aging promise a quantitative approach to defining a neural pacemaker system for aging changes. It is unknown whether the variations in estrogens or other aging-related hormonal changes act on the hypothalamus directly or indirectly, via other hormones, neural factors, or metabolites; estradiol-induced growth factors occur in the brain (Sirbasku, 1978). Hypothalamic aging promises a deep set of problems at molecular, cellular, and physiological levels.

Glial hyperactivity in the arcuate nucleus of the hypothalamus during reproductive aging was found to be prevented by long-term ovariectomy in female rats and in C57BL/6J mice (Schipper et al., 1981). Conversely, glial hyperactivity can be precociously induced in young rats by sustained exposure to estradiol (Brawer et al., 1978). Although the glial morphological changes apparently follow the same formal rules as those leading to the presence or absence of estrous cycles, it remains unknown whether the associated arcuate neurons affected are critically involved in the regulation of cycles. It is also unknown if exposure to estradiol (exogenous or endogenous) causes death of neurons in the hypothalamus, or degeneration of axon terminals and dendrites. Many other changes that follow intense exposure of young rodents to estradiol are also very similar to those of spontaneous aging (Finch et al., 1984). Taken together, these phenomena can be described as an ovary-dependent, estradiol-induced neuroendocrine aging syndrome (Finch et al., 1984).

Exposure to steroids during development influences reproductive aging in

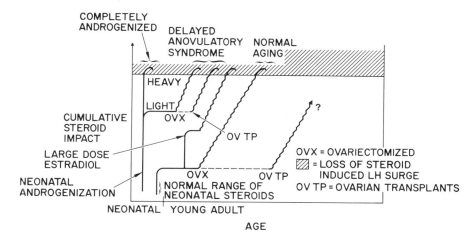

FIGURE 4. A scheme for the cumulative impact of steroids on reproductive development and aging. In this diagram, it is assumed that exposure to a certain (threshold) amount of steroid (probably estradiol) results in the loss of estrous cycles and of the ability of the hypothalamus to produce an LH surge in response to a short-term (2- to 3-day) elevation of estradiol E_2 (see Figure 2). Alteration of hypothalamic functions by E_2 or other aromatizable steroids, such as testosterone, is a normal mechanism in sexual differentiation of the hypothalamus by which males lose the ability to produce an LH surge. Perinatal exposure to steroids, even if subthreshold, can influence the age when cycles are lost in the adult. According to this scheme, the E_2 surges of successive estrous cycles add cumulatively to the perinatal exposure. The primary events of reproductive aging at a hypothalamic level are thus steroid-dependent. For further details, see Finch et al. (1984).

rodents. If neonatal rodents are given large doses of estrogens, or testosterone and other steroids that can be converted by brain enzymes to estrogens ("aromatizable" steroids), then the treated rodents lack cycles as adults, as shown by the failure to respond to rising estradiol with the normal LH surge. In many aspects, then, ovarian function resembles the constant metestrus of normal, aging, old rodents.

Neonatal exposure to steroids is a normal trigger leading to the sexual "differentiation" of the hypothalamus and other brain loci. The active hormones leading to the male-type rodent brain are (surprisingly) estrogens, rather than androgens. Under the influence of estrogens, during development the preoptic region of the hypothalamus acquires altered synaptic and neuronal density (Arnold and Gorski, 1984). However, if small doses of steroids are given to neonatal rodents, these rodents go through puberty and exhibit a number of apparently normal cycles, but then prematurely enter constant metestrus (Swanson and van der Werff ten Bosch, 1964; Harlan and Gorski, 1978; Mobbs et al., 1985). Ovariectomy postpones the loss of cyclicity (Harlan and Gorski, 1977). Therefore, it appears that steroid exposure during cycles in adult rodents can sum with a "background" effect from steroid influences during development. A major question concerns whether the same hypothalamic loci are affected by steroids during development and adult life.

Variations in endogenous steroids between individuals may also influence their patterns of aging. For example, a female pup flanked *in utero* by two male pups has a higher steroid level than does a female flanked by two females (vom Saal and Bronson, 1980a) and exhibits more aggressive adult behavior (vom Saal and Bronson, 1978) as well as longer estrous cycles (vom Saal and Bronson, 1980b). Of great interest is that the sex of the fetal neighbor *in utero* also influences the age when fertility is lost (vom Saal and Moyer, 1985) and age changes in behavioral responses to sex steroids (Rines and vom Saal, 1984). Thus, prenatal exposure to steroids may influence later events of aging (Figure 4). We propose that there is a lifelong continuum of steroid influences on the mammalian brain, in which steroid exposure during development can influence the subsequent amount of steroid exposure needed to cause specific age changes (Finch *et al.*, 1980, 1984). Such mechanisms may also account for the great phenotypic variability seen in reproductive aging patterns of inbred mice, in which some individuals cease cycling 6 months or more before others (Figure 3).

Considered together, these studies suggest that the aging of some hypothalamic loci critical to ovarian cycles is not strictly *time-dependent, but is ovary-dependent* and may be controlled (accelerated or delayed) by exposure to ovarian steroids (Figure 4). These estrogen effects may be an adult analogue of the neonatal masculinization process, although it is not clear that the same loci are affected. Age changes in the rodent hippocampus, including neuronal locus, also result from steroid effects, since they are retarded by adrenalectomy (Landfield *et al.*, 1978) and accelerated by corticosterone treatment (Finch and Landfield, 1985; Sapolsky *et al.*, 1985). It is thus possible that many age-related changes in the brain cells are not intrinsic, but rather result from interactions with endogenous factors.

Another ramification of the ovary–hypothalamus–pituitary interactions is the dependency of ovarian oocyte loss on active pituitary function. Hypophysectomy (pituitary ablation) of neonatal rodents greatly retards the loss of ovarian oocytes (Jones and Krohn, 1961b). Thus, hypothalamic aging depends on ovarian function, and ovarian aging depends on hypothalamic–pituitary function. This fundamental reciprocity between ovarian and neuroendocrine aging processes points to the interdependency of age changes in cascading interactions. It seems likely that immune–neuroendocrine interactions during aging may be similarly interconnected.

A further consequence of these interactions may be the increase in abnormal fetal development with maternal age. Considerable evidence shows that when estrous cycles are prolonged in rodents, thereby delaying ovulation, fetal abnormalities also increase substantially, as monitored by the incidence of lethal malformations and karyotypic abnormalities by midterm in the developing fetus. Abnormal development is also associated with experimentally induced delays of ovulation (Butcher *et al.*, 1969; Kram and Schneider, 1978), as well as spontaneous delays found in the prolonged cycles of aging rats (Fugo and Butcher, 1971). In women, delay of ovulation to day 15 or later drastically reduces the chance of normal conception (Hertig, 1967). In the present context, the increasing length of the estrous cycle in aging rodents may be postulated to result from hypothalamic aging events, which are expressed ultimately as increased fetal abnormalities during pregnancy in older mothers.

The above examples of age changes that can be manipulated independently of chronological age argue powerfully against intrinsic cell and molecular aging phenomena in *these* systems. It is possible that "intrinsic" aging processes do occur in some loci, but it seems unlikely that intrinsic aging changes are widely distributed. In the future, the long-standing designation of processes or changes as "age-dependent" may be less meaningful and may be more accurately replaced by "event-dependent" changes that occur during chronologic time. Biological time, as applied to some aging processes, may therefore also be regarded as a dependent variable.

Temporal Plasticity in Teleost Senescence

The temporal plasticity of some age-related phenomena in mammals also is observed in other vertebrates. As mentioned earlier, in Pacific salmon the adrenal cortex becomes hyperactive during maturation. A syndrome of adrenal steroid toxicosis that occurs is strikingly like Cushing's disease. Death usually follows within 2 weeks of spawning. Castration before maturation in kokanee salmon prevents the adrenal hyperactivity; the castrates continue to grow and survive to at least twice the normal maximum life span (Robertson, 1961). The life span of the European eel, which also usually dies at spawning, can be extended three- to fourfold if eels are prevented from returning to the sea (Lekholm, 1939; Bertin, 1956; Finch *et al.*, 1969). Although there are as yet no correspondingly large manipulations of the mammalian life span (see Sacher, 1977), these examples, when considered with the very similar patterns of aging in short- and long-lived mammals, support the possibility that the genetic program for the timing of aging events is subject to intervention via physiological regulatory molecules in a variety of animals. Such temporal plasticity or dissociability also characterizes some developmental processes. For example, in primates the cellular changes of puberty can be induced at any age over a large time range. Puberty can appear in infants (Tanner, 1969; (Wildt *et al.*, 1980) (precocious puberty) or as late as age 30 or more years in cases of ovarian dysgenesis. Evidently, hormonal target cells in the reproductive system can respond to their hormonal stimuli as soon as they have differentiated, and their responses are not lost if their normal hormonal stimuli are long delayed.

Aging and Heterochrony in Evolution

The temporal plasticity of some developmental and aging processes leads me to consider their similarity to the phenomenon of *heterochrony*, as discussed in evolutionary theory, in which changes in the timing of developmental events are considered to be a major substrate for evolutionary change (Gould, 1977). Comparisons of developmental and adult stages in fossil and extant species give numerous examples of shifts in developmental timing; dissociations of various developmental processes can lead to earlier or later onset of the differentiation of adult characters and greater or lesser extents of growth. Such species variations in timing can be used to classify phylogenetic diversity in development, such as accelerated or retarded sexual maturation with respect to somatic development. Consideration is given to how variations in developmental timing may be a major source of "apparently" new characteristics in evolution. Changes at relatively few genetic loci influencing the developmental timing of specific processes may be involved in these phylogenetic differences.

The above discussions suggest a number of characteristics that are shared by biological aging processes and evolutionary processes:

1. In both aging and evolution, the timing of events during the life span is subject to dissociability. That is, considerable flexibility in the timing of developmental aging events can occur between individuals and species.

2. The signals leading to changes in cell function during aging or the signals that influence the differences in timing of developmental events between species appear to be extrinsic to the "target" cells; extrinsic signals are regulated by neuroendocrine controls or by intercellular contacts.

3. Because the fractions of the life spans in which postnatal development and aging events occur are similar among mammalian species with major variations in life span, the *scaling* mechanisms regulating the timing of developmental and aging events may be relatively invariant in mammalian evolution. It may be considered that the putative scaling mechanisms are controlled by the *same* genetic loci involved in the allometric relationships between longevity, body size, brain size, and metabolic rate. Thus, there may be a set of DNA sequences which have important roles shared in the evolution, development, and aging of mammals.

4. Additionally, I suggest that the genomic mechanisms that determine species differences in the timing of developmental events may involve the same mechanisms as those that control the rates of various aging processes. For example, as described above, the role of steroid interactions with the hypothalamus in sexual differentiation and in aging appears to involve a similar mechanism. As the genetic loci controlling various developmental processes become understood, these analyses may then lead to easier recognition of the controls of aging and may directly indicate the loci influencing heterochronic variations in evolution.

Studies on the eukaryotic genome continue to support theories of hierarchical mechanisms of gene regulation in which repetitive DNA sequences have a major role (Davidson and Britten, 1979). If a relatively few (less than 1000) DNA sequences pleiotropically influence many different gene-dependent processes, then it seems plausible that the heterochronic differences between the development and aging of evolutionarily kindred species (e.g., those of class Mammalia) could result from alterations of such genetic controls. In the future, when the mechanisms of gene regulation in eukaryotes are better understood, it may be possible to identify which features of DNA are involved in heterochrony. Some possibilities include specific base sequences, the organization of specific repetitive and nonrepetitive sequences, or the stoichiometry among specific repetitive sequences in relation to structural genes. In summary, elucidation of how genetic structures influence species differences in the timing of developmental processes (heterochrony) may provide major insights into the control of aging processes as well.

References

Adelman, R. C. (1970) An age dependent modification of enzyme regulation. *J. Biol. Chem.* **245**:1032–1036.

Adelman, R. C. (1977) Macromolecular metabolism during aging. In: *Handbook of the Biology of Aging*, C. E. Finch and L. Hayflick (eds.). Van Nostrand, Princeton, N.J., pp. 63–72.

Adelman, R. C., G. W. Britton, S. Rotenberg, L. Ceci, and K. Karoly (1978) Endocrine regulation of enzyme activity in aging animals of different genotypes. In: *Genetic Effects on Aging*, D. Bergsma and D. E. Harrison (eds.). Original Article Series Vol. 14, No. 1. Liss, New York, pp. 355–364.

Andres, R., and J. Tobin (1977) Endocrine systems and aging. In: *Handbook of the Biology of Aging*, C. E. Finch and L. Hayflick (eds.). Van Nostrand, Princeton, N.J., pp. 357–378.

Arnold, A. P., and R. A. Gorski (1984) Gonadal steroid induction of structural sex differences in the central nervous system. *Annu. Rev. Neurosci.* **7**:413–442.

Aschheim, P. (1965) Resultats fournis par la greffe heterochrone des ovarie dans l'etude de la regulation hypothalamus-hypophyso-ovarienne de la ratte senile. *Gerontologia* **10**:65–75.

Aschheim, P. (1976) Aging in the hypothalamic-hypophyseal-ovarian axis in the rat. In: *Hypothalamus, Pituitary, and Aging*, A. V. Everitt and J. A. Burgess (eds.). Thomas, Springfield, Ill., pp. 376–416.

Barraclough, C. A., and P. M. Wise (1982) The role of catecholamines in the regulation of pituitary luteinizing hormone and follicle stimulating hormone secretion. *Endocr. Rev.* **3**:91–119.

Bell, E., L. F. Marek, D. S. Levinstone, C. Merrill, S. Sher, I. T. Young, and M. Eden (1978) Loss of division potential in vitro: Aging or differentiation? *Science* **202**:1158–1163.

Bertin, L. (1956) *Eels: A Biological Study*. Cleaver-Hume Press, London.

Bertrand, H. A., E. J. Masoro, and B. P. Yu (1980) Maintenance of glucagon-promoted lipolysis in adipocytes by food restriction. *Endocrinology* **107**:591–595.

Brawer, J. R., F. Naftolin, J. Martin, and C. Sonnenschein (1978) Effects of a single injection of estradiol valerate on the hypothalamic arcuate nucleus and on reproductive function in the female rat. *Endocrinology* **103**:501–512.

Brawer, J. R., K. B. Ruf, and F. Naftolin (1980) Effects of estradiol-induced lesions of the arcuate nucleus on gonadotropin release in response to preoptic stimulation in the rat. *Neuroendocrinology* **30**:144–149.

Brody, H. (1976) An examination of cerebral cortex and brainstem aging. In: *Neurobiology of Aging*, R. D. Terry and S. Gershon (eds.). Raven Press, New York, pp. 177–181.

Bulbrook, R. D., J. L. Hayward, and C. C. Spicer (1971) Relation between urinary androgen and corticoid excretion and subsequent breast cancer. *Lancet* **2**:395–398.

Butcher, R. L., J. D. Blue, and N. W. Fugo (1969) Overripeness and the mammalian ova. III. Fetal development at midgestation and at term. *Fertil. Steril.* **20**:223–231.

Chaconas, G., and C. E. Finch (1973) The effect of aging on RNA/DNA ratios in brain regions of the C57BL/6J male mouse. *J. Neurochem.* **21**:1469–1473.

Colman, P. D., B. B. Kaplan, H. H. Osterburg, and C. E. Finch (1980) Brain Poly(A)RNA during aging: Stability of yield and sequence complexity in two rat strains. *J. Neurochem.* **34**:335–345.

Cooper, R. L., S. J. Brandt, M. Linnoila, and R. F. Walker (1979) Induced ovulation in aged female rats by L-dopa implants into the medial preoptic area. *Neuroendocrinology* **28**:234–240.

Davidson, E. H., and R. J. Britten (1979) Regulation of gene expression: possible role of repetitive sequences. *Science* **204**:1052–1059.

DeLong, R., and L. Poplin (1977) On the etiology of aging. *J. Theor. Biol.* **67**:111–120.

Denckla, W. D. (1974) Role of the pituitary and thyroid glands in the decline of minimal O_2 consumption with age. *J. Clin. Invest.* **53**:572–581.

Edelmann, P., and J. Gallant (1977) On the translational error theory of aging. *Proc. Natl. Acad. Sci. USA* **74**:3396–3398.

Everett, J. W. (1939) Spontaneous persistent estrous in a strain of albino rats. *Endocrinology* **25**:123–127.

Fabris, N. (1982) Neuroendocrine-immune network in aging. In: *Developmental Immunology: Clinical Problems and Aging*, Academic Press, New York, pp. 291–298.

Felicio, L. S., J. F. Nelson, R. C. Gosden, and C. E. Finch (1983) Long-term ovariectomy delays the loss of ovulating cycling potential in aging mice. *Proc. Natl. Acad. Sci. USA* **80**:6076–6080.

Felicio, L. S., J. F. Nelson, and C. E. Finch (1986) Prolongation of cessation of estrous cycles in aging C57BL/6J mice and differentially regulated events. *Bio. Repro.,* **34**:849–858.

Finch, C. E. (1971) Comparative biology of senescence: Some evolutionary and developmental considerations. In: *Animal Models for Biomedical Research, No. IV.* National Academy of Sciences, Washington, D.C., pp. 47–67.

Finch, C. E. (1972) Cellular pacemakers of ageing in mammals. In: *Proc. 1st European Conference on Cell Differentiation (Nice, 1971).* R. Harris and D. Viza (eds.). Munksgaard, Copenhagen, pp. 123–126.

Finch, C. E. (1973a) Retardation of hair growth, a phenomenon of senescence in C57BL/6J male mice. *J. Gerontol.* **28**:13–17.

Finch, C. E. (1973b) Catecholamine metabolism in the brains of ageing, male mice. *Brain Res.* **52**:261–276.

Finch, C. E. (1976) The regulation of physiological changes during mammalian aging. *Q. Rev. Biol.* **51**:49–83.

Finch, C. E. (1979) Neuroendocrine mechanisms and aging. *Fed. Proc.* **38**:178–183.

Finch, C. E., (1987) Neural and endocrine determinants of senescence. In: *Modern Biological Theories of Aging,* H. Warner (ed.) Raven, N.Y., pp. 261–306.

Finch, C. E., and K. Flurkey (1977) The molecular biology of estrogen replacement therapy. *Contemp. Obstet. Gynecol.* **9**:97–107.

Finch, C. E., and P. W. Landfield (1985) Neuroendocrine and autonomic functions in aging mammals. In: *Handbook of the Biology of Aging,* 2nd ed., C. E. Finch and E. L. Schneider (eds.). Van Nostrand, New York, N. Y., pp. 79–90.

Finch, C. E., J. R. Foster, and A. E. Mirsky (1969) Ageing and the regulation of cell activities during exposure to cold. *J. Gen. Physiol.* **54**:690–712.

Finch, C. E., L. S. Felicio, K. Flurkey, D. M. Gee, C. Mobbs, J. F. Nelson, and H. H. Osterburg (1980) Studies on ovarian-hypothalamic-pituitary interactions during reproductive aging in C57BL/6J mice. *Peptides* **1** (Suppl. 1): pp. 163–175.

Finch, C. E., L. S. Felicio, C. V. Mobbs, and J. F. Nelson (1984) Ovarian and steroidal influences on neuroendocrine aging processes in female rodents. *Endocr. Rev.* **5**:467–497.

Flatmark, T. (1967) Multiple molecular forms of bovine heart cytochrome C. V. A comparative study of their physicochemical properties and their reactions in biological systems. *J. Biol. Chem.* **242**:2454–2459.

Frantz, A. G., and M. T. Rabkin (1965) Effects of estrogen and sex difference on human growth hormone. *J. Clin. Endocrinol.* **25**:1470–1490.

Fugo, N. W., and R. L. Butcher (1971) Effects of prolonged estrous cycles on reproduction in aged rats. *Fertil. Steril.* **22**:98–101.

Gee, D. M., K. Flurkey, and C. E. Finch (1983) Aging and the regulation of luteinizing hormone in C57BL/6J mice: Impaired elevations after ovariectomy and spontaneous elevations at advanced ages. *Biol. Reprod.* **28**:598–607.

Gershon, H., and D. Gershon (1973) Altered enzyme molecules in senescent organisms: Mouse muscle aldolase. *Mech. Ageing Dev.* **2**:33–41.

Goldberg, A. L., and A. C. St. John (1976) Intracellular protein degradation in mammalian cells, part 2. *Annu. Rev. Biochem.* **45**:747–803.

Goldstein, A., J. A. Hooper, R. S. Schulof, G. H. Cohen, G. B. Thurman, M. C. McDaniel, A. White, and M. Dardenne (1974) Thymosin and the immunopathology of aging. *Fed. Proc.* **33**:2053–2056.

Gordon, H. A., E. Bruckner-Kardoss, and B. S. Wostman (1966) Aging in germfree mice: Life tables and lesions observed at natural death. *J. Gerontol.* **21**:380–387.

Gosden, R. C., S. C. Fyfe, L. S. Felicio, J. F. Nelson, and C. E. Finch (1983) Imminent oocyte exhaustion and reduced follicular recruitment mark the transition to acyclicity in aging C57BL/6J mice. *Biol. Reprod.* **28**:255–260.

Gould, S. J. (1977) *Ontogeny and Phylogeny*. Harvard University Press, Cambridge, Mass.

Harlan, R. E., and R. A. Gorski (1977) Steroid regulation of luteinizing hormone secretion in normal and androgenized rats at different ages. *Endocrinology* **101**:741–749.

Harlan, R. E., and R. A. Gorski (1978) Effects of postpubertal ovarian steroids on reproductive function and sexual differentiation of lightly androgenized rats. *Endocrinology* **102**:1716–1724.

Harley, C. B., J. W. Pollard, J. W. Chamberlain, C. P. Stanners, and S. Goldstein (1980) Protein synthetic errors do not increase during aging of cultured human fibroblasts. *Proc. Natl. Acad. Sci. USA* **77**:1885–1889.

Harman, S. M., and G. B. Talbert (1970) The effect of maternal age on ovulation, corpora lutea of pregnancy, and implantation failure in mice. *J. Reprod. Fertil.* **23**:33–39.

Hart, R. W., and R. B. Setlow (1974) Correlation between deoxyribonucleic acid excision-repair and life-span in a number of mammalian species. *Proc. Natl. Acad. Sci. USA* **71**:2169–2173.

Hayflick, L. (1977) The cellular basis for biological aging. In: Handbook of the Biology of Aging, C. E. Finch and L. Hayflick (eds.). Van Nostrand, Princeton, Ch. 7, pp. 159–188.

Hendry, L. B., and F. H. Witham (1979) Stereochemical recognition in nucleic acid-amino acid interactions and its implications in biological coding: A model approach. *Perspect. Biol. Med.* **22**:333–345.

Hendry, L. B., F. H. Witham, and O. L. Chapman (1977) Gene regulation: The involvement of stereochemical recognition in DNA-small molecule interactions. *Perspect. Biol. Med.* **21**:120–130.

Hertig, A. T. (1967) Human trophoblast; normal and abnormal: A plea for the study of the normal so as to understand the abnormal. *Am. J. Clin. Pathol.* **47**:249–268.

Hirsch, G. P., R. A. Popp, M. C. Francis, B. S. Bradshars, and E. G. Bailiff (1980) Species comparison of protein synthesis accuracy. In: *Aging, Cancer, and Cell Membranes*, C. Brek, C. M. Fenoglio, and D. W. King (eds.). Thieme-Stratton, New York, 7:142–159.

Holland, J. J., D. Kohne, and M. V. Doyle (1973) Analysis of virus replication in ageing human fibroblast cultures. *Nature* **245**:316–318.

Jerne, N. (1976) The immune system: A web of V-domains. *Harvey Lec.* **70**:pp. 93–110.

Jones, E. C., and P. L. Krohn (1961a) The relationships between age, numbers of oocytes, and fertility in virgin and multiparous mice. *J. Endocrinol.* **21**:469–496.

Jones, E. C., and P. L. Krohn (1961b) The effect of hypophysectomy on age changes in the ovaries of mice. J. Endocrinol. **20**:497–508.

Kram, D., and E. L. Schneider (1978) An effect of reproductive aging: Increased risk of genetically abnormal offspring. In: *Aging and Reproduction*, E. L. Schneider (ed.). Raven Press, New York, pp. 237–270.

Landfield, P., J. Waymore, and E. Lynch (1978) Hippocampal aging and adrenocorticoids: Quantitative correlations. *Science* **202**:1098–1102.

Lekholm, G. C. (1939) *En Alderstigen Ål*. Kring Karnan. Halsingborgs Museum Arsskrift.

Levin, P., J. K. Janda, J. A. Joseph, D. K. Ingram, and G. S. Roth (1981) Dietary restriction retards the age-associated loss of rat striatal dopaminergic receptors. *Science* **214**:561–562.

Lewis, V. M., J. J. Twomey, P. Bealmear, G. Goldstein, and R. A. Good (1978) Age, thymic involution, and circulating thymic hormone activity. *J. Clin. Endocrinol. Metab.* **47**:145–150.

Lindstedt, S. L., and W. A. Calder, III (1981) Body size, physiological time and longevity of homeothermic mammals. *Q. Rev. Biol.* **56**:1–16.

Lofstrom, A. (1977) Catecholamine turnover alterations in discrete areas of the median eminence of the 4-and 5-day cyclic rat. *Brain Res.* **120**:113–131.

Makman, M. G., H. S. Ahn, L. J. Thal, N. S. Sharpless, B. Dvorkin, S. G. Horowitz, and M. Rosenfeld (1978) Biogenic amine-stimulated adenylate cyclase and spiroperidol binding sites in rabbit brain: Evidence for selective loss of receptors during aging. *Adv. Exp. Med. Biol.* **113**:pp. 211–230.

Mann, D. M. A., and P. O. Yates (1979) The effects of ageing on pigmented nerve cells of the human locus coeruleus and substantia nigra. *Acta Neuropathol.* **47**:93–97.

Mann, D. M. A., P. O. Yates, and J. E. Stamp (1978) The relationship between lipofuscin pigment and ageing in the human nervous system. *J. Neurol. Sci.* **37**:83–93.

Marx, J. L. (1985) The immune system "belongs in the body." *Science* **227**:1190–1192.

Masters, P. M., J. L. Bada, and J. S. Zigler, Jr. (1975) Aspartic acid racemization in heavy molecular weight crystallins and water-insoluble protein from normal human lens and cataracts. *Proc. Natl. Acad. Sci. USA* **72**:1204–1208.

Masters-Helfman, P., and J. L. Bada (1975) Aspartic acid racemization in tooth enamel from living humans. *Proc. Natl. Acad. Sci. USA* **72**:2891–2894.

McKerrow, J. H. (1979) Non-enzymatic, post-translational amino acid modifications in ageing. A brief review. *Mech. Ageing Dev.* **10**:371–377.

Meites, J., H. H. Huang, and J. W. Simpkins (1978) Recent studies on neuroendocrine control of reproductive senescence in rats. In: *Reproduction and Aging*, E. L. Schneider (ed.). Raven Press, New York, pp. 213–236.

Meltzer, S. J. (1906) The factors of safety in animal structure and animal economy. *Harvey Lec.* **1906–1907**:139–169.

Mobbs, C. V., D. M. Gee, and C. E. Finch (1984a) Reproductive senescence in female C57BL/6J mice: Ovarian impairments and neuroendocrine impairments that are partially reversible and delayable by ovariectomy. *Endocrinology* **115**:1653–1662.

Mobbs, C. V., K. Flurkey, D. M. Gee, K. Yamamoto, Y. N. Sinha, and C. E. Finch (1984b) Estradiol-induced anovulatory syndrome in female C57BL/6J mice: Age-like neuroendocrine, but not ovarian, impairments. *Biol. Reprod.* **30**:556–563.

Mobbs, C. V., L. S. Kannegietor, and C. E. Finch (1985) Delayed anovulatory syndrome induced by estradiol in female C57BL/6J mice: Age-like neuroendocrine but not ovarian impairment. *Biol. Reprod.* **32**:1010–1017.

Moore, R. Y. (1978) Central neural control of circadian rhythms. In: *Frontiers in Neuroendocrinology*, Vol. 5, W. F. Ganong and L. Martini (eds.). Raven Press, New York, pp. 185–206.

Morgan, D. G., J. O. Marcusson, P. Nyberg, P. Wester, B. Winblad, M. N. Gordon, and C. E. Finch (1987) Divergent changes in D-1 and D-2 dopamine binding sites in human basal ganglia during normal aging. *Neurobiol. Aging* **6** (in press).

Nelson, J. F., L. S. Felicio, and C. E. Finch (1980) Ovarian hormones and the etiology of reproductive aging in mice. In: *Aging—Its Chemistry*, A. A. Dietz (ed.). Am. Soc. Clin. Chem., Washington, D.C., pp. 64–81.

Nelson, J. F., L. S. Felicio, H. H. Osterburg, and C. E. Finch (1981) Altered profiles of estradiol and progesterone associated with prolonged estrous cycles and persistent vaginal cornification in aging C57BL/6J mice. *Biol. Reprod.* **24**:784–794.

Oertel, G. W., and P. Benes (1972) The effects of steroids on glucose–6-phosphate dehydrogenase. *J. Steroid Biochem.* **3**:493–496.

Orentreich, N., J. L. Brind, R. L. Rizer, and J. H. Vogelman (1984) Age changes and sex differences in serum dehydroepiandrosterone sulfate concentrations throughout adulthood. *J. Clin. Endocrinol. Metab.* **59**:551–555.

Orgel, L. E. (1963) The maintenance of accuracy of protein synthesis and its relevance to ageing. *Proc. Natl. Acad. Sci. USA* **49**:517–521.

Parker, J., and J. D. Friesen (1980) "Two out of three" codon reading leading to mistranslation in vivo. *Mol. Gen. Genet.* **177**:439.

Parker, R. J., B. A. Berkowitz, C.-H. Lee, and W. D. Denckla (1978) Vascular relaxation, aging, and thyroid hormones. *Mech. Ageing Dev.* **8**:397–405.

Peng, M.-T., and H.-O. Huang (1972) Aging of hypothalamic-pituitary-ovarian function in the rat. *Fertil. Steril.* **23**:535–542.

Piantanelli, L., M. Muzzioli, and N. Fabris (1980) Thymus-endocrine interactions during aging. *Akta. Gerontol.* **10**:199–201.

Popp, R. A., E. G. Bailiff, G. P. Hirsch, and R. A. Conard (1976) Errors in human hemoglobin as a function of age. In: *Interdisciplinary Topics in Gerontology*, Vol. 9, R. G. Cutler and H. P. von Hahn (eds.). Karger, Basel, pp. 209–218.

Quadri, S. K., G. S. Kledzik, and J. Meites (1973) Reinitiation of estrous cycles in old constant-estrous rats by central-acting drugs. *Neuroendocrinology* **11**:248–255.

Reiter, E. O., and M. M. Grumbach (1982) Neuroendocrine control mechanisms and the onset of puberty. *Annu. Rev. Physiol.* **44**:595–613.

Rines, S. P., and F. S. vom Saal (1984) Fetal effects on sexual behavior and aggression in young and old female mice treated with estrogen and testosterone. *Horm. Behav.* **18**:117–129.

Robertson, O. H. (1961) Prolongation of the lifespan of kokanee salmon (*O. nerka kennerlyi*) by castration before beginning development. *Proc. Natl. Acad. Sci. USA* **47**:609–621.

Robinson, A. B. (1974) Evolution and the distribution of glutaminyl and asparaginyl residues in proteins. *Proc. Natl. Acad. Sci. USA* **71**:885–888.

Robinson, A. B., J. H. McKerrow, and P. Cary (1970) Controlled deamidation of peptides and proteins: An experimental hazard and a possible biological timer. *Proc. Natl. Acad. Sci. USA* **66**:753–757.

Roth, G. S., and J. N. Livingston (1976) Reductions in glucocorticoid inhibition of glucose oxidation and presumptive glucocorticoid receptor content in rat adipocytes during aging. *Endocrinology* **99**:831–839.

Rothstein, M. (1983) Enzymes, enzyme alteration, and protein turnover. *Rev. Biol. Res. Aging* **1**:305–314.

Sacher, G. A. (1975) Maturation and longevity in relation to cranial capacity in hominid evolution. In: *Primate Functional Morphology and Evolution*, R. Tuttle (ed.). Mouton, The Hague, pp. 417–441.

Sacher, G. A. (1976) Evaluation of the entropy and information terms governing mammalian longevity. In: *Interdisciplinary Topics in Gerontology*, Vol. 9, R. G. Cutler (ed.). Karger, Basel, pp. 69–82.

Sacher, G. (1977) Life table modification and life prolongation. In: *Handbook of the Biology of Aging*, C. E. Finch and L. Hayflick (eds.). Van Nostrand, Princeton, N.J., pp. 582–638.

Sacher, G. A., and E. F. Staffeldt (1974) Relationship of gestation time to brain weight for placental mammals: Implications for the theory of vertebrate growth. *Am. Nat.* **108**:593–615.

Samuels, L. T. (1956) Effect of aging on the steroid metabolism as reflected in plasma levels. In: *Hormones and the Aging Process*, E. T. Engle and G. Pincus (eds.). Academic Press, New York, pp. 21–32.

Sapolsky, R. M., L. C. Krey, and B. S. McEwen (1985) Prolonged glucocorticoid exposure reduces hippocampal neuron number: Implications for aging. *J. Neurosci.* **5**:1221–1226.

Saunders, J. W. (1966) Death in embryonic systems. *Science* **154**:604–612.

Schipper, H., J. R. Brawer, J. F. Nelson, L. S. Felicio, and C. E. Finch (1981) The role of the gonads in the histologic aging of the hypothalamic arcuate nucleus. *Biol. Reprod.* **25**:413–419.

Schwartz, A. G. (1979) Inhibition of spontaneous breast cancer formation in female C3H(A^{v}y/a) mice by long-term treatment with dehydroepiandrosterone. *Cancer Res.* **39**:1129–1132.

Schwartz, A. G., and A. Perantoni (1975) Protective effect of dehydroepiandrosterone against aflatoxin B and 7,12-dimethylbenz(a)anthracene induced cytotoxicity and transformation in cultured cells. *Cancer Res.* **35**:2482–2487.

Severson, J. A., and C. E. Finch (1980) Reduced dopaminergic binding during aging in the rodent striatum. *Brain Res.* **192**:147–162.

Severson, J. A., J. Marcusson, B. Winblad, and C. E. Finch (1982) Age-correlated loss of dopaminergic binding sites in human basal ganglia. *J. Neurochem.* **39**:1623–1631.

Shaskan, E. G. (1977) Brain regional spermidine and spermine levels in relationship to RNA and DNA in aging rat brain. *J. Neurochem.* **28**:509–516.

Shimazu, T. (1980) Changes in neural regulation of liver metabolism during aging. In: *Neural Regulatory Mechanisms During Aging*, R. E. Adelman, J. Roberts, G. T. Baker, S. I. Baskin, and V. J. Cristofale (eds.). Liss, New York, pp. 159–185.

Shimazu, T., H. Matsushita, and K. Ishikawa (1978) Hypothalamic control of liver glycogen metabolism in adult and aged rats. *Brain Res.* **144**:343.

Simms, H. S. (1946) Logarithmic increase in mortality as a manifestation of ageing. *J. Gerontol.* **1**:13–25.
Simpkins, J. W., G. P. Mueller, H. H. Huang, and J. Meites (1977) Evidence for depressed catecholamine and enhanced serotonin metabolism in aging male rats: Possible relation to gonadotropin secretion. *Endocrinology* **100**:1672–1678.
Sirbasku, D. A. (1978) Estrogen induction of growth factors specific for hormone-responsive mammary, pituitary, and kidney tumor cells. *Proc. Natl. Acad. Sci. USA* **75**:3786–3790.
Sonntag, W. E., R. W. Steger, L. J. Forman, and J. Meites (1980) Decreased pulsatile release of growth hormone in old male rats. *Endocrinology* **107**:1875–1879.
Strehler, B. L. (1977) *Time, Cells, and Aging,* 2nd ed. Academic Press, New York.
Swanson, H. E., and J. J. van der Werff ten Bosch (1964) The early-androgen syndrome; its development and the response to hemi-spaying. *Acta Endocrinol. (Copenhagen)* **45**:1–12.
Szilard, L. (1959) On the nature of the aging process. *Proc. Natl. Acad. Sci. USA* **45**:30–45.
Tanner, J. M. (1969) Growth and endocrinology of the adolescent. In: *Endocrine and Genetic Diseases of Childhood,* L. I. Gardner (ed.). Saunders, Philadelphia, pp. 19–60.
Thimann, K. V., and A. C. Giese (1981) A glance at senescence in plants. In: *Biological Mechanisms in Aging,* R. T. Schimke (ed.). NIH Publ. No. 81-2194, Bethesda, pp. 702–709.
Tice, R. R., and R. B. Setlow (1985) DNA repair and replication in aging organisms and cells. In: *Handbook of the Biology of Aging,* 2nd ed., C. E. Finch and E. L. Schneider (eds.). Van Nostrand, Princeton, N.J., pp. 173–224.
Tomkins, G. A., E. J. Stanbridge, and L. Hayflick (1974) Viral probes of aging in the human diploid cell strain WI-38. *Proc. Soc. Exp. Biol. Med.* **146**:385–390.
Van Heukelem, W. F. (1978) Aging in lower mammals. In: *Biology of Aging,* J. A. Behnke, C. E. Finch, and G. Moment (eds.). Plenum Press, New York, pp. 115–130.
vom Saal, F. S., and F. H. Bronson (1978) In utero proximity of female mouse fetuses to males: Effect on reproductive performance during later life. *Biol. Reprod.* **19**:842–853.
vom Saal, F. S., and F. H. Bronson (1980a) Sexual characteristics of adult female mice are correlated with their blood testosterone levels during prenatal development. *Science* **208**:597–599.
vom Saal, F. S., and F. H. Bronson (1980b) Variation in the length of the estrous cycle in mice due to former intrauterine proximity to male fetuses. *Biol. Reprod.* **22**:777–780.
vom Saal, F. S., and C. L. Moyer (1985) Prenatal effects in reproduction capacity during aging in female mice. *Biol. Reprod.* **32**:1116–1126.
Weindruch, R., S. R. S. Gottesman, and R. L. Walford (1982) Modification of age-related immune decline in mice dietarily restricted from or after midadulthood. *Proc. Natl. Acad. Sci. USA* **79**:898–902.
Weiner, R. I., and W. F. Ganong (1978) Role of monoamines and histamine in regulation of anterior pituitary secretion. *Physiol. Rev.* **58**:905–976.
Wildt, L., G. Marshall, and E. Knobil (1980) Experimental induction of puberty in the infantile female rhesus monkey. *Science* **207**:1373–1375.
Wilson, D. E., M. E. Hall, and G. C. Stone (1978) Test of some aging hypotheses using two-dimensional protein mapping. *Gerontology* **24**:426–433.
Wise, P. M. (1982) Norepinephrine and dopamine activity in microdissected brain areas of the middle-aged and young rat on proestrus. *Biol. Reprod.* **27**:562–574.
Witham, F. H., L. B. Hendry, and O. L. Chapman (1978) Chirality and stereochemical recognition in DNA-phytochrome interactions: A model approach. *Origins Life* **9**:7–15.
Yamaji, T., and H. Ibayashi (1969) Plasma dehydroepiandrosterone sulfate in normal and pathological conditions. *J. Clin. Endocrinol. Metab.* **29**:273–278.
Yen, T. T., J. V. Allen, D. V. Pearson, J. M. Acton, and M. M. Greenburgh (1977) Prevention of obesity in $A^v y/a$ mice by dehydroepiandrosterone. *Lipids* **12**:409–413.

IV

Networks, Neural Organization, and Behavior

Donald O. Walter

Self-organization is interesting to us partly because of patterns it can bring about. Of course, "interest" and "pattern" are defined by the individual, whose own brain, as "pattern recognizer," exemplifies an outcome of the well-known dialectic of heredity with environment. There is a long history of popular and scientific interest in pattern formation in many fields. This section includes three studies on patterns of structure or function in central nervous systems. Such patterns could be fruitfully studied only after Ramón y Cajal in 1909 established the "neuron doctrine," i.e., that the vertebrate brain is not a syncytium of cells connected by continuous protoplasmic bridges but, rather, a vast collection of topologically closed, membranously separated cells and, thus, that any cooperation between them requires intercellular communication (of a type not then specified).

Anatomical relationships between different brain regions or between individual cells in those different regions, e.g., among groups of sensory, associative, and motor neurons, are needed for brain function, and von der Malsburg (Chapter 14) contributes a causal narrative explaining the reliable growth and interconnection (within every normally developed individual mammal) of a behaviorally useful visual system. Stent (Chapter 13) shows how "doctrinally pure" neurons with thresholds, communicating only by unidirectionally transmitted, all-or-nothing impulses, can produce useful rhythmic movement patterns, especially in invertebrate central nervous systems.

Fifty years after Ramón y Cajal's demonstration, Bullock (1959) needed to remind overzealous reductionists that Ramón y Cajal did not imply any "physiological neuron doctrine." Bullock argued that nondiscrete, nonthreshold coop-

---

DONALD O. WALTER • Clinical Electrophysiology Laboratory, Neuropsychiatric Institute and Hospital, University of California, Los Angeles, California 90024.

eration among brain cells in vertebrates (especially mammals) is an equally plausible mode of functioning, leading to a type of patterning he has recently called "the neural throng" (Bullock, 1980). To limit intracerebral communications to threshold-operated pathways would be rather like limiting interpersonal communication to telephone pathways—inhibited and unnatural. The idea of nonthreshold transfer of information between cerebral neurons has so far been neither affirmed nor denied experimentally, but such hypotheses can now be tested. Casual application of ("conservative") Occam-like thought-shaving might suggest that if you have not proved that you need it, then do not assume it. Instead, I believe that such "Occamism" would impoverish brain research by disallowing useful, unconventional hypotheses (see also Walter, 1981).

Arbib (Chapter 15) presents a mechanism-neutral description of high points of studies of vertebrate central nervous function from two complementary viewpoints. The first, standard viewpoint is reductive, i.e., the "bottom-up" approach, which tries to determine the "atomic" constituents (constituents indivisible for present purposes) of the system under study. This level becomes the "bottom" level for theorizing. In this approach, properties of these atomistic elements are then investigated and attempts are made to resynthesize the system's global functions by combining those microscopic properties.

The other viewpoint exploited by Arbib—the "top-down" approach—is less popular with brain theorists. This approach begins with some or all global system functions attributed to the system (the "top"), then attempts are made to determine "natural" (and eventually, identifiable) subsystems that could combine to produce the complex behavior or high-level property. (I warn strongly against any facile presumption that *every* top-level concept is ready for reduction all the way to neurons.)

In brain research, both of these approaches have been exploited for many years—the top-down one for more than a century, at least in a literary way, in the attempts by clinical neurologists (and, more recently, neuropsychologists), to understand brain functioning through intensive study of the defects produced by injury or disease. Indeed, these studies have become more refined, drawing on case material consisting of accidentally injured humans, of humans whose brains developed unusually, of those whose brains have been affected by some disease process, or of injured or diseased animals. A third approach to the study of nervous systems, which might be called "defectology," and described as the "middle, upward, and downward" method, has emerged almost without being noticed by taxonomists of scientific methods. When a small wound is defined in a human patient, or produced in an animal, one can test both the upward fit of the ideas about the lesion into assumed integrated properties of behavior and the downward fit into the functions attributed to smaller subsystems of the brain—perhaps, ultimately, to brain cells. (In fact, middle-level functional defects can serve as a valuable resource for constraining overexuberant theorizing by requiring that proposed linkages between top and bottom levels pass through observed middle-level effects.) Arbib integrates results from this middle approach into his theorizing without defining it as a separate viewpoint.

Although the suggestive utility of reciprocal analogies between a nervous system and a bare, isolated digital computer may have been exhausted, metaphorical appeal to *networks* of *program-containing* computing devices may still help both fields. Bellman and Roosta (Chapter 12) show how a clever network may be designed to adjust its own connections in order to maximize effectiveness at the present task and possibly to reconnect—to do the same at the next task.

All four chapters discuss patterns arising by expression of genomes ("genomes" in a metaphorical sense, in Bellman and Roosta's chapter), within a single, complicated system. The meaning of these patterns, and some of the processes by which they may have come about, can be studied only in terms of a larger, multigenerational evolutionary vision, such as that presented in Section II.

References

Bullock, T. H. (1959) Neuron doctrine and electrophysiology. *Science* **129**:997–1002.
Bullock, T. H. (1980) Reassessment of neural connectivity and its specification. In: *Information Processing in the Nervous System*, H. M. Pinsker and W. D. Willis, Jr. (eds.). Raven Press, New York. pp. 199–220.
Ramón y Cajal, S. (1909) *Histologie du système Nerveux de l'Homme et des Vertébrés.* 2 volumes. Maloine, Paris.
Walter, D. O. (1981) Computer analysis of the synthesizing brain. In: *Recent Advances in EEG and EMG Data Processing*, N. Yamaguchi and K. Fujisawa (eds.). Elsevier, North-Holland, Amsterdam, pp. 3–15.

12

On a Class of Self-Organizing Communication Networks

Richard Bellman[†] and Ramin Roosta

ABSTRACT

Technological networks can be designed to be self-organizing. For example, the method of dynamic programming can be applied to allow communication networks to adjust themselves for adequate reliability of communication at minimum cost, or for maximum reliability under given resources. —THE EDITOR

Growing demand for computing power and the high rate of technological advancements in the electronics industry have demonstrated the need for faster algorithms and better execution of these algorithms. As a result, special-purpose machines were developed and have proven to be faster and more efficient than their general-purpose predecessors. But general-purpose machines contain most, if not all, of the components of the special-purpose machines, and with better system organization as well as automatic system organization they should perhaps diminish the need for multiplying special-purpose designs. How can this economy of design be achieved?

Self-organizing systems are still in their infancy, and the need for further

---

RICHARD BELLMAN • Department of Biomathematics, University of Southern California, Los Angeles, California 90089. **RAMIN ROOSTA** • Department of Electrical and Computer Engineering, California State University, Northridge, California 91330. †Shortly after completing this chapter, Richard Bellman died. His courage through a long illness was an inspiration to his many friends.

research is apparent. There are now several classes of self-organizing systems. In some, an external signal determines the organization; others organize according to an internal purpose; still others have both features. We consider here only the class of systems in which an external signal determines the organization, and show that dynamic programming can be used to treat the mathematical problems that arise. We assume that we have a complex network, and that a path of maximum probability of getting through the network has been chosen. A method to determine this path was first studied by Christofides (1975), and further studies were conducted by Roosta (1982).

Minimum Cost

We first consider the case in which a message goes into the network through a most reliable path. Here we must determine what resources to use in order to obtain a desired reliability. Let us assume a network as shown in Figure 1, in which processing takes place at the nodes and the only purpose of the links is transmission. As soon as a message comes in, it specifies the reliability it needs, which is to be achieved at a minimum cost. There are several ways to represent this problem. For example, we can assume that there is a certain amount of money (resource) assigned in advance to each node and each link in the network that tells us about the reliability. (How the reliability can be improved depends upon the nature of the node and the link—matters not discussed here.)

In a slightly more elaborate case, the message itself tells us how our resources are to be allocated. (At the nodes the resources could be computers, men, and so on.) Let us assume a required reliability equal to r (a probability). If we assign an amount of money X_k to the kth node or link, then the reliability of each node and link depends upon the amount of money allocated, $P_k(X_k)$. If n is the number of links and nodes, then the reliability is specified by

$$r = \prod_{k=1}^{n} P_k(X_k) \qquad (1)$$

We are left with a mathematical problem of ensuring the required reliability, r, while minimizing the cost. Therefore we also have

$$\text{MIN} \sum_{k=1}^{n} X_k \qquad (2)$$

The function dP_k could be determined by investigating the structure of each node and link.

We can handle the above problem by dynamic programming in the following way. Let us call the minimum cost function

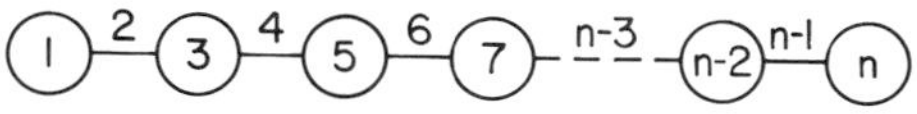

12. Self-Organizing Communication Networks

FIGURE 1. Generalized network with processing nodes and transmission links.

$$f_n(r) = \text{MIN} \sum_{k=1}^{n} X_k \qquad (3)$$

If we assign an amount X_n to the final node or link, we have the following equation:

$$f_n(r) = \mathop{\text{MIN}}_{X_n \geq 0,\ P_n(X_n) \geq r}\left[X_n + \frac{f_{n-1}(r)}{P_n(X_n)} \right] \qquad (4)$$

for $n \geq 2$, where

$$f_1(r) = X_1 \qquad (5)$$

when

$$P_1(X_1) = r \qquad (6)$$

Because r is a probability, we have $0 \leq r \leq 1$.

In general, nodes and links require different types of resources, but this constraint may be easily handled by the method outlined above.

Maximum Reliability

Here we consider the case in which the message tells us what resources are available in order to achieve maximum reliability. Let us assume, for simplicity, that we have only one type of resource, e.g., money, but for a more detailed investigation we could consider multiple kinds of resources, such as money, men, computers, and so on. We now have the problem of how to assign a given amount of money, X, in order to maximize reliability. Mathematically,

$$\text{MAX} \prod_{k=1}^{n} P_k(X_k) \qquad (7)$$

subject to

$$\sum_{k=1}^{n} X_k = X \qquad (8)$$

Once again this problem can be easily handled by dynamic programming. Let us define

$$g_n(X) = \text{MAX} \sum_{k=1}^{n} P_k(X_k) \tag{9}$$

subject to the constraint

$$\sum_{k=1}^{n} X_k = X \tag{10}$$

We obtain the following functional equation:

$$g_n(X) = \underset{0 \leq X_n \leq X}{\text{MAX}} [P_n(X_n) g_{n-1}(X - X_n)] \tag{11}$$

for $n \geq 2$, where

$$g_1(X) = P_1 \tag{12}$$

All the problems considered above can be shown to be equivalent under mild assumptions concerning the probabilities.

In a self-organizing system the organizing stimulus is sometimes external and sometimes internal. Here we have considered only cases in which the stimulus was external. A more interesting example of self-organizing systems is the self-organizing computer that organizes itself according to the task it is working on; in that case the stimulus is internal. Any self-organizing system must involve a combination of software and hardware and will be, in general, quite complex. We have emphasized here only the aspect of reliability of such systems, but the mathematical approach of dynamic programming is very general and can be applied in other, more complicated situations.

References

Christofides, N. (1975) *Graph Theory, an Algorithmic Approach.* Academic Press, New York.
Roosta, M. (1982) Routing through a network with maximum reliability. *J. Math. Anal.* **88**(2):341–347.

13

Neural Circuits for Generating Rhythmic Movements

Gunther S. Stent

ABSTRACT

Breathing, walking, and swimming rhythms (and even heartbeat in subvertebrates) are generated in central nervous systems, often with little regard to sensory inputs. This chapter concerns models and mechanisms for central rhythm generation in a single organism, the leech. Endogenous rhythms in membrane polarization occur in some individual neurons, especially in systems involving chronic, rather than episodic, movement rhythms, which do not require cycle-by-cycle modulation by sensory input. In contrast, oscillations at the network level seem to be used for functions such as walking or swimming, in which quick modulation may be needed. Closed, self-reexciting chains of neurons are one form of network oscillator. Resetting, or restoration, is required for termination of impulse generation by these networks. This can be brought about by activating an inhibitory cell or by reciprocal or recurrent inhibition. Leech swimming, for example, is governed by a network oscillator that operates by recurrent inhibition.

Rhythmic movements in other animals, such as vertebrates, may or may not be generated by the same mechanisms as those found in the leech. Network analysis is much harder in more complex animals, but not many possibilities for rhythm generation exist. Thus, it seems reasonable to expect that the circuits discussed herein might prove to be of general applicability in the entire animal kingdom. —THE EDITOR

The rhythmic movements of vertebrate and invertebrate animals are usually generated by neural elements wholly within the central nervous system (Bullock, 1961). This conclusion is based on the finding that in nearly every analyzed case, a motor neuron activity pattern closely resembling that driving the movement in the intact animal continues to be produced in an isolated preparation deprived

---

GUNTHER S. STENT • Department of Molecular Biology, University of California, Berkeley, California 94720.

of all phasic sensory input. The basic source of such a motor rhythm must, therefore, be a *central nervous oscillator* composed of elements capable of generating an oscillatory activity pattern. The types of rhythmic movements now known to be driven by central nervous oscillators include not only autonomic rhythms, such as heartbeat and breathing, but also voluntary rhythms, such as walking and swimming (Pearson, 1972; Grillner, 1973, 1974, 1975; Huber, 1975; Kristan and Calabrese, 1976; Thompson and Stent, 1976a,b,c; Fentress, 1976). These findings must not be taken to mean that sensory feedback plays no role at all in the realization of rhythmic movements. On the contrary, in most cases the basic rhythm generated by the central oscillator is subject to influence by sensory feedback provided by proprioceptors, which serves to modulate both the period and the amplitude of the rhythm. In this discussion, however, I shall ignore the role of sensory feedback in the generation of rhythmic movements and address the problem of the nature and mode of operation of central nervous oscillators.

As in most other domains of contemporary neurobiology, the study of central nervous oscillators demands a close interplay between theory and observation. Unless the design of theoretically possible oscillatory neural circuits is constantly restrained by observational data, the models can easily stray too far from reality. And unless the design of experiments relevant to the generation of motor rhythms is guided by theoretically sound models, the neurophysiological data that accumulate can easily become too complex for interpretation (see Chapter 18). Accordingly, my presentation consists of a theoretical part, in which a few realistic models of central nervous oscillators are discussed, and an observational part, in which findings relevant to rhythmic movements generated by just one such oscillator—particularly the identification of its neural components—are considered in light of one of the theoretical models. My example is the swimming rhythm of the leech, a member of the annelid phylum of segmented worms, whose comparatively simple nervous system has made possible some recent progress toward the identification of the cellular basis of oscillatory nervous activity (Stent et al., 1978, 1979). More extensive treatments of this subject can be found in two reviews (Friesen and Stent, 1978; Kristan, 1980).

Models and Mechanisms for Central Rhythm Generation

Current models that explain how neurons of the central nervous system can produce rhythmic activity fall into two general classes: *endogenous polarization rhythms* that depend on special oscillatory properties of individual neurons and *network oscillations* that depend on oscillatory properties that arise from the connections linking a set of neurons having endogenously stable polarization levels.

Endogenous Polarization Rhythms

Evidence for the existence of neurons capable of endogenous generation of rhythmic impulse bursts in the absence of any rhythmic synaptic input was first provided in molluscan nervous systems (Strumwasser, 1967; Alving, 1968). The membrane potential of such neurons spontaneously oscillates between a depolarized and a repolarized phase, with the cell producing an impulse burst during the depolarized phase. The ionic basis of these endogenous polarization rhythms has been elucidated for some neurons (Faber and Klee, 1972; Junge and Stephens, 1973; Meech and Standen, 1975; Barker and Gainer, 1975; Smith et al., 1975; Eckert and Lux, 1976). According to these studies, the endogenous cellular polarization rhythm is attributable to an underlying cyclic variation in the intracellular concentration of free calcium ions. This cyclic variation, in turn, arises from the action of two opposing processes: (1) the continuous sequestration, or extrusion, of intracellular calcium ions and (2) their influx from the extracellular medium via a special, voltage-dependent calcium ion channel whose conductance rises with membrane depolarization. Moreover, the membrane of the oscillatory cell also contains a special potassium ion channel, whose conductance rises and falls with the intracellular calcium ion concentration. These periodic variations in intracellular calcium ion concentrations occur in a frequency domain that is low (i.e., on the order of hundreds of milliseconds), as compared to the frequency domain of the polarization cycle of ordinary action potentials (i.e., on the order of milliseconds).

The depolarization phase of each endogenous polarization cycle is initiated by a gradual fall in the intracellular calcium ion concentration due to calcium sequestration or extrusion, while the voltage-dependent calcium ion channel is closed. This decline in intracellular calcium ion concentration results in a conductance decrease of the calcium-dependent potassium ion channel and, hence, in depolarization of the cell. Once depolarization due to this calcium-mediated decrease in potassium ion conductance reaches action potential threshold, the ordinary kinds of voltage-dependent sodium and potassium ion channels also present in the oscillator cell membrane produce an impulse train. The repolarized phase of the cycle is initiated by a conductance increase of the voltage-dependent calcium ion channel, resulting in a gradual rise in intracellular calcium ion concentration, and its attendant gradual increase in conductance of the calcium-dependent potassium ion channel. The impulse train ceases as soon as repolarization falls below the action potential threshold. When full repolarization of the cell is achieved and the voltage-dependent calcium ion channel is once again closed, the depolarized phase of the next cycle is ready to begin upon the gradual fall of the intracellular calcium ion concentration. The production of an impulse train during the depolarized phase is not required for the maintenance of the polarization rhythm, which continues even under conditions that prevent action potential generation (Watanabe et al., 1967; Strumwasser, 1971; Barker and Gainer, 1975).

The cycle period of the endogenous polarization rhythm can be increased or decreased, respectively, by steady passage of hyperpolarizing or depolarizing current into the cell (Frazier et al., 1967; Kandel et al., 1976), inasmuch as flow of hyperpolarizing current delays and flow of depolarizing current hastens the onset of the depolarized phase. Thus, the possibility exists for physiological control of the period of the endogenous rhythm, by setting the level of tonic inhibitory or excitatory input to the endogenously oscillating cell. Moreover, transient passage of current into the cell can shift the phase of its endogenous polarization rhythm Kandel, 1967; Strumwasser, 1971; Kater and Kaneko, 1972; Thompson and Stent, 1976a,b,c). In particular, evocation of a premature impulse burst by transient passage of depolarizing current during the repolarized phase of the cycle delays the onset of the next depolarized phase. This onset is delayed also by transient passage of strong hyperpolarizing current near the end of the repolarized phase. After the delayed onset of the next depolarized phase, the cell will enter the following depolarized phase after lapse of the normal period. Hence, the cycle periods of an ensemble of endogenous oscillator cells can be phase-locked by interconnecting them via a set of excitatory or inhibitory synaptic links. In this way, a complex, multiphasic, rhythmic movement, consisting of more than simply two complementary, on-off phases, can be generated by a network of endogenous oscillator cells, in which the time of onset of the depolarized, active phase forms a progression of phase angles (Thompson and Stent, 1976a,b,c; Calabrese, 1977).

It would appear that neurons with endogenous polarization rhythms find their main employ in the generation of chronic (rather than episodic) rhythmic movements that do not require cycle-by-cycle modulation by sensory input. The heartbeat of the leech provides an instance of such a chronic motor routine driven by endogenous neuronal polarization rhythms (Stent et al., 1979). The visceral rhythms of vertebrates, e.g., that of cardiac muscle, are also based on endogenous polarization rhythms, albeit on endogenous polarization rhythms of the muscle fibers rather than of their controlling neurons (Trautwein, 1973).

Self-Excitatory Networks

In contrast to the chronic rhythms generated by cells with endogenous polarization rhythms, episodic movements, such as walking or swimming, for which cycle-by-cycle sensory modulation may be needed, appear to be better served by network oscillators. The component neurons of network oscillators need not possess any special ion conductance channels that give rise to endogenous oscillations in membrane potential. But, as shall be seen, such network neurons may nevertheless have to be endowed with some special properties to enable their interconnections to produce an activity rhythm.

One type of oscillatory network owes its activity rhythm to self-excitation resulting from neuronal loops with positive feedback. Such networks consist of two or more neurons linked by mutually excitatory connections whose net gain is positive, causing the cells to drive each other to produce impulses at progressively higher frequencies. In order to oscillate, self-excitatory networks must also incorporate some restorative feature that terminates impulse production and repolarizes the network cells as soon as a critical impulse frequency has been attained. In this way the cells are periodically obliged to begin anew their self-excitatory drive to progressively higher impulse frequencies and greater membrane depolarization. Thus, the oscillatory cycle of the cells of self-excitatory networks consists of an active phase of gradually increasing membrane depolarization and impulse frequency, and an inactive phase during which the transient

13. Neural Circuits for Rhythmic Movements

impulse termination process has repolarized the membrane. Such systems can, therefore, generate only a single pair of complementary, on-off phases of a duty cycle.

Whereas the synaptic mechanisms that can give rise to self-excitatory neuronal loops with positive feedback are not difficult to envisage, the mechanisms underlying the restorative termination of impulse production are more elusive. One suggested impulse termination process for networks with self-excitation is the accumulation of refractoriness engendered by each impulse, or adaptation, and its imminent increase in the impulse-generation threshold (Wilson, 1966). Oscillatory model circuits based on this principle, representing two electrically coupled, self-excitatory neurons, have been constructed by means of electronic "neuromime" analogues (Wilson and Waldron, 1968; Lewis, 1968). The neuromimes of these model circuits produced oscillations with cycle periods in the 1-sec range, during which the cell analogue pair gave rise to concurrent impulse bursts. A delicate adjustment of the system parameters, however, was found to be necessary to stabilize these oscillations and to prevent the network from drifting into nonoscillatory, or steady-state, impulse production. No actual case of an oscillatory neuronal network operating in the manner envisaged by this model is known.

Another possible impulse termination process for self-excitatory networks is the activation of an inhibitory cell (Bradley *et al.*, 1975). A simple realization of such a system consists of three cells, A, B, and C, of which A and B are linked by reciprocally excitatory connections and C is provided with excitatory inputs by cells A and B (Figure 1a). Cell C, in turn, is linked via inhibitory connections to cells A and B. Cell C has a high threshold for impulse initiation, which is reached only when cell C receives a high level of excitatory input due to high-frequency impulse activity in cells A and B. This circuit generates a rhythm of concurrent impulse bursts in cells A and B, provided that: (1) the gain of the feedback loop between cells A and B is positive; (2) the eventual activation of cell C causes substantial repolarization of cells A and B; and (3) cells A and B have a source of tonic excitation to ensure that after repolarization they will drive each other again to higher impulse frequencies. The period of the oscillation depends on the system parameters that govern the impulse activity time of cell C and the recovery time of cells A and B for inhibition (both of which times provide the "inertia" of the oscillator), as well as on the rate of impulse acceleration of cells A and B. Figure 1c shows the output of an electronic analog circuit consisting of three neuromime elements representing cells A, B, and C, connected according to the scheme of Figure 1a. As can be seen, the analog circuit does indeed generate stable oscillations. Cells A and B produce concurrent, rhythmic bursts of accelerating impulses, with a cycle period of about 0.5 sec. The summation of the accelerating excitatory synaptic potentials provided by cells A and B to cell C eventually depolarizes cell C to action potential threshold. The single impulse thereupon produced by cell C and its powerful inhibitory synaptic effect repolarizes cells A and B and terminates their impulse burst. This model has been proposed to account for the oscillator in the mammalian central nervous system that drives the contractile rhythm of the diaphragm in breathing (Bradley *et al.* 1975).

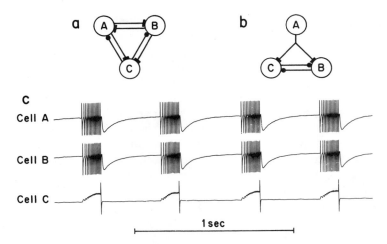

FIGURE 1. Self-excitatory and reciprocally inhibitory networks. In this and all subsequent circuit diagrams, T junctions indicate excitatory and filled circles inhibitory synaptic connections. (a) Self-excitatory network, in which accelerating impulse production in cells A and B is eventually terminated by activation of the inhibitory cell C. (b) Reciprocal inhibition network, in which cells B and C are driven to produce alternating impulse bursts by the tonically active cell A. (c) Output of three electronic neuromime elements connected according to the circuit of panel (a), with the following system parameters: rise time of synaptic potentials: 5 msec; exponential decay time constant of synaptic potentials, A to B and B to A: 6 msec, A or B to C: 30 msec, C to A or B: 80 msec; free-running impulse frequency of cells A and B: 30 Hz. (W. O. Friesen, unpublished experiments.)

Reciprocal Inhibition Networks

Another type of oscillatory network owes its activity rhythm to reciprocally inhibitory rather than self-excitatory loops. In fact, the very first proposal for the neuronal generation of alternating rhythmic movements was of that type—the reciprocal inhibition network proposed by McDougall (1903). The essential elements of McDougall's model are illustrated in terms of modern concepts and terminology by the three-neuron network of Figure 1b. Cell A of this network is tonically active, providing excitation to cells B and C, which are connected by reciprocally inhibitory synapses. Cells B and C will produce alternating impulse bursts if there exists some restorative, hysteretic process by which the inhibitory effect of one of these cells on the other decreases as a consequence of past activity. One such process, proposed by McDougall, as well as by Brown in his equivalent "half-center" model a few years later (Brown, 1911, 1912, 1914), is *fatigue* of the inhibitory synapses, which causes the strength of synaptic transmission to decline with cumulative synaptic use. Other plausible hysteretic processes are *adaptation*, which causes a neuron to respond to constant excitation with a declining impulse frequency, and *postinhibitory rebound*, which causes a transient reduction of the impulse threshold level as the result of past inhibition. No actual neural oscillator depending on synaptic fatigue has been found, nor has such an oscillator been shown to be theoretically feasible by electronic analog or digital computer simulation techniques. By contrast, modeling studies have shown that reciprocal inhibition networks will oscillate if they incorporate adaptation or postinhibitory rebound as restorative processes (Reiss, 1962; Harmon, 1964; Harmon and Lewis, 1966; Lewis, 1968; Wilson and Waldron, 1968; Perkel and Mulloney, 1974; Warshaw and Hartline, 1976).

There is no straightforward relationship between cycle period and tonic excitation level in reciprocal inhibition networks, since both increases (Wilson and Waldron, 1968; Perkel and Mulloney, 1974) and decreases (Reiss, 1962; Harmon, 1964) of cycle period with increasing excitation have been found. The explanation for these opposite variations in the period lies in the dependence of system state variables on excitation levels. Suppose, for instance, that cells B and C in the scheme of Figure 1b have the property of adaptation, and that cell A is producing impulses at a sufficiently high rate to produce stable oscillations. In that case, increases in the impulse frequency of cell A will raise not only the impulse frequency of the currently active cell and its level of inhibition of the currently inactive cell but also the level of inhibitory input necessary to keep the inactive cell below its action potential threshold. Because of these opposite tendencies, the cycle period of oscillators with reciprocal inhibition tends to be relatively insensitive to excitation levels.

Recurrent Cyclic Inhibition Networks

In addition to requiring a hysteretic feature (such as synaptic fatigue, adaptation, or rebound), and a delicate adjustment of system parameters, the reciprocal inhibition network of Figure 1b has one further theoretical limitation as a general rhythm generator: it is limited to the production of biphasic rhythms. Hence, additional network elements are required for the generation of polyphasic motor routines, such as the rhythmic movement of articulated limbs or the metachronal contractile wave of serially homologous muscles found in vertebrate and invertebrate locomotion. As was apparently first realized by Székely (1965, 1967), introduction of such additional elements into the network not only provides for a multiphasic rhythm but also dispenses with the need for hysteresis by opening up the possibility for another source of rhythm generation, namely, recurrent cyclic inhibition. The oscillatory dynamics of a network with recurrent cyclic inhibition can be readily fathomed for its simplest realization, shown schematically in Figure 2a. This network consists of an inhibitory ring formed by three tonically excited neurons, A, B, and C, each of which makes inhibitory synaptic contact with, and receives inhibitory synaptic input from, one other cell. If, as indicated in Figure 2a, cell C happens to be in a depolarized, impulse-generating state, its postsynaptic cell, B, must be in a hyperpolarized, inactive state, while its presynaptic cell, A, is recovering from past inhibition. As soon as cell A has recovered from inhibition and reached its impulse generation threshold, cell C becomes inhibited, thus disinhibiting cell B and allowing the latter to enter its recovery phase. Once cell B has recovered, it inhibits cell A, thus allowing cell C to begin recovery. Once cell C has recovered, so that cell B enters its inactive phase and cell A its recovery phase, one cycle of the oscillation has been completed. If the time required for recovery from inhibition of each cell is R, and if

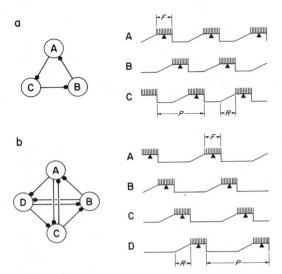

FIGURE 2. Simple networks with recurrent cyclic inhibition. (a) Three-cell network. (b) Four-cell network. The traces to the right of both circuit diagrams represent the membrane potential and impulse burst activity (F) in individual cells, as determined by theoretical analysis of the network. A triangle points to the midpoint, or middle spike, of an impulse burst.

the time required for establishing inhibition is small compared to R, then the period of the oscillator cycle is evidently equal to $3R$. This oscillatory network gives rise to three activity phases separated by phase angles of 120°, and the cycle phases of the three cells progress in a sense opposite to that of the inhibitory connections forming the ring.

Analytical study of a variety of cyclic inhibition networks (Adam, 1968; Kling and Székely, 1968; Dunin-Barkovskii, 1970; Pozin and Shulpin, 1970) has shown that they produce stable oscillations over a broad range of system parameters and can generate as many different cycle phases as the number of cells they contain. Any such ring containing an *odd* number, N, of cells linked in such a manner that every cell receives direct inhibition from one other cell will oscillate with a cycle period

$$P = NR \qquad (1)$$

In this N-membered ring, one cell is always in its recovery phase, while the remaining $N-1$ cells form an alternating sequence of active and inactive phases. By contrast, simple rings containing an *even* number of cells do not oscillate, since they can assume one of two stable states under which either all the even-numbered or all the odd-numbered cells are in the active phase, and no cells are in the recovery phase.

This consideration makes clear why two-cell networks do not oscillate without a source of hysteresis. However, in cases of rings containing four or more cells, topologically more complex networks can be formed, in which recurrent cyclic inhibition does produce an oscillatory activity pattern, even if the number of cells in the ring is even (Kling and Székely, 1968). The simplest (symmetric) realization of such a network is an ensemble of four cells, each of which makes inhibitory contact with, and receives inhibition from, two other cells (Figure 2b).

In this four-cell network, one cell (e.g., cell A) is in the active phase, the two cells subject to inhibition by that cell (i.e., cells C and D) are in the inactive phase, and the cell subject to inhibition by these two cells (i.e., cell B) is in the recovery phase. The network gives rise to four activity phases separated by phase angles of 90°, with the period of the oscillatory cycle being equal to $4R$. The theoretically predicted mode of operation of these networks has been confirmed by means of electronic analog circuits (Kling and Székely, 1968; Friesen and Stent, 1977).

A further theoretical advantage of recurrent cyclic inhibition networks as generators of rhythmic movements is that their cycle period can be made to vary simply by varying the recovery time R. When the nerve cell membrane responds to voltage transients as a passive element, the value of R is given by

$$R = \rho_i C_i \ln[(V_I - V_E)/(V_T - V_E)] \tag{2}$$

where ρ_i and C_i are, respectively, the input resistance and capacitance of the impulse initiation zone of the cell, V_I is the membrane potential after occurrence of the last inhibitory synaptic potential, V_E is the steady-state potential to which tonic excitation would depolarize the cell in the absence of inhibition, and V_T is the threshold potential for impulse generation. Thus, the value of R, and hence the period P of the rhythm, can be shortened or lengthened simply by increasing or decreasing the level of tonic excitation, and hence the value of V_E. This theoretical prediction was confirmed by means of a neuromime analog of the three-cell network of Figure 2a (Friesen and Stent, 1977).

The recurrent cyclic inhibition networks considered up to this point possess the feature of phase-constancy, in that corresponding features of the duty cycles of the oscillator cells, such as their impulse burst midpoints, maintain a fixed phase relation, despite variations in R and, hence, in the length of the period P. However, the phase relations of the component features of many rhythmic movements are not, in fact, invariant (Pearson, 1972; Kristan et al., 1974a; Grillner, 1975; Kristan and Calabrese, 1976). Instead, they depend on the period in a manner such that the duty cycle of the generating oscillator can be inferred to consist of a variable time sector, whose changes in duration are responsible for the variations in the period, and a constant time sector, which has the same length regardless of the length of the period. In order to account for this character of rhythmic movements, the three-cell network of Figure 2a can be modified to include a fixed, period-independent impulse conduction time H in the inhibitory connection from cell A to cell C. The introduction of this additional delay element increases the cycle period from $3R$ to $3R + 2H$. Moreover, in the general case of an N-membered ring with N inhibitory connections, of which M embody a conduction time H, the cycle period has the value

$$P = NR + 2MH \tag{3}$$

Thus, the variable time sector of this period evidently comprises NR, and the constant time sector, $2MH$. These predictions have been confirmed by means of neuromime analog circuits (Friesen and Stent, 1977).

An Identified Neural Network Generating a Rhythmic Movement

The central nervous system of the leech includes two large ganglia, or "brains," one in the head and the other in the tail. Head and tail brains are linked by a ventral nerve cord consisting of a chain of 21 segmental ganglia and their connectives (Figure 3a). Each segmental ganglion contains the cell bodies of some 200 pairs of bilaterally symmetrical neurons and innervates, via two bilateral pairs of segmental nerves, one of the 21 abdominal body segments lying between the head and the tail. The gross anatomy of the iterated segmental ganglia is sufficiently stereotyped from segment to segment and sufficiently invariant from leech to leech, that a large portion of the cell bodies of the central nervous system can be reproducibly identified. It is possible to penetrate these cell bodies with microelectrodes and record action potentials and excitatory and inhibitory synaptic potentials arising from synaptic connections with other neurons.

The leech swims by undulating its extended and flattened body in the dorsoventral plane, forming a body wave that travels rearward, from head to tail (Figures 3b and 4a). The moving crests of the body wave are produced by progressively phase-delayed contractile rhythms of the ventral body wall of successive segments; the moving troughs by similar, but antiphasic, contractile rhythms of the dorsal body wall. As was noted by Leonardo da Vinci (1938) (Figure 4b), the forces exerted against the water by these changes in body form provide the propulsion that drives the leech forward through its fluid medium. The period of

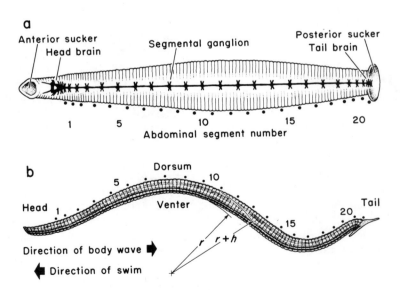

FIGURE 3. (a) Schematic view of the segmental body plan of the leech and of its nerve cord, from the ventral aspect. The skin of most abdominal segments is divided into five annuli. (b) Side view of a leech during the swimming movement. The body wave forms a crest in the 8th and a trough in the 16th abdominal segment. If r and $r + h$ are the radii of curvature of ventral and dorsal body walls at a wave crest, then the length ratio of contracted ventral to distended dorsal longitudinal muscles in the 8th segment is $1-(1 + h-r)$, or equal to 0.8 for the body wave shown here. (c) Dorsal aspect of

the segmental contractile rhythm ranges from about 400 msec for fast swimming to about 2000 msec for slow swimming (Kristan et al., 1974a).

The periodic changes in length of the dorsal and ventral body wall segments are produced by the phasic local contraction of *longitudinal* muscles embedded in the body wall, which, in turn, are innervated by an ensemble of excitatory and inhibitory motor neurons in the corresponding segmental ganglion. It is the rhythmic impulse activity of this motor neuron ensemble that drives the local contraction and distension of the segmental musculature (Kristan et al., 1974a,b; Ort et al., 1974). These motor neurons are located on the dorsal aspect of the segmental ganglion and are designated according to the numerical system indicated in Figure 3c. During swimming, these motor neurons produce impulse bursts in four phase angles of approximately 0°, 90°, 180°, and 270°, as shown in Figure 3d. Inasmuch as the time taken for the body wave to travel from head to tail is about equal to the swim period (so that the body of the swimming leech forms one spatial wavelength), the impulse burst phase of each of these motor

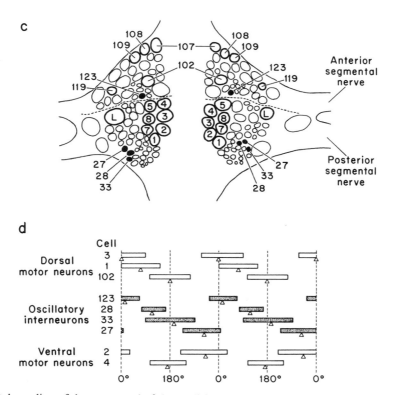

a segmental ganglion of the nerve cord of the medicinal leech, *Hirudo medicinalis*, showing the cell bodies of identified motor neurons (heavy outline) and of interneurons (solid black) related to the generation of the swimming rhythm. The cells are numbered according to the system of Ort et al. (1974). (d) Phase diagram of the activity cycles of excitatory motor neurons (cells 3 and 4), of inhibitory motor neurons (cells 1, 102, and 2), and of oscillatory interneurons of a segmental ganglion of the leech nerve cord during a swimming episode of an isolated preparation. The impulse burst midpoint of cell 3 has been arbitrarily assigned the phase angle 0°.

neurons leads that of its serial homologue in the next posterior segmental ganglion by about 20° (Kristan *et al.*, 1974b).

The motor neurons of an isolated leech ventral nerve cord, deprived of all sensory input from the body wall, can exhibit sustained episodes of swimming activity. Hence, the basic swimming rhythm is produced by a central nervous oscillator whose multiphasic activity pattern is generated independently of any proprioceptive feedback (Kristan and Calabrese, 1976). Four bilateral pairs of

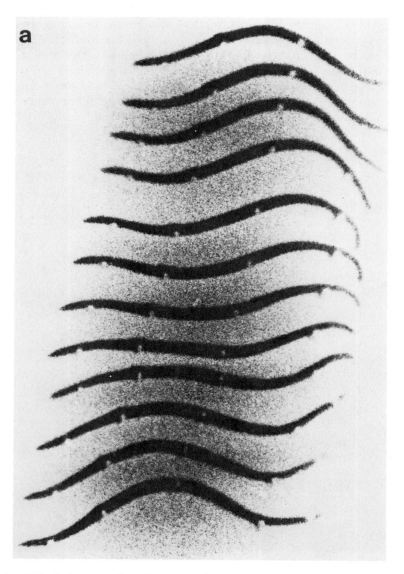

FIGURE 4. (a) The body wave of a swimming leech, as seen in this composite print of successive frames of a cinematographic record of a free-swimming specimen, with white reference beads attached to the 1st, 5th, 10th, and 15th abdominal segments. The right-to-left horizontal displacement of the

interneurons were identified as the component elements of this central swimming oscillator in each of the segmental ganglia (Friesen *et al.*, 1976, 1978). During swimming episodes, these interneurons produce impulse bursts in a phase progression similar to that of the motor neuron activity cycles, as shown in Figure 3d. The oscillatory interneurons impose the swimming rhythm on the motor neurons via a set of identified excitatory and inhibitory connections (Friesen *et al.*, 1976; Poon *et al.*, 1978) (Figure 5a). Since the leech swim oscillator interneurons

animal depicts its true progress in the water. The time occupied by this episode, which corresponds to one cycle period, is about 400 msec. (B) da Vinci's study of aquatic locomotion by generation of a body wave. The third item from the bottom represents a swimming leech.

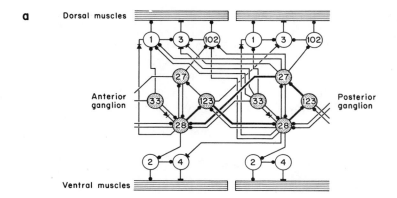

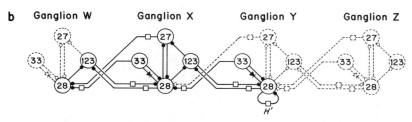

FIGURE 5. (a) Summary circuit diagram of identified synaptic connections between interneurons (shown as shaded circles), motor neurons (shown as plain circles), and longitudinal muscles responsible for the swimming rhythm. Meaning of symbols: T joint = excitatory synapse; filled circle = inhibitory synapse; diode = rectifying electrical junction. The connections forming the basic five-membered, recurrent cyclic inhibition ring are shown as heavy lines. (b) Partial electronic analog model of the network of neuromimes connected according to the circuit shown in (a). The circuit diagram schematizes the oscillatory interneurons of four ganglia W, X, Y, and Z, representing the 1st, 5th, 9th, and last of an isolated chain of 13 ganglia. Cells represented by neuromimes and their modeled connections are shown in solid lines; cells and connections omitted from the model circuit are

do not possess an endogenous polarization rhythm, their impulse burst activity must derive from their assembly into an oscillatory network. This network consists of both intraganglionic and interganglionic synaptic connections of serial homologues of the four oscillatory interneurons, as shown in Figure 5a. The axons of three interneurons project forward along the nerve cord and make inhibitory connections in several more anterior ganglia with serial homologues of one or both cells with which they connect in their own ganglion. The axon of the fourth interneuron projects rearward and makes inhibitory connections in several more posterior ganglia with serial homologues of a cell with which it does *not* connect in its own ganglion.

Analysis of the interneuronal system of Figure 5a has shown that it is an example of a network that owes its oscillatory character to the mechanism of recurrent cyclic inhibition. This network is topologically too complex to permit immediate recognition of its oscillatory features and simple explanation of how it generates the observed swimming rhythm. It is evident at once, however, that the network includes a five-membered recurrent cyclic inhibitory ring formed by

13. Neural Circuits for Rhythmic Movements 259

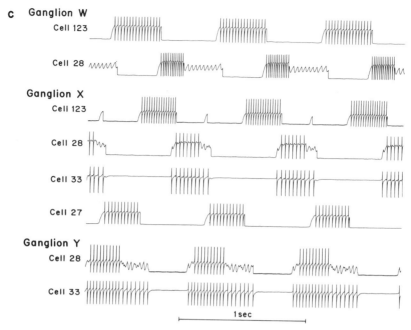

shown in dashed lines. The boxes designate connections with impulse conduction delays of 80 msec. The self-inhibitory phantom connection of cell 28 of ganglion Y incorporating a transmission delay $H' = 250$ msec replaces the presence of cell 123 of ganglion Y and of cells 33 and 28 of ganglion Z. The impulse transmission delays were modeled by means of shift registers. Sufficient tonic excitation was provided to each interneuron analog to produce an impulse frequency of about 80 Hz at the height of its active phase. The details of this model circuit and the justification for use of the phantom connection in place of the two posterior cells are described in Friesen and Stent (1977). (c) Impulse bursts generated by the electronic analog model shown in (b).

two cells of an anterior ganglion and three cells of a posterior ganglion (Figure 5a). Two of the connections of that ring, namely, the interganglionic connections leading from an anterior cell to a posterior cell and from a posterior cell to an anterior cell, incorporate a fixed conduction delay H, attributable to the time taken by impulses generated by an interneuron in one ganglion to reach the synaptic terminals in another ganglion. In a first approximation, therefore, the central swim oscillator can be thought of as an intersegmental network of interlocking five-membered recurrent cyclic inhibition rings with two fixed delay lines. In the absence of any other connections, this system would generate a crude version of the swimming rhythm. Two interneurons in each ganglion would produce antiphasic impulse bursts whose cycle period, according to equation (3), would be $P = 5R + 4H$ and, depending on the level of tonic excitation, could vary over a broad range, according to equation (2). The activity cycles of serial homologues of these two interneurons in successive ganglia of the ventral nerve cord would show a rostrocaudal phase lead of about 20°. The actually identified, topologically more complex network of Figure 5a can then be viewed as an elaboration of the basic five-membered intersegmental ring, in the sense that the additional cell and the additional connections create a set of subsidiary rings that generate the observed four-phased, segmental duty cycle. The cycle period of this network can be shown to depend on two parameters: (1) the intersegmental travel time of impulses conducted from ganglion to ganglion in the axons of the oscillatory interneurons and (2) the recovery time taken by each interneuron to reach action potential threshold upon its release from inhibition (Friesen and Stent, 1977).

Figure 5c presents the output of a partial electronic analog model (Figure 5b) of the full swim oscillator network. The model consists of eight neuromime elements, of which four represent the four oscillatory interneurons of one ganglion (ganglion X) embedded in a chain of 13 ganglia, of which ganglia W and Z are the front- and rearmost, and ganglia X and Y are the fifth and ninth within the chain, respectively. The four other neuromimes represent two cells of ganglion W and two cells of ganglion Y. As can be seen, the model oscillator runs with a realistic system cycle period of about 840 msec, reproduces for the four interneurons of ganglion X a good approximation of the observed interneuronal impulse burst relations shown in Figure 3d, and gives rise to an appropriate rostrocaudal phase progression of the cycle phases of the homologues of these cells in ganglia W, X, and Y. Nevertheless, the schematic of Figure 5a is probably an incomplete version of the actual circuit diagram. As shown by Weeks (1980), a single ganglion of an isolated leech nerve cord can produce the basic swimming rhythm even after its oscillatory interneurons have been surgically deprived of their *inter*ganglionic connections. Since the *intra*ganglionic connections shown in Figure 5a would not by themselves suffice for oscillatory activity in the absence of the known *inter*ganglionic connections, it would appear that there also exist some hitherto unidentified *intra*ganglionic elements responsible for additional rings within the concatenated cyclical interneuronal network.

Inasmuch as the identified neural circuit I have presented here pertains only to the nervous system of an annelid worm, one may ask whether these findings are generally applicable to central nervous oscillators generating rhythmic movements in animals of other species and phyla, particularly in the vertebrates. This question is not easy to answer at this time, because detailed cellular network analyses have thus far been possible only in a very few neurophysiologically favorable preparations. Nevertheless, it is significant that the mechanisms according to which these invertebrate circuits are now thought to generate their oscillations—endogenous rhythmic polarization, reciprocal inhibition, and recurrent cyclic inhibition—were all first proposed to account for generation of rhythmic movements in vertebrates (McDougall, 1903; Brown, 1911, 1912, 1914; Székely, 1965; Bradley et al., 1975). Moreover, the pattern of motor neuron activity in rhythmic movements of vertebrates is not necessarily more complex than the corresponding pattern in analogous movements of invertebrates. The very much greater number of neurons in the central nervous systems of vertebrates does not necessarily imply a greater complexity of the central oscillators generating their rhythmic movements; it may only place greater obstacles in the way of identifying the underlying neuronal circuitry. In any case, it is worthy of note that the current list of fundamentally different and theoretically plausible types of neuronal oscillators is not only quite short but also of long standing. On these grounds it seems reasonable to expect that the circuits discussed here will prove to be of general applicability to the generation of rhythmic movements in the entire animal kingdom.

ACKNOWLEDGMENTS. I thank W. B. Kristan, Jr. for his helpful suggestions and criticism. My research summarized in this chapter was supported in part by NIH Grant No. NS-12818 from the National Institute of Neurological and Communicative Disorders and Stroke and in part by NSF Grant BNS74-24637.

References

Adam, A. (1968) Simulation of rhythmic nervous activities. II. Mathematical models for the function of networks with cyclic inhibition. *Kybernetik* **5**:103–109.
Alving, B. O. (1968) Spontaneous activity in isolated somata of *Aplysia* pacemaker neurons. *J. Gen. Physiol.* **45**:29–45.
Barker, J. L., and H. Gainer (1975) Studies on bursting pacemaker potential activity in molluscan neurons. I. Membrane properties and ionic contributions. *Brain Res.* **84**:461–477.
Bradley, G. W., C. von Euler, I. Marttila, and B. Roos (1975) A model of the central and reflex inhibition of inspiration in the cat. *Biol. Cybern.* **19**:105–116.
Brown, T. G. (1911) The intrinsic factors in the act of progression in the mammal. *Proc. R. Soc. Ser. B* **84**:308–319.
Brown, T. G. (1912) The factors in rhythmic activity of the nervous system. *Proc. R. Soc. London Ser. B* **85**:278–289.
Brown, T. G. (1914) On the nature of the fundamental activity of the nervous centres; together with an analysis of the conditioning of rhythmic activity in progression, and a theory of the evolution of function in the nervous system. *J. Physiol. (London)* **48**:18–46.
Bullock, T. H. (1961) The origins of patterned nervous discharge. *Behaviour* **17**:48–59.
Calabrese, R. L. (1977) The neural control of alternative heartbeat coordination states in the leech. *J. Comp. Physiol.* **122**:111–143.
da Vinci, L. (1938) *The Notebooks*, Vol. 1, E. MacCurdy (translator). Reynal & Hitchcock, New York.
Dunin-Barkovskii, V. L. (1970) Fluctuations in the level of activity in simple closed neurone chains. *Biofizika* **15**:374–378.
Eckert, R., and H. D. Lux (1976) A voltage-sensitive persistent calcium conductance in neuronal somata of *Helix*. *J. Physiol. (London)* **254**:129–151.
Faber, D. S., and M. R. Klee (1972) Membrane characteristics of bursting pacemaker neurones in *Aplysia*. *Nature New Biol.* **240**:29–31.
Fentress, J. C. (ed.) (1976) *Simpler Networks and Behavior*. Sinauer Associates, Sunderland, Mass.
Frazier, W. T. E., E. R. Kandel, I. Kupfermann, R. Waziri, and R. E. Coggeshall (1967) Morphological and functional properties of identified neurons in the abdominal ganglion of *Aplysia californica*. *J. Neurophysiol.* **30**:1288–1351.
Friesen, W. O., and G. S. Stent (1977) Generation of a locomotory rhythm by a neural network with recurrent cyclic inhibition. *Biol. Cybern.* **28**:27–40.
Friesen, W. O., and G. S. Stent (1978) Neural circuits for generating rhythmic movements. *Annu. Rev. Biophys. Bioeng.* **7**:37–61.
Friesen, W. O., M. Poon, and G. S. Stent (1976) An oscillatory neuronal circuit generating a locomotory rhythm. *Proc. Natl. Acad. Sci. USA* **73**:3734–3738.
Friesen, W. O., M. Poon, and G. S. Stent (1978) Neuronal control of swimming in the medicinal leech. IV. Identification of a network of oscillatory interneurones. *J. Exp. Biol.* **75**:25–43.
Grillner, S. (1973) Locomotion in the spinal cat. In: *Control of Posture and Locomotion*, R. B. Stein, K. G. Pearson, R. G. Smith, and J. B. Redford (eds.). Plenum Press, New York, pp. 515–535.
Grillner, S. (1974) On the generation of locomotion in the spinal dogfish. *Exp. Brain Res.* **20**:459–470.

Grillner, S. (1975) Locomotion in vertebrates—Control mechanisms and reflex interactions. *Physiol. Rev.* **55**:247–304.

Harmon, L. D. (1964) Neuromimes: Action of a reciprocally inhibitory pair. *Science* **146**:1323–1325.

Harmon, L. D., and E. R. Lewis (1966) Neural modeling. *Physiol. Rev.* **46**:513–591.

Huber, F. (1975) Principles of motor-coordination in cyclically recurring behavior in insects. In: *Simple Nervous Systems*, P. N. R. Usherwood and D. R. Newth (eds.). Arnold, London, pp. 381–414.

Junge, D., and C. L. Stephens (1973) Cyclical variation of potassium conductance in a burst-generating neurone in *Aplysia*. *J. Physiol. (London)* **235**:155–181.

Kandel, E. R. (1967) Cellular studies of learning. In: *The Neurosciences: A Study Program*, G. C. Quarton, T. Milnechuk, and F. O. Schmitt (eds.). Rockefeller University Press, New York, pp. 666–689.

Kandel, E. R., T. J. Carew, and J. Koester (1976) Principles relating the biophysical properties of neurons and their patterns of interconnections to behavior. In: *Electrobiology of Nerve, Synapse and Muscle*, J. B. Reuben, D. P. Purpura, M. V. L. Bennett, and E. R. Kandel (eds.). Raven Press, New York, pp. 187–215.

Kater, S. B., and C. R. S. Kaneko (1972) An endogenously bursting neuron in the gastropod mollusc, *Helisoma trivolvis*. *J. Comp. Physiol.* **79**:1–14.

Kling, U., and G. Székely (1968) Simulation of rhythmic nervous activities. I. Function of networks with cyclic inhibitions. *Kybernetik* **5**:89–103.

Kristan, W. B., Jr. (1980) Generation of rhythmic motor patterns. In: *Information Processing in the Nervous System*, H. M. Pinsker and W. D. Willis, Jr. (eds.). Raven Press, New York, pp. 241–261.

Kristan, W. B., Jr., and R. L. Calabrese (1976) Rhythmic swimming activity in neurons of the isolated nerve cord of the leech. *J. Exp. Biol.* **65**:643–668.

Kristan, W. B., Jr., G. S. Stent, and C. A. Ort (1974a) Neuronal control of swimming in the medicinal leech. I. Dynamics of the swimming rhythm. *J. Comp. Physiol.* **94**:97–119.

Kristan, W. B., Jr., G. S. Stent, and C. A. Ort (1974b) Neuronal control of swimming in the medicinal leech. III. Impulse patterns of the motor neurons. *J. Comp. Physiol.* **94**:155–176.

Lewis, E. R. (1968) Using electronic circuits to model simple neuroelectric interactions. *Proc. IEEE* **56**:931–949.

McDougall, W. (1903) The nature of inhibitory processes within the nervous system. *Brain* **26**:153–191.

Meech, R. W., and N. B. Standen (1975) Potassium activation in *Helix aspersa* under voltage clamp: A component mediated by calcium influx. *J. Physiol. (London)* **249**:211–239.

Ort, C. A., W. B. Kristan, Jr., and G. S. Stent (1974) Neuronal control of swimming in the medicinal leech. II. Identification and connection of motor neurons. *J. Comp. Physiol.* **94**:121–154.

Pearson, K. G. (1972) Central programming and reflex control of walking in the cockroach. *J. Exp. Biol.* **56**:173–193.

Perkel, D. H, and B. Mulloney (1974) Motor pattern production in reciprocally inhibitory neurons exhibiting postinhibitory rebound. *Science* **185**:181–183.

Poon, M., W. O. Friesen, and G. S. Stent (1978) Neuronal control of swimming in the medicinal leech. V. Connections between the oscillatory interneurons and the motor neurons. *J. Exp. Biol.* **75**:45–63.

Pozin, N. V., and Y. A. Shulpin (1970) Analysis of the work of auto-oscillatory neurone functions. *Biofizika* **15**:156–163.

Reiss, R. F. (1962) A theory and simulation of rhythmic behavior due to reciprocal inhibition in small nerve nets. *Am. Fed. Inf. Process. Soc. Proc. Spring Computer Conference* **21**:171–194.

Smith, T. G., Jr., J. L. Barker, and H. Gainer (1975) Requirements for bursting pacemaker activity in molluscan neurons. *Nature* **253**:450–452.

Stent, G. S., W. B. Kristan, Jr., W. O. Friesen, C. A. Ort, M. Poon, and R. L. Calabrese (1978) Neuronal generation of the leech swimming movement. *Science* **200**:1348–1357.

Stent, G. S., W. J. Thompson, and R. L. Calabrese (1979) Neural control of heartbeat in the leech and in some other invertebrates. *Physiol. Rev.* **200**:101–136.

Strumwasser, F. (1967) Types of information stored in single neurons. In: *Invertebrate Nervous Systems*. C. A. G. Wiersma (ed.). University of Chicago Press, Chicago, pp. 291–319.

Strumwasser, F. (1971) The cellular basis of behavior in *Aplysia*. *J. Psychiatr. Res.* **8**:237–289.

Székely, G. (1965) Logical networks for controlling limb movements in Urodela. *Acta Physiol. Acad. Sci. Hung.* **27**:285–289.

Székely, G. (1967) Development of limb movements: Embryological, physiological and model studies. In: *Ciba Foundation Symposium on Growth of the Nervous System*, G. E. W. Wolstenholme and M. O'Conner (eds.). Little, Brown, Boston, pp. 77–93.

Thompson, W. J., and G. S. Stent (1976a) Neuronal control of heartbeat in the medicinal leech. I. Generation of the vascular constriction rhythm by heart motor neurons. *J. Comp. Physiol.* **111**:261–279.

Thompson, W. J., and G. S. Stent (1976b) Neural control of heartbeat in the medicinal leech. II. Intersegmental coordination of heart motor neuron activity by heart interneurons. *J. Comp. Physiol.* **111**:281–307.

Thompson, W. J., and G. S. Stent (1976c) Neural control of heartbeat in the medicinal leech. III. Synaptic relations of the heart interneurons. *J. Comp. Physiol.* **111**:309–333.

Trautwein, W. (1973) Membrane currents in cardiac muscle fibers. *Physiol. Rev.* **53**:793–835.

Warshaw, H. S., and D. K. Hartline (1976) Simulation of network activity in stomatogastric ganglion of the spiny lobster, *Panulirus*. *Brain Res.* **110**:259–272.

Watanabe, A., S. Obara, and T. Akiyama (1967) Pacemaker potentials for the periodic burst discharge in the heart ganglion of a stomatopod, *Squilla oratoria*. *J. Gen. Physiol.* **50**:839–862.

Weeks, J. C. (1980) The roles of identified interneurons in initiating and generating the swimming motor patterns of leeches. Ph.D. thesis, University of California, San Diego.

Wilson, D. M. (1966) Central nervous mechanisms for the generation of rhythmic behavior in arthropods. *Symp. Soc. Exp. Biol.* **20**:199–228.

Wilson, D. M., and I. Waldron (1968) Models for the generation of the motor output pattern in flying locusts. *Proc. IEEE* **56**:1058–1064.

14

Ordered Retinotectal Projections and Brain Organization

Christoph von der Malsburg

ABSTRACT

The analogy of brain to computer has been sporadically popular. This chapter pursues instead the analogy between brain function and cooperative effects. As an example, the cooperative theory for the establishment of topologically ordered fiber projections is discussed, according to which fibers growing from the retina to the optic tectum retain or recover the geometric relations that they had in the retina by sensing signals transmitted via the optic axons to the tectal cells. By comparison between a retinal cell's signal and that which it senses in tectal cells it contacts, its synapses are either reinforced (eventually to become permanent) or extinguished (in which case the retinal cell's projection withdraws from that tectal cell). This proposal has been tested by extensive computer simulations, which correctly describe normal and experimentally perturbed development in many different situations by means of a single algorithm and one set of parameter values.

The above simulation results illustrate four features common to cooperative systems: (1) the processes are based on systems with a large number of microelements in an initially undifferentiated state; (2) they contain self-amplifying fluctuations; (3) the fluctuations may compete in Darwinian fashion; and (4) fluctuations may cooperate by enhancing the "fitness" of other fluctuations. The result is the emergence of ordered modes, in which all interactions have come to a global equilibrium. The system of ordered modes constitutes a new "macro"-level of complexity. The relationship between micro-level and macro-level is not trivial. No deterministic dynamics can, in general, be formulated for the ordered modes.

The chapter continues on the assumption, prevailing today, that the brain activity relevant for thought processes conforms to the scheme of cooperative phenomena. This conceptual framework is shown to be natural to the discussion of a number of important issues on brain function: the autonomous nature of organization in the brain (in distinction to the computer, which requires a programmer); the perceived unity of our thought processes in a system composed of myriads of elements; causality and determinism, discussed as nonissues; and perception as an *active* process, rather than a passive intake of information.   —THE EDITOR

CHRISTOPH VON DER MALSBURG • Max Planck Institute for Biophysical Chemistry, D-3400 Göttingen, West Germany.

The Significance of Retinotopy

The design and behavior of computers have had a very strong influence on thinking about the brain. The astonishing potential of such machines inspired hope for the possibility of explaining mind on the basis of matter. Other aspects of the analogizing have had less beneficial effects: interpretation of nerve cells as logical switching elements, analogous to the electronic gates of the computer, has never led to successful theories for brain science.

A recently proposed new kind of theory emphasizes a different aspect of brain function. The essential idea behind these theories is that of "cooperative effects," or "relaxation processes." In contrast to the older focus on switching elements, inspired by the computer analogy, these new attempts concentrate more on the ordered, organized behavior of multitudes of elements. This chapter introduces the principles behind such theories, using a particular case: the way in which topological projections of nerve fibers are organized in the embryo. The question to be posed is: How, during embryonic development, can the nerve fibers of a projection establish topological order?

There are various reasons for considering this problem as an important example. Today the dominant general principle of organization asserted by brain morphologists is that of topologically ordered fiber projections. Topological projections can be characterized by the property that neighboring cells of a projecting (presynaptic) sheet connect to neighboring points in the target (postsynaptic) sheet. Such topological projections are found between the sensory surfaces and various levels of the central nervous system. In the human visual system alone there may be more than 20 such projections. The special case of the retinotopic projection between eye and optic tectum, a system that is easily accessible for experiments, has been studied—mainly in amphibians and fish—with great intensity for several decades (for review see Gaze, 1978; Willshaw and von der Malsburg, 1979; Fraser and Hunt, 1980). One of the main sources of motivation for these experiments may have been the hope of elucidating principles of brain organization from a system that is numerically complex (some retinotopic projections contain millions of fibers), yet conceptually simple: a topological projection can be described in simple geometric terms. From the efforts of many neuroscientists we have enough data to allow for a selection among different theories—a rare situation in the brain sciences.

In this chapter I examine the wealth of experimental data on retinotopic projections in terms of a theory published some years ago (Willshaw and von der Malsburg, 1976, 1979; von der Malsburg and Willshaw, 1977). A closely related theory, developed independently, is contained in Fraser and Hunt (1980).

The Cooperative Retinotopy Theory

The recently proposed theory of retinotopic projection considers a growth process of retinal fiber terminals on the tectal surface. In this process the fibers reestablish in the tectum the neighborhood relationships that they had within the retina. The process is organized by signals. The nature of these signals is not yet clear. Nervous activity and chemical markers have been proposed. Recent experiments (Harris, 1980; Meyer, 1983) point to the significance of nervous activity. The issue is not important for the argument of this chapter. Signals originate in the retina and are transmitted by fibers to the tectum. At each tectal location, signals coming from different retinal cells are superposed. The basic idea is that a fiber can "sense," through the particular signal mixture its contacts encounter in the tectum, the presence of other fibers and can selectively grow where neighboring fibers (i.e., fibers that are neighboring in the retina) are present with their contacts. It is essential to this process that the neighborhood relationships of retinal cells express themselves in correlations between the cells' signals. A second prerequisite is that a signal transmitted to a tectal location can be detected also in neighboring points of the tectum. A concrete realization of the theory meeting these requirements follows.

The retinal cells in this model spontaneously create signals. These signals are correlated over some distance by excitatory short-range connections (and possibly longer-range inhibitory connections) within the retina. (In the case of chemical signals, excitatory connections are to be replaced by diffusion paths.) The correlations between the signals express neighborhood-relationships within the retina.

Retinal cells project fibers to the tectum (which they find by a mechanism not discussed here) and establish contacts (synapses) with tectal cells. The signals reach the tectum and are transmitted to tectal cells via the contacts. A synapse is characterized by a number, called its weight. The greater the weight, the stronger is the influence of the presynaptic signal on the signal in the postsynaptic cell. Signals are exchanged between neighboring tectal cells, as is the case for neighboring retinal cells.

Fibers modify their synapses in response to the signal distribution over the tectum in the following way. Each synapse makes a comparison between the signals of its presynaptic and its postsynaptic side. The comparison is evaluated in terms of a correlation, which is 1 for identical signals and 0 for unrelated signals. Each synaptic weight is allowed to grow at a rate that is correlated positively with its correlation. There is a sum rule: that is, the sum of the weights of all the synapses made by one retinal fiber is held constant (by subtractive terms in the rate equations for the synaptic weights). Consequently, a synapse can grow only at the expense of other synapses made by the same retinal fiber. New synapses are created near established ones, and unsuccessful synapses (those whose weight has fallen below a certain threshold) are retracted.

Plausibility arguments easily show that the mechanisms just described lead to a retinotopic projection. A single synapse is self-reinforcing: by transferring the signal carried by its fiber to a tectal cell, it increases the correlation of that cell's signal to the one in its fiber and thereby increases its own rate of growth. Two retinal fibers coming from neighboring cells and connecting to a single tectal cell contribute to each other's growth, because they carry strongly correlated signals. Likewise, two synapses from one retinal cell (or from two neighboring cells) projecting to two neighboring tectal cells support each other, because the tectal cells communicate and the signal transferred by one synapse reaches the postsynaptic cell of the other. On the other hand, if a fiber makes two tectal synapses that are far apart (in comparison to the tectal communication length), the synapses cannot

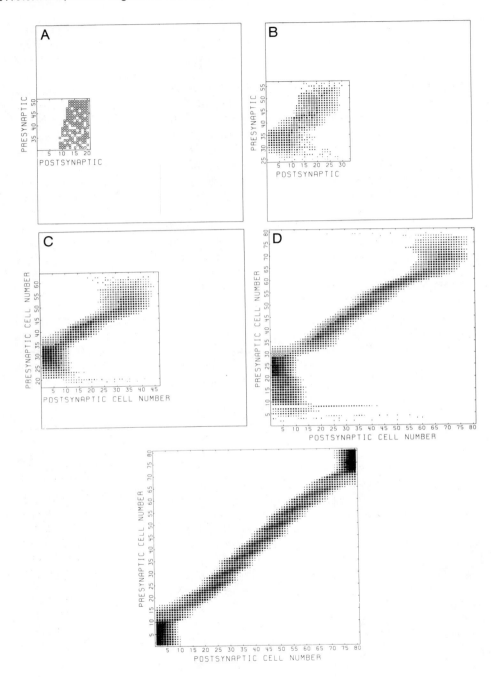

FIGURE 1. Simulation of normal development of a retinotectal projection according to cooperative retinotopy theory. Retina and tectum are one-dimensional. The two chains are growing during the process. Each square in the matrix signifies a synapse between corresponding presynaptic (retinal) and postsynaptic (tectal) cells. Weights of synapses are represented by area of squares. Initial condition is (A). In (B)–(D), open squares correspond to shrinking synapses.

cooperate and, because of the constant sum rule, will compete instead. Similarly, synapses made by fibers from different retinal sites and carrying rather uncorrelated signals interfere with each other when contacting the same tectal cell (or closely neighboring cells).

In summary, the situations most favored by the proposed growth mechanism are those in which fibers from small retinal regions are concentrated into small tectal regions and in which small tectal regions receive fibers from small retinal regions only. If these situations prevail throughout, the projection will be retinotopic.

The mechanism described thus far acts on a microscopic scale. As such, it has no preference for one orientation of retinotopic projection (e.g., the nasal retina projecting to caudal tectum) over another. But, in fact, orientation is nearly invariant; it must be imposed on the system by an additional mechanism. Only a very weak influence is needed in principle for determining the orientation of the projection, and this influence is very helpful in preventing the developing projection from being trapped in a "local optimum." It is not yet clear by which additional mechanism the orientation is actually imposed. [The mechanism may vary from case to case. Chemical markers which are rigidly preestablished in retina and tectum may play an important role, as discussed in Fraser and Hunt (1980) and Whitelaw and Cowan (1981).] For the sake of argument, we assume that there is an initial random distribution of synapses over the tectal surface, subject to a slight restriction: fibers from the rim of the retina are precluded from forming synapses in the part of the tectum that lies opposite to their destined target region. Thus, the initial projection assumed here could be described as one having very imprecise retinotopy.

Simulations

The adequacy of the ideas outlined above to account for the development of retinotopic projections is most easily demonstrated with the help of numerical simulations. An extensive set of simulations has been described elsewhere (Willshaw and von der Malsburg, 1976, 1979; von der Malsburg and Willshaw, 1977; von der Malsburg, 1979) and is summarized here. The simulations accounted correctly for observations of normal development and a large number of different experiments, all with a single algorithm and one set of parameters. Most situations may be treated by one-dimensional calculations that economize computer capacity and are conveniently displayed. The theory has, however, also been tested in two-dimensional simulations (Willshaw and von der Malsburg, 1976; von der Malsburg, 1979).

The normal development of retinotectal projections is complicated by the fact that maturation, i.e., the ability of retinal cells to put out axons and of tectal cells to receive synaptic contacts, spreads out in waves over the two structures.

(B)–(E) correspond to iteration steps 200, 400, 900, 3000, respectively. Note the distortion and global movement of the projection in (B) and (C). These are due to the fact that low-numbered retinal fibers have to colonize low-numbered tectal territory and push away synapses that have settled there earlier. The signals used had linear transport properties (diffusing molecules). Details of simulation are given in Willshaw and von der Malsburg (1979).

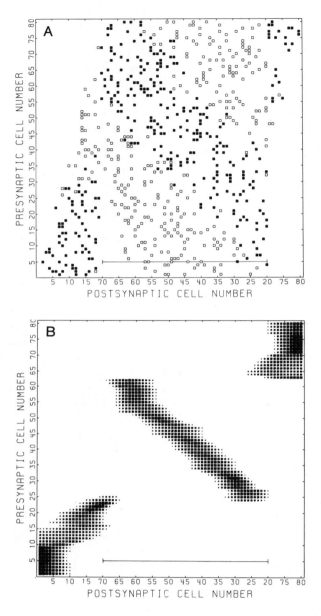

FIGURE 2. Simulation of a graft rotation experiment. After a projection as in Figure 1E had formed, all synapses were severed, a stretch of postsynaptic cells (numbers 20 to 70) was cut out and reversed in order (together with their previously acquired signals), and a new initial distribution of synapses was created. The latter was asymmetric to favor one of the two possible orientations of projection.

(A) shows the system immediately after these manipulations. The extent of the graft is indicated by a bar in the lower part of the matrix. In the simulations, the restriction of growth within the graft to a reversed projection is the effect of surviving signals. With nonpersistent (electrical) signals, a reversal of the projection within the graft could be the effect of a more precise orientation mechanism.

(B) shows the stationary state reached after 1600 iterations. A second simulation, in which signal strengths were allowed to smooth and decay before the arrival of regenerating synapses, ended in a completely continuous projection similar to Figure 1E (Willshaw and von der Malsburg, 1979).

14. Ordered Retinotectal Projections

In the retina, the wave starts in a central region and propagates radially outward in all directions. In the tectum, in the clawed toad *Xenopus laevis* and the goldfish at least, the wave starts at the rostral (front) end and runs in a curved fashion over the tectum to the caudal (rear) end. At each moment only a small fraction of all retinal fibers are entering the tectum, and only a small portion of the tectal cells are ready to be innervated by those fibers. Thus arises a difficulty, in that the fibers originating in the central retina, although first contacting the rostral part of the tectum, later have to shift connections to the central part of the tectum (for a precise description and references, see Easter and Stuermer, 1984). Figure 1 shows how a simulation deals successfully with that difficulty in the developmental case.

Many experiments performed to examine the mechanism of development of retinotectal projections are of the regeneration type: after completion of normal development, the visual system is perturbed by manipulation, the animal is allowed to survive (typically for some months), and the resulting projection is examined. This discussion is restricted to two experiments with seemingly contradictory results.

The first is of the "mismatch" type: the optic nerve is severed and half of the retina is destroyed or removed. The remaining half of the retina regenerates a projection to the tectum that at first covers only the part of the tectum that originally received the projection from the surviving part of the retina. After several months, the half retinal projection has expanded to fill the entire tectum with a continuous retinotopic projection. This feature of the growth mechanism has been called "systems matching" and is accounted for correctly by the present theory (Willshaw and von der Malsburg, 1979).

The second experiment resorts to graft rotation: the optic nerve is destroyed and a square piece of tectal tissue is excised and grafted back in a rotated orientation. About half of the animals with regenerated optic nerve have a completely continuous projection; in the other animals, part of the projection is rotated along with the graft. This latter result seems to contradict the systems matching feature, since a discontinuous projection is observed. Depending on initial conditions of regeneration, both of the possible outcomes of the graft rotation experiment have been reproduced by the computer model (Figure 2). The rotated graft may be stable in the sense that it forms a local optimum.

These and other simulations have been shown to account for all documented facts and peculiarities regarding experimentally manipulated as well as normal retinotopic projections. It may therefore be permissible to discuss retinotopic organization entirely in terms of the cooperative theory.

Typification of Self-Organization in General

The great effort to identify the particular process by which the nervous system makes and maintains retinotopic projections will achieve full yield only if we can extract from that example some invariant principles that may be applied elsewhere. To that end I shall now discuss more abstract aspects of the example. (This part of the discussion is closely related to the contribution by Haken, Chapter 21.)

Self-organization, in all systems capable of the process, may be typified by the five characteristics listed below. The general requirements are given in italics; specifics refer to the retinotopy mechanism.

1. *There is a system consisting of a large number of microscopic elements. The system initially is in a relatively undifferentiated state.* In the example, the initial state is a disordered retinotectal projection. In the extreme case, every retinal cell would connect to every tectal cell with equal strength.

2. *There are self-amplifying fluctuations in (i.e., deviations from) the undifferentiated state.* In the present example, fluctuations in synaptic density produce their own growth by the injection of "their" signal.

3. *Some limitation of resource forces competition among fluctuations and selection of the fittest (i.e., the most vigorously growing) at the expense of others.* Synapses emanating from the same retinal cell are coupled by the constant sum rule, representing a resource limit. Synapses from different retinal locations converging on one tectal cell interfere with each other.

4. *Fluctuations cooperate. The presence of a fluctuation can enhance the fitness of some of the others, in spite of the overall competition for resources in the field. (In many systems the "fitness" of a fluctuation is identical with the degree of cooperativity with other fluctuations.)* A synapse grows much faster if it is helped by others introducing similar markers into the same tectal neighborhood.

5. *Whole systems of cooperatively interacting fluctuations emerge as ordered, differentiated states, or ordered modes—the order often extending over a wide area.* Ordered modes are arrangements of synapses in the form of retinotopic projections.

Ordered modes at one level of organization may be regarded as primitive objects on another (higher) level of complexity. A new dynamics may be formulated in terms of the degrees of freedom of the ordered modes. Originally all theories of retinotopy were formulated on the macroscopic level.

In any partially ordered state, microscopic variables are dominated by forces tending to bring them into conformity with a long-range order. (In Chapter 21, Haken calls this the "slaving principle.") For example, fiber number 8 in Figure 1 lags behind in the ordering process and is dominated by forces adapting the fiber to the ordered mode constituted by the diagonal band of synapses in the matrix.

Many systems contain, in the unorganized state, symmetries that are no longer present after self-organization. Only a complete set of the (mutually exclusive) ordered modes, taken as a whole, contains the original symmetry. In each particular case symmetry is broken spontaneously. In the example, final maps of both orientations would be produced with equal probability, if it were not for the asymmetry introduced artificially by omitting certain synapses from the initial distribution.

The bifurcations leading to spontaneous symmetry-breaking are extremely sensitive to asymmetric external influences, often called "ordering fields." The mechanism determining the final orientation of the map may in principle be extremely subtle—just strong enough to dominate any other random or systematic asymmetries.

In summary, self-organizing systems comprise two levels of complexity. On the microscopic level there are atomistic objects and interactions. Their large number, however, gives rise to a large number of microstates and configurations that are analytically intractable from a detailed kinetic viewpoint. However, the same system can also be characterized on a more macroscopic level constituted by a smaller number of complex objects with fewer (complex) interactions. The smaller number of macroscopic objects and interactions may now be accessible to detailed analysis. The relationship between these two levels usually is very complicated. In the case described here it was established by computer simulations and also in a mathematically rigorous way (Häussler and von der Malsburg, 1983).

Retinotopic Projections Exemplifying Principles of Brain Organization

Spatially ordered projections are a general feature of brain organization. Given the general utility of the broad principles of self-organization outlined above, it is reasonable to ask whether these principles can be broadly applied to organization of the brain and to the emergence of thought processes.

Specificity of the central structure (tectum) in the retinotopy mechanism is induced from the periphery (retina) in the form of signals. There is evidence that a similar induction takes place during other processes of central differentiation, such as those leading to "barrels" in the central representation of rodent and feline whiskers (Van der Loos and Dörfl, 1978) or to ocularity domains. (For a discussion, see von der Malsburg, 1979, in which the organization of ocularity domains is shown to be a direct consequence of the retinotopy mechanism.) It would not be surprising if induction from the periphery of fine specializations in central structures turned out to be a general principle of development, at least for the neocortex. It has been argued that the neocortex, apart from the definition of large functional areas, is an intrinsically homogeneous structure that is ready to process whatever input it receives (for review see Creutzfeldt, 1977). It remains to be seen whether or not it becomes finely specialized according to the principles discussed above.

It is interesting to reflect on the way the retinotopic specialization of a tectal location might be said to be programmed genetically. The main genetic program provides general mechanisms: transport of signals along fibers and the behavior of terminal branches of a fiber in reaction to signals. More specific parts of the genetic program are responsible for guiding fibers to the tectum and for providing the asymmetry in initial or boundary conditions that (reliably!) determines the final orientation of the map.

The quantity of genetic information required is independent of the number of retinotectal fibers. The bulk of the information in the fiber specificity of the final projection is created by interactions among cells. (For example, before the establishment of a projection, a tectal cell is not specialized to receive contact from a particular retinal point.) Any comparison between a naively calculated information content of the brain's microanatomy and the quantity of information contained in the genome is therefore impossible.

Geometry plays a role in the proposed retinotopy mechanism. Geometric continuity is imposed on the process when signals are exchanged between neighboring cells in retina and tectum, i.e., by structuring the topology of transport paths. If a hypothetical "retina" has a pattern of connections with no simple geometrical structure, and if the "tectum" is interconnected in a corresponding way, the mechanism that produces retinotopy may also produce a projection in which pairs of cells connected in the retina project to pairs of cells connected in the tectum. In other words, the mechanism could detect an isomorphy between two structures in terms of inner connections and respond to it with a projection pattern (for applications see von der Malsburg, 1981, 1985a).

The brain is often viewed from the seemingly dichotomous aspects of structure and function. The model discussed here addresses the ontogeny of structure. The important processes are supposed to take place in the course of hours or days. Some thought processes, on the other hand, can take place in fractions of a second, and one may ask whether our paradigm is relevant to them. Synaptic plasticity, which during ontogeny involves the sculpting and movement of fiber terminals, may later take the form of a fast functional modulation of immobile synapses. It is conceivable that mechanisms quite analogous to those that establish retinotopy during ontogeny are also acting during thought processes, e.g., in recall or in the recognition of an object (von der Malsburg, 1981, 1985a).

Ordered Modes as Fundamental Objects of Brain Organization

Does the abstract idea of self-organization of ordered modes as a basic brain process help us find new approaches to the urgent questions about the brain? In the fields of brain theory and artificial intelligence, there is a trend toward believing so. (For a review, see Chapter 15 by Arbib.) Compared to the brain theories prevailing 30 years ago, the ordered mode picture represents a radical change. In the 1950s, conjectures about the brain were dominated by the analogy with switching processes in the computer. The elements of a computer change states in (theoretically) discrete steps, in response to the few signals arriving at their inputs; i.e., decisions are made on a rather local basis (local in terms of the complete data relevant for the computing task). Due to these two aspects of its function—discontinuity and locality—the computer is not an independent system. It requires a "well-organized" mind, a programmer, to make sure that the individual actions intermesh in such a manner as to give rise to meaningful composite action (for a more complete discussion see von der Malsburg, 1985b).

In contrast to a computer, a self-organizing system has an inherent tendency to establish stationary states. It can do so because the state of each element is only slightly modified by the signals arriving at it each moment. Large changes of state need times that are much longer than the signal transmission times. Thus, an element can change its state drastically only after having had the opportunity to communicate with a large part of the network. In addition, a signal can cause lasting effects only if it is validated by other signals acting in the same direction. This cooperativity between signals is created and enhanced by synaptic plasticity of the kind described above. A particular configuration of variables can be stationary only if all interactions within the network reproduce it from moment to moment. Thus, the tendency for global order is inherent in self-organizing systems, whereas in the computer it is the special result of special programs. (Of course, computers can, with the help of special and complicated programs, simulate self-organizing networks, as was described above.)

For a long time the computer has acted as the basic analogy for brain organization and has deeply influenced our thinking. Abandoning it now in favor of the ordered mode concept opens up new vistas. One of the great puzzles of the brain is the mind-body problem. The categories that we use to describe the phenomena displayed by the thinking mind on the one hand, and by the nervous system observed with electrode and microscope on the other, are quite different. How can such divergent phenomena be unified in one scientific theory? A percept or concept that we have in our mind, such as the visual image of an object, unites a large number of elementary perceptions. Yet, at the same time, the concept gives us the feeling of coherence, order, and *unity*; it is a kind of primitive object itself, but on a much higher level than that of the elementary perceptions.

Again and again it has been proposed that high-level objects are represented by physical elements—"cardinal cells" (Barlow, 1972). However, the concept of the cardinal cell has never solved any problem, because it only increases the number of required elements in the brain, thereby aggravating the original problem, that of explaining a perceived unity. A mechanism is needed to tie together hosts of more elementary objects to form objects on a higher level. In its basic outline, that problem is solved by the scheme of self-organization sketched above, although it surely is no trivial problem to work out in more precise terms the nature of the processes required (von der Malsburg, 1981, 1985a).

Consider for a moment that the primitive objects of a phenomenology of mind can be identified with the "ordered modes," or stationary states, of a self-organizing system. These are established only after the exchange of a very large—in mathematical idealization, an infinite—number of signals. Consequently, the ordered modes are states of a very special kind. When discussing the structure of ordered modes, causality (in the sense that one part of the configuration is the result of another part) is no longer a meaningful concept, because each local configuration is as much a cause as it is an effect. There is an infinite number of different transients leading to the same ordered mode, and it becomes impractical to develop theories about them.

The point may be illustrated by referring to the process of perception. In the context of artificial intelligence or of the psychology of perception, an often-discussed question is whether perception is a "top-down" or a "bottom-up" process. "Bottom-up" refers to the process of interpretation that starts from elements excited rather directly by sense organs and ends by constructing a symbolic representation of objects. The representation makes use of elements that are more directly connected to other modalities, e.g., language. A "top-down" process is one that begins with a guess on the symbolic level. The guess is either validated or discarded by performing tests on the sensory elements. Even these "top elements" correspond to the lower level of the scheme discussed here. The upper level in my scheme is the ordered mode, in which top and bottom elements are assembled and "relaxed" into a stationary configuration that gives way as much as possible to cooperative interactions among the elements. The top-down versus bottom-up question then involves (nonstationary) transients and consequently cannot be answered meaningfully in any general theory of steady states or equilibrium structures.

I view thought processes as successions of nearly stationary states concatenated by discontinuous transients. The self-organization scheme implies certain peculiarities in the way states take over from each other. As noted earlier, very weak influences may have large effects if they act as ordering fields. Let me again use perception as an illustration. A vanishingly small percentage of the fibers entering a sensory area of neocortex represents direct sensory input. This fact suggests (although it does not prove) that the sensory input acts as a kind of ordering field, breaking the symmetry of the central mechanism to the extent that the strong interactions inside are nudged off equilibrium. The state of the system then falls into the ordered mode suggested by the weak input, so that most of the "information" inherent in that ordered mode may well have been created internally.

It is often debated whether or not the brain is a deterministic machine. Sometimes this question is raised in connection with existence and the nature of free will. In the framework discussed here the question gets a peculiar answer. In what way do ideas arise and follow each other in the flow of thoughts? Let us consider thoughts as ordered modes of some kind. We have seen above that ordered modes sometimes behave in a probabilistic way, even if the microscopic events are governed by deterministic laws. By considering only ordered mode phenomena, chance cannot be reduced; it cannot be traced back to the microscopic causal events. It is pointless to discuss the phenomenology of microscopic events if one is interested in macroscopic phenomena. The two levels do not have a simple relationship; each level of phenomena deserves and requires its own descriptive language.

The great experimental and theoretical effort that has been invested over decades in the study of the retinotopic projection was inspired by the hope of learning something general about brain organization. I believe that this hope was justified and that the lessons learned are of much deeper significance than could have been expected at the outset. Cooperative phenomena provide a paradigm for analysis of ongoing nervous function as well as for development of nervous structure.

References

Barlow, H. B. (1972) Single units and sensation: A neuron doctrine for perceptual psychology? *Perception* **1**:371–394.

Creutzfeldt, O. D. (1977) Generality of the functional structure of the neocortex. *Naturwissenschaften* **64**:507–517.

Easter, S. S., Jr., and C. A. O. Stuermer (1984) An evaluation of the hypothesis of shifting terminals in goldfish optic tectum. *J. Neurosci.* **4**:1052–1063.

Fraser, S. E., and R. K. Hunt (1980) Retinotectal specificity: Models and experiments in search of a mapping function. *Annu. Rev. Neurosci.* **3**:319–352.

Gaze, R. M. (1978) The problem of specificity in the formation of nerve connections. In: *Specificity of Embryological Interactions*, D. Garrod (ed.). Chapman & Hall, London.

Harris, W. A. (1980) The effects of eliminating impulse activity on the development of the retinotectal projection in salamanders. *J. Comp. Neurol.* **194**:303–317.

Häussler, A. F., and C. von der Malsburg (1983) Development of retinotopic projections: An analytical treatment. *J. Theor. Neurobiol.* **2**:47–73.

Meyer, R. L. (1983) Tetrodotoxin inhibits the formation of refined retinotopography in goldfish. *Dev. Brain Res.* **6**:293–298.

Van der Loos, H., and J. Dörfl (1978) Does the skin tell the somatosensory cortex how to construct a map of the periphery? *Neurosci. Lett.* **7**:23–30.

von der Malsburg, C. (1979) Development of ocularity domains and growth behaviour of axon terminals. *Biol. Cybern.* **32**:49–62.

von der Malsburg, C. (1981) The correlation theory of brain function. Internal Report 81-2, Max-Planck Institute for Biophysical Chemistry, Göttingen.

von der Malsburg, C. (1985a) Nervous structures with dynamical links. *Ber. Bunsenges. Phys. Chem.* **89**:703–710.

von der Malsburg, C. (1985b) Algorithms, brain and organization. In: *Dynamical Systems and Cellular Automata*, J. Demongeot, E. Golès, and M. Tchuente (eds.). Academic Press, New York, pp. 235–246.

von der Malsburg, C., and D. J. Willshaw (1977) How to label nerve cells so that they can interconnect in an ordered fashion. *Proc. Natl. Acad. Sci. USA* **74**:5176–5178.

Whitelaw, V. A., and J. E. Cowan (1981) Specificity and plasticity of retinotectal connections: A computational model. *J. Neurosci.* **1**:1369–1387.

Willshaw, D. J., and C. von der Malsburg (1976) How patterned neural connections can be set up by self-organization. *Proc. R. Soc. London Ser. B* **194**:431–445.

Willshaw, D. J., and C. von der Malsburg (1979) A marker induction mechanism for the establishment of ordered neural mappings: Its application to the retinotectal problem. *Philos. Trans. R. Soc. London Ser. B* **287**:203–243.

15

A View of Brain Theory

Michael A. Arbib

ABSTRACT

"Top-down" brain theory (based upon functional analysis of cognitive processes) is distinguished from "bottom-up" brain theory (as might be based on analysis of the dynamics of neural nets). "Cooperative computation" is proposed as a way of providing a style of analysis for studying the interactions of neural subsystems at various levels.

The section on "Interacting Schemas for Motor Control" provides a top-down analysis of perception and the control of movement in the action-perception cycle. Perceptual "schemas" are introduced as the building blocks for the representation of the perceived environment, and motor schemas serve as control systems to be coordinated into programs for the control of movement. Next, two approaches to the design of machine vision systems are contrasted. The examples exhibit many of the insights to be gained from a top-down analysis but show that such an approach does not guarantee a unique functional analysis of the problem at hand. An algorithm for computing optic flow using the style of cooperative computation is presented. This model has not been validated yet by data from neurophysiology, but seems to be very much "in the style of the brain" and offers interesting insights into the evolution of hierarchical neural structures.

Two established neural models that have developed through a rich interaction between theory and experiment are presented. One emphasizes the possible role of the cerebellum in parametric tuning of motor schemas; the other represents interaction between tectum and pretectum in visuomotor coordination in frogs and toads. A connection between these neural models and the top-down analysis of cognition is described. —THE EDITOR

Brain Theory: "Bottom-Up" and "Top-Down"

"Brain theory" makes systematic use of mathematical analysis and computer simulation to elucidate the interactions of the components of the brain and ways they can subserve such diverse functions as perception, memory, and the control

---

MICHAEL A. ARBIB • Department of Computer Science, University of Southern California, Los Angeles, California 90089.

of movement. As such it differs from the mind theory of cognitive psychology, which seeks to analyze properties of the mind with little concern for how these properties are played out over the structure of the brain. In this chapter, I provide a view of brain theory informed by two viewpoints: the need for a healthy interaction between cognitive studies (top-down) and neuroscience (bottom-up) and the emerging utility of an approach to brain theory that emphasizes the "cooperative computation" of a multitude of subsystems (see also von der Malsburg, Chapter 14, this volume).

Brain theory should confront the bottom-up analyses of neural modeling not only with biological control theory but also with the top-down analyses of artificial intelligence and cognitive psychology (Arbib, 1975, 1978; Marr and Poggio, 1977a). In bottom-up analyses, we take components of known function and explore ways of putting them together to synthesize more and more complex systems. In top-down analyses, we start from some complex functional behavior that interests us and try to determine the natural subsystems into which we can decompose a system that performs in the specified way. Progress in brain theory will depend on the cyclic interaction of these two methods. In advocating a brain theory of this type, I suggest that many experiments in the laboratory of the neuroanatomists and the neurophysiologists should be related to evolving theories of high-level brain function. At the same time, cognitive analysis must constrain the subsystems posited by exploring whether or not they can be mapped into the circuitry of actual brain regions.

The top-down approach complements bottom-up studies, for one cannot simply wait until one knows what all the neurons are and how they are connected to simulate the complete system. Borges (1975) tells of a country that prided itself on the excellence of its cartography. As years went by, the cartographers produced maps of greater and greater accuracy, until finally they achieved the ultimate, full-scale map. And Borges wryly notes that there are places in the Western deserts where even today you can see tattered fragments of the map (presumably pegged to the place they represent). We need a guide to understand a new territory, but any map that provides no simplifications and pointers to distinctive features to aid our exploration does not help us. In the same way, a model that simply duplicates the brain is no more illuminating than the brain itself. We need theory to process data efficiently and to present the facts in an illuminating and insightful way; and we need detailed studies by neurologists to explore how brain lesions impair behavior (both transiently and permanently), as well as studies by neurophysiologists and neuroanatomists to examine circuit, cell, and synapse. Brain theory, properly conceived, can contribute to the design of further experiments to help shape the overall understanding of how portions of the brain interact to make us what we are.

Intermediate Units in the Brain: Between Top and Bottom

Rather than modeling individual activity in millions or more neurons, we can hope to understand much about brain function in terms of the interaction of spatial patterns distributed across a relatively small number of neural arrays. The layered structure of the brain is one of its distinctive features, and may well play an important role in helping us analyze the way in which sensory information—visual, somatic, or other—can be used to control behavior. Another unit of complexity, intermediate between the single neuron and the brain, is the module. One of the earliest modules came from the anatomical study of the reticular formation by Scheibel and Scheibel (1958), who observed that the major neurons of the reticular formation had dendrites parallel to one another and orthogonal to the axons that ran up and down along the head–tail axis. They suggested that nearby neurons could be aggregated into "poker chips" orthogonal to the head–tail axis, with the neurons within a module being roughly uniform in their sampling of the traffic up and down the reticular formation, as well as in their sampling of the peripheral input.

This analysis of the reticular formation, in terms of the interactions between a relatively small number of modules, was used by Kilmer *et al.* (1969) in their RETIC model of the reticular formation. This model was interesting not only because it was related to one of the earliest module concepts within neuroanatomy but also because it showed how a neural system could achieve some overall behavior without executive control. Kilmer and colleagues suggested that the reticular formation had the task of committing the organism to some overall mode of behavior. Each individual module of RETIC used its sample of inputs to make an initial determination of the relative desirability of the different modes. Different modules were coupled to a sample of their neighbors in such a way that the traffic between the modules proved sufficient for them to reach a consensus, in which the majority gave top priority to a single mode, thus committing the organism overall. More about this style of cooperative computation is discussed later.

Another form of intermediate unit is the "column"—an anatomic structure suggested first by the studies of Mountcastle (1957) and Powell and Mountcastle (1959) on somatosensory cortex and later by the work of Hubel and Wiesel (1974) on visual cortex. Moving up and down through the layers of sensory cerebral cortex, all the neurons are responsive to roughly the same stimuli from the external world, and a small displacement will find neurons with roughly the same features, but if we go farther we will move on to another column of cells describable by different features. This suggests, then, that much of the analysis of cortex can be conducted in terms of the interaction between columns, with the analysis of individual neurons playing a more restricted role to explain the dynamics of the column units (Szentágothai and Arbib, 1975; Szentágothai, 1978; Mountcastle, 1978).

Brain theory will progress both by computer simulation and by mathematical analysis. Such mathematical analyses must not only provide formal descriptions of systems but also must prove theorems about their behavior. There are mathematical analyses of properties of general classes of systems in relation to the Hodgkin–Huxley equation (reviewed by Rinzel, 1978), in studies of cooperative computation (as outlined below), and in control theory. On the other hand, there are many cases in which our symbolic representation of systems and their interactions does not take a form that lends itself easily to mathematical analysis.

Then, we turn to the computer to conduct neural simulations and cognitive modeling experiments in the style of artificial intelligence (AI) to gain insight into the capabilities of a system so represented. It may well be that we shall see the growth of a delicate interaction between mathematical analysis and simulation as we use mathematics to determine what is a sufficiently wide sample of different conditions in which to simulate a system to get a proper appreciation of the full range of its behavior.

Cooperative Computation

Time and again, we find that modern studies in brain theory must concern themselves with the integration of the activity of a multitude of subsystems. This brings us to the key question of cooperative computation: How is it that local interaction of a number of systems can be integrated to yield some overall result? The study of cooperative phenomena has its roots in the statistical mechanics of physics, whereby the individual motions of billions of atoms are averaged to produce reliable thermodynamic descriptions of the behavior of a gas or a liquid as a whole. For example, in studying ferromagnets, we seek to understand how atomic magnets can cooperate to yield global magnetism through the mass effects of local interactions. Cragg and Temperley (1954) were perhaps the first to suggest analogies between cortical activity and domain formation in ferromagnets. A number of interesting models were developed in the 1970s, starting with the studies of Harth et al. (1970) and those of Wilson and Cowan (1973), and continuing with the studies of Amari and Arbib (1977), Shaw (1978), and Amari (1980). For example, if we imagine visual cortex to contain a great variety of cells, tagged not only for visual direction but also for depth in the visual field, then we can imagine the process of recognizing regions in the visual field at different depths to be one of suppressing all neural activity except that corresponding to the depths within a given direction. This process of segmentation may have much in common with the process of domain formation in magnets (Julesz, 1971).

The question of how local interactions may yield global function may be studied discretely or continuously. The classic discrete study is von Neumann's (1951) use of tessellation automata to model self-reproduction. These structures comprise nets of regularly spaced automata, each connected to a few neighbors, with the next state of each unit being determined by the present state of units in its neighborhood. The analysis of such nets is purely combinatorial: given a program for local cellular interaction, we check the details of the program to determine if they do indeed yield some desired pattern formation. [See Arbib (1972b) and Ede (1978) for the use of such models in embryology.] In contrast, the continuous approach approximates a tissue of cells by functions varying continuously over the tissue, and uses techniques from differential equations, stability, and statistics. This approach to biological systems goes back to Turing's (1952) paper on morphogenesis, which was foreshadowed by Rashevsky (1948). Turing's paper introduced the use of reaction–diffusion equations into the study of pattern formation.

Turing studied a ring of cells. Within any one cell, substances could engage in chemical reactions; each morphogen could also diffuse between cells. One might think that such interactions would yield identical chemical equilibria in all the cells. To the contrary, Turing was able to show that, even with linear diffusion equations, the system would eventually be structured with standing waves of chemical concentrations, thus providing the substrate for the expression of biological pattern. When someone once asked whether his model would explain the stripes of the zebra, Turing replied, "The stripes are easy, it's the horse part that I have trouble with!" The important point here is that this model shows how local interactions give rise to global pattern. [See Katchalsky et al. (1974), Kopell (1978), and Fife (1979), and various papers in Amari and Arbib (1982) for further information on reaction–diffusion equations and related topics.] Grossberg (1978) has explored analogies between reaction–diffusion equations and neural processes, including the problem of patterning an array of synapses in neural learning. Haken (1978) seeks to provide a unified mathematical framework in which a number of these cooperative phenomena can be viewed. In particular, he relates the reaction–diffusion problem to the mathematics used to look at such phase transitions as the formation of a coherent pulse of light in a laser and the order–disorder transition in a magnet.

In top-down brain theory and AI, we find a discrete style of analysis of cooperative computation. Such AI projects as HEARSAY (Erman and Lesser, 1980) and VISIONS (Hanson and Riseman, 1978a) use a number of interacting knowledge sources to converge upon a perceptual analysis of some sensory input—an acoustic signal encoding a sentence in the first case, and a color photograph of an outdoor scene in the second. The conceptual structure thus created seems to hold promise for helping us understand how different regions of the brain interact—as in language behavior (Arbib and Caplan, 1979)—at the level of analysis of the neurologist concerned with brain lesions, rather than that of the neurophysiologist and neuroanatomist who can trace a few cells at a time.

When we turn to the analysis of neural networks *per se*, we find that the continuous style of analysis of cooperative computation is playing an increasing role. We have already mentioned that the formation of the underlying tissues has been studied in terms of reaction–diffusion equations. Now I shall briefly discuss schemes that explain how one part of the brain can be wired up to another, an analysis of the effects of early environment on the modification of feature detectors via synaptic plasticity, and models of cooperativity in the mature function of the nervous system. Certain of these concepts will be discussed in more detail later.

Retinotopy and Population Encoding

When light is focused by the lens upon the retina, a very small solid angle in the external world can affect each individual receptor. But as we move back through the layers of the retina and along the optic tract into different regions of the brain, neighboring neurons communicate and affect neurons farther along the pathway. Thus, the activity of neurons away from the periphery may be influenced by 20° or more of the visual field. Yet, in moving across layers of neurons in visual cortex, tectum (midbrain visual region), or lateral geniculate (thalamic visual region), there is a lawful direction across the surface corresponding to up and down in the visual field, and another corresponding to left and right. This property of preserving the spatial structure from the retina as we back up into the brain is called *retinotopy*.

Just as retinotopy refers to a variety of maps of the retina within the brain (Allman, 1977), so *somatotopy* denotes maps of the body surface—whether sensory maps of the tactile and other stimuli to the skin, or motor maps of the distribution of contraction of the musculature around the body (Brodal, 1969). In such layered structures, it may be inappropriate to regard a single neuron as conveying a vital message for the brain. Rather, it is the pattern of firing distributed across a whole array of neurons that robustly encodes vital information about the world (Erickson, 1974). If this is so, the occasional misfiring of an individual neuron poses little problem, because the receptive field that it samples overlaps the receptive field of hundreds or even thousands of other neurons in its vicinity in the same layer of the brain, yielding a natural redundancy and stability. This natural redundancy and stability, induced by the way in which layers of neurons represent spatial properties of the world, seems to solve the problem of reliable computation in the presence of noise, which so bothered von Neumann (1956).

The maps within a neural layer are not simple point-by-point transmissions of arrays of stimulation from the periphery. Rather, they involve sophisticated transformations. For example, Lettvin *et al.* (1959) identified several classes of ganglion cells in the retina of the frog, including cells that seemed most responsive to the presence of small wiggling objects in their receptive field, and cells that responded best to the passage of a large, dark object across their larger receptive field. Moreover, they found that these different types of cells distributed their messages to the tectum in such a way that each cell type projected to a different depth, with each projection retinotopic, and that corresponding points in the arrays were on top of one another.

In the mammalian retina, Kuffler (1953) found that the ganglion cells did not respond to these "frog-relevant" stimuli, but rather served to enhance contrast, whereas Hubel and Wiesel (1962) found "edge detectors" in visual cortex of cats and monkeys, which seemed to respond best to the movement of edges with a specific orientation in their receptive field. In this way, we find that the input to the brain is arrayed in maps of distinctive features. The suggestion is that these maps provide the input variables for controlling the animal's behavior. Before developing this idea further in a specific situation, I provide a brief survey of cooperative computation in retinotopic arrays.

Retinotectal Connections

Fibers from the retina reach the visual midbrain—the tectum—and there form an orderly map. Sperry (1951) showed that this retinotopy of the tectum would, at least in the frog, survive rotation of the eyeball after section of the optic tract. After such an operation, nerve fibers growing out from the retina would still find the original tectal locus. This result might suggest that each fiber bears with it a unique "address" and goes directly to the target point on the tectum. However, later experiments (reviewed by Gaze, 1970) showed, for example, that if half a retina were allowed to innervate a tectum, then the map would expand to cover the whole tectum, whereas if a whole retina innervated half a tectum, then the map would be compressed. In other words, the fibers in some sense had to sort out their relative positions in using available space, rather than simply reaching toward a prespecified target.

There are now a number of models that explain this phenomenon, not in terms of an overall global organization principle, but rather by local interactions of a few fibers and the portion of tectum upon which they terminate. These models include the arrow model of Hope et al. (1976), the marker model of von der Malsburg and Willshaw (1977), and the branch-arrow marker model of Overton and Arbib (1982). (See also Chapter 14 by von der Malsburg who, in common with this section, argues for the broader implications of such models for brain theory.)

Cortical Feature Detectors

Many cells in visual cortex are tuned as edge detectors, and it is known that the binocularity of this tuning can be modified by decoupling the input from the two eyes (Hubel and Wiesel, 1965). Moreover, early visual experience can drastically change the population of feature detectors (Blakemore and Cooper, 1970; Hirsch and Spinelli, 1971). This result suggests that the growth of the visual system in the absence of patterned stimulation can at most merely sketch the feature detectors, and that interaction with the world is required, either to express fully the normal situation, or to adapt an array of detectors to the peculiarities of a given environment.

These findings have challenged us to come up with models in which there are mechanisms of local synaptic change within a neuron based on correlation between presynaptic and postsynaptic activity, on reinforcement signals, or on a combination of the two (Grossberg, 1970; von der Malsburg, 1973; Grossberg and Levine, 1975; Kohonen and Oja, 1976; Amari, 1977). In particular, when local synaptic change is coupled with inhibitory interaction between neurons, a group of randomly connected model neurons can eventually differentiate functionally to produce edge detectors for distinct orientations. However, while certain cells in areas 17 and 18 of visual cortex become well tuned as simple cells, other cells in areas 18 and 19 become well tuned as hypercomplex cells. Furthermore, complex cells arise in all three areas. It thus requires a more subtle theory than any proposed to date to understand what it is about the precursive cellular geometry that provides preconditions for different patterns of learning.

Experiments by Spinelli and Jensen (1979) show that early visual experience may actually increase the area of cortex allocated to features represented by a given area of the periphery, contrary to a view that feature detector changes may simply involve atrophy of those neurons that are seldom active. Amari (1980) has modeled topographic organization of two nerve fields connected by modifiable excitatory connections and has indeed proved that a part of the presynaptic field that is frequently stimulated is mapped on a large area of the postsynaptic field (see Kohonen, 1982, for a related model).

Optic Flow and Stereopsis

Given two frames of visual information, we may ask how local features in one frame are matched with the correct features in the other. If the two frames are taken in temporal succession, then the stimulus-matching problem is that of computing optic flow; if the two frames come from two simultaneous vantage points (the left and right eyes, for example), then the stimulus-matching problem is that of stereopsis.

Many algorithms start with an initial set of hypotheses about matching and then use local interactions to change these hypotheses slowly, until a coherent segmentation of the image is obtained on the basis of common motion or common depth. Most models of stereopsis have been based purely on cooperative computation (Sperling, 1970; Julesz, 1971; Dev, 1975; Marr and Poggio, 1977b); however, Marr and Poggio (1979) have offered an alternative model that uses channels of different spatial frequency to obtain first a coarse match, which can then be refined without cooperative computation. Frisby and Mayhew (1980) have offered psychophysical evidence for a version of the model that does require cooperative computation between the channels of different spatial frequencies. A specific cooperative algorithm for the computation of optic flow will be described later.

The Continuity of Development and Function

We have seen a mathematical commonality that linked reaction–diffusion models of basic pattern generation in biological tissues; the formation of projections from one region of brain to another, as in the connection of retina to tectum; the tuning of connections to a cell within a tissue, as in the formation of feature detectors in visual cortex; and the actual function of a brain region, as in the computation of optic flow or stereopsis. This all suggests that many developments in brain theory will result from looking for a commonality of underlying mechanisms between neuroembryology (formation of connections) and adult function. To summarize this conclusion, "The brain is a somatotopically organized, distributed, layered computer" (Arbib, 1972a). Nonetheless, it must not be thought that there is any single method to be "plugged in" to solve all problems of brain theory. The remainder of this chapter will illustrate the diversities as well as the unities.

Interacting Schemas for Motor Control

Control theory has taught us how to represent a continuous system of simultaneously active subsystems linked by message-bearing pathways through complicated patterns of feedback and feedforward. In contrast, computation theory has taught us how to divide a complicated pattern into algorithms describable by flow diagrams, whose boxes correspond to the discrete activation of various data transfers, tests, and operations, and whose lines represent transfer of control from one box to another. In control, then, we have continually active systems in constant intercommunication; in algorithmic computation we have activation of one subsystem after another, with the pattern of activation delicately determined by tests of current data. I suggest that "programs in the brain" might be viewed as combining properties of both control block diagrams and computer flow diagrams. I use the term "schema" to designate the units of control from which these programs are built.

The Action–Perception Cycle

The notion of the action–perception cycle (Neisser, 1976; Arbib, 1981) emphasizes that the current environment does not, in general, determine the behavior of the animal. Feeling hungry, we go to the kitchen to get food from the refrigerator. Our brain models our world (Craik, 1943). Again, many movements are explorations of the world around us. In short, we perceive so that we may plan our actions appropriately; but in acting we provide ourselves with new opportunities to perceive. The cycle of action and perception continues.

The Schema-Assemblage

How is it that, looking around us, we come to recognize objects and their spatial relationships? It seems reasonable to posit that we perceive the world on the basis of our own prior knowledge. We have schemas that are programs or contextual systems that let us recognize a phone, a person, or a mountain from appropriate visual or other stimuli. I suggest that our knowledge of the world is divided into a short-term model, representing our appreciation of our current place in space and time, and a long-term model, representing all that we know both consciously and unconsciously. To a first approximation, especially insofar as it refers to the sensible environment in which we currently find ourselves, the short-term model is construed as an assemblage of activated schemas whose pattern of activations is related to the current state of the environment. By contrast, long-term memory is the distillation of experience (some of it genetic) represented by the repertoire of schemas available for activation. Of course, this account begs many questions as to how schemas are coded in neural terms and how they are related to each other. (This is a top-down analysis, and much remains to be done to determine the extent to which schemas and other processes described here are directly instantiated in neural processes.) Clearly, retinotopy provides a fine framework for understanding the disposition of those schemas that represent objects within our current central and peripheral visual field. (But when we try to account for our current awareness of objects in other parts of the house or in the environment beyond, we come to deep theoretical problems, which I am happy to ignore here.)

Perceptual and Motor Schemas

The present theory posits that a schema is like a neural network and may thus be in a state of greater or lesser activity. It is perceptual to the extent that its activation can vary on the basis of cues from both peripheral stimuli and internal context. It is motor to the extent that, being activated, it can determine an appropriate course of action. Just as a controller may have an identification procedure to tune its parameters to provide appropriate control signals to the controlled system, so, in general, will a motor schema need to be linked to a perceptual schema so that the interaction with the environment is based on a proper appreciation of the nature of that environment. Thus, a schema for grasping an object can be guided by visual perception of not only the position of the object but also its size and orientation. A motor schema, in this sense, is akin to what the Russian school, founded by Bernstein (1968), has called a "synergy" [a different usage of the term from that of Sherrington (1910), who used the term for the coactivation of a set of muscles within a single movement, rather than the overall patterning of some temporally distributed behavior].

Coordinated Control Programs

I now offer two examples of a coordinated control program, combining the features of both conventional control block diagrams and computer flow dia-

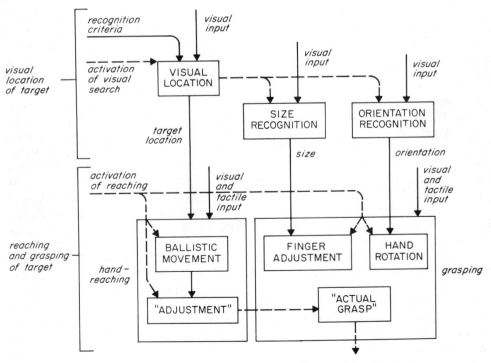

FIGURE 1. A coordinated control program for grasping an object.

15. A View of Brain Theory

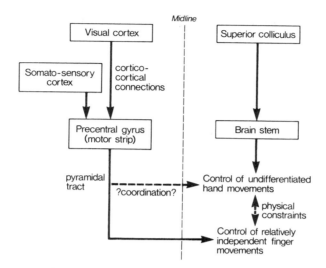

FIGURE 2. A model suggesting how a pyramidal pathway can differentiate extrapyramidal control of reaching movements.

grams. Figure 1 is a diagram of a possible program for the control of reaching toward a visually sensed object. At the top of the diagram are the perceptual schemas that recognize the object and locate it on the basis of a description and those that in turn determine the size and orientation of the object. Jeannerod and Biguer (1982) observed that when a person reaches for an object he already begins to shape his hand to the size and orientation of the object during the reaching movement. Thus, in the motor schemas, the dashed line indicates simultaneous activation of both reaching and shaping of the hand. If we regard the shaping of the hand as part of the grasp schema, we have the interesting fact that the completion of this subtask does not initiate the next task of grasping. It must, in fact, wait for an activation—probably based upon feeling the object touch the hand—consequent upon the successful completion of the reaching movement. [Further developments of these ideas are reported by Arbib *et al.* (1986) and by Iberall *et al.* (1986).]

Figure 2 is based on the studies of Brinkman and Kuypers (1972) and Haaxma and Kuypers (1974). They showed that finely coordinated, visually guided behavior in monkeys involved the cooperative computation of two different systems. A pathway involving the brain stem controls the undifferentiated hand movements akin to the simple grasping discussed in the previous example. A pathway from visual cortex to precentral gyrus and thence directly via the pyramidal tract to motor neurons controls the distal musculature and is responsible for the control of relatively independent finger movements. With interruption either of the corticocortical connections or of the pyramidal tract, the animal was unable to shape its hand in such a way as to dislodge a pellet from a groove whose orientation could be visually determined. Instead, the animal could reach for the pellet, but without preshaping the hand, and would then move its hand back and forth under tactile control until by chance the pellet was dislodged. At that time, the tactile feedback sufficed to allow the animal to grasp the pellet efficiently and bring it to its mouth.

Two Theories of Vision

Earlier I suggested that a task of perceptual systems was to update an assemblage of representations of particular objects in some spatial relationship. This problem is of intense interest to many workers in the branch of AI known as machine vision or computer vision, irrespective of any question of the use to be made of the representation of visual input. This section examines two approaches to machine vision, those of Hanson and Riseman (1978a) and of Marr (1978), and considers the extent to which they can be viewed as cognitive models rather than simply as programs to get machines to emulate certain aspects of vision. In fact, Hanson and Riseman were designing a real computer system to do something useful, while Marr provided a top-down specification of visual systems, whether in brain or machine. The Hanson–Riseman approach provides valuable cues for brain theory that supplement, and occasionally challenge, those offered by Marr. [The fields of both machine vision and neural analysis of vision have been actively explored in recent years. See Arbib and Hanson (1987) for an up-to-date collection of interdisciplinary papers.]

Within a visual system, "bottom-up" processes are those that proceed by more and more elaborate processing of the peripheral signal without reference to knowledge of objects in the world; top-down processes proceed from knowledge of objects to an interpretation of lower-level patterns. (In the rest of this section I use the terms "bottom-up" and "top-down" in a somewhat different sense from that of my earlier comments on styles of brain theory.) Both the Hanson–Riseman and the Marr models are in agreement that there must be two stages of bottom-up processing. The first stage applies a variety of local processors to produce a map that highlights areas likely to be the most information-bearing places within the visual input. They also agree that a second stage must build upon this initial feature map—what Marr calls the "primal sketch"—to yield a representation that provides valuable information about the position and shape of objects, without yet calling upon any interpretation based on other knowledge of objects that might actually be in the world. At the next stage the two models diverge quite drastically, with Hanson and Riseman advocating a pattern of segmentation, whereas Marr advocates what he calls the $2^1/_2$ dimensional sketch. (This $2^1/_2$-D sketch provides the depth and orientation of each patch of surface in the visual field. In fact, as of 1985, no machine vision system has been able to compute a *fine-grained* $2^1/_2$-D sketch.) Having obtained these intermediate representations, both theories similarly invoke high-level information to come up with hypotheses about what objects in the world could be responsible for the observed visual pattern.

The general scheme, then, is bottom-up processing through several levels of representation until world knowledge can be invoked to generate hypotheses;

these hypotheses then act top-down to verify or disqualify themselves by determining whether or not other data from the visual image are compatible with those data that evoked the hypothesis in the first place. This same overall system organization can be seen in AI studies of speech understanding, as in the HEARSAY system (Erman and Lesser, 1980). The stage at which the two theories differ the most is discussed below.

The Hanson–Riseman Approach

As mentioned above, the Hanson–Riseman approach uses a pattern of segmentation as its intermediate representation. An attempt is made to segment the image into different regions, using one of two methods. One is edge-finding, based on discontinuities in color or texture or depth that could signal a break between two surfaces. The other is region-growing, i.e., aggregating areas of similar visual stimulus by finding clusterings in the feature space and then mapping representative symbolic labels back upon the image to determine a partition of the visual field. The variety of shapes and illuminations in the world are such that it proves virtually impossible, at least with current techniques, to come up with regions that are in 1:1 correspondence with surfaces of distinct objects. The pattern of light and shade in a tree can break it into a number of chromatically distinct regions. A highlight may make it impossible to see an edge separating one region from another. Shadows and highlights may themselves be treated as distinct regions rather than features lying upon a given surface. In fact, there is a hierarchical problem of grouping texture elements—consider leaves, clumps of leaves and branches, trees on a hillside, and so on. Although it is true that more sophisticated, bottom-up processing can be designed to take into account various processes of color change under highlighting and shadowing, to allow merging of regions that would be separated on a crude analysis, it nonetheless seems fair to posit that total segmentation cannot be done without invoking real-world knowledge. Note, however, that segmentation can be improved by competitive cooperation among different segmentation processes, such as those based on edge-finding and those based on region-growing.

Once approximate segmentation is completed, the Hanson–Riseman approach calls various processes into play to make hypotheses about the objects whose surfaces contain the regions. The schemas which represent objects or other visual regions must thus contain the necessary routines to determine whether that which they represent is present within the scene. For example, if a region is blue and is near the top of a picture taken outdoors with a level camera, then a reasonable hypothesis is that the region is sky. If it is green and near the bottom of the picture in an outdoor scene, then it is a plausible hypothesis that it is grass. Other cues may suggest the presence of bushes, trees, houses, windows, cars, and so on. Such hypotheses can then be checked by determining whether or not the region can be merged with other regions that satisfy the bounds on their spectral attributes and other features, and if the resulting posited surfaces are appropriate in terms of shape (with extra processing required to account for occlusion effects and size). A camera model can be invoked to infer the size of an object from an estimate of its distance along the ground plane when other depth information, e.g., from stereo disparity or a range finder, is unavailable. (For a review of the current status of this work, see Hanson et al., 1985.)

The Marr Theory

In Marr's theory [masterfully reviewed in his posthumously published opus, Marr (1982)], the intermediate representation, after the "primal sketch," is based not on segmentation but on a depth analysis that seeks to assign to each point of the scene an estimate of its orientation in space. Recall my earlier comments on stereopsis; other cues come from surface highlights (Horn, 1974) and motion (Ullman, 1979). The resultant, still somewhat hypothetical, $2^1/_2$-D sketch is like a bas-relief, in which the shape is determined at each point, but there is no symbolic representation of separation into distinct objects. Further computations are then designed to find axes of symmetry for different bulges in the $2^1/_2$-D sketch, and the resultant axes are used as stick figures that might provide access to a hypothetical data base of different objects known to the system. In other words, the 3-D representation here is not based on surface patches but on "body-centered coordinates" of the represented object, providing a sort of stick figure skeleton, with a specification of cross sections to be swept up and down these axes to flesh out the 3-D object. Programs for such data-base access remain a topic for future research. It is open to question whether or not the $2^1/_2$-D sketch can in fact be constructed with sufficient accuracy to drive the process of axis inference well enough to allow reliable retrieval of hypotheses.

These two approaches provide a useful base for understanding the visual system, but certainly do not stand alone. Data are often noisy, so that inference of the $2^1/_2$-D sketch, or the segmentation sketch, is unreliable and can at best suggest hypotheses rather than lead to the selection of a unique hypothesis. Marr's approach posits a uniquely 3-D representation of objects, rather than accepting the perhaps more plausible view, as in, e.g., the theory of frames by Minsky (1975), that our knowledge of an object is often a synthesis of views from a number of different perspectives. Although Marr has downplayed cooperative computation in his latest stereopsis algorithm, I believe that the proper development of a theory of vision systems, synthesizing and building upon features of many different approaches (sampled in Arbib and Hanson, 1987), will involve cooperative computation among a multitude of processes.

In a feature-rich environment, there are always more features available than can be taken into account in a reasonable processing time. It is thus necessary to initiate processes that extract certain salient features; but the system must be so designed that the use of these features does not preclude taking into account other features. In the Hanson–Riseman approach, a process initiated on the basis of feature measure cues could then be rigorously checked by invoking other processes that could take size or shape into account. This type of interaction of multiple knowledge sources is, it seems to me, the style of the brain, with its incessant interaction of hundreds of continuously active brain regions.

Computing the Optic Flow

We have seen that machine vision research postulates high-level systems to build upon representations initially determined at lower levels (as in the primal sketch), utilizing perceptual schemas to recognize objects within the environment. Gibson (1955, 1966, 1977) most forcefully made clear to psychologists that there was a great deal of information that could be picked up by low-level systems and that, moreover, this information could be of great use to an animal or to an organism even without invocation of high-level processes of object recognition. For example, if, as we walk forward, we recognize that a tree appears to be getting bigger, we can infer that the tree is getting closer. What Gibson emphasized, and others such as Lee (1974; Lee and Lishman, 1977) have since developed, is that object recognition is not necessary to make such inferences. In particular, the "optic flow"—the vector field representing the velocity on the retina of points corresponding to particular points in the environment—is rich enough to support the inference of where collisions may occur within the environment and, moreover, the time until contact.

A problem often glossed over in Gibson's writings is the computation of the optic flow from the changing retinal input. Our studies to date have been "in the style of the brain," but they have not been related to actual neural circuitry. Rather than now asking how neurons might pick up the optic flow on the basis of continuously changing retinal input, I shall simply offer an algorithm (the MATCH algorithm: Prager, 1979; Prager and Arbib, 1983; for other algorithms, see, e.g., Ullman, 1979), played out over a number of interacting layers, each of which involves parallel interaction of local processes, where the retinal input is in the form of two successive snapshots and the problem is to match corresponding features in these two frames. (Mathematically, the problem is the same as that of stereopsis, discussed earlier. However, whereas there are only two eyes, there may be many successive moments in time, so that the initial algorithm for matching a successive pair of frames can be vastly improved when the cumulative effect of a whole sequence can be exploited.)

The problem is posed in Figure 3, where we see four features extracted from Frame 1, shown as circles, and four features from Frame 2, represented as crosses. The *stimulus-matching problem* is to try to match features in the two frames that correspond to a single feature in the external world. Figure 3a shows an assignment that seems far less likely to be correct than that shown in Figure 3b. Lacking other information, we would prefer the latter stimulus-matching because the world tends to be made up of surfaces, with nearby points on the same surface being displaced similar amounts. [This use of the plausible hypothesis that our visual world is made up of relatively few connected regions to drive a stimulus-matching process was enunciated, for stereopsis, by Arbib *et al.* (1974).] This algorithm, then, will make use of two consistency conditions:

- *Feature Matching:* Where possible, the optic flow vector attached to a feature in Frame 1 will come close to bringing it in correspondence with a similar feature in Frame 2.

- *Local Smoothness:* Because nearby features will tend to be projections of points on the same surface, their optic flow vectors should be similar.

In developing an algorithm "in the style of the brain," I assume that there is a retinotopic array of local processors, which can make initial estimates of the local optic flow but which will then pass messages back and forth to their neighbors in an iterative process to converge eventually upon a global estimate of the flow. That interactions are necessary to obtain a correct global estimate is shown in Figure 4, where we see a local receptive field for which the most plausible estimate of the optic flow is greatly at variance with the correct global pattern. Our MATCH algorithm is then as shown in Figure 5. We fix two frames and seek to solve the matching problem for them. An initial assignment of optic flow vectors might be made simply on the basis of nearest match. The algorithm then proceeds through successive iterations, with the local estimate for the optic flow vector assigned to each feature of Frame 1 being updated at each iteration.

Consider, for example, the Frame 1 feature A of Figure 5 and the position B, which is the current hypothesis as to the location of the matching stimulus in Frame 2. If feature matching were the sole criterion, the new optic flow would be given by the wavy arrow that matches A to the feature in Frame 2 closest to the prior estimate, namely B. On the other hand, if only local smoothness were taken into account, the new optic flow vector assigned to A would be the average of the optic flow vectors of features within a certain neighborhood. The MATCH algorithm updates the estimate at each iteration by making the new optic flow estimate a linear combination of the feature-matching update and the local smoothness update, as indicated by the dashed arrow emanating from A in Figure 5. The algorithm works quite well in giving a reliable estimate of optic flow within 20 iterations.

If we take advantage of the availability of a whole sequence of frames, rather than just two, then we can obtain an increasingly accurate estimate of the optic

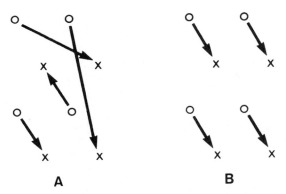

FIGURE 3. A stimulus-matching problem. In a world made up of surfaces, nearby features are likely to have similar displacements. Thus, the flow of (B) is far more likely to be correct than that of (A).

15. A View of Brain Theory

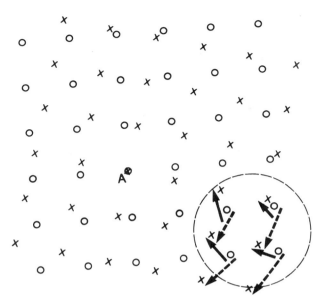

FIGURE 4. Stimulus-matching problem illustrating the need for interaction "in the style of the brain," to obtain a correct global estimate of optic flow. Frame 1 comprises the dots indicated by circles. Frame 2 is obtained by rotating the array about the pivot at A to place the dots in the positions indicated by crosses. The dashed circle at lower right is the receptive field of a local processor. The solid arrows indicate the best local estimate of the optic flow, the dashed arrows show the actual pairing of features (the global pattern) under rotation about A.

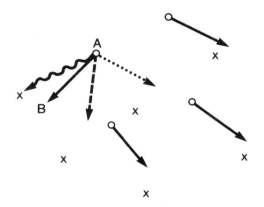

FIGURE 5. The MATCH algorithm for estimating optic flow. The circles indicate features in Frame 1; the crosses, features in Frame 2; and the solid arrows, the current estimate of the optic flow—the head of the arrow shows the posited position in Frame 2 of the feature corresponding to the Frame 1 feature at the tail of the arrow. The feature-matching consistency conditions (see text) alone would adjust A's optic flow to the wavy arrow pointing to the Frame 2 feature nearest to B (the current estimate of A's Frame 2 position). The "local smoothness" condition (see text) would yield the dotted arrow—the average of the optic flow of the neighbors. The relaxation algorithm actually yields the dashed arrow as a weighted combination of these two estimates.

flow, with fewer iterations to match each new frame as it is introduced. For example, if, having matched Frame n to Frame $n + 1$, we try to match Frame $n + 1$ to $n + 2$, it is reasonable to assume that—to a first approximation—the optic flow advances a feature by roughly the same amount in the two frames. If we thus use the repetition of the previous displacement, rather than a nearest neighbor match, to initialize the optic flow computation of the two new frames, we find from simulations that only 4 or 5 iterations, rather than the original 20, are required and that the match on real images is definitely improved.

The algorithm just described is based on two consistency conditions: feature matching and local smoothness. It is instructive to note where these constraints break down. If one object is moving in front of another object, then points on the rear object's surface will be either occluded or disoccluded during this movement, depending on whether the front object is tending to cover or uncover the object behind it. Thus, if we look at the current estimate of the optic flow and find places where the flow vector does not terminate near a similar feature to that from which it starts, we have a good indication of an occluding edge. On the other hand, the local smoothness will also break down at an edge, for the two objects on either side of the edge will, in general, be moving differentially with respect to the organism. Thus, we can design edge-finding algorithms that can actually use the breakdown of our consistency conditions to find edges in two different ways—on the basis of occlusion/disocclusion and on the basis of optic flow discontinuity. To the extent that the estimate of edges by these two processes is consistent, we have the cooperative determination of surfaces within the image. To the extent that good edge estimates become available, the original basic algorithm can be refined

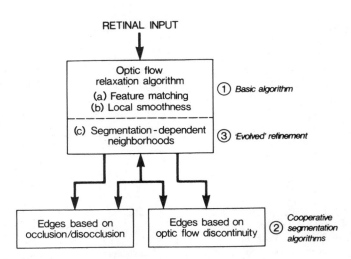

FIGURE 6. Evolutionary design of optic flow algorithm. (1) The basic optic flow relaxation algorithm uses the consistency conditions of feature matching and local smoothness. (2) The resultant optic flow estimate permits the hypothesization of edges based on occlusion/disocclusion cues and on optic flow discontinuity. (3) The resultant edge hypotheses can be used to refine the computation of optic flow by dynamically adjusting the neighborhoods used in employing the consistency conditions.

(Figure 6). (This extension of the algorithm has not yet been implemented.) Now, instead of having bleeding across edges, we can dynamically change the neighborhood of a point, so that the matching of features or the conformity with neighboring flow can be based almost entirely upon features on the same side of the hypothesized boundary. (But not completely, for at any time the edges will themselves be confirmed with limited confidence and, thus, may be subject to later change.)

Figure 6 represents an "evolutionary design process." The basic algorithm provides new information which can then be exploited in the design of the cooperative segmentation algorithms, but once the segmentation information is available, the original algorithm can be refined by the introduction of segmentation-dependent neighborhoods. This idea is not simply an interesting engineering speculation; rather, it gives us some very real insight into the evolution of the brain. Basic systems provide the substrate upon which "higher-level" systems may evolve. These higher-level systems then enrich the environment of the lower systems, which may then evolve to exploit the new sources of information. Although it is still useful, to a first approximation, to talk of low-level and high-level systems, there is no longer any univocal flow of information. We are here very close again to the classical Jacksonian notion of levels (Jackson, 1874, 1878–1879).

A Model of the Cerebellum

To see how the top-down analysis of interactions between perception and movement that was presented earlier can be related to details of neural circuitry, I shall now examine a model of the cerebellum (Arbib et al., 1974; Boylls, 1975, 1976). The model brings together the notion of a motor schema and the notion of maps as control surfaces. It is important in that it exhibits neural layers acting as control surfaces representing levels of activation for the coordination of muscles, complementing our study of retinotopic representations of visual input. (For a current view of research on the cerebellum, see Ito, 1984.)

I have suggested that the problem of motor control is one of sequencing and coordinating motor schemas, rather than directly controlling the vast number of degrees of freedom offered by the independent activity of all the motor units. We must not only activate the appropriate schemas but must tune them. To understand this notion of tuning we need an important concept from modern control theory—that of the *identification algorithm*. In the familiar realm of feedback control theory, a controller compares feedback signals from the controlled system with a statement of the desired performance of the system to determine control signals that will move the controlled system into ever greater conformity with the given plan. The appropriate choice of control signal must depend upon having a reasonably accurate model of the controlled system—for example, the appropriate thrust to apply must depend upon an estimate of the weight of the object that is to be moved. However, there are many cases in which the controlled system will change over time in such a way that no *a priori* estimate of the system's parameters can be reliably made. To that end, it is a useful practice to interpose an identification algorithm that can update a parametric description of the controlled system in such a way that the observed response of the system to its control signals comes into greater and greater conformity with that projected on the basis of the parametric description. If a controller is equipped with an identification algorithm, if the controlled system belongs to a class whose parameters the algorithm is designed to identify, and if, finally, the changes in parameters of the controlled system are not too rapid, then in fact the combination of controller and identification algorithm provides an adaptive control system, which is able to function effectively despite continual changes in the environment.

So far, our analysis has been top-down. Turning now to neurophysiological data, we shall use as a working model the cerebellar function in locomotion of the high-decerebrate cat (Shik *et al.*, 1966). Sherrington (1910) noticed that stimulation of Deiters nucleus in the standing animal would lead to extension of all the limbs. Orlovsky (1972) found that in the high-decerebrate cat, stimulation of Deiters nucleus during locomotion would not affect extension during the swing phase of walking but would yield increased extension during the support phase. Since the locomotory motor schema has been shown to be present even in the "spinal" cat [as discussed in the classical work by Sherrington (1910) and in more recent publications (Herman *et al.*, 1976)], it seems reasonable to view the system in which the cerebellum and Deiters nucleus are involved as providing an identification algorithm for the parametric adjustment of the spinal schema. We now turn to Boylls' (1975, 1976) model, which shows how the adjustment of those parameters might be computed within the cerebellar environs.

The only output of cerebellar cortex is via the Purkinje cells, which provide inhibitory input to the cerebellar nuclei (Eccles *et al.*, 1967). Each Purkinje cell has two input systems. One input is via a single climbing fiber that ramifies and synapses extensively over the Purkinje cell's dendritic tree. The other input system is via the mossy fibers, which activate granule cells whose axons rise up into the layer of Purkinje cell dendrites. These dendrites are deployed in planes, with all the dendritic trees parallel to one another. The axons of the granule cells enter the dendritic layers and form T's, whose crossbars run parallel to one another at right angles to the planes of the Purkinje dendritic trees. (There are also various kinds of interneurons in cerebellar cortex, but I shall not model these here, concentrating instead on the basic cerebellar circuit.)

The climbing fiber input to a Purkinje cell is so strong that when a climbing fiber is fired, the Purkinje cell responds with a sharp burst of four or five spikes, known as the climbing fiber response (CFR). Many investigators have thought that the "secret" of the climbing fiber is to be found in this sharp series of bursts, but the true role of the climbing fiber input may be to suppress Purkinje cell activity for as much as 100 msec. Such suppression has been found to follow the CFR (Murphy and Sabah, 1970).

The overall architecture of Boylls' model, distributed over an array of interacting control surfaces, is shown in Figure 7, which is an anatomical template of circuitry ubiquitous in cerebellar transactions. That is, specific labels could be given to, say, the "brain stem output nucleus" as the red nucleus or Deiters

15. A View of Brain Theory

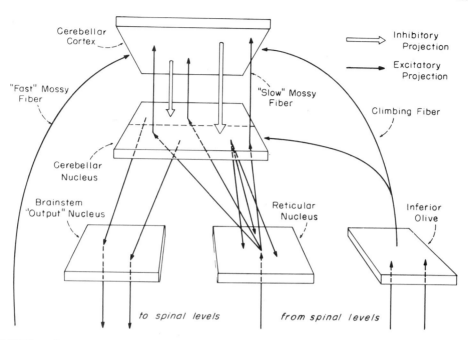

FIGURE 7. Generalized schematic of the interacting control surfaces involved in tuning of motor schemas by the cerebellum and related nuclei.

nucleus; the "reticular nucleus" could be reticularis tegmenti pontis or paramedian; and so on. From this architecture we gather that the output from the cerebellar nuclei via the brain stem output nucleus results from the interaction between cerebellar cortical inhibition supplied by the Purkinje cells and by drives from the reticular nucleus. Tsukahara (1972) demonstrated the possibility of intense reverberation between the reticular and cerebellar nuclei following removal of Purkinje inhibition, and Brodal and Szikla (1972), among others, have demonstrated the anatomical substrate for such loops. The somatotopy is indicated in Figure 8. We can thus postulate that there will be explosively excitatory driving of the cerebellar nucleus by reticulocerebellar reverberation, unless it is blocked by Purkinje inhibition.

The output of cerebellar tuning is expressed as a spatiotemporal neuronal activity pattern in a cerebellar nucleus that can then be transmitted via the brain stem nuclei to spinal levels. A careful analysis of the anatomy enabled Boylls to predict that the agonists of a motor schema would be represented along a sagittal strip of cerebellar cortex, while its antagonists would lie orthogonal to that strip (in the mediolateral plane). Applications of this formula to cortical topography of the anterior lobe, as developed by Voogdt (1969) and Oscarsson (1973), allowed Boylls to identify particular cortical regions as associated with particular types of hindlimb–forelimb, flexor–extensor synergistic groupings. This result led to conclusions that are experimentally testable.

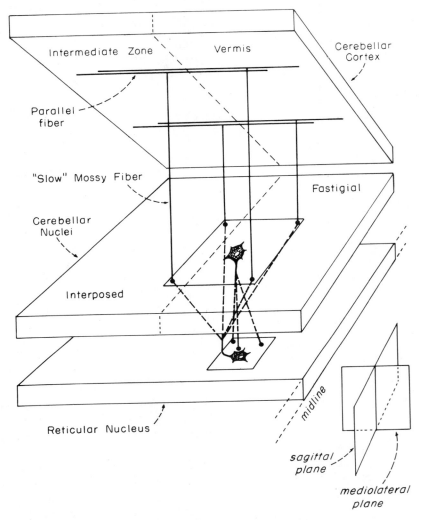

FIGURE 8. Anatomical template of cerebelloreticular reverberatory loop.

The Boylls model suggests that activity within the cerebellar nucleus is initiated through topically precise climbing fiber activity; the mechanism involves their direct cerebellar nuclear activation, coupled with the suppression of the target Purkinje cell activity in the cortex via the above-mentioned inactivation response. Once activity is installed in corticonuclear interactions via climbing fiber intervention, the underlying reverberatory excitation helps to retain or store it. At the same time this activity is transmitted to cerebellar cortex by mossy fibers, eventually altering the inhibitory pattern in the nuclear region surrounding the active locus. The relevant pathways involving mossy fibers and the corticonuclear projection are schematized in Figure 9. The spread of parallel fibers indicated in Figure 9 yields a form of lateral inhibition that provides spatial sculpting in a manner dependent on the elaborate geometry of cerebellar cortex and corticonuclear projections. Mossy inputs of various types tune the resultant patterns to the demand of the periphery, and the program is read out at the spinal cord level, as appropriate.

Testing of the various hypotheses has required computer simulation of this neuronal apparatus. Simulation results corroborated the conjecture that cerebellar-related circuitry could support the short-term storage of motor schema param-

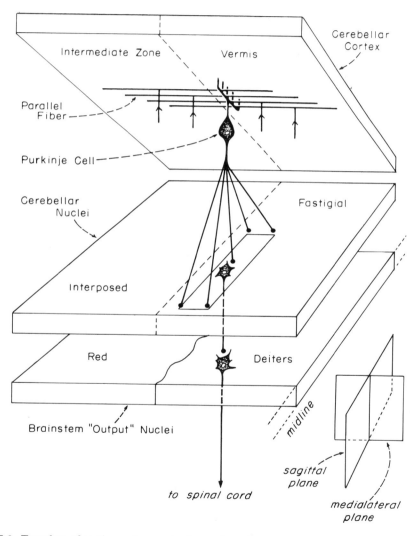

FIGURE 9. Template of corticonuclear projection and cerebellar outflow. For simplicity of the diagram, the granule cells are omitted.

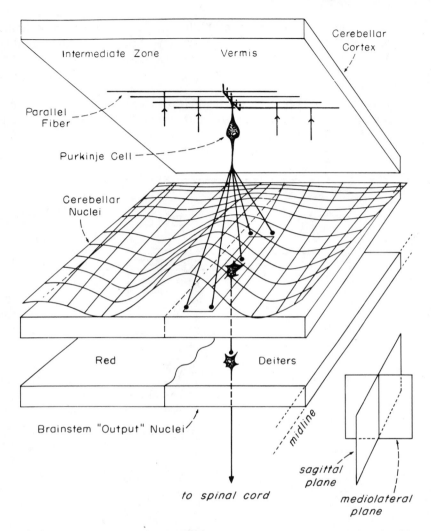

FIGURE 10. Activation pattern stored in cerebellar neuronal interactions via climbing fiber activity represents the parameters for a particular motor schema, which may activate the musculature.

eters initiated (and periodically refreshed) by climbing fiber activity. Figure 10 suggests a typical nuclear activation pattern so introduced.

Modeling Frog Visuomotor Coordination

My own group has chosen visuomotor coordination in the frog and toad as a setting in which the top-down and bottom-up approaches to brain theory may interact. After discussing the basic models of prey selection and of prey-predator discrimination, I shall discuss how new behavioral experiments are being used to develop a top-down analysis, which, we hope, will allow us to extend our current modeling of the animal's behavior to models that take context more fully into account.

Lettvin *et al.* (1959) asked, "What does the frog's eye tell the frog's brain?" Didday (1970, 1976) and I asked, "What does the frog's eye tell the frog?" It is one thing to say that the human monitoring a cell through a microelectrode can correlate the cell's activity with some feature of the external world; it is quite another to say that the neural circuitry within the brain of the animal can actually make use of that information in determining behavior (Perkel and Bullock, 1968). We sought to ask, then, how the "bug detector" information from the retina might be used to guide the animal's activity. Our basic perspective was formed by the behavioral studies of Ingle (1976), who noted that a frog will orient and snap with the tongue at a small, wiggling stimulus within a certain range about the animal. If two stimuli are simultaneously presented within the snapping range, the animal will usually snap at just one of them—but there are cases in which it will snap at neither or will snap in between at the average fly.

Pure top-down analysis of prey selection might specify the task as the command: "Develop a procedure for finding the greatest element in an array of elements." Unless constrained by the requirement to "use local, parallel computations," this imperative might be realized on a computer simply by scanning a list of values to find the maximum. The no-fly and average-fly effects could be handled by subroutines that would detect when the largest entries of the list bore some designated relationship. However, this serial process would not be interesting as a brain model, whatever its utility as a computational summary of the behavior. We thus asked how this process of selection could be implemented through the interaction of neurons rather than through the supervention of some executive program. The model that we finally developed involved an array of neurons modulated by another array of inhibitory neurons in such a way that peaks of activity in the first layer would compete through the second layer. In general, the highest peak on the first layer would finally suppress all other peaks and emerge from the system to control motor activity. However, in some cases two peaks of similar amplitude would hold each other below the threshold for action.

Recent advances in neurophysiology have led to better identification of different cell types, and the original model of prey selection, which was thought to correspond to activities of cells located in the tectum, is now conceived of as resulting from interaction between cells in the tectum and pretectum (Figure 11). In particular, the detailed anatomy of the tectum was modeled as an array of columns by Lara (1980) (see also Lara *et al.*, 1982). Ingle (1975) observed that presentation of a flylike stimulus to a frog for 0.3 sec will rarely elicit a response, whereas presentation for 0.6 sec usually will. If a stimulus is presented for 0.3 sec, taken away for several seconds, and then presented again for 0.3 sec, this second short presentation will elicit a response. The response is history dependent. This short-term memory has been modeled, and it has been shown that a plausible geometry of the tectal column will yield this facilitation effect.

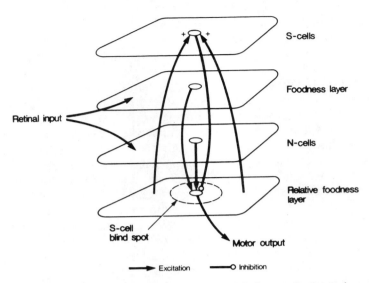

FIGURE 11. Tectal-pretectal interaction in model of prey selection in frog.

In the model of prey selection, efforts must be made to determine the processes that can modulate an array of activity on the tectal map that serves as control surface for the snapping response. This effort moves one step beyond a simple stimulus–response model, in that we are now asking how, given a structured stimulus, one part of it can be selected for response. If an object is small and moves in a certain way, it is prey to be snapped at; if it is large and moves in a certain way, it is a predator to be jumped away from. How does the animal make this discrimination? Ewert (1976) has shown that if the pretectum–thalamus region of the toad (toads are similar enough to frogs for the present purpose) is ablated, the animal will snap at any moving object, no matter how large it is. This result has suggested that the tectum is not, as was first thought, a device to guide snapping at flies, but is rather a device to guide snapping at moving objects. It is then the task of other brain regions to modulate this tectal activity, differentiating the process of recognition to make finer discriminations. Ewert and von Seelen (1974) developed a model of filters for both tectum and pretectum of toad, with inhibitory modulation of tectal activity when pretectal cells signaled the presence of "antiworms." Cervantes-Perez *et al.* (1984) synthesized—in a way consistent with current behavioral, anatomical, and physiological data—our original model of prey selection with Ewert's work on prey–predator discrimination, using an 8×8 array of tectal columns in interaction with a simple array of pretectal elements.

The above models of frog and toad behavior treat the environment as made up of a number of objects. The animal in essence has to choose the object of its response and to determine whether to respond to that object as prey or as predator. How do we bring "context" into the world of the animal? We have devised rough models of an AI kind that are programs rather than simulations of neural nets and that will yield the behavior. Unlike the serial list approach to prey selection mentioned briefly above, these models are structured in terms of interacting processes, so that they can be used as plausible hypotheses about the interaction between brain regions. This is in the spirit of many workers in AI who use psychological observations to design their models of how to achieve some intelligent behavior and then structure the models in terms of concurrently active, interacting processes. Lesion studies impose further neural constraints on the model. We thus hope to see in the near future not only the fuller articulation of the top-down model on the basis of recent behavioral experiments but also continued physiological modeling and experimentation.

Ingle (1976) observed that when confronted with a predator stimulus, a frog chooses a direction of escape that is a compromise between the forward direction and the direction directly away from the stimulus. When Ingle interposed a barrier in the preferred escape path of the frog, he found that the frog would now tend to jump to one side or other of the barrier. A plausible interpretation of these results is that the animal would come as close to the preferred route as it could while avoiding an obstacle in the way. Here, then, we see *context* at work: the animal is no longer simply responding to the aversive stimulus but, rather, is integrating the spatial structure of the world around it in choosing its response.

In more recent work, Collett and colleagues have not only observed the animal's behavior with respect to barriers but have also studied toads faced with a chasm between them and a worm (Collett, 1979; Lock and Collett, 1980). He finds, as might be expected from our (usually unexamined) ideas about animal behavior, that if the chasm is narrow, the animal will jump across; if it is wide but shallow, the animal will step down and walk across; but if the chasm is both deep and wide, the toad will simply turn away. For models of this, see Arbib and House (1983) and Lara et al. (1984).

In summary, in the study of visuomotor coordination in frog and toad, we can carry out subtle behavioral studies to bring manageable cognitive aspects into the analysis, at the same time finding interesting subproblems in which detailed neural circuit analysis is possible.

A Style for the Brain

The very richness of current research on the brain guarantees that any single view must be incomplete. Most neuroscience is experimental and emphasizes the cellular and subcellular levels with special emphasis on chemical mechanisms. By contrast, while stressing the need for interaction with experiments, this chapter is focused on theory, with special emphasis on models that relate neural circuitry (bottom-up) to the analysis of cognitive processes in terms of interacting functions (top-down). Our examples of cognitive processes pertain to vision and the control of movement [see Arbib and Caplan (1979), Lavorel and Arbib (1981), Arbib (1982a,b), and and especially Arbib et al. (1987) for a parallel attempt to relate cooperative computation to studies of brain and language]. Within this chosen area, we have sampled a number of fruitful models with varying degrees of neural veridicality—optic flow, cerebellar tuning of motor schemas, visuomotor coordination in frog and toad. In closing, I shall try to extract a general lesson from such models. But first, I will return to my previous top-down analysis to put these models into perspective. We begin to have neural models of individual schemas, but little understanding of the way in which coordinated control programs are constructed, embodied, or executed in neural tissue. Studies of neural circuitry in breathing and rhythmic locomotion (see Stent, Chapter 13 this volume), provide a useful first sequence of steps but do not address the problems of coordinated phasing in and out of diverse activities in a goal-dependent way. The awareness of such shortcomings is not intended as a criticism of the current state of our science but, rather, as a spur to the further articulation of top-down analyses to better stretch the scope of neural modeling.

In summarizing my view of cooperative computation as "a style for the brain," I do not wish to reiterate the case for a cooperative analysis of neural interactions in layers of neurons, whether in development, learning, or perceptual-motor function—a case made earlier, and argued persuasively by von der Malsburg in his discussion of the ontogenesis of retinotopy as a paradigm for organization in the brain (Chapter 14). Nor do I want to say more about the discrete form of cooperative computation as embodied in such AI systems as VISIONS and HEARSAY. Instead, I want to recall an intermediate level of cooperative computation that seems to provide general insight into layer-by-layer interactions within the brain.

Our algorithm for optic flow first developed a basic, low-level system of cooperative computation of optic flow within a retinotopic array (the MATCH algorithm) and then developed (hypothetically) high-level systems to conduct segmentation on the basis of the output of this array; the higher-level systems enrich the environment of the basic systems, which can then be adjusted to exploit the new sources of information. The distinction between low and high levels becomes blurred—the low-level systems provide the necessary data to initiate high-level hypotheses, but these hypotheses are needed to constrain the lower-level processes. This is very much the approach to visual systems espoused by Hanson and Riseman (1978a) and by Marr (1978). *There is no longer a simple one-way flow of information* but, rather, a "coming to equilibrium" of multiple systems. The equilibrium is of course dynamic, because the retinal input to the system changes with time. The system must possess something of the "adiabatic approximation" of Haken (1978, Chapter 7)—it must be able to adjust to significant changes in the world at a rate faster than that at which those changes occur. In the same fashion, the interacting layers of the cerebellar model must adjust the tuning parameters of the motor schemas rapidly enough to keep pace with changes in the environment. This is what makes the system adaptive, with perceptual processes (identification) intimately intertwined with control.

Again, we saw that the tectum of frog and toad could be regarded as a basic array for the control of snapping movements, while the pretectum apparently serves to differentiate the recognition of moving objects so that snapping would not be directed at large objects. In the same way, the work of Brinkman and Kuypers (1972) and Haaxma and Kuypers (1974) (see Figure 2) stresses that we should not view the pyramidal and extrapyramidal pathways as alternative paths for motor control. Rather, the pyramidal pathway serves to differentiate and refine distal movements that ride atop movements of proximal musculature under extrapyramidal control.

The monkey's basic, undifferentiated reaching movements can work quite well with visual input from the retina directed through the superior colliculus (the mammalian analogue of the tectum), even in the absence of visual cortex. In fact, primates can exercise quite elaborate control of movement in the absence of visual cortex. In his paper "What the frog's eye tells the monkey's brain," Humphrey (1970) built on the argument that a monkey without visual cortex should have at least as much visual ability as a frog. But monkeys had hitherto appeared to be blind when they had lost visual cortex. Humphrey argued that the monkeys had not been taught to pay attention to the visual cues they have, and after 2 years he was able to get a monkey without visual cortex to grab at moving objects and to use changes in luminance—such as an open door—for navigation, even though delicate processes of pattern recognition were never regained.

Thus, we see the development of a paradigm that can guide us in the mathematical analysis and computer simulation of interacting brain regions as we come to address an ever richer array of cognitive processes.

ACKNOWLEDGMENTS. Portions of this chapter were presented in two talks: "Perceptual Structures and Distributed Motor Control" delivered at the School of Social Sciences, University of California, Irvine, on January 29, 1980; and "The Theoretical and Experimental Bases of Brain Theory" delivered in the Department of Biomedical Engineering at the University of Southern California on February 14, 1980. I thank Ken Wexler and his colleagues and George Moore for their hospitality on these occasions. Preparation of the manuscript was supported in part by NIH Grant NS-14971-02. The manuscript was essentially completed in 1980, save for the addition of current references in February 1985.

References

Allman, J. (1977) Evolution of the visual system in the early primates. In: *Progress in Psychobiology and Physiological Psychology.* Vol. 7, Academic Press, New York, pp. 1–53.
Amari, S. (1977) Neural theory of association and concept-formation. *Biol. Cybern.* **26**:77–87.
Amari, S. (1980) Topographic organization of nerve fields. *Bull. Math. Biol.* **42**:339–364.
Amari, S., and M. A. Arbib (1977) Competition and cooperation in neural nets. In: *Systems Neuroscience.* J. Metzler (ed.). Academic Press, New York, pp. 119–165.
Amari, S., and M. A. Arbib (1982) *Competition and Cooperation in Neural Nets, Lecture Notes in Biomathematics,* **45**. Springer-Verlag, Berlin.
Arbib, M. A. (1972a) *The Metaphorical Brain: An Introduction to Cybernetics as Artificial Intelligence and Brain Theory.* Wiley-Interscience, New York.
Arbib, M. A. (1972b) Automata theory in the context of theoretical embryology. In: *Foundations of Mathematical Biology.* Vol. II, R. Rosen (ed.). Academic Press, New York, pp. 142–215.
Arbib, M. A. (1975) Artificial intelligence and brain theory: Unities and diversities. *Ann. Biomed. Eng.* **3**:238–274.
Arbib, M. A. (1978) Segmentation, schemas and cooperative computation. In: *Studies in Mathematical Biology Part 1: Cellular Behavior and the Development of Pattern.* S. Levin (ed.). Math. Assoc. of America, Washington, D.C., pp. 118–155.
Arbib, M. A. (1981) Perceptual structures and distributed motor control. In: *Handbook of Physiology,*

The Nervous System, II. Motor Control, V. B. Brooks (ed.). American Physiological Society, Bethesda, pp. 1449–1480.

Arbib, M. A. (1982a) From artificial intelligence to neurolinguistics. In: *Neural Models of Language Processes.* M. A. Arbib, D. Caplan, and J. C. Marshall (eds.). Academic Press, New York, pp. 77–94.

Arbib, M. A. (1982b) Perceptual-motor processes and the neural basis of language. In: *Neural Models of Language Processes.* M. A. Arbib, D. Caplan, and J. C. Marshall (eds.). Academic Press, New York, pp. 531–551.

Arbib, M. A., and D. Caplan (1979) Neurolinguistics must be computational. *Behav. Brain Sci.* 2:449–483.

Arbib, M. A., and A. R. Hanson (eds.) (1987) *Vision, Brain, and Cooperative Computation.* MIT Press, Cambridge, Mass.

Arbib, M. A., and D. H. House (1983) Depth and detours: Towards neural models. In: *Proc. 2nd Workshop on Visuomotor Coordination in Frog and Toad,* R. Lara and M. A. Arbib (eds.). COINS Technical Report 83-19, University of Massachusetts, Amherst.

Arbib, M. A., C. C. Boylls, and P. Dev (1974) Neural models of spatial perception and the control of movement. In: *Cybernetics and Bionics.* W. D. Keidel, W. Handler, and M. Spreng (eds.). Oldenbourg, Munich, pp. 216–231.

Arib, M. A., E. J. Conklin, and J. C. Hill (1986) *From Schema Theory to Language,* Oxford University Press, New York.

Arbib, M. A., T. Iberall, and D. Lyons (1985) Coordinated control programs for movement of the hand, *Exp. Brain Res. Suppl.* 10:111–129.

Bernstein, N. A. (1968) *The Coordination and Regulation of Movements.* Pergamon Press, Elmsford, N.Y.

Blakemore, C., and G. Cooper (1970) Development of the brain depends on the visual environment. *Nature* 228:477–478.

Borges, J. L. (1975) Of exactitude in science. In: *A Universal History of Infamy.* Penguin Books, New York, p. 131.

Boylls, C. C. (1975) A Theory of Cerebellar Function with Applications to Locomotion, I. The Physiological Role of Climbing Fiber Inputs in Anterior Lobe Operation. Computer and Information Science Department Technical Report 75C-6, University of Massachusetts, Amherst.

Boylls, C. C. (1976) A Theory of Cerebellar Function with Applications to Locomotion, II. The Relation of Anterior Lobe Climbing Fiber Function to Locomotor Behavior in the Cat. Computer and Information Science Department Technical Report 76-1, University of Massachusetts, Amherst.

Brinkman, J., and H. G. J. M. Kuypers (1972) Split-brain monkeys: Cerebral control of ipsilateral and contralateral arm, hand, and finger movements. *Science* 176:536–539.

Brodal, A. (1969) *Neurological Anatomy.* Oxford University Press, London.

Brodal, A., and G. Szikla (1972) The termination of the brachium conjunctivum descendens in the nucleus reticularis tegmenti pontis: An experimental anatomical study in the cat. *Brain Res.* 39:337–351.

Cervantes-Perez, F., R. Lara, and M. A. Arbib (1984) A neural model of interactions subserving prey–predator discrimination and size preference in anuran Amphibia. *J. Theor. Biol.* 113:117–152.

Collett, T. S. (1979) A toad's devious approach to its prey: A study of some complex uses of depth vision. *J. Comp. Physiol.* A131:179–189.

Cragg, B. G., and H. N. V. Temperley (1954) The organization of neurones: A cooperative analogy. *Electroencephalogr. Clin. Neurophysiol.* 6:85–92.

Craik, K. J. W. (1943) *The Nature of Explanation.* Cambridge University Press, London.

Dev, P. (1975) Computer simulation of a dynamic visual perception model. *Int. J. Man-Mach. Stud.* 7:511–528.

Didday, R. L. (1970) The Simulation and Modelling of Distributed Information Processing in the Frog Visual System. Ph.D. thesis, Stanford University.

Didday, R. L. (1976) A model of visuomotor mechanisms in the frog optic tectum. *Math. Biosci.* 30:169–180.

Eccles, J. C., M. Ito, and J. Szentágothai (1967) *The Cerebellum as a Neuronal Machine.* Springer-Verlag, Berlin.

Ede, D. A. (1978) *An Introduction to Developmental Biology.* Halsted Press, New York.

Erickson, R. P. (1974) Parallel "population" neural coding in feature extraction. In: *The Neurosci-*

ences: Third Study Program. F. O. Schmitt and F. G. Worden (eds.). MIT Press, Cambridge, Mass. pp. 155–169.

Erman, L., and V. R. Lesser (1980) The HEARSAY II system: A tutorial. In: *Trends in Speech Recognition,* W. A. Lea (ed.). Prentice-Hall, Englewood Cliffs, N.J., pp. 361–381.

Ewert, J.-P. (1976) The visual system of the toad: Behavioral and physiological studies on a pattern recognition system. In: *The Amphibian Visual System: A Multidisciplinary Approach,* K. V. Fite (ed.). Academic Press, New York, pp. 141–202.

Ewert, J.-P., and W. von Seelen (1974) Neurobiologie und System-Theorie eines visuellen Muster-Erkennungsmechanismus bei Kroten. *Kybernetik* **14**:167–183.

Fife, P. C. (1979) *Mathematical Aspects of Reacting and Diffusing Systems. Springer Lecture Notes in Biomathematics* **28**. Springer-Verlag, Berlin.

Frisby, J. P., and J. E. W. Mayhew (1980) Spatial frequency tuned channels: Implications for structure and function from psychophysical and computational studies of stereopsis. *Philos. Trans. R. Soc. London Ser. B* **290**:95–116.

Gaze, R. M. (1970) *The Formation of Nerve Connections.* Academic Press, New York.

Gibson, J. J. (1955) The optical expansion-pattern in aerial location. *Am. J. Psychol.* **68**:480–484.

Gibson, J. J. (1966) *The Senses Considered as Perceptual Systems.* Allen & Unwin, London.

Gibson, J. J. (1977) The theory of affordances. In: *Perceiving, Acting and Knowing.* R. E. Shaw and J. Bransford (eds.). L. Erlbaum Assoc., Hillsdale, N.J.

Grossberg, S. (1970) Neural pattern discrimination. *J. Theor. Biol.* **27**:291–337.

Grossberg, S. (1978) Communication, theory and development. *Prog. Theor. Biol.* **5**:183–232.

Grossberg, S., and D. Levine (1975) Some developmental and attentional biases in the contrast enhancement and short term memory of recurrent neural networks. *J. Theor. Biol.* **53**:341–380.

Haaxma, R., and H. Kuypers (1974) Role of occipito-frontal cortico-cortical connections in visual guidance of relatively independent hand and finger movements in rhesus monkeys. *Brain Res.* **71**:361–366.

Haken, H. (1978) *Synergetics: An Introduction.* 2nd ed. Springer-Verlag, Berlin.

Hanson, A. R., and E. M. Riseman (1978a) VISIONS: A computer system for interpreting scenes. In: *Computer Vision Systems,* A. R. Hanson and E. M. Riseman (eds.). Academic Press, New York, pp. 129–163.

Hanson, A. R., and E. M. Riseman (eds.) (1978b) *Computer Vision Systems.* Academic Press, New York.

Hanson, A. R., and E. M. Riseman (1987) A methodology for the development of general knowledge-based vision systems. In: *Vision, Brain and Cooperative Computation,* M. A. Arbib and A. R. Hanson (eds.). MIT Press, Cambridge, Mass., pp. 285–328.

Harth, E. M., T. J. Csermely, B. Beek, and R. D. Lindsay (1970) Brain functions and neural dynamics. *J. Theor. Biol.* **26**:93–120.

Herman, R. M., S. Grillner, P. S. G. Stein, and D. G. Stuart (eds.) (1976) *Neural Control of Locomotion.* Plenum Press, New York.

Hirsch, H. V. B., and D. N. Spinelli (1971) Modification of the distribution of receptive field orientation in cats by selective visual exposure during development. *Exp. Brain Res.* **12**:509–527.

Hope, R. A., B. J. Hammond, and R. M. Gaze (1976) The arrow-model of retinotectal connections. *Proc. R. Soc. London Ser. B* **194**:447–466.

Horn, B. K. P. (1974) Determining lightness from an image. *Computer Graphics and Image Understanding* **3**:277–299.

Hubel, D. H., and T. N. Wiesel (1962) Receptive fields, binocular interaction and functional architecture in the cat's visual cortex. *J. Physiol. (London)* **160**:106–154.

Hubel, D. H., and T. N. Wiesel (1965) Binocular interaction in striate cortex of kittens reared with artificial squint. *J. Neurophysiol.* **28**:1041–1059.

Hubel, D. H., and T. N. Wiesel (1974) Sequence regularity and geometry of orientation columns in the monkey striate cortex. *J. Comp. Neurol.* **158**:267–294.

Humphrey, N. K. (1970) What the frog's eye tells the monkey's brain. In: *Subcortical Visual Systems,* D. Ingle and G. E. Schneider (eds.). Karger, Basel, pp. 324–337.

Iberall, T., G. Bingham, and M. A. Arbib (1986) Opposition space as a structuring concept for the analysis of skilled hand movements. *Exp. Brain Res. Ser.* **15**:158–173.

Ingle, D. (1975) Focal attention in the frog: Behavioral and physiological correlates. *Science* **188**:1033–1035.

Ingle, D. (1976) Spatial vision in anurans. In: *The Amphibian Visual System*, K. V. Fite (ed.). Academic Press, New York, pp. 119-141.
Ito, M. (1984) *The Cerebellum and Neural Control*. Raven Press, New York.
Jackson, J. H. (1874) On the nature of the duality of the brain. *Med. Press Circ.* **1**:19-63.
Jackson, J. H. (1878-1879) On affections of speech from disease of the brain. *Brain* **1**:304-330, **2**:203-222, 323-356.
Jeannerod, M., and B. Biguer (1982) Visuomotor mechanisms in reaching within extra-personal space. In: *Advances in the Analysis of Visual Behavior*, D. J. Ingle, R. J. W. Mansfield, and M. A. Goodale (eds.). MIT Press, Cambridge, Mass.
Julesz, B. (1971) *Foundations of Cyclopean Perception*. University of Chicago Press, Chicago.
Katchalsky, A., V. Rowland, and R. Blumenthal (1974) *Dynamic Patterns of Brain Cell Assemblies*. MIT Press, Cambridge, Mass.
Kilmer, W. L., W. S. McCulloch, and J. Blum (1969) A model of the vertebrate central command system. *Int. J. Man-Mach. Stud.* **1**:279-309.
Kohonen, T. (1982) A simple paradigm for the self-organized formation of structured feature maps. In: *Competition and Cooperation in Neural Nets*, S. Amari and M. A. Arbib (eds.). Springer-Verlag, Berlin, pp. 248-266.
Kohonen, T., and E. Oja (1976) Fast adaptive formation of orthogonalizing filters and associative memory in recurrent networks of neuron-like elements. *Biol. Cybern.* **21**:85-95.
Kopell, N. (1978) Reaction-diffusion equations and pattern formation. In: *Studies in Mathematical Biology Part 1: Cellular Behavior and the Development of Pattern*. S. A. Levin (ed.). Mathematical Association of America, Washington, D.C., pp. 191-205.
Kuffler, S. W. (1953) Discharge patterns and functional organization of mammalian retina. *J. Neurophysiol.* **16**:37-68.
Lara, R. (1980) The amphibian visual system: Modelling the tectal column. *Cognition and Brain Theory* **3**:90-100.
Lara, R., M. A. Arbib, and A. S. Cromerty (1982) The role of the tectal column in facilitation of amphibian prey-catching behavior. *J. Neurosci.* **2**:521-530.
Lara, R., M. Carmona, F. Daza, and A. Cruz (1984) A global model of the neural mechanisms responsible for visuomotor coordination in toads. *J. Theor. Biol.* **110**:587-618.
Lavorel, P. M., and M. A. Arbib (1981) Towards a theory of language performance: Neurolinguistics, perceptual-motor processes, and cooperative computation. *Theor. Ling.* **8**:3-28.
Lee, D. N. (1974) Visual information during locomotion. In: *Perception: Essays in Honor of James J. Gibson*, R. B. MacLeod and H. L. Pick, Jr. (eds.). Cornell University Press, Ithaca, N.Y., pp. 250-267.
Lee, D. N., and J. R. Lishman (1977) Visual control of locomotion. *Scand. J. Psychol.* **18**:224-230.
Lettvin, J. Y., H. Maturana, W. S. McCulloch, and W. H. Pitts (1959) What the frog's eye tells the frog's brain. *Proc. IRE* **47**:1940-1951.
Lock, A., and T. Collett (1980) The three-dimensional world of a toad. *Proc. R. Soc. London Ser. B* **206**:481-487.
Marr, D. (1978) Representing visual information. In: *Computer Vision Systems*. A. R. Hanson and E. M. Riseman (eds.). Academic Press, New York, pp. 61-80.
Marr, D. (1982) *Vision: A Computational Investigation into the Human Representation and Processing of Visual Information*. Freeman, San Francisco.
Marr, D., and T. Poggio (1977a) From understanding computation to understanding neural circuitry. *Neurosci. Res. Progr. Bull.* **15**:470-488.
Marr, D., and T. Poggio (1977b) Cooperative computation of stereo disparity. *Science* **194**:283-287.
Marr, D., and T. Poggio (1979) A theory of human stereopsis. *Proc. R. Soc. London Ser. B* **204**:301-328.
Minsky, M. L. (1975) A framework for representing knowledge. In: *The Psychology of Computer Vision*, P. H. Winston (ed.). McGraw-Hill, New York, pp. 211-277.
Mountcastle, V. B. (1957) Modality and topographic properties of single neurons of cat's somatic sensory cortex. *J. Neurophysiol.* **20**:408-434.
Mountcastle, V. B. (1978) An organizing principle for cerebral function: The unit module and the distributed system. In: *The Mindful Brain*, G. M. Edelman and V. B. Mountcastle MIT Press, Cambridge, Mass, pp. 7-50.

Murphy, J. T., and N. H. Sabah (1970) The inhibitory effect of climbing fiber activation on cerebellar Purkinje cells. *Brain Res.* **19**:486–490.

Neisser, U. (1976) *Cognition and Reality.* Freeman, San Francisco.

Orlovsky, G. N. (1972) The effect of different descending systems on flexor and extensor activity during locomotion. *Brain Res.* **40**:359–372.

Oscarsson, O. (1973) Functional organization of spinocerebellar paths. In: *Handbook of Sensory Physiology*, Vol. II, A. Iggo (ed.). Springer-Verlag, Berlin, pp. 339–380.

Overton, K. J., and M. A. Arbib (1982) The extended branch-arrow model of the formation of retinotectal connections. *Biol. Cybern.* **45**:157–175.

Perkel, D. H., and T. H. Bullock (1968) Neural coding. *Neurosci. Res. Progr. Bull.* **6**(3):223–348.

Powell, T. P. S., and V. B. Mountcastle (1959) Some aspects of the functional organization of the cortex of the postcentral gyrus of the monkey: A correlation of findings obtained in a single unit analysis with cytoarchitecture. *Bull. Johns Hopkins Hosp.* **105**:133–162.

Prager, J. M. (1979) Segmentation of Static and Dynamic Scenes. Ph.D. thesis, Computer and Information Science Department, University of Massachusetts, Amherst.

Prager, J. M., and M. A. Arbib (1983) Computing the optic flow: The MATCH algorithm and prediction. *Computer Vision, Graphics, and Image Processing* **24**:271–304.

Rashevsky, N. (1948) *Mathematical Biophysics*, 2nd ed. University of Chicago Press, Chicago.

Rinzel, J. (1978) Integration and propagation of neuroelectric signals. In: *Studies in Mathematical Biology Part 1: Cellular Behavior and the Development of Pattern*, S. A. Levin (ed.). Mathematical Association of America, Washington, D.C., pp. 1–66.

Scheibel, M. E., and A. B. Scheibel (1958) Structural substrates for integrative patterns in the brain stem reticular core. In: *Reticular Formation of the Brain*, H. H. Jasper, L. Proctor, R. Knighton, W. Noshey, R. Costello (eds.). Little, Brown, Boston, pp. 31–68.

Shaw, G. L. (1978) Space-time correlations of neuronal firing related to memory storage capacity. *Brain Res. Bull.* **3**:107–113.

Sherrington, C. S. (1910) Flexion-reflex of the limb, crossed extension-reflex, and reflex stepping and standing. *J. Physiol. (London)* **40**:28–121.

Shik, M. L., F. V. Severin, and G. N. Orlovsky (1966) Control of walking and running by means of electrical stimulation of the mid-brain. *Biophysics* **11**:756–765.

Sperling, G. (1970) Binocular vision: A physical and neural theory. *Am. J. Psychol.* **83**:463–534.

Sperry, R. W. (1951) Mechanisms of neural maturation. In: *Handbook of Experimental Psychology*, S. S. Stevens (ed.). Wiley, New York, pp. 236–280.

Spinelli, D. N., and F. E. Jensen (1979) Plasticity: The mirror of experience. *Science* **203**:75–78.

Szentágothai, J. (1978) The neuron network of the cerebral cortex: A functional interpretation. *Proc. R. Soc. London Ser. B* **201**:219–248.

Szentágothai, J., and M. A. Arbib (1975) *Conceptual Models of Neural Organization.* MIT Press, Cambridge, Mass.

Tsukahara, N. (1972) The properties of the cerebello-pontine reverberating circuit. *Brain Res.* **40**:67–71.

Turing, A. M. (1952) The chemical basis of morphogenesis. *Philos. Trans. R. Soc. London* **237**:37–72.

Ullman, S. (1979) *The Interpretation of Visual Motion.* MIT Press, Cambridge, Mass.

von der Malsburg, C. (1973) Self-organization of orientation-sensitive cells in the striate cortex. *Kybernetik* **14**:85–100.

von der Malsburg, C., and D. J. Willshaw (1977) How to label nerve cells so that they can interconnect in an orderly fashion. *Proc. Natl. Acad. Sci. USA* **74**:5176–5178.

von Neumann, J. (1951) The general and logical theory of automata. In: *Cerebral Mechanisms in Behavior: The Hixon Symposium*, L A. Jeffress (ed.). Wiley, New York, pp. 1–32.

von Neumann, J. (1956) Probabilistic logics and the synthesis of reliable organisms from unreliable components. In: *Automata Studies*, C. E. Shannon and J. McCarthy (eds.) Princeton University Press, pp. 43–98.

Voogdt, J. (1969) The importance of fiber connections in the comparative anatomy of the mammalian cerebellum. In: *Neurobiology of Cerebellar Evolution and Development*, R. Llinás (ed.). American Medical Association, Chicago, pp. 493–514.

Wilson, W. R., and J. D. Cowan (1973) A mathematical theory of the functional dynamics of cortical and thalamic nervous tissue. *Kybernetik* **13**:55–80.

V

Epistemology of Self-Organization

Gregory B. Yates

The general phenomenon of self-organization has been described in preceding sections. It is thus natural to ask whether a general theory of self-organization is emerging or whether phenomena of self-organization are fundamentally inexplicable. Attempts to resolve these questions raise a host of issues that lie in the realm of epistemology. Epistemological questions must be addressed in theories of self-organization because they cannot be avoided.

The antiquity and universality of creation myths reveal the depth of human fascination with self-organization, and reasonably explicit discussions of the subject may be traced through millenia in the works of philosophers and theologians. Epistemology concerns the properties of observers. Relevant debate about these properties has ranged between two poles, loosely identified with the respective views of Plato and Aristotle. In Platonic thought the mind of an observer either possesses, or has transcendental access to, information independent of that imposed upon it by the senses. In Aristotelian thought the mind of the observer is viewed as a blank tablet upon which information about the natural world is inscribed by a general inductive process. The physical and biological sciences, because their subject is the natural world, have assumed an Aristotelian perspective, and the increasing success of physics and biology over the past two centuries has often been construed as an implicit validation of the Aristotelian view of the observer. But in the last century, there has also been a striking resurgence of interest in the structure and process of the observer—brought on largely by the exigencies of quantum physics. The demands of explaining self-organization and complexity now raise fresh questions about the role of observer properties in the explanation of natural phenomena. The chapters in this section are arranged roughly in order of their increasing concern with these properties.

GREGORY B. YATES • Crump Institute for Medical Engineering, University of California, Los Angeles, California 90024.

Whether the theories of biology may be reduced to those of physics is addressed by Ayala, who shows that they involve several different, but often-confused types of reduction. For example, the question of whether life processes can be explained without recourse to vitalistic forces is of a different type than that of whether the theories of biology are logically implied by those of physics. Accepting these distinctions, it is possible to make clear statements about the probable future relationship of the sciences. Ayala concludes that complete epistemological reduction of biology to physics is unlikely but that such reduction nevertheless remains a reasonable goal of scientific work.

Pattee focuses on two major classes of self-organizing systems: the statistically unstable systems, such as fluids, versus the information-dependent ones, such as cell differentiation. He shows that complementary views of instability underlie theories concerning the two classes. Events that have no observable cause (instabilities) may be viewed as the result either of hidden causes or of pure chance. The effects of this choice of views cascade to produce two types of self-organization theories characteristic of biology and physics, respectively. These two theory types are simultaneously valid but formally incompatible. Pattee argues that both views are essential in any general theory of biological self-organization.

By introducing the choice of view as an important aspect of self-organization theory, Pattee implicitly raises questions about the properties of the one who chooses. Stent makes these questions much more explicit. First he rejects the programmatic behavior of computers as a metaphor for developmental neurobiology and suggests the historical phenomena of developing ecosystems as an alternative. He then shows that the analysis of historical phenomena is formally very similar to hermeneutics—the interpretation of sacred texts. But in hermeneutics the validity of an interpretation is discussable only by those who bring the same preunderstanding to the subject, so that an objectively valid theory of biological self-organization is likely to be impossible. The properties of the observer become central in Stent's chapter.

The writers of this section seem to agree that formulation of a general and useful theory of self-organization is beset by epistemological difficulties and by the inevitable interdependence between theories and their human interpreters.

16

Biological Reductionism
The Problems and Some Answers

Francisco J. Ayala

ABSTRACT

The problems of biological reductionism, that is, questions of the relationship between whole organisms and their parts, fall into three domains: ontological, methodological, and epistemological. Ontological reductionism occurs when an organism is found to be exhaustively composed of the same components as inorganic matter. Methodological reductionism is the claim that the best strategy of research is to study living phenomena at the lowest levels of complexity. Epistemological reductionism results if the laws and theories of biology can be derived as special cases of the laws and theories of the physical sciences.

The mechanism versus vitalism issue belongs to the ontological domain, as does the issue of emergent properties. All evidence indicates that organisms are exhaustively composed of nonliving atoms and that life processes can be explained without recourse to any substantive nonmaterial entity. However, atoms in association do display properties not normally included among those of isolated atoms, and in practice there is no way to ascertain what properties one object may show in conjunction with any other.

Methodological reductionism seems justified in a moderate form: often the best strategy of research is an alternation between analysis and synthesis. This moderate position finds support from epistemological reductionists, who admit the heuristic value of nonreductionistic experiments, and from opponents of epistemological reductionism, who nevertheless think that organisms should be studied at all levels of integration.

Epistemological reduction of one theory to another takes the form of a deductive argument in which one of the premises is the primary theory and the conclusion is the secondary theory. This operation requires that the disparate terms of the two theories be connected by suitable definitions. These conditions have been largely satisfied in the reduction of thermodynamics to statistical mechanics, the reduction of chemical valence theory to the physics of orbital electrons, and the reduction of some of genetics to molecular biology. If the reduction of one science to another is not possible at present, then claims that such a reduction will be possible in the future carry little weight. In any case, there is an unresolved residue accompanying any attempt at epistemological reductionism, and the

---

FRANCISCO J. AYALA • Department of Genetics, University of California, Davis, California 95616.

terms and patterns of explanation in some sciences seem wholly unconnectable with those of other sciences. For these reasons it seems unlikely that complete epistemological reduction of biology to physics will ever be possible. Nevertheless, epistemological reductions are successful forms of scientific explanation and remain a reasonable goal of scientific work. —The Editor

Organisms are complex self-organizing entities made up of parts: organs, tissues, cells, organelles, and ultimately molecules and atoms. One question that arises concerns the relationship between the whole and its component parts. The issue at stake is sometimes called "the question of reduction" or "the problem of reductionism." Few if any questions in the philosophy of science have received more attention and been more actively debated than the question of reduction (e.g., Koestler and Smythies, 1969; Ayala and Dobzhansky, 1974). The debates, however, often involve several different issues, not always properly distinguished. Issues about the relationship between organisms and their physical components, or between biology and the physical sciences, arise in at least three domains, which may be called "ontological," "methodological," and "epistemological." I shall identify the issues raised in each domain and then consider each domain in turn.

Reductionistic questions arise, first, in what may be called the ontological, the structural, or the constitutive domain. The issue here is whether or not physicochemical entities and processes underlie all living phenomena. Are organisms constituted of the same components as those making up inorganic matter? Or, do organisms consist of other entities besides molecules and atoms? Two other questions are related to the previous ones: Are organisms nothing else than aggregations of atoms and molecules? Do organisms exhibit properties other than those of their constituent atoms and molecules?

Second, there are reductionist questions that might be called methodological, procedural, or strategical. These questions concern the strategy of research and the acquisition of knowledge—the approaches to be followed in the investigation of living beings. The general question is whether a particular biological problem should always be investigated by studying the underlying (ultimately, physical) processes, or whether it should also be studied at higher levels of organization, such as the cell, the population, and the community.

The third type of reductionistic questions concern issues that may be called epistemological, theoretical, or explanatory. The fundamental issue here is whether or not the theories and laws of biology can be derived from the laws and theories of physics and chemistry. Epistemological reductionism is concerned with the question of whether biology may be ignored as a separate science because it represents simply a special case of physics and chemistry.

Distinction of the various kinds of questions being asked in debates about reductionism is the first step toward solving the issues. Much argumentation and confusion have resulted from failure to identify the question being argued in particular instances. It is typical, for example, to see a reductionist concerned primarily with the ontological question, accusing a self-proclaimed antireductionist of having vitalistic ideas, while the latter may be an epistemologically antireductionist but also in fact an ontological reductionist.

Constitutive Reductionism

In the ontological or constitutive domain, the reductionist–antireductionist controversy in its extreme form resolves into the mechanism-versus-vitalism issue. The mechanist position is that organisms are ultimately made up of the same atoms that make up inorganic matter, and of nothing else. Vitalists argue that organisms are made up not only of material components (atoms, molecules, and aggregations of them) but also of some non-material entity, variously called by different authors, entelechy, vital force, *élan vital*, radial energy, and the like. Aristotle, a philosopher who was also the best biologist of his time, is sometimes said to have been the first systematic proponent of vitalism. The modern mechanism–vitalism controversy dates from the 17th century when René Descartes proposed that animals are nothing else than complex machines. Early in the 20th century, vitalism was defended by philosophers such as Henri Bergson and by some biologists, notably Hans Driesch. At present, this extreme form of vitalism has no distinguished proponents among biologists, and few if any among philosophers.

Vitalism has been excluded from science primarily because it does not meet the requirements of a scientific hypothesis, i.e., it is not subject to the possibility of empirical falsification and therefore leads to no fruitful observations or experiments. Moreover, all evidence indicates that organisms and life processes can be explained without recourse to any substantive nonmaterial entity.

Ontological reductionism claims that organisms are exhaustively composed of nonliving parts; no substance or other residue remains after all atoms making up an organism are taken into account. Ontological reductionism also implies that the laws of physics and chemistry fully apply to all biological processes at the level of atoms and molecules.

Ontological reductionists do not necessarily claim, however, that organisms are nothing but atoms and molecules. The inference that because something consists only of something else, it is nothing but this "something else" is an erroneous inference, called by philosophers the "nothing but" fallacy. Organisms consist exhaustively of atoms and molecules, but it does not follow that they are nothing but heaps of atoms and molecules. A steam engine may consist only of iron and other materials, but it is something else than iron and the other components. Similarly an electronic computer is not only a pile of semiconductors, wires, plastic, and other materials. Organisms are made up of atoms and molecules, but they are highly complex patterns, and patterns of patterns, of these atoms and molecules. Living processes are highly complex, highly special, and highly improbable patterns of physical and chemical processes.

A much-debated reductionist question that belongs in the ontological domain is whether organisms exhibit "emergent" *properties*, or whether their properties are simply those of their physical components. For example, are the functional properties of the kidney simply the properties of the chemical constituents of that organ? The question of emergent properties is not exclusive to biology, but arises for all complex systems with respect to their parts. The general formulation of this question is whether the properties of a certain kind of object are simply the properties of other kinds of objects, namely, their component parts, organized in certain ways.

Whether complex systems exhibit emergent properties is largely a spurious issue that can be solved as a matter of definition. Consider the following question: Are the properties of common salt, sodium chloride, simply the properties of sodium and chlorine when they are associated according to the formula NaCl? If among the properties of sodium and chlorine we include their association into table salt and the properties of the latter, the answer is "yes." In general, if among the properties of an object we include the properties that the object has when associated with other objects, it follows that the properties of complex systems, including organisms, are also the properties of their component parts. This is simply a definitional maneuver that contributes little to understanding the relationships between complex systems and their parts.

In common practice we do not include among the properties of an object all the properties of the systems resulting from its association with any other objects. There is a good reason for that. No matter how exhaustively an object is studied in isolation, there is usually no way to ascertain all the properties that it may have in association with any other object. Among the properties of hydrogen we do not usually include the properties of water, of ethyl alcohol, of proteins, and of humans. Nor do we include among the properties of iron those of the steam engine.

The question of emergent properties may also be formulated in a somewhat different manner. Can the properties of complex systems be *inferred* from knowledge of the properties that their component parts have in isolation? For example, can the properties of benzene be predicted from knowledge about oxygen, hydrogen, and carbon? Or, at a higher level of complexity, can the behavior of a cheetah chasing a deer be predicted from knowledge about the atoms and molecules making up these animals? Formulated in this manner, the issue of emergent properties is an epistemological question, not an ontological one; we are now asking whether the laws and theories accounting for the behavior of complex systems can be derived as logical consequences from the laws and theories that explain the behavior of their component parts. Epistemological questions of reductionism will be discussed later.

The Study of Organisms

The outstanding characteristic of living beings is a complexity of organization recognized in their common name, "organisms." There is a hierarchy of levels of complexity that runs from atoms, through molecules, macromolecules, organelles, cells, tissues, multicellular organisms, populations, and communities. Some biological disciplines focus on one or a few of these levels of complexity of organization. Cytology is the study of cells, histology is the study of tissues, ecol-

16. Biological Reductionism

ogy comprises studies of populations and communities. Yet the biological disciplines are distinguished more by the kinds of questions asked and the kinds of answers sought than by the levels of organization that are investigated.

Methodological reductionism is the claim that the best strategy of research is to study living phenomena at increasingly lower levels of complexity and, ultimately, at the level of atoms and molecules. For example, genetics should seek to understand heredity ultimately in terms of the behavior and structure of DNA, RNA, enzymes, and other macromolecules, rather than in terms of whole organisms, which is the level at which the Mendelian laws of inheritance are formulated. Methodological reductionism has its counterpart in what may be called methodological compositionism (Simpson, 1964). This is the claim that in order to understand organisms, we must explain their organization—how organisms and groups of organisms come to be organized and what functions the organization serves. According to methodological compositionism, organisms and groups of organisms must be studied as wholes, not only in their component parts.

Methodological reductionism in its extreme form would be the claim that biological research should be conducted only at the level of the physicochemical component parts and processes. Research at other levels is, allegedly, not worth pursuing or is at best of only provisional value, since biological phenomena must ultimately be understood at the molecular and atomic levels. Methodological compositionism in its extreme form makes the opposite claim, namely, that the only biological research worth pursuing is that at the level of whole organisms, populations, and communities. Research at lower levels of organization may, allegedly, be good physics or good chemistry, but it has no biological significance.

It is unlikely that any scientist would thoughtfully sponsor the extreme forms of either compositionism or reductionism advanced in the previous paragraph. Extreme methodological reductionism would imply the unreasonable claim that genetic investigations should not have been undertaken until the discovery of DNA as the hereditary material or that a moratorium should be declared in ecology until we can investigate the physicochemical processes underlying ecological interactions. Similarly, extreme methodological compositionism would imply that understanding the structure of DNA or the enzymatic processes involved in its replication is of no significance to the study of heredity or that the investigation of physicochemical reactions in the transmission of nerve impulses is of no interest to the understanding of animal behavior.

A moderate position of methodological reductionism points to the success of the analytical method in science and to the obvious fact that the understanding of living processes at any level of organization is much advanced by knowledge of the underlying processes. Moderate methodological reductionists would claim that the best strategy of research is to investigate any given biological phenomenon at increasingly lower levels of organization as this becomes possible and, ultimately, at the level of atoms and molecules.

The positive claims of moderate methodological reductionists are legitimate. The analytical method is of great heuristic value; much is often learned about a phenomenon through the investigation of its component elements and processes. In biology, the most impressive achievements of the last few decades are those of molecular biology. But there is little justification for any exclusionist claim that research should always proceed by investigation of lower levels of integration. The only criterion of validity of a research strategy is its success. Compositionist as well as reductionist approaches, synthetic as well as analytic methods of investigation, are justified if they further our understanding of a phenomenon—if they increase knowledge. Reductionist and compositionist approaches to the study of a biological problem are complementary; often the best strategy of research is an alternation between analysis and synthesis.

Investigation of a biological phenomenon in terms of its significance at higher levels of complexity often contributes to the understanding of the phenomenon itself; compositionist investigations are also heuristic. For example, it is doubtful that the structure and functions of DNA would have been known as readily as they were if there had been no previous knowledge of Mendelian genetics. The problem of the specificity of the immune response of antibodies proved refractory to a satisfactory solution as long as antibodies, antigens, and their structures alone were taken into consideration. The natural selection theory of antibody function emerged only when antibodies were considered in their organismic milieu. Although the idea of clonal selection was first logically inadequate and quite vague, it had an enormous heuristic value in helping to understand how the specificity of antibodies comes about (Edelman, 1974).

Methodological reductionism and compositionism are sometimes based on convictions about the possibility of epistemological reduction. A methodological reductionist might claim that research should be pursued at increasingly lower levels of organization because he is convinced that ultimately all biological phenomena will be explained by the laws and theories of the physical sciences and thus that the only knowledge of lasting value is that acquired at the level of atoms and molecules. A methodological compositionist might claim that epistemological reductionism is impossible and, therefore, that full understanding of a biological phenomenon requires knowledge of its significance for higher levels of integration.

Methodological and epistemological reductionism are nevertheless separate issues. Methodological reductionism is concerned with the strategies of research and the acquisition of knowledge. Epistemological reductionism deals with the organization of knowledge and the logical connections between theories. An epistemological reductionist, for example, might nevertheless accept compositionist approaches to research because of their heuristic value. Similarly, epistemological antireductionists often claim that biological research should be pursued at all levels of integration of the living systems, including the atomic and molecular levels.

Biology and the Physical Sciences

When philosophers of science speak of reductionism, they generally refer neither to ontological nor to methodological issues but, rather, to epistemological reduction. In biology the question of epistemological (theoretical, explanatory) reduction is whether or not the laws and theories of biology can be shown to be derived as special cases from the laws and theories of the physical sciences.

Science seeks to discover patterns of relationships among many kinds of phenomena in such a way that a small number of principles explain a large number of propositions concerning those phenomena. Science advances by developing gradually more comprehensive theories, i.e., by showing that theories and laws that had hitherto appeared as unrelated can in fact be integrated in a single theory of greater generality. For example, the theory of heredity proposed by Mendel can explain, with respect to many kinds of organisms, diverse observations such as the proportions in which traits are transmitted from parents to offspring, why progenies exhibit some traits inherited from one parent and some from the other parent, and why the offspring may exhibit traits not present in their parents. The discovery that the behavior of chromosomes during meiosis is connected with the Mendelian principles made possible the explanation of many additional observations concerning heredity, for example, why certain traits are inherited independently from each other, whereas other traits are transmitted together more often than not. Further discoveries have made possible the development of a unified theory of inheritance, which has great generality and explains many diverse observations, including the distinctness of individuals, the adaptive nature of organisms and their traits, and the discreteness of species.

The connection among theories has sometimes been established by showing that the tenets of a theory or branch of science can be explained by the tenets of another theory or branch of science of greater generality. The less general theory (or branch of science), called the secondary theory, is then said to have been reduced to the more general or primary theory. Epistemological reduction of one branch of science to another takes place when the theories or experimental laws of a branch of science are shown to be special cases of the theories and laws formulated in some other branch of science. The integration of diverse scientific theories and laws into more comprehensive ones simplifies science and extends the explanatory power of scientific principles, and thus conforms to the goals of science. Epistemological reductions are of great value to science because, as special cases of the integration of theories, they greatly contribute to the advance of scientific knowledge.

The reduction of a theory or even of a whole branch of science to another has been repeatedly accomplished in the history of science (Nagel, 1961; Popper, 1974). One of the most impressive examples is the reduction of thermodynamics to statistical mechanics made possible by the discovery that the temperature of a gas reflects the mean kinetic energy of its molecules. Several branches of physics and astronomy have been to a large extent unified by their reduction to a few theories of great generality, such as quantum mechanics and relativity. A large sector of chemistry was reduced to physics after it was shown that the valence of an element bears a simple relation to the number of electrons in the outer orbit of the atom. Parts of genetics were to some extent reduced to chemistry after discovery of the structure, replication mode, and action of the hereditary material, DNA.

The impressive successes of these reductions, and in particular the spectacular achievements of molecular biology, have led some authors to claim that the ideal of science is to reduce all natural sciences, including biology, to a comprehensive physical theory that would provide a common set of principles of maximum generality capable of explaining all observations about natural phenomena.

Some authors have gone so far as to claim that the only biological research worth pursuing is that contributing to the explanation of biological phenomena in physicochemical terms.

Nagel (1961) has formulated the two conditions that are necessary and jointly sufficient to effect the reduction of one theory or branch of science to another. These are the condition of *derivability* and the condition of *connectability*. Epistemological reduction takes place when the experimental laws and theories of a branch of science are shown to be special cases of the experimental laws or theories of some other branch of science. The condition of derivability simply states that in order to reduce a branch of science to another, it is necessary to show that the laws and theories of the secondary science can be derived as logical consequences from the laws and theories of the primary science.

No terms can appear in the conclusion of a demonstrative argument that do not appear in the premises. The reduction of one theory to another takes the form of a deductive argument in which one of the premises is the primary theory and the conclusion is the secondary theory. For the deduction to be logically valid, there must be another premise that establishes the connection between the terms of the primary theory and the terms of the secondary theory. This is the condition of connectability. Generally the experimental laws and theories of a branch of science contain distinctive terms that do not appear in other branches of science. To accomplish an epistemological reduction, it is necessary that suitable connections be established between the terms of the secondary science and those used in the primary science. This result may be accomplished by redefining the terms of the secondary science using terms of the primary science. The reduction of thermodynamics to statistical mechanics required the definition of terms such as "temperature" by means of terms such as "kinetic energy." The reduction of theories or experimental laws of genetics to physicochemistry requires that terms such as "gene" and "chromosome" be defined by means of terms such as "hydrogen bond," "nucleotide," "deoxyribonucleic acid," "histone protein," and the like. Whenever the conditions of connectability and derivability are satisfied, the epistemological reduction of a theory to another becomes logically feasible. If all the experimental laws and theories of a branch of science can be reduced to experimental laws and theories of another branch of science, the former science will have been completely reduced to the latter.

It bears repetition that epistemological reduction is not a question of whether the *properties* of a certain kind of objects, such as organisms, result from the properties of other kind of objects, such as their component parts. That problem is an issue of ontological reductionism, which can be solved by a convention as to what is to be included among the properties of component parts. The reduction of one science to another is rather a matter of deriving a set of *propositions* from another such set. Scientific laws and theories consist of propositions about the natural world. The question of epistemological reduction can only be settled by concrete investigation of the logical consequences of propositions—not by discussions about the nature of things or their properties. It is of course a legitimate epistemological question to ask whether *statements* concerning the properties of organisms can be deduced logically from statements concerning the properties of their physical components.

16. Biological Reductionism

It follows from the previous comments that questions of epistemological reduction can only be properly answered by concrete reference to the actual state of development of the disciplines involved. Certain parts of chemistry were reduced to physics after the modern theory of atomic structure was advanced half a century ago. That reduction could not have been accomplished before such a development. If the reduction of one science to another is not possible at the present stage of development of two given disciplines, claims that such a reduction will be possible in the future carry little weight, because such claims depend on the hypothetical development of as yet nonexisting theories.

There are some extreme positions on epistemological reductionism that can be quickly discounted. Some substantive vitalists have claimed that, in principle, biology is not reducible to the physical sciences because living phenomena are the manifestation of nonmaterial principles, such as vital forces, entelechies, and so on. Epistemological antireductionism is then predicated on ontological antireductionism. But vitalism is not an empirical hypothesis because it does not lend itself to the possibility of empirical falsification. Moreover, the origin, structure, and functions of organisms can be explained without recourse to nonmaterial components or principles.

At the other end of the spectrum is the claim that the epistemological reduction of biology to the physical sciences is possible and, indeed, is the most important task of biologists at present. The impressive successes of molecular biology during recent decades have moved some people to claim that the only worthy and truly scientific biological investigations are those leading to the explanation of biological phenomena in terms of the underlying physicochemical components and processes. Nevertheless, the epistemological reduction of biology to the physical sciences is not yet possible. In the current stage of scientific development, a great many biological terms, such as organ, species, consciousness, mating propensity, fitness, competition, predator, and many others, cannot adequately be defined in physicochemical terms. Nor is there any class of statements and hypotheses in physics and chemistry from which every biological law could logically be derived. Neither the condition of connectability nor the condition of derivability—the two necessary conditions for epistemological reduction—is satisfied.

A moderate reductionist position is probably not uncommon among biologists. It is claimed that, although the reduction of biology to physicochemistry cannot be effected at present, it is possible in principle and is indeed a goal to be actively sought. The factual reduction of biology to the physical sciences is made contingent upon further progress in the biological or physical sciences or both. This moderate form of epistemological reductionism is often based on convictions about ontological reductionism. It is generally accepted by biologists that living beings are exhaustively made up of physical components. It does not follow, however, that organisms are *nothing but* physical systems. Ontological reductionism does not entail epistemological reductionism. From the fact that organisms are exhaustively composed of atoms and molecules, it does not logically follow that the behavior of organisms can be exhaustively explained by the laws advanced to explain the behavior of atoms and molecules.

The claim that the eduction of biology to physicochemistry will eventually be possible is contingent upon unspecified, and at present unspecifiable, scientific advances. It is, therefore, a position that cannot be convincingly argued. Moreover, there are reasons to believe that complete reduction of biology to physics will never be possible. Popper has shown that no major case of epistemological reduction (including such model cases as the reduction of thermodynamics to statistical mechanics) "has ever been *completely* successful: there is almost always an unresolved residue left by even the most successful attempts at reduction" (Popper, 1974, p. 260; see also Hull, 1974). It does not follow, however, that biologists should not attempt to reduce their theories to those of the physical sciences, whenever such an undertaking seems likely to be successful. On the contrary, epistemological reductions are successful forms of scientific explanation. A great deal is learned from epistemological reductions even when they are unsuccessful or incomplete, because much understanding is gained by the partial success, and valuable insight is gained from the partial failure (Popper, 1974). In biology, the reduction of Mendelian genetics to molecular genetics has been far from completely successful (Hull, 1974). Yet there can be little doubt that much has been learned from what has been accomplished up to the present.

Some authors have claimed that the reduction of biology to physics and chemistry is impossible in principle because biological disciplines have patterns of explanation that do not occur in the physical sciences (Simpson, 1964; Ayala, 1968). Historical explanations, which are sometimes described as distinctively biological and social, play a role in evolutionary theory, although this theory is primarily concerned with causal explanations of evolutionary processes. Historical patterns of explanation, however, occur in some physical sciences such as astronomy and geology. On the other hand, as I have proposed elsewhere (Ayala, 1970), teleological patterns of explanation are appropriate in biology but appear to be neither necessary nor appropriate in the explanation of natural physical phenomena.

References

Ayala, F. J. (1968) Biology as an autonomous science. *Am. Sci.* **56**:207–221.
Ayala, F. J. (1970) Teleological explanations in evolutionary biology. *Philos. Sci.* **37**:1–15.
Ayala, F. J., and T. Dobzhansky, eds. (1974) *Studies in the Philosophy of Biology.* Macmillan & Co., London, and University of California Press, Berkeley.
Edelman, G. M. (1974) The problem of molecular recognition by a selective system. In: *Studies in the Philosophy of Biology,* F. J. Ayala and T. Dobzhansky (eds.). Macmillan & Co., London, pp. 45–56.
Hull, D. (1974) *Philosophy of Biological Science.* Prentice-Hall, Englewood Cliffs, N.J.
Koestler, A., and J. R. Smythies (1969) *Beyond Reductionism.* Hutchinson, London.
Nagel, E. (1961) *The Structure of Science.* Harcourt, Brace & World, New York.
Popper, K. R. (1974) Scientific reduction and the essential incompleteness of all science. In: *Studies in the Philosophy of Biology,* F. J. Ayala and T. Dobzhansky (eds.). Macmillan & Co., London pp. 259–284.
Simpson, G. G. (1964) *This View of Life.* Harcourt, Brace & World, New York.

17

Instabilities and Information in Biological Self-Organization

Howard H. Pattee

ABSTRACT

Two classes of self-organizing systems have received much attention, the *statistically unstable systems* that spontaneously generate new dynamical modes, and *information-dependent systems* in which non-statistical constraints harness the dynamics. Theories of statistically unstable systems are described in the language of physics and physical chemistry, and they depend strongly on the fundamental laws of nature and only weakly on the initial conditions. By contrast, the information-dependent systems are described largely by special initial conditions and constraints, and they depend only weakly, if at all, on the fundamental laws. This results in statistically unstable theories being described by rate-dependent equations, while the information-dependent systems are described by rate-independent (nonintegrable) constraints.

It is argued that an adequate theory of biological self-organization requires that these two complementary modes of description be functionally related, since the key process in morphogenesis is the harnessing of cellular dynamics by the informational constraints of the gene. This could arise if the triggering role of fluctuations could be displaced by informational constraints in the control of the dynamical behavior. However, the spontaneous replacement of chance fluctuations by deterministic informational codes is itself a serious problem of self-organization. At present the only approach requires complementary modes of description for the molecular informational constraints and for the macroscopic dynamical behavior that they harness. —The Editor

What do we expect to learn from a theory of biological self-organization? What types of observables or events do we begin with, and what regularities or laws would we accept as explanations of the self-organizing consequences of these events? Is a theory of biological self-organization fundamentally different from a theory of evolution, a theory of development, or a theory of thermodynamics; or

HOWARD H. PATTEE • Department of Systems Science, State University of New York, Binghamton, New York 13850.

is self-organization only the result of special types of behavior, such as the singularities found in dynamical systems, or the genetic shuffling found in evolutionary systems? If we survey the many approaches to the problem taken by the other authors of this volume, we might conclude that everyone sees self-organizing behavior in terms of his or her own discipline's theoretical framework. This is to be expected. We all use what we know to better comprehend what we do not know, as Newton used his theory of gravitation better to comprehend God. Nevertheless, even if all disciplines have something to contribute to our comprehension of self-organizing behavior, we still must be able to distinguish this "special type" of behavior from the "normal" events associated with each discipline, if self-organization is to have a distinct meaning.

This chapter focuses on two classes of self-organizing system that have received the most attention: the *statistically unstable systems* and the *information-dependent systems*. The question to be addressed is: What do these two types of systems have to do with each other? Reading the chapters in this book, we find that instability and information are nearly disjoint subjects that are not even discussed in the same language. The chapters on instabilities contain the language of physics or physical chemistry and are particularly sensitive to the basic laws of nature, whereas the chapters focusing on informational concepts contain biological language and are not concerned with physical laws at all. Of course this might simply reflect the differences of the two disciplines of physics and biology. It is also true that authors of papers on instabilities tend to use relatively simple nonliving chemical systems as examples, whereas those writing papers involving information discuss organs as complex as the brain of humans. Again, the differences in approach might simply reflect this enormous difference in levels of organization chosen for study.

These reasons for different approaches are understandable, but ignoring either dynamical instability or symbolic information evades a fundamental aspect of biological self-organization. Although it is true that dynamical theory and symbolic information are not associated in our normal way of thinking, they are epistemologically complementary concepts that are nevertheless both essential for a general theory of biological self-organization. Moreover, instabilities are the most favorable condition of a dynamical physical system for the origin of nondynamical informational constraints, and the evolution of self-organizing strategies at all levels of biology requires the complementary interplay of dynamical (rate-dependent) regimes with instabilities and nondynamic (rate-independent, nonintegrable) informational constraints (Pattee, 1971). Finally, after discussing these points, I shall comment on the limitations of self-organization theories in terms of epistemological and methodological reductionism (see Ayala, Chapter 16, this volume).

The Necessity of Stability

One epistemological requirement of scientific explanation of events is some form of homomorphism between the behavior of the theory as a model or simulation and the behavior of corresponding events or measurements. Or, as Hertz (1899) expressed it, we have an explanation "... when the necessary consequents of the images in thought are always the images of the necessary consequents in the nature of things pictured." This classical epistemological requirement for a correspondence between object and image implies the basic concept of stability—that all sufficiently small changes in the images or descriptions of events represent some corresponding, small changes in the events themselves. Again, it is hard to improve on the classical formulations. Poincaré (1952) called a system stable when "small causes produce small effects." Although Newton had no formal methods for testing the stability of his laws of motion, he clearly understood the profound importance of stability for an explanation, since he argued that one of God's primary functions was to maintain the stability of the planetary orbits against the perturbations of other planets. Deterministic laws of motion alone do not ensure stability.

When Laplace proved that Newton's equations were inherently stable, the universe became a great deterministic machine that required no divine intervention once it was created. The only remaining problem, as Laplace (1951) saw it, was the practical impossibility of the human mind acquiring a complete and precise set of initial conditions for all the bodies in the universe. The best he could do was develop the theory of probability so that we could deal rationally with our remaining ignorance of initial conditions. From this classical, objective epistemology of Newton and Laplace, the idea of self-organization made little sense. Either God intervened locally to keep motions stable, or any unpredictable behavior could be attributed only to our ignorance of details. In either case, there was no apparent desire for additional theories of self-organization or emergent behavior.

The Necessity of Instability

Logically, any theory of self-organization must accept the precondition of a disorganized subsystem, or a partially disorganized system, since an inherently totally organized system leaves no room for more organization without either redundancy or contradiction. Thus, in the ideal Laplacian universe where every microscopic initial condition is precisely given, we have a totally organized system with an inexorably determined past and future. The well-known escape from this complete determinism, which Laplace recognized, is the condition of human ignorance of initial conditions. This condition is the basis of statistical mechanics; but the most general consequence of this condition was only an increasing disorganization or entropy in the course of time.

The main contribution of Prigogine's school has been to find a new description of how instabilities in these statistical systems may result in entirely new structures from chaotic initial conditions. However, both the ontological and epistemological status of instability remain a fundamental and still controversial problem. One may briefly, and somewhat naively, describe the problem as follows: Our knowledge of the world depends upon our ordering of experience by images, models, descriptions, theories, and so on, which we try to make as clear and unambiguous as possible. To test our theories we predict the consequences from observed initial conditions using the rules of the theory and then look for the corresponding consequences in our later observations. If this correspondence holds over a wide-enough range of initial conditions, and if the theory has other coherent logical and aesthetic properties, which we find difficult to define, then we may feel that we understand or have described some laws of nature, or some laws of knowledge, depending on our metaphysics.

Now, although we may observe unstable behavior both directly in events or in our mathematical models of events, we still have a problem in establishing their correspondence by the same criteria we use to establish the correspondence between predictions and measurements of stable behavior. This results from the fact that we recognize events as unstable only because they appear to have no observable deterministic cause, and we recognize descriptions of events as unstable only because the descriptions fail to completely define those events. Poincaré (1952) defined instabilities in the same way that he defined chance events, i.e., as observable events that have no observable cause. In other words, Poincaré's definition of an unstable event implies that it can be described only by a probabilistic model that is logically incompatible with a deterministic model of the same event. The question, then, of whether instabilities are fundamentally the result of chance or determinism in an objective sense or whether they result from the failure of our *descriptions* of events is unresolved and perhaps unresolvable, in any scientific or empirical sense.

In any case we can see that self-organizing behavior must involve something more than stable, deterministic trajectories of classical theories, or the assumption of ignorance of initial conditions, which leads to the stable, deterministic distributions of statistical mechanics. Even though these laws of physics are a foundation for all organization, including self-organization, we recognize in the self-organizing behavior of both nonliving and living systems many entirely new forms and patterns that are not simply the perturbations of stable systems or the probabilistic behavior of unstable systems. The novelty and persistence of emergent forms characteristic of living systems do not fit our definition of either stable or unstable behavior. Therefore, we may expect theories of self-organization also to require complementary subjective or functional modes of description.

Thom's (1975) catastrophe theory of self-organization may at first appear to have little relation to these concepts, since his starting point is pure mathematics. However, Thom's concept of form is also expressed in terms of stability; whatever remains stable under a small perturbation has the same form. Change of form or morphogenesis consequently involves instability. The mathematical concept of "structural stability," as it is called, is technically complex, but can naturally be associated with two complementary modes of description—a control-space (subjective) description and a corresponding, state-space (objective) description. The essence of structural stability is that a gradual change in the control-space description induces a corresponding, gradual change in the state space description ("small causes produce small effects"). In the other case, where a gradual change in the control-space description induces a sudden, discontinuous change in the state-space description (bifurcation), there is a change of form, or what Thom calls a catastrophe and we call a form of self-organizing behavior. Of course, many other epistemological interpretations of the formalism are possible since the model, or "method," as Thom calls it, is essentially mathematical. However, it is fundamental to the method that two complementary structures are necessary for developing the concept of catastrophe.

The Necessity of Complementarity

The modern concept of complementarity is associated with Bohr and the development of quantum theory (Bohr, 1928, 1963; Jammer, 1974; d'Espagnat, 1976), but Bohr believed that the concept was of much more general epistemological significance. Although it was the theory of the electron that forced recognition of complementary modes of description, Bohr felt that complementarity "bears a deep-going analogy to the general difficulty in the formation of human ideas, inherent in the distinction between subject and object." Bohr's definitions of the complementarity principle were not formal or precise, and they have generated much controversy. This is to be expected of any epistemological principle that claims both universal applicability as well as some empirical necessity. It is generally accepted, however, that the principle includes at least two components: first, that to account for or to explain an observed event, two distinct modes of description or representations are necessary, and second, that these two modes of description are incompatible, both in the logical sense that a contradiction would arise if the two descriptions were combined into one formal structure and in the conceptual sense that trying to combine the meanings of both descriptions into one image leads to confusion. Although my concept of complementarity was greatly influenced by reading Bohr and his interpreters, I do not wish to defend or attack his epistemology. I simply find no alternative but to accept multiple, formally incompatible descriptions as a satisfactory explanation of many types of biological events (Pattee, 1979).

Perhaps the most fundamental epistemological complementarity arises in our perception of events as either deterministic or chance. The Laplacian ideal of determinism is certainly more than a rational hypothesis about point masses and universal gravitation. One of the most easily observed beliefs of a 5-or 6-year-old child is the child's assumption that every event has a cause or that events could not be capricious (Piaget, 1927). Of course, the "causes" that children see are usually animistic or moralistic, but in any case, the idea of determinism is very primitive and does not easily die out in the course of intellectual development. For example, a mature Wigner (1964) characterizes his acceptance of explanation as the feeling that "events could not be otherwise," and most of us are at least emotionally sympathetic with Einstein's belief that "God does not play dice."

The concept of chance comes developmentally with experience, but is never assimilated into our thought with the clarity of the concept of determinism. In fact, gamblers and physicists alike behave as if chance is only determinism disguised by ignorance. The two types of classical theory illustrate this contrast: the so-called microscopic, deterministic descriptions based on the Laplacian ideal of total knowledge and the complementary macroscopic, statistical descriptions based on some predefined ignorance. The former theories are usually pictured as laws of nature that are inexorable in every detail, whereas the latter theories assume some alternative behaviors that are ascribed to chance. The deterministic laws of nature are reversible (time-symmetric) and give predictions that are crucially dependent on knowledge of initial conditions, whereas statistical laws are irreversible and give predictions that are more or less independent of knowledge of initial conditions.

Much of what we call nonliving organization is explained by one or the other of these types of description. Nevertheless, attempts to use both types of theory to explain organization have presented conceptual and formal difficulties since the time of Boltzmann. Prigogine and his collaborators have discovered from their formal attempts to define nonequilibrium entropy that a complementarity principle appears inescapable. The description of deterministic, reversible trajectories is incompatible with a simultaneous description of entropy (Prigogine, 1978; Misra, 1978). Even without the mathematical formalism, most of us will conceptually agree with Planck (1960) that "it is clear to everybody that there must be an unfathomable gulf between a probability, however small, and an absolute impossibility," i.e., determinism. Therefore, whether one looks at the principle of complementarity as an evasion or as a solution of the problem of determinism and chance, we have very little choice at present but to use complementary models for explaining self-organization in biological systems, in which chance and determinism play such interdependent roles. However, in my biologically oriented epistemology, I am going to suggest that chance is displaced in some optimal sense by informational constraints that efficiently control the higher levels of dynamical behavior.

The Nature of Dissipative Structures

We have seen that deterministic descriptions are not adequate for explaining self-organizing behavior even at the prebiological levels. The instabilities in deterministic dynamics may be regarded as escape hatches, which in effect leave the behavior of the system undefined in some regions and, hence, subject to unknown or chance events. But chance serves only as an *escape* from classical determinism; it is not a theory of self-organization. The basic contribution of Prigogine was to find an alternative description for the system behavior that exhibits a new structure that is *stable* with respect to the chance events of the previous level. In this way, the instability can serve as a source of chaos in the deterministic, microscopic description and also as a source of new order in the statistical, macroscopic description. However, even though these new modes of behavior effectively introduce a history into physical description, the selection of alternative modes is left to chance.

The physics of this situation is described in terms of far-from-equilibrium thermodynamics, in which instability produces amplification of the fluctuation or chance behavior at the microscopic level and which is then stabilized in the form of new organizations or dissipative structures at the macroscopic level. Although the formalism describing this behavior becomes very complex, it is the epistemological basis of the approach that has the most significance for our theories of self-organization. Classical epistemology, as we said, assigns determinism and objectivity the primary role in theory, with chance and the subjective observer only accepted as unavoidable perturbations in God's universal mechanism. Modern physics, especially quantum theory, forced us to assign a more fundamental role to probability as well as to the role of the observer and the process of measurement. Explanation came to mean not simply reduction of statistical observations to deterministic microscopic events but an effective procedure for correlating our descriptions of observations with our descriptions of laws. Prigogine's epistemological approach to dissipative structures requires this same complementarity between laws and observers. He associates irreversibility, or a direction of time with the necessary epistemological conditions for observation, thereby achieving a consistent definition of nonequilibrium entropy and the instabilities that allow dissipative structures (Prigogine, 1979).

However, this association of instability with the conditions for observation produces an apparent paradox, which is fundamental for theories of self-organization. Recall that the classical concept of an instability associates it with chance events—Poincaré's "observable events that have no observable cause." This means that lack of information is associated with chance and instability. Thus, flipping a coin is a chance event only if we do not measure the initial conditions accurately enough. Or, in other words, instability is interpreted classically as a lack of knowledge of the system. But the modern view requires instability as a *condition* for measurement. In this sense, observation and knowledge seem to require a system that is complex enough to display instability in some sense. What is the difference, then, between the instabilities that we associate with loss of knowledge and the instabilities that we associate with the acquisition of knowledge? Or, in terms of theories of self-organization, what is the difference between the organizations that develop from loss of information and the organizations that develop from acquiring new information? One answer is that a loss of information occurs when systems acquire alternative behaviors (bifurcations), while a gain of information occurs when alternatives are reduced (selection).

The Nature of Symbolic Information

These questions bring us directly back to the fundamental difference between the physicist's approach and the biologist's approach to a theory of self-organization, for despite the new epistemology of physics that gives more weight to probabilistic descriptions and the requirements of observation, there is still an enormous gap between the types of self-organization found in the dissipative structures of macroscopic chemical systems and even the simplest living cells. There is also a serious discrepancy in the idea that any hierarchy of levels of statistical, dissipative structures could ever, by itself, lead to biological self-organization, since the latter is clearly instructed and controlled by individual molecules of nucleic acids and individual enzymes. Statistical mechanics can play no more role in describing these individual molecules than in describing a computer program.

The basic epistemological distinction I wish to make is between the organizations that are constrained by symbolic information and those that develop through chance. Physical theory can be expected to describe self-organizing behavior only insofar as the laws of nature and of statistics are responsible for the behavior; but when the self-organizing behavior is under the constraints of a symbolic information system, then the history of the system dominates our description. The concept of physical laws has meaning only for universal and inexorable regularities, i.e., for systems in which our information about the system can change only with respect to our knowledge of initial conditions. In other words, the form of the laws of nature—even the statistical laws—must be expressed as invariant relative to the information the observer may have about the state of the system. Physical laws do not change in time and have no symbolic memory of past events. On the contrary, observers and symbol systems are characterized by their response to selected past events, which are recorded as memory.

A basic discrepancy, then, between the physicist's and biologist's approach to self-organization is that the physicist's theory recognizes no symbolic restrictions and no historical regularities, whereas the biologist's theory assumes genetic symbol systems with more than 3 billion years of selected historical structures. The two approaches to self-organizing theories—the instability theories and the information-dependent theories—reflect these two complementary views toward symbols and matter, the instability theories emphasizing fluctuations and ignoring symbolic constraints, and the information-dependent theories ignoring physical laws and emphasizing genetic instructions in their respective formulations. One can find this complementarity sharply distinguished in our two modes of describing computers. On the one hand, the basic electronic gates and memory devices that form the hardware require a description in the language of solid-state physics with no reference to syntax or symbols, whereas the software description is entirely symbolic using programming languages that have no reference to physics. Of course, this same complementarity in our descriptions is required for even the simplest symbolic behavior. The "hardware" description of a pencil has nothing to do with the function of a written message, and the chemical description of a gene has nothing to do with the function of the enzyme whose synthesis it instructs.

We can therefore recognize that the physicist's instability-based concepts and the biologist's information-based concepts of self-organization are also closely related to the two sides of the structure-function complementarity. The concept of function and the concepts of symbol and memory are not a part of physical theory itself. It is significant, however, that when the physicist tries to extend his descriptions to the process of measurement, he cannot avoid the concepts of symbol and function, since measurement is defined as a functional activity that produces symbolic output. For this reason the majority of physical scientists simply ignore the problem of measurement or place it in the biological world, often at the level of the conscious observer. Attempts at unified physical theories of measurement, especially in quantum theory, have not been satisfactory and remain controversial (e.g., see Jammer, 1974; d'Espagnat, 1976). On the other hand, most biological scientists tacitly assume a classical reductionism they expect will ultimately explain biological activity in terms of physical theory.

When it comes to theories of the brain, cognitive activity, and consciousness, there is not only controversy over what constitutes an explanation, but also basic disagreement on what it is that we are trying to explain. I therefore find it useful to pay more attention to primitive symbol systems where we may expect the matter–symbol relationship to be less intricate, and where the fundamental complementarity of physical instabilities and symbolic information can be more easily explored. My use of the concept of information is strictly limited to semantic information, i.e., to information that is characterized by its meaning, value, or function. Of course, semantic information is not as well defined as is the structural information in communication theory or complexity theory; but the problem of value and function is obviously the central issue in biological organization. My concept of symbol system is essentially the concept of a languagelike set of rules (e.g., codes, lexical constraints, and grammars) that are necessary conditions for executing or interpreting symbolic information, instructions, or programs (cf. Stent, Chapter 18, this volume).

Instabilities and Information

Since the concept of symbolic information is not as well defined as the concept of instability, we need to consider the primitive conditions for symbolic behavior in more detail. The simple concept of a symbol is that it is something that stands for something else by reason of a relation, but it is implicit in this concept that the relationship of symbol to referent is somewhat exceptional. In other words, it is not a physical law. We do not call the electron a symbol for a proton because there is a relationship of attraction between them, nor is the symbol's relationship to its referent the result of statistical laws. We do not call the temperature of a gas a symbol for the velocity distribution of its molecules. In natural languages we say that the symbol–referent relation is a convention. But what does a convention correspond to at the most primitive levels? At the level of the genetic code there is no evidence of a physical or chemical basis for the particular relations between codons and their amino acids. Crick first called this type of relation a frozen accident, and Monod has generalized this biological arbitrariness as a principle of gratuity. To achieve such an arbitrary or conventional aspect of the symbol–referent relation, the physical system in which the symbol vehicles are to exist must exhibit some form of instability. We could also say that a totally deterministic, stable description of the world in which small changes inexorably produce corresponding small effects does not allow the arbitrariness necessary to generate new symbol–referent relations.

Moreover, once created, a symbol system must persist under the same instabilities through which it came to exist, since these instabilities are an inherent property of the underlying physical system. These sound very much like the conditions for Prigogine's dissipative structures, which are created by fluctuations in an unstable thermodynamic regime but which are subsequently stabilized against the same level of fluctuations by the macroscopic coherence of energy flow through the system. However, there is a basic discrepancy between the characteristics of dissipative structures and the characteristics of symbols or informational structures. Dissipative structures are dynamic, i.e., they depend crucially on the *rate* of matter and energy flow in the system (e.g., reaction and diffusion rates). By contrast, symbol systems exist as *rate-independent* (nonintegrable) constraints. More precisely, the symbol–referent relationship does not depend, within wide limits, on the rate of reading or writing or on the rate of energy or matter flow in the symbol-manipulating hardware. On the other hand, the effect or meaning of symbols functioning as instructions is exerted through the *selective* control of rates. For example, the rate of reading or translating a gene does not affect the determination of which protein is produced. However, the synthesis of the protein as instructed by the gene is accomplished through the selective control, by enzymes, of the rates of individual reactions. At the other extreme of symbol-system evolution, we have the example of computers where the rate of reading a program or the rate of computation does not affect the result or what is being computed. On the other hand, the program, as instructions, is actually selectively controlling the rate at which electrons flow in the machine.

A second difference between dissipative structures and symbols is in their size. Dissipative structures occur only when the size of the system exceeds some critical value that is significantly larger than the fluctuating elements that trigger the instability that creates them. This restriction may also be considered as a statistical requirement for a large number of elements to allow stabilization of the new structure. Symbol vehicles, whether bases in nucleic acids, synaptic transmitters, or gate voltages in a computer, are generally not large relative to the size of the organizations they control, nor are symbols essentially statistical in their structure or behavior. A symbol is a localized, discrete structure that triggers an action, usually involving a more complex system than the symbol itself. Also, symbolic inputs are generally amplified in some sense. In other words, symbols act as relatively simple, individual, nondynamical (nonintegrable) constraints on a larger dynamic system.

This suggests that with respect to their relative size, discreteness, nondynamical behavior, and triggering action on larger dynamical systems, symbols act at the same level as the fluctuations that generate dissipative structures. But clearly symbols are completely unlike fluctuations in other respects. Symbol systems are themselves exceptionally stable, and if we are to understand the origin of symbolic behavior at the fluctuation level or molecular level of organization, then we must explain how symbol systems could stabilize themselves without depending on the statistics or averages of macroscopic organization. How can we expect high reliability in molecular information structures that are embedded in a noisy thermal environment?

There are several possible answers to this question, but quantitative results based on theory are still very difficult to produce. The problem was first discussed by Schrödinger (1944), who recognized that the quantum dynamical stationary state in a covalently bonded macromolecule ("aperiodic crystal") effectively isolated its primary structure from thermal fluctuations. Schrödinger also suggested the analogy of a true "clockworks" at the molecular level, based on quantum dynamical order—and not the statistical order of classical clocks.

London (1961) also speculated that individual enzyme molecules may function in some form of quantum superfluid state that allows the advantage of stationary states along with the possibility of internal motions isolated from thermal fluctuations. In a discussion of the physical basis of coding and reliability in biological macromolecules, I proposed that the correlation of specificity and catalytic rate control in the enzyme required a quantum mechanical nonintegrable (rate-independent) constraint formally analogous to a measurement process (Pattee, 1968). However, no quantum mechanical models of a quantitative relationship of specificity and catalytic power exist at this time.

We should also mention von Neumann's (1966) ideas on the logical requirements for reliable self-replication. Although he gave no proof, or even a precise statement of the problem, von Neumann conjectured that in order to sustain heritable mutations without losing the general self-replicative property, there must be a separate *coded description* of the "self" that is being replicated along with the general translation and synthesis mechanism that reads and executes the description (the "universal constructor").

The most fully developed study of the reliability requirements for hereditary propagation were begun by Eigen and developed into an extensive theory of self-organization at the chemical kinetic level (Eigen and Schuster, 1977). Given that some error in information-processing including the template replication of polymer sequences, is physically inevitable, Eigen and Schuster show how cooperation between replicating sequences, each with error-restricted information capacity, can lead to hierarchical systems of greater and greater capacity (see Schuster and Sigmund, Chapter 5, this volume). Their dynamical treatment of this problem requires a special assumption or special boundary condition that I would associate with arbitrary measurement constraints rather than with laws of nature. This in no way weakens their mathematical arguments; however, it bears directly on the question of epistemological reductionism. The special assumption is the existence of specific catalysts that are coordinated to form a primitive code. I want to make it clear that there is nothing wrong with this assumption—indeed, this may actually be the way life began. All I claim is that a code is not reducible to physical laws in any explanatory sense, i.e., without basically revising the concept of explanation. The practical question of origins is simply whether any such functional code had a reasonable probability of occurring by *chance* under primitive earth conditions. The epistemological question is whether the logical concept of codes in general is derivable from only physical laws or whether the concept of code also requires a complementary functional mode of description. In my earlier discussions of the relationship of specific catalysts to the problem of measurement, I concluded that symbolic information can only originate from processes epistemologically equivalent to measurements (Pattee, 1968, 1979). Measurement may occur over an enormous range of levels of biological organization, from the specific catalysis of enzymes to natural selection processes that are the ultimate origin of symbolic information. The artificial measurements of physicists fall somewhere in this hierarchy. Syntactical constraints or codes are analogous to the constraints embodied in measuring devices. Without such coherent constraints, neither informational nor measurement processes would have any function or meaning; yet these constraints are arbitrary in the sense that many physically distinguishable syntactical constraints (i.e., different languages and measuring devices) may produce indistinguishable meanings or results. This redundancy or degeneracy is found in all levels of symbol systems from the genetic code to human languages and contributes to the stability and reliability of symbolic information. One is tempted to contrast Poincaré's concept of instability, i.e., distinguishable events for which we find no distinguishable antecedents, with this degeneracy characteristic of symbol vehicles, i.e., distinguishable antecedents for which we may find no distinguishable consequences. However, this is too simple a comparison, for even in the most reliable symbol systems some unstable ambiguities in function must exist with respect to perturbations in symbol vehicle structure. I shall return to these epistemological issues in the last section, but my main purpose here is to suggest how dynamics, instabilities, dissipative structures, and symbolic information are related in biological self-organization.

I begin with the hypothesis that the elemental basis for the symbol–referent relationship is the individual specific catalytic polymer in which the folded shape of the molecule is arbitrarily or "gratuitously" coupled to control a specific dynamical rate. This is the only general type of coupling that provides the necessary conditions for a symbol–referent relationship, although it is by no means sufficient. The recognition site or substrate binding site is not related to the catalytic site simply by dynamical or statistical laws. Rather, it requires the particular rate-independent constraints of the folded polymer to determine what molecules are recognized and what bonds are catalyzed. Given this type of specific catalyst, a true code relationship between structure and dynamics is logically possible, along with self-replication and Darwinian natural selection. The question is, can we predict some general organizational consequences of this complementary view of dynamics and information? We expect all stable dynamical behavior to be largely autonomous; that is, since stable dynamics generate no alternatives, there is no need for informational control except in constraining internal boundary conditions. Of course in a stable dynamical regime there still are fluctuations (variations) in these boundary conditions, e.g., modifying protein sequences with or without selection. However, if the limits of dynamical stability are exceeded through excessive competition or new interactions, the informational constraints become dominant in choosing a new stable dynamical structure. This should result in a sudden, large, phenotypic change incommensurate with any structural information measure that could be observed and produce two distinguishable patterns of genotypic and phenotypic change. Under stable phenotypic dynamics there may be observable structural changes in the gene with little corresponding phenotypic change. This would superficially appear as selective neutrality. However, under unstable phenotypic dynamics, a major evolutionary change, like speciation, could result from only minor genetic change. This situation would of course appear to be even more complex, because dynamical instabilities will evolve at all levels of the organizational hierarchy (see Gould, Chapter 6, this volume).

A second consequence of this complementary view of self-organization may appear in our approach to the evolution of symbol systems themselves. Since symbolic information at all levels, from nucleic acids to natural languages, is so obviously effective for instruction and control of dynamical systems, we usually jump to the conclusion that this capability is an intrinsic and autonomous property of the symbol system alone. From our view of the symbol–matter relationship, however, there is virtually no meaning to symbols outside the context of a complex dynamical organization around which the symbolic constraints have evolved. It is useless to search for the meaning in symbol strings without the complementary knowledge of the dynamic context, especially since the symbolic constraints are most significant near dynamical instabilities.

From this point of view the relationship of the present design of computers and their programming languages represents a bizarre extreme. The hardware is designed to have no stable dynamics at all. The gates and memories are a dense maze of instabilities that require explicit, detailed programs of informational constraints for every state transition. Programming languages therefore have no natural grammars. Machine language is conceptually vacuous, whereas high-level languages that try to mimic natural languages are generally compiled or translated to machine code only at the cost of speed and efficiency. The benefit of this design, of course, is universality of computation. The central epistemological problem of computer simulations of such complex systems as the brain is the ability to distinguish the part of the computer simulating the brain's dynamics from the part simulating the brain's information, since in the universal computer, all dynamics are simulated by information. This raises the deeper question of whether or not we can determine if the brain itself uses a dynamical mode for its representations or if all knowledge is restricted to informational constraints.

Epistemological Limitations

The conclusion that it is useless to search for meaning in symbols without complementary knowledge of the dynamics being constrained by the symbols raises the classical issue of what it would mean to "know" the dynamics. We can mention only briefly here two major schools of thought: that of the information processors who believe that to know the dynamics means representing the dynamics with yet another symbol system (e.g., Newell and Simon, 1972) and that of the subjectivists who believe that to know the dynamics is tacit or ineffable, i.e., that it cannot be represented by any symbol system (e.g., Polanyi, 1958). In light of these opposing concepts of knowing, we may reconsider the question of whether symbols can be epistemologically reduced to dynamics, i.e., whether codes can be derived from physical laws. To information processors the reduction of symbols to dynamics is not logically possible, since to them, knowing means reducing dynamics to symbols. To a subjectivist the reduction of symbols to dynamics would place knowing entirely in the realm of the ineffable, which, although acceptable epistemologically, is methodologically impotent.

I do not believe that any of these arguments, including my own, are likely to be convincing at the level of the nervous system, because of the complexity of the hierarchical organization of both its dynamical and symbolic modes. There is no question that information processing at many coded symbolic levels goes on in the brain. It is also obvious that much of what goes on in the brain is unconscious and inaccessible to objective analysis. However, it is largely because of these difficulties that studying the relationship of symbol systems to dynamics at the molecular level may uncover useful concepts and organizational principles. At least at this level we can say that the symbolic instructions of the gene go only as far as the primary sequence of the proteins. From there, thus constrained, the dynamical laws take over. To the gene, these dynamics are ineffable.

References

Bohr, N. (1928) The Como Lecture, reprinted in *Nature* **121**:580.
Bohr, N. (1963) *On Atomic Physics and Human Knowledge*. Wiley-Interscience, New York.
d'Espagnat, B. (1976) *Conceptual Foundations of Quantum Mechanics*. Benjamin, New York.
Eigen, M., and P. Schuster (1977) The hypercycle: A principle of natural self organization. *Naturwissenschaften* **64**:541–565; **65**:7–41, 341–369.
Hertz, H. (1899) *The Principles of Mechanics* (translated by D. E. Jones and T. E. Walley), London.
Jammer, M. (1974) *The Philosophy of Quantum Mechanics*. Wiley, New York.
Laplace, P. S. (1951) *A Philosophical Essay on Probabilities*. Dover, New York.
London, F. (1961) *Superfluids*, 2nd ed., Vol. 1. Dover, New York, P. 8.
Misra, B. (1978) Nonequilibrium entropy, Lyapounov variables, and ergodic properties of classical systems. *Proc. Natl. Acad. Sci. USA* **75**:1627–1631.
Newell, A., and H. A. Simon (1972) *Human Problem Solving*. Prentice-Hall, Englewood Cliffs, N.J.
Pattee, H. H. (1968) The physical basis of coding and reliability in biological evolution. In: *Towards a Theoretical Biology*, Vol. 1, C. H. Waddington (ed.). Univ. of Edinburgh Press, Edinburgh, pp. 69–93.
Pattee, H. H. (1971) Physical theories of biological coordination. *Q. Rev. Biophys.* **4**:255–276.
Pattee, H. H. (1979) The complementarity principle and the origin of macromolecular information. *BioSystems* **11**:217–226.
Piaget, J. (1927) *The Child's Conception of Physical Causality*. Routledge & Kegan Paul, London.
Planck, M. (1960) *A Survey of Physical Theory*. Dover, New York, p. 64.
Poincaré, H. (1952) *Science and Method*, F. Maitland (tr.). Dover, New York.
Polanyi, M. (1958) *Personal Knowledge*. Routledge & Kegan Paul, London.
Prigogine, I. (1978) Time, structure and fluctuations. *Science*, **201**:777–785.
Prigogine, I. (1979) Discussion in "A Question of Physics," P. Buckley and D. Peat (eds.). University of Toronto Press, Toronto, p. 74.
Schrödinger, E. (1944) *What Is Life* Cambridge University Press, London.
Thom, R. (1975) *Structural Stability and Morphogenesis*. Benjamin, New York.
von Neumann, J. (1966) *Theory of Self-Reproducing Automata*, edited and completed by A. W. Burks. University of Illinois Press, Urbana.
Wigner, E. P. (1964) Events, laws and invariance principles. *Science* **145**:995–999.

18

Programmatic Phenomena, Hermeneutics, and Neurobiology

Gunther S. Stent

ABSTRACT

Contemporary neurobiology presents a rare case in which philosophical attention might further scientific progress. Semantic confusion about the term "program" has led to the suggestion that genes embody a program for the development of the nervous system. It is unlikely, and perhaps impossible, that the events of neural development are isomorphic with the structure of any program. Development of the nervous system is not a programmatic, but a historical phenomenon—like the development of an ecosystem—in which events follow a well-defined sequence in the absence of any program. The analysis of historical phenomena bears a strong epistemological affinity to the activity of hermeneutics—a term originally applied to the interpretation of sacred texts. To avoid logical dilemma, hermeneutics requires in the interpreter a preunderstanding resulting from experience and intuition. The fact of variation in preunderstanding puts in doubt the attainability of objectively valid explanations, particularly in the "soft" sciences that address phenomena of great complexity. The more complex a phenomenon, the more hermeneutic preunderstanding is required by an explanation and the less likely it is that the explanation will have the aura of objective truth. Psychoanalytic theory classically demonstrates this weakness: no critical tests of the theory are possible because the failure of any prediction based on the theory can almost always be retrodictively rationalized by modifying slightly one's preunderstanding of the phenomenon. Much of neurobiology displays this character. The student of a complex neural network must bring considerable preunderstanding to the task of interpreting its function, so the explanations that are advanced may remain beyond the reach of objective validation. —The Editor

Practicing scientists are wont to regard the philosophy of science as an activity of has-beens or parasitic ne'er-do-wells—a subject suitable at most for discussion over brandy after dinner. That view is not wholly unjustified, since many lines of scientific work can be successfully pursued without any clear understanding of

GUNTHER S. STENT • Department of Molecular Biology, University of California, Berkeley, California 94720.

their epistemological basis. Moreover, in the two centuries since Kant, during which philosophy has existed as a distinct professional calling, there have been very few known instances in which professional philosophers of science have actually made a productive contribution to scientific progress at the cutting edge of research. Nonetheless, there *have been cases in which explicit* philosophical considerations did play a crucial role in a fundamental scientific advance, such as in the development of relativity theory and of quantum mechanics. But here the philosophical groundwork was done, not by professional philosophers, but by Einstein and by Bohr and Heisenberg. It is my belief that contemporary neurobiology happens to present another one of those rare cases in which some philosophical attention (or "intellectual hygiene," as my teacher Andre Lwoff called it) would be of some benefit for further progress. Two aspects of neurobiology in particular seem in want of conceptual clarification, namely, the development of the nervous system and the functional analysis of its networks.

What Is a Programmatic Phenomenon?

In the mid-1960s, at the time of the triumphant culmination of molecular biological research in the cracking of the genetic code and the elucidation of the mechanism of protein synthesis, research projects began to be formulated to wed the disciplines of genetics and developmental neurobiology. Thus, the idea gained currency that the deep biological problem posed by the metazoan nervous system, namely, how its cellular components and their precise interconnections arise during ontogeny, could, and even should, be approached by focusing on genes. In particular, the notion arose that the structure and function of the nervous system, and hence the behavior of an animal, is "specified" by its genes. Admittedly, it cannot be the case that the genes really embody enough information to permit explicit specification of a neuron-by-neuron circuit diagram of the nervous system (Brindley, 1969). But, all the same, the circuit might somehow be *implicit* in the genome, since purely quantitative information-theoretical arguments, such as those advanced by Horridge (1968), against genetic specification of the nervous system are clearly invalid (Stent, 1978). No, the problem with the notion of genetic specification is other than information-theoretical, for even if the genes did embody a neuron-by-neuron circuit diagram, the existence of an agency that reads the diagram in carrying out the assembly of the component parts of a neuronal Healthkit would still transcend our comprehension. So a seemingly more reasonable view of the nature of the genetic specification of the nervous system would be that the genes embody, not a circuit diagram, but a *program* for the development of the nervous system (Brenner, 1973, 1974). But this view is rooted in a semantic confusion about the concept of "program." Once that confusion is cleared up, it becomes evident that development of the nervous system is unlikely to be a programmatic phenomenon.

Development belongs to that large class of regular phenomena that share the property that a particular set of antecedents generally leads, via a more or less invariant sequence of intermediate steps, to a particular set of consequences. However, of the large class of regular phenomena, programmatic phenomena form only a small subset, almost all the members of which are associated with human activity. For membership of a phenomenon in the subset of programmatic phenomena, it is a necessary condition that, in addition to the phenomenon itself, there exists a "program," whose structure is isomorphic with, i.e., can be brought into one-to-one correspondence with, the phenomenon. For instance, the onstage events associated with a performance of "Hamlet," a regular phenomenon, are programmatic since there exists Shakespeare's text with which the actions of the performers are isomorphic. But the no less regular offstage events, such as the actions of house staff and audience, are mainly nonprogrammatic, since their regularity is merely the automatic consequence of the contextual situation of the performance. Or, the operation of a digital computer has programmatic aspects, insofar as there exists a program, or set of instructions separate from the hardware, whose structure is isomorphic with the sequence of operations performed by the machine. However, in the example of computer programs the demand for isomorphism has to allow for the possibility that the structure of the program is actually more elaborate than that of the phenomenon. Here the program often calls for one of two or more alternative operations at various stages of the process, depending on the result of earlier computations (and hence on the initial state). In such cases the phenomenon is evidently isomorphic with only part of the program.

One of the very few regular phenomena independent of human activity that can be said to have a programmatic component is the formation of proteins. Here the assembly of amino acids into a polypeptide chain of a particular primary structure is programmatic, because there exists a stretch of DNA polynucleotide chain—the gene—whose nucleotide base sequence is isomorphic with the sequence of events that unfold at the ribosomal assembly site. However, the subsequent folding of the completed polypeptide chain into its specific tertiary structure lacks programmatic character, since the three-dimensional conformation of the molecule is the automatic consequence of its contextual situation and has no isomorphic correspondent in the DNA. This example of the formation of proteins can serve also to clarify the distinction made earlier between the embodiment by the genes of a neuron-by-neuron circuit diagram of the nervous system on the one hand and a program for its development on the other. In the case of proteins, the genes evidently do not embody an explicit atom-by-atom specification of spatial coordinates of the tertiary structure of proteins but, merely, a program for assembly of their primary structure from ready-made amino acid building blocks.

When we extend these considerations to the regular phenomenon of development we see that its programmatic aspect is confined mainly to the assembly of polypeptide chains (and of various species of RNA). But as for the overall phenomenon, it is most unlikely—and no credible hypothesis has yet been advanced how this *could* be the case—that the sequence of its events is isomorphic with the structure of any second thing, especially not with the structure of the genome. The fact that mutation of a gene leads to an altered neurologic phenotype shows that genes are part of the causal antecedents of the adult organism, but does not in any way indicate that the mutant gene is part of a program for development of the nervous system.

But are not polemics about the meaning of words such as "program" just a waste of time for those who want to get on with the job of finding out how the nervous system develops? As Woodger (1952) showed in his Tarner Lectures *Biology and Language*, which were published shortly before Watson and Crick's discovery of the DNA double helix and Benzer's reform of the gene concept, semantic confusion about fundamental terms, such as "gene," "genotype," "phenotype," and "determination," had become the bane of classical genetics. It would be well to avoid reconstituting that confusion in the context of developmental biology and to remember Woodger's advice that "an understanding of the pitfalls to which a too naive use of language exposes us is as necessary as some understanding of the artifacts which accompany the use of microscopical techniques" (Woodger, 1952, p. 6).

The general notion of genetic specification of the nervous system not only is defective at the conceptual level but also represents a misinterpretation of the knowledge already available from developmental studies, including those that have resorted to the genetic approach. As Székely (1979) has pointed out, we already know enough about its mode of establishment to make it most unlikely that neuronal circuitry is, in fact, prespecified; rather, all indications point to stochastic processes as underlying the apparent regularity of neural development. That is to say, development of the nervous system, from fertilized egg to mature brain, is not a programmatic but a historical phenomenon under which one thing simply leads to another. To illustrate the difference between programmatic specification and stochastic history as alternative accounts of regular phenomena, we may consider the establishment of ecological communities upon colonization of islands (Simberloff, 1974) or growth of secondary forests (Whittaker, 1970). Both of these examples are regular phenomena, in the sense that a more or less predictable ecological structure arises via a stereotyped pattern of intermediate steps in which the relative abundances of various types of flora and fauna follow a well-defined sequence. The regularity of these phenomena is obviously not the consequence of an ecological program encoded in the genome of the participating taxa. Rather, it arises via a historical cascade of complex stochastic interactions between various biota (in which genes play an important role, of course) and the world as it is.

Cerebral Hermeneutics

But to fathom the complex interactions that produce historical phenomena, it is necessary to understand the context in which they are embedded. And upon recognizing the importance of contextual relationships to development, we move into a domain of phenomenological analysis to which conventional scientific methodology is no longer fully applicable. That domain bears a strong epistemological affinity to the scholarly activity called "hermeneutics." This designation was originally given by theologians to the theory of interpretation of sacred texts, especially of the Bible. The name is derived from that of Hermes, the divine messenger. In his capacity as an information channel linking gods and men, Hermes must "interpret," or make explicit in terms that ordinary mortals can understand, the implicit meaning that is hidden in the gods' messages. In recent years, scholars have applied the term hermeneutics also to the interpretation of secular texts, since there may be implicit meanings hidden even in the literary creations of ordinary men that need to be made explicit to their fellow mortals. But hidden meanings pose a procedural difficulty for textual interpretation, because one must understand the context in which implicit meaning is embedded before one can make it explicit. In other words, one must know what the whole text means before one can uncover hidden meanings in any of its parts. Here we face a logical dilemma, a vicious hermeneutic circle. On the one hand, the words and sentences of which a text is composed have no meaning until one knows the meaning of the text as a whole. On the other hand, one only can come to know the meaning of the whole text through understanding its parts. To break this vicious circle—which comes first, the chicken or the egg?—hermeneutics invokes the doctrine of *preunderstanding*. As set forth by Rudolf Bultmann, hermeneutic preunderstanding, or *Vorverstandnis*, represents the lifetime of experience and insights that the subject must bring to the task of interpreting a particular text. Accordingly, understanding the Bible requires possession of those experiences and insights that its author presupposed in the audience to which his message of salvation is addressed. [A lucid overview of the philosophical aspects of hermeneutics, including a discussion of the role of preunderstanding and of the differentiation between understanding and explanation central to hermeneutic thought, can be found in Coreth (1969)].

In assessing the epistemological status of hermeneutic studies, we may ask to what extent the concept of objective validity is applicable to their results. An objectively valid interpretation would presumably be one that has made explicit the "true" meaning hidden in the text, i.e., the meaning intended by the author. But here we encounter two difficulties. First, the author may not have been—in fact, according to the teachings of analytical psychology, most likely *was* not—consciously aware of the (subconsciously) intended meaning of his own text. Therefore, the outcome of the only operational test of the validity of an interpretation, namely, asking the author: "Is this your intended meaning?" (or discovering the author's own explicit statement of the meaning of his text), does not provide an objective criterion of interpretative truth. What is needed in addition is an (also interpretative) exploration of the author's subconscious. Second, to be eligible for even attempting a true interpretation in the first place, the interpreter must possess just those experiences and insights that the author presupposed (consciously or subconsciously) in the audience to which his text is addressed. But those experiences and insights, i.e., the interpreter's preunderstanding, are necessarily based on his own subjective historical, social, and personal background. Hence, agreement regarding the validity of an interpretation could be reached only among persons who happen to bring the same preunderstanding to the text. Thus, because of the conceptual lack of an operational test for truth on the one hand and the necessarily subjective nature of preunderstanding on the other, there cannot be such a thing as an objectively valid interpretation. It is this evident unattainability of universal and eternal truth in interpretation that makes hermeneutics different from science, for which the belief in the attainability of objectively valid explanations of the world is metaphysical bedrock.

To what extent is this belief actually justified? According to such writers as

Thomas Kuhn, Imre Lakatos, and Paul Feyerabend, it is not really justified because resort is made also to subjective notions equivalent to hermeneutic preunderstanding in the search for scientific explanations of phenomena. Thus, in judging the objective validity of these explanations, we must try to assess the degree to which preunderstanding enters into their development. Such an assessment can help us understand why the belief in the attainability of objectively valid explanations does seem to be more appropriate in the "hard" physical sciences than in the "soft" human sciences, such as economics, sociology, and psychology.

One of the main reasons for this epistemological difference between the hard and soft sciences is that the phenomena which the soft sciences seek to explain are much more complex than those addressed by the hard sciences. And the more complex the ensemble of events that scientists isolate conceptually for their attention, the more hermeneutic preunderstanding must they bring to the phenomenon before they can break it down into meaningful atomic components that are to be governed by the causal connections of their eventual explanations. Accordingly, the less likely it is that their explanations will have the aura of objective truth. By way of comparing a pair of extreme examples—one very hard, the other very soft—we may consider mechanics and psychoanalysis. There is an aura of objective truth about the laws of classical mechanics, because the phenomena that mechanics consider significant, such as steel balls rolling down inclines, are of low complexity. Because of that low complexity it is possible to adduce critical observations or experiments about rolling steel balls. By contrast, there is no comparable aura of truth about the propositions of analytical psychology, because the phenomena of the human psyche that it attends are very complex. Here there are no critical observations or experiments, because the failure of any prediction based on psychoanalytic theory can almost always be explained away retrodictively by considering additional factors or by modifying slightly one's preunderstanding of the phenomenon. Hence, in psychoanalysis a counterfactual prediction rarely qualifies as negative evidence against the theory that generated it. As is well known, Freud failed to appreciate this fundamental epistemological limitation of his discipline. He thought at first that he had founded the physics of the mind. But as it turned out, he had founded its hermeneutics. As pointed out by Habermas (1968), psychoanalysis consists of the hermeneutic interpretation of the complex text that is provided by his subject to the analyst. This misunderstanding by its founding father still remains at the root of the ambiguous relationships of psychoanalysis to the sciences. [For a discussion of the hermeneutic character of the human sciences, see Bauman (1978).]

Neurobiology covers a broad range on this hardness-softness scale. At its hard end, neurobiology is represented by cellular electrophysiology, whose phenomena, although more complex than those associated with rolling steel balls, can still be accounted for in terms of explanations that are susceptible to seemingly objective proof. But at its soft end, neurobiology is represented by the study of the function of large and complicated neural networks. The output of these networks comprises phenomena whose complexity approaches that of the human psyche, in fact *includes* the human psyche. Hence, at that soft end, neurobiology takes on some of the characteristics of hermeneutics; the student of a complex neural network must bring considerable preunderstanding to the system as a whole before attempting to interpret the function of any of its parts. Accordingly, the explanations that are advanced about complex neural systems may remain beyond the reach of objective validation. Or, in cerebral hermeneutics, to paraphrase my former employer, Governor Edmund G. Brown, Jr., we may have to be satisfied with less.

References

Bauman, Z. (1978) *Hermeneutics and Social Science.* Columbia University Press, New York.
Brenner, S. (1973) The genetics of behavior. *Br. Med. Bull.* **29**:269–271.
Brenner, S. (1974) The genetics of *Caenorhabditis elegans. Genetics* **77**:71–94.
Brindley, G. S. (1969) Nerve net models of plausible size that perform many simple learning tasks. *Proc. R. Soc. London Ser. B* **174**:173–191.
Coreth, E. (1969) *Grundlagen der Hermeneutik.* Herder, Freiburg.
Habermas, J. (1968) *Erkenntnis und Interesse.* Suhrkamp, Frankfurt am Main.
Horridge, G. A. (1968) *Interneurons.* Freeman, San Francisco.
Simberloff, D. S. (1974) Equilibrium theory of island biography and ecology. *Annu. Rev. Ecol. Syst.* **5**:161–182.
Stent, G. S. (1978) *Paradoxes of Progress.* Freeman, San Francisco, pp. 169–189.
Székely, G. (1979) Order and plasticity in the nervous system. *Trends Neurosci.* **October**:245–248.
Whittaker, R. H. (1970) *Communities and Ecosystems.* Macmillan Co., New York.
Woodger, J. H. (1952) *Biology and Language.* Cambridge University Press, London.

VI

Control Theory View of Self-Organization

F. Eugene Yates

In the following chapters, Stear provides an advanced and synoptic commentary on the current stage of development of modern control theory, and Tomović highlights the peculiarities, from a control viewpoint, of three biological processes: self-reproduction, motor control, and gene expression. He notes what is missing from numerical control theory and claims that this lack renders it nearly useless in describing biological control. Newer extensions of control theory in the realm of nonmetric control may be more helpful, he suggests, but they are not ready yet to be of service to biologists. Both Stear and Tomović, after careful consideration, find that technological control and biological control do not resemble each other closely enough for current engineering theories to apply clearly to both. (See also Yates, 1982.)

Bellman (1967) commented, "One might describe control theory as the care and feeding of systems." And, he added,

> It should be constantly kept in mind that the mathematical system is never more than a projection of the real system on a conceptual axis.... Intuitively, we can consider a system to be a set of interacting components subject to various inputs and producing various outputs.... [Models of systems] are designed to aid our thinking. This means that we must be careful not to take them too seriously. What will be a useful representation in one case can be quite misleading in another.... Once the mathematical die has been cast, equations assume a life of their own and can easily end by becoming master rather than servant. In the same vein, it cannot be too strongly emphasized that real systems possess many different conceptual and mathematical realizations.

Mathematical control theory makes no important distinction between regulation and control: regulation is treated as just a special case of control, in which

---

F. EUGENE YATES • Crump Institute for Medical Engineering, University of California, Los Angeles, California 90024.

the controller's command to the plant is, for example, "Keep this output constant, within these limits." But in the world of practical machines, control and regulation are not the same, and in nature itself, control scarcely exists: the only known controllers reside within nervous systems of terrestrial animals, and more recently in the artifacts (machines) created by the nervous systems of man. In contrast, regulation is everywhere, as in the dynamic regulation of the size of a star, based on the balancing of gravitational collapse tendencies against thermal and thermonuclear expansion tendencies.

Practical, technological control consists of directing or commanding a system by means of external arrangements of physical components (the controller) connected to that system so that the system's behavior is brought within the bounds and to the trajectories set by the desire of a human being. The controller is an active attachment, requiring an energy source.

In contrast, regulation arises from the physical structure of the plant and of the regulator. Regulation does not require the pentad of measurement, feedback, amplification, computation, and decision, as controllers do. Nor does it require any explicit criteria such as reference values or specifications of optimality from outside the structure of the plant or regulator itself. Regulation may be static or dynamic, and in either case often involves self-regulation, in which no particular device is identifiable as a regulator except the factory-plant itself. If an external component—the regulator (governor, compensator)—is attached, the additional part does not explicitly amplify, compute, or decide. Regulators may be either active or passive devices.

Typically, a problem in controller design involves some functioning system (a plant), whose dynamics or rules of behavior are at least partially known and mathematized by a *deus ex machina* residing outside system (i.e., by the engineer himself). The plant always obeys physical laws. But the design also involves a criterion, the expression of a value judgment, originating in the engineer or in his client, outside the plant. The engineer makes arrangements, theoretical or practical, that sample the plant's performance and pass that information to a device (controller) especially constructed to evaluate the actual performance in comparison to some nominated ideal or criterion. The controller resides outside the plant system being controlled, in the sense that it is not constructed out of parts of the plant itself, though it is often parasitic on the plant for supply of power, because its power requirement is usually very small compared to that of the plant itself. Attaching the controller's inputs to sample the plant's outputs supposedly does not alter the dynamics of the plant, but the controller's outputs are connected to some of the plant's inputs, often after power amplification. The control inputs may alter plant behavior either by providing corrective information or signals to be processed by the usual, unchanged dynamical arrangements of the plant structure, or by modifying plant structure or parameters to achieve new dynamics. In either case there is a new system: the plant-controller. If the plant dynamics, and the controller functions of measurement, estimation, filtering, computation, decision, evaluation, amplification, feedback, and so on, can be modeled (i.e., mathematized), then the properties of the total equation set can be explored, and abstract optimal controllers can be designed for an abstract model of a plant.

VI. Control Theory View of Self-Organization

Modern control theory depends heavily on *in numero* modeling, i.e., on simulations running on digital computers.

For at least 20 years we have had computer programs that simulate self-organization. These are pattern generators, and in some cases they suggest simple rules out of which complex structures might arise. They provide metaphors for self-organization in physical systems, and they are entertaining. At their best, they suggest possibilities for biological organization; at their worst, they suggest that humans are machines. But powerful as control, communication, and computation sciences are believed to be as a basis for affecting some political, economic, and social systems, they have not yet contributed much to our understanding of biological systems.

In contrast, principles of dynamic regulation, as this term is understood in engineering practice, apply to nearly every physiological system. Instead of feedback, we find circular causalities, in which there is no compelling basis for labeling one pathway of influence as "forward" and another as "backward." We find that distributed physical-chemical parameters of system structure, rather than "programs," determine the performance. Biological systems usually show dynamic regulation rather than control, nonnumerical processing rather than numerical, circular causality rather than negative feedback, nonlinear stability in motional states rather than linear stability of rest states or trajectories toward them. When we do find negative feedback in biological systems, its presence quickens responses, but it is not usually fundamental for the operation of the system.

The concept of dynamic regulation was well understood by Adam Smith, and the idea appears in his *The Wealth of Nations* (1776). The line of thought begun by Smith and carried through in greater detail in the early 1800s by David Ricardo included the now-famous idea of an "invisible hand" regulating the marketplace, i.e., the assurance that *laissez-faire* is the best policy. These ideas originated from the hope that there is some kind of compensatory lawfulness in economic processes. (Modern views are more sophisticated, but dynamic regulation still predominates.) Hardin (1959) points out that Darwin had similar notions about the evolutionary process, viz., if a mutation moves an organism from the well-adapted mean of its population, natural selection is likely to prune the outliers and stabilize the distribution around that mean. (Of course, as we now know from the technological manipulation of microorganisms for industrial processes, a very strong selection pressure can achieve a new mean state determined by the mutants, but in nature, Darwin's regulatory view probably applies more often, as in the spontaneous abortion of "defective" fetuses or fertilized eggs.)

If control is so rare in nature, why include here a section on the control theory view of self-organizing systems? It is the powerful, heuristic property of control theory that justifies its detailed presentation in the context of analysis of self-organizing systems. It invokes the idea of intention and a great problem in biology is to account for the seemingly intentional behaviors of all living systems. The exchanges 20 years ago among Rosenbleuth, Wiener, and Bigelow, on one hand, and Richard Taylor, on the other (see Buckley, 1968), highlighted the problem of purpose and teleology in life sciences and technology. A mechanistic conception of purposefulness, and the distinction between purposeful and nonpurposeful behavior, engaged mathematicians, engineers, and biologists, but the issues were not satisfactorily settled. Wiener (1948) gives an account of the rise of the feedback metaphor as the basis of control theory in the introduction to the original edition of *Cybernetics: Or Control and Communication in the Animal and the Machine.*

A sample of the current theoretical state of the art of control can be found in the excellent paper by Landau (1981), in which he proposes a unified approach to "model reference adaptive controllers and stochastic self-tuning regulators." His presentation shows both the strengths and the weaknesses of modern control theory. A chief weakness is the inescapable requirement for a reference model. These always arise as the result of an intelligence operating outside the system to be controlled.

It is the challenge of a scientific approach to self-organizing systems that the "plans" and "programs" emerge from plausible dynamic regulatory mechanisms that are properties of the plant itself at each stage of development. Only when that challenge is met will vitalism be dead.

References

Bellman, R. (1967) *Introduction to the Mathematical Theory of Control Processes.* Academic Press, New York, pp. 2, 4.
Buckley, W. (1968) *Modern Systems Research for the Behavioral Scientist—A Source Book,* Part V, Section A. Aldine, Chicago, pp. 221–242 (exchanges among Rosenbleuth, Wiener, and Bigelow with Richard Taylor, four articles).
Hardin, G. (1959) *Nature and Man's Fate.* Holt, Rinehart & Winston, New York.
Landau, I. D. (1981) Model reference adaptive controllers and stochastic self-tuning regulators—A unified approach. *J. Dyn. Syst. Meas. Control* **103**:404–416.
Smith, A. (1776) *The Wealth of Nations.* (See, e.g., the 1966 edition published by Kelley, New York.)
Wiener, N. (1948) *Cybernetics: On Control and Communication in the Animal and the Machine.* Wiley, New York.
Yates, F. E. (1982) Systems analysis of hormone action: Principles and strategies. In: R. G. Goldberger and K. Yamamoto (eds.). *Biological Regulation and Development,* Vol. 3A, Plenum Press, New York, pp. 25–97.

19

Control Paradigms and Self-Organization in Living Systems

Edwin B. Stear

ABSTRACT

This chapter describes control of systems characterized by numerically valued variables. It opens with a qualitative description of the fundamental concepts behind open-loop and feedback control methods. The chief aim of feedback control is to cause a system's output to maintain some desired relation to an input, in spite of disturbances or deviations in plant dynamics.

Classical linear control theory is quantitatively described, using the standard, frequency-domain ($j\omega$) notation and Fourier transform method. The open-loop and closed-loop (feedback) equations are each shown for a continuous, linear system with time-invariant parameters and structure and single input with single output. In some cases it is necessary to achieve independent control of each of multiple disturbances or to control more than a single plant output. Or there may be more than a single control input available with which to control the plant's outputs. One may then employ multiloop control or more complex control arrangements for multivariable control.

Classical linear control concepts have been extended in modern optimal control theory. The new approach is far more general, and it emphasizes the state-space description of dynamic systems. It is general enough to encompass time-varying and nonlinear conditions. Using the general formulation of the deterministic, optimal control problem, open-loop control is reexamined from the point of view of the Pontryagin Maximum Principle. Some similarities between the optimal control problem and Hamiltonian mechanics are pointed out. It is concluded that optimal control theory in this form represents a significant conceptual generalization of the variational theory of classical mechanics to a much wider class of problems. In this sense, optimal control theory overlaps and transcends ordinary physical theory and cannot properly be considered to be a simple consequence of the known physical laws of mechanics.

The optimal control approach is next generalized to include stochastic processes appearing as unpredictable perturbations. The mathematical difficulty increases, but such difficulties do not detract from the underlying rich structure of optimal control theory and its substantial connections with and differences from the variational theory of mechanics.

The equations of stochastic optimal control do not arise in statistical mechanics because standard

---

EDWIN B. STEAR • Washington Technology Center, University of Washington, Seattle, Washington 98195. This chapter was written while the author was Chief Scientist of the United States Air Force.

statistical mechanics does not deal with concepts that involve estimating the state of dynamic systems for purposes of control. Neither does physical theory appear to deal with any problems analogous to or embedded in the stochastic optimal control problem. Again, one is forced to the conclusion that control theory cannot, in general, be considered to be a part of, or a simple extension of, known physical theory.

Finally, this chapter addresses control in living systems, and asks whether or not the principles of technological control, as outlined in the first part, apply. Living systems are dominated by regulatory processes. The history of our discovery of some of these processes is summarized. Enzyme-catalyzed reactions and feedback control of enzyme levels by genetic repression-derepression constitute interesting sample cases for control theory analysis. The analysis shows that control theory provides a basis for showing what experiments and data are necessary to validate the claims of the biochemists that the closed, inhibitory pathways they discover (usually *in vitro*) actually have physiological relevance. This validation has yet to be made. It is concluded that some of the concepts of classical control theory do illuminate some biochemical control processes, but that the richness of modern, optimal control theory cannot be brought to bear on the richness of modern genetic control of protein biosynthesis. —The Editor

---

This chapter examines the role of control paradigms in living systems and some of their possible relationships to the self-organization of such systems. It delineates both the power and the limitations of control theoretic views of general systems behavior—with special emphasis on the evolution and behavior of living systems—and establishes the relationship of such views to several other useful views of living systems.

That control paradigms are useful in describing and characterizing a great many of the processes and functions occurring in living systems *in qualitative terms* is now well established. To become convinced of this, one need only read the literature on any aspect of living systems to find many references to the concepts of feedback and feedback control. Such references range from the level of genetic processes involving DNA and RNA, through cellular processes involving enzyme (protein)-mediated reactions, to the functioning of organs and organ systems involving hormones and/or neuronal processes, and even to the functioning of individual members of species and societies of species involving higher-level neuronal processes.

This widespread qualitative use of control concepts to characterize the behavior of living systems illustrates the general pervasiveness and universality of control concepts—whether they are applied to living systems; to the control of nonliving systems such as chemical plants, machines, robots, and so on; or to mixed systems, such as man–machine systems in which humans control the machines. However, like all widely applicable concepts, control concepts require elaboration and specialization when applied quantitatively to particular situations. For example, in mechanics one must specify the masses or mass distributions of the bodies involved, the interconnections among bodies, and the quantitative nature of the relevant force fields on the bodies in order to analyze quantitatively the motions of the bodies. With similar aims in the case of coupled enzyme-mediated reactions, one must specify the chemical species (including enzymes) involved, their chemical interactions, and the rate equations and constants for all reactions to define the time course of the concentrations of the chemical species involved. Analogously, in the case of control systems one must quantitatively specify the dynamics of the relevant variables of the systems being controlled, the variables being used by the controllers to effect the desired control actions, and the structure of the controllers and their dynamics in order to fully and quantitatively understand the overall dynamic behavior of the control systems, their fundamental properties, and their performance. Also, the sets of variables for the systems being controlled must be specified no more and no less than required to make the responses of the set unique. And, insofar as it is possible, given the state of knowledge of the systems being controlled, the variables and their dynamics should be selected so as to be dynamically isomorphic to the actual system being controlled.

In this chapter, concepts of control are characterized in both qualitative and quantitative terms. In both cases, the fundamental limitations and capabilities of control processes are addressed. The qualitative discussion is included because it has universal applicability and makes the concepts available to those who are not mathematically prepared for the full description. The quantitative discussion is for those who are better prepared. It establishes certain connections of optimal control theory with the variational theory of mechanics and the theory of statistical mechanics. This discussion of the fundamental concepts and theory of control is then used to examine the possible roles of control paradigms in living systems, including the self-organization of such systems.

The Fundamental Concepts of Control

The fundamental concepts of control (Truxal, 1955; D'Azzo and Houpis, 1960; Horowitz, 1963) can be conveniently described and discussed qualitatively in terms of the block diagrams shown in Figure 1. These block diagrams of the two basic types of control systems illustrate the major components of control systems, the major kinds of signals or variables of interest, and their interconnections. (It is assumed that the only connections or interactions between components are those shown in the block diagrams.)

As suggested by the block diagrams, *all control systems*—whether open-loop or feedback—contain a dynamical system labeled "plant," whose output variable or variables, labeled c, are to be controlled in some desired fashion by means of an input variable or variables, labeled u. In addition, the "plant" is normally subject to unpredictable disturbances, labeled d, which originate in the plant's operating environment and often cause its output to deviate unacceptably from the desired behavior unless appropriate action is taken to control such effects. Finally, there is the important and commonly occurring situation—not easily illustrated by means of block diagrams of the kind shown—in which the dynamic characteristics of the plant's response, c, to the control input, u, and/or the disturbance, d, are either not known exactly or vary in an unpredictable but bounded manner. These deviations from the known, nominal, dynamic characteristics of the plant can also cause the plant's output to deviate unacceptably from its

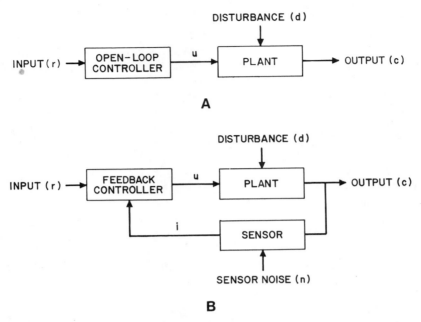

FIGURE 1. Basic single-input, single-output control system block diagrams. See text for definition of symbols. (A) Open-loop control system. (B) Feedback control system.

desired behavior. Such unpredictable variations in the plant's dynamic characteristics can arise from either parameter variations within a given plant structure or variations in the plant structure itself (or both). The latter case is less commonly encountered but is undoubtedly more important in the context of self-organizing systems. The plant, its dynamic characteristics, the disturbances to which it is subjected, and the expected variations of the plant dynamics must all be specified or determined through appropriate modeling based on physics, chemistry, endocrinology, neurophysiology, or other appropriate sciences.

As further suggested by the block diagrams, *all control systems* also have an input variable or variables, labeled r, to which the plant output(s) c are to be related in a desired and specified manner. The primary purpose of a control system is to achieve such a desired relationship. The exact nature of the specified relationship between r and c can vary widely, depending on the specific application. In the simplest case, it might be desired to have c simply follow or track r over a frequency range (time scale) of interest, with a special case occurring when r is constant and it is desired to keep c at some constant value that depends only on the value of r itself. In the general case, c is specified to have some *physically realizable* dynamic relationship or dependence on r. The phrase "physically realizable" is italicized here to emphasize the fact that, in general, it is not possible to realize an arbitrary dynamic relationship between r and c, because of limiting characteristics of the "plant" dynamics, such as input or internal variable satu-

19. Control Paradigms and Self-Organization in Living Systems

ration, rate limiting of the input or internal variables, plant instability, and so on. More will be said about these limitations later.

For a given plant, the simplest and most direct, generic way of establishing a desired dynamical relationship between r and c is illustrated in Figure 1A. One simply inserts a properly designed dynamical system—called an open-loop controller—in series (cascade) with the plant. If the dynamical characteristics of the plant do not vary from their nominal values, and if there are no disturbances in the plant, then it is possible—within the (previously discussed) limits imposed by considerations of physical realizability—to select appropriate dynamical characteristics for the open-loop controller so that the desired relationship between r and c is realized. Moreover, selection of these appropriate dynamical characteristics for the open-loop controller depends *only* on the nominal dynamical characteristics of the plant and the desired relationship between r and c.

However, despite their useful, fundamental ability to achieve a desired overall dynamic response from r to c as described above, open-loop control systems have an inherent limitation, which is obvious from their topological structure as shown in Figure 1A: they have no mechanism for reducing undesirable effects that might arise because of unpredictable internal variations in the plant dynamics or unpredictable external disturbances. To develop a mechanism for exercising control of such unpredictable circumstances, it would be necessary to sense or measure such variations as they manifest themselves in the course of time and to vary the control input u in a compensatory manner. Open-loop controllers cannot do this, because they respond *only* to the input r through dynamical characteristics, which depend *only* on the (predictable) nominal dynamical characteristics of the plant (as described above). Thus, open-loop control is limited to those cases in which unpredictable plant disturbances produce negligible variations in the plant output c from that desired in response to r.

These limitations of open-loop control can be overcome by means of the fundamental concept of feedback control as illustrated in Figure 1B. Here, undesirable deviations of the plant output arising from unpredictable internal or external fluctuations are sensed directly (in real time) by an output sensor. The output, i, of this sensor is then sent to a feedback controller that uses the information contained in i to generate appropriate compensating variations in the control input u.

Even in this very qualitative discussion, it is possible to discern several important general requirements that must be satisfied if such feedback control is to be effective. First, if the variations in c are to be kept acceptably small in the face of large disturbances or deviations in the plant dynamics, then small, sensed variations in c must produce (correspondingly) large compensating variations in the control input u. That is, the "gain"[1] of the feedback controller from i to u must be large; there must be amplification. Second, the time scale in which the sensor–feedback controller combination responds to sensed variations in c must be comparable to (and generally smaller than) the time scale of the disturbances or deviations in plant dynamics that are to be controlled. Finally, the sensor

---

[1] The concept of "gain" will be formally defined and characterized in quantitative terms later.

noise, n, must be comparable to or smaller than the (already small) acceptable level of deviations in c; otherwise the sensor noise itself will produce unacceptably large variations in c.

This latter effect is a direct consequence of the previous requirement that the sensor–feedback controller must produce the appropriate control inputs to keep the sensed variations in c acceptably small. It simply expresses the fact that small errors made in sensing variations in c will be treated by the feedback controller as if they were real variations in c resulting from disturbances or deviations in the plant dynamics. The feedback controller will produce a compensating variation in c to offset these sensing errors caused by the sensor noise. Thus, if sensor noise is too large, it can easily become a limiting factor in the performance achievable by feedback control.

Although it is now evident that feedback control can be used to reduce undesired effects, it is not yet clear whether or not it can also simultaneously achieve a desired dynamic response from r to c. That it can do both simultaneously—again within limits imposed by considerations of physical realizability—is seen as follows. If the plant output, c, is responding to disturbances and/or to deviations in the plant dynamics in the same direction as the desired response, then the feedback controller simply reduces the control input u from the nominal value it would have in the no-disturbance, no-deviation-in-plant-dynamics case. On the other hand, if c is responding to disturbances and/or to deviations in plant dynamics in a direction opposite to the desired response, then the feedback controller increases the control input from its nominal value to overcome the disturbances (and/or effects due to deviations in plant dynamics) and to produce the desired response. The key point here is that a *single output is being controlled,* and *this can be done with a single input*. In general, the simultaneous achievement of a desired overall dynamical relationship between r and c and a reduction of the effects of unpredictable disturbances and/or deviations in plant dynamics requires a larger range of control input values than would be the case in the absence of disturbances or plant deviations. This fact has important implications concerning the saturation effects that can be allowed in the control input to the plant.

The Classical Linear Control Theory

A deeper insight into the fundamental control concepts discussed in the previous section can be provided through the quantitative methods of classical linear control theory (Truxal, 1955; D'Azzo and Houpis, 1960; Horowitz, 1963). An analysis of both open-loop control systems and feedback control systems using linear quantitative methods is given below. The analysis addresses both the capabilities and the limitations of the two basic types of control systems.

The classical linear control theory is emphasized in this section because, in addition to being both historically important and very effective in past and current engineering applications, it also is the simplest and most direct known quantitative approach that encompasses all aspects of the important fundamental control concepts discussed in the previous section. However, despite the emphasis on the linear case in this section, it is very important to remember that *all the fundamental concepts of control discussed above—including capabilities, limitations, and requirements for effectiveness—are applicable in general and are in no way dependent on assumptions of linearity for their validity.*

In the classical linear control theory, all dynamic components (i.e., plants, controllers, sensors, and so on) are assumed to be time-invariant (with respect to parameter values and structure) linear systems. As a result, they can be quantitatively characterized in terms of their time-domain response to input sine wave signals at all frequencies of interest.[2] In a complementary fashion, all the input signals can be represented in terms of their frequency content through Fourier analysis (or Wiener's generalized Fourier analysis, if required.)[3] Because of the assumed linearity of the components, the response of any component to a sum of two input signals (from zero initial conditions) is simply the sum of the responses of the component to each of the two input signals taken separately (i.e., superposition holds). Thus, one can analyze overall control system characteristics at each frequency of interest (i.e., one frequency at a time) with the assurance that responses due to inputs that are sums of sine waves (which form results from the Fourier analysis of the input signals regardless of the shape) are simply the sums of the individual responses for each frequency component in the input. This consequence of linearity results in a great conceptual and operational simplification in the analysis of systems and is responsible for the popularity of linear analysis (when it is valid, i.e., when nonlinear and/or time-varying effects are negligible).

With the assumptions of linearity and time-invariance described above, the block diagrams in Figure 1 can be specialized and simplified as shown in Figure 2. Here, $P(j\omega)$ denotes the *steady-state* response of the plant (in the frequency domain) to a simple harmonic control input $U(j\omega)$ of unit amplitude and of frequency ω of the form $u(t) = e^{j\omega t}$ in the time domain. The complex quantity $P(j\omega)$ is commonly called the frequency response of the plant. Similarly, $C(j\omega)$, $H(j\omega)$, and $M(j\omega)$ represent the frequency responses of control elements, and $S(j\omega)$ represents the frequency response of the sensor. The signals shown also are denoted in terms of their frequency content, with $R(j\omega)$ denoting the complex amplitude of $r(t)$ at frequency ω obtained by resolving $r(t)$ into its frequency components by means of generalized Fourier analysis,[4] $U(j\omega)$ denoting the complex amplitude of $u(t)$ at frequency ω, and so on. With this representation and notation, the result of the action of any component on its input, at any frequency ω,

---

[2] This response can be found by taking the (generalized) Fourier transform of the time response of the system provolked by an impulsive (technically a Dirac impulse) input occuring at time equal to zero.

[3] Methods based on the Laplace transform also can be used, but frequently-based (i.e., Fourier transreader needs to recognize that for linear conditions there are various formal, exact relationships between a time-domain funtion, e.g., $\mu(t)$, and its frequency-domain representation, e.g., $V(j\omega)$. Frequency-domain representations have certain mathematical convenience and are used for that reason chiefly. (See also footnote 4.)

[4] That is, $R(j\omega) = \int r(t) e^{-j\omega t}\, dt$, and $r(t) = \dfrac{1}{2\pi} \int R(j\omega) e^{j\omega t} d\omega$.

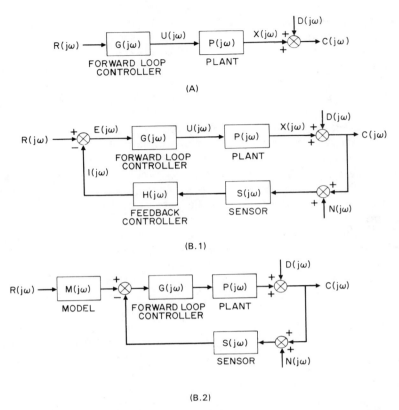

FIGURE 2. Basic linear control system block diagrams. (A) Linear open-loop control system. (B.1) Linear feedback control system. (B.2) Alternative configuration for a linear feedback control system.

can be determined by simply multiplying the input component amplitude by the appropriate frequency response function. Thus, it follows from Figure 2A that

$$U(j\omega) = G(j\omega)R(j\omega) \tag{1}$$

and

$$X(j\omega) = P(j\omega)U(j\omega) = P(j\omega)G(j\omega)R(j\omega) \tag{2}$$

The circles in Figure 2 represent summing or differencing devices, which sum or difference the incoming signals depending on the + or − notation shown. Thus, it also follows from Figure 2A that

$$C(j\omega) = X(j\omega) + D(j\omega)$$
$$= P(j\omega)G(j\omega)R(j\omega) + D(j\omega) \quad \text{(open-loop control)} \tag{3}$$

19. Control Paradigms and Self-Organization in Living Systems

Equation (3) expresses the complete quantitative representation of the linear open-loop control system shown in Figure 2A. It will be used later to explore quantitatively the capabilities and limitations of linear open-loop control.

An analogous quantitative representation of the linear feedback control system shown in Figure 2B.1 can be obtained through the sequence of steps given below:

$$C(j\omega) = X(j\omega) + D(j\omega) = P(j\omega)G(j\omega)E(j\omega) + D(j\omega) \quad (4)$$
$$E(j\omega) = R(j\omega) - I(j\omega)$$
$$= R(j\omega) - H(j\omega)S(j\omega)[C(j\omega) + N(j\omega)] \quad (5)$$

Eliminating $E(j\omega)$ by substituting Eq. (5) into Eq. (4) and rearranging gives the desired result:

$$C(j\omega) = \left[\frac{P(j\omega)G(j\omega)}{1 + P(j\omega)G(j\omega)H(j\omega)S(j\omega)}\right] R(j\omega)$$
$$- \left[\frac{P(j\omega)G(j\omega)H(j\omega)S(j\omega)}{1 + P(j\omega)G(j\omega)H(j\omega)S(j\omega)}\right] N(j\omega) \quad \text{(feedback control)}$$
$$+ \left[\frac{1}{1 + P(j\omega)G(j\omega)H(j\omega)S(j\omega)}\right] D(j\omega) \quad (6)$$

The corresponding result for the system shown in Figure 2B.2 is

$$C(j\omega) = M(j\omega) \left[\frac{P(j\omega)G(j\omega)}{1 + P(j\omega)G(j\omega)S(j\omega)}\right] R(j\omega)$$
$$- \left[\frac{P(j\omega)G(j\omega)S(j\omega)}{1 + P(j\omega)G(j\omega)S(j\omega)}\right] N(j\omega) \quad \text{(feedback control)}$$
$$+ \left[\frac{1}{1 + P(j\omega)G(j\omega)S(j\omega)}\right] D(j\omega) \quad (7)$$

The expressions in Eqs. (6) and (7) are identical if $G(j\omega)$ in Eq. (7) is replaced by $G(j\omega)H(j\omega)$ and if $M(j\omega)$ is taken to be $H^{-1}(j\omega)$. Thus, these two configurations of linear feedback control are equivalent, and from now on only the configuration shown in Figure 2B.1 will be discussed. The two configurations were shown in Figure 2B to establish the significant fact that seemingly different configurations may, in fact, be equivalent with regard to their control action if one makes the appropriate selection of the controller elements' dynamic characteristics.

Equations (3) and (6) can be used to establish quantitatively several important facts concerning the capabilities and limitations of open-loop control systems and feedback control systems, which have already been discussed in qualitative terms. First, if there are no disturbances [i.e., $D(j\omega) = 0$], and if the plant dynamics $P(j\omega)$ do not vary from their nominal value, denoted by $P_0(j\omega)$, then it follows from Eq. (3) that

$$C = P_0 GR = T_0 R; \quad T_0 \triangleq P_0 G \quad \text{(definition)} \quad (8)$$

where the ($j\omega$) notation has been suppressed to simplify notation. The quantity T will be used to denote the overall system response from R to C. In the usual case, in which the dynamics of a component can be described by a finite-order, ordinary linear differential equation of the form

$$a_n d^n x/dt^n + \cdots + a_1 dx/dt - a_0 x = b_m d^m u/dt^m + \cdots + b_0 u \qquad (9)$$

it is easily shown that the plant frequency response function $P(j\omega)$ is of the form

$$P(\rho) = \frac{b_m \rho^m + \cdots + b_1 \rho + b_0}{a_n \rho^n + \cdots + a_1 \rho + a_0}; \qquad \rho = j\omega \qquad (10)$$

The *plant* is stable if the denominator of $P(\rho)$ has no roots with positive or zero real parts, but T_0 is stable if and only if *both* P_0 and G are stable. It is seen from Eq. (10) that a *desired stable linear overall dynamic relationship* T_0^5 between R and C can be achieved using an open-loop control system simply by selecting $G(j\omega) = T_0(j\omega)/P_0(j\omega)$ at all frequencies of interest, providing that the plant is stable and the numerator of $P(\rho)$ also has no roots with zero or positive real parts.[6] However, examination of Eq. (3) establishes a serious limitation of linear open-loop control systems, namely, *any disturbances of the plant show up directly and without attentuation in the plant output*. Furthermore, it is also clear that *variations in the plant dynamics* (i.e., variations in P) *produce a directly proportional change in the overall dynamic relationship, T, between R and C*. A useful and intuitively satisfying measure of the sensitivity of T to changes in P is given by the following (dimensionless) ratio:

$$S_P^T = \frac{\Delta T/T}{\Delta P/P} \qquad (11)$$

where $\Delta T = T - T_0$ and $\Delta P = P - P_0$. This ratio is called the sensitivity function for the control system (Horowitz, 1963). The use of ratios in the definitions of S_P^T serves to normalize the magnitudes of the changes in the variables and provides a sensitivity measure that can be used to compare the effectiveness of the control of plant variations in different applications in which the magnitudes of the variables and their variations may differ widely. Evaluating S_P^T for the open-loop control case gives the result

$$S_P^T = \frac{(T - T_0)/T}{(P - P_0)/P} = \frac{(PG - P_0 G)/PG}{(P - P_0)/P} = 1 \qquad (12)$$

---

[5]Normally, only stable systems are desired because unstable systems have internal variables that become unbounded as time approaches infinity, and hence becomes nonlinear, blow up, burn up, break up, or otherwise behave badly.

[6]Systems for which one or more of the roots of the numerator of their corresponding frequency response function have positive real parts are called nonminimum phase systems. These systems occur in a variety of applications, including biological systems.

19. Control Paradigms and Self-Organization in Living Systems

for all frequencies. Equation (12) quantitatively expresses the fact, previously discussed, that open-loop control systems *have no capability to reduce the effects of unpredictable variations in the plant dynamics*.

The analogous quantitative relationships for feedback control systems can be obtained as follows. First, it is seen from direct examination of Eq. (6) that the effect of unpredictable disturbances on the plant output is given by the expression

$$C\Big|_{\substack{R=0 \\ N=0}} = \left(\frac{1}{1+PGHS}\right) D = \left(\frac{1}{1+L}\right) D \tag{13}$$

Thus, the effects of disturbances are attenuated by the factor $1/(1+L)$, where L is defined as $PGHS$. This dimensionless quantity "L" is commonly referred to as the *loop gain* of the feedback control system, and it is *the* important measure of the ability of the feedback control loop to reduce the effects of plant disturbances or plant noise.

The sensitivity of feedback control systems to unpredictable variations in the plant dynamics is also determined solely by the loop gain as the following computation shows. From Eq. (6) it is seen that

$$T = \frac{C}{R}\Big|_{\substack{D=0 \\ N=0}} = \frac{PG}{1+PGHS} \tag{14}$$

and, hence,

$$S_P^T = \frac{\left(\dfrac{PG}{1+PGHS} - \dfrac{P_0 G}{1+P_0 GHS}\right) \Big/ \left(\dfrac{PG}{1+PGHS}\right)}{(P-P_0)/P} \tag{15}$$

$$= \frac{1}{1+P_0 GHS} = \frac{1}{1+L_0}$$

Thus, the effects of unpredictable plant variations on T are also reduced by the factor $1/(1+L_0)$. It is seen from Eqs. (13) and (15) that the loop gain L_0 must be large at all frequencies for which D and/or ΔP are large, if the effects of external disturbances at the plant, and/or of variations in plant dynamics, are to be substantially attenuated. Furthermore, the feedback loop must be responsive over the same range of frequencies (i.e., same time scale) at which significant disturbances and/or variations of the plant dynamics occur.

The above discussion suggests that the effects of both external disturbances and variations in the plant dynamics could be made arbitrarily small by making L_0 arbitrarily large over the frequency range of interest. In general, however, it is not possible to make L_0 arbitrarily large because of other effects that limit the loop gain that can be achieved. For example, from Eq. (6),

$$C\Big|_{\substack{R=0 \\ D=0 \\ P=P_0}} = -\left(\frac{L_0}{1+L_0}\right) N \approx -N; \quad \text{if } |L_0| \gg 1 \tag{16}$$

Equation (16) shows that, for very large loop gains, the sensor noise "N" appears unattenuated on the plant output. Furthermore, sensor noise can also lead to large control input rates and even *rate saturation* if the loop gain is too high. To see this effect, consider the following expression for the absolute value of $U(j\omega)$, based on Figure 2B.1:

$$|\dot{U}(j\omega)|\bigg|_{\substack{R=0 \\ D=0}} = \frac{|j\omega G(j\omega)H(j\omega)S(j\omega)|}{|1+L(j\omega)|}|N(j\omega)|$$

$$\approx \frac{|j\omega|}{|P(j\omega)|}|N(j\omega)|; \qquad |L(j\omega)| \gg 1$$

or

$$\approx |j\omega G(j\omega)H(j\omega)S(j\omega)||N(j\omega)|; \qquad ||L(j\omega)|| \ll 1 \qquad (17)$$

Hence, sensor noise ultimately limits the rate at which plant output variations can be reduced.

Another significant factor that limits the achievement of arbitrarily high loop gains in feedback control is the inevitable instability of the feedback control loop. This instability arises from control elements of limited complexity as the loop gain is made increasingly large. This phenomenon can be demonstrated analytically in specific cases by any one of several stability analysis methods. However, it can only be established in general by means of a fairly complex analysis, which will not be given here. For the purpose of this chapter, it is sufficient that the reader be aware of the existence of loop gain limits arising from gain-induced instability. The design of linear feedback control systems involves careful shaping of the loop gain of (minimum complexity) controllers G and H (Figure 2B), as a function of frequency, to achieve a "trade-off" balance between reducing the effects of disturbances or deviations in plant dynamics and obtaining an acceptably low risk of loop instability or of output variation due to sensor noise.

Finally, in addition to reducing the effects of unpredictable plant disturbances or unpredictable variations in the plant dynamics, feedback control can also be used to stabilize unstable plants. This claim may seem paradoxical at first, in view of the gain-induced instability of feedback loops discussed above. However, when you recall that in the former case we are considering stability of the whole feedback control loop, but that in the latter case we address only the stability of the plant, the seeming paradox disappears. Gain-induced instability as a gain-limiting factor in feedback control systems applies in both cases. For unstable plants, the feedback loop must be active, i.e., $|GH|$ must be definitely greater than zero and have an appropriate phase response over the frequency range at which the plant instability occurs. The added requirement of stabilizing an unstable plant has the effect of putting lower limits on the magnitude and phase of the loop gain over an appropriate range of frequencies.

Multiloop and Multivariable Control

In some cases, it is necessary to achieve independent control of each of the multiple disturbances or to control more than a single plant output (Truxal, 1955; D'Azzo and Houpis, 1960; Horowitz, 1963; Stear, 1975), or there may be more than a single control input available with which to control the plant's outputs. One may then employ more than a single feedback control loop, and additional control concepts may be required. As an example, consider the system shown in Figure 3, which involves two independent disturbances to the plant. In the linear case [see Eq. (6)],

$$C = \frac{P_1 G_2^1 G_1}{1 + L_1} R + \left(\frac{P_1}{1 + L_1}\right)\left(\frac{1}{1 + L_2}\right) D_2 + \left(\frac{1}{1 + L_1}\right) D_1 \\ - \left(\frac{P_1}{1 + L_1}\right)\left(\frac{L_2}{1 + L_2}\right) N_2 - \left(\frac{L_1}{1 + L_1}\right) N_1 \quad (18)$$

where $L_1 = G_2^1 G_1 H_1 S_1 P_1$, $L_2 = P_2 G_2 H_2 S_2$, and $G_2^1 = P_2 G_2 / (1 + L_2)$, by definition. Here L_2 and L_1 are selected to achieve the desired reduction in the response of C to D_2 and D_1, respectively, and G_1 is selected to achieve the desired response of C to R. Again, L_1 is limited by the effect of N_1 on C and U_1, and L_2 is limited by the effect of N_2 on C in a way completely analogous to the single-loop case.

An example of the multivariable case is illustrated in Figure 4. Here, the system (inside the dashed rectangle) is input cross-coupled through P_{12} and output cross-coupled through K_{12}. If the cross-coupling is negligible, then the multivari-

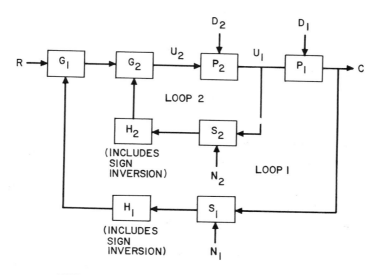

FIGURE 3. Multiloop control system, single variable.

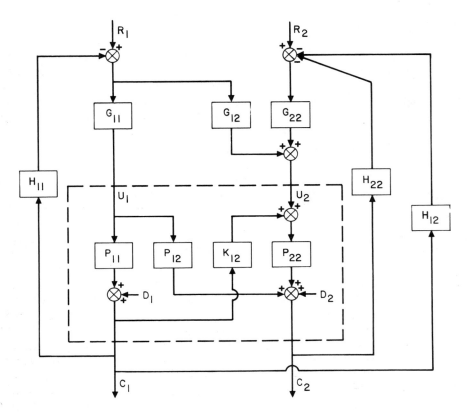

FIGURE 4. A multivariable control system.

able plant can be considered simply as two independent single-output plants that can be controlled independently through G_{11} and H_{11} and through G_{22} and H_{22}, respectively, as discussed earlier for the single-input, single-output case. However, when there is significant cross-coupling, the input U_1 and the disturbance D_1 at plant P_{11} can lead to undesired effects on C_2. The control of such effects requires the concepts of input decoupling and output decoupling. Input decoupling can be controlled through incorporation of an additional feed-forward element G_{12} in the multivariable controller and can be theoretically eliminated altogether if G_{12} is selected to satisfy the relationship $G_{11}P_{12} + G_{12}P_{22} = 0$ at the frequencies of interest. Similarly, output decoupling can be controlled through incorporation of the feedback control element H_{12} and eliminated altogether if H_{12} is selected in order to satisfy the relationship $K_{12} - H_{12}G_{22} = 0$ at all frequencies of interest.

Except for the new concepts of input and output decoupling, no other fundamental concepts arise in consideration of either multiloop or multivariable control.

Modern Optimal Control Concepts

During the 1960s, the classical control concepts discussed in the previous sections were reinterpreted, augmented, and generalized in several significant ways (Bryson and Ho, 1969; Jazwinski, 1970). These advances were accomplished through the development of a new approach to control system design and analysis based on the calculus of variations and its extensions to problems involving inequality constraints and stochastic processes. This new approach has had far-reaching consequences, many of which are still being explored and developed. For example, it provided a new, unified, and much more general approach to specifying the desired performance of control systems through introduction of the concept of a *scalar performance measure,* which is to be minimized (or maximized if appropriate) by an "optimal" control input to the plant. Through its consistent emphasis on the state-space description of dynamic systems, it provided a new conceptual and analytical framework that was general enough to encompass time-varying and nonlinear conditions in addition to the linear, time-invariant systems to which the classical theory was limited. Moreover, the state-space description provided a unified treatment of both the single-input, single-output case and multiple-input, multiple-output (i.e., multivariable) cases. This general treatment was lacking (or was awkward) in the classical linear theory. Finally, the new approach provided an analytical framework close in spirit and viewpoint to that represented by the variational theory of mechanics. From this viewpoint it is possible to discern various parallels, analogies, and divergences between this new control theory and the general variational theory of physics.

I shall now present the modern theory and concepts of optimal control in general terms. Their connections to the classical control concepts described above will be outlined, and their analogies to the variational theory of classical mechanics will be noted. Also, some important limitations will be pointed out. For the most part, the discussion will be carried out in analytical, as opposed to descriptive, terms; results will be simply stated without derivations or proofs, which can be found in the references cited (Bryson and Ho, 1969; Jazwinski, 1970).

Formulation of the Problem

In *deterministic optimal control* theory, one is usually given a dynamical plant to be controlled, which can be described by means of a state-space equation of the form[7]

$$\dot{\mathbf{x}} = f(\mathbf{x}, \mathbf{u}, t); \quad \mathbf{x}(t_0) \text{ given} \qquad (19)$$

where $\mathbf{x}(t)$ is an n-vector representing the state of the system at time t and $\mathbf{u}(t)$ is an m-vector representing the values of the m control inputs at time t (or better,

---

[7] In spatially distributed systems, Eq. (19) must be replaced by an appropriate partial differential equation. Only systems with a finite dimensional state space will be discussed here.

at state **x**). The system outputs are some functions of the system state, and, possibly, of time independently, of the form

$$\mathbf{c}(t) = h(\mathbf{x}(t),t) \tag{20}$$

where $\mathbf{c}(t)$ is a p-vector representing the relevant or measurable plant outputs. Saturation effects on either the input variables or the state variables can be represented by inequalities of the form

$$|\mathbf{u}_i(t)| \leq a_i$$
$$|\mathbf{x}_i(t)| \leq b_i \tag{21}$$

or, for more general constraints, by vector inequalities of the form

$$\gamma(\mathbf{x},\mathbf{u},t) \leq 0$$
$$\Sigma(\mathbf{x},t) \leq 0 \tag{22}$$

where γ and Σ are vector-valued functions of appropriate dimensions. Finally, it is assumed that the desired control action can be characterized as the one that minimizes a *scalar performance measure* or index of the form

$$J = \phi(\mathbf{c}(t_f),t_f) + \int_{t_0}^{t_f} L(\mathbf{c}(t),\mathbf{u}(t),t)\, dt \tag{23}$$

subject to the plant dynamics represented by Eqs. (19) and (20) and the constraints represented by Eqs. (22). Here $[t_0,t_f]$ is the interval over which the control action is to take place and $L(\cdot,\cdot,t)$ is a performance measure reflecting the cost of control $\mathbf{u}(t)$ and achieved output value $\mathbf{c}(t)$ per unit time. Equation (23) is a special case of a more general form for the performance index, given by

$$J = \phi(\mathbf{x}(t_f),t_f) + \int_{t_0}^{t_f} L(\mathbf{x}(t),\mathbf{u}(t),t)\, dt \tag{24}$$

which can be obtained by substituting $h(\mathbf{x}(t),t)$ for $\mathbf{c}(t)$ in Eq. (23). Equation (24) is the form usually used.

There are several important points to be noted concerning the above formulation of the deterministic optimal control problem. First, it accommodates both nonlinear and linear plants, i.e., nonlinear and linear functions $f(\cdot)$, with equal ease. Second, it accommodates both multiple control inputs and multiple plant outputs as well as the single-input, single-output case. Finally, in contrast to the classical linear theory, it allows the direct incorporation of limiting nonlinear effects, such as input and state variable saturation through inequalities of the form given in Eqs. (21) and (22).

The optimal control problem as formulated above differs slightly from the normal variational theory of classical mechanics in that (1) the loss function

$L\{\mathbf{x}(t),\mathbf{u}(t),t\}$ need not be the Lagrangian (i.e., the kinetic energy minus the potential energy) of a physical system and (2) plant dynamic constraints of the form given in Eq. (19), which involve variables **u** not directly representing coordinates of particles or bodies with mass, do not occur in variational mechanics. Later, it will be seen that the variational theory of classical mechanics can be regarded as a special case of optimal control theory, providing an appropriate identification of variables and functions is made. Moreover, when this appropriate identification is made, the inequality constraints in Eqs. (22) are directly analogous to non-holonomic constraints as defined in the theory of classical mechanics.

Open-Loop Control

Given the above formulation of the optimal control problem, for the case that Eqs. (22) reduce to $\gamma(\mathbf{u}(t),t) \leq 0$, McShane (1939) and Pontryagin et al. (1962) have shown that the desired control action [i.e., the minimizing control time history $\mathbf{u}^\circ(t)$] satisfies the following necessary conditions, referred to as the Pontryagin Maximum Principle,[8] namely,

$$H(\mathbf{x}^\circ(t), \boldsymbol{\lambda}^\circ(t), \mathbf{u}^\circ(t), t) \leq H(\mathbf{x}^\circ(t), \boldsymbol{\lambda}^\circ(t), \mathbf{u}(t), t); \quad t_0 \leq t \leq t_f \quad (25)$$

for all $\mathbf{u}(t)$ satisfying the constraints $\boldsymbol{\lambda}(\mathbf{u}(t),t) \leq 0$ where

$$H(\mathbf{x},\boldsymbol{\lambda},\mathbf{u},t) = \boldsymbol{\lambda}^T f(\mathbf{x},\mathbf{u},t) - L(\mathbf{x},\mathbf{u},t) \quad (26)$$

and

$$\begin{aligned}\dot{\mathbf{x}} &= (\partial H/\partial \boldsymbol{\lambda})^T = f(\mathbf{x},\mathbf{u},t) \\ \dot{\boldsymbol{\lambda}} &= -(\partial H/\partial \mathbf{x})^T = -(\partial f/\partial \mathbf{x})^T \boldsymbol{\lambda} + (\partial L/\partial \mathbf{x})^T\end{aligned} \quad (27)$$

and where the boundary conditions on Eqs. (27) are given by $\mathbf{x}(t_0) = \mathbf{x}_0$, the given initial state of the plant, and $\boldsymbol{\lambda}(t_f) = -(\partial \phi/\partial \mathbf{x})^T|_{t=t_f}$. The superscript T indicates transpose; the superscript o denotes the desired or optimal condition.

In general, to find $\mathbf{u}^\circ(t)$ it is necessary to solve Eqs. (25) and (27) simultaneously, using the given boundary conditions for $\mathbf{x}(t)$ and $\boldsymbol{\lambda}^\circ(t)$. That is, one first finds $\mathbf{u}^\circ(t)$ as a function of $\mathbf{x}^\circ(t)$ and $\boldsymbol{\lambda}^\circ(t)$ by maximizing $H(\cdot)$ over **u** to get $\mathbf{u}^\circ(t) = \mathbf{u}^\circ(\mathbf{x}^\circ(t), \boldsymbol{\lambda}^\circ(t), t)$ and then substitutes this into Eqs. (27) to get the pair of (Hamilton-like) equations

$$\begin{aligned}\dot{\mathbf{x}}^\circ &= f(\mathbf{x}^\circ, \mathbf{u}^\circ(\mathbf{x}^\circ, \boldsymbol{\lambda}^\circ, t), t) \\ \dot{\boldsymbol{\lambda}}^\circ &= \{-(\partial f/\partial \mathbf{x})^T \boldsymbol{\lambda}^\circ + (\partial L/\partial \mathbf{x})^T\}|_{\mathbf{u}^\circ = \mathbf{u}^\circ(\mathbf{x}^\circ, \boldsymbol{\lambda}^\circ, t)}\end{aligned} \quad (28)$$

These coupled differential equations must then be solved (integrated), subject to the boundary conditions.

---

[8] It is called the maximum principle because Pontryagin defined his Hamiltonian to be the negative of that used above.

The resulting solutions for $\mathbf{x}^\circ(t)$ and $\lambda^\circ(t)$ are then substituted in $\mathbf{u}^\circ(t) = \mathbf{u}^\circ(\mathbf{x}^\circ(t),\lambda^\circ(t),t)$ to find $\mathbf{u}^\circ(t)$. Unfortunately, because the boundary condition on $\mathbf{x}$ is given at the initial time, t_0, while that on λ is given at the final time, t_f, Eqs. (28) represent a two-point boundary value problem, which can only be solved iteratively, except in the linear case. This requirement for solving a nonlinear two-point boundary value problem in order to find $\mathbf{u}^\circ(t)$ has greatly complicated the application of optimal control theory to nonlinear problems. However, difficulties with computation are no different here than they are with nonliner problems in the general sciences. Hence, they should not be used to denigrate optimal control theory any more than they are used to denigrate general scientific theory.

Equations (27) bear a strong resemblance to Hamilton's canonical equations of motion in the theory of classical mechanics, except for the occurrence of the control variables $\mathbf{u}(t)$. If the control variables and state variable are unconstrained, i.e., no constraints of the form given in Eqs. (22) are present, then they can be made to have an essentially identical form to Hamilton's equations. If there are no constraints, then Eq. (25) implies that $\partial H/\partial \mathbf{u}|_{u=u^\circ} = 0$. In this case, let

$$H^\circ(\mathbf{x}^\circ,\lambda^\circ,t) = H(\mathbf{x}^\circ,\lambda,\mathbf{u}^\circ(\mathbf{x}^\circ,\lambda^\circ,t),t) \quad \text{(definition)} \tag{29}$$

Then, it follows that

$$\begin{aligned}\partial H^\circ/\partial \lambda^\circ &= \partial H/\partial \lambda^\circ + (\partial H/\partial \mathbf{u})(\partial \mathbf{u}/\partial \lambda^\circ) = \partial H/\partial \lambda^\circ \\ \partial H^\circ/\partial \mathbf{x}^\circ &= \partial H/\partial \mathbf{x}^\circ + (\partial H/\partial \mathbf{u})(\partial \mathbf{u}/\partial \mathbf{x}^\circ) = \partial H/\partial \mathbf{x}^\circ\end{aligned} \tag{30}$$

$$\begin{aligned}\dot{\mathbf{x}}^\circ &= (\partial H^\circ/\partial \lambda^\circ)^T \\ \dot{\lambda} &= -(\partial H^\circ/\partial \mathbf{x}^\circ)^T\end{aligned} \tag{31}$$

In this case, $H^\circ(\mathbf{x}^\circ,\lambda^\circ,t)$ can be considered as representing the "Hamiltonian" function for the optimal control problem, and Eqs. (31) can be considered to be the corresponding Hamilton canonical equations for the optimal motion $\mathbf{x}^\circ(t)$ of the system. Clearly, the components λ_i° of λ° then play the role of "momentum" variables corresponding to the state variable components $\mathbf{x}^\circ_i$ of the state vector $\mathbf{x}^\circ$ of the controlled system. These momentumlike variables, however, do not represent real momenta of real particles. Rather, they are induced by the requirement for optimal control of the system.

If Eq. (19) is of the special form $\dot{\mathbf{x}} = \mathbf{u}$, and if $L(\mathbf{x},\mathbf{u},t) = L(\mathbf{x},\dot{\mathbf{x}},t)$ is the Lagrangian of a physical system, then the above optimal control equations lead directly to Hamilton's equations of motion for the system. Similarly, in this special case, the inequalities in Eqs. (22) represent nonholonomic constraints in the physical system. Thus, one may conclude from this discussion that the variational theory of classical mechanics is a special case of optimal control theory or, perhaps more historically accurate, that optimal control theory represents a significant conceptual generalization of the variational theory of classical mechanics to a much wider class of problems. In this sense, *optimal control theory transcends ordinary physical theory and cannot properly be considered to be a simple consequence of the known physical laws of mechanics.*

Feedback Control with Perfect Knowledge of the State

Other connections between optimal control theory and the variational theory of mechanics will emerge later. At this point it is more appropriate to address the relationship of optimal control to the fundamental control concepts discussed in previous sections. The optimal control theory discussed above dealt only with the problem of finding an optimal control time history $\mathbf{u}°(t)$ to achieve the desired control action of minimizing the performance index J over the set of all those possible control time histories, $\mathbf{u}(t)$, satisfying the state variable and control variable constraints. It did *not* address the important issue of achieving such an optimal control action in the presence of unpredictable plant disturbances and/or unpredictable variations in the plant dynamics. Thus, *it must be considered as an open-loop control theory only.* Because feedback control must be used to reduce undesired effects of unpredictable plant disturbances and/or unpredictable variations in the plant dynamics, a theory of optimal feedback control is obviously required if optimal control theory is to be a comprehensive generalization of the earlier classical theory of control, as claimed earlier.

One useful way of introducing the concept of feedback into the optimal control theory discussed above is simply to determine the optimal control input $\mathbf{u}°$ at any time t in terms of the state $\mathbf{x}(t)$ of the system at time t. The rationale for this seemingly *ad hoc* approach is based on the following observation. If the state of the system is somehow suddenly and inadvertently disturbed at time t from the value that it would otherwise have had in the absence of such a disturbance, and if no other disturbances act on the system in the future, then the desired control action for minimizing J in the presence of such a disturbance would be achieved by solving for the optimal control as a function of the disturbed state $\mathbf{x}(t)$ at time t. If this were done for every time t and every possible disturbed state $\mathbf{x}(t)$, then the resulting feedback control $\mathbf{u}(\mathbf{x}(t),t)$ would optimally reduce the undesired effects of such a sudden disturbance and minimize J no matter when the disturbance occurred or what its magnitude was. The required feedback control function $u(\mathbf{x}(t),t)$ can be computed, in principle, from the above equations as follows.

Consider the integration of Eqs. (28) backward in time from t_f, using the boundary conditions $\mathbf{x}(t_f) = \mathbf{x}_f$ and $\lambda(t_f) = \partial\phi(\mathbf{x},t_f)/\partial\mathbf{x}|_{\mathbf{x}=\mathbf{x}_f}$. Then the resulting solutions[9] will be functions of $\mathbf{x}_f$ of the form $\mathbf{x}(t) = X(t;\mathbf{x}_f)$, $\lambda(t) = \Lambda(t;\mathbf{x}_f)$, which will, for each $\mathbf{x}(t)$, satisfy Eqs. (28) and the terminal boundary condition $\mathbf{x}(t_f) = \partial\phi/\partial\mathbf{x}|_{\mathbf{x}=\mathbf{x}_f}$. In the usual case there are no conjugate points on the optimal control trajectory $\mathbf{x}°(t),\mathbf{u}°(t)$. Thus, the function $X(t;\mathbf{x}_f)$ will be one-to-one for all $\mathbf{x}_f$ in the neighborhood of $\mathbf{x}°(t_f)$. Hence, the equation $\mathbf{x}(t) = X(t;\mathbf{x}_f)$ can be solved for $\mathbf{x}_f$ in terms of $\mathbf{x}(t)$ for each t, giving $\mathbf{x}_f = X^{-1}(t,\mathbf{x}(t))$. Substituting this result into the

---

[9] It is assumed that the optimal control problem is such that these solutions exist and are unique.

expression $\lambda(t) = \Lambda(t;\mathbf{x}_f)$ gives $\lambda(t)$ as a function of $\mathbf{x}(t)$ in the form $\mathbf{x}(t) = \Lambda(t;X^{-1}(t,\mathbf{x}(t)))$. Substituting this in turn into the expression $\mathbf{u}(t) = \mathbf{u}°(\mathbf{x}(t),\lambda(t),t)$ gives the optimal control $\mathbf{u}°(t)$ as a function of the current state in the form $\mathbf{u} = \mathbf{u}°(\mathbf{x}(t),t) \equiv \mathbf{u}°(\mathbf{x}(t),\Lambda(t;X^{-1}(t,\mathbf{x}(t))),t)$ for all $\mathbf{x}(t)$ in a neighborhood of $\mathbf{x}°(t)$. This is the optimal feedback control law sought. In the nonlinear case the above computation is very complex and normally cannot be accomplished in closed form.

A considerably different and very instructive approach to the problem of optimal feedback control is to use a corresponding Hamilton–Jacobi-like theory. Moreover, introduction of such a theory provides another natural connection between optimal feedback control and the variational theory of mechanics. In the Hamilton–Jacobi theory of optimal control, attention is focused on the value of the scalar performance index J along optimal trajectories. It is obvious from the previous discussion of the optimal control problem that the minimum value of J depends only on the initial time t_0 and the initial state $\mathbf{x}(t_0)$. This is because the optimal control $\mathbf{u}°(t)$ only depends on t_0 and $\mathbf{x}(t_0)$ through the solution of Eqs. (28) as described above, and the optimal trajectory $\mathbf{x}°(t)$ depends only on $\mathbf{x}(t_0)$ and $\mathbf{u}°(t)$ through Eq. (19). Finally, since the minimum value of J is totally determined by $\mathbf{u}°(t)$ and $\mathbf{x}°(t)$ for $t_0 \leq t \leq t_f$, it is clear that this minimum is a function only of t_0 and $\mathbf{x}(t_0)$.

Let the minimum value of J be denoted by $S(t,\mathbf{x})$ for arbitrary initial times t and initial states $\mathbf{x}$. Then, under reasonable conditions it can be shown that $S(t,\mathbf{x})$ satisfies the Hamilton–Jacobi–Bellman equation

$$(\partial S/\partial t) = H°[\mathbf{x}, -(\partial S/\partial \mathbf{x})^T, t] \tag{32}$$

where $H°(\mathbf{x},\lambda,t)$ is as defined by Eq. (29), i.e., $H°(\mathbf{x},\lambda,t)$ is by definition min $H(\mathbf{x},\lambda,\mathbf{u},t)$. Furthermore, it can be shown that the boundary condition for Eq. (32) is given by the expression

$$S(t_f,\mathbf{x}_f) = \phi(\mathbf{x}_f,t_f) \tag{33}$$

The vector variable λ is related to $S(t,\mathbf{x})$ by

$$\lambda^T = -\partial S/\partial \mathbf{x} \tag{34}$$

and it follows that the optimal feedback control is given by the expression

$$\mathbf{u}°(\mathbf{x},t) = \mathbf{u}°[\mathbf{x}, -(\partial S/\partial \mathbf{x})^T, t] \tag{35}$$

When solving for $\mathbf{u}°(\mathbf{x},t)$ with Eqs. (28), it is seen that the Hamilton–Jacobi–Bellman equation must also be integrated backward in time, because its boundary condition is given at the final time t_f. This is a distinguishing, characteristic property of optimal control in general, and it is directly related to the fact that *current control actions have future effects on the performance index being minimized* that must be accounted for in determining the optimal control.

In general, the solution of Eq. (32) is impossible in closed form and one must resort to numerical computational methods. These can also be very difficult and demanding of computer resources. Moreover, solution of Eq. (32) is much more difficult than was finding an open-loop optimal control input $\mathbf{u}°(t)$, in that it requires the solution of Eqs. (28) for all t and all $\mathbf{x}(t)$ of interest. However, in the important special case in which disturbances are small, it is possible to simplify greatly all computations and to reduce the solution of Eq. (32) to the solution of a matrix Ricatti equation. In this case of small disturbances, the plant dynamics given by Eq. (19) can be linearized and the performance index given by Eq. (23) can be expanded to quadratic terms about the optimal open-loop control $\mathbf{u}°(t)$ and trajectory to obtain $\mathbf{x}°(t)$:

$$\dot{\bar{\mathbf{x}}} = f_x(t)\bar{\mathbf{x}} + f_u(t)\bar{\mathbf{u}} \tag{36}$$

and

$$\delta^2 J = J° + (\tfrac{1}{2})\bar{\mathbf{x}}^T \phi_{xx}(t_f)\bar{\mathbf{x}}(t_f) + (\tfrac{1}{2}) \int_{t_0}^{t_f} [\bar{\mathbf{x}}^T(t)L_{xx}(t)\bar{\mathbf{x}}(t) + \bar{\mathbf{x}}^T(t)L_{xu}(t)\bar{\mathbf{u}}(t)$$
$$+ \bar{\mathbf{u}}^T(t)L_{ux}(t)\bar{\mathbf{x}}(t) + \bar{\mathbf{u}}^T(t)L_{uu}(t)\bar{\mathbf{u}}(t)] \, dt \tag{37}$$

where

$f_x(t)$ is defined as $(\partial f/\partial \mathbf{x})|\mathbf{x}=\mathbf{x}°(t), \mathbf{u}=\mathbf{u}°(t)$
$f_u(t)$ is defined as $(\partial f/\partial \mathbf{u})|\mathbf{x}=\mathbf{x}°(t), \mathbf{u}=\mathbf{u}°(t)$
$L_{xu}(t)$ is defined as $\dfrac{\partial}{\partial \mathbf{u}}((\partial L/\partial \mathbf{x})^T)|\mathbf{x}=\mathbf{x}°(t), \mathbf{u}=\mathbf{u}°(t)$

and so on, and where $\bar{\mathbf{x}}$ is defined as $\mathbf{x} - \mathbf{x}°$ and $\bar{\mathbf{u}}$ is defined as $\mathbf{u} - \mathbf{u}°$. Moreover, it is clear that minimizing J under this assumption[10] of small disturbances is equivalent to minimizing $\delta^2 J$ subject to Eq. (34). This is the so-called *second variation problem*, which has been long understood to be the problem of interest in the optimal feedback control of the undesired effects of small plant disturbances.

Applying the general theory discussed above to the second variation problem leads to the following equations in addition to Eq. (37):

$$\overline{H}(\bar{\mathbf{x}}, \bar{\lambda}, \bar{\mathbf{u}}, t) = \lambda^T\{f_x(t)\bar{\mathbf{x}} + f_u(t)\bar{\mathbf{u}}\} - \tfrac{1}{2}[\bar{\mathbf{x}}^T L_{xx}(t)\bar{\mathbf{x}} + \bar{\mathbf{u}}^T L_{uu}(t)\bar{\mathbf{u}}] \tag{38}$$

$$\dot{\bar{\lambda}} = -f_x^T(t)\bar{\lambda} + L_{xx}(t)\bar{\mathbf{x}} \tag{39}$$

where it has been assumed for convenience (and without loss of generality) that $L_{xu}(t) \equiv 0$ and $L_{ux}(t) \equiv 0$. The optimal control $\bar{\mathbf{u}}°$ satisfies the condition that

---

[10] The first-order term δJ in the expansion of J (above) vanishes along $\mathbf{u}°(t)$, $\mathbf{x}°(t)$, because $\mathbf{u}°(t)$, $\mathbf{x}°(t)$ are the optimal control and corresponding trajectory which minimizes J.

$(\partial \overline{H}/\partial \overline{\mathbf{u}}) = 0$ or $\overline{\mathbf{u}}^\circ = L_{uu}^{-1} f_u^T \lambda$. The Hamiltonian $\overline{H}^\circ$ is given by

$$H^\circ(\mathbf{x},\lambda,t) = \lambda^T f_x \overline{\mathbf{x}} + \lambda^T f_u L_{uu}^{-1} f_u^T \lambda - \tfrac{1}{2}[\overline{\mathbf{x}}^T L_{xx} \overline{\mathbf{x}} + \overline{\lambda}^T f_u L_{uu}^{-1} f_u^{T} \overline{\lambda}] \quad (40)$$
$$= \lambda^T f_x \overline{\mathbf{x}} + (\tfrac{1}{2})\overline{\lambda}^T f_u^T L_{uu}^{-1} f_u \overline{\lambda} - (\tfrac{1}{2})\overline{\mathbf{x}}^T L_{xx} \overline{\mathbf{x}}$$

and Eq. (36) becomes

$$\dot{\overline{\mathbf{x}}} = f_x \overline{\mathbf{x}} + f_u L_{uu}^{-1} f_u^T \overline{\lambda} \quad (41)$$

The Hamilton–Jacobi–Bellman equation in this case is given by

$$(\partial \overline{S}/\partial t) = -(\partial \overline{S}/\partial \overline{\mathbf{x}}) f_x \overline{\mathbf{x}} + \tfrac{1}{2}(\partial \overline{S}/\partial \overline{\mathbf{x}}) f_u L_{uu}^{-1} f_u^T (\partial \overline{S}/\partial \overline{\mathbf{x}})^T - \frac{1}{2}\overline{\mathbf{x}}^T L_{xx} \overline{\mathbf{x}} \quad (42)$$

subject to the boundary condition $\overline{S}(\mathbf{x},t_f) = (\tfrac{1}{2})\overline{\mathbf{x}}^T S_{xx}(X^\circ(t_f),t_f)\overline{\mathbf{x}}$. Equation (42) can be solved directly by assuming that $S(\mathbf{x},t)$ is of the form

$$\overline{S}(\mathbf{x},t) = (\tfrac{1}{2})\overline{\mathbf{x}}^T P(t) \overline{\mathbf{x}} \quad (43)$$

where P is assumed to be a positive definite symmetric matrix. Substituting Eq. (43) for Eq. (42) and symmetrizing the resulting term $(\partial \overline{S}/\partial \overline{\mathbf{x}}) f_x \overline{\mathbf{x}}$ leads to

$$(\tfrac{1}{2})\overline{\mathbf{x}}^T \dot{P} \overline{\mathbf{x}} = -(\tfrac{1}{2})(\overline{\mathbf{x}}^T P f_x \overline{\mathbf{x}} + \overline{\mathbf{x}}^T f_x^T P \overline{\mathbf{x}}) + \tfrac{1}{2}\overline{\mathbf{x}}^T (P f_u L_{uu}^{-1} f_u^T P - L_{xx}) \overline{\mathbf{x}} \quad (44)$$

Since $\overline{\mathbf{x}}$ is arbitrary, Eq. (44) will be satisfied if, and only if, the following matrix Ricatti equation for $P(t)$ is satisfied:

$$\dot{P} = -P f_x - f_x^T P + P f_u L_{uu}^{-1} f_u^T P - L_{xx} \quad (45)$$

and P is obtained by integrating Eq. (45) backward in time subject to the boundary condition $P(t_f) = S_{xx}(\mathbf{x}_0(t_f),t_f)$. The optimal (linear) feedback control is given by the expression

$$\mathbf{u}^\circ(\mathbf{x},t) = -L_{uu}^{-1}(t) f_u^T(t) P(t) \mathbf{x} \quad (46)$$

By definition, this is $-K(t)\overline{\mathbf{x}}$.

The above treatment of the effect of small plant disturbances via the second variation yields a feedback control that achieves an optimal balance (in the sense of minimizing J) between increases in J due to the effect of such disturbances on the state $\mathbf{x}$ and increases in J due to the use of additional control action $\mathbf{u}$ to reduce the effect of such disturbances. As seen from Eqs. (45) and (46), the resulting feedback gain $K(t)$ depends directly on the weighting matrices $L_{xx}(t)$ and $L_{uu}(t)$, which quantitatively characterize the increases in J due to $\overline{\mathbf{x}}$ and $\overline{\mathbf{u}}$, respectively. Moreover, it can be seen from Eq. (46) that the feedback gain $K(t)$ "increases" as the ratio $L_{uu}^{-1} L_{xx}$ "increases." Thus, the gain $K(t)$ can be increased even further by replacement of L_{xx} and L_{uu} with other matrices Q and R, respectively, for which the product $R^{-1}Q$ is larger than $L_{uu}^{-1}L_{xx}$. Although this change would increase J

19. Control Paradigms and Self-Organization in Living Systems

above the value it would have for the optimal feedback control, it would achieve an even greater reduction in the deviation of the state $\mathbf{x}(t)$ from its undisturbed open-loop optimal value $\mathbf{x}^o(t)$. Such a reduction would lead, of course, to correspondingly larger feedback control inputs $\mathbf{u}(t)$, which would have to be accommodated through use of more capable control actuation devices. There is a cost for control!

Stochastic Processes

The above discussion of optimal control in the presence of unpredictable plant disturbances was based on the assumption that the plant found its state perturbed by a temporary disturbance at time t and that it was undisturbed from t onward. It was then argued that the feedback control should minimize J starting from the perturbed state $(\mathbf{x}^o + \mathbf{x})$ at time t. This choice led to the optimal nonlinear feedback control law given by Eq. (35) for minimizing the effect of such a disturbance on J no matter when it might occur or what its magnitude might be. A more general case occurs when the state of the plant is continuously perturbed by an unpredictable *stochastic process*, $\mathbf{w}(t)$:

$$\dot{\mathbf{x}} = f(\mathbf{x},\mathbf{u},t) + g(\mathbf{x},t)\mathbf{w}(t) \qquad (47)$$

where $\mathbf{w}(t)$ is a vector-valued Gaussian stochastic process with zero mean and covariance $E[\mathbf{w}^T(t)\mathbf{w}(\tau)] = Q(t)\delta(t - \tau)$.[11] In this case, $\mathbf{x}(t)$ becomes a stochastic process and, hence, J becomes a *stochastic variable*. As a result, any attempt to minimize J in the usual way does not make sense mathematically. We try to find a feedback control law that would minimize the expected value of J, denoted $\langle J \rangle$:

$$\langle J \rangle = E\left\{ \phi(\mathbf{x}(t_f),t_f) + \int_{t_0}^{t_f} L(\mathbf{x}(t),\mathbf{u}(t),t)\, dt \right\} \qquad (48)$$

In this case, it can be shown that the resulting optimal feedback control law $\mathbf{u}^o(\mathbf{x},t)$ is again given by Eq. (35), but this time $S(\mathbf{x},t)$ satisfies the equation

$$(\partial S/\partial t) + (\tfrac{1}{2})\mathrm{TR}[(\partial^2 S/\partial \mathbf{x}^2)GQG^T] = H^o[\mathbf{x},\, -(\partial S/\partial \mathbf{x})^T, t] \qquad (49)$$

where $H^o(\mathbf{x},\lambda,t)$ is again as defined by Eq. (29) and where TR[ ] denotes the trace operator (i.e., sum of the diagonal elements) for matrices. Equation (49) differs from the Hamilton–Jacobi–Bellman equation (Eq. 32) only through the term $(\tfrac{1}{2})\mathrm{TR}[(\partial S/\partial \mathbf{x}^2)GQG^T]$, and it reduces to Eq. (32) when $Q \equiv 0$. The occurrence of this term is essentially due to rectification (square-law nonlinearity) effects on the disturbance $\mathbf{w}(t)$.

In the special case that the plant is linear and ϕ and L are quadratic functions

---

[11] Here $\mathbf{w}(t)$ plays the same role that D played in the earlier sections. The mean of $\mathbf{w}(t)$ is predictable and, hence, unpredictable processes must have zero mean values for the formulation to make sense.

of all variables, it can be shown that the resulting linear optimal feedback control law is *identical* to that given before for a temporary inadvertent disturbance.

The comments made earlier regarding the difficulties in solving the Hamilton–Jacobi–Bellman equation also apply to the problem of solving Eq. (49). However, also as before, such difficulties in application of this theory do not in any way detract from the underlying rich structure of optimal control theory and its substantial connections with and differences from the variational theory of mechanics.

Optimal Feedback Control with Sensor Noise

A critical implicit assumption made in the above treatment of optimal feedback control theory was that *perfect measurements of the total state vector* $\mathbf{x}(t)$ were available for use by the optimal feedback control law as given by Eq. (35). However, in most control problems, only one or at most a few functions of the state variables can be measured or sensed directly for use by a feedback controller, as indicated by Eq. (20). For example, in the discussion of single-input, single-output systems in the previous section, only a single output variable c was available to the feedback controller (see Figures 1 and 2). Moreover, in most cases measurements are corrupted by significant amounts of sensor noise, which must be suppressed to an appropriate degree by the feedback controller, as was explained in previous sections. From these considerations, it is clear that the optimal feedback control theory discussed above must be further generalized to allow for imperfect measurements of the system state. Otherwise it cannot be considered to be a comprehensive theory of control incorporating all the fundamental concepts of the classical control theory. Fortunately, some of the appropriate generalizations have been developed, and they are summarized below.

Consider the case that all measurements are perfect (i.e., made without sensor noise), although only certain combinations of the state variables are available for measurement, and that plant dynamics and output structure as given by Eqs. (19) and (20) are linear and jointly satisfy a certain reasonable condition known as observability. It is then possible to construct a physically realizable dynamic system (called an observer) that reconstructs all the state variables from the output combinations available to within an error that can be made to vanish arbitrarily rapidly. The output of this observer (i.e., the reconstructed states) can then be used directly in the feedback control law in place of the actual states and thereby achieve the desired optimal feedback control.

The assumption of observability was always implicitly made in the classical linear control theory discussed above. If the plant–output measurement combination is not observable, then some plant state combinations cannot be controlled by means of feedback using the given set of measurements, because the motions are simply not observable in the measurements available to the feedback controller. If control of these unobservable state combinations is required, additional sensors must be used to make them observable. The observer theory can be extended to cover appropriate nonlinear plant–output measurement combinations as well, but this case is less well understood because of the obvious analytic difficulties. The general subject of observers for state reconstruction will not be discussed further here because the assumption of no measurement errors (i.e., no sensor noise) is very restrictive and is not often satisfied sufficiently well in practice anyway.

Stochastic Optimal Control

The remaining major generalization to be discussed is optimal control in the presence of unpredictable measurement errors (i.e., sensor noise). This problem (often called the stochastic optimal control problem) is far more complex than any of those discussed above. As a result, its structure and the nature of its solution are not nearly as well understood as are the previous examples, except for the linear dynamics, Gaussian process, quadratic performance index case. In fact, the general nonlinear, non-Gaussian process case is still a subject of active research. Furthermore, it will be seen that this case—which is in some sense the most general of all optimal control problems—*does not appear to have any direct connections or analogies with the variational theory of mechanics or even with statistical mechanics* (to which it might be expected to be connected). Rather, it appears to be a problem that falls outside the realm of theoretical physics as it is understood today. More will be said on this subject later.

The most studied version of the general stochastic optimal control problem can be formulated as given by Eqs. (47), (48), and (20), except that Eq. (20) is modified to include a sensor noise term $\mathbf{n}(t)$ in the form

$$\mathbf{c}(t) = h(\mathbf{x},t) + \mathbf{n}(t) \tag{50}$$

where $\mathbf{n}(t)$ is a zero mean, Gaussian, vector-valued white noise process with $E[\mathbf{n}(t)\mathbf{n}^T(\tau)] = R(t)\delta(t - \tau)$. As before, the problem is to find an optimal feedback control that minimizes $\langle J \rangle$ as in Eq. (48), but this time all knowledge of the state required by the feedback controller must be obtained through the noisy measurements given by Eq. (50). Other versions in which the processes $\mathbf{w}(t)$ and $\mathbf{n}(t)$ are allowed to be non-Gaussian have also been studied, but they will not be discussed here because, although they lead to different equations for the optimal solution, they do not lead to any fundamentally different viewpoints or approaches to the problem.

A reasonable *ad hoc* approach to the problem of dealing with sensor noise is first to find an optimal estimate $\hat{\mathbf{x}}(t)$ of the state $\mathbf{x}(t)$ for each t, and then use this estimate in place of the real state $\mathbf{x}$ in the optimal control law $\mathbf{u}^o(\mathbf{x},t)$ in Eq. (35). This approach breaks the general stochastic optimal control problem into two parts, which are solved separately and then combined to obtain a final feedback control law. The problem of finding $\mathbf{u}^o(\mathbf{x},t)$ under the assumption of direct sensing of each component of the system state without sensor errors was discussed above. The remaining problem of finding an optimal estimate of $\mathbf{x}(t)$ given the current and past noisy measurements $\mathbf{c}(\tau)$ for $t_0 \leq \tau \leq t$ is widely known as the *optimal nonlinear filtering* problem. It can be resolved, in principle, as follows.

It is clear from Eq. (47) that $\mathbf{x}(t)$, being the solution of a stochastic differential equation [i.e., one with an unpredictable stochastic input and a stochastic initial condition, $\mathbf{x}(t_0)$], is a stochastic vector with some probability density $p(\mathbf{x}(t),t)$.[12] Equations of the general form given in Eq. (47) are encountered in the theory of statistical physics, and it is well known from this theory that $p(\mathbf{x},t)$ satisfies the Fokker–Planck equation

$$(\partial p/\partial t) = -\Sigma(\partial[pf_i]/\partial x_i) + (½)\Sigma(\partial^2[p(gQg^T)_{ij}]/\partial x_i \partial x_j) \quad (51)$$

and the initial condition $p(\mathbf{x},t_0) = p_{x(t_0)}(\mathbf{x})$. Here, $p_{x(t_0)}(\mathbf{x})$ is the probability density function of the initial plant state $\mathbf{x}(t_0)$, which is assumed to be given.

In the absence of any knowledge of $\mathbf{x}(t)$ beyond the fact that it satisfies Eq. (47), $p(\mathbf{x},t)$ represents the most complete quantitative information available concerning the actual value of $\mathbf{x}(t)$. However, in the optimal control problem, which is the case of interest here, the noisy measurements $\mathbf{c}(\tau)$ for $t_0 \leq \tau \leq t$ provide additional information concerning $\mathbf{x}(t)$, which must be accounted for through appropriate modifications to $p(\mathbf{x},t)$. Specifically, $p(\mathbf{x},t)$ must be replaced by the conditional probability of $\mathbf{x}(t)$, given the measurements $\mathbf{c}(\tau)$ for $t_0 \leq \tau \leq t$. This conditional probability density will be denoted by $p(\mathbf{x},t|C_t)$, where $C_t = \{\mathbf{c}(\tau) | t_0 \leq \tau \leq t\}$ denotes the whole set of observed or measured values $\mathbf{c}(\tau)$ for each τ in the interval $[t_0,t]$.

The conditional density $p(\mathbf{x},t|C_t)$ formally satisfies the following stochastic partial differential equation:

$$(\partial p/\partial t) = -\Sigma(\partial[pf_i]/\partial x_i) + (½) \sum_{ij} (\partial^2[p(gQg^T)_{ij}]/\partial x_i \partial x_j)$$
$$+ (h - \hat{h})^T R^{-1}(t)(\mathbf{c}(t) - \hat{h}) \quad (52)$$

where $\hat{h}$ denotes the conditional expected value of h, i.e.,

$$\hat{h} = \int_{-\infty}^{\infty} h(\mathbf{x},t) p(\mathbf{x},t|C_t) \, d\mathbf{x} \quad (53)$$

Where $R^{-1}(t) \to 0$, Eq. (52) reduces to the Fokker–Planck equation, i.e., Eq. (51). This reduction simply expresses the fact that since $R(t) \to \infty$ [and hence, $R^{-1}(t) \to 0$], the measurements become increasingly noisier and contain less and less reliable information concerning $x(t)$. This limiting result is important for consistency of the theory, but it is not very surprising. Given the conditional density $p(\mathbf{x},t|C_t)$, the question remains as to just what statistics of $p(\cdot)$ should be used in the feedback control law. Various statistics proposed include the conditional mean, the conditional mode, and the conditional median. The choice depends, of course, on one's criterion of good control performance. For a variety of reasonable criteria, however, the conditional mean is the best statistic. That is, one

---

[12] It is assumed here that the functions $f(\cdot)$ and $g(\cdot)$ in Eq. (47) satisfy the rather mild conditions that guarantee the existence of unique solutions with densities. Also, at this point, $p(\mathbf{x},t)$ clearly depends on $\mathbf{u}(\tau)$ for $t_0 \leq \tau \leq t$, but this dependence is suppressed for convenience in the notation.

should take as the estimate $\hat{\mathbf{x}}(t)$ of $\mathbf{x}(t)$ the value of

$$\hat{\mathbf{x}}(t) = \int_{-\infty}^{\infty} \mathbf{x} p(\mathbf{x},t | C_t) \, dt \tag{54}$$

This procedure is widely advocated.

To compute $\hat{\mathbf{x}}(t)$ in real time, it is necessary to solve Eqs. (52) and (53) in real time as the noisy measurements $c(t)$ become available and then evaluate $\hat{\mathbf{x}}(t)$ using Eq. (54). With available computers, this requirement is an all but impossible task in practical nonlinear problems of interest. However, many approximate solutions based on various types of series expansions have been proposed, and their performance has been evaluated in some simple cases. The real-time computational burden normally limits the number of terms that can be used, and most of the time one is content to use only second-order to third-order expansions with appropriate *ad hoc* corrections.

Such approximate solutions will not be discussed further here; however, one important case is worthy of further discussion. That is, when Eqs. (47) and (50) are linear, both $p(\mathbf{x}(t),t)$ and $p(\mathbf{x}(t), | C_t)$ are Gaussian. Then, one needs only to determine the first and second statistical moments as a function of time in order to specify the probabilities completely. For instance, for $p(\mathbf{x}(t), | C_t)$, the conditional mean or expected value $\hat{\mathbf{x}}(t)$ of $\mathbf{x}(t)$ and its conditional covariance matrix $\Sigma(t)$ satisfy the equations

$$\dot{\hat{\mathbf{x}}} = F(t)\hat{\mathbf{x}} + \Sigma(t)H^T(t)R^{-1}(t)[\mathbf{c}(t) - H(t)\hat{\mathbf{x}}] \tag{55}$$
$$\dot{\Sigma} = F(t)\Sigma + \Sigma F^T(t) + G(t)Q(t)G^T(t) - \Sigma H^T(t)R^{-1}(t)H(t)\Sigma \tag{56}$$

where $F(t)$, $G(t)$, and $H(t)$ are the appropriate matrix representations of $f(\cdot)$, $g(\cdot)$, and $h(\cdot)$ in the linear case. These equations represent a feasible real-time computation in most cases and have been used widely, either directly for linear problems or as approximate solutions based on linearization techniques for nonlinear problems.

There is one other important result in the linear case of Eqs. (47) and (50) if the functions $\phi(\cdot)$ and $L(\cdot)$ in Eq. (48) are quadratic in their arguments. This result, often called the *certainty-equivalence principle*, asserts that the *ad hoc* approach to the optimal control problem in the presence of sensor noise (i.e., when the overall problem is broken into two parts separately solved and the results combined to obtain a feedback control law) actually produces *the optimal stochastic feedback control*. That is, one can determine an optimal linear feedback control law by the procedure leading to Eq. (46). That procedure consists of independently estimating $\mathbf{x}(t)$, using Eqs. (55) and (56) with an appropriate simple correction to account for the suppressed value of $\mathbf{u}(t)$, and combining the results in the form

$$\mathbf{u}^\circ(t) = -K(t)\hat{\mathbf{x}}(t) \tag{57}$$

to obtain *the optimal* feedback control solution for this special case.

Unfortunately, the certainty-equivalence principle does not hold in the general case. As a result, the *ad hoc* approach discussed above usually fails to give the optimal solution whenever the plant and/or measurement equations are nonlinear, the performance index is nonquadratic, or the disturbances and sensor noise are non-Gaussian. This failure is associated with the occurrence of the control variable in the plant dynamic equation (47). If feedback control is used in any form, then $\mathbf{x}(t)$ will depend on *all* of the past values of $\mathbf{x}(\tau)$ for $t_0 \leq \tau \leq t$; hence, $\mathbf{x}(t)$ will *not be a Markovian process*. However, the derivation of the equations for conditional density $p(\mathbf{x},t|C_t)$ used to find $\hat{\mathbf{x}}(t)$ for the control requires that $\mathbf{x}(t)$ be Markovian. Thus, it is not surprising to find that the *ad hoc* approach fails in the general case. Perhaps it should have been surprising to find that it does give the correct result in the linear, quadratic-performance, Gaussian stochastic process.

A rigorous proof of the certainty-equivalence principle is not trivial even in the linear, quadratic, Gaussian case because of the non-Markovian nature of $\mathbf{x}(t)$. Unfortunately, the observation that the state $\mathbf{x}(t)$ is not Markovian in the presence of feedback control forces the use of an even more general mathematical framework and theory for the stochastic optimal control problem than has been introduced above! A proper formulation and solution of the general stochastic optimal control problem will require a considerable generalization of both deterministic optimization and statistical estimation theories to infinite-dimensional function spaces followed by a suitable synthesis of the two. This generalization and synthesis have yet to be accomplished and represent an open and very challenging problem in the ongoing development of control theory.

Equations of the type given in Eq. (52) do not arise in the theory of statistical mechanics, because statistical mechanics does not deal with concepts that involve estimating the state of dynamic systems for purposes of control. Neither does physical theory appear to deal with any problems analogous to or embedded in the stochastic optimal control problem. *Again one is forced to conclude that control theory cannot generally be regarded as a part of, or a simple extension of, known physical theory!*

Control Processes in Living Systems

It is now almost certain that the first control processes to appear on earth were those that emerged billions of years ago in the course of evolution of complex living systems from primitive life forms (Gordon, 1968). However, the first recognition of the possible occurrence of control processes in living systems apparently dates back only to the 18th century when Lavoisier suggested, "Regulation may consist in governed exchanges of substance."[13] A century later, Claude Bernard carried Lavoisier's idea further when he discovered (by chemical

---

[13]Use of the term "regulation" here is consistent with its earlier definition as a special kind of feedback control process for which the goal is to maintain a plant output quantity or quantities near a desired constant output level.

analysis) the constancy or stability of the composition of plasma and interstitial fluids, which he called the internal environment. In 1929, Cannon suggested the term "homeostasis" to characterize this stability of the internal environment (Cannon, 1929, 1939). Subsequently, Cannon's definition of "homeostasis" was broadened to designate either the constancy or stability of the body's internal environment, or the processes by which this constancy or stability is maintained. From this broadened definition, it is but a short step to *the interpretation of homeostasis as a special kind of feedback control process known as a regulatory control process.*

Much earlier, the increasing awareness by humans of their direct involvement in a variety of control processes and their interest in the automation of these processes led to the invention and use of a great diversity of automatic control devices. The first was probably the invention some 5000 years ago of the float-operated sluice gate that automatically controlled the water level in irrigation systems. This device was followed much later by Watt's very significant invention of the steam engine governor for regulation of engine speed in the presence of load disturbances. Maxwell analyzed the operation of Watt's governor, and his analysis marked the true beginning of control theory. It was followed by Minorsky's analysis of control processes for steering ships (Minorsky, 1922). Minorsky's work provided significant further demonstration of the wide applicability of control processes for achieving stability and regulation of physical variables in the presence of disturbances.

The first definitive statement of all the properties achievable by feedback processes of the type used by Watt and Minorsky (also used in the float-operated sluice gate) was given by Black (1934). The problems imposed by the ultimate instability of feedback control loops, appearing as the loop gain is increased, were greatly clarified by Black and also by the beautiful stability analysis technique developed by Nyquist (1932).

From the 1930s to the present, the theory and application of control processes have grown explosively. Many of the major advances are summarized in the preceding sections of this chapter. This same period was also a time of startling growth in our understanding of the fundamental processes of living systems at all levels of organization. As might be expected—given their parallel development and the existence of certain obvious analogies between the concepts involved—the new theory of control processes and the new understanding of the processes of living systems were thought to have overlapping domains. Wiener's book on cybernetics brought the possibilities for fruitful interactions to the attention of both control theorists and life scientists (Wiener, 1948). As a result, principles of control were discovered in the processes of living systems. The jargon of control theory eventually permeated the scientific literature dealing with molecular biology, endocrinology, and neurosciences. This effect was most obvious in physiology, as illustrated by Guyton's (1971) *Textbook of Medical Physiology.* Physiology came to be considered by some as dealing largely with control processes operative in living systems.

Despite the invasion of the life sciences by control jargon, control theory did not play an important role in the major experiments and discoveries of the rapidly advancing fields of molecular biology, endocrinology, or neurobiology. Rather, it seems merely to have provided an after-the-fact framework for interpreting the discoveries and developments in these fields [see, e.g., Judson's recent history of major discoveries in molecular biology (Judson, 1979)]. Most life scientists have been very reluctant to embrace control concepts and theory. For instance, some have pointed out that words such as "feedback" and "control" do not convey the same meanings to biologists as to control theorists. They have consistently pointed out that the physicochemical control processes in living systems differ markedly from those found in control systems designed by control engineers (Chance, 1961). Others have stated what they believe to be serious limitations of control theory (especially the classical theory with its explicit assumption of linearity), which may prevent it from ever becoming a satisfactory unifying theory of the fundamental integrative processes of living systems.

Thus, control theory has been left in a somewhat ambivalent state with regard to its usefulness in explaining any processes occurring in living systems. The remainder of this chapter examines some control processes operating in living systems in order to delineate both the power and the limitations of control concepts and theory in this context.

Control Processes at the Cellular Level—Enzyme Inhibition

Enzyme-Catalyzed Reactions

Fundamental processes in living systems occur at the molecular and cellular levels; many of them involve enzyme-catalyzed reactions, including the synthesis and breakdown of the enzyme proteins themselves (Lehninger, 1975; Savegeau, 1976). The control characteristics can be examined by considering the simplest, classical model of an enzymatic reaction: there is a single substrate, S, a single enzyme, E, a single product, P, and rate constants, k_i. The reaction proceeds according to the formula

$$S + E \underset{k_2}{\overset{k_1}{\rightleftharpoons}} ES \overset{k_3}{\rightarrow} P + E \tag{58}$$

where S and E combine reversibly to form an enzyme–substrate complex, ES, which can either dissociate to S + E or else irreversibly convert to P + E. If it is assumed that the enzyme–substrate reaction is fast compared to the rate of change of the concentrations [S] and [E] of S and E, respectively (which is the usual case), then the reaction velocity in a constant volume is given by

$$v = d[P]/dt = k_3[E]([S]/(K_m + [S])) = k_3[E]/((K_m/[S]) + 1) \tag{59}$$

where $K_m = (k_2 + k_3)/k_1$. K_m, the Michaelis constant, is not an equilibrium constant for a single reaction but represents the substrate concentration at which half-maximal velocity will be reached in the reaction system of Eq. (58). It is seen from Eq. (59) that at constant substrate concentration, the rate of reaction is directly and *linearly* controlled by the enzyme concentration [E]. This process can be interpreted as an *open-loop* control where the *control variable* is [E] and the *output or controlled variable* is the rate, v, of formation of the product, P (Figure 5). Equation (59) also shows that v depends nonlinearly on the substrate concentration [S]. In this regard, two limiting cases are of interest. First, if [S] $\gg K_m$, then Eq. (59) reduces to

$$v = d[P]/dt = k_3[E] \tag{60}$$

and the controlled variable is independent of substrate concentration. On the other hand, if [S] $\ll K_m$, Eq. (59) reduces to

$$v = d[P]/dt = (k_3/K_m)[E][S] \tag{61}$$

and the controlled variable depends on [E] and [S]. That is, in the first limiting case, the dependence of the controlled variable on the control input is linear; in the second, the dependence is bilinear. In either case, the question remains as to what role should be assigned to variations in the variable [S] in this context. For reasons that will become clear later, I consider variations in [S] to be disturbances acting on the plant, as illustrated in Figure 5. The strong nonlinearity of the plant dynamics (shown in the box in Figure 5) should be noted.

Sequences of Enzyme-Catalyzed Reactions

In general, the synthesis of biological molecules does not proceed as a single-enzyme-catalyzed reaction as in Eq. (58). Rather, it usually occurs as a sequence of enzyme-catalyzed reactions. Well-studied examples include the biosynthesis by some bacteria of tryptophan from chorismate, of isoleucine from threonine, and of pyrimidine nucleotides from aspartic acid and carbamyl phosphate. Specifically, the biosynthesis of isoleucine from threonine involves a sequence of five reactions, catalyzed by five different enzymes (Lehninger, 1975). In each of these reactions, the product of one reaction becomes the substrate for the following reaction (Figure 6). The enzyme threonine deaminase, E_1, catalyzes the first step of the biosynthetic pathway, and it has a special role in the control of this process.

Biosyntheses such as that shown in Figure 6 also can be viewed as open-loop control processes where the concentration of the (active form of the) first enzyme, e.g., threonine deaminase, is taken to be the primary control variable, and the

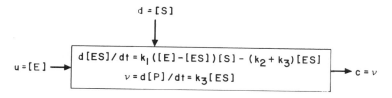

FIGURE 5. Enzyme-controlled reaction characterized as an open-loop control process.

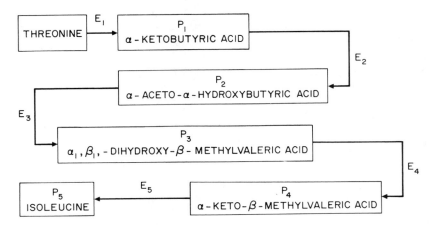

FIGURE 6. Biosynthetic pathway for isoleucine—a typical enzymatic cascade. E_i = enzyme catalyst; P_i = chief products of reaction.

rate of product generation by the final reaction, e.g., rate of isoleucine generation, is taken to be the controlled output variable.[14] As before, variations in the initial substrate concentration, e.g., threonine concentration, are here considered as disturbances acting on the system for control analysis. The concentrations of the products of the individual reactions P_1 through P_5 are the state variables for biosynthesis.

It is also possible to view the above reactions as open-loop control processes. In this alternative view, the substrate concentration becomes the control variable and the enzyme concentration simply becomes a parameter of the process. *There is nothing inherent in control theory that forces the selection of one viewpoint over the other!* In a sense, this egalitarianism is a limitation of control theory that results from its generality. In order to favor a particular viewpoint, more information on the functioning of the process in relation to its environment is required. A useful general rule is to select control variables that are directly influenced by a preceding process, by a parallel process, or by direct feedback effects of other downstream variables that are themselves controlled variables. The remaining variables are then either exogenous variables arising from sources external to the open-loop control process or variables directly dependent on control variables and the exogenous variables. Variations in the former are then taken as disturbances acting on the open-loop control process, whereas the latter become either the internal state variables of the process or output variables from the process. This rule obviously leads to a consistent application of control theory concepts as they were described in earlier sections. Because the synthesis of enzymes is under genetic control by a preceding process, according to this rule one can take concentrations of (the active forms of) appropriate enzymes as con-

---

[14] The choice of rate as the controlled variable is based on the assumption that a fluctuating demand for the product of the biosynthetic pathway must be met.

Double-Substrate Reactions

In most cases, enzymes do not catalyze single-substrate reactions (Lehninger, 1975; Savegeau, 1976). Rather, they usually catalyze reactions involving two substrates. Therefore, the analysis given above for single-substrate reactions has to be generalized. Most two-substrate reactions belong to either of two classes: single-displacement or double-displacement reactions. Single-displacement reactions can be further divided into two forms—random or ordered.

Consider the case of two substrates, A and B, and an enzyme E. In random, single-displacement reactions, either substrate can first combine with the enzyme to yield a binary complex to which the other substrate then binds to form the tertiary complex EAB. EAB then dissociates to form products P and Q, plus the free enzyme. The resulting rate equation, analogous to Eq. (59), for the formation of either P or Q will not be given here. It is again possible to interpret the reaction as open-loop control, in which the enzyme concentration is taken to be the control variable, and variations of the substrate concentration are considered as disturbances acting on the plant.

In ordered, single-displacement reactions, the enzyme E must first combine with substrate A to form the binary complex EA, which then can combine with B to form the tertiary complex EAB. B cannot bind directly to E. The tertiary complex EAB then dissociates as before to form products P and Q plus free enzyme. The rate equation for the ordered, single-displacement reaction is

$$\nu = V_{max}/((K_S^A K_M^B/[A][B]) + (K_M^A/[A]) + (K_M^B/[B]) + 1) \tag{62}$$

where V_{max} is the maximum rate achievable for the given amount of enzyme, K_S^A is the equilibrium constant for the reversible formation of the binary complex EA, and K_M^A and K_M^B are the Michaelis constants for substrates A and B, respectively. In the limiting case where $[A] \to \infty$ or $[B] \to \infty$, this expression reduces to the single-substrate case discussed earlier, where S is taken to be either B or A. Again, this reaction can be interpreted as an open-loop control process in exactly the same way as was done for a random, single-displacement reaction. That is, the variables are assigned in the same way, even though the dynamics are obviously different in the two cases.

Finally, in double-displacement reactions, the enzyme first combines with substrate A to form the binary complex EA, which then dissociates to form product P plus a covalently substituted enzyme E*. Substrate B then combines with E* to form E*B, which finally dissociates to form product Q plus the original enzyme E. The general rate equation for such double-displacement reactions is

$$\nu = V_{max}/((K_M^A/[A]) + (K_M^B/[B]) + 1 \tag{63}$$

where V_{max} and K_M^{Si} are as defined for Eq. (62) where Si denotes either A or B. Again, when either $[A] \to \infty$ or $[B] \to \infty$, this equation reduces to the single-substrate case given in Eq. (59). This reaction can also be properly viewed as an open-loop control process, using the same identification of variables as for the single-displacement reactions.

It is pleasing to find that all double-substrate reactions can be viewed within a single open-loop control system framework, even though the dynamics vary widely. This observation demonstrates again the universality of *control theoretic concepts as systems-level concepts*. Up to this point, however, I have merely shown the affability of control theoretic descriptions; I have not demonstrated any new insights that might be obtained by translating the description of common forms of biochemical reactions into open-loop control language and images. To gain those insights, the reader should now return to the detailed discussion of open-loop control given earlier in this chapter, where the general properties of such controlled systems are specified. The point here is that these are also properties of the common enzymatic reaction forms just presented. From the functional viewpoint of a living cell, it is exactly these control properties that are persuasive. (To an engineer it is not sloppy thinking to anthropomorphize a cell on occasion or, for that matter, to objectify humans on other occasions—for defined purposes only.)

Modulation of Enzyme Reaction Rates

There are several different mechanisms for modulating the activity or concentration of enzymes. These mechanisms can be used to control the rates of reactions and are thus of primary importance in understanding the control processes that may operate within cells. Such mechanisms may be reversible or irreversible. Irreversible mechanisms are used by cells to break down enzymes; some poisons may damage enzymes by binding to them irreversibly.[15] The classical reversible mechanisms include competitive inhibition, uncompetitive inhibition, and noncompetitive inhibition. (Allosteric inhibition is a more elaborate reversible mechanism that occurs in highly specialized, regulatory enzymes. Because of their great importance for biological control, regulatory enzymes are discussed in detail separately in the following section.) The classical reversible mechanisms have been displayed in standard texts of biochemistry since the 1950s. A recent summary can be found in Savegeau (1976), and there is no need to provide much detail here. However, I want to show how traditional "enzyme kinetics" looks to the control engineer. To do that, I must present the standard model reaction rate equations.

In competitive inhibition, the concentration of the active enzyme is modified by an inhibitor substance, I, chemically similar to the normal substrate molecules. Because of similarity to substrate, inhibitor molecules compete with the substrate

---

[15]Irreversible mechanisms are of central interest in the field of toxicology. They provide a potential mechanism for selection in the evolution of enzymes and possibly even for the extinction of some forms of life when exposed to new environments.

molecules for binding at the active site of the enzyme. But the inhibitor molecules are not chemically changed after interaction with the enzyme. Thus, the net effect of the binding of such an inhibitor to an enzymatic active site is to prevent the binding of a normal substrate, thereby inhibiting the catalyzed conversion of substrate to product. The rate equation for competitive inhibition is

$$v = k_3[E][S]/(K_m(1 + ([I]/K_I)) + [S]) \tag{64}$$

where K_I is the equilibrium constant for the formation of the enzyme–inhibitor complex. From Eqs. (64) and (59) it is seen that the slope of a curve relating rate of formation of product, v, to substrate concentration, [S], at low substrate concentrations (i.e., as [S] → 0) is decreased by factor $1/(1 + ([I]/K_I))$. However, the maximum rate of formation of product (i.e., the value of v as [S] → ∞) is unaffected by the inhibitor concentration.

In uncompetitive inhibition, the inhibitor combines reversibly with the enzyme–substrate complex to produce an inactive enzyme–substrate–inhibitor complex that cannot dissociate to form product. The rate equation is

$$v = k_3[E][S]/(K_m + (1 + ([I]/K_{IS})[S])) \tag{65}$$

where K_{IS} is the equilibrium constant for the reaction ES + I ⇌ ESI. In this case, as seen in Eqs. (65) and (59), the slope of the rate versus substrate concentration curve at low substrate concentrations (i.e., as [S] → 0) is unaffected by the inhibitor concentration, but the maximum rate is reduced by the factor $1/(1 + ([I]/K_{IS}))$.

Noncompetitive inhibition occurs when the inhibitor binds reversibly to the enzyme at a site distinct from the substrate binding site and yields two inactive bound forms of the enzyme, namely, both EI and ESI complexes. The rate equation is

$$v = k_3[E][S]/(K_1(1 + ([I]/K_I)) + (1 + ([I]/K_{IS}))[S]) \tag{66}$$

where K_1, K_I, and K_S are the equilibrium constants for the reactions E + S ⇌ ES, E + I ⇌ EI, and ES + I ⇌ ESI, respectively. Equations (66) and (59) show that this case is a combination of the previous two, in that the maximum rate is reduced by the factor $1/(1 + ([I]/K_{IS}))$ and the slope of the rate versus substrate curve at low values of [S] is reduced by the factor $1/(1 + ([I]/K_I))$.

All three models of enzyme inhibition described above have the potential for controlling enzyme-catalyzed reactions or reaction sequences. Although few real enzyme systems obey these idealized rate equations, they serve as a useful heuristic.

From a control viewpoint, competitive inhibition is not very interesting, because control based on competitive inhibition would be active only when substrate concentration is very low. For larger values of substrate concentration, the rate is effectively independent of inhibitor concentration, as illustrated by the curves in Figure 7A.

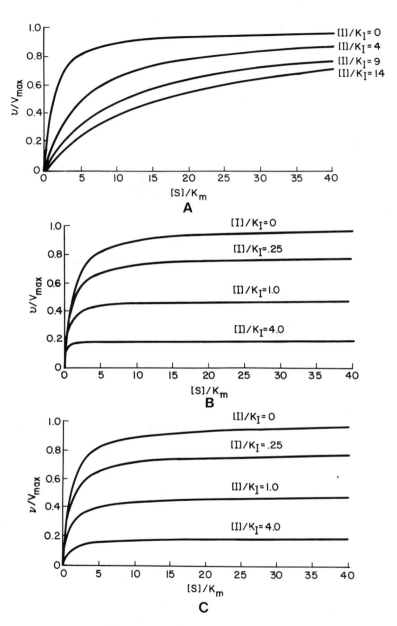

FIGURE 7. Normalized kinetics of enzyme inhibition (symbols defined in text). (A) Rate dependence on inhibitor concentration for competitive inhibition. (B) Rate dependence on inhibitor concentration for uncompetitive inhibition. (C) Rate dependence on inhibitor concentration for noncompetitive inhibition. (D) Rate dependence on inhibitor concentration for a regulatory (allosteric) enzyme.

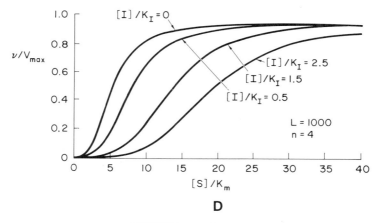

FIGURE 7 (*continued*)

Uncompetitive inhibition has just the opposite characteristics with regard to the range of effective substrate concentrations. It is effective only for large substrate concentrations, as illustrated by the curves in Figure 7B.

The most interesting case for control analysis is noncompetitive inhibition, because the inhibitor concentration affects *both* the initial slope of the rate versus substration curve *and* V_{max}. Hence, it could provide an effective control over the entire range of possible substrate concentrations. This feature is illustrated in Figure 7C.

The curves in Figure 7C bear a strong qualitative resemblance to the characteristic curves of active electronic devices such as pentodes and transistors, where reaction rate is analogous to the plate or collector current, the substrate concentration is analogous to the cathode–anode or emitter–collector voltage, and the inhibitor concentration is analogous to the grid voltage or base current. Also, because noncompetitive inhibition does not require the inhibitor to be sterically similar to the proper substrate, it is much more flexible as a mechanism for achieving control of biochemical reactions.

Regulatory (Allosteric) Enzymes

Control processes involving competitive, uncompetitive, or noncompetitive inhibition as described above do not appear to be very common or important, compared to the many well-established examples of control by regulatory enzymes. Although the history of the discovery of regulatory enzymes is well known, it is worthwhile to review it here, because it illustrates the transition from open-loop to closed-loop (negative feedback) models of biochemical control.

One of the major discoveries of biochemists is that the *end products* of the biosynthetic pathways for amino acids and nucleotides (Figure 6) *normally inhibit the activity of the enzyme involved in the first step in the pathway.* This effect was noted by Novick and Szilard (1954) in the tryptophan biosynthetic pathway, and it was thought by them to be an instance of a *negative feedback loop* operating at the cellular level. Similar effects were also observed by Yates and Pardee (1956) in the pathway that makes pyrimidine nucleotides and by Umbarger (1955) in the isoleucine pathway shown in Figure 6. Changeux (1961) and Gerhart and Pardee (1962) conclusively established that the negative feedback inhibition involved binding sites other than the substrate binding sites and, hence, was not competitive. It was also observed that the inhibition in such pathways was "atypical," in that the rate versus substrate concentration curve for a given level of inhibitor was sigmoidal. This surprising shape (Figure 7D) implied the operation of a mechanism different from those shown in Figure 7A–C. Monod and Jacob (1961) also reported that this feedback inhibition involved inhibitors that were not chemical analogues of the substrates. They designated the mechanism involved as "allosteric inhibition." They also argued, "Since the allosteric effect is not inherently related to any particular structural feature common to substrate and inhibitor, the enzymes subject to this effect must be considered as pure products of selection for efficient regulatory devices." Moreover, they pointed out the significant possibility that such an effect could be used to allow the level of metabolite synthesized in one pathway to control the rate of reaction in another pathway, i.e., it could provide a mechanism for cross-coupling of biosynthetic pathways (including those in which end product inhibition is also operative). The outstanding potential of such a mechanism for integration of biosynthetic pathways is obvious.

Umbarger (1961) preferred to use the term "end product inhibition" rather than "feedback inhibition" when referring to the allosteric inhibition effect described in the preceding paragraph, because "end product inhibition is an operational term that signals the potential importance of an interaction as a physiological mechanism, yet its use does not require proof of that significance." He went on to say, "The assignment of a role as a feedback mechanism should only be made after careful appraisal of such a role in the economy of the cell." This was a very perceptive distinction, because none of the biochemists had actually showed that the proposed "feedback" systems were stable or that they had sufficient loop gain to make the feedback control effective. In fact, *they did not address the question of stability, loop gain, or the actual performance of the feedback loop at all!* Biochemists are not engineers, and the discussion of end product inhibition in the literature of molecular biology provides a good illustration of the fundamentally different interests and viewpoints of molecular biologists and control scientists. Molecular biologists are chiefly interested in discovering the sites of interactions between biologically active molecules and their metabolic pathways and in the detailed biophysical and biochemical mechanisms by which these interactions take place. Their interest seems to wane once the potential existence of feedback loops is established, and they seem content with *qualitative* implications of such existence on cell behavior. On the other hand, control scientists are most interested in the *quantitative* properties and performance achievable by feedback control loops, and they are little interested in the specific mechanisms of how such feedback effects are realized, as long as they exist and can be quantitatively characterized. In a very real sense, these viewpoints are complementary.

Biologists will have to place more emphasis on the control scientist's viewpoint to acquire a full understanding of the physiological significance of end product inhibition (feedback control) in the proper functioning of cells. Otherwise, we shall not really know whether the apparent feedback involving end product inhibition is actually effective in regulating the rate of biosynthesis of end product to meet a varying demand. If further effort were to show that it is not effective (i.e., that loop gains are small), then some alternative other than such regulation must be found to rationalize the widespread existence of end product inhibition in biosynthetic pathways.

Next, allosteric inhibition is examined in considerable detail to show the control issues from an engineering perspective. My example is end product inhibition of threonine deaminase by isoleucine, which suggests the presence of a feedback control loop of the form illustrated in Figure 8, where the enzymes E_i and products P_i are as given in Figure 6. To establish whether or not this putative feedback control loop is in fact effective, it is necessary to analyze the response of the pathway both with and without the feedback (allosteric inhibition). By so doing we can determine the extent to which the feedback loop reduces (1) the effects of disturbances, e.g., fluctuations in the substrate (threonine) concentration or in the demand for the end product, isoleucine, and (2) the effects of changes in the kinetic parameters of the enzymes involved. The necessity for such an analysis follows directly from the earlier discussion of control theory.

The dynamics of the biosynthetic pathway shown in Figure 8 can be described quantitatively for time scales long compared to the enzyme–substrate association time constants. That is, it is assumed in this description that the enzyme–substrate complexes are in the appropriate steady state at a time relative to the enzyme, substrate, and inhibitor concentrations occurring at that time. The rate equations are as follows:

$$\begin{aligned} dP_1/dt &= f_1(E_1,S,P_5) - f_2(E_2,P_1) \\ dP_2/dt &= f_2(E_2,P_1) - f_3(E_3,P_2) \\ dP_3/dt &= f_3(E_3,P_2) - f_4(E_4,P_3) \\ dP_4/dt &= f_4(E_4,P_3) - f_5(E_5,P_4) \\ dP_5/dt &= f_5(E_5,P_4) - D_p \end{aligned} \quad (67)$$

Here, the E_i's, P_i's, and S represent the concentrations of the enzymes, the products, and the initial substrate, respectively; D_p represents the unpredictably fluctuating rate of demand for the end product P_5; and the f_i's represent the dependence of the rate of product formation on the appropriate enzyme, substrate, and

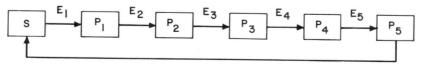

FIGURE 8. Potential feedback control loop in isoleucine biosynthetic pathway.

inhibitor concentrations. For example, the f_i's are the right-hand sides of rate equations of the form given in Eqs. (59), (62), and (63).

The function $f_1(E_1,S,P_5)$ represents the operation of the allosteric enzyme threonine deaminase in the example under discussion. This particular allosteric enzyme from several microorganisms has been purified and studied extensively. The purified enzyme has a number of common characteristics among these microorganisms (Sophianopoulos, 1973; Umbarger, 1978). In the absence of isoleucine[16] (i.e., when the feedback loop is open), it exhibits a hyperbolic substrate saturation curve typical of ordinary Michaelis–Menten kinetics, as given by the right-hand side of Eq. (59), but in the presence of isoleucine it exhibits a sigmoidal saturation curve, interpreted as indicating "cooperativity" between the binding of threonine and isoleucine. The enzyme from *E. coli* exhibits a K_m of 1.6×10^{-3} M at 24°C, and its sigmoidal substrate saturation curve shape appears at a 10^{-6} M isoleucine concentration.[17]

The nonlinearity of the functions f_i precludes a simple, quantitative analytic analysis of the ability of the feedback control loop shown in Figure 8 to reduce the effects of disturbances on the system (e.g., of variations in threonine concentration and variations in the demand for isoleucine) or to reduce the sensitivity of output to variations in the dynamics of the system being controlled (such as variations in the rate characteristics of the various enzymes involved). In such cases one must resort to approximate computer simulations to explore dynamic properties of the feedback control system. However, Savegeau (1976) has developed an analysis procedure for examining such properties *in the steady state* (i.e., at zero frequency), providing the functions f_i can be reasonably approximated by simple power laws,[18] as indicated below.

With the power law assumption (electing the minimum number of interactions), Eqs. (67) assume the following form:

$$
\begin{aligned}
dP_1/dt &= \alpha_1 S^{g10} P_5^{g15} - \beta_1 P_1^{h11} \\
dP_2/dt &= \beta_1 P_1^{h11} - \beta_2^{h22} \\
dP_3/dt &= \beta_2 P_2^{h22} - \beta_3 P_3^{h33} \\
dP_4/dt &= \beta_3 P_3^{h33} - \beta_4 P_4^{h44} \\
dP_5/dt &= \beta_4 P_4^{h44} - \beta_5 P_5^{h55}
\end{aligned}
\qquad (68)
$$

[16] Or in the presence of valine. Valine reverses both the allosteric inhibition by isoleucine and the cooperative binding.

[17] K_m was defined in the discussion of Eq. (59).

[18] The power law formalism incidentally can capture the "chaotic" dynamics of nonlinear mechanics— a matter of current interest in topology. These equations represent extremely abstract accounts of the reaction process, and should be used very cautiously in assessing the dynamic range of the actual reaction system. Over a limited range, the equations are very useful. Detailed modern studies of allosteric enzymes show that the Monod kinetic model, introduced about 20 years ago, does not usually fit the actual case. Instead we see snap-diaphragm kinetics, jump discontinuities, periodic functions, and other kinetic exotica. Nevertheless, as with competitive, uncompetitive, and noncompetitive inhibition, I shall invoke with allosteric inhibition, the classical, oversimplified model of the kinetics. These are quite rich enough to permit me to illustrate what the control viewpoint can add to standard enzyme kinetic analysis, as practiced by biochemists.

19. Control Paradigms and Self-Organization in Living Systems

(Loosely speaking, the exponents are related to the orders of the reactions, and the coefficients α and β are measures or weights for the contributions of the reactions, whether positive or negative.) The power law approximations for the allosteric enzyme are determined from plots similar to the form given in Figure 7D. Here it has been assumed that the demand rate for isoleucine is also given by a power law of the form $\beta_5 P_5^{h55}$. In the steady state, which results from a fixed threonine concentration S, $dP_i/dt = 0$ for all i, and the equations reduce to nonlinear algebraic equations that can, in turn, be reduced to linear algebraic equations by taking logarithms of each term, as shown by Savegeau. Letting $Y_0 = S$, $Y_i = \log P_i$ for $i = 1, \ldots, 5$, $b_1 = \log(\beta_1/\alpha_1)$, and $b_i = \log(\beta_{i-1}/\beta_i)$ for $i = 2, \ldots, 5$, the resulting equations are

$$\begin{aligned}
b_1 &= g_{10} Y_0 - h_{11} Y_1 + g_{15} Y_5 \\
b_2 &= h_{11} Y_1 - h_{22} Y_2 \\
b_3 &= h_{22} Y_2 - h_{33} Y_3 \\
b_4 &= h_{33} Y_3 - h_{44} Y_4 \\
b_5 &= h_{44} Y_4 - h_{55} Y_5
\end{aligned} \qquad (69)$$

and the solution for Y_5 is

$$Y_5 = (g_{10} Y_0 - b_1 - b_2 - b_3 - b_4 - b_5)/(h_{55} - g_{15}) \qquad (70)$$

In the open-loop case, when inhibition is not present, $g_{15} = 0$ and Eq. (70) becomes

$$Y_5 = (g_{10} Y_0 - b_1 - b_2 - b_3 - b_4 - b_5)/h_{55} \qquad (71)$$

Now, from Eqs. (70) and (71), the effectiveness of the feedback loop (in the steady state) can be established as follows.

First, consider the effect of disturbances on the output. Variations in isoleucine concentration due to variations in the substrate (threonine) concentration are given by

$$\frac{\partial Y_5}{\partial Y_0} = \frac{\partial \log P_5}{\partial \log S} = \begin{cases} g_{10}/(h_{55} - g_{15}) & \text{with feedback} \\ g_{10}/h_{55} & \text{open-loop} \end{cases} \qquad (72)$$

Hence, the feedback loop reduces the output variations by the factor $h_{55}/(h_{55} - g_{15})$ or $1/(1 - g_{15}/h_{55}) = 1/(1 + L)$, where $L = g_{15}/h_{55} > 0$ is the (logarithmic) loop gain. This result is consistent with the earlier description of negative feedback from an engineering standpoint.

Similarly, variations in the logarithm of isoleucine concentration, which are due to variations in the dynamics of the plant, and are specified by the logarithms of the ratios of the rate constants, b_i, are given by

$$\frac{\partial Y_5}{\partial b_i} = \begin{cases} 1/(h_{55} - g_{15}) & \text{with feedback} \\ 1/h_{55} & \text{open-loop} \end{cases} \qquad (73)$$

Again, feedback reduces perturbing effects (in the steady state) by the factor $1/(1 + L)$, as expected from the earlier discussions of feedback.

Finally, the percentage variation in the logarithm of isoleucine concentration, due to the percentage variations in the rate of demand for isoleucine, as reflected in changes in h_{55}, is given by

$$(\partial Y_5/Y_5)/(\partial h_{55}/h_{55}) = \begin{cases} -1/(1 + (-g_{15}/h_{55})) & \text{with feedback} \\ -1 & \text{open-loop} \end{cases} \quad (74)$$

Once again the feedback reduces a perturbation effect by the factor $1/(1 + L)$. This example establishes the *primacy* of the loop gain (here $L = g_{15}/h_{55}$) in determining the effectiveness of the feedback control loop. *For a negative feedback control loop to be effective, it is necessary that* $L \gg 1$.

Savegeau has noted that g_{15} for allosteric enzymes typically ranges from -0 to -4 and that $h_{55} = 0.5$ if the enzymatic process creating demands for isoleucine is working at half-maximal rate. In this case, L varies from 1 to 8—a range that clearly establishes the effectiveness of the feedback loop. In biosynthetic pathways of the type under discussion, g_{15} commonly ranges from -2 to -4.

The stability of such biosynthetic feedback loops is very important. The open-loop system representation of isoleucine biosynthesis is stable. This conclusion can be derived from the block diagram of the system shown in Figure 9; each integration has a stabilizing negative feedback term around it and the entire open-loop system is simply a cascade of stabilized integrator subsystems. However, as previously discussed, all real feedback systems eventually become unstable if the loop gain is made sufficiently large. Thus, the obvious question in the current context is the following: Can biosynthetic pathways involving allosteric inhibition by the end product be unstable for the range of loop gains commonly found for such systems? That is, can they be unstable for loop gains ranging from 0 to 9?

The answer depends on the number of intermediate reactions in the pathway and on the rate constants of the intermediate reactions involved. If the rate constants for all the intermediate reactions are identical, then the stability for a given strength of inhibition is determined by the common rate constants for each step

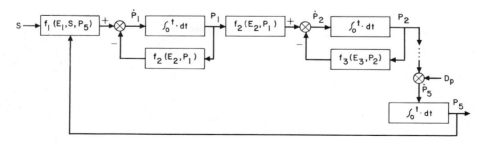

FIGURE 9. Block diagram of feedback control by end product inhibition in the isoleucine biosynthetic pathway.

and by the number of reactions in the pathway. But if the rate constants for some of the reactions are much larger than those of the remaining reactions, stability is determined by those remaining reactions. Because of the poorly developed state of stability analysis for nonlinear systems, it is not possible to provide definitive quantitative conditions for the stability of such systems as a function of loop gain. However, it is possible to examine the stability (or rather the instability) of linearized approximations to such systems around steady-state operating points or to examine the stability of such nonlinear systems by computer simulation. For example, in one such simulation, Savegeau (1976) showed that as the number of intermediate reactions increased from 3 to 10 (with all reactions having identical rate parameters), the value of g_{1n}, and hence, the loop gain, at which the system exhibited borderline stability decreased by a factor of about 5. These simulation results clearly demonstrate the strong dependence of stability on the number of effective reactions in the sequence. Furthermore, they suggest that in the long sequences of reactions, it may be important that some of the intermediate reactions be much faster than others in order to achieve stability at values of loop gain required to make the feedback loops effective. (In this connection, enzyme–enzyme complexes might provide a potential mechanism for achieving higher rates for some of the intermediate reactions.)

As far as I can determine, no one has studied the actual *effectiveness* of feedback control by regulatory enzymes in any of the biosynthetic pathways involved in amino acid biosynthesis. This omission is apparently due to a lack of the required detailed information on *all* of the enzymes involved. From my viewpoint, it would certainly be pleasing to have such a definitive study of at least one of the pathways. If it was found that the loop gain was high enough for effective negative feedback control, and also that the control loop was stable, then, as an engineer, I could understand that the biochemists had indeed characterized the system dynamically. They would understand real traffic patterns. What they have now looks merely like roadways from which they can postulate any dynamics they wish. From an engineering viewpoint, that is too free a game, and not science.

Control Processes at the Genetic Level—Repression of Enzyme Biosynthesis

Perhaps the most significant discovery of molecular biology that pertains directly to feedback control of biosynthetic pathways is the recognition that the end product of such a pathway exerts a negative feedback effect on the genome, expressed through the rate of synthesis of enzymes involved in the pathway. This discovery immediately suggests that isoleucine biosynthesis, for example, may be subject to multiloop feedback control of the kind previously described (Figure 3). Here the inner loop could be closed by the allosteric inhibition of threonine deaminase by isoleucine whereas the outer loop is closed by the negative feedback effect of isoleucine (through the genome) on the rate of synthesis of threonine deaminase (and the other four enzymes in the pathway).

Negative feedback effects of end product on enzyme synthesis are known to exist for many biosynthetic pathways. The most thoroughly understood example is for the synthesis of tryptophan in *E. coli* (Platt, 1980).[19] It is tempting to use the now-classical operon model of Jacob and Monod (1961a), developed for explaining induction and repression of enzyme synthesis in prokaryotic cells, in an analysis of the control characteristics of such processes. The biochemistry is fairly clear in several instances (Platt, 1980; DeRobertis and DeRobertis, 1980), and the Jacob–Monod model has been invoked successfully in a qualitative way (Jacob and Monod, 1961b). The discovery that cyclic AMP is used by *E. coli* as a "symbol" of carbon source limitation (Reznikoff and Abelson, 1980) has revealed some additional pathways for feedback genetic control of enzyme synthesis.

Unfortunately, we do not have enough quantitative data to establish the actual effectiveness (i.e., the loop gain) of genetic feedback loops. I make no attempt at a control theoretic analysis. It is even possible that loop gain is more or less an irrelevant concept for analysis of switching networks of the kind involving genomes. (Loop gain, as I have used it here, is pretty much a continuous-system, near-steady-state concept.)

Regardless of the difficulties in analysis, the rate of reaction at a given substrate concentration is directly and linearly related to enzyme concentration. Thus, end product control of the rate of synthesis of enzymes provides a powerful control over the rate of synthesis of the end product of a biosynthetic pathway.

The time scale for feedback control of enzyme biosynthesis is much longer than that for end product feedback control of enzyme activity by a process such as noncompetitive inhibition. These facts lead to the conjecture that end product inhibition of enzyme activity provides a rapid, fine control of the rate of end product biosynthesis, whereas end product repression of enzyme biosynthesis provides a superposed, slower, and coarser control of the rate of biosynthesis of end product. This conjecture is consistent with standard control theoretic views on the design and capability of multiloop control systems of the type indicated in Figure 3.

Applying Control Theory to Biological Systems

The preceding discussions pertained to classical control theory, but application of optimal control theory to biochemical systems was noticeably absent! No known criterion of optimality has been shown to be generally applicable across biological control systems in the same sense that the principle of least action is known to be operable across much of physics. The desirability of finding such a criterion, if it exists, is obvious. In addition to providing a great unification of

---

[19]The simplest cases occur for unbranched pathways, such as those that synthesize tryptophan and histidine. As might be expected, the details of the mechanisms of such negative feedback effects are currently best understood for these (simplest) cases. On the other hand, branched pathways—such as those that synthesize isoleucine, valine, and leucine (Umbarger, 1978)—are much more complex; they involve common enzymes at branch points of the pathway, and they might require complex cross-coupling and cross-feedback between branches in order to achieve coordinated responses. The previous discussion of isoleucine biosynthesis dealt with only one branch of the pathway.

biological control theory, it would permit one to explore the relationships between biological control systems and physical theory at a much deeper level. But perhaps living systems have so many different suboptimal solutions to the same problems that no general criterion will ever be found.

One of the great difficulties in applying control concepts to biological systems arises because of the fundamental nonlinearities of biological processes (e.g., enzyme kinetics, as discussed above). Further developments in control theory are needed in order to overcome these difficulties. Recent developments in the theory of dynamics of physical systems appear to hold considerable promise in this regard (Abraham and Marsden, 1978). Eventually, an effective control theory may evolve from these mathematical foundations. Biological control processes promise to be a fertile area for the development and application of any such theory, as seen in the use of bilinear control processes (Mohler, 1973) in immunology (Mohler et al., 1980).

Connections with Other Viewpoints

In addition to the control theory, there are many other interesting and useful viewpoints of systems for biological processes. Because these alternative viewpoints appear to have some interesting connections with control theory concepts, it is useful to try to clarify the details of such connections. Such a clarification is attempted below for two alternative viewpoints: Iberall's concepts of homeokinetic physics (Soodak and Iberall), 1978; Iberall and Soodak, Chapter 27, this volume) and Haken's concept of synergetics (Haken, 1977, and Chapter 21, this volume).

In his concept of homeokinetic physics, Iberall perceives all systems in terms of a space–time hierarchy, ranging from the smallest and fastest systems involving subatomic particles and forces at one end, to systems of galaxies at the other (largest and slowest) end (see Chapters 24 and 27). Biological processes and systems occupy several contiguous intermediate levels in this hierarchy. At each level of the hierarchy, identifiable entities (called atomisms) are engaged in persistent, oscillatory motion associated with thermodynamic engine cycles. In this context, the central physiological doctrine of homeostasis as the regulation of the internal variables of the organism is modified to accommodate "homeostasis by dynamic regulators" (the thermodynamic engine cycles) "whose mean operating state" represents "the process of regulation." Beyond their homeokinetic operation at a given hierarchical level, it is further perceived that these atomisms associate through collective forces into a continuumlike ensemble whose essential properties can be usefully described in hydrodynamic terms. Dynamic instability of the continuum occurs following large fluctuations at a sufficiently large spatial scale and sufficiently long time scale to produce superatomisms at the next higher hierarchical level, where the process is repeated. In Iberall's concept of homeokinetic physics, feedback control (or better, dynamic regulation) occurs within and, possibly, between atomisms at a given hierarchical level. Moreover, because the central oscillatory behavior (the thermodynamic engine cycles) involved is inherently nonlinear, he asserts that the relevant feedback control concepts must also centrally involve nonlinear behavior. For this reason he dismisses quantitative *linear* control theory as not being relevant in the homeokinetic context, although he apparently does not dismiss all the fundamental control concepts described earlier, including the variational control theory. Within his viewpoint, regulation occurs within and among entities at a given hierarchical level, which he calls "side–side" physics, whereas control inputs (the "R" variables in my notation here) originate at the next higher level of the hierarchy, which he calls "up–down" physics.

In Haken's viewpoint, which he calls synergetics, emphasis is placed on techniques for distinguishing, within a given system, between variables that can be usefully considered to be control variables and other variables slaved to (i.e., controlled by) the control variables. Application of his techniques leads to a natural ordering of processes into a hierarchy arranged according to increasing time scales. From a control theory viewpoint, it is clear that Haken's synergetics is concerned with defining and characterizing open-loop control processes as described in preceding sections. In Iberall's terms, he is dealing with up–down physics.

This chapter has provided a comprehensive overview of some important aspects of control systems theory and has illustrated their application to, and effectiveness in, a selected set of cellular and subcellular biological processes. Higher-level processes (e.g., endocrine processes and neuromuscular processes) could also have been used to test the relevance of control concepts in understanding the essential processes supporting living systems. Obviously, however, much remains to be done.

References

Abraham, R., and J. Marsden (1978) *Foundations of Mechanics,* 2nd ed. Menlo Park, Calif.
Black, H. S. (1934) Stabilized feedback amplifiers. *Bell Syst. Tech. J.* **12**:1–19.
Bryson, A. E., Jr., and Y. C. Ho (1969) *Applied Optimal Control.* Ginn (Blaisdell), Boston.
Cannon, W. B. (1929) Organization for physiological homeostasis. *Physiol. Rev.* **9**:399–431.
Cannon, W. B. (1939) *The Wisdom of the Body.* Norton, New York.
Chance, B. (1961) Control characteristics of enzyme systems. *Cold Spring Harbor Symp. Quant. Biol.* **26**:289–299.
Changeux, J. P. (1961) The feedback control mechanism of biosynthetic L-threonine deaminase by L-isoleucine. *Cold Spring Harbor Symp. Quant. Biol.* **26**:313–318.
D'Azzo, J. J., and C. H. Houpis (1960) *Control System Analysis and Synthesis.* McGraw-Hill, New York.
DeRobertis, E. D. P., and E. M. F. DeRobertis (1980) *Cell and Molecular Biology,* 7th ed. Saunders, Philadelphia.
Gerhart, J. C., and A. B. Pardee (1962) The enzymology of control by feedback inhibition. *J. Biol. Chem.* **237**:891–896.
Gordon, M. S. (1968) *Animal Function: Principles and Adaptions.* Macmillan Co., New York.
Guyton, A. C. (1971) *Textbook of Medical Physiology.* Saunders, Philadelphia.
Haken, H. (1977) *Synergetics—An Introduction.* Springer-Verlag, Berlin.
Horowitz, I. M. (1963) *Synthesis of Feedback Systems.* Academic Press, New York.
Jacob, F., and J. Monod (1961a) On the regulation of gene activity. *Cold Spring Harbor Symp. Quant. Biol.* **26**:193–211.

Jacob, F., and J. Monod (1961b) Genetic regulatory mechanisms in the synthesis of proteins. *J. Mol. Biol.* **3**:318–356.

Jazwinski, A. (1970) *Stochastic Processes and Filtering.* Academic Press, New York.

Judson, H. F. (1979) *The Eighth Day of Creation.* Simon & Schuster, New York.

Lehninger, A. L. (1975) *Biochemistry,* 2nd ed. Worth, New York.

McShane, E. J. (1939) On multipliers for Lagrange problems. *Am. J. Math.* **61**:809–819.

Minorsky, N. (1922) Directional stability of automatically steered bodies. *J. Am. Soc. Nav. Eng.* **34**(2).

Mohler, R. R. (1973) *Bilinear Control Processes.* Academic Press, New York.

Mohler, R. R., C. Bruni, and A. Gandolfi (1980) A systems approach to immunology. *Proc. IEEE* **68**:964–990.

Monod, J., and F. Jacob (1961) Teleonomic mechanisms in cellular metabolism, growth, and differentiation. *Cold Spring Harbor Symp. Quant. Biol.* **26**:389–401.

Novick, A., and L. Szilard (1954) Experiments with the chemostat on the rates of amino acid synthesis in bacteria. In: *Dynamics of Growth Processes.* Princeton University Press, Princeton, N.J., pp. 21–32.

Nyquist, H. (1932) Regeneration theory. *Bell Syst. Tech. J.* **11**:126–147.

Platt, T. (1980) Regulation of gene expression in the tryptophan operon of *Escherichia coli.* In: *The Operon,* 2nd ed., J. H. Miller and W. S. Reznikoff (eds.). Cold Spring Harbor Laboratory, Cold Spring Harbor, N.Y., pp. 263–302.

Pontryagin, L. S., V. G. Boltyanskii, R. V. Gamkrelidze, and E. F. Mishchenko (1962) *Mathematical Theory of Optimal Processes.* Wiley–Interscience, New York.

Reznikoff, W. S., and J. N. Abelson (1980) The *Lac* promotor. In: *The Operon,* 2nd ed., J. H. Miller and W. S. Reznikoff (eds.). Cold Spring Harbor Laboratory, Cold Spring Harbor, N.Y., pp. 221–243.

Savegeau, M. (1976) *Biochemical Systems Analysis.* Addison–Wesley, Reading, Mass.

Soodak, H., and A. Iberall (1978) Homeokinetics: A physical science for complex systems. *Science* **201**:579.

Sophianopoulos, A. (1973) Differential conductivity. *Methods Enzymol.* **17**:557–590.

Stear, E. B. (1975) Application of control theory to endocrine regulation and control. *Ann. Biomed. Eng.* **3**:439–455.

Truxal, J. G. (1955) *Automatic Feedback Control Synthesis.* McGraw-Hill, New York.

Umbarger, H. E. (1955) Evidence for a negative-feedback mechanism in the biosynthesis of isoleucine. *Science* **123**:848.

Umbarger, H. E. (1961) Endproduct inhibition of the initial enzyme in a biosynthetic sequence as a mechanism of feedback control. In: *Control Mechanisms in Cellular Processes,* D. M. Bonner (ed.). Ronald Press, New York, pp. 67–85.

Umbarger, H. E. (1978) Amino acid biosynthesis and its regulation. *Annu. Rev. Biochem.* **47**:533–606.

Wiener, N. (1948) *Cybernetics,* 2nd ed. Wiley, New York.

Yates, R. A., and A. B. Pardee (1956) Control of pyrimidine biosynthesis in *Escherichia coli* by a feedback mechanism. *J. Biol. Chem.* **221**:757–770.

20

Control Theory and Self-Reproduction

Rajko Tomović

ABSTRACT

Control theory is a branch of technology. A control task addresses an object to be controlled, some set of controls with at least two possible states, and some limitations on both the system to be controlled and the controller. The designer of the controller introduces his intentions through goal functions, or selection principles. Goal functions are usually expressed numerically, whereas selection principles can be simpler and qualitative, such as "on" or "off." In either case, the designer intends to optimize the performance of the system under control.

Notions of optimality are very anthropomorphic. In applying control theory to biological processes, severe difficulties arise. Examples are given of the nonnumerical nature of skeletal muscle control, of self-reproduction, and of gene expression. In each case, the biological process has characteristics foreign to the domain of technological control theory. These characteristics are highlighted, and it is concluded that the images and concepts of technological control theory fail to take note of important features of biocontrol systems. The features that lie outside the range of modern control theory include the use of both positive and negative feedback relations between the information in a memory and the product of the readout operation at the genome; the increased possibility for richness of control that arises from mapping from a symmetrical memory structure (DNA) to an asymmetric catalytic structure (an enzyme); the merging of structure with function in the microscopic domain; and the acquisition of information by interaction with the environment. Furthermore, biocontrol is of the nonnumerical type. When all these features are considered together, one must conclude that modern control theory so far has had little to offer biologists.   —THE EDITOR

Control theory may be used to illuminate the differences between biological control systems and technological control systems designed by humans. Some considerations are the implications of using microscopic systems to achieve control

RAJKO TOMOVIĆ • Department of Computer Science, University of Belgrade, EOX 816 Belgrade, Yugoslavia.

and natural selection as a criterion of successful operation. Certain aspects of self-reproducing systems cannot yet be effectively represented in terms only of physics or nonlinear dynamics. To begin, consider some conventional principles of modern control theory as a branch of applied mathematics or physics.

Control Principles in Technology

In the most general sense, a control task requires the following three elements:

- An object to be controlled—a dynamical system of some kind (i.e., the "plant").
- A set of controls with at least two distinct members. That is, there must be more than one option available to the controller and a basis for choosing among options.
- Constraints, limitations, or bounds on both the plant and the controller. Not everything is possible; a plant can undergo only limited transformations.

The basic control mechanism of engineering is of the input–output type. The carrier of control information is a time-dependent function, or state-dependent vector function, i.e., $f(t)$ or $f(\mathbf{x})$. This type of control implies the concepts of "states" and "state space." The state space approach assumes that the behavior of a dynamical system can be described by a finite set of measurable attributes of the real object.

In a dynamical system, the plant attributes vary with time ("dynamics" means motion and change). Therefore, there is a one-to-one correspondence between states of the system and an ordered set of time instants or elements of some other linearly ordered set. (Even if the system is continuous, it is possible to think of time as a series of discrete instants—a view encouraged by quantum mechanics.)

The basic issue of any control process is selection of a definite control condition from the set of allowed controls. How is this done? In technological control applications, the designer must know the preferred states of the plant and its trajectories. He knows the "better" controls. This value judgment arises from the many different considerations of low cost, high energy efficiency, safety, stability, getting the job done accurately, and so on. The designer therefore establishes an *ordering relation* for the set of allowed controls. Unless this relation can be introduced, there can be no technological control at all. We must next ask, how can the controller recognize the order (total or partial) in a set of options?

Goal Functions, Selection Principles, and Technological Control

The only way to solve a technological control task is to have either a *goal function*, or a *selection principle*, against which the controls can be checked for "optimality." In standard control engineering the goal function is ordinarily, but not always, given in numerically measurable form. Typically, the mathematical form of the goal function for dynamical systems is the functional. (Functionals are functions of functions, i.e., their arguments are not simply an independent variable, but another function, like a curve, surface, or whatever.) In many real situations, such as those that apply to socioeconomic tasks, health care delivery systems, urban management, and so on, the goal function cannot be fully described mathematically.

There are important differences between optimizations based on a goal function and those based on a selection principle. When a goal function is used, whether in the standard numerical form or in nonnumerical form (discussed later), the set of allowed controls can be ordered, at least partially. Control by a selection principle is much simpler to optimize. At the simplest, the principle may be just "on-off" selection. For example, successful job applicants are selected by virtue of their university degrees, and aspirants for military services must meet a minimal height requirement. Between the simple optimization by a selection rule and various optimizations around goal functions, there is a hierarchy of optimization strategies of increasing complexity and difficulty.

When speaking of the relative simplicity of optimization by selection principle, I refer to the recognition mechanism by which the solutions are divided into acceptable and nonacceptable ones. Division of allowed controls into two or more equivalence classes requires a much simpler recognition mechanism than is needed for the metrics inherent in most goal-function optimizations. In fact, optimization by selection principle amounts to pattern matching and creating Boolean expressions (truth statements). The common feature of these selection-rule optimization mechanisms is absence of numerical data processing. As demonstrated later, this approach seems to be the basis of biological control. Nonnumerical control methods have the great merit of being practically *independent of the number of variables* to be controlled. Consequently, they work well for large system control problems—a class to which biological control processes certainly belong.

What Is the Origin of the Criterion of Optimality?

Notions of optimality are very anthropomorphic. In technological control, the engineer determines the criterion, acting as a *deus ex machina*. Certainly the object to be controlled cannot impose the goals by itself; the control goals must be introduced from a source *external* to the plant. At the macro-level, this is done by the designer, but in microscopic biocontrol there is a need to identify some other source for the criterion. Recombinant DNA work represents a startling merger between criterion functions determined by a human agent and the microscopic biocontrol machinery. The following discussion of biological control refers to the spontaneous control mechanisms observed in living systems that arose without human intervention. The aim of much of modern medicine is to achieve correct macrocontrol based upon human purposes, working with the biocontrol systems of patients with disorders of function.

Biocontrol Principles

The control mechanisms used by living systems differ from the usual forms of technological control in many ways. For example, biocontrol cannot be numerical control, because we have no reason to believe that any living organism contains a measuring device that maps physical variables onto the scale of natural numbers, or some such metric scheme. Therefore, we do not expect to find goal-function optimizations in biocontrol. What about selection-rule optimization? "Natural selection" for "survival of the fittest" suggests itself as a possibility. Unfortunately, there are grave logical difficulties with Darwinian ideas when they are cast in these terms, because we must have an *independent* criterion of fitness, other than survival itself, to avoid a tautology. This well-known difficulty, and others, have led to new views about evolution of life (see Chapter 6 by Gould, in Section II, for a full discussion).

Despite the above philosophical issue, we must still look to *plant–environment interactions* as the source of simple classification mechanisms, which serve as if there were a criterion. There have been interesting discussions by Kimura (1961) and Pattee (1979), among others, that indicate how the environment can act as a source of information for an evolving living system. Their arguments are too detailed for discussion here. Suffice it to say that such matters as temperature limits on bond stability, variations in food availability, or other physicochemical factors may in principle act as mechanisms for separating controls into dichotomy classes. Thus, long before the invention of technological control systems and goal-setting by humans, biocontrol operated on a selection-rule optimization principle that depended upon system–environment interactions.

I shall consider three aspects of biocontrol: (1) the nonnumerical nature of skeletal muscle motor control, (2) self-reproducing systems, and (3) gene expression.

Nonnumerical Nature of Skeletal Muscle Control

The motions of human extremities, as well as skeletal activity in general, can be described by differential equations of mechanics. Consequently, the control of human extremities may be interpreted as a specific case of the general multivariable control problem in state space. For rockets, for example, the controller processes system states to derive new control inputs. If this is properly done, the multivariable mechanical system will follow "desired" trajectories. Such an approach can be called trajectory control. But technologically, trajectory control is based on real-time solutions of differential equations, i.e., on processing of numerical data. Numerical trajectory control, although rationalizable on the grounds of mechanics, has important shortcomings from the biological point of view. Because technological trajectory control depends on integration procedures, an increase in the number of controlled variables necessarily increases delay times and the amount of calculation. The notion that complex skeletal activities like biped gait, manipulation, and the requirements of various sports are numerically controlled in real time is unrealistic on this basis alone. Furthermore, technological trajectory control must be related to a fixed external coordinate system and to fixed initial conditions, precisely because it is the outcome of integration of differential equations. But experience shows that biped locomotions and other skeletal activities can be performed independently of the body position relative to a fixed coordinate frame. Standard trajectory control is also highly sensitive to perturbations arising from the parameter changes that are continuously mapped upon solution variations, whereas motor control in living systems can be startlingly insensitive to (some) parameter changes (Partridge and Benton, 1981).

Because of these shortcomings in numerical methods for control of processes with a large number of variables, an approach to the technological control of skeletal activities based on *nonnumerical procedures* has been developed for use with prosthetic and orthotic devices. The basic idea of this method is pattern recognition. The term "pattern recognition" is used here in a much broader sense than is usual. For instance, *interpretations* of sensory information, like "warm," "cold," "hard," "soft," "slippery," "even," "steep," and so on, are essential for manipulation and locomotion. Using derived sensory information of this type, the designer can substitute something much simpler than the numerical procedures required by negative feedback as it is implemented in most technological control. Once a pattern like "hot" has been recognized, it is possible to match this information directly to muscle control, without use of feedback. In this special open-loop case, the sensory information will produce fast disengagement from the hot object.

The principle of matching sensory patterns directly to patterns of skeletal activities has been developed into a general method for the control of multivariable, anthropomorphic robots. It has been applied to the design of anthropomorphic hands and assistive devices for locomotion (Tomović and McGhee, 1966). Pattern recognition by biological sensors has unique features compared to standard machine methods, because it does not depend on feature extraction, metrics, or clustering. It operates holistically by virtue of special structures, rather than by the processing of numerical data. In this way, complete reflex loops can be developed without using calculations that are sensitive to the magnitude of the control task. The price paid for the advantages of speed and holistic responses is that very specific sensory transducers must exist. They need not, however, be highly reliable! Unfortunately, we do not have many suitable man-made transducers, whereas the biological systems are full of appropriate sensors.

Technological Reproduction and Biological Self-Reproduction

Reproduction of parts from plans, templates, or blueprints is a low-level control task in the sense of "control" as discussed above. This statement is true, however, only if self-reproduction processes are excluded from consideration. If self-reproduction is involved, then the very rich control aspects of the process become prominent.

There is a clear distinction between reproduction and self-reproduction. A technological reproduction process requires a description of the features to be reproduced. The description is located in an external memory. Several different kinds of descriptions may be used for the content of external memory: statements in a natural language or in artificial languages, drawings, templates. There is one-way information flow from memory to the product and no direct feedback from product to memory. In other words, reproduction is a *feedforward process* with no optimization involved. Control (product optimization) is accomplished only at a next higher level, where human factors enter the loop. Consequently, one can speak of control and feedback only at a level superior to the memory–product interaction.

Biological self-reproduction is basically different from the above technological reproduction case. Both feedforward and feedback channels are part of self-reproduction processes *at the implementation level*. A higher level is responsible for adaptive behavior (optimization) matching the memory content to the requirements of a selection-rule operated through interaction with the environment. The existence of a feedback interaction between the product and the memory, in addition to the feedforward channel, is essential for normal functioning of self-reproduction. If the feedback loop ceases to function, the process could lead to uncontrolled reproduction of the memory content. This danger is always present in self-reproduction (e.g., cancer?), where the implementation level is normally under the control of feedback. (Such danger is nonexistent in a standard technological reproduction process.) Self-reproduction starts without involvement of human factors at all, as during the evolution of the terrestrial biosphere (until recently).

Biocontrol and Physical Laws

It is very difficult to reduce the self-reproduction process to terms of thermodynamics and nonlinear dynamics. A structural output (product) may either repress or enhance memory expression in self-reproduction (as discussed below). This fact adds new features to the reproduction process, even though the feedback principle itself invokes no special laws of physics. Furthermore, if factors external to the dynamic process (e.g., humans or environment) introduce ordering relations (weighting factors) into the set of available options, the laws of physics certainly remain as they were, yet attaching weighting factors to responses of dynamic systems may lead to new, higher levels of functional behavior. Thus, a purely methodological reduction of biology to physics will fail to encompass the essential differences between biocontrol of self-reproduction and technological control. Self-reproduction must be viewed as a specific control process with unique properties; it cannot be simply described. Many of its aspects are reducible to laws of physics, but it is a historical process, because the memory–product interaction is exposed to the influence of the environment. Consequently, biological phenomena are reducible to physics *plus the informational action of the environment* (Pattee, 1979). A full description of the evolution of biological systems is not possible unless the complete history of the environment is known. The same laws of physics produce alternative forms of self-reproduction processes in different environmental conditions.

A clue to better understanding of self-reproduction may be found in the study of differences between the control at both the macro-level and micro-level (e.g., gene expression). The scale of the physical support of the control seems to be the limiting factor determining the size and the complexity of feasible control tasks. Unfortunately, this aspect of control theory has been completely neglected in the past. The message of most contributions in mathematical control theory is that the limits to system growth are independent of the dimensions of the state space. In other words, with more equations and more computer power the solution of control tasks of any size and complexity may be possible. This conclusion is wrong. To see why, consider control of gene expression.

Gene Expression and Structural (Micro-level) Control

For biological details of gene expression, the reader should consult the splendid volume edited by Goldberger (1979). The control of gene expression is fundamentally different from technological input–output control mechanisms. At the micro-level the carrier of control information is the *structure* itself; the specific control capability corresponds to each macromolecular structure, and structure and function are nearly the same thing at the micro-level. There is no need for state identification at this level. Furthermore, there is no need for numerical data processing. Control is not generated from outside, but from within the dynamical process itself! Each macromolecular structure may support a specific control task. Finally, the size of the controlled system is not limited by the amount of computation or the dimensionality of the state space. It is no wonder that the molecular biologist has had so little need for technological control theory!

Structural control adds a new factor to self-reproduction (Tomović, 1978). For any kind of reproduction there must be two distinct elements: a memory containing the product description and the output product itself. The new feature at the micro-level is the use of an *internal* alphabet in DNA macromolecules for the product description. The output products in this case are proteins and polypeptides. No external memory is used.

The triplet code of purines and pyrimidines maps a three-letter "word" into a single amino acid in the primary sequence of a polypeptide or protein, or protein subunit. In that mapping, both posttranscriptional and posttranslational processing occur. But in addition to these potential sources of additional information (and noise), it is a remarkable fact that many proteins (e.g., enzymes, receptors, ionophores) have a functional richness that far exceeds that of the nucleic acids. The mapping goes from nearly inert macromolecules to rate-controlling macromolecules that participate in dynamic processes. Proteins have a control power not seen in DNA. Thus, symmetric memory units are transformed by one-to-one mappings into asymmetric products having new control potential. Broken symmetry leads to genesis of control power. Such a process can be formally described in the following way: From the structural point of view, gene expression is a one-to-one mapping between two macromolecular chains:

$$X \to Y \qquad (1)$$

where X is nucleic acid DNA and Y is enzyme protein.

From the control point of view, the mapping (1) involves

$$I_x \to I_y + C_y \qquad (2)$$

where I means the information content in parts of the respective macromolecular chains and C stands for the control capability of Y measured in terms of active sites. There has been an increase in control potential, but C_y is not explicitly represented in I_x. I_x specifies only I_y. C_y arises from the physicochemical field surrounding Y. Processes with increasing control potential certainly must enhance chances for self-reproduction. Furthermore, the product may act catalytically in the copying process that makes it.

Conclusions

The images and concepts of technological control theory fail to take note of the features of biocontrol systems that permit such remarkable functions as gene expression, motor control, and self-replication. The features of biocontrol that stand out are (1) the use of both positive and negative feedback relations between the information in a memory and the product of the readout operation, (2) the increased control potential that arises from mapping from a symmetrical memory structure to an asymmetrical catalytic structure, (3) the merging of structure with function in the microscopic domain, (4) the acquisition of information by interaction with the environment, and (5) the use of nonnumerical, nonmetric control optimizations. When all these features are considered together, one must conclude that modern control theory has little to offer biologists, except at its most abstract level.

It was at the abstract level that von Neumann began the analysis of the logical requirements for a self-reproducing automaton (see, e.g., the discussions by Myhill, 1964, or by Laing, 1979). But real systems confront physical realities as well as logical truths, and the general-purpose Turing machine cannot solve all computable problems in reasonable time. The abstractions of the logicians offer no shortcut to the problem of control in a real world with its vicissitudes. Control as a branch of mathematics has been of less help in understanding biocontrol than has engineering practice, such as the process control experience of the chemical engineer. But even there, the technologist has much to learn from biology. In fact, in antibiotic production, fermentation of beers and wines, and so on, the engineer uses biocontrolled elements in his process chain. So does the new recombinant DNA engineer. We appear to be a long way from imitating biocontrol in the technological world, but we are learning how to use it.

References

Goldberger, R. F. (ed.) (1979) *Biological Regulation and Development*, Vol. 1. Plenum Press, New York.

Kimura, M. (1961) Natural selection as the process of accumulating genetic information in adaptive evolution. *Gen. Res.* **2**:127–140.

Laing, R. (1979) Machines as organisms: An exploration of the relevance of recent results. *BioSystems* **11**:201–215.

Myhill, J. (1964) The abstract theory of self-replication. In: *Views on General Systems Theory*, M. D. Mesarovic (ed.). Wiley, New York, pp. 206–218.

Partridge, L. D., and L. A. Benton (1981) Muscle, the motor. In: *Handbook of Physiology—The Nervous System II*, V. B. Brooks (ed.), pp. 43–106.

Pattee, H. H. (1979) The complementarity principle and the origin of macromolecular information. *BioSystems* **11**:217–226.

Tomović, R. (1978) Some control conditions for self-organization: What the control theorist can learn from biology. *Am. J. Physiol.* **235**:205–209.

Tomović, R., and R. McGhee (1966) A finite state approach to the synthesis of bioengineering control systems. *IEEE Trans. Hum. Factors Electron.* **7**(2): pp. 65–69.

VII

Physics of Self-Organization

F. Eugene Yates

Physics dominates the sciences through the generality, parsimony, and implied causality of its explanations. With only a few general forces (not more than three; perhaps only one), a few principles (e.g., conservation, invariance, symmetry—these three even being interrelated), a few "natural" constants (e.g., C, h, k, G), the logical equivalences of mathematics, and a small number of general theories and concepts, the physicist attempts to explain all the beings and becomings of the universe. Yet there is a peculiarity: in practice, physicists seem to do better in illuminating processes very much faster (e.g., at Planck time, 10^{-43} sec; propagation of light at 3×10^{10} cm/sec) or very much slower (e.g., the unfolding of the cosmos for perhaps 2×10^{10} years) than those at the scale of individual man (10^{-10} sec to 10^2 years). They theorize better concerning spatial or mass domains very much smaller (quarks) or very much larger (supragalactic clusters) than man, or temperatures very much colder [a few tenths of a degree (K)] or very much hotter (10^{15} GeV; 10^{31}°K) than those of the human condition (near 300°K).

The frontier work of physicists prospers at the above extremes, but why do we not have a more fully developed physics of life or man? The difficulty is not that ordinary dynamics are poorly defined at the space-time scales of humans. On the contrary—the detailed processes that we find in man pose no particular mystery after we uncover them. Of course, there are currently interesting physical investigations at terrestrial scales (e.g., acoustical properties of the nonlinear media of earth, atmosphere, and oceans; electromagnetic properties of amorphous substances), but the problem seems to be that a human operates as a system whose degrees of freedom and interactions are mostly internal and whose responses are dominated by action (energy × time) rather than by momentum (mass × velocity). As a result, the human system is slow and "soft" in the sense of being characterized by numerous transports and transformations only loosely

---

F. EUGENE YATES • Crump Institute for Medical Engineering, University of California, Los Angeles, California 90024.

coupled, with greatly delayed internal processing. The transports are predominantly diffusive and convective; only a few are propagative. The transformations are weak exchanges of chemical bonds of no more than a few electron volts per exchange. Given the soft couplings, it becomes difficult to ascribe sharply defined macroscopic output behaviors to any particular set of macroscopic inputs, present or past. Stimulus and response are joined by numerous, delayed internal couplings. To the observer accustomed to the classical physical idea of close connections by forces between causal inputs and their provoked outputs, these nonclassical dynamic properties raise the larger question: What is the status of the physics of complex, supraatomic systems?

The chapters in this section examine the physics and mathematics of complex processes such as morphogenesis. Haken extends the "slaving principle" and the concept of "order parameters" with particular reference to reaction-diffusion systems. He builds from the work of Landau and Turing and goes beyond, using the methods of "synergetics." He chooses the laser as an example of coherence arising from chaotic states, of ordered structures being bootstrapped out of incoherence and maintained by energy and matter fluxes. He notes, though, that biological systems become organized differently from the light in the laser, because biological structures can persist (at least briefly) when the external power (food) is turned off, whereas laser light cannot. Haken recapitulates and examines some of the ideas expressed by other chapters in Sections VII and VIII. He does not, however, treat the peculiar stability characteristics of life in detail.

Landauer goes immediately to the matter of relative stability of complex systems, particularly those capable of information handling. There are only a few ways in which information storage can be made persistent, one of which involves a continuing expenditure of energy. Noise makes a fundamentally important contribution to the selection of stable states (as well as to their destruction). Landauer asks if the effects of variable noise are merely incidental to the evolution of the terrestrial biosphere or if they are essential. Although he leaves the matter open, he strenuously objects to grandiose misuse of the term "self-organization" to describe relatively primitive, physically constrained behaviors, such as the formation of Bénard cells in heated liquids (simple second-order transitions). Biological systems are especially remarkable for their capability of self-repair and self-maintenance to restore local stability. Reproduction accomplishes that with a relatively simple tactic. But what does life require of an environment in order to sustain its variety? Could life be complex in a physically simple or uniform environment? This haunting question is made to seem physical and, in Landauer's skillful treatment, demands a physical answer.

Is Life "Far from Equilibrium"?

"Equilibrium" is a term that refers to processes and ensembles. It can be a macroscopic (local or global) or a microscopic condition, but need not be all simultaneously. We have mechanical, thermal, electrical, chemical, and other equilibria as separate aspects that require consideration. With respect to any process, local equilibrium requires detailed balance and reversibility: we go from microscopic state A to B just as often as we go from B to A, and we do not lose or gain free energy or entropy in the exchange. Nevertheless, a system can be at equilibrium locally (macroscopically), but be far from equilibrium microscopically. The wide range of velocities of molecules in the macroscopic equilibrium state of an ideal gas, from zero to supersonic, is an example; it is the overall *distribution* of velocities that is unchanging at macroscopic equilibrium. Brownian motion provides another example of microscopic disequilibrium in a system that can be at macroscopic equilibrium—see Lavenda (1985).

"Close-to-equilibrium" usually is taken to mean that, although detailed balances are not perfect, and although entropy production is not zero, and although free energy changes are occurring, these deviations are so small in a local region that they can be described by a first-order linear deviator theory for change of local ensemble states. That is, transport coefficients (of viscosity or, more generally, of all the diffusivities) aptly describe the motional changes. Onsager relations effectively hold and specify what different processes may be coupled to each other. Above all, the local state is specifiable through the parameters used in equilibrium thermodynamics (thermostatics). Any relaxation process toward equilibrium smoothly approaches the small-amplitude (linear) deviations governed by coefficients that are independent of the state of the system.

It is not always appreciated that the fluctuations and couplings and energy transformations that may hold an open system off of, though close to, equilibrium depend upon *nonlinear* mechanisms. In close-to-equilibrium, nonlinear thermodynamic cyclic transformations, a "squirt" followed by a relaxation typically makes up the cycle overall. Each cycle has two components—a nonlinear "kick" followed by an approximately linear recovery. The human heartbeat is the archetype.

The assumption of being close to equilibrium locally does *not* mean that it is valid to integrate over the whole macroscopic field to determine the entropy density found by assuming local equilibrium. Macroscopically the thermostatic variables *do change* from region to region. The gradients or flows that occur are described by time-independent transport coefficients that provide measures of the *irreversible* component of entropy change. The macroscopic field processes are dissipative and time irreversible. Some believe that dissipations may arise microscopically as well (Prigogine and Stengers, 1984), but that controversial issue is rejected by most physicists (e.g., Pagels, 1985), who commonly work with models of fully conservative fields and microscopically reversible time.

Far from Equilibrium

Nonlinear phenomena, in chemical systems especially, have been extensively studied from a mathematical viewpoint by Prigogine and his colleagues (Glansdorff and Prigogine, 1971; Nicolis and Prigogine, 1977). They have emphasized the importance of criticality conditions and of fluctuations (Horsthemke and Lefevér, 1977, 1980; Nicolis and Turner, 1979; Kondepudi and Prigogine, 1981).

They treat the problem of dynamic transitions in certain nonlinear systems, extending from rather simple behavior to complex form and function. In their models the driving sources behind the transitions arise either from deterministically increasing the values of some control parameter set or from capturing noisy fluctuations at the particular control parameter values that specify a bifurcation from one dynamical stability regime to another.

These presentations by Prigogine's group rest largely on bifurcation theory as a branch of nonlinear mathematics. They not only emphasize the role of internal fluctuations in morphogenesis, but also present a complementary account of the role of environmental fluctuations. In both cases, they address nonequilibrium transitions in what they call "far-from-equilibrium" systems (Nicolis and Prigogine, 1981). What they seem to mean by that phrase is that the equations or solutions describing the dynamics are in their nonlinear rather than linear ranges; i.e., they describe nonlinear conditions in which dynamics other than damped, decaying, linear relaxational phenomena may appear. But a purely mathematical account of conditions for qualitative changes in system dynamics may lead to confusion, because the phrase "far-from-equilibrium" has different mathematical and physical interpretations that may not always agree. In a common mathematical interpretation, the phrase indicates that some parameter (or set of parameters) to which system dynamics are sensitive has been extended and bifurcations or qualitative changes in dynamic modes occur as the parameter value is increased. (Subsequent decrease may lead to a reverse path on the same trajectory, or there may be hysteresis or other more elaborate behavior.) The physical terms "thermodynamic" and "nonthermodynamic" are applied by Prigogine (inappropriately, I believe) to various mathematical solutions of model equations at various parameter values, according to whether or not damped (linear), decaying behavior is observed. But, as I shall try to argue, the domain of applicability of near-equilibrium thermodynamics in physics can include the dynamics referred to by Prigogine as "nonthermodynamic" in domains "far-from-equilibrium" in his mathematical terminology.

Microscopically, living systems do have some regions, near boundaries or in ionophores, that are, indeed, far from chemical equilibrium, but I suspect that the relevance of Prigogine's work to biology is limited by the fact that life physically operates locally near macroscopic equilibrium in almost every region and under conditions that are near global equilibrium as well, with "low-duty" cycles, as explained below. Life is perhaps better imagined as developing through a series of first-order transitions, thermodynamically close to equilibrium but nonlinear, rather than nonthermodynamically through second-order transitions, far from equilibrium.

The Domain of Thermodynamics

A process is physically "thermodynamic" if the answer to each of the following questions is "yes." Does a continuum description that assumes both local near-equilibrium and a mean disequilibrium *field* theory actually apply? Are

VII. Physics of Self-Organization

transport coefficients of irreversible thermodynamics defined locally? If so, can they be treated as functions of thermodynamic potentials only, and not of time? For example, is viscosity a function of temperature and concentration only? Do the internal conversions of energy operate without "knock," as a gasoline engine does in its range of normal sustained operation?

In the domain of irreversible thermodynamics, spatial gradients and temporal process rates are not extreme, i.e., they are less than those of the molecular fluctuations. *The important point is that overall thermodynamic near-equilibrium holds even if a field is microscopically far from a chemical equilibrium!* As long as a chemical reaction can be accurately described by reaction rate coefficients, even if there are a number of steps in the reaction, the thermodynamic description will hold. The region of breakdown of the thermodynamic description of a field is at shocks, which include chemical reaction shocks when affinity (A/RT) is nearly "infinite" or at least extremely high. In this sense living systems are piecewise continuous fields. They operate between surface regions where there might be some equivalent of chemical shocks or "knocks." (Inside an internal combustion engine that is not knocking, near-equilibrium thermodynamics adequately describe the combustion process. A knock is an explosive shock wave that indicates a departure from the domain of irreversible near-equilibrium thermodynamics. That is the physical far-from-equilibrium condition.)

So, we can rephrase our question to ask: Do biological engines knock? The test is again: Are the transport coefficients for reactions and diffusions, and the parameters of wave propagations, independent of time, and dependent only on thermostatic potentials? If so, then biological systems are, in fact, close to macroscopic equilibrium in their operations, and for them a continuum, irreversible thermodynamic description will apply (except perhaps for their "information" content and for very limited spatial domains, as within membranes and ionophores. Studies of membrane regions must ultimately yield a detailed kinetics of transport processes, to join the thermodynamic phenomena on either side of the discontinuity).

In addition to questions about the "far-from-equilibrium" requirement of Prigogine's dissipative structure theory (with regard to its applicability to biological phenomena), there is another difficulty. Anderson and Stein examine two aspects of symmetry-breaking—one leading to equilibrium phase change and the other to conditions far from equilibrium leading to dissipative structures. They raise the possibility that symmetry-breaking effects in systems far from equilibrium, such as the appearance of vortices, may *not* produce stable dissipative structures after all. In that case, speculations about the origin or permanence of life, cast in the form of interpretations of far-from-equilibrium dissipative structure theory, are limited by the lack of a well-developed theory for such structures.

The concluding chapter by Soodak and Iberall proposes that instead of seeking a general physical theory of emerging, self-organizing systems, we should look to extended thermodynamics of the fluid and solid states for strategic natural principles. They suggest that force systems are capable of generating only a limited number of types of condensed phases. The condensations always involve broken symmetries. All forms of life, including our social systems, fall under this view, which emphasizes the creative, constructive aspects of flow processes, and chemical bonding. To understand the physics especially relevant to life it is necessary to examine a combination of hydrodynamics and chemical reactivity that, within the domain of near-to-equilibrium thermodynamics, leads to condensed phases of matter that have rich, not trivial, behaviors.

Physics captures the strategy of nature as well as the more familiar tactics (which are given by specific laws, initial conditions, and boundary conditions), but has not yet yielded its full content of strategic concepts. The physics of life may not be comprehensible until that is done. The demand for a truly physical biology, rather than for a limited biophysics, may power some of the further development of physics itself. The constructs of irreversible, close-to-equilibrium thermodynamics appear to me to be up to the task, but this view is not endorsed by everyone, and especially not by Prigogine and his co-workers (Prigogine and Stengers, 1984).

Problems of self-organization by living systems inject a useful and dramatic tension into both mathematics and physics, and we are indebted to Turing, Prigogine, Nicolis, Lefevér, Anderson, Haken, Rössler, Ruelle, Takens, Smale, Feigenbaum, Horsthemke, and many others, for their recent illumination of nonequilibrium phenomena and the primacy of fluctuations in morphogenesis and in transitions.

Today we have a more interesting picture of the physical status of biological systems than we have ever drawn before. I believe that there are very strong reasons for believing that terrestrial life has the following physical characteristics:

1. It obeys all the known laws of physics, with their universal constants (both dimensional and dimensionless) and conservations, that are applicable to its scales of time and space. No new fundamental postulates are required to explain the dynamics of life.
2. It operates lawfully in the domain of applicability of near-to-equilibrium, irreversible thermodynamics, so that atomistic interactions and constraints below imply a limited set of possibilities for "governors" and "bureaucracies" above, but each level appears to have "independent" rules, because of the emergence of new structures as (nonholonomic) constraints, at each field level.
3. It has historical, evolutionary (time irreversible), and informational features that are not aspects of standard physics. We must either accept these features as being in some sense extraphysical, or else, as argued here, we must extend normal physics to rationalize these attributes of the biosphere (Blumenfeld, 1981, p. 18).
4. Its communication systems arise out of broad-band noise, being only barely coherent by conventional physical measures (Iberall, 1978, and Chapter 27, this volume). But their $1/f$ noise properties merely represent the effective carrier processes; the systems have tunable, narrow band filters that extract coherent signals from $1/f$ noise.
5. Its dynamics are nonlinear in the couplings (the cubic, bistable characteristic of the snap diaphragm may be the archetypical nonlinearity). They

VII. Physics of Self-Organization

are close to equilibrium (i.e., thermodynamic) in the relaxations (e.g., the macroscopic motions are mostly trajectories in slow manifolds).

6. It does not possess any unusual "antientropic" tendencies. [It is entertaining but gratuitous to claim life feeds on negentropy for its survival. Life has no more order (by entropic measures) than does a rock (Blumenfeld, 1981, pp. 8–14). It is the quality, the specialness, of the order, not its quantity, that is remarkable.]

7. Its genetic code is arbitrary—a frozen accident. The nucleotide sequences of prebiological macromolecules were very likely chemically arbitrary or only slightly biased by chemical constraints (see Orgel, Chapter 4, this volume). However, after a first two-stranded molecule arose by a random process of low probability, such that on one strand matrix synthesis had time to be completed before the spontaneous polymer hydrolysis, then

> the situation changed abruptly. The nucleotide sequence existing in such long-living two-strand polymers is now meaningful. This meaning is quite obvious: in a stable two-strand molecule capable of reduplication this sequence exists, but other, in principle allowed sequences, do not. Due to specific properties of the two-strand structure (diminished hydrolytic rate, matrix synthesis) the concentration of unistrand polymer molecules, having this, now specific, sequence, will increase. These unistrand polynucleotides are now in dynamic equilibrium not only with monomers, but also with stable two-strand structures having the same nucleotide sequence. The two-strand polymer concentration will increase on account of monomers due to the reduplication process.... Thus, owing to *the memorizing of the random choice* ... a meaningful ordering has arisen, a system capable of creating meaningful information has been made.... The physical properties of polymer molecules having this sequence may not differ from those of other polymers. This sequence has a meaning just because a self-reproducing polymer system with this sequence has accidentally arisen. The meaning of ordering is, thus, a biological category, i.e., it is determined by all the history of the formation of a system, by its evolution. It does not mean that the concept of 'meaning' is somewhat mystical and does not obey the laws of physics. All properties of meaningful objects can be logically derived from their physical characteristics, provided we know their history, their evolution.... For biological systems with a meaningful ordering it is the quality and not the quantity of information that is most important. [Blumenfeld, 1981, pp. 16–18]

8. Its systems are dissipative structures (of the first kind) in the sense that they are maintained off equilibrium by dissipation of energy which enters from the outside. However, they are stabilized mainly by kinetic means: transition to the equilibrium state which would involve destruction of the already existing structure would require that a potential barrier be overcome. Living systems are not dissipative structures of the second kind—those whose deviations from equilibrium are extremely great. Such structures may not be stable (see Anderson and Stein, Chapter 23, this volume).

If these views are correct, and life does indeed lie not far from the reach of standard physical models, then we can be optimistic about the possibilities for small generalizations of existing physics as a basis for comprehending life in its extraordinary variety. I believe that the generalization needed is the joining of nonlinear mechanics to irreversible thermodynamics in the near-equilibrium domain. Here Onsager relations and linear deviator terms in stress tensors apply for the relaxations—but not for the nonlinear, soft escapements tapping the dissipations, that recurrently bootstrap the system off equilibrium and give it stable life.

We now need to decide what aspects of standard physics to keep (surely symmetries, conservations, invariances) and what to discard (perhaps the classical—mechanical initial and boundary conditions). We need to go from models of smooth, simple coordinate systems to more elaborate pictures of intrinsic geometries, in order to preserve in our abstractions the invariances over transformations that are observed in morphogenetic processes within individuals and across species. We need new renormalizations to keep the physics of life simple. It is not yet necessary to abandon the Assumption of Simplicity in Nature—even if the mathematics of our models is not simple.

References

Blumenfeld, L. A. (1981) *Problems of Biological Physics*. Springer-Verlag, Berlin.
Glansdorff, P., and I. Prigogine (1971) *Thermodynamics of Structure, Stability and Fluctuations*. Wiley, New York.
Horsthemke, W., and R. Lefevér (1977) Phase transition induced by external noise. *Phys. Lett.* **64A**:19.
Horsthemke, W., and R. Lefevér (1980) A perturbation expansion for external wide band Markovian noise: Application to transitions induced by Ornstein-Uhlenbeck noise. *Z. Phys. B*. **40**:241–247.
Iberall, A. (1978) A field and circuit thermodynamics for integrative physiology: Power and communicational spectroscopy in biology. *Am. J. Physiol.* **3**: R3–R19.
Kondepudi, D. K., and I. Prigogine (1981) Sensitivity of nonequilibrium systems. *Physica* A **107**:1–24.
Lavenda, B. H. (1985) Brownian motion. *Sci. Am.* **252**:70–85.
Nicolis, G., and I. Prigogine (1977) *Self-Organization in NonEquilibrium Systems*. Wiley, New York.
Nicolis, G., and I. Prigogine (1981) Symmetry breaking and pattern selection in far-from-equilibrium systems. *Proc. Natl. Acad. Sci. USA* **78**:659.
Nicolis, G., and J. W. Turner (1979) Effect of fluctuations on bifurcation phenomena. *Ann. N. Y. Acad. Sci.* **316**:251.
Pagels, H. R. (1985) Is the irreversibility we see a fundamental property of nature? [Review of *Order Out of Chaos*, I. Prigogue and I. Stengers, Bantam Books, New York, 1984]. *Physics Today*, Jan., pp. 97–99.
Prigogine, I., and I. Stengers (1984) *Order out of Chaos*. Bantam Books, New York.

21

Synergetics
An Approach to Self-Organization

Hermann Haken

ABSTRACT

Synergetics is offered as a physical construct for understanding self-organization. It is a mathematical–physical way of studying how collections of subsystems (such as atoms, cells, animals) can produce structures and patterns by self-organization. The construct is applicable to all kinds of matter. The great generality of thermodynamics pertains to thermal equilibrium or, in irreversible thermodynamics, to systems driven only slightly from that equilibrium. When systems are driven very far from equilibrium, however, new things can happen, and the classic concepts of thermodynamics are no longer adequate. Ordered structures can arise out of formerly chaotic states and can be maintained by energy and matter fluxes passing through the system. The laser is used as an example to exhibit the general concepts and principles of these new "instability" states that can possess ever-increasing order.

The concept of "slaving" is fundamental in synergetics. Long-lasting quantities may enslave short-lasting quantities. Symmetry-breaking or selection of bistable state alternatives also may be important in nonequilibrium systems. Because of the slaving principle, there can be an enormous reduction in the number of degrees of freedom of complex systems. Order parameters emerge as representative of such higher-ordered enslaving principles. At critical instability points, nonequilibrium systems "test" various configurations or collective motions by fluctuations. These configured dynamic "modes" can act as order parameters and enslave all other modes of the system. This structuring of order parameters can occur within an entire hierarchy of instability points.

There are two possible outcomes of such phenomena: completely chaotic motion that can decay to a structureless state when power is turned off; or, typical of biological systems, structure that is maintained even when metabolic processes stop. Function can be latent in form.

A mathematical schema is provided and applied to a "morphogenetic" example.
—THE EDITOR

Synergetics is a new field of interdisciplinary research (Haken, 1970a; Haken and Graham, 1971) that studies how systems composed of many subsystems can pro-

HERMANN HAKEN • Institute for Theoretical Physics and Synergetics, University of Stuttgart, 7000 Stuttgart 80, West Germany.

duce macroscopic spatial, temporal, or functional structures in a self-organized way. The subsystems may be atoms, molecules, cells, animals, humans, or computers. Despite the diversity of these elements, striking analogies in the macroscopic behavior of systems become apparent when these systems dramatically change their macroscopic order or structure (Haken, 1977). Many underlying principles of synergetics were first developed in the realm of physics (Haken and Sauermann, 1963; Haken, 1964; Graham and Haken, 1968, 1970); I shall use these as a starting point in this chapter, which contains an outline of some ideas of traditional physics, including thermodynamics.

A Glance at Thermodynamics

There is a consensus today that the fundamental laws of physics apply not only to physics and chemistry but also to biology. These laws are particularly well formulated within such fields as mechanics, electrodynamics, quantum mechanics, and statistical physics. Other branches of physics seem to offer general laws, but applications to biology may not be possible. At first sight, thermodynamics is attractive as a basis for physical biology, because its laws apply to all kinds of matter and it requires only a few, general concepts, such as temperature, heat, pressure, and entropy. These concepts refer to macroscopic quantities. Boltzmann, however, showed that these macroscopic concepts can be derived from a microscopic level description by means of statistical mechanics to deal with the motions or behaviors of ensembles of individual atoms or molecules, and how their behavior leads to the properties of macroscopic bodies composed of atoms or molecules. The application of statistical physics to thermodynamics may reveal important analogies to biology. However, these analogies must be taken with a grain of salt, because thermodynamics is not sufficient to treat many important features of biological systems.

Statistical physics deals with ensembles of large numbers of molecules. It would make no sense to list all the positions and velocities of individual molecules or to follow the paths of all individual molecules. Such an effort would reveal nothing about the macroscopic state, which requires taking certain averages based on statistical assumptions. The macroscopic state of an ensemble is described by concepts (e.g., entropy, temperature) that are alien to individual atoms or molecules. A *single* molecule does not have a temperature or an entropy. This fact sheds light on one aspect of reductionism: although it is possible to deduce macroscopic laws from microworld phenomena, we recognize the necessity of simultaneously introducing *new concepts referring to new qualities* at the higher hierarchical level, that of the macroworld.

The generality of classical thermodynamics requires that systems be in thermal equilibrium. But systems can be driven away from thermal equilibrium by pumping energy and matter into (or through) them. If systems are driven only slightly from thermal equilibrium, the laws of irreversible thermodynamics apply; for these conditions, concepts such as entropy production are central. However, when systems are driven very far from thermal equilibrium, entirely new things

21. Synergetics

can happen, and the classical concepts of thermodynamics, such as entropy, are no longer adequate to describe the newly arising phenomena. In some far-from-equilibrium situations, ordered structures can arise out of formerly chaotic states; these new structures are maintained by fluxes of energy and matter passing through the systems. (See Section VII for more details.) Here we deal with "partially structured" systems—those that are partially disordered, but also partially ordered. Being driven away from thermal equilibrium, a system may pass through a number of "instabilities" in which its macroscopic state changes dramatically, and progressively more ordered states appear. One such system, the laser, provides a number of general principles.

The Laser as a Synergetic System

The laser is more than merely a new kind of lamp. Some of its features directly lead to the question of self-organization. In addition, the predictions of laser theory were experimentally verified in detail, so that the theory and the general concepts (e.g., slaving principle) on which it is based are truly operational.

A laser consists of a rod of laser-active material, i.e., a material that contains atoms that can emit light. Two mirrors at the end-faces of the rod generate a preferential direction in which laser light is eventually emitted. The laser atoms are excited from the outside, perhaps by another lamp. The most interesting feature of the laser is the following: when laser atoms are excited only weakly, each emits an individual light wave with random phase, as in a common lamp. For instance, in a gas-discharge lamp the light field looks very much like "spaghetti." However, as the atoms are increasingly excited, *very suddenly a totally different kind of light is emitted*: a practically infinitely long sinusoidal wave (Figure 1). The structural change of the emitted light can be interpreted as follows. In the ordinary lamp, electrons of individual atoms make their optical transitions independently of each other, but in the laser the electrons make their optical transitions cooperatively. To relate this process to self-organization, consider a simple model: Several people stand at a channel filled with water (Figure 2). Each one has a bar that can be pushed at will into the water. The action of these people represents the behavior of individual atoms, whereas the water represents the behavior of the light field. A situation corresponding to that of the ordinary lamp

FIGURE 1. The transition from a disordered to an ordered state in the laser. The electric field strength of the emitted light is plotted versus time. The left part of the figure shows a light field consisting of individual wave tracks caused by uncorrelated emission of atoms. The right part shows the completely sinusoidal field emitted by a laser; the emission of light by atoms is now completely correlated.

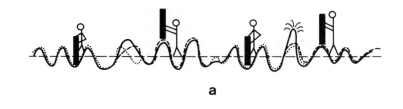

a

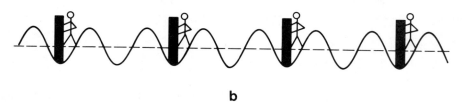

b

FIGURE 2. An anthropomorphic interpretation of the laser process. In both cases people are standing at a channel filled with water. The ones pushing bars into the water symbolize the action of the light-emitting atoms; the water symbolizes the state of the light field.

(a) The people push their bars into the water independently of each other, creating an entirely irregular motion of the water surface. This action corresponds to that of light emission from an ordinary lamp. (b) The people push their bars into the water in a fully coordinated, regular way. This action symbolizes the behavior of laser light emission.

results when the people push their bars into the water independently of each other: an entirely irregular motion of the water surface is produced. However, if they instead push their bars into the water cooperatively and entirely regularly, i.e., in a well-ordered way, their action is analogous to the activity of laser atoms. Their cooperative activity can be easily brought about if an external boss gives orders to them. But in the laser, there is nobody to give orders. Thus, the regular, cooperative behavior of the laser atoms is an act of *self-organization*.

To illuminate the mechanism responsible for the self-organization of the laser, I present the laser equations in a simplified version to exhibit the essentials. The laser field is described by the light-field amplitude $E(t)$. In laser theory this amplitude obeys Eq. (1) (see, e.g., Haken, 1970b):

$$\dot{E}(t) = -\kappa E(t) - \sum_\mu p_\mu(t) + F \qquad (1)$$

The first term on the right side of Eq. (1) describes the decay of the field due to energy losses, for example, at the mirrors. The second term describes the generation of the field by oscillating dipole moments, which represent the electrons of the atoms, where p_μ is the dipole moment of the atom with label μ. F is a force describing random fluctuations exerted by the mirrors on the light field. In turn, the dipole moments p_μ also change during the laser process, as described by

$$\dot{p}_\mu(t) = -\gamma p_\mu(t) + E(t)D_\mu(t) + F_\mu \qquad (2)$$

This decay of a dipole moment because of interactions with its surround is described by the first term on the right side, $-\gamma p_\mu(t)$, in which γ is the decay constant. The second term on the right describes how the dipole moment is driven by the electric field strength E. D_μ represents the difference between the occupation numbers of the upper and lower levels of the atom μ. F_μ is any random force acting on the atom μ.

Let us assume that the decay constant γ in Eq. (2) is much larger than κ in Eq. (1). That is, the field strength is long-lived, but dipole moments are short-lived. According to Eqs. (1) and (2), we can therefore say that E decays slowly and p_μ decays quickly. The fast decay of p_μ means that the system quickly relaxes to being driven by the second term, i.e., to being driven by E. The equilibration is fast enough that p_μ is proportional to the instantaneous value of E: $p_\mu(t) \propto E(t)$. We describe this situation by saying that the dipole moment is "slaved" by the field strength. The concept of slaving plays a fundamental role in synergetics. Long-lived quantities may enslave short-lived (or quickly adapting) quantities. In this context, "slaving" is a technical term and does not imply any ethical position, even when it is applied to disciplines, e.g., sociology, other than physics.

This argument goes under many names and has been given many distinct mathematical expressions. Sometimes it is called "adiabatic elimination of fast-relaxing variables," a generalization of the procedure whereby a system containing some variables that equilibrate much faster than others may be described by considering only the more slowly relaxing ones, because the slow-relaxing variables allow the fast-relaxing ones to return to equilibrium. Hence, the system is a moving equilibrium in the fast-relaxing variables, and those variables can be eliminated from the dynamics.

The mathematical treatment of this principle has been varied. In some cases it is equivalent to a "slow manifold/fast foliation" decomposition, or a "two-time" procedure. Ralph Abraham suggests that the most general expression of this principle is the theory of invariant manifolds, e.g., the stable manifold master theorem of Duistermaat (based on the methods of Perron). (See Abraham and Marsden, 1978, p. 529.) This theorem implies that an analogous decomposition can be made in any dynamical system with a *spectral gap*, i.e., a splitting of its tangent bundle into two components in such a way that all the flow in one component is faster than a given exponential flow, and all the flow in the other is slower than some other exponential flow.

The above argument, which has been proved very useful in the development of laser theory (see, e.g., Haken and Sauermann, 1963; Haken, 1970b), is a *special case* of the slaving principle of synergetics (Haken, 1977; Haken and Wunderlin, 1982; Haken, 1983). This theorem contains the theorems on invariant manifolds as special cases, since it also treats the neighborhood of the center manifold including fluctuations, i.e., it applies to stochastic differential equations as well. In addition, by the introduction of finite-bandwidth excitations, the assumption of the spectral gap can be avoided.

We can also show that D_μ is slaved by the field strength. This result allows us to express both p_μ and D_μ by E, from which we get

$$\dot{E}(t) = (G - K) E(t) - CE^3(t) + F_{\text{tot}} \tag{3}$$

G is a constant depending on the power input to the laser, C is another constant, and F_{tot} is the total random fluctuating forces acting on the field. For a lamp, $G < \kappa$, but for a laser, $G > \kappa$. This equation can best be interpreted by a mechanical model in which we identify E with the coordinate q of a particle and interpret Eq. (3) as describing particle motion in a potential field $V(q)$ (Figure 3):

$$m\ddot{q} + \dot{q} = -\frac{\partial V}{\partial q} \qquad (4)$$

where m is the mass of the particle. The motion of this particle is the same as that of a ball within one or two valleys between hills. When m becomes very small, Eq. (4) coincides with Eq. (3).

The behavior of laser fields using this mechanical model (Haken, 1964, 1970c) is repeated here because the context is new. When G is smaller than κ, i.e., for weak pump power, the dashed curve of Figure 3 applies. The fictitious particle, which moves in the corresponding potential field, relaxes to its equilibrium position at q_0. The action of the fluctuating force F can be visualized as a team of soccer players who kick their soccer ball at random. Thus, the fictitious particle will stay close to $q = 0$, around which point it undergoes random motion. When G is bigger than κ, a qualitatively new situation occurs. As shown by the solid line of Figure 3, the former position $q = 0$ has become unstable and is replaced by two stable positions. However, because the ball (or fictitious particle) can occupy only one position, it must choose between the left or right position. In physical terms, the *symmetry is broken*. From studies of systems in thermal equilibrium it is known that symmetry-breaking is a rather universal phenomenon. Its importance has been stressed in particular by Anderson (1972). In this case we deal with a system far from thermal equilibrium, and we shall see that symmetry-breaking also plays a fundamental role in the behavior of such nonequilibrium systems. Symmetry-breaking can be found in very complex systems—even in the most

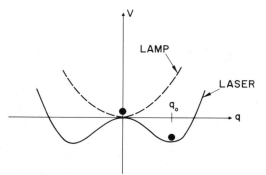

FIGURE 3. Behavior of the field amplitude of the laser interpreted by analogy to particle coordinate. The particle coordinate or field amplitude appears as the abscissa. The potential function V plotted as the ordinate can be interpreted as mountains with one or two valleys in between, over which the ball moves under the action of gravity and an additional friction force. The dashed line describes a situation for low power input into the laser ($G < \kappa$) (see text). When we neglect fluctuations, the fictitious particle comes to rest at $q = 0$. The fluctuating force that represents individual acts of light emission by atoms causes a random motion around $q = q_0$. With this motion the wave tracks on the left part of Figure 1 are connected. At sufficiently high pump power ($G > \kappa$), the solid curve applies. The particle can now choose between two minima, and a fixed amplitude evolves. As a consequence of this fixed amplitude, a sinusoidal wave, depicted in the right part of Figure 1, evolves. The fluctuations have now only a minor effect on the sinusoidal wave.

complex one, the brain (Figure 4). In this context, the notion "symmetry-breaking" is used rather loosely. More precisely, the situation is "bistable." The system (brain) must choose between one of the two states; it cannot select both states simultaneously.

The laser transition has many other features common to phase transitions of systems in thermal equilibrium, but I must refer the reader to the more special literature for that (Haken, 1977). Here I stress another point. When the equilibrium position q_0 is plotted as a function of $G - \kappa$, we find the situation depicted in Figure 5. The former state $q = 0$ is replaced by *two branches*; in other words, *bifurcation* occurs. The laser is one of the simplest examples of bifurcation theory—an important branch of mathematics (see Chapter 29). Note, however, bifurcation theory neglects an aspect of reality whose importance is well known from phase transition theory—fluctuations. Fluctuations play a decisive role in transition regions; they are especially responsible for the way in which symmetry

FIGURE 4. An illustration of "symmetry-breaking" in perception. This figure contains two equivalent perceptions, but the brain must decide which one to select. To this end further information is necessary. This information can be: "consider the white part as foreground." In that case we perceive a vase, and the symmetry is broken. On the other hand, when we receive the information: "consider the black parts as foreground," we perceive two faces. This example shows quite clearly that "symmetry-breaking" can appear even in very complex systems such as the brain, and one may derive a theory of pattern recognition as a sequence of symmetry-breaking events.

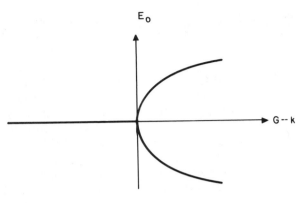

FIGURE 5. Illustration of the phenomenon of bifurcation. In this figure we plot the pump power minus the energy loss, $G - \kappa$, along the abscissa and the equilibrium positions, E_0, along the ordinate. For $G < \kappa$, we have the equilibrium position $E_0 = 0$; for $G > \kappa$ two branches evolve. (This description is somewhat oversimplified because in our representation E_0 is considered a real quantity, whereas in the practical laser case E_0 is a complex quantity and the bifurcation scheme is more complicated.)

is broken. Furthermore, fluctuations are responsible for transient phenomena, i.e., when a system goes from state $q = 0$ to state q_0.

Another important conclusion can be drawn from the above considerations. Because a laser consists of many atoms, it has very many degrees of freedom. However, as is apparent from Eq. (3), the essential behavior of the laser is described by a single E, i.e., by a single degree of freedom. *Because of the "slaving principle," the system undergoes an enormous reduction of the number of degrees of freedom.* Only one degree of freedom shows up macroscopically in the laser light field, and through it we can describe the state of the total system. Because E both describes the macroscopic order and simultaneously "gives orders" to "slave" the atoms, E is an *order parameter*. [Incidentally, this development generalizes Landau's order parameter concept (Landau and Lifshitz, 1959) beyond systems in thermal equilibrium to those far from it. Furthermore, order parameters need not be discrete quantities. They can belong to a whole continuum; in other words, a "single" order parameter can be a slowly varying function of space and time. This property has been shown for the continuous mode laser (Graham and Haken, 1970).] Below threshold ($G < \kappa$), E is zero (aside from fluctuations); it cannot then give orders, and the dipole moments are just driven by fluctuations incorporated in the fluctuating forces F_μ. Above threshold, the establishment of macroscopic order together with symmetry-breaking allows the storage and processing of information. Denoting the two different stable states by a and b, we see that the system can now store information by being either in state a or in state b (Figure 6).

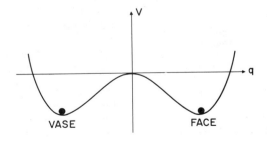

FIGURE 6. Symmetry-breaking as a basis for the storage of information. In this example two states are possible. In complex systems those states may be associated with quite different macroscopic states or with quite different information contents. In the present case, for instance, the left side could be identified with "vase"; the right side with "face."

Coupling of ordered systems allows the performance of logical processes, as in computer elements. For the sake of simple illustrations of the concepts of "order parameter" and "slaving," I have discussed only the single-mode laser. This device can be realized experimentally (and was used to check the theoretical predictions). But laser physics can be much richer. For instance, multimode laser action is possible under certain conditions (see, e.g., Haken, 1970b). The different laser modes behave similarly to biological species: they may compete so that only one survives; Darwin's principle of the "survival of the fittest" thus finds a realization even in the inanimate world. Under other physical conditions, different laser modes are sustained by different atoms and therefore can coexist. In this way the inanimate world provides a model system of the action of "ecological niches." In addition to these effects, the laser can undergo further transitions to still more complicated ordered behavior, such as specific oscillations. The concepts of "slaving" and "order parameter" allow us to deal also with these more complex phenomena.

A General Approach to the Treatment of Synergetic Systems

The laser is a typical example of a large class of systems that can produce order on macroscopic scales through self-organization: in physics, the spontaneous occurrence of macroscopic patterns of velocity fields in fluids and plasmas; in chemistry, macroscopic oscillations or spatial patterns of reacting systems (e.g., Nicolis and Prigogine, 1977); in biology, models of population dynamics, morphogenesis (e.g., Gierer and Meinhardt, 1974), and evolution (Eigen, 1971; Eigen and Schuster, 1977, 1978a,b). In the approach of synergetics the same principles lie behind the occurrence of new macroscopic structures in all of these fields (Haken, 1977). When we change the external conditions, e.g., the pump power input into a laser or the concentration of certain chemical substances in reaction–diffusion systems, the macroscopic state of the system changes drastically. At those instability points the system tests various configurations or collective motions of its parts by fluctuations. The size (amplitude) of some of the new collective configurations tends to increase progressively, whereas other configurations relax rapidly to certain equilibrium values. The increasing configurations ("modes") play exactly the same role as the field strength of the laser. These modes act as *order parameters* and are capable of *slaving* all other modes of the system. Therefore, in general, the behavior of the total system is governed by only a few order parameters that prescribe the newly evolving order of the system. When external conditions are changed further, a whole hierarchy of such instability points can be reached. For example, a homogeneous and quiescent state of a system can be replaced by a static but spatially inhomogeneous structure. Then, at the next stage of parameter extension, this structure is replaced by a new spatial structure that can oscillate. In this way, increasingly complicated structures and dynamics can be realized as the system is driven further from equilibrium. Some typical examples are shown in Figure 7.

In most systems studied in this way the motion eventually becomes chaotic, i.e., entirely irregular. An example of such chaotic motion is shown in Figure 7. Chaotic motion is becoming a lively research subject in mathematics, physics, and biology. (For further discussion see Chapters 29 and 30.) In principle one can take two attitudes toward this new kind of dynamic phenomenon when trying to explain biological processes. One view is that there are intrinsic laws of chaotic motion not yet discovered; but, after being discovered, they would be of great relevance to biological processes. For instance, understanding chaotic motion could mean that chaotic motion represents a highly logical process that can be decoded by adequate concepts and algorithms. On the other hand, biological systems may not be exclusively dynamic systems. In the dynamic systems described above, the quiescent, structureless state is quickly restored when the energy flux is switched off. In contrast, biological systems retain their structure for some time, even when metabolic processes stop (as in freezing or dehydration). Quite obviously, in biological systems there is an interplay between function and structure; functions are embodied in structures, whereas structures are the substrate on which functions can be performed, and they even specify which kind of functions must be performed. Structures, in this sense, are informational. By use of solid components, biological systems may avoid chaotic processes. If solid structures were used as primers, evolution could proceed to build still more complicated structures. They, in turn, allow increasingly complicated processes, which are needed to construct increasingly complicated structures. We observe a self-consistent field situation already known in physics from order parameters and sub-

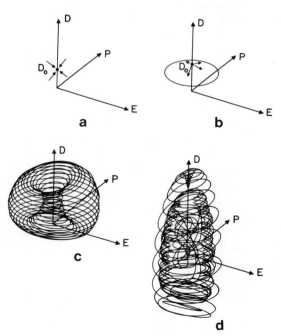

FIGURE 7. Instability hierarchy of the laser. Here the system is always exactly the same, but we change the external conditions, such as pump power or losses. Equations (1) and (2) of the text are only an approximation of the full laser equations. The three coordinates are the electric field strength, polarization, and inversion. The true physical meaning of these quantities is unimportant here. What is important, however, is the qualitative change of behavior of the system. Part (a) applies to the case of small pump power. The system is in a stable quiescent state (as long as we neglect the effects of noise). In part (b) the pump power has been increased. As oscillation of the system sets in (compare also Figure 1, right part), in the terminology of dynamic systems theory, a limit cycle appears. In part (c) the pump power was further increased. The double periodic motion that appears is called a torus in dynamic systems theory. Part (d) finally shows the situation at high pump power and high losses. In this case an entirely irregular, chaotic motion appears.

systems. While order parameters (e.g., field strengths) determine the behavior of subsystems (e.g., atoms), the subsystems collectively determine the order parameter. There is a circular causality.

More explicit examples of the interplay between function and structure are found in current theories of morphogenesis, which assume that the production, interaction, and diffusion of activating and inhibiting substances lead to the formation of morphogenetic fields that enable appropriate processes to prevail for that stage. At the next stage, however, it is assumed that the activators and inhibitors are capable of switching on genes that cause the differentiation of cells. These cells then form a rigid structure.

The interplay between function and structure is also seen in brain models in which an initially blank network is assumed and in which structures are formed by receiving information (which is a process). Hebb's synapse and the formation of *clusters* of neurons represent this type of model, as does von der Malsburg's model on the growth of nerve fibers connecting retina and tectum (see Chapter 14). During these processes, phenomena such as symmetry-breaking again play an important role. With respect to brain models, different possibilities can be tested:

1. The neural network is initially globally homogeneous and orderly, but the homogeneity is broken by certain input signals.
2. The network is globally homogeneous, but has locally a random structure that would facilitate symmetry-breaking by incoming signals.
3. The network is initially globally structured, before any signals have arrived.

How do order parameters and slaving apply to these processes? When new structures evolve, governed by order parameters, several order parameters may occur that can either compete or cooperate. Competition or cooperation determines the behavior of all individual parts of the system. In this way, order parameters can be attributed to incoming signals (or patterns), and whole groups of cells are switched accordingly. Through these concepts the behaviors of many classes of complex systems become accessible to detailed analysis. Recently I found a class of order parameter equations that describe games; connections could be made with the theory of evolutionary games (Schuster and Sigmund, Chapter 5, this volume).

Through this kind of analysis we can understand how self-organizing systems transform energy of low quality into energy of high quality. (In the laser example, the random light of the lamp is transformed into *coherent* laser light.) We get a feeling for how microscopic processes can become correlated to produce (persistent) macroscopic features. These kinds of transformations seem to be an important feature of life. Complex *microscopic processes* become correlated in a self-organized way and *lead to coherent action on macroscopic scales*, such as locomotion, formation of organs, pattern recognition, and, eventually, thinking.

An Outline of the Mathematical Approach

The process of self-organization can be modeled in various ways, particularly by evolution equations or by stochastic equations, such as the master equations. The approach based on evolution equations (Haken, 1977, 1978a,b) is particularly transparent:

We begin by describing the behavior of subsystems of a total system by a set of variables

$$q_1, q_2, q_3, \ldots \tag{5}$$

which may be lumped together into a state vector

$$\mathbf{q} = (q_1, q_2, q_3, \ldots) \tag{6}$$

In the case of continuously extended media, $\mathbf{q}$ can also depend on the vectorial space coordinate $\mathbf{x}$. The evolution of the system is described by stochastic nonlinear partial differential equations of the following type:

$$\dot{\mathbf{q}} = \mathbf{N}(\mathbf{q}, \nabla, \mathbf{x}, \alpha) + F \tag{7}$$

N is a nonlinear function of space coordinate $\mathbf{x}$, state vector $\mathbf{q}$, and differential operators ∇, which act on $\mathbf{q}$. In addition, N depends on control parameters α, which describe the impact of the environment on the system. F is a stochastic force responsible for "partial disorder." Equation (7) describes a general class of equations, such as nonlinear wave equations, reaction–diffusion equations, the Navier–Stokes equations, and so on. To simplify the present representation, we neglect F and refer the reader to the literature which shows how F can be included in the analysis (Haken, 1977, 1978a,b).

We assume that we have found a stable solution $\mathbf{q}_0(\mathbf{x},t)$ to Eq. (7) for a specific control parameter value α_0. We then alter α and study the stability of the newly developing solution. To this end we make a linear stability analysis using the hypothesis

$$\mathbf{q} = \mathbf{q}_0 + \mathbf{u} \tag{8}$$

where $\mathbf{u}$ is the deviation from the stable solution. Inserting Eq. (8) into Eq. (7) and linearizing the resulting equation with respect to $\mathbf{u}$, we obtain

$$\dot{\mathbf{u}} = L\mathbf{u} \tag{9}$$

Because N is autonomous, i.e., it does not depend explicitly on time, the time dependence of $\mathbf{q}_0$ determines that of L. Now consider the following cases:

(a) $\mathbf{q}_0$, and thus L, are time-independent. In this case the solutions of (9) are

$$\mathbf{u} = e^{\rho t}\mathbf{v} \tag{10}$$

where $\mathbf{v}$ is a time-independent vector.

(b) q_0, and thus L, are periodic. According to Floquet's theorem, the solutions of (9) then are

$$\mathbf{u} = e^{\rho t}\mathbf{v}(t) \tag{11}$$

where $\mathbf{v}(t)$ is *periodic*.

(c) q_0, and thus L, are quasiperiodic, i.e., they depend on several frequencies $\omega_1, \omega_2, \ldots$

There are large classes of L's in which the solution of Eq. (9) is

$$\mathbf{u} = e^{\rho t}\mathbf{v}(t) \tag{12}$$

where $\mathbf{v}(t)$ is a *quasiperiodic* function.

To solve the fully nonlinear equation (7), hypothesize for case (a), that

$$\mathbf{q}(t) = \mathbf{q}_0 + \xi_\mu(t)\mathbf{v}_\mu \tag{13}$$

where the ξ_μ are the amplitudes of the modes v_μ.

For case (b),

$$\mathbf{q}(t) = \mathbf{q}_0(t + \phi(t)) + \sum_\mu \xi_\mu(t)\mathbf{v}_\mu(t + \phi(t)) \tag{14}$$

where $\phi(t)$ is the phase of the (now periodic) functions v_μ. I have also treated case (c), but the hypothesis on $\mathbf{q}$ there is rather involved. Inserting (13), (14), or the still more general "Ansatz" of case (c) into (7) allows us to derive equations for the amplitudes ξ and the phase $\phi(t)$. The equations then take the form

$$\dot{\xi}_\mu = \rho_\mu \xi_\mu + g_\mu(\xi,\phi), \quad \dot{\phi}_i = \omega_i + \lambda_i(\xi,\phi) \tag{15}$$

In the above equations we must distinguish between those cases in which the real part (Re) of ρ_μ is positive and those in which the real part is negative. I have devised a general procedure that exactly eliminates all variables connected with ρ_μ for which Re $\rho_\mu < 0$ ("slaving principle"). We are then left with a reduced set of equations for the *order parameters* ξ_μ, where Re $\rho_\mu > 0$. If we include stochastic forces, F, the resulting order parameter equations contain effective stochastic forces F, causing critical fluctuations of the order parameter close to critical points of α. The importance of these phenomena is well known in physics; I expect that they are equally important in biological systems.

If we ignore stochastic forces (as does dynamic systems theory in its present condition), interesting relationships can be established between the order parameter concept and dynamic systems theory. If case (a) applies and ρ_μ of the order parameters is real, the space spanned by means of the order parameters corresponds to the "slow manifold" of an attracting (or repelling) point. On the other hand, in cases (b) and (c) our approach goes considerably beyond the "slow manifold" or the "center manifold theorem." It treats the bifurcation of limit cycles and tori into other tori (or limit cycles in the case of mode locking), and it includes stochastic forces. If the spectrum of linearization $\{\rho\}$ is continuous around Re $\rho \approx 0$, (13) or (14) must be replaced by sums over wave packets such that the *order parameters* ξ_μ become slowly varying functions of space, in complete analogy to the Ginzburg–Landau theory. The resulting order parameter equations (derived in Haken, 1977) can be considered as generalizations of those

In contrast, we expect a discrete spectrum in biology because of the presence of finite boundaries. However, the essential information about spatial structure is carried by the v_μ's. The ξ's determine the superposition of the v_μ's and, in this way, also the evolving pattern. In many cases the resulting equations of the order parameters fall into only a few classes, which means that seemingly quite different systems behave in very much the same way close to the critical points at which q_0 becomes unstable. In this sense the transitions discussed above become "mechanism-independent." Such behavior is strongly reminiscent of the equilibrium phase transition of physics (occurrence of ferromagnetism, superconductivity, and so on). For this reason, these transitions are often called "nonequilibrium phase transitions." However, there is an important difference between equilibrium and nonequilibrium phase transition: in equilibrium phase transitions, infinitely extended systems are considered, and the geometrical shape of the objects is thus irrelevant; but in nonequilibrium systems the geometry (i.e., boundary conditions) is crucial in the selection of the evolving spatial structures. (For further details see Haken, 1977.) An illustration is the evolution of morphogenetic fields using the Gierer–Meinhardt model. The whole procedure can be generalized to situations in which, by *adding new subsystems*, suddenly the macroscopic state can be changed dramatically. (This is self-organization through increase of number of components.)

An Application to Morphogenesis

During the development of animals, organs are formed out of ensembles of originally undifferentiated cells. The individual cells do not initially carry all the specific information about the kind of cell into which they will differentiate. If at an early enough state of development cells are transplanted to a new location in the developing organism, they form that part which corresponds to their *new* position and not to their old one. (Later this situation changes.) There must be means by which cells get information about their position within the ensemble of cells. According to current ideas, this information is provided by a "morphogenetic field" (see Chapters 8 and 9). Turing's original proposal may serve as a possible mechanism for such a field: The cells may produce chemical substances that interact with each other and diffuse. By the interplay of these processes the spatial symmetry (i.e., the homogeneous distribution of chemicals) can be "broken," and an inhomogeneous distribution of chemicals results. The notion of spatial "symmetry-breaking" used in this way is quite different from the notion of "symmetry-breaking" invoked above, where symmetric breaking referred to the *order parameter*.

21. Synergetics

Haken and Olbrich (1978) based their explicit mathematical treatment of morphogenesis on model equations proposed by Gierer and Meinhardt (1974). To model, for instance, morphogenetic processes in hydra, these authors assume the cellular production of an activator and an inhibitor substance, which activate or inhibit the formation of the head. Such substances were isolated by Schaller (1980) for head and foot of hydra. The equations of Gierer and Meinhardt are given below to give the reader a feeling for their character, but many quite different chemical mechanisms may lead to precisely the same macroscopic patterns. From any observed pattern one may not draw many conclusions about underlying microscopic mechanisms.

In the Gierer–Meinhardt equations the concentrations of activator and inhibitor substances are denoted by a and h, respectively. The time variation of a, da/dt, is produced by the following effects: a constant production rate ρ, a decay rate $-\mu a$, and spatial diffusion $D_a \nabla^2 a$, where ∇^2 is the Laplace operator and D_a the diffusion constant. An autocatalytic production rate is assumed, $\approx a^2$, which is limited, however, by the inhibitor, so that the effective rate is taken as a^2/h (in dimensionless variables). Thus, the overall rate equation for a reads

$$\frac{da}{dt} = \rho - \mu a + D_a \nabla^2 a + \frac{a^2}{h^2} \tag{16}$$

The rate equation for h is derived under the following assumptions: h is produced by a bimolecular process using a, a^2; it decays according to $-h$; and it diffuses according to $D_h \nabla^2 h$, so that the overall rate equation for h reads

$$\frac{dh}{dt} = -h + D_h \nabla^2 h + a^2 \tag{17}$$

Equations (16) and (17) are typical examples of "reaction–diffusion" equations. We have solved such equations using the concepts and methods described above.

For small production rate ρ in (16), homogeneous, time-independent distributions of a and h result; $a = a^\circ$ and $h = h^\circ$ so that a°, h° correspond to our former $\mathbf{q}_0$. When we increase ρ above a critical value, ρ_c, distributions become possible. The linearized equations together with boundary conditions (e.g., "nonflux" boundary conditions) fix the "unstable" modes $\mathbf{u}$. For rectangular boundaries, the $\mathbf{u}$'s are given by sine functions; for two-dimensional circular geometry we find Bessel functions. The amplitudes of these functions serve as order parameters that obey the corresponding nonlinear equations. We have solved these equations analytically. Figures 8 and 9 show some results (Haken and Olbrich, 1978). In Figure 9 the cooperation of three order parameters leads to the structure depicted. Depending on initial conditions, fluctuations, and boundary conditions, quite different structures may evolve; for example, in a circular geometry structures of the form of Figure 8 evolve. Berding and Haken (1982) have also performed calculations in spherical geometry that lead to models of the formation of living cells in the developmental stages of morula and blastula for multicellular embryos.

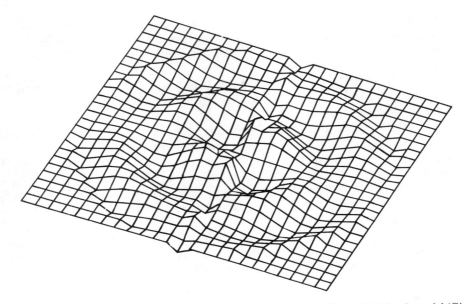

FIGURE 8. An analytical solution (Haken and Olbrich, 1978) of the Gierer–Meinhardt model (Gierer and Meinhardt, 1974). The activator concentration (ordinate) develops within two-dimensional sets of cells (which are taken to be homogeneously distributed). We have used nonflux boundary conditions with circular geometry.

Such models can be only a first step toward an understanding of morphogenesis, but they demonstrate how the cooperation of cells (via chemical communication) may lead to spatial patterns.

Concluding Remarks

This outline of basic ideas developed in the physical construct called synergetics demonstrates that a large variety of processes of macroscopic pattern for-

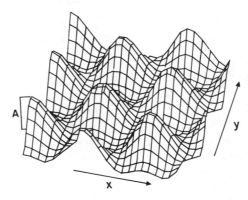

FIGURE 9. The same problem as treated in Figure 8, but with periodic boundary conditions. Note the entirely different pattern of the activator concentration.

mation by self-organization can be understood by the same mechanism, operating close to instability points where the dynamics are governed by few order parameters describing the macroscopic features of the system. Through these mechanisms still more complex processes can be created and described. Competition or cooperation of order parameters can lead to structural and behavioral richness as well as to more complicated cases in which order parameters "play games." This scheme can be generalized to even more complex phenomena if the order parameters of one hierarchical level act as control parameters at a different hierarchical level. This approach allows us to draw entirely new pictures of self-organization processes that are not described merely by means of subsystems but, rather, by functional entities of subsystems carring out specific collective processes.

ACKNOWLEDGMENT. I wish to thank Philip Anderson for his valuable comments on the manuscript.

References

Abraham, R. H, and J. E. Marsden (1978) *Foundations of Mechanics,* 2nd ed. Cummings, Menlo Park, Calif.
Anderson, P. W. (1972) More is different: Broken symmetry and the nature of the hierarchical structure of science. *Science* **177**:393.
Berding, C., and H. Haken (1982) Pattern formation in morphogenesis. *J. Math. Biol.* **14**:133–151.
Eigen, M. (1971) Selforganization of matter and the evolution of biological macromolecules. *Naturwissenschaften* **58**:465–523.
Eigen, M., and P. Schuster (1977) The hypercycle: Part A. *Naturwissenschaften* **64**:541–565.
Eigen, M., and P. Schuster (1978a) The hypercycle: Part B. *Naturwissenschaften* **65**:7–41.
Eigen, M., and P. Schuster (1978b) The hypercycle: Part C. *Naturwissenschaften* **65**:341–369.
Gierer, A., and H. Meinhardt (1974) Biological pattern formation involving lateral inhibition. *Lect. Math. Life Sci.* **7**:163–183.
Graham, R., and H. Haken (1968) Quantum theory of light propagation in a fluctuating laser-active medium. *Z. Phys.* **213**:420–450.
Graham, R., and H. Haken (1970) Laser light—First example of a second-order phase transition far away from thermal equilibrium. *Z. Phys.* **273**:31–46.
Haken, H. (1964) A nonlinear theory of laser noise and coherence. *Z. Phys.* **181**:96–124.
Haken, H. (1970a) Lectures at Stuttgart University (unpublished).
Haken, H. (1970b) Laser theory. In: *Encyclopedia of Physics,* Vol. XXV/2c, L. Genzel (ed.). Springer-Verlag, Berlin.
Haken, H. (1970c) Laserlicht—ein neues Beispiel für eine Phasenumwandlung? In: *Festkörperprobleme X,* O. Madelung (ed.). Pergamon/Vieweg, Brunswick, pp. 62–73.
Haken, H. (1977) *Synergetics—An Introduction: Nonequilibrium Phase Transitions and Self-Organization in Physics, Chemistry and Biology.* Springer-Verlag, Berlin.
Haken, H. (1978a) Nonequilibrium phase transitions and bifurcation of limit cycles and multiperiodic flows. *Z. Phys. B* **29**:61–66.
Haken, H. (1978b) Nonequilibrium phase transitions and bifurcation of limit cycles and multiperiodic flows in continuous media. *Z. Phys. B* **30**:423–428.
Haken, H. (1983) *Advanced Synergetics.* Springer-Verlag, Berlin.
Haken, H., and R. Graham (1971) Synergetik: Die Lehre vom Zusammenwirken in Wissenschaft und Technik. *Umschau* **6**:191.
Haken, H., and H. Olbrich (1978) Analytical treatment of pattern formation in the Gierer-Meinhardt model of morphogenesis. *J. Math. Biol.* **6**:317–331.
Haken, H., and H. Sauermann (1963) Frequency shifts of laser modes in solid state and gaseous lasers. *Z. Phys.* **176**:47–62.

Haken, H., and A. Wunderlin (1982) Slaving principle for stochastic differential equations with additive and multiplicative noise and for discrete noisy maps. *Z. Phys. B* **47**:179–187.

Landau, L. D., and I. M. Lifshitz (1959) *Course of Theoretical Physics*, Vol. 5. Pergamon Press, Elmsford, N.Y.

Nicolis, G., and I. Prigogine (1977) *Self-Organization in Non-Equilibrium Systems.* Wiley, New York.

Schaller, C. (1980) Talk given at the Tagung der Ges. deutscher Naturforsch und Arzte, Hamburg.

22

Role of Relative Stability in Self-Repair and Self-Maintenance

Rolf Landauer

ABSTRACT

This chapter opens with a discussion of relative stability, from the point of view of the physics of information storage. Bistable elements that store information are of two kinds: (1) static devices with performance resembling that of particles occupying either one of several energy (potential) wells in a field; (2) dynamic systems consisting of structures that by energy conversions and dissipations hold one of several possible steady states. The latter, dynamic type is examined here.

The relative stability of a locally stable state of a dynamic system depends strongly on the character of the *noise* along paths between two (or more) such states. No amount of study of the possible terminal states, by themselves, will suffice to discover the more probable one. To do that, we must take account of the fluctuations in the unlikely intervening states between those that are locally stable. This account does not require that a distinction be made between "internal" and "external" noise.

Because of fluctuations, a multistable system may be buffeted or agitated out of noisy regions of its state space, then "condense" into a low-noise "cold trap" region, where it remains. Simplistic schemes based only on "information" or "entropy production" cannot suffice as accounts of relative dynamical stability in the presence of fluctuations nor of the constructive aspects of noise for self-organizing systems such as life.

"Self-organization" is in some respects an unfortunate term, because it hints at an extraphysical "*élan vital*," which we, as physical reductionists, should drive out of our explanations. Even worse, the term "self-organizing" is sometimes applied to Bénard cells in hydrodynamic fields exposed to strong thermal gradients, but considering that under the constraints these "structures" are the only physically allowed pattern of behavior, do they really represent "self-organization" any more than does an electron circulating about an atom in a quantum state? Surely the notion of "self-organization" has often been used too grandiosely.

Experience with technological machinery shows that it always tends to wear out. True, a scheme of maintenance can sustain its operations for long or very long epochs. But can a technological machine be made self-maintaining, so that it persists with form and function intact, without the intervention of people acting as *deus ex machina*? How does life accomplish this without help from humans? Two schemes of maintenance or adaptation are examined: (1) slow and careful modification of parts or blueprints and (2) reproduction, variation, and descent with modification. The scheme used

---

ROLF LANDAUER • IBM T. J. Watson Research Center, Yorktown Heights, New York 10598.

by living systems to ensure persistence (especially at the species level of organization) is of the latter kind. The role of random mutations in that scheme resembles that of nucleation in the physical case of first-order phase transitions. The fluctuation leads from one locally stable state to another, different stable state. Mutations do *not* resemble physical second-order phase transitions that require that a symmetry-breaking threshold be exceeded. After the threshold is exceeded, the more symmetrical original state is no longer stable—*any* fluctuation then drives the system away from it. Though such bifurcations are of great interest in the mathematical field of qualitative dynamics, it is not clear that they pertain to the unfolding of the evolution of the terrestrial biosphere.

The chapter closes with two related questions: (1) Will a system close to but not at (thermal) equilibrium evolve, and if so, does the level of complexity that can be achieved depend on the degree of departure from equilibrium? (2) Does simplicity of environment (e.g., uniformity) prevent evolution. It is suggested that complex systems require for their continuation or origination large deviations from equilibrium, and nonuniform, rich environments. —The Editor

---

Interest in self-organization arises largely from the hope that we can characterize, in some general way, the setting that leads to the origin of life and the unfoldings of evolution. It is presumptuous for physical scientists to assume that their background, without a hard confrontation with many dirty biochemical details, will enable them to say much about how "we" came about. It is equally presumptuous, however, for biologists to assume that their knowledge of our terrestrial life form qualifies them to speculate about the generality of this phenomenon and the frequency with which it arises in the universe. Such a presumption is only a slight extension of the "bug-eyed monster" concept of extraterrestrial life. As a computer technologist I can at best speculate about a metatheory, describing some of the aspects of a more detailed theory. In particular, I stress the essential role of a stochastic view of relative stability in an eventual detailed theory. Therefore, I turn first to a brief discussion of relative stability.

Computers are built of bistable information-holding elements, which, through linkages of varying complexity, influence each other. As part of a continuing attempt to understand the ultimate limitations imposed by the laws of physics on information handling (Landauer and Woo, 1973; Landauer, 1976), we have asked how small can we make the bistable elements and still have them be sufficiently immune to fluctuations to be useful. Bistable elements can be of two kinds. First of all, there are devices that resemble a particle in a potential field with two (or perhaps more) adjacent potential wells; a stabilizing potential barrier between the wells prevents the particle from jumping too often from one to the other. In practice, of course, if the information-bearing degree of freedom is massive enough, e.g., as for a hole in a punched card, we can dispense with the stabilizing potential and rely only on the slowness of the information deterioration process. Such devices are stabilized by static forces and do not require energy to hold information. In addition, however, we have a dynamic class: "The second class of devices consists of *structures* which are in a steady (time invariant) state, but in a *dissipative* one, while holding on to information" (Landauer, 1961) (italics added for emphasis). Gas jets and arc lights constitute well-known illustrations of systems of this latter sort. Under the same external conditions, i.e., rate of flow for the gas jet, electrode spacing and voltage for the arc light, these systems can be either in a quiescent or else in an ignited state. External intervention, e.g., a match for the gas jet, can be used intentionally to modulate the system temporarily and switch it from one state to another. Just as in the case of the bistable wells, however, noise also can cause the system to switch.

For an asymmetrical system, such as a gas jet or arc light, in which the two locally stable states are very dissimilar, we can expect that the system, if left alone long enough under steady-state noise, will settle into one of the two states, with a probability close to one (certainty). (The parameters of the dynamic system can, of course, be carefully adjusted to make the two probabilities comparable or equal.) I have analyzed this type of problem in a series of papers describing both the steady-state distribution reached after a long time, as well as the relaxation time that describes the rate with which the steady-state distribution is reached (Landauer and Woo, 1973; Landauer, 1975, 1978a,b, 1979a,b). Similar points were made independently and early by Stratonovich (1963, 1967). A key point in this work is that the variation of the noise along the paths between the locally stable states is crucial to the determination of relative stability, i.e., to the determination of the probable state. This point was stated two decades ago (Landauer, 1962) and has, in recent years, been widely echoed in many papers, of which I can cite here only a small sample (Horsthemke and Malek-Mansour, 1976; de la Rubia and Velarde, 1978; Schenzle and Brand, 1979; Lindenberg et al., 1983).

This recent literature invokes a new terminology—"multiplicative noise" and "external noise"—to describe noise that is a function of the state of the system. (Strictly speaking, the question at issue is more complex than the presence or absence of variation in the noise. For example, nonuniform noise causes no complications in a damped Hamiltonian system, which obeys an Einstein relation for a fixed temperature and in which the noise variation reflects only the nonuniformities of the dissipative parameters of the system.) All this literature, however, leads us to the same conclusion: considerations confined to an analysis of the two terminal states to be compared cannot tell us which is the more likely one; to do that the fluctuations in the unlikely intervening states must also be taken into account. A distinction between "internal" and "external" noise does not seem unambiguous or fundamental; it depends on where we decide to draw the boundaries of our system. All noise, in fact, simply represents degrees of freedom that are not followed closely enough to be included explicitly in the equations of motion under consideration.

The concern with relative stability, in systems far from equilibrium, has now become widespread in many fields, including open chemical reactions (Janssen, 1974; Matheson et al., 1975; Schlögl, 1980), control systems (Lindsey, 1969), bistable laser excited systems (Roy et al., 1980; Zardecki, 1980; Drummond et al., 1980; Hanggi et al., 1980; Sczaniecki, 1980; Mandel, 1980; Hasegawa et al., 1980; Bonifacio et al., 1981; Englund et al., 1981), and social systems (Weidlich and Haag, 1980). Other applications and many additional citations will be found in the reviews by Haken (1975, 1977). The probability distribution in such multistable systems can be regarded as a competition between the coherent (noiseless) dynamics of the systems, driving it toward certain preferred states, and noise, which permits the system to get away from these states. That viewpoint makes it quite obvious that the variation of the noise along the paths does matter. Indeed, if the noise variation is strong enough, it can determine the points of local stability, i.e., the system will be agitated out of the noisy parts of the system's state space, and will "condense" into the low-noise "cold traps" (Horsthemke and Lefevér, 1977; Landauer, 1978a). Whether or not the effects of variable noise in the emergence and evolution of biological systems are an incidental complication or play a crucial role is not clear to me. But, as I have emphasized (Landauer, 1978b), as long as this variation is present we cannot expect simple "magical" arguments based on a comparison of entropy, information, entropy production, and so on, between two states of local stability, to tell us which is the more likely one toward which the system will evolve.

I am not very happy with the expression "self-organization" used in connection with the appearance of life or the tendency of biological forms to become more complex. It carries with it too much of a suggestion of an *"élan vital"*—a driving force toward complexity. If, however, we have learned anything from the Darwinian view of evolution and from its modern detailed elaborations in terms of genetics and molecular biology, it is that there is no such force, but only experience with random variations—with the concomitant ability for some variations to sustain themselves. Any attempt to draw an analogy, say, between the appearance of life and the Bénard instability, seems wrong. The Bénard instability appears in a thin fluid layer in a gravitational field, when the temperature drop across the layer exceeds a threshold. Once this threshold is exceeded, however, the original spatially uniform fluid, exhibiting only a temperature gradient and no circulating flow, is not even metastable, it is unstable. A fluid layer put temporarily in this homogeneous state will be forced out of it upon the appearance of almost any fluctuation. By contrast, the appearance of life, or the "forward motion" of evolution, carries the system from one possible state to another, i.e., between two states, both of which are locally stable under the same external conditions, much as in the asymmetrical multistable systems I have already discussed. Indeed, we can ask whether or not the circulating cells of flow in the Bénard instability, *under conditions in which they are the only allowed pattern of behavior*, represent "self-organization" any more than does an electron, circulating about an atom in a quantum state.

Machinery, once it is brought into existence, tends to wear out and deteriorate. A gasoline engine is an example of a multistable system: with the ignition on, it can either be running or be stationary. It is possible—as in the case of maintenance or repair of the automobile—to offset the wear-out and deterioration mechanisms to a large extent and thus lengthen the life of the machinery. The machinery can be equipped with automatic adjustments to compensate for wear, or for environmental fluctuations, such as in temperature. The machine can also be equipped with redundant devices to permit it to keep functioning in the presence of failing components, as is done with dual hydraulic brake systems. All this, however, only reduces the rate for functional failure of the machine and does not totally eliminate it. Thus, if we consider our device in a large-enough phase space, which describes not only the multistability inherent in its basic kinetics but also the wear and oxidation of parts, then the original states of local stability slowly become unstable with respect to the troublesome degrees of freedom.

We can now go on to invoke forms of self-repair and self-maintenance to restore local stability. Maintenance can be preventive, as in the use of fluorescent bulbs in a large office building, which require systematic replacement only after an extended time interval. There we count on the presence of enough fluorescent fixtures in any one space, so that the premature dimming of one bulb will not affect the basic functions of the space. Alternatively, we can invoke detection methods for faulty parts, which then direct the replacement of a failing or failed part. If we associate a complete description of the machinery with the machinery, then this self-repair capability can be pushed far. Indeed, how far it has to be pushed is a function of the complexity of the environment. In an environment containing parts stores, repair shops, and high-level fault detection devices in the form of people, our cars have just about reached the status where they can be kept operating indefinitely. An extraterrestrial observer might well observe that our houses, for example, are self-maintaining, although admittedly the self-maintenance involves a symbiotic parasitic relationship with smaller occupying entities. We would be inclined to object. After a hard weekend of repair work, or after paying the plumber, we are not inclined to think of the house as self-repairing. The detached observer objects: "You are not essential. I have observed that healthy houses that lose their parasites soon manage to attract new ones." To avoid such semantic debates it would be best to avoid the emotional and goal-oriented noun "self" and instead simply discuss the stability of forms of behavior. For simplicity, however, and in consistency with other chapters in this volume, I shall continue to use the noun.

To prolong the life span of a piece of self-repairing machinery that cannot take advantage of parasitic relationships, we need machinery that not only replaces all the direct functional parts of the unit but also can handle repair and diagnostic apparatus by itself. This arrangement is possible in principle, but I am not sure that it has been demonstrated in full detail. (Of course, no system can be designed to cope with all possible failure modes, including those that can cause very extended damage.) We also need to cope with the deterioration of the blueprints and of the repair directions. That, however, is trivial. We have several copies in digital form, which we periodically compare to each other. When one copy has changed, it is discarded and replaced by a new restandardized version. We can also use more subtle forms of redundancy (Swanson, 1960), instead of simple duplication. If we allow a long-enough period, then we shall eventually see enough simultaneous fluctuations, so that even the information protected by redundancy is destroyed. The rate of information loss, however, can easily be made minute (Swanson, 1960), e.g., by restandardizing sufficiently often.

Can such a long-lived, self-repairing system also modify itself and its blueprints? Obviously the system cannot afford to try out random and major changes—"lethal" changes will be lethal. But if the system can monitor its own performance and tries out changes in a sufficiently careful way, allowing for fallback positions, there is no reason it should not be possible to evolve to a status that is adapted to cope more effectively with its environment or with a wider range of environments. Can we expect to find nonreproductive life forms of this sort, eventually, in the universe? Our complex self-repairing and self-modifying system is far too complex to have come into existence spontaneously; it had to be built. On the other hand, it is quite possible that a life form that is initially reproductive, and evolves enough additional self-repair and self-modification capability (neo-Lamarckism), eventually can come to the point at which an attempt to discard the original reproductive apparatus is one of the successful attempts at self-modification. Human civilization, regarded as a single entity, is an example of a self-modifying, nonreproductive organism.

Nature has given us a simpler recipe for evolution by utilizing reproduction. We have many copies of a system, or of closely related systems, and can try out variations at random. If the variations are lethal or deleterious, they will either disappear immediately or else be at a disadvantage. A successful mutation, taking us to a new stable ecological balance, is very much like a nucleation event in a *first-order* phase transition. In both cases we go from one locally stable state to another, more favored one. In both cases we require an initial fluctuation that takes us in a very particular direction toward the new state, but in neither case is it required to take us all the way there. In the phase transition, a critical nucleus must be formed, characterizing the new phase in its structure, and large enough so that its volume gain in free energy, over the original phase, outweighs the penalty due to the surface or interface energy. In the case of the organism the fluctuation need not change the entire organism but only the genetic pattern. Not any fluctuation in the genetic structure will do; the genetic structure must be modified completely enough to be that of the new successful form, or must at least take us far enough toward that form to be within later genetic self-repair capabilities. This requirement for specifically directed fluctuations of adequate size is much like the nucleation process or like the mechanism for noise-activated transitions in an active dissipative system, out of one locally stable state into another more favored state. In all these cases the fluctuations must take us far enough so that we can leave the subsequent progress toward the new state to the coherent (i.e., noiseless) dynamics of the system. Mutations are unrelated to *second-order* phase transitions, or transitions in more general systems exhibiting a symmetry-breaking threshold. In the latter, as I have already pointed out, once the threshold is exceeded, the original and more symmetrical state is no longer stable; *any* fluctuation will drive the system away from it. Analogies between nucleation theory and biological phenomena are not new; they were invoked, for example, in a theory of cancer (Fisher and Hollomon, 1951). A more modern variation on this theme, emphasizing the role of nonuniform noise, is presented by Lefevér and Horsthemke (1979).

The biological state space is, of course, tremendously complex, with a myriad of locally stable states, most of which will never be realized. Furthermore, within a locally stable state there can be subordinate kinetics, characteristic of that state. Thus, for example, the development of human civilization has taken place substantially within the "valley" corresponding to a single genetic pool, without the need for further biological evolution. In our cultural and technological evolution, we can find exploration attempts similar to those in biology, resulting occasionally in acceptance of a modified form, e.g., in the invention of wheels and nations. Thus, the single biological valley can be broken up into cultural subunits (Kuhn, 1976, 1978). Although the total number of locally stable states (possible ecologies for terrestrial conditions) is tremendous, in this world with finite resources it must be a limited number. Thus, there is a maximum complexity achievable by evolution.

We can come to the same conclusion by simply pointing out that in a world with a limited amount of matter and energy there is a limit to the amount of genetic information that can be provided. Clearly the more complex life forms include those that understand something about the structure of the adjacent stable states and can modify themselves. It is also worth emphasizing another way in which the stochastic multistable systems underlying life are much more complex than some of the simplest systems analyzed by physicists and chemists. Many of the latter, after they have been left alone long enough under the influence of noise, reach a steady-state probability distribution characterized by detailed balance, i.e., transitions from state A to state B are balanced directly by the inverse transition, rather than through some more circuitous route. Most biological phenomena, and most of our everyday technical machinery, are more complex than that. Clocks, with remarkable predictability, go on to 11:59 from 11:58, and rarely go the other way.

Let us now speculate about the relationship of the environment to the complexity of the evolving organisms. My colleague, Charles Bennett, has put forth the following question. Consider a fluid layer, subject to a modest temperature gradient, say, 1°C, and thus not far from equilibrium. It is almost certain that such a modest departure from equilibrium will drastically slow up evolution. But will it prevent the appearance of life altogether, or just severely limit the extent of the resulting evolution? If we allow enough time, can we expect to find a civilization that has invented thermal engines that take advantage of the small temperature drop? We are really posing two important and somewhat separate questions here. One concerns the need for deviations from thermal equilibrium. Does the origin of life require a minimal deviation from equilibrium? If not, does evolution saturate at a level of complexity that depends on the deviation from equilibrium? After all, as Bennett points out, the speed with which operations can be carried out depends on the extent of the deviations from equilibrium, and if this speed is inadequate, we cannot carry out the self-repair or reproduction of a complex organism fast enough to prevent its deterioration. A closely related point is made in the "Note added in proof" in Landauer (1976).

The second question relates to the uniformity and simplicity of the environment. Don Glaser has attempted experiments in which a culture of microorganisms is kept under very carefully controlled steady-state conditions; he proposes to find retrograde evolution under such conditions. Organisms do not need their full genetic baggage to cope with such a constant and controlled environment and can reproduce more rapidly if they discard the unnecessary complexity. Thus, a predictable and limited environment removes the need for sophisticated systems and, of course, also removes some of the effects and devices that the organism might use. It is hard to visualize any complex system built out of a single chemical element at a temperature at which the element is gaseous or liquid. Some minimal richness in the surroundings, in terms of a choice of available materials, in topography as manifested by interfaces between gas, liquid, and solid phases; temperature and light variations; and so on; seems needed.

The two aspects of the "Bennett" question are, of course, not really independent. A rich and varied environment is, by definition, not in thermal equilibrium. On the other hand, strong deviations from equilibrium create their own inhomogeneities, e.g., turbulence.

We have hinted that Bennett's question has a negative answer, that the civilization with thermal engines is too complex to be supported by such a small deviation from equilibrium, in such a simple environment. At this point, however, it is worth stressing one of Bennett's key points: one cannot depend on intuition to guess at the possible forms of complex self-stabilizing behavior. After all, if there were a fairly direct design method for such things, evolution would have found it by now, and utilized it.

Postscript added in 1985: A paper which overlaps this one, to some extent, but has a higher ratio of equation to words, and a higher ratio of physics to speculation, is in existence (Landauer, 1983).

A set of sociological comments regarding my colleagues is given in Landauer (1981).

Postscript added in 1987: See also Landauer (1987).

ACKNOWLEDGMENTS. The author is particularly indebted, for his stimulation, to two sources. One is H. Kuhn (1976, 1978); the other is my local colleague, Charles Bennett, with his unpublished notions about the "depth" of computations and the relationship of that concept to the questions discussed here.

References

Bonifacio, R., L. A. Lugiato, J. D. Farina, and L. M. Narducci (1981) Long time evolution for a one-dimensional Fokker-Planck process: application to absorptive optical bistability. *IEEE J. Quantum Electron.* **17**:357–365.

de la Rubia, J., and M. G. Velarde (1978) Further evidence of a phase transition induced by external noise. *Phys. Lett.* **69A**:304–306.

Drummond, P. D., K. J. McNeil, and D. F. Walls (1980) Non-equilibrium transitions in sub/second harmonic generation. I. Semiclassical theory. *Opt. Acta* **27**:321–335.

Englund, J. C., W. C. Schieve, W. Zurek, and R. F. Gragg (1981) Fluctuations and transitions in the absorptive optical bistability. In: *Optical Bistability*, C. M. Bowden, M. Cliftan, and H. R. Robl (eds.). Plenum Press, New York, p. 315.

Fisher, J. C., and J. H. Hollomon (1951) A hypothesis for the origin of cancer foci. *Cancer* **4**:916–918.

Haken, H. (1975) Cooperative phenomena in systems far from thermal equilibrium and in nonphysical systems. *Rev. Mod. Phys.* **47**:67–121.

Haken, H. (1977) *Synergetics—An Introduction.* Springer-Verlag, Berlin.

Hanggi, P., A. R. Bulsara, and R. Janda (1980) Spectrum and dynamic-response function of transmitted light in the absorptive optical bistability. *Phys. Rev. A* **22**:671–683.

Hasegawa, H., T. Nakagomi, M. Mabuchi, and K. Kondo (1980) Nonequilibrium thermodynamics of lasing and bistable optical systems. *J. Stat. Phys.* **23**:281–313.

Horsthemke, W., and R. Lefevér (1977) Phase transition induced by external noise. *Phys. Lett.* **64A**:19–21.

Horsthemke, W., and M. Malek-Mansour (1976) The influence of external noise on non-equilibrium phase transitions. *Z. Phys. B* **24**:307–313.

Janssen, H. K. (1974) Stochastisches Reaktionsmodell für einen Nichtgleichgewichts-Phasenübergang. *Z. Phys.* **270**:67–73.

Kuhn, H. (1976) Evolution biologischer Information. *Ber. Bunsenges. Phys. Chem.* **80**:1209–1223.

Kuhn, H. (1978) Modellvorstellungen zur Entstehung des Lebens (I) and (II). *Phys. Bl.* **34**:208–217, 255–263.

Landauer, R. (1961) Irreversibility and heat generation in the computing process. *IBM J. Res. Dev.* **5**:183–191.

Landauer, R. (1962) Fluctuations in bistable tunnel diode circuits. *J. Appl. Phys.* **33**:2209–2216.

Landauer, R. (1975) Inadequacy of entropy and entropy derivatives in characterizing the steady state. *Phys. Rev. A* **12**:636–638.

Landauer, R. (1976) Fundamental limitations in the computational process. *Ber. Bunsenges. Phys. Chem.* **80**:1048–1059.

Landauer, R. (1978a) Distribution function peaks generated by noise. *Phys. Lett.* **68A**:15–16.

Landauer, R. (1978b) Stability in the dissipative steady state. *Phys. Today* **31**:23–30.

Landauer, R. (1979a) The role of fluctuations in multistable systems and in the transition to multistability. In: *Bifurcation Theory and Applications in Scientific Disciplines*, O. Gurel and O. E. Rössler (eds.). N.Y. Acad. Sci., New York, pp. 433–452.

Landauer, R. (1979b) Relative stability in the dissipative steady state. In: *The Maximum Entropy Formalism*, R. D. Levine and M. Tribus (eds.). MIT Press, Cambridge, Mass., pp. 321–337.

Landauer, R. (1981) Nonlinearity, multistability, and fluctuations: Reviewing the reviewers. *Am. J. Physiol.* **241**:R107–R113.

Landauer, R. (1983) Stability and relative stability in nonlinear driven systems. *Helv. Phys. Acta* **56**:847–861.

Landauer, R. (1987) Computation: A fundamental physical view. *Phys. Scr.* **35**:88–95.

Landauer, R., and J. W. F. Woo (1973) Cooperative phenomena in data processing. In: *Synergetics*, H. Haken (ed.). Teubner, Stuttgart, pp. 97–123.

Lefevér, R., and W. Horsthemke (1979) Bistability in fluctuating environments: Implications in tumor immunology. *Bull. Math. Biol.* **41**:469–490.

Lindenberg, K., K. E. Shuler, V. Seshadri, and B. J. West (1983) Langevin equations with multiplicative noise: Theory and applications to physical processes. In: *Probabilistic Analysis and Related Topics*, Vol. 3, A. T. Bharucha-Reid (ed.). Academic Press, New York, pp. 81–125.

Lindsey, W. C. (1969) Nonlinear analysis of generalized tracking systems. *Proc. IEEE* **57**:1705–1722.

Mandel, P. (1980) Fluctuations in laser theories. *Phys. Rev. A* **21**:2020–2033.

Matheson, I., D. F. Walls, and C. W. Gardiner (1975) Stochastic models of first-order nonequilibrium phase transitions in chemical reactions. *J. Stat. Phys.* **12**:21–34.

Roy, R., R. Short, J. Durnin, and L. Mandel (1980) First-passage-time distributions under the influence of quantum fluctuations in a laser. *Phys. Rev. Lett.* **45**:1486–1490.

Schenzle, A., and H. Brand (1979) Multiplicative stochastic processes in statistical physics. *Phys. Rev. A* **20**:1628–1647.

Schlögl, F. (1980) Stochastic measures in nonequilibrium thermodynamics. *Phys. Rep.* **62**:267–380.

Sczaniecki, L. (1980) Quantum theory of subharmonic lasers: Non-equilibrium phase transition of the first order. *Opt. Acta* **27**:251–261.

Stratonovich, R. L. (1963) *Topics in The Theory of Random Noise*, Vol. I. Gordon & Breach, New York.

Stratonovich, R. L. (1967) *Topics in the Theory of Random Noise*, Vol. II. Gordon & Breach, New York.

Swanson, J. A. (1960) Physical versus logical coupling in memory systems. *IBM J. Res. Dev.* **4**:305–310.

Weidlich, W., and G. Haag (1980) Migration behaviour of mixed population in a town. *Coll. Phen.* **3**:89–102.

Zardecki, A. (1980) Time-dependent fluctuations in optical bistability. *Phys. Rev. A* **22**:1664–1671.

23

Broken Symmetry, Emergent Properties, Dissipative Structures, Life
Are They Related?

Philip W. Anderson and Daniel L. Stein

ABSTRACT
The authors compare symmetry-breaking in thermodynamic equilibrium systems (leading to phase change) and in systems far from equilibrium (leading to dissipative structures). They conclude that the only similarity between the two is their ability to lead to the emergent property of spatial variation from a homogeneous background. There is a well-developed theory for the equilibrium case involving the order parameter concept, which leads to a strong correlation of the order parameter over macroscopic distances in the broken symmetry phase (as exists, for example, in a ferromagnetic domain). This correlation endows the structure with a self-scaled stability, rigidity, autonomy, or permanence. In contrast, the authors assert that there is no developed theory of dissipative structures (despite claims to the contrary) and that perhaps there are no stable dissipative structures at all! Symmetry-breaking effects such as vortices and convection cells in fluids—effects that result from dynamic instability bifurcations—are considered to be unstable and transitory, rather than stable dissipative structures.

Thus, the authors do not believe that speculation about dissipative structures and their broken symmetries can, at present, be relevant to questions of the origin and persistence of life.
—The Editor

The more that theoretical physicists penetrate the ultimate secrets of the microscopic nature of the universe, the more the grand design seems to be one of ultimate simplicity and ultimate symmetry. Because all the interesting parts of the

PHILIP W. ANDERSON and DANIEL L. STEIN • Department of Physics, Joseph Henry Laboratories, Princeton University, Princeton, New Jersey 08544.

universe—at least those of interest to us, such as our own bodies—are markedly complex and *unsymmetric*, the first, correct conclusion one draws from this statement is that the deep probing of the nature of matter (on which physicists expend great effort and greater sums of money) is becoming more and more irrelevant to *us*. But that is not really an adequate retreat for any scientist who hopes to achieve the ultimate goal of science, which we take to be real understanding of the nature of the world around us from first principles. It is essential to explain the real world in terms of the ultimately simpler constituents of which it is made. In fact, scientists must thank their stars that the world becomes simpler as each underlying level is discovered—the opposite case would make their task difficult indeed.

The simplicity to which we refer is, of course, that implied by the recent success of elementary particle theorists in reducing the equations of the fundamental constituents of matter to perfectly symmetrical ones, in which all constituents initially enter in exactly the same way and in which all the interactions themselves are derived from a principle which *itself* is a manifestation of an especially perfect kind of symmetry. But those not acquainted with these developments need not fear that what we say will depend on them in any way. We wish merely to make the point that there is a sharp and accurate analogy between the breaking-up of this ultimate symmetry to give the complex spectrum of interactions and particles we actually know and the more visible complexities we shall shortly discuss.

During the past 20 years or so there has gradually arisen a set of concepts relating the ways in which complexity in nature arises from simplicity. Some of these concepts are quite rigorously and soundly based in the theoretical physics of large and complex systems, whereas others extend all the way to the speculative fringe between physics and philosophy.

The most basic question with which such a conceptual structure might hope to deal would be placing life within the context of physics in some meaningful way: to relate the emergence of life itself from inanimate matter to some general principle of physics. Can we understand the existence or even the origin of life in some purely physical context? We approach the answer through four questions.

Clearly, we are trying to view life as an *emergent property*, a property of a complex system not contained in its parts. So we start from the very simple question of whether or not such properties exist:

1. Can properties emerge from a more complex system if they are not present in the simpler substrate from which the complex system is formed?

The most rigorously based, physics-oriented description of the growth of complexity out of simplicity is called the theory of *broken symmetry*, and it gives an unequivocal "yes" answer to this question. In equilibrium systems containing large numbers of atoms, new properties, such as rigidity or superconductivity, and new stable entities or structures, such as quantized vortex lines, can emerge that are not just nonexistent, but even meaningless on the atomic level.

Unfortunately, the emergent properties we are most seriously interested in are not these simple ones of equilibrium systems. Specifically, we need to know whether or not life, and then consciousness, can arise from inanimate matter; and the one unequivocal thing we know about life is that it always dissipates energy and creates entropy in order to maintain its structure. So we come to a second question:

2. Are there emergent properties in dissipative systems driven far from equilibrium?

The answer is yes: dynamic instabilities such as turbulence and convection are common in nature and their source is well understood mathematically. When they occur, these phenomena exhibit striking broken symmetry effects, which very much resemble the equilibrium structures that exist in condensed matter systems. These have been called "dissipative structures." Examples are convection cells or vortices in turbulent fluids, but these seem always very unstable and transitory. Can they explain life, which is very stable and permanent (at least on atomic time scales)?

3. Is there a theory of *dissipative structures* comparable to that of equilibrium structures, explaining the existence of new, stable properties and entities in such systems?

Contrary to statements in a number of books and articles in this field, we believe that there is *no such theory*, and it even may be that there are no such structures as they are implied to exist by Prigogine, Haken, and their collaborators (Glansdorff and Prigogine, 1971; Nicolis and Prigogine, 1977; Haken, 1977). What does exist in this field is rather different from Prigogine's speculations and is the subject of intense experimental and theoretical investigation at this time.

The statement that life is "stable" is, of course, not valid if we look at long-enough time scales: on long time scales, life has the character of a rather slowly growing, dynamic instability. But one has an intuitive feeling that living systems have an extraordinarily great ability to ignore perturbations and changes in boundary conditions, i.e., to be autonomous and *rigid* in some sense, whereas our technological experience leads us to expect that to stabilize dynamic systems requires great efforts directed at avoiding external noise and providing correct boundary conditions. Perhaps a better word entirely to describe the dynamics of life is *autonomy* rather than *stability*, the latter being, as Leslie Orgel has pointed out to me, more a pun than a rigorous statement.

4. Can we see our way clear to a physical theory of the origin of life that follows these general lines?

The answer to the fourth deep question is already evident. It is "no," because there is no theory of dissipative structures. The best extant theoretical speculations about the origin of life, those of Eigen (Eigen, 1971; Eigen and Schuster, 1979), are only tenuously related to the idea of dissipative structures and instead are *sui generis* to the structure of living matter. Still, it may be that they contain deep problems of the same sort that destroy the conventional ideas about dissipative structures. We are setting out to study this question in detail.

The above is our basic outline. Now we will set it all out in more specific and detailed terms.

Initially, we must consider some things about the real physics: What is "broken symmetry"—more properly called "spontaneously broken symmetry"—and why does it occur? The answers to both questions are so simple that we almost miss their depth and generality. First, what is it? Space has many symmetries—it is isotropic, homogeneous, and unaware of the sign of time, at the very least. Correspondingly, the equations that control the behavior of all particles and systems of particles moving in space have all these symmetries. But nature is not symmetric; "*Nature abhors symmetry.*" Most phases of matter are not symmetric; the crystals of which all rocks are made, for instance, are neither homogeneous nor isotropic, as Dr. Johnson forcefully pointed out. Molecular liquids are often not isotropic but form liquid crystals (Figure 1); magnets, such as iron, or rust, which is antiferromagnetic, are not invariant under time-reversal. Superfluids

FIGURE 1. (A) Nematic liquid crystal in the disordered state. The line segments represent the rodlike molecules of the nematic. Averaging molecular orientations over macroscopic distances yields zero. (B) For a suitable choice of thermodynamic parameters, the nematic enters the ordered state, with the appearance of a macroscopic order parameter (the director $\vec{D}$). The system is no longer isotropic, but has chosen a special direction; rotational symmetry has been broken.

break one of the hidden symmetries of matter, the so-called gauge symmetry, allowing the phase of quantum wavefunctions to be arbitrary and related to the laws of conservation of charge and number of particles.

Why does symmetry-breaking occur spontaneously? Fluctuations, quantum or classical, favor symmetry. Gases and liquids are homogeneous; magnets at high temperatures lose their magnetism. Potential energy, on the other hand, always prefers special arrangements: atoms like to be at specific distances from each other; spins like to be parallel or antiparallel; and so on. Thus, we define *spontaneously broken symmetry:*

> *Definition.* Although the equations describing the state of a natural system are symmetric, the state itself is *not*, because the state can become unstable with respect to the formation of special relationships among its component atoms, molecules, or electrons.

So far the idea of spontaneous symmetry-breaking is purely descriptive; it becomes meaningful when we find that it relates and explains many apparently different and unrelated phenomena. Initially, the concept was introduced by Landau and Lifschitz (1969) to solve a series of problems related to the nature and meaning of thermodynamic phase transitions, but it also connects and explains many other properties of broken symmetry phases. The single most important idea in Landau's whole theory was that of the *order parameter.*

According to Landau, loss of symmetry requires a new thermodynamic parameter (the order parameter, η) whose value is zero in the symmetric phase. It appears, for instance, in the magnetization of a ferromagnet (Figure 2), and the magnitude of the order parameter η measures the degree of broken symmetry. It is a quantitative measure of the loss of symmetry.

The canonical order parameter $\vec{M}$ is in a ferromagnet, which is the mean moment $\langle \mu_i \rangle$ on a given atom. Others are:

1. Director $\vec{D}$ of the nematic liquid crystal
2. Amplitude ρ_G of the density wave in a crystal
3. Mean pair field in a superconductor $\langle \psi(r) \rangle$

This idea has many implications. For instance, the appearance of a wholly new thermodynamic variable is a necessary condition for a continuous (so-called second-order) phase transition. The transition can occur discontinuously, and often does, but it need not—see $M(T)$ in Figure 2, and for contrast, $\rho_G(T)$ in Figure 3 for a typical crystal. Also, because there is an extra parameter, the free energy and all the thermodynamic properties can never be the same mathematical

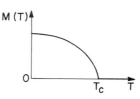

FIGURE 2. Variation of magnetization M with temperature T in a simple ferromagnet. This is a typical second-order phase transition, in which the order parameter grows continuously from zero as T is lowered below a critical temperature T_c.

functions in the two phases of different symmetry, so the phases are always separated by a sharp phase transition—unlike, for instance, that between liquid water and steam.

The thermodynamic consequences that flow merely from the amplitude of η are sufficient justification for the broken symmetry concept. Even more important consequences follow from another property of the order parameter that Landau never formalized but sometimes used, viz., *the order parameter is a quantity that always has a phase, and the free energy $F(T, |\eta|)$ is a function of its magnitude $|\eta|$ but must not be so of its "phase" or direction, because of the existence of the original symmetry.* For instance, the energy may not depend on the direction of the director because space is isotropic; it may not depend on the orientation or position of a crystal, or on the phase, ϕ, of the superfluid wave function. Another way to say this is that the order parameter has a space within which it is free to move without changing the energy. (In quantum-mechanical terminology, the ground state is highly degenerate in the broken symmetric phase—a condition that in a way is a remnant of the original symmetry of the Hamiltonian, which remains unchanged. This condition is connected to some of the dynamical consequences of broken symmetry, such as Goldstone modes and the Higgs phenomenon, as will be discussed below.)

Another property of η is obvious if we see it as a physical thermodynamic parameter; η may vary over macroscopic distances in the sample, and $\eta(r)$ may be defined locally, just as we can define a local temperature or pressure in a sample not too far from equilibrium, if they do not vary too rapidly.

From these characteristics of η we can rationalize three major emergent properties of spontaneously broken symmetry:

1. Generalized rigidity
2. New dynamics
3. Order parameter singularities and their role in dissipative processes

All these properties are very interesting, because most of the important characteristics of solids depend upon them. Here we discuss only the first—the simplest and most general.

Again we use the idea that η is a physical thermodynamic parameter to which we can by one means or another apply a force. This claim is clear in the case of $\vec{D}$—which couples to boundary orientation—or in the case of $\vec{M}$, or of crystal orientation θ, ϕ; but it can be a little more esoteric for "hidden" order parameters

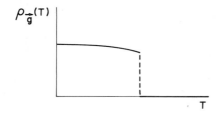

FIGURE 3. In a first-order transition, such as the liquid-to-solid crystal transition shown here, the order parameter will exhibit a discontinuous jump at the transition with an associated release (or absorption) of latent heat.

like sublattice magnetization in the antiferromagnet, or ψ, the superfluid order parameter. Nonetheless, it is always possible to grasp η at any point in the system. While F is not a function of the phase angles of η, it is naturally a function of the gradient of these phases, because otherwise arbitrarily large relative fluctuations of the phase would destroy the existence of the order parameter. Thus, we must have

$$F = F(|\eta|, |\nabla\eta|^2, \ldots)$$

and

$$\frac{\partial^2 F}{\partial(\nabla\eta^2)} > 0$$

There is a positive stiffness for variations of η. This result is enough to ensure that if we exert a force on $\eta(r)$ at one end of a sample, then $\eta(r')$ will respond at the other end. We can essentially use η as a crankshaft to transmit forces from one point to another, i.e., to exert action at a distance (Figure 4). This rigidity is a true *emergent property*; none of the forces between actual particles are capable of action at a distance. It implies that the two ends cannot be decoupled completely without destroying the molecular order over a whole region between them.

Rigidity of solids, then, is a model for a wide class of other rigidity properties, including permanent magnetism, ferroelectricity, and superconductivity and superfluidity. These last two have, since the discovery of the Josephson effect, been understood to be the phase rigidity of the order parameter $\psi(r)$ (Anderson, 1966).

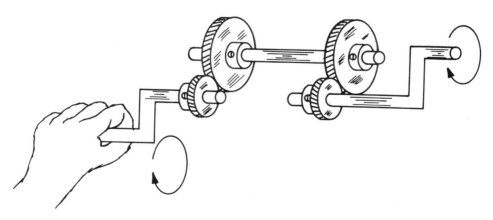

FIGURE 4. Illustration (somewhat schematic) of generalized rigidity. An external force (the crank) couples to the order parameter at one end of the system, represented as a gear. A change in the order parameter at any point in the ordered system is transmitted to all other parts of the system (first gear turns the second gear). The second gear turns the second crank; a force has been transmitted from one end of the system to the other via the order parameter.

The other two major emergent properties listed above are also consequences of this phase freedom in broken symmetry systems. The phase freedom underlies the existence of long-wavelength collective motions of the order parameter, such as phonons and spin waves, which are the models for the Goldstone and Higgs phenomena of elementary particle physics; it also accounts for the existence and classification of singularities and textures of the order parameter. The singularities give the possible order parameter fields that are allowed when we permit lower-dimensional regions to be excluded from the order parameter field $\eta(r)$—including vortex lines and dislocations, domain boundaries, singular points, and so on. Broken symmetry gives rise to the appearance of new length scales that did not exist in the symmetric phase.

We return now to the main theme of our discussion--that there does *not* exist a corresponding theory of the dissipative case. First, we describe the kinds of experiments that lead to very similar types of broken symmetry in the dissipative case. The canonical example is the Bénard instability; a layer of fluid heated from below, which, once a critical heating rate is exceeded, exhibits very regular-appearing "rolls" of convection, arising spontaneously with a rather fixed size or wavelength (Figure 5). Other examples abound, such as the Couette instability of a viscous fluid between rotating cylinders (Figure 6) or even the laser exhibiting a periodic wave of excitation density (Figure 7).

Clearly, all these systems exhibit spontaneously broken symmetry in the simple sense. In each case the sign is arbitrary, and an initially homogeneous state changes into an inhomogeneous one. The initial transition is often continuous, as in the typical second-order transition, and it has often been suggested that there is some kind of deep analogy between these two types of systems. There is indeed one mathematical aspect in which there is at least a similarity—both are examples of dynamic instabilities, for which there exists a general mathematical theory ("catastrophe theory"), described by Thom (1975), that has been elaborated in recent years by many mathematicians (e.g., Ruelle and Takens, 1971). But the thermodynamic phase transitions invariably present only the simplest kind of catastrophe, the so-called "bifurcation," and the simplest type of state, the so-called "fixed point," whereas the other dynamical instabilities seem always to evolve—even oversimplified mathematical models of them evolve—toward more and more complex behavior leading eventually to completely chaotic behavior. The evolution of chaos in such systems has been beautifully described by Gollub and Swinney (1975) and by Abraham and Marsden (1967). Some of the beautiful work in following the successive instabilities from classical to steady

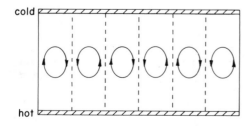

FIGURE 5. The Bénard instability in rectangular geometry. A layer of fluid between two horizontal rectangular plates is heated from below. When a sufficient thermal gradient is reached between top and bottom plates, convection arises in the form of rolls. In this cutaway edge-on view, the arrows represent the fluid velocity.

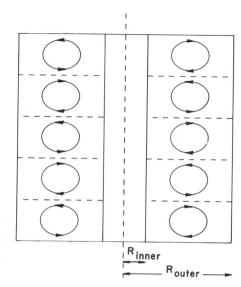

FIGURE 6. Couette flow: A fluid is placed between two cylinders with different rotational velocities about their axes. When the velocity gradient exceeds a critical value, rolls of vortices form. In this view the cylinder is cut along its length.

rolls, to singly-periodic dynamics, to multiply-periodic dynamics, and finally to total chaos are described in the last section (Section IX) of this book.

Experimentally the situation is more complex. Ahlers and Behringer (1978) particularly have shown that even the complicated behavior seen by Gollub and Swinney (1975) and predicted by the mathematicians may be an artifact of an overconstrained system heavily influenced by its boundary conditions: they find finer-scale chaos or near-chaos even in the apparently quiescent region of the Bénard system. This finding is inevitable, and dissipative structures in a real, physical, open system unconstrained by artificial boundary conditions will inevitably be chaotic and unstable (Anderson, 1980; Stein, 1980). [For instance, the laser can be persuaded to oscillate in a single mode only with the utmost artificiality and difficulty. This result depends on the proper placement of endplates or mirrors so that here broken symmetry is strongly dependent on externally applied boundary conditions. Lasers occurring naturally in nature (e.g., from astrophysical sources) seem to show no mode selection.]

Prigogine and his school have made a series of attempts to build an analogy between these systems and the Landau free energy and its dependence on an order parameter, which leads to the important properties of equilibrium broken symmetry systems. The attempt is to generalize the principle of maximum entropy

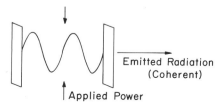

FIGURE 7. In a laser, a standing wave of excitation density is set up between two end plates, or mirrors, resulting in emission of a beam of coherent radiation.

production, which holds near equilibrium in steady-state dissipative systems, and to find some kind of dissipation function whose extremum determines the state. As far as we can see, in the few cases in which this idea can be given concrete meaning, it is simply incorrect. In any case, it is clearly out of context in relation to the observed chaotic behavior of real dissipative systems.

Thus, we conclude that there is no analogy between the stability, rigidity, and other emergent properties of equilibrium broken symmetry systems and the properties of dissipative systems driven far from equilibrium. The latter types of systems have never been observed to exhibit the rigidity, stability, and permanence that characterize the thermodynamically stable broken symmetry systems, nor has any mathematical reason been found why they should. [One driven system that might have exhibited broken symmetry but failed to do so is described in Anderson (1980) and Stein (1980).] Unfortunately, many authors have chosen to use such systems as the laser and the Bénard instability as models for the nature and origin of life itself, as an emergent property of inanimate matter. It is indeed an obvious fact, noted since Schrödinger (1945), that life succeeds in maintaining its stability and integrity, and the identity of its genetic material, at the cost of increasing the rate of entropy production of the world as a whole. It is, at least in that sense, a stable "dissipative structure"—i.e., an existence proof by example.

Turing (1952) long ago observed that a fertile source of dynamic instabilities was the autocatalytic chemical reaction in which reaction products serve as catalysts as well. The base-pairing mechanism of DNA is an obvious and good example. Eigen in particular has tried to develop a theory of autocatalytic instabilities in the primeval soup as a detailed explanation of the origin of life (Eigen, 1971; Eigen and Schuster, 1979). It is a glorious picture to imagine the growth of an "order parameter in molecular information space," driven by a dynamic autocatalytic instability and self-stabilized in some mystical way by the magical power of Darwinian evolution. This may be in fact the way it happened—one can hardly assume it did not! But there are reasons to be skeptical of the claim that we have found the full story. Why should dynamic *instability* be the general rule in all dissipative systems except this supremely important one? We are attempting a computer simulation of a model of the origin of information-carrying macromolecules that is already producing quite interesting results in terms of the spontaneous generation of complex molecules.

In the simple picture we are using, we begin with a "soup" of monomers of two different varieties, A and B, and an externally applied energy flux that drives the system toward formation of strings of monomers, according to a simple set of rules for lengthening and shortening chains. This process relies on temperature cycling. In the low-temperature phase, two strands (or a strand and monomers) attach weakly via A–B attraction (as in hydrogen bonding between a purine–pyrimidine pair). Although the pair is held together in this fashion, stronger bonding may take place at the ends or in other regions between two adjacent, previously unattached strands. In the high-temperature phase, the hydrogen bonds break, but the stronger bonds along the length of the strands do not, and the newly created (or lengthened) strands separate until the next cycle. There is a slightly higher probability for strong bonding between dissimilar monomers than between similar ones. There is also a certain chance that, in the high-temperature phase, a strong bond may be broken (and a strand thereby shortened) because, for example, of interaction with an energetic cosmic ray. In addition to these "birth" and "death" rates (more accurately, lengthening and shortening processes), there is also a small error probability; that is, in the low-temperature phase, an A monomer may mistakenly hydrogen bond to another A, rather than to a B as it should. These birth, death, and error rates thus form a complete prescription for building many lengthy strands, starting from a sea of lone monomers and a single strand of two or three monomers.

We wish to see if, from this very simple picture, a polymer with nontrivial information content will be selected from the near-infinite number of possibilities, selection (if it exists) being implicit in the strong nonlinearity of the problem. Clearly, if most chains are of the form ABABABAB ... (or AAAAA ... or BBBBB ...), nothing very interesting has happened. On the other hand, if many chains with irregular sequences, such as ABBABABAAABBA..., are formed but no pattern appears to predominate, again little of interest has occurred.

In looking for patterns that may predominate, we have found it most useful to search for "triplets," by which we mean the following. Suppose we are given the strand

$$\underbrace{AB}_{2} \; \underbrace{BABABA}_{6} \; \underbrace{A}_{1} \; \underbrace{AB}_{2} \; \underbrace{BA}_{2}$$

Whenever two like monomers appear adjacent, we draw squiggly lines separating them, as pictured. We then count the respective lengths of the purely alternating sequences that make up the polymer (these are the numbers that appear in the example above). A "triplet" is then the triplet of lengths of three adjacent alternating sequences; in the example above, we have a (2, 6, 1), (6, 1, 2), and (1, 2, 2). Note that the mirror image of the polymer would give the same result. We therefore wish to see if certain triplets make up the bulk of most long polymers. This seems to us to be more useful than trying to select an entire polymer itself as the prototype of what should be selected.

Our preliminary results indicate that for certain choices of bonding probabilities, selection of a number of triplets occurs and, in fact, can be quite strong (as well as persistent over many cycles, which is a requirement if we are to say selection has occurred). It is also amusing, and somewhat unexpected, that a small error probability is necessary for selection to occur in the cases studied so far.

Many questions remain unanswered, the most prominent of which is, how does one assign a meaningful information content to a polymer? So far we have only discussed necessary, but not sufficient, conditions for symmetry breaking in "information space" to occur. One would guess that, in some sense, structure and function are intimately related. In the real world DNA serves as a blueprint for manufacture of proteins, some of which act as enzymes in replication and other processes governing the DNA molecule itself. Is there any way in which this process can be seen in the simple model presented here? This is one of the most fundamental problems in understanding the origin of life, a not so subtle variant of the "chicken and egg" problem (Eigen, 1971; Eigen and Schuster, 1979). We are not attempting to answer this problem at this stage, but rather the somewhat less ambitious problem of whether or not one can relate the issues of symmetry-breaking discussed earlier to the problem of the origin of life (specifically, a primitive genetic code in this instance), and in what context this is possible and meaningful.

This preliminary work has been extended in a series of publications (Anderson, 1983; Stein and Anderson, 1984; Stein, 1984; Rokhsar et al., 1985) using some additional concepts, especially a special ansatz about Darwinian selection, and we feel that the beginnings of a real theory of origins may be at hand. While the newer work has a considerably more sophisticated technical and philosophical basis, none of the above ideas has been seriously abrogated. We might add that, to our thinking, there is no serious disagreement at this time, except perhaps as to emphasis, between us, on the one hand, and the Eigen–Schuster group, on the other, and that similar thinking has been described by H. Kuhn. In particular, at present the discovery of the catalytic properties of RNA itself (Cech et al., and others) has made RNA the overwhelming favorite for the earliest true biomolecule.

We conclude by iterating our main point. Because we understand the process in all details, we still believe in the reality of emergent properties, i.e., in the ability of complex physical systems to exhibit characteristics unrelated to those of their constituents. But we do not believe that stable "dissipative structures" maintained by dynamic driving forces can be shown to exist in any inanimate system, and thus we do not see how speculations about such structures and their broken symmetry can yet be relevant to the still open question of the origin and nature of life.

ACKNOWLEDGMENTS. The work at Princeton University was supported in part by National Science Foundation Grant DMR 78-03015 and in part by U.S. Office of Naval Research Grant N00014-77-C-0711.

References

Abraham, R., and J. Marsden (1967) *Foundations of Mechanics.* Benjamin, New York.
Ahlers, G., and R. P. Behringer (1978) Evolution of turbulence from the Rayleigh-Bénard instability. *Phys. Rev. Lett.* **40**:712–716.
Anderson, P. W. (1966) Considerations on the flow of superfluid helium. *Rev. Mod. Phys.* **38**:298–310.
Anderson, P. W. (1981) In: *Order and Fluctuations in Equilibrium and Nonequilibrium Statistical Mechanics,* G. Nicolis, G. Dewel, and P. Turner (eds.) Can broken symmetry occur in driven systems? Wiley, New York, pp. 289–297.
Anderson, P. W. (1982) *A Suggested Model for Pre-Biotic Evolution: The Use of Chaos.* In press.
Anderson, P. W. (1983) A suggested model for pre-biotic evolution: The use of chaos, *Proc. Natl. Acad. Sci. USA* **80**:3386.

Eigen, M. (1971) Self-organization of matter and the evolution of biological macromolecules. *Naturwissenschaften* **58**:465–523.
Eigen, M., and P. Schuster (1979) *The Hypercycle: A Principle of Natural Self-Organization.* Springer-Verlag, Berlin.
Glansdorff, P., and I. Prigogine (1971) *Thermodynamic Theory of Structure, Stability and Fluctuations.* Wiley, New York.
Gollub, J., and H. L. Swinney (1975) Onset of turbulence in a rotating fluid. *Phys. Rev. Lett.* **35**:927–930.
Haken, H. (1977) *Synergetics: An Introduction.* Springer-Verlag, Berlin.
Landau, L. D., and E. M. Lifschitz (1969) *Statistical Physics.* Pergamon Press, Elmsford, N.Y.
Nicolis, G., and I. Prigogine (1977) *Self-Organization in Non-Equilibrium Systems.* Wiley, New York.
Rokhsar, D. S., P. W. Anderson, and D. L. Stein (1985) Self organization in prebiological systems: A model for the origin of genetic information. To be published in *J. Mol. Evol.* **23**:119–126.
Ruelle, D., and F. Takens (1971) On the nature of turbulence. *Commun. Math. Phys.* **20**:167–192.
Schrödinger, E. (1945) *What is Life?* Cambridge University Press, London.
Stein, D. L. (1980) Dissipative structures, broken symmetry, and the theory of equilibrium phase transitions. *J. Chem. Phys.* **72**:2869–2874.
Stein, D. L. (1984) A model for the origin of biological information. *Int. J. Quant. Chem.* **11**:73–86.
Stein, D. L. and P. W. Anderson (1984) A model for the origin of biological catalysis, *Proc. Natl. Acad. Sci. USA* **81**:1751–1753.
Thom, R. (1975) *Structural Stability and Morphogenesis.* Benjamin, New York.
Turing, A. M. (1952) The chemical basis of morphogenesis. *Philos. Trans. R. Soc. London B* **237**:37–72.

24

Thermodynamics and Complex Systems

Harry Soodak and Arthur S. Iberall

ABSTRACT

This chapter describes a generalized thermodynamic construct competent to deal with simple or complex natural field systems, at all levels of organization. The construct applies to cosmic, galactic, stellar, planetary, chemical, biological, and social systems and has the capability to deal not only with the ongoing dynamics of these fields, but also with their slower evolution. To understand motion and change in these field systems, it is necessary to distinguish between fluid processes, which can develop patterns, and condensation processes, which can create more permanent forms. Both processes break symmetry. Although the symmetry-breaking involved in fluid processes is well known, the symmetry-breaking occurring in condensation of matter (self-organization of form) is more obscure. Three properties characterize condensed matter: rigidity, an elastic limit, and flow. Flow processes involved in matter condensation may be either external or internal. Organization of form occurs by an external, in part radial, flow process that brings together atomistic constituents. The constituents develop an elastic limit by giving up an energy of binding. The cooperative binding also enhances the rigidity of the field. The field can be stressed by local processes up to the elastic limit without appreciable change in form. At stresses beyond that limit, the form of the system can be degraded by induced, new flow processes, arising either within the previously bound atomistic constituents or outside.

Complexity of systems may be ascribed to associational (bulk) viscosity, which expresses internalized, fluid mechanical and chemical dissipative processes. The measure of the cooperative binding which leads to condensation of form is given by a flow criticality condition, described by a generalized Reynolds number. The way is pointed out to apply these very general ideas in the study of global geophysical phenomena and of human cultures. —THE EDITOR

Structure and Function in the Universe

Structure and process occur at all material scales, from elementary particles to the cosmos. Quarks bond to form protons and neutrons, which bond to form

HARRY SOODAK • Department of Physics, The City College of New York, New York, New York 10031. **ARTHUR S. IBERALL** • Department of Oral Biology, University of California, Los Angeles, California 90024.

nuclei, which bond with electrons to form atoms, which bond to each other to form molecules. Atoms, ions, and molecules in larger numbers bond into a considerable variety of organized matter forms—clouds of gas, liquids and solids of various sizes, from dust to planetary subsystems, to planets, on to stars and galaxies. The living cell is an aggregate of atoms and molecules in a bonded structure of liquid–solid form. Cells bond to form colonies, organelles, organs, and organisms. Organisms of a given species bond to form breeding groups and societies. Human organisms even bond to form polities and civilizations. Structure and process are causally linked: process is guided and constrained by structure; structure is laid down, maintained, changed, and degraded by process. Thus, one symmetry-breaking mechanism can beget another in a historical sequence that constitutes the evolutionary history of the universe.

The structures and processes at any level are determined by the atomistic units, by the forces that bind them, and by dissipative mechanisms. Atomic models postulating hard spheres with no mutual attractions can account for the gaseous state and for flow fields described by Navier–Stokes equations, involving shear viscosity and thermal conductivity. The addition of a van der Waals attraction introduces the richness of binding and leads to condensation into the liquid or solid states and to multiphase systems that may have associational (i.e., bulk) viscosity as well as flow (i.e., shear) viscosity. Adding the effect of gravitational attraction and nuclear reactions leads to the formation of galaxies, stars, clouds of dust, and planets. The presence of various atomic types, such as Si, C, H, N, O, P and S and their associated compounds, leads further to the complexities of geophysics and geochemistry on planetary bodies and to the living state of matter—cells, organisms, and societies.

A Generalized Thermodynamic View

In our view, one set of thermodynamic principles applies to all levels of organized unit activity, including those of the living state. At each level stereotypic activities within the individual atomistic units and among units as they interact in small numbers, determine the dynamic behavior of a system formed from many units. The kinetic behavior is described in terms of the microscopic coordinates of the units. The macroscopic behavior of the system is governed by laws representable as continuum or field equations in terms of macroscopic coordinates which are integrals or averages over appropriate microscopic quantities. These macroscopic coordinates and their interrelations are to be regarded in a dual fashion: they are emergent properties, arising from the kinetic behavior; and they represent set measures or constraints on the kinetic behavior. Thus, the micro- and macro-levels are mutually linked. The kinetics at the micro-level determines the continuum field mechanics and is the source of internal *macroscopic* fluctuations. In turn, that kinetics is constrained by the macrostate. Finally, the behavior of the macrosystem, as described by its field equations, is constrained by boundary conditions from outside the system. These outside constraints (which may be fixed or variable) often originate from a higher-level system of which the macroscopic system is itself simply one of the atomistic units.

We therefore regard thermodynamics as addressing systems at any level, including their relation to the levels immediately below and above them.

Complexity

Complexity begins with associational (bulk) dissipation processes. A system composed of only hard-sphere atomic or atomistic units with no attraction forces (e.g., a near-ideal gas), is therefore not a complex system. Yet it does reveal a first level of symmetry-breaking. The kinetic behavior of such a simple system consists of a repeated set of elastic collisions characterized by an emergent mean path and mean time. This hydrodynamic field, as described by the Navier-Stokes equation set, invoking only shear viscosity and thermal conductivity, then manifests a variety of flow patterns determined by constraints or boundary conditions from a higher level. The flow patterns cannot give rise to bound, condensed forms. Nevertheless, the birth and death of these flow patterns are a consequence of a first level of symmetry-breaking.

The addition of attractive forces between the hard-sphere units augments the external diffusivities, but also leads to associational effects and processes among small groups of units, to associational disequilibrium and damping, to the formation of liquid and solid forms (gels, plastics, memory-processing associations, ion chain polymers, living cells), and to the beginning of complex behavior.

As the internal activities within each unit and the modes of interaction within and among units become richer or more complex, so does the system composed of such units and its behavior. The description of the kinetics of such complex systems then must be extended to include both a stereotypic description of the internal activities within the units and the interaction among small groups of units. If form and function within the field are maintained, then the local kinetics consists of a repeated set of the various types of interaction, along with the set of internal behavior patterns or modes that occur within each unit. The resultant may be described in terms of a "factory" and "factory day." The factory consists of the repetitive activities within each atomistic unit, and the factory day is the time required for a cycle or ring of all of those repetitive activities. At the end of each factory day the unit is substantially ready to repeat the activities for the next factory day, thereby assuring autonomy of the unit and of the field system. Because the dissipative energy of internal activities is stored or in transit during any or all segments of the factory day, the factory day may be used as one measure of an overall internal delay time.

A single overall measure of the complexity of a system of units is the ratio of the internal delay time (the factory day measure) to the translational delay time of interaction between units. This measure is on the order of magnitude of the bulk-to-shear-viscosity ratio. It is clear, however, that a full treatment of complexity requires a greater number of measures (not just a single shear and single bulk viscosity) to describe both interactional dissipation and internalized dissipation. Interactional dissipations comprise the translational or shear processes; internalized dissipations represent the associational or bulk processes, including chemical reactivities. Complex systems tend to display a cascade spectrum of many relaxation processes. The field equations for complex systems involve a comparably large number of macroscopic coordinates and transport coefficients describing the various dissipative mechanisms at different scales.

Structure Formation

If a system governed by continuum field equations is isolated within a given region of space, then the dissipative mechanisms drive it inexorably to thermodynamic equilibrium among its interacting units. The situation is changed, however, if the system is stressed by externally applied forces and fluxes. For small stresses the system achieves a quasi-steady state described simply in terms of gradients, potentials, and dissipative fluxes. The system is then on what Prigogine (1978, 1980) calls the "thermodynamic branch." When the stresses exceed certain critical values, a symmetry-breaking can occur because of flow or associational effects (first or second kinds of symmetry-breaking), or both. This bifurcation leads to spatial or temporal patterning of flow (as in eddies) or of association (as in chemical patterns). These patterns are called "dissipative structures" by Prigogine. They signal the possible presence of forms which may condense into stable structures by the combined effects of binding forces and associational dissipation. If such stable structures appear, they represent the formation of multiphases, of liquids and solids, living cells, galaxies, stars, planets, and other such widely repetitive units. In our view, the emergence of form, whether as evanescent flow pattern or more solid state, is signaled by stress criticality conditions. The actual formation of stable structures is completed by the action of binding and dissipation.

As a first example of flow criticality, consider a Navier-Stokes fluid with only shear viscosity. In shear flow criticality is reached when the Reynolds number, Re, is on the order of unity. A Reynolds number of this magnitude signals the instability of laminar sheet flow and the emergence of self-generated turbulent eddies. In the absence of binding forces, these eddies simply remain flow patterns with no formation of stable structure. For flow past a sphere, the condition Re = 1 specifies departure from laminar flow (and thus from the Stokes drag formula) at distances from the sphere on the order of the sphere size. For flow through pipes, the condition Re = 1 represents the scale of the boundary layer of transition between laminar and turbulent flow, which is the locus of eddy initiation. Specifically, the Reynolds number is

$$\text{Re} = LV\rho/\mu \tag{1}$$

where ρ is mass density, μ is shear viscosity, L is sphere size or boundary layer size, and V is the far-field flow speed which stresses the fluid. (A similar criticality condition arises in Bénard thermal instability, where the measure of criticality involves the dimensionless Rayleigh number, which includes effects of dissipation by shear viscosity and thermal conductivity.)

The shear viscosity is related to macro and micro properties of the fluid through the relation

$$\mu \approx \beta \tau_{tr} = \rho c^2 \tau_{tr} \qquad (2)$$

where β is the fluid bulk modulus, $C = \sqrt{\beta/\rho}$ is the sonic speed, and τ_{tr} is the mean time between atomic collisions. The mean path δ between collisions is then

$$\delta \approx c\tau_{tr} \qquad (3)$$

For given values of the mechanical properties ρ and β, the value of the thermodynamic transport coefficient, μ, specifies the values of δ and τ_{tr}, the space and time scales of the atomistic fluctuations.

The macro–micro aspect of the Reynolds number may be exhibited in the following fashion: the relationship

$$\mu/\rho = c^2 \tau_{tr} \qquad (4)$$

is an expression for the momentum diffusivity (the diffusion coefficient for momentum transport by shear viscosity). The time required for momentum to *diffuse* a distance L is on the order of $L^2\rho/\mu$. The time required for momentum to be *convected* into a region of extent L is L/V, where V is the flow speed. The Reynolds number is then the ratio of diffusive time to convective time. A Reynolds number which is large compared to unity means that the diffusive time is not short enough to handle the incoming energy, and a new process must appear.

A richer situation exists when the fluid has an associational or bulk viscosity, λ, in addition to shear viscosity. In this case, additional time delays (τ_{bulk}) of an associational nature are present as well as those represented by the translational interactional delay of shear. These extra delays, called into play by the imposition of time varying bulk stress (e.g., a density change), extend the relaxation time required to achieve thermodynamic equilibrium by energy partitioning or by a process related to achieving configurational equilibrium. (These additional time delays arise from relaxations within the atomistic constituents or from the associations among them.) In a simplified treatment we let a single overall relaxational time, τ, stand for the total delay process:

$$\tau \approx \tau_{tr} + \tau_{bulk} \qquad (5)$$

τ governs the absorption of sound waves for fluids having bulk viscosity. In an acoustic wave of frequency f, the density is high for half a cycle (of duration $1/2f$) and low for the next half cycle. Small absorption may be expected at low frequency when

$$\tau \ll 1/2f \qquad (6)$$

because the displacement motion within the degrees of freedom can keep pace with the slow change. Small absorption also may be expected at high frequency, when

$$1/2f \ll \tau \qquad (7)$$

because displacements within the degrees of freedom are relatively unaffected by such rapid changes, responding instead mainly to the average density of the fluid. Thus, maximum absorption may be expected at a frequency given by

$$1/2f = \tau \qquad (8)$$

or

$$f\tau \approx 1/2 \qquad (9)$$

See Herzfeld and Litovitz (1959) for more precise results.

The maximum absorption coefficient can be expressed in the form of a critical value of unity for an acoustic Reynolds number (Greenspan in Herzfeld and Litovitz, 1959). This may be defined by

$$\text{Re(ac)} = \frac{(\ell/2)\rho c}{\mu + \lambda} \qquad (10)$$

where the length scale is half a wavelength ($\ell/2$), the speed is the sonic speed C, and the denominator is the sum of the two viscosities, shear (μ) and bulk (λ). It then may be seen that

$$\text{Re(ac)} = 1/2f\tau \qquad (11)$$

and that the absorption peak occurs at

$$\text{Re(ac)} = 1 \qquad (12)$$

The critical value unity for the acoustic Reynolds number represents a maximum disequilibrium generated in some internal degrees of freedom. *This criticality condition is clearly an appropriate gateway to the creation of new structure.* We conjecture that this criticality condition holds for all Reynolds numbers dominated by bulk viscosity

$$\text{Re} = LV\rho/\lambda \qquad (13)$$

The actual formation of a stable, associational structure depends on two additional conditions. First, binding energy must be transformed (usually by

24. Thermodynamics and Complex Systems

internal dissipation within the forming structure) and eliminated from the structure. We have illustrated that process in Chapter 27, indicating that the elimination has to be rapid. Second, the binding energy must be large compared to the energy of interaction between the structure and external agents. Otherwise the structure might not form or, if formed, might be disrupted easily.

In summary, the creation of a new form depends on a criticality condition given by the Reynolds number dominated by associational viscosity. Actual creation of the form will be initiated at criticality and completed if binding energy can be dissipated. Binding confers both additional rigidity and an elastic limit (e.g., an energy density per unit volume) below which external stresses do not change the basic structure. These ideas now can be applied in general to the solid state.[1]

Application of thermodynamics to solid or associated state matter must provide a uniform description of both small and large strain processes and their time-dependent displacement yields. That description requires a coordination of the three properties (as processes): rigidity, elastic limit, and flow. Flow may be either momentum flow or internalized flow. Assembly can be seen as flow. Form is generated by a radial, or aggregating, flow in which atomistic units carry matter and energy into a condensation or nucleation region (see also Chapter 27). A considerable fraction of that energy is then given up to create binding into a cooperative system. It is assumed that there are no other energetic fluctuations in the vicinity to fracture or evaporate the bonds. That is, an asymmetry between the energy of binding and of unbinding appears.

The solid state can be detected by the increased resistance to external stress, which depends on the rigidity. Disassembly will depend on flow process yields beyond the elastic limit among or within atomistic units. We now point the way to apply these thermodynamic constructs to the study of geophysics and of cultures.

Extension to Geophysics

The above rudimentary account of the properties of materials is competent to deal with the thermodynamic physics of the earth. It is not known exactly how the earth's constituents assembled as an aggregating flow process.[2] It is probable that they accreted into distinct layers, each with its own rigidity. It is known that

---

[1] As an illustration of a system just barely reaching the solid state of form, a good-quality latex rubber is possibly the stretchiest elastomer known. It may exhibit a bulk modulus of about 100,000–200,000 psi, a Young's modulus of about 50–100 psi, a Poisson's ratio of about 0.499$^+$, and a strain at the elastic limit of about 25%. To appreciate fully how near it is to liquidlike characteristics, consider that it can stretch almost 1000% before breaking. But its true elastic limit (to which it can be stretched repeatedly without taking set) is only 25%.

[2] One current, humorous description: "By chance a few bodies grow especially large and eat up their neighbors rather like tadpoles growing from frog spawn" (Smith, 1976).

some layers, such as the mantle and the asthenosphere, flow under thermal-mechanical stresses and give rise to plastic–elastic characteristics of the surface of the earth, its plate dynamics, and its interactions with the contiguous film of oceans.

As far as is now known, the first 0.8 billion years of the earth's history (4.6–3.8 billion years before present) was devoted largely to assembly and settling out of major plastic–elastic and fluid processes in independent phases (Windley, 1976; Iberall and Cardon, 1980). From about 3.8 to 2.5 billion years before present, most of the sequestering of surface materials occurred. Plates probably formed, and vertical plate motion (rocking) began. The ocean basins filled. From that period on, it appears that there was a rough constancy of mean sea level and a semistationary spectrum of sea level fluctuations or changes in height in a range up to several hundreds of meters. These fluctuations relate largely to vertical displacements and movements of plates. They arise from horizontal motions of plates, subductions and collisions, erosion of plates, midoceanic rift production of surface, and thrust faults. (See, e.g., Uyeda, 1984.) The fluctuational processes, in general, are "slow" relaxations in ocean level followed by "rapid" emergence of land.

The major "periods," in millions of years (My), associated with these fluctuations (for the past 2.5 billion years) and their possible causal processes are:

- 200- to 300-My relaxations, probably resulting from a turnover in mantle convection cells. (Mantle viscosity, for example, lies in the 10^{19} to 10^{23} poise range. This might be compared to a value on the order of 10^{13} poise for a glass, 10^{-2} poise for water, or 10^{-4} poise for a gas.)
- 40- to 60-My relaxations, associated with horizontal plate movements. (This relaxation, for example, is represented by the latest phase of the making and positioning of current continents.)
- 3- to 5-My relaxations, caused by continental erosion, making the unbalancing sedimentary deposits on plates around continental margins. (These materials are statically unbalancing because of the density differences of major chemical rock materials. The chemical thermodynamics of crust and mantle, both oceanic and continental, relate to two major constituents, the basalts and the granites. The erosion time scale can be computed from the hydrological cycle. In 100 My, that cycle can completely erode a continent to ocean level. Thus, the observed erosions proceed only to the creation of mechanical—tectonic—plate instabilities.)
- 100,000-year relaxations, related to weather patterns, temperature fluctuations, and the attendant sequestering of water by glaciation.
- 200- to 400-year relaxations, related to weather processes not yet well understood and associated with the smallest glacial fluctuations.

These many and varied process times and scales set the requirements for a thermodynamic account of geophysics.

24. Thermodynamics and Complex Systems

Extension to Society

We offer here a sketch of a thermodynamic approach to a study of human society. We select as our system an individual culture. Our approach has three interrelated aspects:

1. The organized activities of the society must be described in terms of macroscopic coordinates.
2. That description must be related to the microscopic properties and kinetics of the atomistic units, the individual persons making up the society.
3. There must be an account of the outside constraints, fluxes, and forces imposed on the culture, both from the geophysical environment and from the societal level above. The societal level above consists of neighboring, interacting cultures (making up a polity), as well as the genetic species development occurring on the evolutionary scale. In addition, because cultures are born, live, and die, there also must be an account of historical processes on time scales between the lifetime of a culture and the evolutionary scale for the development of the human species.

We define a culture as organized (spatially coordinated) fragments of one or more breeding groups, who—in the human case—also share a common memorizable and transmissible heritage of customs, language use, and tools (in toto, making up an epigenetic potential). A culture is a complex system consisting of varied atomistic units which are themselves complex systems. They are "factories" as described above. The genetic variety of the living units in a culture is determined partly by their peculiar chemical potential (that is, the genetic code). This genetic potential represents the possibilities inherent in the chemically determined genetic pool, affected both by evolutionary development of the species and by immigration and emigration. Phenotypic variety among individuals also is determined by the genetic variety, as well as by the culture itself and by the environment. The factory day activities, the modes and patterns of external and internal individual actions, depend on the genotypic and phenotypic distribution and, in addition, are guided and constrained by the culture.

The factory day actions comprise the "kinetics" of a culture. These (nonexhaustively) include person-to-person behavior patterns of various types—within a family, with outsiders, between different age groups, between sexes, or with man's exclusive division of labor, among occupational groups. They include the activities required for physiological and psychological maintenance day by day, season by season, and life-stage by life-stage. Finally, they include economic and political perceptions and actions. Most of these kinetic activities are culture dependent, and a strong relation between the macro and micro levels of society is evident. It appears that there is no single factory day time scale for societal kinetics. Just as in a liquid, in society there is a nested sequence of relaxation time scales. Nor was there a single factory day time scale for continental processes, as previously noted. Nevertheless, for each nearly autonomous operational scale there is a dominant "factory day".

The primary biological factory day is the earth day. Seasonal variations among interacting members of an ecosystem define another factory day—the earth year. There are natural factory days within any species consisting of generation time (approximately 20–30 years for us) and of an individual lifetime for each species, because each unit is itself a complex system that is born, develops into maturity, begets a new generation, becomes old, and dies. There are also intermediate time scales, such as multiple days, months, and years, associated with biological, economic, social, and political activities (e.g., 3- to 6-year business cycle; average time scales for political rulers; and for wars).

The microscopic coordinate variables that describe the kinetics of individual units in human cultures fall into five major groups. They appear in the magnitudes and fluctuations in: demographics, materials usage, energy usage, economics, and human action. Demographic variables refer to population and its change, its distribution measures by age, sex, and race, and death and reproduction of the units. Material variables refer to food, clothing, shelter, and other tools used by society. Energy variables refer to food, to physiological modes of activity, and to additional processes or systems manipulated by man—animal, chemical, and nuclear. Economic variables describe the rules and perceptions by which an individual trades goods and services. In such interactions "value-in-trade" is the transfer measure. Action (energy–time product) variables describe the processes, rules, and perceptions by which a person relates socially and professionally to others and to self—the things that people do to maintain self and social relationships. They involve sharing, teaching, learning, and invoking values and skills. To describe the human being in modern society from a physical viewpoint, the classical momentum and energy variables that account for the exchanges between very simple atoms and molecules, must be expanded into value-in-trade and action variables.

In our generalized thermodynamic model of society, the internal microprocesses that emerge as the action variables make up the components of the associational (bulk) viscosity and create an internal stress that extends beyond the normal hydrostatic pressure. We call this internalized stress component the "social pressure." It wells forth from the interiors of the atomistic units and influences their total cooperative, cultural process (Iberall *et al.*, 1980).

The macroscopic coordinates in terms of which the societal macrosystem must be described, fall into the same five groups already listed as microscopic variables. A full view of a society is then to be obtained by studying and describing all five classes of variables. It is necessary to seek out the factory day of the society and the modes of interactions with surrounding cultures comprising the level above. It is also necessary to try to describe the dynamic modes at all the process time scales between a societal factory day and the factory day of the individual units. In our view the primary societal factory day begins at the time scale of three generation times, or about 70 to 90 years, and extends as far as the life of a civilization, about 300 to 500 years. (Iberall and Wilkinson, 1987).

The study of society is a complex task that has been aided by contributions from genetics, biology, psychology, sociology, anthropology, economics, historical study, and evolutionary theory. We believe that the physical, thermodynamic view sketched here is a useful and necessary strategy for an extended study of society. According to this view, standard demographic theory is necessarily incomplete because it does not take into account the interactions among all five coordinate groups of variables. (The same criticism applies to standard economic theory.) Although current treatments of demography do address birth rate, death rate, immigration, emigration, and predation, they do not take into account the full thermodynamic nature of birth rate. To a large extent birth rate is determined by *choice* on the part of the individuals. But changes in population depend on the average choice function of a generation. Economic activity is pervasively linked both as cause and effect to the other four components. Choice, in turn, depends on perceptions arising from cultural activities in all of the five coordinate compartments within the culture. On the global scale, in which cultures are bound into polities, which are in turn bound into a field ecumene, the major large-scale modal process that emerges, obstinately escaping free will, is an alternation of trade and war.

Elaboration of some of these themes appears in Chapter 28 and elsewhere (Iberall *et al.*, 1980; Iberall, 1985).

References

Herzfeld, K., and T. Litovitz (1959) *Absorption and Dispersion of Ultrasonic Waves*. Academic Press, New York.

Iberall, A. (1985) Outlining social physics for modern societies—locating culture, economics, and politics. The enlightenment reconsidered. *Proc. Acad. Sci. (USA)* **82**:5582–5584.

Iberall, A., and S. Cardon (1980) Contributions to a thermodynamic model of earth systems. Third Quarterly Report of Gen. Tech. Serv., Inc., to NASA Headquarters, Washington, D.C. (November). Contract NASW-3378.

Iberall, A., H. Soodak, and C. Arensberg (1980) Homeokinetic physics of societies—a new discipline: Autonomous groups, cultures, polities. In: *Perspectives in Biomechanics*, Vol. 1, Part A, H. Reul, D. Ghista, and G. Rau (eds.). Harwood, New York, pp. 433–528.

Iberall, A., and D. Wilkinson (1987) Dynamic foundations for complex systems. In: *Exploring Long Cycles,* G. Modelski (ed.). Lynne Rienner, Boulder, CO, pp. 16–55.

Prigogine, I. (1978) Time, structure, and fluctuations. *Science* **201**:777–785.

Prigogine, I. (1980) In: *From Being to Becoming: Time and Complexity in the Physical Sciences*. W. H. Freeman, San Francisco, pp. 84–94.

Smith, J. (1976) Development of the earth-moon system with implications for the geology of the early earth. In: *The Early History of the Earth*, B. Windley (ed.). Wiley, New York, pp. 3–19.

Uyeda, S. (1984) Subduction zones: Their diversity, mechanism and human impacts. *GeoJournal* **8.4**:381–406.

Windley, B. (ed.) (1976) *The Early History of the Earth*. Wiley, New York.

VIII

Extensions of Physical Views of Self-Organization

F. Eugene Yates

Contributions of physical sciences to life sciences began early, culminating in the work of Helmholtz by the middle of the last century. In 1867 William James, 25 years old, attended a meeting in Germany and wrote the following enthusiastic reaction to the work of Helmholtz and Wundt:

> Perhaps the time has come for psychology to become a science—some measurements have already been made in the region lying between the physical changes in the nerves and the appearance of consciousness. [James, reprinted 1950]

But 25 years later he reconsidered:

> [There is only] a string of raw facts; a little gossip and wrangle about opinions; a little classification and generalization on the mere descriptive level.... We don't even know the terms between which the elementary laws would obtain if we had them. This is no science, it is the hope of a science. [See Joravsky, 1982]

We still respect James' resistance to pseudosolutions

> in which inconsistencies cease from troubling and logic is at rest. It may be a constitutional infirmity, but I can take no comfort in such devices for making a luxury of intellectual defeat. They are but spiritual chloroform. Better live on the ragged edge, better gnaw the file forever! [James, reprinted 1950]

Since Helmholtz, biology has continued to be refreshed by small tactical contributions from physics, chiefly in the form of experimental methods and simple conceptual models. Occasionally, grander strategic views have shaped progress in the life sciences, often with prompting from the wings, as by Leo Szilard or Erwin Schrödinger. And scientists from physics or chemistry were very much on stage

---

F. EUGENE YATES • Crump Institute for Medical Engineering, University of California, Los Angeles, California 90024.

at the beginning of the drama of modern molecular biology (e.g., Elsasser, Pauling, Crick, Delbrück). But, as noted in the introduction to Section VII, and by Iberall and Soodak in Chapter 27, the phenomena of life, human beings, and society continue to lie outside the ordinary professional concerns of the community of physicists, even though in retrospect we see that the interaction between the two great sciences was continual. The interaction could be seen in the bold gesture made by E. U. Condon to open the pages of the *Reviews of Modern Physics* to the "Biophysics Study Program"; in the works of Britton Chance or Aaron Katchalsky; in the publication of the four volumes entitled *Towards a Theoretical Biology,* edited by C. H. Waddington (1968, 1969, 1970, 1972); or in *Theoretical and Mathematical Biology,* edited by Watterman and Morowitz (1965). Mathematics also participated, led by D'Arcy Thompson, by Lotka and Rashevsky, and, more recently, by the appearance of *Stabilité Structurelle et Morphogénèse: Essai d'une Théorie Générale des Modèles* (Thom, 1972).

The penetration of physical ideas into biology, particularly in this century, can be seen in studies on the origins of life, the dynamic encoding of genetic information, regulation and control processes, sensory processes, and language and communication. But biologists have not been satisfied; they want more. Is there not some extension of "normal" physics that can yield strategic principles underlying the behavior of complex systems, such as the varieties found in the terrestrial biosphere or in the social and political activities of human beings?

The four chapters in this section touch on thermodynamics, statistical mechanics, stability theory, information theory, nonlinear mathematics, hierarchy theory, fluid mechanics, and quantum mechanics—all applied to the dynamics of life, especially to the phenomenon of self-organization that appears so richly in biology.

In Chapter 25 Caianiello reminds us of the structure latent in natural languages and examines the enrichment of systems, in an informational sense, by the introduction of rules for the creation of levels. He uses the surprising example of monetary systems to illustrate his analysis. Although he describes equilibrium phenomena, he hopes that the approach might permit the analysis of more complex dynamical and interactive systems and situations.

Musès is the only author in either Section VII or VIII on physics to invoke quantum mechanical ideas explicitly to explain living matter and its behavior. (Of course, the laser metaphor of Haken in Chapter 21 has quantum mechanical aspects, but their details were not essential to his points.) He complains that modern science has not explained as much as it sometimes intimates and that the origin of living systems from lifeless matter still must be regarded with amazement. After reviewing the early history of quantum mechanics, he turns to the nature of the vacuum and some of the virtual transitions and minimum energy now thought to be present there. He suggests that the minimum energy of the vacuum might be a substrate for macroscopic effects in chemical systems, including biochemical systems. To him the relationship between mental events, neurophysiological events, and physical events may lie at the lowest possible levels of physical organization of matter. This view is in marked contrast to those in other chapters of Sections VII and VIII, in which the authors invoke statistical thermodynamic fluctuations, but not quantum mechanical fluctuations, to produce or explain complex macroscopic dynamics.

Modern quantum mechanics has a variety of interpretations, even within the domain of professional physicists. Among them are the "Copenhagen" interpretation, the Einstein interpretation, the Wheeler interpretation, and the conservative or neutralist position (which is equivalent to saying that one should just compute with the mathematical relationships of quantum mechanics and forget the interpretations). None of these interpretations has been proved wrong yet, and they greatly enrich discussions of modern physics (see Epilogue). But such diversity at the foundations produces in some people a sense of unease, of incompleteness in the science, even though its power remains undiminished. The unease is stronger when quantum mechanical metaphors or concepts are drawn upon to explain aspects of life or (hardest case) consciousness, as in Chapter 26. Yet the question dogs us: can quantum fluctuations ever be amplified so as to propagate through neural networks?

We are beginning to see some successes in simple quantum chemical studies on molecular determinants for drug action (Weinstein *et al.*, 1981). The theoretical methods of quantum chemistry can reveal the relationship between structure and reactivity of certain molecules, and this relation takes on special interest in the "recognition" of ligands (such as hormones) by their cellular receptors. The recognition requirements are computed in terms of electrostatic potentials, polarizabilities, and proton affinities—all properties calculated from molecular wavefunctions. An optimal distance (3.5 Å) is nominated for the intermolecular interactions. Electrostatic orientation vectors and dipole vectors together give an orientation vector representing all kinds of interactions. Used in this way, classical quantum mechanics certainly provides an important adjunct to molecular biology. Similar approaches are being tried to account for spontaneous protein-folding, which leads to the formation of tertiary, three-dimensional structures from one-dimensional, linear strings of amino acids. Such a process is an extremely weak force system. In all this, quantum mechanics appears in its most conventional guise as a source of algorithms for computation. Important invariances among classes of molecules, including drugs, become apparent. But psyche does not appear. Is there a legitimate, strictly quantum mechanical approach to the understanding of minds? Below is an argument saying "no!"

The nonzero scaled value of Planck's constant, of which we are reminded by Musès, has been used imaginatively and, I think, effectively by Elsasser and Jakobsson (see Jakobsson, 1972) to show that there is a maximum size of a mappable (black) box, whose internal operations are fully deterministic in the classical physical sense. That maximum size or complexity is very small, in terms of the number of interacting elements. The irreducible complexity arises at as few as 70 interacting elements if there is just one box on which to experiment. If there are m identical boxes, the situation is only slightly improved, because the condition for full determination becomes

$$2^n \hbar / E \geq m t_f$$

where n is the number of deterministically interacting components, $\hbar$ is $h/2\pi$ (h is Planck's constant), E is the difference between (two) energy states that are switched by an interaction, t_f is the (average) lifetime of a component before it wears out and ceases to act deterministically, and m is the number of identical systems that must be prepared in order to identify the system experimentally.

This result depends on the quantum nature of the universe, which sets a maximum rate (infinity machines are not allowed) at which tests of the connectivity of the elements can be made. The number of elements allowed for a mappable box is small, because the number of possible functional connections increases exponentially as the number of components is increased linearly, whereas the maximum rate at which a test can be made is always $2E/\hbar$ sec^{-1} (see Jakobsson, 1972, for details). If this reasoning is correct, then it must be concluded, given the almost incredible connectivity of the human nervous system, that it can never be known in principle whether or not the nervous system acts deterministically, probabilistically, or with a nonphysical "will" of its own. $\Delta E \Delta t \geq \hbar/2$ forbids much, including proving or disproving the autonomy of the mind.

Iberall and Soodak hold the view that we should not expect or require a physical *theory* of life or societies, but rather that physics yields *strategic principles* that would permit us to frame questions about life and society in a way that would give answers possessing the elegance of physical explanations: power resting on few principles. Building on Chapter 24, they sketch here a strategic physics and invoke the concepts of competing force systems and generalized criticality conditions (in the form of a generalized Reynolds number). They offer a few competing processes to explain particular, material self-organizations and the appearance of order parameters and linguistic behavior in complex systems. They then speculate on the origin of life, using these constructs. Applications are made to demographic processes and to the study of civilizations.

The authors of this section intend to present purely physical constructs. Perhaps these chapters should not have been separated from those of Section VII— all eight chapters, 21 through 28, examine self-organizing systems, or life, from aspects of physical or mathematical sciences. Perhaps the reader will find in both sections adumbrations of a profound physical biology to come.

References

Jakobsson, E. (1972) An alternative approach to generalized complementarity. *J. Theor. Biol.* **37**:93–103.
James, W. (reprinted 1950) *The Principles of Psychology*. Dover, New York.
Joravsky, D. (1982) Body, mind and machine. *New York Review of Books* **xxix**, No. 16, Oct. 21, pp. 14–47.
Thom, R. (1972) *Stabilité Structurelle et Morphogénèse: Essai d'une Théorie Générale des Modèles*. Benjamin, New York.
Waddington, C. H. (ed.) (1968, 1969, 1970, 1972) *Towards a Theoretical Biology*: 1, *Prolegomenon*; 2, *Sketches*; 3, *Drafts*; 4, *Essays*. Aldine, Chicago.
Watterman, T. H., and H. J. Morowitz (eds.) (1965) *Theoretical and Mathematical Biology*. Ginn (Blaisdell), Boston.
Weinstein, H., R. Osman, S. Topiol, and J. P. Green (1981) Quantum chemical studies on molecular determinants for drug action. *Ann. N.Y. Acad. Sci.* **367**:434–451.

25

A Thermodynamic Approach to Self-Organizing Systems

Eduardo R. Caianiello

ABSTRACT

Starting from models of neural systems and natural languages, the author was led to a study of "structure" in general systems. Simplest among these are (decimal, binary, and so on) counting system and monetary systems, which appear as archetypes of hierarchical modular systems (HMSs).

HMSs are characterized by *levels*, determined by a *module* that specifies how many elements of a given level *cluster* to form an element of the *next higher* level. To each level a *value* is assigned (which for counting and monetary systems happens to coincide with the module).

A careful use of *entropy* (which bypasses, without mentioning, the many traps that lurk when the entropies of Shannon, Boltzmann, Gibbs, Kullback, Reny, and others, are not well discriminated) provides a *rationale* for the process of level formation as an efficient tool to deal with *complexity*, and a link with information theory.

A formal analogy with thermodynamics can then be established and put to use. The environment, or universe, of an HMS, interacts with it by fixing its "average value" computed over all levels; this must not change when module and number of levels are altered by processes of rearrangement, i.e., "self-organization" of the system. The methods of physics permit then to determine the *population* of each level and to establish some relevant properties. HMS can change in two ways: by *evolution* and by *revolution* (the latter term denotes change of module); it acts as a *template* for any other HMS with which it interacts, because equilibrium is possible only when structure is the same.

Quantitative results fit remarkably well the known data on Zipf's law, monetary systems, military commands, population distributions, and a vast number of other situations. —THE EDITOR

This chapter defines a very specific model as narrowly as possible. Its appearance is one step in a chain of reasoning that started in the late 1950s. At that time I was introduced to cybernetics through collaboration with Norbert Wiener, and I

EDUARDO R. CAIANIELLO • Faculty of Sciences, University of Salerno, 84100 Salerno, Italy.

was concerned with models of neural activity—sets of discontinuous, highly nonlinear equations that described the behavior of assemblies of neurons, as well as the comparatively slow changes in their interconnections that I took as being responsible for memory. The same approach led me to a parallel study of natural languages—those that are written or spoken by humans—because they express "thought," the product of a nervous system in articulated form. A complete understanding of a real language presents problems no less difficult than those posed by the study of the brain itself. It soon became evident that writing equations for the behavior of individual neurons was not sufficient for understanding how nervous structures could actually function, any more than an equation of atomic physics can tell us about all the qualities of wood or iron. I realized that considering only properties of strings of letters or symbols would not lead very far. Thus, I began to study the structure of specific languages, such as Italian or English—not just of nearly arbitrary sequences of letters, as is often done.

I needed some set of notions about "structure." Not all ways of combining letters give acceptable syllables; not all combinations of words give acceptable sentences. Some *structural rules* evidently exist that, although we may not know them explicitly, drastically reduce the number of combinations of letters that form meaningful sequences in a natural language. The next step seemed to lie not in developing more models of specific neuronal assemblies, or of more detailed procedures for generating languages, but rather in trying to gain some understanding of what a "structure" can be. The first model I tried that displays "structure" at work in its barest essence worked with unexpected accuracy. This model addresses instances such as monetary systems, populations on a territory (Caianiello *et al.*, 1984), and so on, and gives "laws" that are observed, apparently, in all countries. Zipf's law, some properties of military chains of command, and many other hierarchical organizations also could be expressed. This model is proposed as a tool for the understanding of "structure," which I take to be the major scientific issue of our time.

Some systems exhibiting a hierarchical structure (Caianiello, 1977) can change within specified "equivalence classes," so as to adjust to the "universe"—their environment. They are "self-organizing."The simplest cases are "modular" systems (e.g., the decimal system for numbers). A thermodynamics can be built for them. Their development in time alternates between phases of "evolution" and "revolution." A key point that emerges is that *discreteness* ensures *stability*. This issue is, I am convinced, very profound.

One example will fix ideas and render most of the following intuitive: given a number system (e.g., decimal), consider the system formed by all possible sequences of digits. No connection with the operations of arithmetic is implied. This discussion is restricted to systems composed of a *discrete*, however large, number of elements.

This chapter is confined to a particular class of systems called the *hierarchical modular systems (HMSs)*. Each HMS has structure, and behaves much like a number system having arithmetic base $M(=10)$.

A structured system may be subject to two types of changes over time. For

25. A Thermodynamic Approach

example, suppose the system is isomorphic with a set of numbers. Of the first type are changes of these numbers as elements (i.e., "evolution"); the second type of change pertains to alteration of the structure itself. A structural alteration corresponds here to a change in the base of the number system (to provide isomorphism with the new situation of the system). I call this a "revolution."

This chapter is focused mainly on revolutionary change, and the use of the term "self-organizing" is restricted to systems that can spontaneously change their structure to meet new situations. (They may also evolve.)

Structure emerges as a discrete quantification of a continuum. In the model under discussion, a change of structure will be a change among integer values, so that a criterion of stability is provided to protect self-organizing systems against arbitrarily small perturbations. (The rounding-off of a number to the nearest integer will be understood without further mention, as a necessary approximation in the application of theory, e.g., $\sqrt{10} = 3$, when it denotes the number of cells, coins, or people.)

Definitions

Bose–Einstein and Boltzmann Counting

Elements will be regarded as identical or nondistinguishable when, for a collection, their individual identity is irrelevant. Thus, any two groupings into a same state consisting of the same number, n, e.g., of infantrymen to form a squad, or of $100 bills to form a given total (if one is not interested in forgery and serial numbers), are indistinguishable from the point of view of a commanding officer or a tax collector, considered here as physical "observers." I pose no *a priori* restriction on the values that n, the number of identical elements of any sort, may assume; given n, the number W of *different* states that can be formed with them is $W = n + 1$ ($n = 0$ is included). This is the counting of Bose–Einstein statistics. (Not treated here are situations in which Gentile or Fermi counting may be appropriate.)

Whenever it is important to keep track of nonidentical elements, e.g., coins of different denominations, Boltzmann counting will apply.

Information

The function of hierarchical levels in a system can be examined through the example of a monetary system. What is the reason that all such systems (in other than primitive societies) are quantified into discrete units of different denominations and values, rather than being "continuous"? In this discussion I assume at the start an indefinite number of coins, or tokens, of the same unit value. (Because I use quantities that are familiar in thermodynamics, my notation and symbols conform to standard physical usage.)

Figure 1 shows the relationship between the number of different possible states (sums of money), W, and the corresponding entropy or information, S,

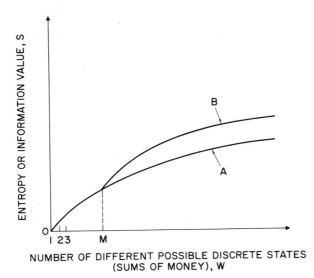

FIGURE 1. Classical, statistical-mechanical relationship between complexity of a system, as measured by the number, W, of possible microstates, and its Boltzmann entropy (or information), S. The two curves, A and B, are described by Eqs. (1), (2), and (3)(see text).

where

$$S = K \log W = K \log (1 + n) \tag{1}$$

The logarithmic growth of S (nearly linear for small n, increasingly slower for higher n) well typifies our experience of increasing awkwardness in handling a system whenever n grows, whatever we may want to do or know about it. Doubling n nearly doubles S for small n, but for large n enormous increase is necessary to double the information.

Clustering and Level Formation

Suppose the system is restructured into a new one, in which identical "clusters" of coins of identical unit denomination, each containing M elements, are formed. A cluster is a bag of unit-denomination coins (or equivalently, a single, second type of coin of M-unit denomination). We can now form any amount that exceeds $M - 1$ in different ways. Suppose we use n_1 coins of unit denomination and n_2 coins of the second type. The latter type being distinguishable from the former, we have the following number of combinations that define all the different microstates of the same total:

$$W = (1 + n_1)(1 + n_2) \tag{2}$$

and thus

$$S = K \log (1 + n_1) + K \log (1 + n_2) \tag{3}$$

As soon as the new (M-unit) coins are used, curve A of Figure 1 changes into curve B, which has a kink, and separates from A to form a new logarithmic arc that is again nearly linear for small n_2. The introduction of a new value, or level n_2, keeps the information nearly linear within a broader range of W. Another way of looking at curve B is to imagine the second arc as deriving from a contraction by a scaling factor $1/M$ of the subjacent portion of the horizontal axis. We may then want

$$S = K \log [(1 + n_1)(1 + n_2)] = K \log (1 + n_1 + Mn_2) \qquad (4)$$

which occurs only if $n_1 = M - 1$; that is, when we do not use unit coins unless absolutely necessary to create an amount less than M, the cluster value.

Thus, we can keep information growing linearly in any system of identical elements, provided the system is organized into levels, the elements of each level being identical clusters of elements of the level below. Such clusters can be treated as, or replaced by, new elements, identical among themselves but not with those of levels below or above. Clustering emerges as a typical process of level and structure formation. Furthermore, a hierarchical arrangement of elements, clusters, clusters of clusters, and so on, appears as a rather natural mechanism to achieve this effect. An element, by becoming a part of a cluster of the next higher level, *loses* many of its features and *acquires* the new function of a member of the cluster.

Hierarchical Systems (HS)

Consider a system composed of elements of h levels [$h = 0, 1, 2, \ldots, L (\leq \infty)$]. The elements at each level are indistinguishable among themselves, but distinguishable from those at other levels. Let there be n_h elements in level h. The total number of elements, N, is then

$$N = \sum_{h=0}^{L} n_h \qquad (5)$$

Equation (5) describes a partition π of the elements of the system.

Attach now to each level h an (integer) value v_h. If we assume for simplicity only one value function, then the total value, V, of the system is

$$V = \sum_{h=0}^{L} n_h v_h \qquad (6)$$

The average value, $\langle v \rangle$, of an element of the system is then

$$\langle v \rangle = \sum_{h=0}^{L} n_h v_h \Big/ \sum_{h=0}^{L} n_h \qquad (7)$$

A self-organizing system organizes itself by the partitioning process into a hierarchical structure. We now require that the value function be such that

$$v_h/v_{h-1} = \text{integer} > 1, \text{ any } h; \quad v_0 = 1 \text{ (in suitable units)} \tag{8}$$

If, in particular, this requirement takes the following form:

$$v_h/v_{h-1} = v_{h-1}/v_{h-2} = \cdots = v_1/1 = M \tag{9}$$

we call the system hierarchical and modular, and M is the module or base of the system and of the corresponding partition.

Hierarchical Partitions

We are now concerned with relationships among different partitions of a given self-organizing system. A partition π_b is a refinement of a partition π_a if it has more levels than π_a and retains *all* the levels of π_a, which stay invariant under π_b.

A *hierarchical modular partition* $\pi_b^{(\text{mod})}$ is therefore a refinement of a previous $\pi_a^{(\text{mod})}$ if and only if

$$M_b = M_a^{1/p}; \quad p \text{ integer} > 1 \tag{10}$$

From now on our interest will be confined to HMSs, and for brevity, using Eq. (10), I shall call any $\pi_b^{(\text{mod})}$ a *p-refinement* of $\pi_a^{(\text{mod})}$.

Hierarchical Modular Systems

It is important now to make a clear-cut distinction between two different issues. The first pertains to the assignments of specific values to the levels of an HS or of an HMS (in the latter case to M^h). These assignments, as well as the choice of number of levels, $L + 1$, are here assumed to be given *a priori*. The second issue relates to the distribution n_h of the N elements of the system among its levels, $h = 0, 1, 2, \ldots, L$, once their number and values are known. This distribution will reflect the way the system *adapts* to an external requirement or constraint. General properties may be derived from a study of the second issue in typical instances, using models.

As long as a self-organizing HS is isolated, it has no reason (from what has been said thus far) to prefer any particular choice of n_h. If, however, it interacts with an environment, a preferred distribution may occur. Assume that the system, which has some specific values, is free to change that assignment by any refinement of its original hierarchical partition. This principle claims that the universe does not care which partition the system chooses for organizing itself. The interaction of environment with the system is of a global nature, and requires only that the value of some *mean* quantity of the system be fixed by that interaction so that the average value of an element of an HMS stays invariant under any p-refinement of the HMS:

$$M \to M^{1/p} \tag{11}$$

Then

$$L \to pL \tag{12}$$

and for $n_n \equiv n_h^{(1)}$,

$$\langle v \rangle = \frac{\sum_{h=0}^{L} n_h^{(1)} M^h}{\sum_{h=0}^{L} n_h^{(1)}} + \frac{\sum_{h=0}^{pL} n_h^{(p)} M^{h/p}}{\sum_{h=0}^{pL} n_h^{(p)}} \tag{13}$$

Note that an HMS possesses a natural invariance under modular changes, that of the ratio of the maximum number of states to the module

$$\frac{M^{L+1}}{M} = \frac{(M^{(1/p)(pL+1)})}{M^{1/p}} = M^L \tag{14}$$

We must find $n_h^{(p)}$ so that, for given L and M, $\langle v \rangle$ stays the same for any p; this will be

$$n_h^{(p)} = N \frac{1 - M^{-1/2p}}{1 - M^{-1/2p - L/2p}} M^{-h/2p} \tag{15}$$

That is, Eq. (15) is the distribution law that secures the wanted invariance of $\langle v \rangle$ under p-refinements. In the following discussion we need not consider values of p other than 1. There is no way of knowing a "past history" of levels. Note that, in particular

$$n_h = n_{h-1}/\sqrt{M} \quad \text{and} \quad \langle v \rangle = \sqrt{W} = M^{L/2} \tag{16}$$

Thermodynamics of Hierarchical Modular Systems

Given an HMS with a distribution law (N assumed constant),

$$p_h = \frac{n_h}{N} = \frac{1 - M^{-1/2}}{1 - M^{-1/2 - L/2}} M^{-h/2} \quad (h = 0, 1, \ldots, L) \tag{17}$$

We can regard p_h as expressing a frequency, or a probability scheme, with which we may associate an information, or entropy, in the standard way:

$$S = -K \sum_{h=0}^{L} p_h \log p_h \qquad (18)$$

A change of variables will now clarify the form, without altering the content. Let

$$\beta = \frac{1}{KT}; \quad \frac{E_0}{L} = \epsilon_0; \quad \epsilon_h = \epsilon_0 h; \quad M = \varepsilon^{2\beta\epsilon_0} \qquad (19)$$

and

$$Z = \sum_{h=0}^{L} e^{-\beta \epsilon_h} = \frac{1 - M^{-(L+1)/2}}{1 - M^{-1/2}}; \quad \psi = \log Z \qquad (20)$$

Then p_h can be rewritten in the familiar form

$$p_h = \frac{1}{Z} e^{-\beta \epsilon_h} \qquad (21)$$

and the entropy becomes

$$S = K\psi + (1/T) \langle \epsilon \rangle \qquad (22)$$

where

$$\langle \epsilon \rangle = \sum_{h=0}^{L} p_h \epsilon_h = \epsilon_0 \sum_{h=0}^{L} h p_h = -\frac{\partial \psi}{\partial \beta} \qquad (23)$$

$\langle \epsilon \rangle$ corresponds thermodynamically to the average energy. Here it designates, to within a factor ϵ_0, the average order (or string length) of a level. For instance, if we attribute to each word, conceived of as a string of h code letters, an energy ϵ_h, this energy would become proportional to h. Thus, we recover from our premises a concept that has been used on occasion by mathematical linguists. Finally, note that

$$\log \langle v \rangle = \frac{E_0}{KT} \qquad (24)$$

The quantities $\epsilon_h = \epsilon_0 h$ that correspond to the usual energies are *not*, however, connected with the value, but only with the order of the levels. To have found a formal definition of energy, however remote our starting point, is significant, because then one can develop for any such HMS the formalism of thermodynamics. For a full treatment of this point see Rothstein (1951, 1962), Jaynes (1957), Tribus (1961), and Caianiello (1977).

By analogy with physics, two different HMSs, 1 and 2, will be said to be in equilibrium (zeroth law of thermodynamics) if they have the same "temperature" T [or β, see Eq. (19)]. *But we must now add the additional requirement that $\langle v \rangle$ be the same for both*:

$$T^{(1)} = T^{(2)} \tag{25}$$
$$E_0^{(1)} = \epsilon_0^{(1)} L^{(1)} = E_0^{(2)} = \epsilon_0^{(2)} L^{(2)} \tag{26}$$

If $\epsilon_0^{(1)} = \epsilon_0^{(2)}$, then the two systems, besides having the same "temperature," have also the same number of levels.

Some General Considerations

Invariance within an Equivalence Class

To provide a general formulation of the method so far applied only to a subclass (N = const) of HMSs, I take as initial data a description of the levels and of the value functions attached to them (these define the "structure" of the system). In so doing, one is naturally led to consider transformations, such as the p-refinements for HMSs, that change the system into an "equivalent" one with respect to its interactions with its universe (as measured by average values) and its structure (e.g., modularity must be retained). The specification of such transformations can be regarded as part of the initial data: one defines an *equivalence class* within which a self-organizing system can freely move.

Transformations within an equivalence class can change the structure of the system during its development. The transformations will be forced by external influences or by the growth either of the number of individuals in a population model, or of the volume of trade in an economic model. To some extent transformations will be specific for each system.

Our attention has been focused on finding a general criterion that might allow the determination of the population of each level, once the initial data are given. To do so I have invoked a principle of invariance of response under allowed structural transformations. Response or interaction is measured through mean values; allowed transformations are the p-refinements. Again, the universe does not know and does not care about which state a self-organizing system chooses to settle in, within an equivalence class.

Mathematical Linguistics

The present research stems from work done in mathematical linguistics (Caianiello and Capocelli, 1971) aiming at an inductive study of the hierarchical organization of a language. I studied HMSs only as a first step to formulate questions. However, it was interesting to find a connection between "energy" and "word length," such as was postulated by Mandelbrot (1954) in his attempt to explain the so-called "Zipf's law" (Zipf, 1949). Proceeding as he does, I would immediately obtain the same result for *any* HMS, not only for somewhat idealized languages. This I refrain from doing, because I feel that Herdan's (1962) criticism should be taken seriously.

Human Society and Military Structures

Some consideration of human social organization (Jacques, 1976) is appropriate to this subject. I shall take as the only basic factor common to any form of society the necessity that members *communicate* among themselves in order to act socially. Communication takes time, in amounts that increase with the complexity of the task to be agreed upon or commanded. Rational communication (here undefined) becomes less efficient the larger the group with which an individual has to communicate. Aside from one-level societies, which are anarchic and are found only in very small groupings, all others are hierarchical (though not necessarily authoritarian). Therefore, the crudest model is that of an HMS that treats all individuals and tasks as equal. The module M denotes the number of individuals (constant by assumption) who are communicating with, or controlled by, an element of the next higher level.

This model may be compared with the chain of command typical of military organizations. Within my experience, $M \cong 10$ can be very nearly taken as the module (unless otherwise required for technical reasons). But then one is immediately struck by the systematic appearance of the number 3. There are often three leaders (one sergeant and two corporals) for a squad of ten soldiers; three squads to a platoon; three platoons plus one squad to a company; and so on. $\sqrt{10} \cong 3$ is just what an HMS would require in this case.

In our model *a hierarchical system interacting with another acts as a template for it: structure forces structure upon the environment.*

I am far from believing that HMSs describe everything, any more than harmonic motion can describe all motion. I have tried only to establish a method, whose essence is the use of *conservation* and *invariance principles* in an (assumed) ignorance of specific dynamical laws, as is customary in physics. I could multiply examples in which modularity is in fact respected, as well as the distribution law I find for it [e.g., rabbits in a particular experiment, $M = 12$; city of Athens, $M = 7$ (Viriakis, 1972); some anthropological data (Carneiro, 1969); perhaps the columnar organization of the cerebral cortex]. However, it is more relevant at this stage to search instead for systems with structures and distributions radically *different* from those of HMSs.

It can be shown that the k-convolution of HMSs yields immediately γ-distributions of order k, such as have been introduced to describe asymptotically most diverse situations, from the distribution of animal species into niches to hadron jet or galaxy clustering.

Some seemingly evident particulars have not been addressed, e.g., that level formation is a consequence of entropy maximization, that it is attributable to the increased cooperativity of structured systems (army versus crowd), or that it comes from a balance between organizing and disorganizing agents, as happens with all equilibria. These various statements are not contradictory, but I have avoided them as a matter of method, since it is not yet clear to what extent the claims are merely specific and to what extent more generally true.

Monetary Systems

HMSs are perhaps the crudest possible models of self-organizing systems, especially under the (unnecessary) restriction that the number of elements stay constant under refinements. It is therefore rather surprising to find a realistic situation described by them quite satisfactorily. The evidence was first provided in an interesting study by Hentsch (1973, 1975, 1983) in which a penetrating, empirical analysis was made of monetary circulation in various countries of the world to find some regularities. Similar results were then seen to hold for nearly all countries for which data were available. They all reduce, in our notation, to the law

$$n_h v_h \propto \sqrt{v_h} \qquad (27)$$

which exactly coincides with our $n_h^{(p)} = \propto M^{-h/2p}$, when $p = 1$ and $v_h = M^h$. Various fractional powers of 10 were examined by Hentsch (1973), in particular

$$10^{1/3} \rightarrow 1, 2.15, 4.64, 10, 21.5, 46.4, 100, \ldots \qquad (28)$$

This value of the module M, after rounding off to the nearest integer, should be familiar to the reader. Figures 2 through 5 represent the situation in several countries, which have quite different structures and cultures.

The explanation of this behavior, once it is ascertained empirically that a monetary system is an HMS (M need not, of course, be $10^{1/3}$), becomes quite obvious from my perspective. A monetary system is not just a game; it is a system that interacts with the universe of all that can be exchanged with money. The principle of invariance under p-refinements here means simply that the average value of the monetary token must equal the average value, or price, of anything that may be bought or sold, from a needle to a skyscraper. This universe, I repeat, does not know or care which module a country may choose for its monetary system.

I refer the reader to Caianiello et al. (1980, 1982) for a more detailed study of this subject as well as for an explanation of the numerical determination $M = 10^{1/3}$ empirically found for the module of monetary systems. The result of this analysis is that, from Eqs. (6) and (27), $V_h = n_h v_h = n_0 \sqrt{v_h}$, and so $\log V_h = \log n_0 + \frac{1}{2} \log v_h$.

The coefficient $\frac{1}{2}$ should be the same for all countries that have modular monetary systems. Some (like Japan) do not; in others, not all values of the sequence are present. Such instances bring about deviations from the linear plot. The deviations would disappear if modularity were established. Of course, devia-

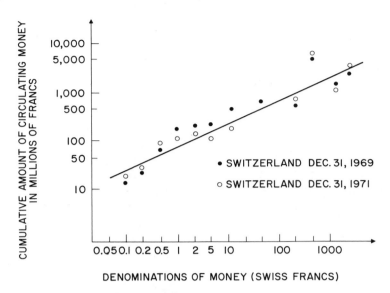

FIGURE 2. Relationship between denominations of money (Swiss francs) and the cumulative amount of money accounted for. Two important phenomena may be observed: the 200-franc piece is the cause of the considerable increase of circulating 500-franc pieces; moreover, the monetary circulation tends to the canonical distribution (except for transient anomalous effects such as the hoarding of silver coins). Data include silver coins of 0.5, 1.2, and 5 francs. The abscissa and ordinate are, respectively, denominations and cumulative amounts in circulation in all the following figures (3 through 5).

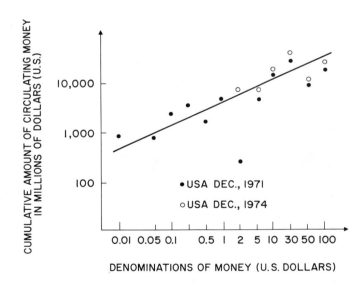

FIGURE 3. United States data (see legend of Figure 2). The circulating money for different denominations agrees fairly well with the canonical distribution (the agreement is better with the 1974 data), except for the $2 bill, which has an anomalous behavior. (The United States public rejected this denomination strongly.)

25. A Thermodynamic Approach

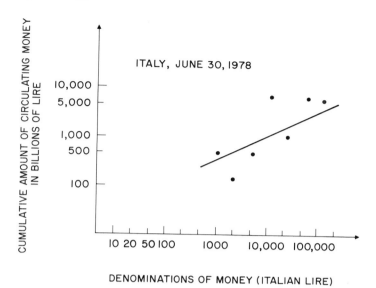

FIGURE 4. Data for Italy (see legend of Figure 2). The differences between the data and the canonical distribution (given by the straight line) are higher than those of other countries.

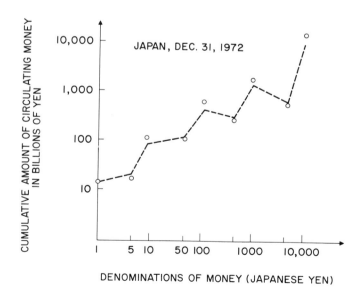

FIGURE 5. Data for Japan (see legend of Figure 2). The saw-toothed behavior occurred because of the lack of 2-, 20-, 200-, and 2000-yen denominations. This deficiency is compensated for by the corresponding increase of the circulating money of 1-, 10-, 100-, and 1000-yen denominations.

tions may have other causes. The distribution law was violated somewhat in Switzerland in 1969, when some small tokens were coined in silver (and hence hoarded out of circulation), but it went back to linear form in 1971, when precious metal was no longer used.

Only equilibrium phenomena have been described in these pages, but this approach might also permit the analysis of more complex, dynamical and interactive situations.

ACKNOWLEDGMENT. This research was supported by CNR Grant 79.00968.11.

References

Caianiello, E. R. (1977) Some remarks on organization and structure. *Biol. Cybern.* **26**:151.
Caianiello, E. R., and R. Capocelli (1971) On form and language: The Procrustes algorithm for feature extraction. *Kybernetik* **8**:223.
Caianiello, E. R., G. Scarpetta, and G. Simoncelli (1980) Sulla leggeydi distribuzione delle Monete. *Rass. Econ.* **44(4)**:771–794.
Caianiello, E. R., G. Scarpetta, and G. Simoncelli (1982) A systemic study of monetary systems. *Int. J. Gen. Sys.* **8**:81.
Caianiello, E. R., M. Marinaro, G. Scarpetta, and G. Simoncelli (1984) The population distribution as a hierarchical modular system. In: *Analisi e Controllo dei sistemi complessi*. Assoc. Merid. diMecc.
Carneiro, R. (1969) *Trans. N.Y. Acad. Sci.* **31**:1013.
Hentsch, J. C. (1973, 1975, 1983) *J. Soc. Stat. (Paris)* 4.
Herdan, G. (1962) *The Calculus of Linguistic Observations*. Mouton, The Hague.
Jacques, E. (1976) *A General Theory of Bureaucracy*. Heinemann, London.
Jaynes, E. T. (1957) Information theory and statistical mechanics. *Phys. Rev.* **106**:620.
Mandelbrot, B. (1954) Structure formelle des textes et communication. *Word* **10**:1–27.
Rothstein, J. (1951) Information, measurement, and quantum mechanics. *Science* **114**:171–175.
Rothstein, J. (1962) Discussion: Information and organization as the language of the operational viewpoint. *Philos. Sci.* **29**:406.
Tribus, M. (1961) Information theory as a basis for thermostatics and thermodynamics. *J. Appl. Mech.* **28**:1–8.
Viriakis, T. (1972) *Ekistics* **199**:503.
Zipf, G. K. (1949) *Human Behaviour and the Principle of Least Effort*. Addison-Wesley, Reading, Mass.

26

Interfaces between Quantum Physics and Bioenergetics

Charles Arthur Musès

ABSTRACT

The historically weak philosophical underpinnings of the concept of abiogenesis—the origination of living organisms from lifeless matter—calls to mind circular proofs of the existence of God, which proved neither side. Blind credulity is a problem for science, and we must guard against it. Some of it is a fault of scientists themselves, who believe they have explained away more than they have. Just what, for example, is the state of affairs about our knowledge of matter? The early history of quantum mechanics is reviewed in this light, and some of the elements of the theory are presented, in order to establish that a vacuum is not devoid of physical attributes, but possesses some minimum energy. That minimum energy can be a sort of substrate for macroscopic effects in chemical systems, including biochemical systems. Psychological phenomena clearly have some physical referent because they can produce physiological effects, for example, psychosomatically induced symptoms or physiological phenomena under hypnotic induction or as reactions to psychological stress. Mental events, neurophysiological events, and physical events have a connectedness that may be traceable to the very "lowest" levels of physical organization of matter, though as quantum physics has shown, not the least complex or subtle. It is suggested that certain atoms in functional molecules (such as magnesium in chlorophyll) occupy positions of control that could be subjected to microfluctuations. Those fluctuations of zero-point vacuum energies are addressed and their triggering potential for more macroturbulent or bifurcational "emergent" fields pointed out. It is becoming more clearly understood that turbulent, bifurcational, and attractor-type fields are microgenerated from singularities. The recent history of science reveals that our explanations depend on microstructure and microprocesses. We suggest that the control of biological phenomena may be explained at levels of organization even more primal than those currently addressed by molecular biology, thus leading to an ultimate reductionism of maximal parsimony and profundity in which the deepest physics and the deepest psychology would have to meet, as Eugene Wigner prophetically dreamed they might—leading to a revolutionary scientific view of reality otherwise unattainable that began with the quantum leap taken by Max Planck (1900) that ushered in the fateful twentieth century.

CHARLES ARTHUR MUSÈS • Editorial and Research Offices, Mathematics and Morphology Research Center, 1052 Santa Fe Ave., Albany, California 94706.

The subject of molecular and submolecular bioenergetics opens up a wealth of fascinating findings and problems increasing rapidly in number, and though volumes can and have been written on the topic, the prime problems persist, and even ramify with deepening knowledge [see Ingram's (1969) still valuable commentary].

Because it has enjoyed for more than a century a large share of publicity as a specious speculation, let us first take a closer look at an extreme form of mechanical (including thermodynamic) reductionism: abiogenesis—the notion that living organisms could originate from lifeless matter. Thus, a modern dictionary of biology (Steen, 1971) logically equates abiogenesis and spontaneous generation as synonyms, cross-referencing the two terms as such. Abiogenesis is the extremal opposite of hylozoism—a sophisticated form of animism postulating that lifeless matter is actually nonexistent. Thus, ironically enough, abiogenesis turns out to be a form of the old theological *creatio ex nihilo* in atheistic guise, both opposed by the rather cogent *ex nihilo, nihil—out of nothing can come only nothing*. More explicitly, if life and mind are not included to begin with in the definition of matter, they cannot manifest out of matter; and if they are included in the definition of matter, then one is a hylozoist and not an abiogenesist.

Using words like "prebiotic" is a practice that tends to mislead more than help because such tactics only evade the primal issue of how and where the *bios* was in the first place, or worse, produce vicious circles. An idea emerging today is that what we call "nonliving matter" may not be appropriately so designated, if we forsake inadequate nineteenth century concepts of matter and take account of the new and profound discoveries that are emerging from quantum physics, both in theory and in laboratory observation.

We need a sharper awareness of the current limits of what we do *not* as yet know. That awareness should furnish the necessary humility for us to proceed with a salutary form of caution—unbiased observation, ready not to force prejudgments but to change tentative hypotheses, if need be, because of challenging new facts. Such an attitude can diminish the boundaries of the as yet unknown, but one must admit that we all have notable difficulties in foregoing cherished beliefs, even in the face of inspiring facts. With this caveat, which has served and will still serve to explain how scientific enquiry is and has been obstructed, to use Charles Sanders Peirce's apt phrase, let us reexamine some changes in the views of the nature of matter, and hence of mechanics, that occurred during those three remarkable decades, from 1900 to 1930.

If electrons were to lose even the slightest energy in their orbitals, all atoms would quickly collapse and the world as we know it would cease to exist. Because that catastrophe obviously does not occur, Bohr included in his early atomic theory of 1913 the still-arresting assertion that matter under certain conditions could move without loss of energy. A whole new era of nonentropic and negentropic thermodynamics or energetics had begun.

Max Planck's quantum theory of radiation had already been verified experimentally. Rayleigh's and Stefan's laws connecting radiation and absolute temperature at either end of the frequency spectrum were now satisfyingly combined

26. Quantum Physics and Bioenergetics Interfaces

in Planck's radiation equation, which sprang out of his fundamental assumption of quantized action (energy multiplied by time) and, hence, quantized energy. That assumption, as one studies the development of Planck's thought, seems to have arisen from the idea of applying Maxwell's and Gibbs' statistical mechanics to radiation and its associated phenomena. Niels Bohr then applied Planck's approach to the hydrogen atom, and described consequences that fit laboratory observations that nonquantized approaches to matter and energy could not explain.

An interesting implication from Planck's equations, first perceived and made explicit by de Broglie in 1923, is that a guiding wave, as he termed it, is associated with every moving mass. Actually, both Bohr's and de Broglie's developments were prophesied by the insights published in W. R. Hamilton's *General Dynamics* in 1834, in which the deep analogy between particle dynamics and optics or radiation dynamics was first demonstrated. In 1926 Erwin Schrödinger, following up de Broglie's suggestion, published his theory of the wave nature of matter. In confirmation of quantum theory, the wavelike structure of matter finally was demonstrated in the laboratory by Clinton Davisson, who provided some then-amazing electron micrographs of the interference patterns formed when an electron (a "material particle")[1] traverses a thin sheet (crystal lattice) of similarly constituted matter. Here was matter being refracted by matter, exactly as light.

Combining Schrödinger's and Hamilton's work, we can generalize classical to quantum dynamics, using Hamilton's quaternions (i_1, i_2, i_3), a form of hypernumber defined by $i_1^2 = i_2^2 = i_3^2 = -1$ and $i_b i_a = -i_a i_b$. Hence, we can write

$$\sqrt{H - (V + T)} = (\hbar/\sqrt{2m})(\partial/\partial i_1 x_1 + \partial/\partial i_2 x_2 + \partial/\partial i_3 x_3) \tag{1}$$

Squaring, we have

$$H - (V + T) = -\hbar^2/2m(\partial^2/\partial x_1^2 + \partial^2/\partial x_2^2 + \partial^2/\partial x_3^2) = -(\hbar^2/2m)\nabla \tag{2}$$

where $\hbar = h/2\pi$ is the angular momentum form of Planck's constant h; m is mass; H is the Hamiltonian operator; and $x_{1,2,3}$ are positions along three orthogonal directions in space. V is a potential energy function related to the three spatial coordinates, and T is the nonquantum kinetic energy contribution. In classical systems H is the total energy, and $H - (V + T)$ is zero. But in quantum mechanics, this difference is not zero because Planck's constant, although very small, is not zero. The equation (not hitherto published in this form) explicitly shows this nonzero difference, which vanishes in classical formulations where h tends to zero, and hence $\hbar$ also since $\hbar = h/2\pi$. The contribution to the energy from quantum effects drops to zero in the prequantum view, whereby the left-hand side of the above equations must be identically zero.

---

[1] Actually, just as much a "matter wave," matter and light fusing into one ultimate notion in quantum physics, the substratum of both remaining as yet inaccessible except in terms of recondite theory of the structure of the "vacuum" as voiced by P. Dirac, for example, and more recently by A. Salam and S. Weinberg. See also Musès (1980).

The Physically Effective Vacuum

When Paul Dirac (1928) applied the concept of retarded potentials [previously used by P. Gerber (1898, 1902, 1910) in the equation for the perihelion shift of Mercury in what later was "relativity theory"] to dynamics and optics, he found that he was forced to posit particle spin as well as "antimatter" and a sea of negative energy in the vacuum of space. Dirac's antimatter was confirmed experimentally by C. D. Anderson's discovery of the positron in 1932. Later it was observed that a particle and its antiparticle mutually "annihilate" when they collide, releasing a corresponding amount of radiation. Conversely, two beams of radiation can fuse to create a particle–antiparticle pair. From these advances made by quantum physics, a consummating observational datum emerged: the zero-point energy of the vacuum. It is known that the quantum mechanical equation for a simple harmonic oscillator shows a residual energy in the ground state at absolute zero. But the theory goes further and predicts also a (minimum) intrinsic energy for the vacuum state. Such energies have been observed experimentally in the anomalous Lamb effect, first observed by Willis E. Lamb, Jr.

Actually, the inherent energy of the vacuum is much higher than the tiny spurts of it seen in the Lamb effect, which is bound to be small because it is merely a polarization arising from electron–positron pairs being created and almost instantly annihilated. Because of the Uncertainty Principle, such particle pairs exist observationally only for a brief time given by the small number $(h/4\pi)\,\Delta E$, where ΔE is the variation of the radiation energy involved in such creation. The intrinsic energy of the vacuum even causes deviations from Coulomb's law between two static electric charges, as shown by Uehling in 1935.

Import for Psychobiology

The above experimental and theoretical aspects of quantum physics suggest a possibility for psychobiology: if the physical effectiveness of a vacuum is needed for a better understanding of ordinary matter, it would be needed even more to explain entities considerably further removed from ordinary physical structure, such as emotion, symbolic thought, and motivation. All these psychological terms have physical referents that are effective psychophysiologically, as the reactions inducible by placebos (a form of suggestion) and hypnosis show.

A very important factor in the effectiveness of hypnotic and related suggestions seems to be the "firmness" (obtainable via third-party protocols) of the subject's emotional conviction of the worth of the procedure. Apparently, fear also is a potent factor in the production of psychophysiological effects (e.g., those produced by the present priests of *vodun* in Dahomey or by the Australian aborigine "bone-pointers" of former days). In our culture Hans Selye (1950) explored such psychophysiological phenomena, with emphasis on the psychic factor. He calls a "stressor" any agent, physical or psychic, capable of causing the physiological alarm reaction. This reaction grows into a general stress syndrome if unchecked. Resistance to a stressor requires a systemic "effort" on the part of the organism.

Selye reported striking examples of the stress syndrome in men under continuous, severe emotional stress: "The basic reaction pattern is always the same, irrespective of the agent used to produce the stress." (More recent work has shown that the regularities he observed are not as invariant as he supposed on the basis of his techniques. Yet his point remains.) To Selye's researches we can add the testimony of similar observations before him, including multiple observed psychosomatic effects. See Scott (1961). Indeed, the brilliant philosopher-emperor Flavius Claudius Julianus wrote in his well-known (though less read) Oration to Hera, c. 360: "That all diseases, or at least the majority of them, occur from a wrongful changing and action of the mind will not, I believe, be denied by any true physician."

Process control in living systems occurs at the molecular and submolecular levels. Single atoms could be in controlling positions; if so, by virtue of these atoms the system might be amenable to quantum triggering effects such as vacuum polarization or London forces (Musès, 1970). Single atoms of magnesium, iron, vanadium, copper, and iodine occupy key positions in chlorophyll, hemoglobin, sea squirt blood, crustacean blood, and the vertebrate thyroid hormone, respectively. Manganese is necessary for RNA function, and chromium ion is important in the glucose tolerance factor that helps regulate insulin function.

We may now assume that the functions of aminoacyl tRNA synthetases and of recognition reactions in general involve: (1) characteristic distribution of patterns of negatively and positively charged regions on a molecular surface; (2) the patterns of hydrophilic and hydrophobic ionic groups on the molecular surface; and (3) the flexibility of a given molecular region. All these factors can be affected by extremely small trigger fluctuations, which are amplifiable thereafter through cascade effects.

The great specificity of some enzymes for their substrates rivals that of some antibodies for their antigens. Again, we have a situation possibly subject to control by microfluctuations. To extrapolate the idea further, neurobiological substrates of psychic phenomena may involve fluctuational phenomena at a very low level of physical organization. Indeed, they may even involve fluctuations of the zero-point energy of the vacuum, which I am here postulating as a psychophysiological transducer.[2]

---

[2]The vacuum is defined as devoid of both matter and radiation and hence formally nonphysical by definition, since the domain of physics is completed by matter and/or radiation. Although quite formally nonphysical, the vacuum is clearly physically effective, as quantum-physical observations show; e.g., the so-called anomalous Lamb effect produces energy on the order of 1060 Hz. This opens a new pathway of investigation, that of the *physically effective*, which is not necessarily to be equated with the physical per se. See also Demys (1972). Stunningly reinforcing the concepts of this note and this chapter are the newly published findings by Arnold Scheibel and Joyce Kovelman (1984) of UCLA's Brain Research Institute—a research project that began in 1975. As Scheibel reported to Gladwin Hill (1985) after over 13,000 confirmed micro-observations: "One direction our research will be going in the months ahead is a study of some of the molecules on brain nerve surfaces—molecules recognized as playing key roles in guidance and alignment of cells during the development of the nervous system soon after conception." Again, the emphasis of the leading edge of research substantiates the herein proposed theory of microcausality at work in processes that extend into molecular and hence quantum-biological domains.

Much of our understanding of physiology consists of increasing knowledge of microstructure and microprocess. Part of the history of physics, as quantum field theory abundantly shows, has been in the same fertile direction. If history, including the most recent, is any guide at all, then the future of control science in general and psychobiology in particular may not lie on global or large-scale levels, as useful as those are in ecosystem problems (Casti, 1979), but on very profound microlevels. When high chemical specificity is required, it appears that simple gross models of the fluid dynamic type, for example, could not serve adequately. Thus, one of the most useful of all the global parameters of a statistical mechanical field is entropy (or negentropy), which proves on analysis to depend largely upon availability and accessibility of microstates.

Macrophenomena turn out to be effects of microarrangements and microphenomena, although they may manifest lawful behavior on their own terms, too, and micro-singularities can create new macro-fields. At the start of a microcausal chain are uncharted regions of energy transductions through space itself that play a fundamental role in the maintenance of the bonds of matter and, hence, at least remotely, in the maintenance of organic structure and biochemical function. Could it be that the ultimate controls for physiological and psychobiological processes may well lie in those regions? Biological controls might then share the same relationships with our radiation-created matter as does the sea of negative energy at the foundation of the physical world. Here is a reductionism that may be at last viable logically and biologically as well. It is actually not reductionism in the crude sense at all since it "reduces" to deeper levels of actually *greater* profundity—the now-leading emphasis of quantum physics (see, e.g., Halzen and Martin, 1984). One need mention only superstring theory.

Such a primal basis on the microlevel has profound implications for theoretical biology and in particular for genetic theory and how genera and species proliferate by long and minutely connected sequences of orchestrated genetic (DNA/RNA) changes.

As far back as a 1962 lecture published three years later (Musès, 1965), I posited repeated organism/environment interactions leading, by microprocesses utilizing quantum effects if need be, to consequential RNA and/or DNA shifts by inverse transcriptase processes and appropriate plasmid insertions. I later pointed out (Musès, 1972) that RNA/DNA plasmids (open or not) were the benign prototype of the virus, the general principle here being that parasitic forms (e.g., viruses) are by necessity later developments of originally nonparasitic forms of similar phenotype. Given the grave logical hiatuses in the Darwin–Wallace and neo-Darwinian apologisms for the observed fact of long sequences of coordinated physiological–anatomical changes both within and between genera of fauna and flora, a new and general theory of evolution must emerge; and the line of reasoning presented in this chapter points the way to the micro-underpinnings of such a theory. If it is to be viable, it must be more than a repatched and tired Darwinism, mixed to suit taste with a naive randomism that would demonstrably require more than given geological time to produce the extraordinary taxonomy we in fact observe (see Berlinski, 1979, and references therein). We have had enough Darwinian and neo-Darwinian hogwash that takes us out through the same door wherein we went, explaining nothing about the actual origin of species, let alone phyla and orders. It was proposed some time ago (Musès, 1965) that the only scientific answer to these questions lies in the custom-tailored process (neuron-hormonally regulated) of allowing RNA-structure and the environment to interact, finally, through reverse transcriptase, writing these changes in terms of interpolated codon plasmids (the benign counterpart of viruses) in the DNA-genome.

We suggest that evolutionary change, whether proceeding toward increased insight and awareness, or merely toward parasitism, is governed by micropsychophysiological processes resulting in the long, orchestrated chains of genetic change that are observed in fact to occur. The neuropeptide system is the key to evolution.

Summary

I am in full agreement with William James, whom I regard as one of the few great perennial writers on science. For the same reasons I am opposed to that form of obstruction of inquiry which, naively identifying its own preassumptions (and presumptions) with "science," proceeds to call any who challenge its views "unscientific." William James would heartily agree that such tactics degenerate scientific discussion into a vulgar display of mere name calling and muddying waters that should be kept clear in the interests of fruitful interchange.

In the interests of such clarity, let me very briefly and unequivocally summarize what I am saying in this chapter.

First, I am not "invoking" quantum physics, either as a demon or as a mere metaphor. I am simply saying that in view of the laboratory findings that overthrew billiard-ball versions of atomic physics (whether the balls be small, large, simple, or structured, it matters not in this context) and logically and empirically necessitated quantum physics to avoid fatal self-contradictions in the older views—that in view of this history, not to use quantum physics to the hilt in biology would be self-defeating.

Unfortunately, however, as Eugene Wigner has stressed to me several times in conversation and has scattered through his writings, far too many biologists, psychologists, and sociologists still remain concept-bound to outmoded versions of physics that advanced physicists—those who know most about the nature of matter—have long since rejected as inadequate theories.

I am not "complaining" about "modern science" at all, but about any clique whatsoever that presumes to speak for all of science when, on examination, its own views tend to be largely descriptive and also verbally overcomplicated in obscuring a paucity of explanatory power. In other words, I am for calling a spade a spade and letting facts invalidate any theory that is invalid, whether it be highly favored or not by any steering group. I am for science, not dogmatism.

I do not at all regard "the origination of living systems from lifeless matter ... with amazement," but rather insist that such a statement is itself fundamentally misconceived. Matter has not yet been shown lifeless, nor has the nature of the interface between molecules and the awareness so characteristic of life been adequately discussed. Indeed, such an interface is not even discussable in the sweep-the-psyche-under-the-rug theories, whose proponents, however, are psychically energized to fight "to the death" to defend their *ideas*—a term that interestingly finds no place in their theories, thus rendering them under their own assumptions walking self-delusions. Reductionism, thus misused, itself becomes a gigantic *reductio ad absurdum*.

Phenomenologically, of course, there is a necessary interface between phenomena like attention, resentment, planning, conceiving, and hoping on the one hand and like momentum and position on the other. All other physical phenomena can be deduced from these latter two, whose product is action (mass × velocity × distance), the *lingua franca* of quantum theory, and also the physical dimension of Planck's universal constant $h = 6.6262 \times 10^{-27}$ erg-seconds, since energy × time has the same physical dimension as momentum × space.

To me the situation is decidedly better and not at all "worse" (as Yates seems to feel) if the highest physics we know—quantum field theory—is used to shed light on biology and psychodynamics. Indeed, such use precisely characterizes the evolution of science in this century. I am merely insisting that biologists become increasingly aware of what their own best physicists are telling them: namely, that "matter" at its roots contains tremendous unsolved problems. Whether this flux at the foundations causes us malaise or not, it is what nature is telling us loud and clear. And we had better be Jamesian enough to live with it.

Finally, I heartily agree with Yates' citing the work of E. Jakobsson. But the manifest conclusion from that work is not at all that there is no legitimate quantum approach to the study of mind and life. Indeed, without the quantum approach, our current physicochemical understanding of organic molecules would be impossible! Rather, the conclusion to be drawn is that there is a level of organization below which we cannot in principle determine. Yet as I have shown (Musès, 1970), such indeterministic energy fluctuations can be marshalled and deployed by target-seeking processes. Indeed, that is what intelligent life does constantly with its environment. The prime example of such processes remains indubitably the psychic process of awareness itself, the literal terminus *à quo* of every statement ever made by any scientist. Would it not be totally improbable that that prime target-seeking process should *not* be involved with the deployment of the abundantly existing random energies for its own ends? In fact, the most concentrated use of such energies would finally have to be the case in which entropy is kept low enough for life to emerge in a physical manifestation, since the harnessing of random energies is a negentropic process par excellence, and thus involves time itself. In this connection see Musès (1983, 1985).

Finally, we do not prove the autonomy of our minds ultimately by any measurements, quantum or otherwise, because "measurement" already implies mind; and any such "proof" would be a mere vicious circle foredoomed to failure. We all cut that Gordian knot quite naturally, however, by *self-awareness*, of which each person is provided with an immediate and incontrovertible demonstration. The core of the matter is that I am saying biology, and *a fortiori* psychology, require and include a physics of the so-called "vacuum"; and that a physics, deliberately and on an ad hoc basis defined without life and mind, can by no valid logic include *them*. Those who wish to deny this in the face of their own self-awareness (with which they perforce must begin their every statement) are left with an enormous burden of both proof and artificiality, and they have nothing but their own inverse credulity—parading as skepticism—to blame.

References

Berlinski, D. (1979) Philosophical aspects of molecular biology. *J. Philos.* **69**:333.
Bohr, N. (1913) *Philos. Mag.* **26**:476f, 857f.
Casti, J. (1979) *Connectivity, Complexity and Catastrophe in Large-Scale Systems*. Wiley, New York.
de Broglie, L. (1923) *Nature* **112**:540.
Demys, K. (1972) CM (consciousness–matter) quantum theory: The beginning of a quantum psychology. *J. Study Consc.* **5**:239–242.
Dirac, P. A. M. (1928) *Proc. R. Soc. London Ser.* **117**:610; **118**:654.
Gerber, C. L. P. (1898) Die Räumliche und Zeitliche Ausbreitung der Gravitation. *Z. Math.* **43**:93–104.
Gerber, C. L. P. (1902) *Die Fortpflanzungsgeschwindigkeit der Gravitation*. Hendess, Stargard in Pommern.
Gerber, C. L. P. (1910) *Gravitation und Electrizität*. Hendess, Stargard.
Halzen, F., and A. D. Martin (1984) *Quarks and Leptons*. Wiley, New York.
Hill, G. (1985) Where schizophrenia comes from. *California* **10**:72.
Ingram, P. (1969) *Biological and Biochemical Applications of Spin Resonance*. Hilger, London.
Musès, C. (1965) Systemic stability and cybernetic control. In: *Cybernetics of Neural Processes*, E. R. Caianiello (ed.). National Research Council of Italy, Rome, p. 190, note 24.
Musès, C. (1970) On the modification of random fluctuations by a target-seeking process utilising random energies. *Int. J. Bio-Med. Comput.* **1**:75–80.
Musès, C. (1972) RNA and DNA in biological evolution and cancer. *J. Study Consc.* **5**:292–294.
Musès, C. (1980) Hypernumbers and quantum field theory. *Appl. Math. Comput.* **6**:63–94.
Musès, C. (1983) Hypernumbers and time operators. *Appl. Math. Comput.* **12**:139–167.
Musès, C. (1985) *Destiny and Control (Chronotopology)*. Kluwer–Nijhoff, Boston.
Planck, M. (1900) *Verh. Dtsch. Phys. Ges.* **2**:237.
Scheibel, A., and J. Kovelman (1984) A neurohistological correlate of schizophrenia. *J. Biol. Psychiatry* 1601f.
Schrödinger, E. (1926) *Ann. Phys.* **79**:361f, 489f.
Scott, M. (1961) *Hypnosis in Skin and Allergic Diseases*. Thomas, Springfield, Ill.
Selye, H. (1950) *The Stress of Life*. McGraw–Hill, New York.
Steen, E. B. (1971) *Dictionary of Biology*. Barnes & Noble (Harper & Row), New York.

27

A Physics for Complex Systems

Arthur S. Iberall and Harry Soodak

ABSTRACT

The fundamental problem for a science of complex systems is to explain how diverse forms and evolutionary processes arise from the operation of the few principles and materials required for a physical description of nature. A theory is offered here to account for emergent properties in complex systems. Its central theme is that the emergence of structure as well as process is a stability transition in which new form arises because changing parameters have made the less structured state unstable. A generalization of the Reynolds number concept of hydrodynamics is suggested as a criterion for emergence, and a number of examples of its application are presented including social theory (see also Iberall, Chapter 28).

Biological systems employ language and information. Information is usually regarded as being static, but catalysis is offered here as a dynamical basis for information and language, applicable to all complex systems. Such systems transform internal energy through internal, time-delayed, dissipative processes; they deploy it in various action modes in response to command-control algorithms that emerge from dynamic regulators. The behavior of the system is modulated catalytically. Zipf's rank-ordering law of word usage is suggested as the way to characterize linguistic catalysis.

In the authors' view, complex systems achieve stability by coupling aggregative to degradative processes. Dissipation (of free energy) is seen to guarantee both emergence and degradation of local forms, as well as to describe the overall nature of the global process. As a final application of their theory, the authors address the origin of life as a self-organizing process. —THE EDITOR

Introduction

A Physics for Complexity

Modern physical theory is no longer faced with great difficulty in dealing with most of the nested hierarchical systems of nature—of cosmos, galaxies, many of

---

ARTHUR S. IBERALL • Department of Oral Biology, University of California, Los Angeles, California 90024. HARRY SOODAK • Department of Physics, The City College of New York, New York, New York 10031.

the galactic fragments or subsystems, atoms-ions-molecules, or even leptons and quarks and a vacuum. But the principles of physics do not seem able to confront the organization and phenomena of such systems as are represented by life, man, mind, and society explicitly. Do these systems and their rich phenomena fall outside the legitimate bounds of physical theory? We doubt it. There is no evidence that biological or social systems require materials or forces beyond those identified by physics. Yet most biologists and social scientists have trouble connecting basic physics to a theory of the structure and operation of systems characterized by autonomous organogenesis, evolution, adaptation, nearly invariant self-reproduction, historical irregularity, goal-directed behavior and their social analogues. For physics to be a universal science, gaps between physical principles and applied systems analysis in all existing domains must be bridged. Principles must be extended strategically to link basic physics to theories of organization and self-organization of both simple and complex systems. Such strategic principles should support the lateral reduction of a problem to standard physics. Further, both nature and the constructs we fashion in mind should utilize the same principles.

By developing and using such strategic principles, we hope to bring physics to bear on organic and social complexity and to revitalize Aristotle's conception of physics (McKeon, 1941) as a general science of motion and change, including the history and evolution of all natural systems. This generality of physical theory was very much a part of the outlook of physics until the 19th century. Building on Newton's 17th century world-machine image (see Randall, 1940), scholars of the Enlightenment period proposed mechanistic models and metaphors from physics to include life, man, mind, and society. This grand scope disappeared from physics in the 19th century, however, largely because Newtonian mechanics appeared to be inadequate for the task. Actually, before or without its extension by statistical physics, thermodynamics, electrodynamics, quantum field theory, and relativity, physics really could not explain either simple or complex systems at their many scales. Is physics now complete or mature enough to deal with the complexity in biology and social sciences? We believe the answer is "yes." Here we present ten strategic physical principles of organization that perhaps may constitute the competence of physical science to account for processes in complex systems. The general schema that we believe is required to extend normal physics of simple systems to a physics of complex systems we denote as homeokinetic physics. Such a physics relates to those notions required to maintain form and function within complex field systems at the many levels at which they are found. In an earlier article we developed five relevant propositions about that physics (Iberall and Soodak, 1978).

Principles, Materials, and Forces Are Few

A fundamental science of organization and self-organization must begin with modern physics itself. The existence of such a body of thought is the primary foundation for our explanatory construct—Principle 0. (Landau and Lifshitz,

1951–1979, for example, offer a technical introduction to the concepts of physical theory.) *All physical nature operates with only a few principles, but they acquire many forms and are expressed in a variety of emergent processes—Principle 1a.* ("Nature is not economical of structures—only of principles," Abdus Salam.)

Our scientific knowledge of the material constituents of the universe has developed from early Greek conceptions of the one substance, the few, or the many (see Toulmin and Goodfield, 1965), through the elements of Mendeleev's atomic tables, to the nuclear constituents of the Bohr atom, and now to the more elementary particles. There are still uncertainties about particular details: for example, we do not know all the forms of biochemical macromolecules that function as "informational" messengers in living organisms; we are not sure that we understand all the material processes associated with the formation and operation of galaxies. Nevertheless, it is clear that *the basic materials of physical systems throughout the entire universe are few—Principle 1b.*

Natural systems are acted on by only a few forces—Principle 1c. In the Greek classical period, Aristotle vaguely described the action of causal agents; in the 17th century Newton formally defined force and identified one major force system (gravitation). A unified form of the electromagnetic forces was described subsequently, in the 19th century. Two additional forces have been defined in the 20th century to account for processes at the nuclear level of matter organization: the weak nuclear force that governs, for example, spontaneous β decay in radioactive nuclei, and the strong nuclear force that governs the close-packed binding of protons and neutrons. Two significant changes in the concept of force also have been made. Einstein generalized Newton's law of inertia and described gravitation geometrically: a particle follows a geodesic path in space–time curved by the presence of matter (the theory of general relativity). Maxwell's electromagnetic field theory has been generalized to a more encompassing gauge field theory. This theory ascribes the force between particles to a fluctuating exchange of gauge field quanta between them and has already led to a unification of electromagnetic and weak nuclear forces.

The diversity of all natural systems emerges from these few forces and materials. *The deep problem for any physics of complex systems is to show how historical and evolutionary processes, with their diverse morphologies, arise from the operation of these fundamental elements.*

Atomisms, Fields, and Field Transport Processes

Systems manifest themselves as alternations of atomistic and continuum field levels (Iberall, 1972; Soodak and Iberall, 1978a). At some spatial–temporal scale of observation a system acts as a continuum, but on a finer scale the apparent continuum is an atomic collection, such as the molecules of a fluid or the cells of an organism. Much of the behavior of the system, particularly its capability to create pattern and formal structure, lies in the interaction between the kinetics of its atomistic levels and the continuumlike mechanics of its fields.

At the fine-scale level, a system consists of a number of discrete, rather similar or identical physical entities ("atomisms"). The entities interact repetitively among themselves in a fluctuating fashion (local kinetic processes). According to the principles of statistical mechanics, these kinetic interactions at an atomistic level are characterized by conserved physical quantities (e.g., momentum, mass, energy, electric charge) that become distributed or partitioned throughout the population of atomisms.

The relationship between the atomistic and the field levels of a system can be seen in the equilibrium thermodynamic description of a one-constituent, electrically neutral gas. The atomisms of the gas (molecules) interact according to conservations of mass, energy, and momentum, leading to sharing of the conserved quantities among the population. The field description of the gas is based on the local spatial averages of its atomistic measures: the local spatial average of mass is *density* ρ, of energy is *specific energy* e or *temperature* T, and of momentum is *pressure* p. Such three independent measures are functionally connected in the (macroscopic) *equation of state* for the gas. Summing over the local atomistic fluctuations, we obtain a correlated expression

$$f(p, \rho, T) = 0 \tag{1}$$

which, for a near-ideal gas, for example, is

$$\frac{p}{\rho T} = \text{const} \tag{2}$$

The existence of such a relationship among macroscopic field variables, however, does not describe *dynamics* at the field level; the state description is thermostatic rather than thermodynamic. To describe dynamics within the field it is necessary to account for the transport processes that are characteristic of accommodations or relaxations of the local measures in any near-continuum field and to show how they can be related to the atomistic level. These accommodations between differing local neighborhoods or states are known as *equations of change*.

Transport processes between differing field regions are of three kinds: *diffusion*, *wave propagation*, and *convection* (Iberall, 1976). Diffusion and wave propagation act locally; convection operates at field scale. The atomistic basis for these processes is simple to visualize using, for example, the motions among translational degrees of freedom of atomisms in a fluid field.[1]

---

[1] Imagine the field to be locally represented by a regular small boxed array of atoms a near-continuum array. They have an average distance of separation related to the mean free path of motion. The instantaneous atomistic motion will consist of movements away from this idealized picture, but if we examine the array intermittently at a period near the equilibrium relaxational time scale (a small number of fluctuating relaxation times), we can see the regular boxed array transformed into another, similar regular array within the same box. (Both the original state and the final state have to represent near-equilibrium stereotypes.) In the transformed array the identities of some of the original individuals may be changed, but the mean positional relations are unchanged. Only state exchanges of a few neighboring individuals, e.g., in position or momentum, have occurred. At the near-equilibrium time scale, only two general kinds of local exchanges can occur in the array. A few border atoms within

This decomposition into three kinds of processes suggests a basis for field dynamics in general; the constructive argument is applicable to other kinds of degrees of freedom, including chemical—atomic or nuclear—transformation and transport among more than one type of mass species and, in solids, diffusive and propagative transports that are associated with phonons rather than with the mass particles themselves. The same kinds of processes also can take place among atomisms at any level of organization. *Physical nature operates with only a few types of field processes: diffusion, wave propagation, convection—Principle 2.*

The Emergence of Form and Its Diversity

The scale at which a force operates is determined by the form with which it is associated; a force scaled to a particular form is a force system. Nuclear forces, gravity, electromagnetic forces, hydrostatic pressure, chemical potentials, and social bonding forces, including a "social pressure" (Iberall et al., 1980), are examples of force systems. *Principle 3: A form emerges from interaction between two (or more) force systems. This process of emergence of form may be described as follows: matter and energy "ingather" (an absorption process), are tied together coherently for an epoch, and then are released (an emission process). We refer to this asymmetric ingathering and release as a hop. The ingathering and release phases of form are unsymmetrical. In composite the pair of processes represents a hopping Brownian motion, which we regard as a hierarchical version of Einstein's account of ordinary Brownian motion.*

The coherence-for-an-epoch that represents the emergent form occurs because forces and subassemblies of matter are available to interact at many scales. Force and form join locally and create forms of greater size and time scales. As scale increases, forms vary: they may exist as atomic nuclei, atoms, ions, molecules, crystallites or organelles, solid-state matter or cells, inanimate or living subsystems, organisms, societies, lithospheres, hydrospheres, and atmospheres. With each scaling-up, the new force systems that emerge remain based on the primitive physical forces, which are still present but transformed in their realization. ("Now the smallest of particles of matter may cohere by the strongest attractions and compose bigger particles of weaker virtue; and many of these may cohere and compose bigger particles whose virtue is still weaker and so on for diverse successions . . ." Isaac Newton.)

Even the forces of psychology and sociology are built on these fundamental forces. For example, an interpersonal exchange force has to be based ultimately on electromagnetic forces (see Iberall and McCulloch, 1969). The electromagnetic aspects, externally are acoustic or mechanically vibratory, but internally sensory–

---

the box can be exchanged with atoms outside the array—this process is a model for *incoherent exchange* or *diffusion*. There also might be a relay motion along any line: an atom enters the array, sequentially displacing other atoms until an atom at another edge leaves the array. This model represents *coherent* exchange or *wave propagation*. These two local processes, one incoherent and one coherent exhaust the local possibilities. The only other process that might occur is the displacement of the entire array, which represents the nonlocal process of *convection* in the field.

linguistic communications are effected through neural and neuroendocrine (electric and electrochemical) means. The interpersonal force varies with time, instead of with distance of separation, as potentials do, and it is not usefully measured in dynes. (It might be characterized, nevertheless, by bonding energy. For example, a considerable fraction of the metabolic energy expended by the partners in a marriage is associated with actions that express the bonding between the two people.) Psychologists, ethologists, social scientists, and even physicists may be reluctant to accept the proposition that there might be an electromagnetic basis for social force and a physical basis for social pressure. Perhaps neurophysiologists and neuroendocrinologists will be more sympathetic. In any case, it is possible, in at least a crude sense, to identify the linkages in social systems (Iberall et al., 1980; Iberall, 1984). The connections are only moderately more tortuous than those of the chemical potential differentials between solvent and solutes in a multicomponent solution (Soodak and Iberall, 1978b, 1979). Social interaction is more complex because of the greater internal complexity of the individual human atomisms, who possess an elaborate memory function, a history (a dependence on experience), and a more hierarchically structured chemistry.

Explaining Emergence

A specific emergent form is a solution to a generalized stability problem, by spontaneous application of physical law involving a fluctuation–dissipation process operating on whatever atomistic materials are at hand. Spontaneous natural designs pose the same conceptual problems as do human designs of man-made systems. The fluctuation–dissipation principle varies according to the scale and medium of the subsystems involved, but its essence is this: small fluctuations produced by lower levels of organization may be quickly dissipated, or they may be amplified and stabilized within a higher-ordered organization. Form thus may emerge from a relatively homogeneous underlying medium, as, for example, in the formation of dew droplets from water vapor, of galaxies within the cosmos, or in the birth of a hurricane. *Principle 4: Emergence is a stability transition— new patterns or forms arise because changes of forces and scales make the existing patterns or forms unstable.*

What is the driving force that is sufficient to create the instability out of which new pattern or form emerges? To answer we must look to the scale of the upper, continuum level: the mechanism that forces the instability is the imposition of sufficient driving force "from above." A flow of *authority* extends from the upper scale to the atomisms. A step in a hierarchy is thereby established. Forces impressed from above create *order parameters* and impose requirements on the atomisms. The atomisms interact to adjust to the demands of the order parameters.

Diverse possibilities of atomistic response is the necessary condition for morphological diversity. In this view, form arises from an interaction: a macroscopic field dominated by force systems interacts with its own atomistic microstructure to create a new middle form of organization. The homogeneous field may then be said to be fractured by symmetry-breaking into new (larger) collections of subatomisms. Diversity of form emerges; it is not already present in the ground-level atomisms. The atomisms may be permanent, whereas the new patterns or forms, in their own independent existence, are born, live, and die.

In hydrodynamics (the dynamics of mobile atomisms), emergence of pattern occurs in a field with underlying atomistic fluctuations when a critical value of a flow parameter is reached because of some change in scale or magnitude imposed from outside—as when more energy is injected. Derived first from Reynolds' study of pressure-driven fluid flow in a pipe (extended later to flow around a sphere or cylinder) the critical parameter that governs the emergence of intermediate flow patterns is now known as the Reynolds number, i.e., $\text{Re} = V/(\nu/L)$ where V is a characteristic velocity in a flow field, ν is the kinematic viscosity of the fluid, and L is some characteristic (length) dimension of the field. The numerator represents a convective velocity, sweeping matter into the field, whereas the denominator is a diffusional velocity, the rate of transport of momentum. With the scale suitably selected, if $\text{Re} < 1$, the flow is *laminar*, without global structure.[2] But if $\text{Re} > 1$, the homogeneous laminar solution is no longer physically stable. Mathematically, a bifurcation appears in the solutions to the governing equations. Physically, the bifurcation corresponds to a transition in the field's ability to handle the imposed stress. (In the neighborhood of $\text{Re} = 1$, the situation is uncertain and unstable.)

We wish to expand this criterion to a general condition for emergence of form as well as flow pattern, starting from a generalization of Reynolds number

$$\text{Re} = \frac{V(\text{convection})}{V(\text{diffusion})} \qquad (3)$$

The issue addressed by this generalized parameter is whether or not the energy associated with the global convective velocity (numerator) that sweeps into the field can be absorbed into the internal energy at the atomistic level by some diffusive process (denominator). If not, the field becomes unstable, and some new larger-scale, structured form or pattern emerges. Convective field processes and local diffusive transports compete. A diffusive process may be momentum diffusivity (as in the kinematic viscosity of the fluid flow case), or some other dominant mode of diffusion (e.g., electrical, thermal, chemical).

The single concept embodied in the generalized Reynolds number gives us useful insight into many different cases of emergence. We present several examples below.

Fluid Flow. The critical Reynolds number criterion ($\text{Re} > 1$) for the change in stability of hydrodynamic flow fields is well known. Much of the history of 20th century fluid dynamics, associated with Prandtl, Blasius, von Kármán, G. I. Taylor, Oseen, and others, is a series of successes in explaining what happens in various boundary layers as Re passes critical values. The onset of Rayleigh–Bénard convection or the initiation of cellular vorticity as Taylor cells are critical

---

[2] For a pipe, the characteristic length scale for which the critical Reynolds number is unity is the boundary layer, as it is for flow near a flat plate. See, for example, Landau and Lifshitz (1959).

transitions determined by the increasing Reynolds (or the related Rayleigh) numbers. The theme is common: the emergence of new patterns at critical values of parameters is a stability transition. The old solution becomes unstable to underlying fluctuations. Modern hydrodynamic stability theory (Ruelle and Takens, 1971; Ristel, 1975; Gollub et al., 1980) suggests that driving any continuum field into the domain beyond its first instability leads to a number of branching instabilities at other critical (bifurcation) points. At some intermediate bifurcation, for example, quantized patterns capable of some autonomy are possible; and at higher levels of stress and repeated bifurcations the field system can exhibit such complicated behavior that some feel is described appropriately by the term "chaos."

Our purpose here is to extend the generalized Reynolds number criterion beyond the second-order transitions of hydrodynamics. As suggested by Landau and Lifshitz (see also Haken, 1977, or Chapter 21, this volume), the concept of an *order parameter* is useful for considering hydrodynamic instabilities in flow fields and the phenomena of *phase transitions*. We shall attempt here to apply the generalized Reynolds number criterion, as an order parameter, to the phenomenon *of matter condensation*. For this example, we shall use the very simple case of a first-order phase transition, the condensation of vapor into a liquid droplet (or equivalently to condensation into the solid phase).

Matter Condensation. Condensation of an initially homogeneous vapor into a two-phase system of vapor and liquid droplets occurs in nonideal gases, commonly illustrated as a van der Waals gas. The van der Waals attraction force is required to destabilize the low-density vapor phase, and particle repulsion is required to limit the high-density liquid phase. The competition between these forces is described by the van der Waals isotherm (plotted graphically as pressure p versus density ρ) at temperatures below the critical point temperature. These isotherms have two horizontal tangents, $dp/d\rho = 0$, at densities ρ_v^* and ρ_l^* leading to the unstable intermediate region of negative $dp/d\rho$ in the density range between ρ_v^* and ρ_l^*. Thermodynamic equilibrium at any temperature exists at the vapor and liquid densities, ρ_v^e and ρ_l^e, corresponding to the horizontal constant pressure line (of constant p versus changing ρ) according to Maxwell's equal area rule. This rule emerges from the requirement that the chemical potentials of a saturated vapor and liquid in contact with the vapor across a liquid surface be equal. The surface may be flat, or that of a liquid drop of large enough radius for surface tension to play a negligible role.

Ordinarily, condensation of a near-saturation vapor occurs at nucleation centers, such as dust particles or ions, that play the role of catalysts. A dust particle allows vapor molecules to condense upon it into a droplet of finite size. This growth in size avoids the surface tension that can destabilize a liquid by increasing the pressure (by twice the surface tension divided by the drop radius) and thereby increasing the chemical potential. An ion acts as a nucleation center for polar molecules (e.g., water) by providing an attraction center, resulting in capture of incident vapor molecules into a droplet of increasing size. Macroscopically, the ionic electrical field lowers the chemical potential of the liquid. A stable droplet can form and grow only by eliminating the latent heat associated with the phase change. Thus, to form a drop requires the simultaneous parallel processes of matter inflow and heat outflow as in the standard psychrometric wet bulb process (Iberall and Soodak, 1978).

The full hydrodynamic nature of the condensation process is exposed by considering condensation of an initially homogeneous vapor containing no nucleation centers. If the vapor is near saturation, with density near ρ_v^e where $dp/d\rho$ is positive, condensation does not take place. Local density fluctuations are rapidly propagated away at sonic speed c, and thereby dissipated, simply leading to normal fluctuation and dissipation. However, as the density (or pressure) is increased above saturation to near the horizontal tangent value of ρ_v^* of the van der Waals isotherm, fluctuations result in local regions in the unstable density range greater than ρ_v^* where $dp/d\rho$ is negative. Because such a density leads to a decreased pressure, the result is a contraction of this region into still higher densities and still lower pressures with an increasing rate of contraction. A true implosion process thus takes place, leading to contraction speeds on the order of the thermal speed C (as in a shock wave). The size of such a density fluctuation is only that of several mean-free paths and so the hydrodynamic implosion process is scaled by a Reynolds number value of the order of unity [as in the more usual acoustic Reynolds number situation[3] (Soodak and Iberall, Chapter 24)].

Reynolds numbers scale several different but related flow processes. For example, in matter condensation, the order parameter that drives the instability is the thermodynamic potential difference (or density difference). It leads to the implosive flow process, scaled by the hydromechanical Reynolds number, resulting (after heat elimination) in droplet formation and a segregated two-phase system. In the pure hydrodynamic flow field, the order parameter that drives the flow instability is the ordinary flow Reynolds number that leads to vorticity production and dissipation and to turbulence. Turbulence consists then of a two-phase flow field: (1) a laminar field within the boundary layer adjacent to a solid boundary, and (2) the mainstream vorticity field. The relationship between the two cases may be made tighter by recasting the flow Reynolds number order parameter in the form of a pressure difference or gradient that drives the flow field. Although the Reynolds number represents the same concept in both cases, the difference between the two lies in the fact that matter condensation occurs from an initially homogeneous field, whereas the flow separation into two phases stems from a nonhomogeneous boundary-dominated condition that effectively destroys the field homogeneity.

Chemical Patterns. Turing, Prigogine, Goodwin, Chance, Belousov, and Zhabotinsky, and many chemical engineers (see, e.g., Gmitro and Scriven, 1969) have indicated that autocatalytic chemical reactions can lead to temporal oscillations and that reaction–diffusion systems can lead to flow patterns. Within such systems, Prigogine has demonstrated the existence of a chemical stability criterion that can be regarded as a chemical "Reynolds number" (for discussion see Iberall and Cardon, 1980). Many other forms of couplings of chemical processes

---

[3] A view of this kind may be helpful in understanding the condensation of galaxies in the postrecombination era. (The existence of such a problem is described by Soodak in Chapter 1.)

leading to regular fluctuations or oscillations are known (e.g., the chemomechanical and chemothermal couplings of combustion engines). There is, of course, a large chemical engineering literature on spatially quantized combustion processes in fast flow fields. The principles are the same as those presented above.

Social Patterns. The generalized Reynolds number criterion for emergence can even be applied to the nucleation of people into urban settlements in the post-Neolithic period, as discussed in Chapter 28 (see also Iberall and Soodak, 1978).

We believe that each of these accounts of form or pattern formation is reducible to a generalized Reynolds number criterion in terms of physical and chemical diffusivities. The production of new forms in this manner is nature's way of resolving a stability problem. The new emergent forms or designs stem from a fluctuation–dissipation process operating on available materials. As stated earlier (Principle 3), we regard the creation of forms as an extension of the Einsteinian theory of Brownian motion: matter–energy ingathers, is tied together coherently for an epoch, and is then released.

Principle 5: the basic physical laws of nature are expressed in terms of formal force systems scaled to the structures on which they act. The creation and stability of new forms require cooperation of two or more force systems, so that form and force systems entwine upon available material ensembles and create new forms of greater size and time scale (e.g., molecules, cells, organisms, stars, galaxies, and societies). This is the principle of hierarchy.

Simple versus Complex Systems

A simple field system is one whose atomisms are themselves simple in the sense that internal processes within the atomisms are few and have small or zero time delays—on the same order as, or less than, external time delay (the relaxation time of fluctuating interactions with other atomisms) (Soodak and Iberall, 1978a, and Chapter 24). Simple fields and atomisms have very little internal dissipation and little or no memory. Simple systems at any hierarchical level are organized by the external (with respect to their atomisms) sharing and conservation within the field of charge, mass (and number), momentum, and energy.

The atomisms in a *complex system* have many internal degrees of freedom; they undergo dissipative, long-time-delayed internal processes. A measure of such internal dissipation is the associational (bulk) viscosity (Soodak and Iberall, Chapter 24; Herzfeld and Litovitz, 1959). In complex systems the energy conversions lead to a ring, chain, or net of internal atomistic modal actions, rather than merely to the external space–time momentum trajectory characteristic of simple systems. These atomistic modal actions at any moment are influenced by past events, the present trace of which represents memory.

A complex system shows the effects of the internal behavioral modes of its atomisms through external fluctuations (that may organize into global dynamic modes) that extend over a time period (a "factory day") during which the local atomisms complete the cycle or ring of their actions. That cycle of actions appears coupled largely as a Markov chain. Repetitive cycling through its macroscopic modes enables a system to carry out the various (sometimes simultaneously incompatible) activities required for survival and to maintain and regulate its internal environment. To say that a complex system exists is to say that a set of macroscopic processes has emerged and stabilized in action-space by cycling the materials and energies at hand through persistent patterns and forms. We call this concept *homeokinetics*—a term that emphasizes the role of kinetic processes as dynamic regulators of the internal environment of the complex field system (Iberall and McCulloch, 1969; Soodak and Iberall, 1978a).

Principle 6: Whereaas the physics of simple systems deals with the organized space–time motion consequences (including rest states) of simple mean path-relaxation time movements of their atomisms, complex field systems manifest dynamic regulating behavior, describable in an action-space. Action is discretized into modes characteristic of the atomistic species. The physics of complex systems involves three largely independent phases in a grand dynamic pattern—a start-up phase in which complex atomisms assemble to make up the field system, a long life phase in which form and function of the field system are maintained by modal actions, and a degradation or dissolution phase.

Language in Complex Systems

The various action modes of a complex system dissipate energy at comparable rates, but the energy transformations within the atomistic subsystems are different for each of the global dynamic modes. As a mathematical description in action (energy–time) phase space, we can say that a global action mode is a low-dimensional topological *attractor* of the system dynamic; the system quickly settles onto one or another attractor, acts in that mode, and then moves to the basin of another attractor. Changes of mode are effected by switching procedures, consisting of low-energy impulses that carry the system over the barrier from the basin of one attractor to another. Complex systems can be said to use *language* to effect such switchings. In modern biology we speak, for example, of "messages" transported between hypothalamus and pituitary, we say that there is "communication" between cells, or that there is a genetic "code" of bases in nucleic acids for sequences of amino acids in proteins. But what does this mean? In essence we mean that low-energy impulses can evoke microstates in the atomisms of the system or can cause the system to switch from one mode to another. Small impulses that switch the cells or the body from one mode to another can be described functionally in linguistic terms,[4] but physically they represent *catalytic* processes. The generalized catalysis referred to here is very similar to that of the chemist's—a catalyst is a facilitator of a reaction that permits it to take place under milder conditions (e.g., at lower energy) than would be required for the *same rate* of reaction in the absence of a catalyst. (In extension, catalysts in complex systems may also be inhibitors and cause otherwise probable reactions to be slowed, stopped, or prevented.)

Language is low-energy in the sense that it adds or removes little or no energy

---

[4]The linguistic aspects of the hormone system and the relation between linguistics and dynamics in the hormone control of metabolism are discussed in detail by Yates (1982).

to any process it commands. Natural languages in complex systems arise during development as mode-switching, state-evoking regulators of the system. They affect both the development and the operation of the system. Language is created out of whatever physical materials and processes are at hand that can be tied into a catalytic process. Language generally has a significant degree of design tolerance because of redundancy (in choice of channel, in linguistic elements, in dynamic sequence) and of error-checking mechanisms. (See, e.g., Iberall, 1983, for an account of the physical character of language.) Its function is highly robust in the face of noise. Natural languages stand in contrast to so-called formal languages, which are highly structured (i.e., they are nearly hard switches—very intolerant of errors).

Language coordinates the processes of a hierarchical system within each level and between levels, as in the flow of authority downward from higher to lower levels, or in the flow of information about system status from lower to higher levels. *In toto*, this spreading flow of information represents command-control within the system. Only complex systems use language and have command-control. We regard our description to be a physical view of language in complex systems. *Principle 7: A recurring ring of action modes in complex systems involves comparable (equipollent) energetics in each mode, with some small barriers between modes. Command-control systems must exist to relate internal and external events. That command-control is catalytic. The catalytic switchings that negotiate the barriers among action modes may be viewed as linguistic signals.*

Each type of complex system has its own language and uses linguistic elements—words, syntax, semantics—appropriate to its atomisms and its field processes (according to constraints drawn from the processes and materials at hand). The study of any one language is a very difficult task. The study of a general theory of language is even more difficult. Nevertheless, a step toward such a physical theory was taken by Zipf (see Iberall *et al.*, 1978). From his work it appears that all human written languages possess a common element in the distribution function of their word usage, in which the probability of their usage is inversely proportional to the rank order, $1/r$, or to frequency, $1/f$, of usage.

The standard presentation of Zipf's law (Brillouin, 1962) is as a reciprocal rank-ordering law, but a more useful view of the Zipfian distribution in many cases is its log-uniform character, i.e., the cumulative sum in each logarithmic interval (e.g., octave or decade) is approximately the same. The range of phenomena that display a Zipfian-like distribution is wide: the rank order of word frequencies in common languages, the sizes of companies (or nations) in relatively free markets, the rank order of distribution of populations among urban settlements of different sizes, the frequency of trade inversely proportional to distance, and the $1/f$ (reciprocal of frequency) character of working electrical systems whose actions are not tightly constrained—as in the electrical noise of biological membranes of excitable cells, in the turbulence in a pipe at its major coherent power region, and in the acoustic power distribution of ordinary speech patterns. Even a pattern as seemingly highly organized as the music of Bach displays a frequency distribution of acoustic power which exhibits a $1/f$ character. Each of these complex systems is an autonomous thermodynamic system in its own right (the company, the nation, the trading system, the cell, the music, and the musician). Each of these complex systems uses language to communicate within the community of systems in which it is involved. And their linguistic characteristics are all Zipfian-like.

How can the same distribution law cover such diverse phenomena? An inverse square law of force, or a $1/d$ (reciprocal of distance) potential, applicable to all kinds of weakly associated social phenomena, has been suggested (Stewart, 1947, 1948, 1960). In urban science this idea has become associated with place theory, or as a "gravity law" for populations. There is no physical basis whatever for these "laws." However, if such distributions are seen to represent equal power over each logarithmic interval, a rationale begins to emerge, based on concepts of *cost*. These costs may be those of living in population centers of various sizes, of transportation according to distance, of the number of transactions, or of energy associated with the number of units in a message. Language usage tends to be economical—as much is encoded into a particular set of units as possible. Because the evolution of freely competing, autonomous, complex systems requires that they be ready for business at many different frequency scales, each system will tend to distribute linguistic power equally over those bands (scaled logarithmically) until all dynamic process scales required for survival of the system are covered. This arrangement is most cost-effective from the point of view of the language-using system; but paradoxically, from the point of view of *physics*, meaningful language is thus very close to noise. *Principle 8: The distribution function characteristic of catalytic switch modes used as language is that of $1/f$ noise. Communication among complex autonomous units is just barely coherent.*

Birth–Life–Death: The Problem of History

Principle 9: The appropriately scaled physics at each level of organization is thermodynamic in nature (i.e., thermodynamics is the physics of "systems"). Renormalization of the thermodynamic equation set at any level of organization, both as an intellectual or descriptive process and as a natural emergent process, is a necessary historical process by which new organized levels (the organized system of forms and their force systems) emerge from a lower atomistic structure. Renormalization requires that there be some prior physical "construct" at a lower level, to build upon, and further that the construct be metalinguistic in character. The strategic theme is that both the physics of the historical process creating a new hierarchical level and the physics at the new level are thermodynamic; both involve only a renormalization of the physics of the lower level. In this sense, the physics between levels, "up–down" physics, is similar to the physics at a given level, "flatland" physics.

Up to this point, as in Principle 6, we have suggested that when complex systems come to life, they temporarily (for their lifetime) nearly balance out entropic degradations, by *cycling* among a set of internal and external action modes in which energy and materials are drawn from outside. This behavior is as true of inorganic processes as it is of organic. At the most basic thermodynamic level, the Carnot cycle explains what the limits are on an elementary engine that is cycled to do work. The concept of a time-ordered (irreversible) cyclic process is essential to understand generally how a structured functioning is maintained in the face of thermodynamic dissipations. (See Tolman, 1938, for an excellent account of the notion of cyclic processes in statistical mechanical ensembles.)

As an elaboration of these notions (Iberall and McCulloch, 1969; Soodak and Iberall, 1978a), we claim that self-organizing systems use cyclic processes to regulate their inner environments. These cyclic processes transport matter and energy and therefore, in effect, import and export entropy. We call this thermodynamic cyclic conception of the life phase of complex systems *homeokinetics*, by extension of Claude Bernard and Walter Cannon's concept of homeostasis. In the homeostatic view, internal mechanisms regulate the organism so as to return the organism to an equilibrium. Homeokinetics emphasizes the role of kinetic processes, particularly in the form of ongoing periodic thermodynamic engine processes, in carrying out the regulating functions of the organism. Obvious examples can be found—from the level of the biosphere as a whole (the nitrogen cycle, the photosynthetic cycle) to the individual animal (the various metabolic cycles).

A river represents a typical homeokinetic system. A river is not a fixed form, but is a process cycle; it is not a hydraulic element in a fixed channel, but the effect of rainfall in a hydrological cycle. The rainfall causes both water runoff and the appearance of a bedload of transportable material that erodes and maintains the form of the river (see Iberall and Cardon, 1962; Iberall, Chapter 2). Viewing rivers in this way, as an effectively ergodic class of similar cyclical processes, helps to explain the more general phenomena of the heterogeneity or diversity of individual forms, and, ultimately, their death. The key to both individuality and death lies in the fact that force systems at one level emerge by a splitting of lower-level force systems into higher-ordered and weaker forms. Two different component force systems combine to produce reinforcements and resonances in some regions and annulments and dissonances in others. From such competition and cooperativity, functioning form emerges. That original emergence may be very sensitive to initial conditions, but after that continuing history and evolution are relatively stable. Such initial sensitivity allows for the appearance of heterogeneity and individuality. As depicted in Chapter 2, imagine a series of water faucets playing on many similar hills of sand. Water runs downhill in every case, but the pattern of rivulets, their meanders and their number density, will vary from example to example and from time to time. Nevertheless, their distribution statistics will be effectively stationary and similar.

Death of functioning form comes ultimately because the individual system cannot maintain the delicate balance of forces indefinitely.[5] The universal entropy generation by all ongoing processes guarantees that no process can operate in a perfect closed cycle. With each passage around the cycle, some small change in the system (e.g., wear and tear) must occur. The system is initially stable to such

---

[5]Although the stability of a system's processes tends to be robust over most of the operating range, it becomes marginal at the limits of the range.

perturbations, but ultimately the entropic accumulation destabilizes the system's regulatory mechanisms.

On the Historical Process in Which Birth, Life, and Death Appear

Earlier we noted that the themes of complex systems—of life, man, mind, and society—disappeared from physics in the 19th century when social scientists had to confront a considerable degree of nonmechanistic, noisy, even goal-oriented behavior in social phenomena, rather than merely the majestic regularity of motion in gravitationally bound solar systems. Such descriptive methods as the harmonic analysis of Fourier seemed hardly up to dealing with the consequences of the French and Industrial Revolutions. In their search for a science of man, 19th century inheritors of the Newtonian world machine model of the 18th century Enlightenment literally wanted to develop a social physics based on the Newtonian concepts. This effort included such masters of the social sciences, and of history and evolution in the social and biological sciences, as Saint Simon, Comte, Quetelet, Spencer, Darwin, and even Marx and Engels (who were against the so-called vulgar mechanics of Newton, offering dialectical materialism as a substitute for a generalized dynamic process). Unfortunately, the only scientific construct that was immediately available was the gravitational force model that had worked so elegantly for solar system motion (or the concept of an ebb and flow of fluid fields as a model for dialectical materialism, in the case of Engels and Marx).

Comte and those sociologists who immediately succeeded him (Spencer, Ward) elected, as an alternative description of the dynamic process in history, the notion of progress. History would proceed along that path in which a greater complexity, expressed as a perfectibility of man, occurs. Thus, they adopted explanations in terms of evolutionary typologies (e.g., from savage to civilized, from simple biological forms to more complex, from underlying geological stratigraphy to overlay, from lithic to chalcolithic to bronze to iron cultures, from feudalism to capitalism to communism, from childhood fancies to adult repressions). But the concept of progress (and the concept of degradation) could not be derived from the physics of the Enlightenment. It was not until the development of thermodynamics in the 19th century that physics provided a directional arrow to time in terms of dissipative or degradative processes. However, degradative processes hardly seem to be a model for the continued persistence of systems, let alone their progressive development. The obvious question then arises—How can the arrow of time be associated both with an aggregative process (e.g., greater complexity, greater order, greater perfectibility) *and* with a degradative process (the dissipative production of entropy)?

A simple illustration of the connection between aggregative and degradative processes can be seen in the population law for living systems. As originally proposed by Malthus, the time variation of population P may be represented by a first-order reaction rate law:

$$\frac{dP}{dt} = KP \tag{4}$$

Subsequent investigators, such as Verhulst, or Pearl and Reed, or Yule (1925), modified this equation to the "logistic" form

$$\frac{dP}{dt} = KP\left(1 - \frac{P}{P_0}\right) \tag{5}$$

where P_0 is a "carrying capacity" for the populated region. This expression avoids the disasters of an exponentially changing population exploding to infinity or dying out to zero, but it does not explain how population carrying capacity asserts itself dynamically. Why does population growth decrease as P approaches P_0? Does the death rate increase with density? Does the birth rate decline? How? But, most important, the logistic model offers no theory of the regulatory mechanism that maintains the population near some equilibrium, possibly such as P_0. This model is merely empirical, even questionable, curve-fitting. (See, e.g., Yule, 1925, and discussion following; or Kingsland, 1982.)

A better approach would recognize in the Malthusian growth law (Eq. 4) that K is itself a dynamically varying parameter. K is the difference between the birth rate, b, and death rate, d:

$$K = b - d \tag{6}$$

and it can be positive or negative. Instead of logistic saturation, a thermodynamic approach requires the difference between b and d to approach zero, but allows it to approach zero from complex effects associated with variations in either rate. Equilibrium is maintained in the independent but interrelated fluctuating birth and death rates. This statement expresses a thermodynamic approach: change in population is a homogeneous property of a population—*but its birth rate b and death rate d must connect to each other causally.* At thermodynamic steady state, they are equal. At other phases, they have partially independent causalities, but are still coupled. The dynamics of birth and death emerge from the effects of the available potentials and of the equations of change associated with the various other conservations: the fluxes of matter, of action, of energy, and, in advanced societies, of value-in-trade. These other fluxes assert themselves, for example, as perceived pressures on the individuals, typically, to have families of a given size (e.g., a fertility rate of about seven children per woman in 1900, but only two and a half in 1970).

Any living species, as an emergent form, exists as a stable system because it has found a way to link the fluctuations of an aggregative process (birth) with a degradative one (death). The historical process of man itself emerges from that fluctuating linkage. Classical equilibrium thermodynamics gives us only half the process—the degradation demanded by the Second Law. Irreversible thermodynamics permits an account for the other half of all complex systems, the aggregative process by which new form itself is created. It is from study of such processes that we have derived *Principle 9*.

The linkage between the two processes of self-organization and degradation, of life and death, is extremely important to understand, because the same dissipation that ultimately takes life away also makes life possible. A complex system couples its aggregative tendencies to its degradations by many cyclic internal processes. Stable complexity requires dissipative processes (especially their large capacity for internalization, as expressed in their bulk viscosity measure), but the dissipation must also appear externally in the shear viscosity measure of the ensemble. This coupling is true of the individual organism, which joins its various metabolic cycles to structure-building and action modes, and it is also true on the species level, where birth and death are coupled in cyclic processes that maintain the species.

The fact that the two viscosities, shear and bulk, are both positive guarantees that energy transformations will be dissipative, but also assures the emergence as well as the degradation of local forms during the overall dissipative global process.[6]

Origin of Life as a Self-Organizing Process

The astronomer Sandage stated more than a decade ago that the cosmological problem is the problem of the three clocks, representing (1) the inequality in which cosmos starts up before galaxy, which in turn starts up before star (e.g., before our sun); (2) the start-up of our planet; and (3) the geological time scale of the lithosphere, e.g., the oldest rocks. Those issues have been reasonably settled, resulting now in a fairly coherent belief in a start-up age for the cosmos (15×10^9 years), old galaxies ($> 10 \times 10^9$ years), on down to our sun (5×10^9 years) and earth (4.5×10^9 years). The fundamental mode of thinking in settling these problems was ordered typology, conceptually not much different from the method of the would-be social scientist of the Enlightenment, who, confronted by the limited physics of the Newtonian world-machine, tried to order the historical development of social (and biological) forms. The difference, of course, is that the physicist can attempt to connect his space–time typological ordering by the principles of physics.

Table 1 lists our guesses about the ordering of salient earth processes—the

---

[6]The general significance of the bulk viscosity measure of time-delayed action in complex systems and in the hierarchy of such systems in nature can be glimpsed in the following three quotations:

1. Gal-Or (1975): "Causal links among the thermodynamic, electrodynamic, and cosmological arrows of time are explained within the framework of... general relativistic theory.... The fundamental role played by the expansion (bulk) viscosity is stressed...."

"It is the 'large' system which dominates the evolution of smaller systems, not vice versa!"

2. Iberall (1976): "... time... arises out of the mechanical processes which are locked in, level by level, as resonators, rotators, or vibrators (cosmic expansion, galactic whirl, stellar formation and evolution...)... *and* the fact that the lossy 'Q' of the processes, binding level to level, helps to act as the escapement for the time keeping (i.e., the arrow of time is kept by the isochronous mechanics of each level, locked level to level by its thermodynamic losses)."

"... the living system is a state (using the cell as a prime example) which adjusts in self-regulatory manner the ratio of the bulk to shear viscosity of the boundary (i.e., the admission characteristics) and thereby provides a variable dynamic for major physical chemical processes of life..."

3. Weinberg (1977): "In each cycle (of a cosmological oscillating model)... the entropy per nuclear particle is slightly increased by a... friction known as 'bulk viscosity' as the universe expands and contracts." (See also Sanchez, 1986.)

TABLE 1. Nine Great Process Start-ups on the Earth

Process	Billions of years (eons) ago
Assembly of earth and heating up	4.6
Plastic–elastic puckering and degassing of earth	4.5
First stages of condensation of water	4.4
Formation of elastic plates	4.3
Buckling of the plates into continents	4.2
Start-up of geochemistry (sedimentary geochemistry)	4.1
Start-up of relatively regular weather patterns	4.0
Start-up of prebiotic carbon chemistry	3.9
Start-up of life (biosphere starts to influence the earth's surface by interactions with lithosphere, hydrosphere, and atmosphere)	3.8

start-up of nine major clocks. We have been guided by Hart's very provocative model of the conditions for evolution of planetary atmospheres (Hart, 1978, 1979a,b).[7]

Some of the processes shown in Table 1 began running simultaneously. That simultaneity encouraged evolution. [This opinion agrees with Smith's view of the "cold start" earth (Smith, 1976, 1982).

We also hypothesize the following physical model for the start-up of life. Terrestrial life originated and developed as one of the feasible complex systems whose scale is next above that of simple molecules. Macroscopic heterogeneity in the form of thermodynamic engine cycles can emerge at the scale of a small number of molecular diameters by molecular cooperativity. This cooperativity is seen in the combination of a number of molecules into a chemical complex expressing an internally time-delayed action mode, importing and exporting potentials in exchange with the surrounding medium. (Micelles of thermal proteinoids are illustrative candidates for this stage of prebiotic physics and chemistry.) Because the medium may be homogeneous in the large, a local self-scaling was required. We have in mind not a rapid flow process scaling (which is vorticity dominated), but rather a more complex associational process, more related to phase or conformational change. In our view, processes involving jointly an associational step and a fluid motion step are not rare.

We believe that the opportunities for starting up the physical–chemical processes of life were maximal at condensation centers within heterogeneous reactor beds, and that solid-state precipitation was a first requirement. It probably took the form of such materials as alumina or silica gels, substrates on and within which fluid and solid interaction can occur (as in clays. Bernal originally called

---

[7]Hart's model is still very interesting for its evolutionary details, whether or not our planet started up with an initial oxidizing or reducing atmosphere—the subject, currently, of considerable controversy. For current views, see Brock (1980).

attention to the possible importance of clays in the origination of terrestrial life). Then the interplay of gas (atmosphere) and liquid (oceans and continental runoff) within such porous gel-like geochemical beds at continental margins provided the scaling for reaction cycles. The reactions led up to the prebiotic carbon, hydrogen, nitrogen, and oxygen reactions, with the aid of atmospheric, hydrological, and solar stirrers and impulsive injectors of energy. Such a process of heterogeneous catalysis is far more likely to create complexity than is one that occurs without flow fields in homogeneous catalysis in fluid media.

We conjecture that this kind of process probably took place within the early sedimentation beds 3.8×10^9 years ago. It is the kind of process that is also being proposed to explain the relationship between banded-iron formation and the development of wide continental shelves and enclosed basins adjacent to deeper oceans 2–2.5×10^9 years ago (see Brock, 1980).

Opportunities for emergence of new forms and processes arise when a field is unstable. Although it is never obvious which fluctuation may then be the creative impulse for a new system, after many random events, given the richness of the memory trace of the previous trials, it becomes increasingly likely that some fluctuation will originate a new system. *Principle 10: Start-up emerges as an S-shaped transition from stable near-homogeneous field I (the flatland of lower atomistic structure, which generally will contain heterogeneous regions, such as gas or dust clouds in a galaxy out of which stars form) to stable near-homogeneous field II (the flatland of the higher level, e.g., a local community of stars).* That transition itself passes through a considerable number of logarithmic time intervals whose measure is that of creative impulses in the prior system I. The creative impulses are those fluctuations that lead to the destabilization of system I and to the formation of higher-level system II. The space–time scale of the second system is determined by the actions of the force systems and materials at hand. The theme: systems arise out of the noisy fluctuations of an unstable system, which converge on a new emergent level of stability whose scale is determined by the available materials, forms, and forces at hand.

Summary

We anticipate on the basis of many reviews and discussions that our views on extending the strategies of physics toward the biological and social sciences might be criticized to various degrees on at least four counts. It might be thought that:

1. We are proposing merely a hermeneutic view—an exegesis of physics not based on the purity of experimentally verifiable principles.
2. Our principles are antidialectical and in some manner antibiological.
3. Social and organismic systems are far enough removed from thermodynamic equilibrium so as to invalidate analysis by methods of near-equilibrium thermodynamics.
4. We are advocating a metascience of physics rather than physics itself.

We reject each of these complaints with the following comments:

1. We are proposing a view or construct that offers a common physical science base for all complex systems including biological and social. Homeokinetics may be thought of as "physicalizing" biology and social science (and incidentally as "socializing" physics by providing frankly anthropomorphic ways of characterizing complex inanimate systems, using notions such as command-control, language, and culture). The homeokinetic descriptions of complex systems grew out of the experimental demonstration of an extensive spectrum of metabolic time scales or cycles in the organism (human and mammal) (see Iberall, 1974, which brings a number of papers together). Because we believe that thermodynamic cycles are responsible for an observed spectrum, the homeokinetic view calls attention to the existence and time scales of some underlying near-periodic processes. Discovery of the details of process chains requires study by experts in the system under investigation. Experts such as physicists, earth scientists, meteorologists, and others already apply similar reasoning without calling it homeokinetics. Biologists and sociologists are just starting to use these principles.

2. Our homeokinetic view of all complex systems including society is indeed proposed as a substitute for the dialectics of Marx and Engels.

We know that any serious attempt to treat an individual human and a society of humans as "just" another pair of complex systems to be studied objectively by the same techniques that are used for nonbiological systems is likely to antagonize many biologists. No doubt our "socializing" physics is just as offensive to physicists.

3. It is true that modal macroscopic behaviors in homeokinetic systems arise because the systems are enough removed from static thermodynamic equilibrium to generate process cycles rather than simple transports along slowly varying gradients, such as are seen in creeping flow fields. Nevertheless, the existence of process cycles does not invalidate the idea of local near equilibrium. This description holds in the cases of turbulent eddies, von Kármán vortex streets, Bénard cells, Taylor cells, air mass movements, and liquid droplets in near equilibrium with their vapor. It is also the case in many biological and biochemical process cycles in which the irreversible entropy production of nonequilibrium chemical reactions can be expressed in terms of chemical potentials. By treating the nonequilibrium process scale as a higher level of fluctuation–dissipation (Einsteinian Brownian motion) we renormalize the thermodynamics. Averaged over a few such fluctuation–dissipation cycles, processes at the new scale are indeed close to equilibrium. These issues were clarified by Einstein and Onsager. Such modeling can be performed at any scale.

4. The dynamics of a simple fluid field at all levels between an underlying molecular fluctuational scale and the boundary conditions impressed from above (or without) are well described by a single master equation set, the Navier–Stokes equations (together with the appropriate thermostatic equation of state). On the other hand, complex systems cannot be described by a single master equation set. Instead, they must be dealt with in a piecewise manner, mode by mode, scale by scale, level by level, and by patching connections between modes, scales, and levels. Nevertheless, the fields are piecewise continuous and one can deal with their conservations by a cycle of relaxations.

The principles discussed in this chapter are our attempt to represent the description of a strategic approach to such physical modeling. Further, we believe that nature employs the same strategy in a historical evolution in which systems are born, live, and die.

Whether these ideas are called physical or metaphysical is not important. What matters is whether or not they are or can be useful in addressing the complexities of systems that are generally considered by both physicists and students of the particular complex systems alike to be beyond the reach of standard strategies of physics.

References

Brillouin, L. (1962) *Science and Information Theory*. Academic Press, New York.
Brock, T. (1980) Precambrian evolution. *Nature* **288**:214–215.
Gal-Or, B. (1975) Cosmological origin of irreversibility, time, and time anisotropies. *Found. Phys.* 6:407–426, 623–637.
Gmitro, J., and L. Scriven (1969) A physicochemical basis for pattern and rhythm. In: *Towards a Theoretical Biology*, 2 Sketches, C. Waddington (ed.). Aldine, Chicago, pp. 184–203.
Gollub, J., V. Benson, and J. Steinman (1980) A subharmonic route to turbulent convection. *Ann. N.Y. Acad. Sci.* **351**:22–28.
Haken, H. (1977) *Synergetics—An Introduction: Nonequilibrium Phase Transitions and Self-Organization in Physics, Chemistry and Biology*. Springer-Verlag, Berlin.
Hart, M. (1978) The evolution of the atmosphere of the earth. *Icarus* 33:23–39.
Hart, M. (1979a) Habitable zones about main-sequence stars. *Icarus* 37:251–257.
Hart, M. (1979b) Was the pre-biotic atmosphere of the earth heavily reducing? *Origins Life* 9:261–266.
Herzfeld, K., and T. Litovitz (1959) *Absorption and Dispersion of Ultrasonic Waves*. Academic Press, New York.
Iberall, A. (1972) *Toward a General Science of Viable Systems*. McGraw-Hill, New York.
Iberall, A. (1974) *Bridges in Science—From Physics to Social Science*. Gen. Tech. Serv., Inc., Upper Darby, Pa.
Iberall, A. (1976) *On Nature, Life, Mind and Society, Fragments of a Vigorous Systems Science*. Gen. Tech. Serv., Inc., Upper Darby, Pa.
Iberall, A. (1983) What is 'language' that can facilitate the flow of information. *J. Theor. Biol.* **102**:347–359.
Iberall, A. (1984) Contributions to a physical science for the study of civilization. *J. Soc. Biol. Struct.* 7:259–283.
Iberall, A., and S. Cardon (1962) The physical description of the hydrology of a large land mass pertinent to water supply and pollution control. Four reports by General Technical Services, Inc. to Div. Water Supply and Pollution Control, Public Health Service, Department of Health, Education and Welfare. Contract SAPH 78640.
Iberall, A., and S. Cardon (1980) Task II—Reviewing the work of other investigators. Gen. Tech. Serv., Inc., report to Transp. Syst. Center, U.S. Dept. Transp., Cambridge, Mass. Report No. TSC-1734-II. Part 2 of a study of regional and urban organization.
Iberall, A., and W. McCulloch (1969) The organizing principle of complex living systems. *J. Basic Eng.* **91**:290–294.
Iberall, A., and H. Soodak (1978) Physical basis for complex systems—Some propositions relating levels of organization. *Collect. Phenom.* 3:9–24.
Iberall, A., H. Soodak, and F. Hassler (1978) A field and circuit thermodynamics for integrative physiology II. Power and communicational spectroscopy in biology. *Am. J. Physiol.* **3**: R3–R19.
Iberall, A., H. Soodak, and C. Arensberg (1980) Homeokinetic physics of societies—A new discipline: Autonomous groups, cultures, polities. In: *Perspectives in Biomechanics*, Vol. 1, Part A, H. Reul, D. Ghista, and G. Rau (eds.). Harwood, New York, pp. 432–528.

Kingsland, S. (1982) The refractory model: The logistic curve and the history of population ecology. *Q. Rev. Biol.* **57**:29–52.
Landau, L., and E. Lifshitz (1951–1979) *A Course of Theoretical Physics*, 9 vols. Addison–Wesley, Reading, Mass.
Landau, L., and E. Lifshitz (1959) *Fluid Mechanics*. Pergamon Press, Elmsford, N.Y.
McKeon, R. (1941) *The Basic Works of Aristotle*. Random House, New York.
Randall, J. (1940) *The Making of the Modern Mind*. Houghton Mifflin, Boston.
Ristel, T. (1975) *Fluctuations, Instabilities, and Phase Transitions*. Plenum Press, New York.
Ruelle, D., and F. Takens (1971) The transition to turbulence. *Commun. Math. Phys.* **20**:167–192.
Sanchez, N. (1986) *From Black Holes to Quantum Gravity*. Taylor and Francis, Philadelphia, Pa.
Smith, J. (1976) Development of the earth–moon system with implications for the geology of the early earth. In: *The Early History of the Earth*, B. Windley (ed.). Wiley, New York, pp. 3–19.
Smith, J. (1982) The first 800 million years of earth's history. *Philos. Trans. R. Soc. London,* A. **301**: 401–422.
Soodak, H., and A. Iberall (1978a) Homeokinetics: A physical science for complex systems. *Science* **201**:579–582.
Soodak, H., and A. Iberall (1978b) Osmosis, diffusion, convection: A brief tour. *Am. J. Physiol.* **4**:R3–R17.
Soodak, H., and A. Iberall (1979) More on osmosis and diffusion. *Am. J. Physiol.* **6**:R114–R122.
Stewart, J. (1947) Empirical mathematical rules concerning the distribution and equilibration of population. *Geogr. Rev.* **37**:461.
Stewart, J. (1948) Concerning social physics. *Sci. Am.* **178**:20–23.
Stewart, J. (1960) The development of a social physics. *Am. J. Physics.* **18**:239–243.
Tolman, R. (1938) *The Principles of Statistical Mechanics*. Oxford University Press, London.
Toulmin, S., and J. Goodfield (1965) *The Architecture of Matter*. Harper & Row, New York.
Weinberg, S. (1977) *The First Three Minutes*. Basic Books, New York.
Yates, F. E. (1982) Systems analysis of hormone action: Principles and strategies. In: *Biological Regulation and Development*, Vol. IIIA, R. F. Goldberger and K. R. Yamamoto (eds.). Plenum Press, New York, pp. 25–97.
Yule, G. (1925) The growth of population and the factors which control it. *J. R. Statist. Soc.* **88**:1–62.

28

A Physics for Studies of Civilization[1]

Arthur S. Iberall

ABSTRACT
This chapter applies the general physical outlook described in Chapters 24 and 27 to human social systems. Social systems, like other complex systems, are characterized by conservations that determine their dynamics over a wide range of time scales. In addition to the usual physical driving forces, human social systems also introduce a number of distinct factors, such as epigenetic memory storage, transmission of information (e.g., tool use, technology), and transfer of social values. They also introduce two new conservations: (1) reproduction of number and (2) value-in-trade. These new factors are given an extended physical interpretation.
The author uses this extension of physics to explain the emergence of settled civilizations, which is seen by him as a stability transition similar to that of matter condensation. Following a first condensation to fixed settlements, a second transition to urban civilization occurs—through the rise of trade, which he interprets as the onset of a macroscopic convection process. Trade and war emerge as the dominant large-scale processes in the ecumene of civilizations. —THE EDITOR

This chapter applies the physical outlook described in Chapters 24 and 27 to the analysis of civilizations. It is a new attempt at a social physics, a social thermodynamics, in which the dynamics of the behavior of an ensemble of interacting individuals, comprising in common a social-physical field, is characterized by identifying certain basic conservations. A physical approach to society was proposed during the Enlightenment, as is evident from the writings of Saint-Simon, Comte, and Quetelet. In this century some hope for a social physics was expressed

[1] The theses outlined here may be examined against the background of events involved in the transition to civilization in the Nile valley. An introduction to suitable source material may be found in Wendorf and Schild (1976, 1980) and Hoffman (1979). The appearance of the latter sources originally prompted the writing of this essay. The Egyptian prehistory is not as well known, in a popular sense, as is the Tigris–Euphrates prehistory. Nevertheless, it poses the same range of problems in the emergence of civilization. (For more details, see Iberall, 1984.)

ARTHUR S. IBERALL • Department of Oral Biology, University of California, Los Angeles, California 90024.

most strongly by Henry Adams and J. Stewart. Current opinion in the fields of sociology, anthropology, history, and physics, however, emergent in particular after Marx and Engels, seems at odds with this hope. A vivid example of the distance that many social scientists wish to put between their field of study and the natural sciences is explicit in Quigley's (1961) *The Evolution of Civilizations*.

This chapter demonstrates the ways in which certain deep physical concepts, particularly those of conservation principles, also govern human social activity and the birth, life, and death of civilizations. Comparing the dynamics of humans with those of atoms or molecules in a statistical thermodynamic ensemble does not trivialize man's endeavors; rather it illuminates and explains them. My philosophical stance is one of those offered by Bunge (1977)—moderate reductionism. Ayala has called this reductionism "epistemological" (see Chapter 16). In this moderate epistemological reduction, I identify levels in the hierarchical systems of nature and seek the application of physical laws to each level. Thus, it is a "lateral" rather than a "vertical" reduction of the type visualized by Comte and Spencer. In my approach these laws appear as statements about local physical conservations; it is based on a method for unification of explanation of the operation of complex systems. Homeokinetic physics (Chapter 27) assumes that throughout the universe there will be local differences in structure or form, but not in the principle that a limited number of conservations underlie motion and change in ensembles and fields, whether in simple *or* in complex systems. It is my premise that only a description by such conservations can come close to being fundamental and not merely *ad hoc*. In any system of atomistic participants, the physics of motion and change within the ensemble is expressed in terms of conservations. (Hirschfelder *et al.*, 1964) may be used to examine a prototype of such description in the atomistic field of molecules. I draw my justification for extending normal physics to larger ordered entities (cells, people, stars, galaxies), whereby they comprise atomisms and near-continuum fields, from Einstein's theory of Brownian motion. In that theory, he indicated how larger ordered entities establish and equilibrate their energies within a sea of underlying atomisms.

Background

Atomisms

In dealing with complex field systems, we are usually concerned with systems engaged in fluidlike motion in which the atomistic participants interact and move relative to each other (Soodak and Iberall, 1978; Iberall *et al.*, 1980; Iberall and Soodak, Chapter 27). I treat human social organizations as an example of such fluidlike systems in which individual humans are the atomisms, each having also a fluidlike system in its interior. "Atomism" is used both for the doctrine of atomisticlike entities and for the entities themselves. The concept of atomisms has been found in Western thought since Greek times (e.g., Democritus, Lucretius, and even Aristotle, who argued against their existence), and particularly from the time of the Enlightenment (e.g., Galileo, Boyle, Newton, Dalton, Avogadro, and Mendeleev). A living creature is an atomism among the organisms with which it

28. A Physics for Studies of Civilization

associates, breeds, and interacts. For the purposes of physical analysis of social systems, we emphasize certain properties that are involved in the interactions among the atomisms. (In this approach, the ways in which people are alike are of concern, not the ways in which they differ.)

Simple versus Complex Systems and Their Conservations

In a simple system (one in which the atomisms quickly equipartition kinetic energy among their internal motions) three quantities are conserved in each atomistic interaction—mass, energy, and momentum (the product of mass and velocity). In a complex system new processes emerge from actions delayed within the atomistic interiors. The delays arise because of complex fluidlike processes in these atomistic interiors. Delays modify the external appearances of basic conservations in the following three ways:

1. Chemical change may appear, i.e., atomisms may be transformed among themselves. The appearance of new forms requires that the observer identify conservations for each individual new type of atomism (e.g., for each different mass species) that may emerge and persist. In chemical change the participating atomisms making up the field may retain their identity as their internal ingredients turn over, or they may transform from one type of mass species to another. In the field of an ensemble of living organisms, the atomistic members of a species (of organisms) turn over material chemically, but do not transform their type. Persons beget persons, but remain persons. (I am not dealing here with much slower evolutionary processes.)

2. In the case of species of living systems generally, in which reproductive processes involving birth, growth, life, death, and dissolution may occur, conservation of population number or density emerges as a renormalized conservation. If the appearance of a reproduction form were transient (e.g., the appearance of perhaps only one or two generations of stars) or if their continued appearance in time and place did not become independent of initial start-up conditions, one would not view the process of reproduction as embodied as a renormalized conservation. This new conservation, in the case of long-persistent living species, is expressed by the fact that generation begets generation, and the length of the generation time becomes the minimum time that characterizes the interactions. (Note that the size of the entire population making up each "generation" need not be exactly constant for the conservation principle to hold with sufficient generality at the generational scale to sustain a physical approach to characterization of a social system. Thermodynamic steady state in the renormalized social-biological set would imply no change in number. Thus, actual changes are dynamic perturbations near equilibrium.)

3. The physical conservation of momentum transfer between participants requires conceptual extension when there are significant time delays in the action within atomisms. In simple systems the salient relevant time scale (out of a sequence of subordinate processes) for momentum transfer is the average time between interactive collisions, and is called the relaxation time (more appropriately for our purposes here the principal relaxation time). In complex systems the

principal relaxation time is the total time scale (which I shall call a "factory day") over which all processes within the atomistic interiors complete a cycle, thereby reaching an "equilibrium" in the sense that the system is then ready to repeat another similar day starting from approximately the same initial conditions. That cycle of processes is made up of action modes. Such a ring of actions modes, indexed by i with action H_i characteristic of each mode, may be derived from momentum by a rule resembling Bohr's early quantum theory ($\oint pdq = \Sigma_i H_i = H$, where p is the instantaneous momentum integrated over all displacements dq for the entire factory day $\oint$, to give the daily action H).

Factory Day

The factory day is characteristic not only of living systems but also of all complex atomistic systems that persist and that express much of their action internally. For an individual human atomism the factory day largely, but not completely, coincides with the earth's day. For the species, the principal factory day is the generation time. The factory day of the individual commonly begins with awakening from sleep, voiding, and looking for food. The cycle is finally closed again with sleep. As in the organism's interior (which itself is a factory involving cellular and organ atomisms at lower levels), so in human society do various associative linkages among parts create action modes. Associative bondings are necessary in order to hunt, sense, perceive, move, to reproduce, to farm, to nurture, and to build shelters or altars.

Ensemble Physics in Brief

The science of ensemble physics pertains to (and only to) the quantities that are conserved upon interactions among atomisms and the potential sources of supply for these conservations. Details of individual-to-individual interactions are described by the kinetics related to these conservations. Motion and change in an ensemble are described thermodynamically by summing or averaging over the ensemble for the conservative aspects of the local kinetic interactions to produce a continuum description of the ensemble's internal motion. That continuum description has only those facets that are determined by the conservations. As a result of continuing interactions throughout the ensemble, the conserved quantities are partitioned (shared) among the participants, producing *statistical distribution functions* of the ensemble. These are statements about how the density of matter, momentum (or action), and energy are each distributed throughout the space (or any local region). Distribution functions are macroscopic aspects of underlying conservations.

The *equation of state* of the ensemble relates the macroscopic variables that are ensemble average measures of the essential conservations. The equation of state exists because the fluctuations of the local conserved qualities are coconstrained. Such an equation holds when the entire ensemble is contiguously in equilibrium, but also when it is only near equilibrium locally. At true equilibrium—neglecting body forces—groups of atomisms in different localities will possess the same external kinetic energy measure (temperature) and the same external momentum measure (pressure). If the ensemble is only near equilibrium, there will be *transports* of the conserved variables between regions of the field. The transports are described by *equations of change*. The actual *forcing of flows in a continuum* field results from a gradient difference (in terms of some of the conserved variables) between regions. Detailed kinetic studies are required to model the transports, which may be diffusive, convective, or propagative (wavelike).

In addition to transient differences in the conservation measures between field regions, there may be heterogeneous depots of the conserved quantities within the field or at its boundaries. These may be regarded as the *potentials*. Potentials can be a source of field gradients and act to drive fluxes.

The complete construct of statistical physics for fluidlike material field ensembles consists of two equation sets (equation of state, equations of change), interacting atomistic participants, and boundary conditions that are constraints on the field.

Ensemble Physics Applied to Social Fields

For an ensemble of humans, the distribution functions of the ensemble are implicit in the following statements: The average daily metabolism of adults living a sedentary life is 2000 ± 200 kcal. The average weight of adults is 150 ± 30 lb. (The mass measure permits adjusting metabolic data for sexual dimorphism or variations in stature.) At equilibrium or near equilibrium, the ensemble tends to preserve its number. Just as the kinetic pressure of a gas is augmented by intermolecular forces in a liquid, so the stress tensor, generalized for human interactions, is augmented by an internal "social pressure," dependent on bulk or associational viscosity, that dominates the action (Iberall *et al.*, 1980). The possible action modes number perhaps 20 for humans (Iberall and McCulloch, 1969).

Having specified that people are the atomisms of a social field, that they are energetically endowed, that they maintain mass, that they tend nearly to conserve local number, and that there is a social, ensemble pressure whose minute action fluctuations account both for the larger-scale conservations and the conservations of other cooperative social actions, I must now identify the minimum space and time scales at which field (near-continuum) descriptions actually hold. For example, social behavior at the field level within any community organizes a population of individuals in time at the numerical scale of an earth day. This aggregated behavior may involve a roaming range for a small tribe, of perhaps 25 or more people, minimally over a few miles but up to 25 miles (see Murdock, 1967).

There are also maximum space and time scales for which continuum field results apply. With respect to the evolutionary history of civilizations, the conceivable time scale is that of modern man's 40,000-year cultural and biological history; the entire land surface of the earth explored and occupied by man becomes the spatial field. Thus, there will be a spectrum of processes in the temporal range of a day to 40,000 years and in the spatial range occupied by a small tribe or a few cooperating bands and extending to the ecumene of civilizations occupying a large portion (40 million square miles) of the earth's land surface. If we can identify the various forces or potentials external to any particular social ensemble of concern and also postulate or discover the spectral domains (i.e., the relevant process times as multiples of factory days), then equations of state and equations of change might be developed for each space–time domain of interest. The task is to build the descriptions in each domain, starting at the smallest space–time scale and integrating (or aggregating) to the next scale, until the maximum scale for which the results hold is reached.

Temporal Spectrum: Characteristic Times for Major, Invariant Social Processes

External conditions or forces (e.g., potentials) creating "authoritative" processes determine certain natural periodicities in the social system. Four time scales for recurrent processes are readily identified:

The social process of the day is dominated by earth's day–night variation; it has become internalized as part of the chemical encoding of most biological organisms in the form of 24-hr endogenous rhythms (one of the many "biological clocks").

The social process of the year is dominated by planetary seasonal variation. The material and energy supply in ecological niches of most, if not all, species fluctuates with that period.

The social process of generation turnover is determined by the genetic code of each species, which is carried as an onboard chemical potential. Each species has a generation or maturation time scale; for humans it averages between 20 to 30 years.

The social process of the total life span is also determined by the genetic constitution. (In poorly developed communities, plagues and famines may sharply curtail the life expectancy far below the maximum life span.) The maximum human life span of about 90 years marks the period beyond which very few individuals are found continuing in the social scene. In human societies, the old have been used traditionally as ambulatory, external memories. Social continuity depends on memory and the transfer of information among humans, because people die.

The four scales—day, year, generation, life span—are determined by physical, chemical, or biological features of organisms, and their environments and these act as sources of "authority" (e.g., as order parameters) for social fields.

I suggest two additional, significant time scales that represent authority (natural, not political) imposed on human social fields, although they are more speculative:

1. *The social process associated with the life span of a culture.* The time scale is on the order of 300 to 500 years, an estimate that is in agreement with earlier ones by Mencius and Spengler, and with the time scales found by Blegen's dissection of the levels of Troy. It is the scale at which a small, isolate culture can maintain coherence—at which it can retain a "founder-figure" myth and transmit

its heritage from generation to generation even without the means of an extensive, abstract, written language. It is also the scale over which a large civilization can maintain political coherence. My estimate is that the number of generations over which cultural information might be passed without considerable drift or transformation is about 20. Murdock's (1967) *Ethnographic Atlas* premises that a small group develops cultural independence if it is separated from neighboring groups either by a few hundred miles or by a thousand years (40–50 generations). The life expectancy of a culture is a result of circumstances that often are largely exogenous to the social ensemble, and the details of events for each society will be specific to that particular civilization. Similarly, the specific time and cause of death of an individual contrasts strongly with whatever determines the maximum life span for that species. My estimate of a 20-generation scale for the lifetime of a culture implies that social processes generally remain or can be maintained coherent for that time, even in the presence of other groups in a coupled, interacting ecumene.

2. *The social process associated with independent cultural waves that cross and mix in their passage over a large land mass.* Their time scale is of the order of a few millenia, and their front moves approximately 1 mile/year. That diffusional velocity arises from the penetration of a new 25-mile roaming range at the generation time scale for a tribe or band of people. See, for example, van Doren Stern (1969) for an archeological discussion of that velocity. In a few thousand years a number of such diffusive movements will lead to reformations of social strategies and changes in the character of the cultural values. (A process scale for movement among animals at the atomistic level relates to body size and daily metabolism.[2] The appearance of more integrative cultural scales is uniquely a human process.)

To survive, any complex system must develop an underlying physical program for maintenance of action. If not genetically set, the program will have the attributes of a "strategy." Eventually the strategy may become stereotyped as a form of governance. For example, we may look for the grossest scale at which strategies are reformed. I suggest that such change occurs at the few millennia time scale as a property of the nature of large-scale reformation in the social ensemble, e.g., urban codes, and moral codes, forms of state and of government. (For example, one finds different forms of government discussed by Aristotle and Polybius.)

Consider also the following topological element in the temporal organization of functional fluctuations. At one extreme of time for an individual organism, Iberall and McCulloch (1969) note that the shortest "moment" of organized action (reaction in this case) consists of three response units: 0.1 sec to determine position, 0.1 sec to determine velocity, and 0.1 sec to determine acceleration. For

---

[2] One may note (Garland, 1983) that the daily movement distance of the most active carnivores is about 12 miles/day. While this measure may or may not have a precise meaning, there are other time-distance measures that are related, but depend on species characteristics. Thus, for example, the human roaming range nominally extends out to perhaps 25 miles; the human daily "travel" budget seems characteristically to be about 1 hr, which in walking distance is about 4 miles out and in per day.

our society it has been shown that business activity has a fluctuation time scale of about 3 years (Dewey and Dakin, 1947, or see Montroll and Badger, 1974). Investment and business decisions require a minimum observation time of the yearly balance sheets, but three units of annual decisions (one each year) must pass before a change in course can be justified, planned, and executed. I propose that three or four process unit cycles are characteristic of overall "business" cycles in memory-endowed systems. Similarly, with regard to social change through cultures, I assume that an ecumene cannot change its world views until a number of smaller cultural changes have occurred.

Fluxes and Potentials Involved in Conservations

At each hierarchical level and for each characteristic time domain, the following fluxes and potentials are identifiable in living systems. They express the conservations.

- *Energy flow* (the daily caloric expenditure, approximately 2000 kcal/day for sedentary persons).
- *Matter* (the conservation in the adult organism of carbohydrate, fat, protein, minerals, oxygen, trace metals, and water; for example, for the human this represents approximately 60 g of protein and perhaps a pound of suitable carbohydrate intake per day).
- *Action modes* (the factory day actions or energy–time integrals that are characteristic of the species). Among mammals—as a generalized matrix of actions—these include ingestive, eliminative, sexual, care-giving, care-soliciting, conflictual, imitative, shelter-seeking, and investigatory behaviors (Scott, 1962). Humans display perhaps 20 such action modes (Iberall and McCulloch, 1969).
- *Population*. For systems that grow, live, and die, conservation of the species requires that generation begets generation (individual isolated breeding pools may die out, but the species persists by immigration and emigration. Ethnicity is exported).

The availability of the following potentials serves as boundary condition for those flux interactions characterized by conservations:

- *Temperature*. The solar energy flux, interacting with earth's atmosphere, produces a temperature range that supports the biosphere. (Both the solar flux and the atmospheric temperature are required potentials.)
- *Chemical potential*. The earth, as substrate and depot, provides materials such as foodstuffs both for building materials and for energy.
- *Genetic potential*. An internal chemical potential is carefully transmitted from generation to generation by specific macromolecules in germ cells.
- *Geographic potential*. The lithosphere, hydrosphere, and atmosphere are available as substrate to surround and to support life processes.

- *Epigenetic potential* (an internal potential arising within the human nervous system). The nervous system furnishes various competences for action, such as memory, stimulus-bound responses, in higher mammals cognition, and, in humans, value systems that become central to strategies for both individual and social behavior.[3]

Summary

The general character of dynamical physical laws is largely contained in the statement that potential gradients drive fluxes. This statement is almost tautological because potentials are just storage bins for the flux quantities. I claim that the above potentials act as forcing functions for human social processes of the species and invoke the same dynamics associated with other "fluid mechanical" fields, such as those of concern in nonlinear stability theory (see Gurel and Rössler, 1979; Helleman, 1980). The conditions for emergence of form are not sharply determined, but represent mixed stochastic–deterministic selections from two or more comparably probable paths of evolution for both form and function.

How do potentials come into being, and how do they drive fluxes? How does the cosmos become the nurturing source for galaxies, or galaxies for stars, or the sun for life on earth? No scientist can regard discovery of the origin of the potentials that drive modern society and the fluxes that constitute modern society as a trivial problem. Ultimately we must determine how the epigenetic potential arose from a biophysical construct within the central nervous system of hominids.

The General Physical Construct Specialized for the Study of Civilizations

To apply the potential-flux principle to the study of modern man, I must first present a specialized view of the conservations and potentials. Our Plio-Pleistocene ancestors originated and enlarged their epigenetic potential by the evolution of rapid capability in the nervous system of dealing in abstractions, e.g., tools. In

---

[3] I see the epigenetic potential as consisting of two components, both emergent from transformations in the cortical capacity and arrangement of Plio-Pleistocene hominids. One is the memory function contained in a value system; the other is a tool-creating capacity. Both evolved from the capability to deal in abstractions and languages. The value system includes world images of (and a language description for) *self, interpersonal relationships, nature, society, ritual and institution, other living organisms, technology* (more broadly, culture), *spirituality* (fathers, leaders, gods), and *art forms* (abstract representations designed to attract attention in various sensory modes) (Iberall and Cardon, 1980). Certainly this is the strangest potential from a physical point of view. One notes that it was preceded, among hominids, by a tool-creating potential (more than 2 million years old), before the cultural potential that marks modern man emerged (about 40,000 years ago). That continuing tool-creating potential appears to be the amplification of man's "normal" actions. It creates technology, and—with values—makes politics and economics—trade and war—possible. We identify culture in modern man by his linguistic-like or symbolic artifacts—representations as tools, language—both internal and external, statues, paintings, artifacts that may be used for a great number of abstract functions.

these hominids the cortical nervous system of the primate developed to the point that an epigenetic heritage could be transmitted from generation to generation. I propose that the epigenetic heritage included a rudimentary value system and a particular feature that might be called a technological rate potential. This potential (of the central nervous system) is represented by the capability of each generation to add additional tool-making complexity beyond its ancestral heritage. The increase of this measure seems to have been linear among man's hominid predecessors during the last 2 million years; each hominid made approximately the same incremental change in tool complexity. This assertion is based on a crude estimate of the gain in power-handling capability provided by the various tool assemblages evolved over the past few million years (e.g., eoliths, hand axes, flakes, microliths). With the evolution of modern man, about 40,000 years ago, the increase in power-handling capability changed to a much greater, albeit still linear, rate. I regard the technological rate potential as a separate component of the epigenetic potential, because it apparently began earlier than the first traces of human culture, e.g., the Venus figures, and it represents a specific form of abstraction, more than mere transmission of memories. The later components of an epigenetic heritage, such as the development of religious symbols or perhaps the more primitive recognition of self and others, may have much less to do with changing technology.[4] Some cultures have transmitted a learned heritage from generation to generation with little or no change in their technological rate potential. In Tasmania (a 25,000-square-mile island), the techniques of fishing—a major technological expression—even *regressed* over a period of about 5000 years. (One surmises that such a change was the result of geographic isolation. Smaller island populations, when they become isolated, often disappear entirely, a very drastic result of operation as an even more closed system).

The social life of our early hominid ancestors probably differed little from that of other primate species (Eisenberg et al., 1972). As these prehensile, upright primates changed from fruit-gatherers to hunter-gatherers, however, they "condensed" into bands with a considerable division of labor (Steward, 1936). When they were fully transformed into our sapient subspecies, they were no longer dependent at all on the pecking orders common in many other species. But why did a subsequent series of discontinuous cultural changes occur, through which subpopulations settled for long periods in one place and eventually organized into what we now call urbanized civilizations? From a physical point of view, the only potentials and fluxes involved were those previously named—energy, matter, momentum or action, and population; no new exogenous force of a special character can be invoked. What then caused the rather rapid social and cultural evolution?

Human social evolution has to be a matter of dynamic instability in which a transition from one type of field process to another occurs. Ever since pioneering works in elastic stability by Euler (the buckling of a loaded column), in hydrodynamics by Reynolds (transitions from laminar to turbulent flow), and in

---

[4]For example, Binford (1973) noted that no "patterned 'stylistic' variability has been demonstrated in the archaeological record prior to the upper paleolithic [starting from 40,000–35,000 ybp]." Neanderthal man's view of the world was apparently different from ours.

mechanical orbits by Poincaré and Lyapounov, the subject of mechanical stability has flourished. It has been realized that the stability both depends upon and is transformed by nonlinear dissipative, thermodynamic processes. Recent conferences (see reports by Gurel and Rössler, 1979; Helleman, 1980) have assembled a large amount of material demonstrating the generality of the problem. Elsewhere, Soodak and I have speculated on the application of physical stability principles to the human social transition problem (Iberall and Soodak, 1978).

The First Transition: Horticulture Emerges

I view the discontinuous social change manifested by the appearance of food-producing societies (e.g., from hunting-gathering to horticulture to settled agriculture) as evidence of internal rearrangements, new associations or configurations, and a new phase condensation—as if a gaslike phase of matter were becoming liquidlike or solid state-like. I take this apparent analogy to be deeply physical and an apt homologue for describing the birth, life, and death of civilizations. To illustrate the applicability of the physical ideas of conservation and condensation to societies, I have traced as well as I could the story of social change in 2000- to 6000-year intervals from approximately 40,000–30,000 years before present (ybp), when modern man emerged, to 5000 ybp, the dawn of the truly historical period of records. I suggest that epochs of a few millennia have tended to define a physical relaxation process in which societies modify their dominant life-styles.

The task I face now is to offer a physical theory for the cultural transition, whose story begins from 40,000 to 30,000 ybp, as related to the geophysical boundary conditions and the conservations, fluxes, and potentials previously discussed. At his beginning, modern man apparently lived in hunter-gathering groups, operating in a range appropriate to human size and metabolism (there being moderate differences among animals in such ranges for carnivores and frugivores). This species reproduced and diffused slowly over wide spaces, increasing its overall niche in the earth habitat (Iberall and Wilkinson, 1984). Climatic change and the search for food largely governed the diffusive process. During the first phase of that diffusion small bands of people tended to separate on the large land masses by distances of up to 70 to 100 miles. This distance of separation is estimated from the number of independent cultures that have been identified in the two large land regions of the Americas and Australia (e.g., the latter by Birdsell, 1973). If, as appropriate to his size, man had the typical mammalian metabolism and a potential roaming range of about 25 miles/day, cultures separated on the order of 50 miles would have had little interaction. Cross-cultural studies (Murdock, 1967) suggest that a distance of a few hundred miles between groups was sufficient for almost complete cultural independence. The 70- to 100-mile separation of populations, as empirically found, is highly suggestive of a system of weak force, "gaslike" interactions, in which the human bands in a group or tribe (or settlement) correspond to social molecularities. Thus, the diffusion of an early, small population could be considered nearly a gaslike motion, such as that of a two-dimensional gas. Its approximate mean free path was one roaming range—25 miles; its relaxation time was one generation—25 years. Its diffusional speed, therefore, was approximately 1 mile/year (see van Doren Stern, 1969, for confirmation of this estimate). Information related to cultural processes could not be transmitted faster.

In a second stage, density increased and stronger near-neighbor interactions occurred, with a diffusion and remixing of ethnicity. The increased densification produced van der Waals-like forces. However, with such forces (a cubic instability), a field can remain metastable for an extended period of time. A second process was involved in subsequent phase condensation.

I surmise that decreases in the levels of the required potentials (temperature, water, and food) caused condensation of small bands on fixed centers of population. This claim might seem surprising—one might think that diversity of form and complexity is associated with rich potentials. This is not the case. At high energy, systems tend to move toward a gaseous phase because the low-energy bonds are broken. Remaining degrees of freedom of motion are then equally endowed with energy (equipartitioning). As Einstein showed for the specific heat of matter, it is only at low temperature (low kinetic energy) that energy supports many ordered configurations. A cooperative coordinative phenomenon then occurs: order is imposed and authority is established. Thus, matter in condensation, in liquid and solid states, always exhibits more diversity and complexity of order than it does in the gaseous state. Most animal species probably respond to lowered potentials by some sort of similar condensation, and in extreme cases of impoverishment of potentials there will be very great changes in group behavior.[5] The internal genetic potential usually determines the particular manifestations. In the long term (a few millions of years), hominids, with more epigenetic and tool-making (technological) potentials than those of other animals, adapted by two means: genetic and cultural (with technological change a subclass of the cultural).

As Sahlins (1972) has pointed out, preagricultural hunter-gatherer societies did not live a more energetic life pattern than did agricultural societies. Instead, they lived at sparser density and with fewer possessions. Life at higher densities involves a greater variety of processes and a greater yield from action. Given potentials within the earth substrate (e.g., sunlight or the chemical potentials) can be used for multiple purposes. After condensation the interactional processes are faster, more intense. (Walking speed, for example, is faster in big city settlements than it is in small ones.) The more significant condensations occurred in the populations associated with river valleys—as part of the invariable drive to search for reliable water supplies. Among the physical–chemical potentials, adequate oxygen pressure, which varies with altitude, and the availability of water seem to be the most critical (followed by suitable temperature). The nature of the social phase condensation, however, depends on the amplifying capability of the technological potential. Associated with the two chief potentials—water supplies and technology (tools)—came changes in modes of living, improvement in the use of water resources, and localized social development through domestication of

---

[5]Such behavior is seldom beneficial to the species, although when taken aboard within the genetic potential—in the form of hibernation, sporulation, or desiccation—such transformation can have survival value.

plants and animals. Under the influence of these two changing potentials, a condensation of population into a stable agricultural community could occur.

In Eurasia the great change in water potential occurred with the withdrawal of the glaciers at the end of the last ice age about 12,000 to 10,000 ybp. At the melting front there were grasses, grazers, predators, new permanent water supplies in the form of river valleys, well-watered mountain flanks, lakes, and springs. In Africa, there were wet–dry alternations largely concomitant with the glacial changes. In the alternation the social transitions varied between temporary condensations and evaporations. Nomadic life-styles developed, or people gathered around water. If the supply dried up, they scattered to find new sources. Thus, the self-sufficient agricultural village, well located with respect to water supplies, became a new feature of the landscape, emerging in the Nile valley and elsewhere in Africa, in the Tigris–Euphrates valley, in the Americas, in the Asian steppes, in the Indus valley, and in China. Some of these developments were independent; others, derivative. The diffusive nature of the spread of agriculture over Europe at the millennial time scale has been carefully documented (Cavalli-Sforza, 1974).

The transition from hunter-gatherer to agrarian society, although profound, was not difficult for humans. To accomplish the change, they first had to adapt their behavior to that of a species they wished to domesticate (e.g., as nomads following migratory herds). It was then necessary for them to put selection pressure on the reproduction of the chosen species to accelerate their adaptation toward human requirements. Results were produced on a time scale much shorter than that of random, evolutionary natural selection. Human epigenetic processes have a time scale of perhaps 1000 to 2000 years and are about 100 to 1000 times faster than genetic evolutionary processes at the species level. Symbioses are common in the biological world, and domestication is hardly a novel invention. But humans brought to it an epigenetic memory of time and place and of sequential ordering, as tradition, that must have proved useful and stabilizing. All these are general aspects of the transition toward condensation.

Prior to the appearance of the first major civilizations, there was deterioration of environmental conditions, a pressure for technological innovation, regionalization of social groups, and a mosaic of types of cultures (Hoffman, 1979). These cultures were sympatric societies, occupying roughly the same niche, though slightly separated geographically, and having different tool traditions. Such a field has a latent tendency toward condensation; two or more atomistic, fluidlike assemblages coexisted. A more organized (precipitated) group, already condensed or nearly condensed, would have superior influence and capacity to force the other groups into opposition, to drive them to disperse, or to absorb them into a growing condensation. I propose that this type of pattern likely characterized the Mousterian to Cro-Magnon transition 50,000 to 40,000 ybp (see ApSimon, 1980) as well as the much later hunter-gatherer to agricultural transition.

I do not intend to suggest novel social histories to the archeologist and prehistorian who know the actual histories much better than I do, but instead to stress the role of flows and phase transitions in determining social field stability.

R. B. Lee has discussed such stability from a Marxian perspective, commenting on the relative stability of an agricultural way of life as opposed to a hunter-gatherer way of life in a current African case (Lee and DeVore, 1968). But rather than argue about the transition from the mysticism of dialectical materialism, as he does, I wish to call attention to stability issues as studied within physics.

Civilization, a Second Transition: The Appearance of a New Conservation[6]

Given the long-term stability of the social forms of hunter-gatherer bands and agricultural village settlements, why was there any further transition leading to civilizations? Village settlements obtained greater permanence of food supply through agriculture and provided greatly enhanced free energy as stores or sur-

---

[6] In Chapter 27, Soodak and I have offered a general stability criterion for the transition from a fluidlike state to a second fluidlike or formlike (solid) state. It has advantages over our earlier attempt (Iberall and Soodak, 1978) that incidentally applied the criterion to transition from mobile (fluidlike) hunter-gatherer societies to fixed trading (urbanized) societies. The essence of the transition argument as applied to a societal phase change, is that the transition criterion

$$Re_{cr} = 1$$

(where D = characteristic size scale, ν = dominant diffusivity, ν/D = a characteristic diffusive velocity) can be unfolded as follows:

$$Re = \frac{V \text{ (convective)}}{V \text{ (diffusive)}} = \frac{V}{\nu/D}$$

$$D = D_0 \sqrt{N}$$

where D = diameter of a trading constellation; D_0 = diameter associated with a single settlement, e.g., $D_0 \sim 40$ miles, twice the magnitude of a daily walking range that might be associated with the settlement; and N = the number of such settlements bound in trading.

In comparison to the translational momentum diffusivity ν_0 associated with the human atomism within a hunter-gatherer society, measured say by $\nu_0/D_0 \sim 1$ mile/year, the trading group diffusivity is augmented by its internal action

$$\nu = \nu_0(1 + \tau_{int}/\tau_{ext})$$

As we showed in Chapter 24, form criticality is based, not on the external time delay but on that time scale augmented by the internal time delay. That time delay, loosely speaking, is represented by the generation time. With reference to the daily translational scale, the ratio of the time scales is about 7000 days/day. While human forced marches could be of the order of 25 miles/day, the transportation (convective) velocity in a simple post-Neolithic society, walking between settlements to conduct trade, would be more like a fraction of that velocity, e.g., 5 miles/day. Thus,

$$Re_{cr} = 1 = \frac{\sqrt{N} \times 365 \times 5}{7000}$$

This criterion requires that N be on the order of 16 communities. The $D = D_0 \sqrt{N}$ relation suggests that the trading constellation is bound up in a region on the order of a few hundred miles. Comparing these schematic results with the first trading constellations known (Mellaart, 1965; Hamblin, 1973), reasonable agreement is found.

pluses for further evolution. Surpluses permitted an increase in population and population density. Settlement separation became less than a daily roaming range—less than 25 to 50 miles. But mutual needs—for materials, for alliances against other cultures, for security from climatic vicissitudes, for common support against perceived uncertainties—made some form of trading interaction necessary. Because the locations of the settlements were fixed, the materials of trade had to be carried. A fluidlike flow process, a convection, arose with the invention of a fundamentally new conservation, value-in-trade (or value-in-exchange). The appearance of trade as a convective process did not provide a new condensation because the settlements already represented condensations. Trade provided instead a new facilitated mechanism for high-speed movement and change throughout the field. The question then is, how did the field respond to the new internal processes that generated the trade?

There is a requirement for social cohesion in any population center. People must recognize each other and have a basis for bonding. Such a recognition and bonding process is well represented within the agricultural village [it probably limits the size to less than 500, roughly the number of faces that can be recognized by one person (Pfeiffer, 1969)]. What we know of primate social ordering (Eisenberg et al., 1972) or the formation of the more specialized hominid bands (Iberall et al., 1980) suggests possibilities for the kinds of village leadership that may have emerged, dependences on kinship, and the hereditary and appointive positions in the occupations that developed from a division of labor (see Murdock, 1967).

Convective trade inevitably creates the need to let the foreigner in and transforms the social pressure previously described. The conservational exchanges of materials and energies are no longer processes localized within the kinship or cooperative band molecularity, as a person-to-person Brownian motion. Instead they are convections and field diffusions. Thus, the transport processes are intensified and accelerated. Economy of effort soon requires that symbols be recognized (tokens, money) and transported rather than the increasingly burdensome flux of material objects themselves. Thus, storage, ownership, and trade by value-in-exchange develops and evolves. It is another renormalized conservation because at the moment of exchange, nominally for all transactions in society, value (according to the epigenetic value system) is exchanged. Trade and war emerge as the major large-scale processes among (urban) condensation centers of population. Economics and politics are joined.

The Nature of Civilization

What is the essential nature of civilization? It is a source of argument among civilizationists whether religion, agriculture, urban settlement, trade, literacy, or recorded tradition is the immediate causal ingredient for birth of a civilization. Eventually a central ingredient is the establishment of *civitas*, a formal set of objective rules setting forth to the insider or outsider those relations that govern processes and complex forms as hierarchies, heterarchies, or stratified classes in the society. The rules define relationships: ruler–citizen, master–slave, citizen–

citizen, citizen–outsider. The set of rules acts as a political constraint on the flow of authority (Lasswell, 1958) and determines the impedances or conductances of the social flows. Diffusions in a civilization are not determined only narrowly by physical principles, but are extended, facilitated, or impeded by man-made law. Internal police, rules of conduct, external-acting military establishments, class processes, records—all of these become necessities for governing. The foreigner, having been allowed in for trade, also recognizes the alternative possibility of war. War is a way to establish new ownership and new command. Trade and war become the dominant patterns of civilization, opposite sides of the social coin.

Written language, originating perhaps 5500 to 5000 ybp, was at first merely a recording of person-to-person transactions (or of the heroic deeds of new rulers), but by 4400 to 3700 ybp, we find records of codes governing class relations in the urban city-state. Such class relations are powerful elements in the emergence of ensembles of city-states (empires). It would be very impressive if we could find records of such city-state codes dating back another millennium, or even more impressive if such codes were proven to be recorded (instead of merely implied) for the period beginning 10,000 ybp (see the conjectures about writing described by Schmandt-Besserat, 1978). We have to allow the possibility, however, that a number of two-millennia "relaxations" were required for man to make such drastic transitions. The agricultural transition start-up of 10,000 ybp perhaps could not be accompanied by further stability transitions before a few such relaxations had occurred.

The creation of bureaucratic institutionalized forms by which human actions are regulated obviously involves abstractions, and these are created by human minds with great ease. But how is it that the abstractions can have social force and utility? It seems plausible that freedom of abstraction led to *symbols* and the formation of arbitrary linguistic associations (Iberall, 1973). Magic and religion became a value system involving shaman, totem, tabu, and ritual. Subsequently, the epigenetic evolution of agricultural, and later urban, systems required much more complex abstractions. It is hardly accidental that ziggurats in the Tigris–Euphrates valley mark some of the earliest structures in that urban expansion, or that we find evidence of religious formalism in Catal Huyuk (Eliade, 1978) and prehistoric Egypt (Hoffman, 1979). Certainly, explicit, institutionalized religions emerged early in civilizations.

Summary of the Start-up of Civilizations

From the archeological record I conclude that civilizations began when there was extensive trade (convective flow) among population concentrations (condensations). The urban centers held cumulative populations greater than 2500 and were composite groups. The threshold size can be estimated from the absence of complex cultures of smaller population (Murdock, 1967). The estimate is consistent with data about the earliest trading constellations in Anatolia and Armenia. Convective interaction among centers involves contact with strangers and outsiders. It cannot be governed only by tradition or oral heritage accepted by all internal parties in the family or in the local isolated village. An objective symbolism must be invented and externalized as value-in-trade, a symbolic form invented by the human mind that endows every transaction with value. In each transaction, equal value is traded according to the current notions of equality of value. *Thus, the economic conservation is invented.* With it arises a pricing system (and also the problem of inflation, a runaway value system). The economic variable acts as the final renormalized physical conservation in societies.

I have shown that an effective image of societies cannot be created merely as a purely demographic model, a purely economic model, or a psychophysiological or cultural model of human activities. Nor will the engineer's energy model or the chemist's material balance suffice. The modeling of civilizations requires synthesizing all the above models and specifying the driving fluxes, both outside and inside the system. The overarching model is based on hydrodynamic processes and principles.

In summary, I believe that the transition to civilization occurred in two steps. The first step (physically) was a matter condensation, largely for example, to agricultural settlement. Biologically it was a symbiosis among species; culturally it was a social stratification. The second step (again physically) was a hydrodynamic instability transition to the convection of trade. Little biological change was involved (except for a change in the level and character of stress on physiological systems), but the complexity in cultural stratification increased enormously. I have tried to illustrate that whenever a complex system is studied at its own organizational level from a physical point of view, one finds a commonality of operating principles. Accordingly, social phenomena, including the processes of birth, life, and death of civilizations, have fundamental physical attributes.

Death of Civilizations

Why do civilizations die? A reasonable answer lies in the study of the conservations and potentials that make up the description of the regular functioning of social systems. Each person brings anew to the social experience an optimistic view that this is *the* life, that the now of this lifetime is the only one that counts. That optimism is doomed to failure because the social system is unstable thermodynamically.

The young enter into a modern, developed society and join its hierarchical institutions—family, neighborhood, local political community (either urban, migrant, or rural), and national political community. They acquire their epigenetic heritage and gradually take on the roles of the adults they displace and replace. That turnover practically guarantees that there will be no new achievements. The problems of the past are propagated, and new ones are acquired. The conservations that must be satisfied remain the same. Unfortunately, it seems to take a lifetime of 40 to 60 years to learn the total operational wisdom that the past may have acquired. In traditional societies in which escape or movement away is rare or slow, the old can act as stores of memories of how things are best done. But to deal with complex social systems in which the convective currents are large and fast, technological turnover is appreciable, introducing changes daily, weekly, seasonally, or yearly, requires adaptation techniques not yet discovered or learned, because of a limitation of the human mind. Most leaders are thrown into their roles in society after fewer than 20 years of experience past their youth. A political time scale of only a few years is and has been the most common characteristic of man since civilization began. That rapid turnover period is well suited to the agricultural village or the hunter-gatherer society, because in such societies the only conservations that must be satisfied are those of materials, energy, population number, and action modes, particularly as they relate to reproduction. Under these circumstances, the need for longer-range planning or information of an abstract nature about the total value system as a potential can be left to a few people—the leaders, elders, shamans, and priests.

But civilizations must contend with the continued influx of strangers and the choice between trade and aggression, with the requirement for symbolic balance in trade or in social relations. They must show a responsiveness to ever-changing external conditions of opportunity and threat. Each young person must develop an enormous capability to deal with a complex series of interactions. As usual, it is not the entire populace that is concerned with the mastery of such operations, but an elite structure as a new form of priesthood, constituting only a few percent of the population.

But leaders are not queen bees endowed with some royal jelly containing special powers. They are ordinary humans very much like their followers. As young priests of the economic and political market place—whether their societies are feudal, commercial, capitalistic, communistic, socialistic, dictatorial, or anarchical—they cannot and do not learn how to operate a complex society. They are taught stereotypes. They too quickly fall into the generational variation of trade or war, and its historical consequences within the ecumene. Unfortunately, the operation of a society requires understanding and controlling conservational balances for periods of about three generational relaxation times, about a 90-year lifetime—the period that comes close to permitting social equilibrium. Not only must the leaders understand how to strike all the necessary balances for such periods, they must also convince the people, their followers, to satisfy the requirements so posed. Over 1 year they can do it; for 3 years they perhaps can do it. Leader can be replaced by leader. But they cannot impose the kind of rational regulation and control required for longer periods, even if they create a dynasty. Thus, trade and war become the standard political alternation at the generational scale within the ecumene (Iberall, 1973).

Eventually the overall mismanagement of the social system destroys it and the system diffuses into incoherence, perhaps after about 500 years (20 generations worth, a few handfuls of short-term leaders). But the outlook and imagination of societal members and their local institutions are still scaled for the period of 1 year or a few years. As Sorokin (1937) pointed out, when civilizations come apart or reform, the local units largely survive intact and begin the task of putting together a new form. And so it goes for our species. My account might be called a topological theory of pattern coherence within the birth or death of civilizations.

Many might not accept that skill or wisdom are necessarily proportional to age or experience or that an accumulation of complex problems necessarily limits the duration of coherence of a civilization. In inquiring about the development of individual managerial (not scientific) skills on the part of corporate and political managers, I found always the same estimate—that there is a steady growth of skills from the time a person enters the system, at perhaps 25 years of age, to achievement of a modest competence a decade later. Competence increases during the next two decades, reaching a peak at about 55 years. There is a plateau until about 65 years and diminishing competence thereafter. (The declining competences of later ages are largely the result of loss of interest!)

I suggest that a model for civilization involves a start-up phase, a life phase, and a deterioration phase. Elites may act twice. In the start-up phase, their characteristic human actions help to build the system. In the deterioration phase, the same kind of actions accelerate the system's deterioration. The basic issue is not so much the experience–age distribution function for all the leaders, but the limited nature of an individual's judgment. What works in one context, such as at start-up or in a time of troubles, does not necessarily work in another; after we have attempted a gamut of processes, our responses become stereotypic. Polybius early noted this for the operation and alternation among political systems. In a simple hydrodynamic instability there may be only two dynamic paths between which to select; in the electrohydrodynamic instability of the human brain there may be only a few more. The vaunted originality of the human brain has very severe limits. Thus, we find significant engagement in war every generation in the ecumene.

References

ApSimon, A. (1980) The last Neanderthal in France? *Nature* **287**:271–272.
Binford, L. (1973) Interassemblage variability—The Mousterian and the 'functional' argument. In: *The Explanation of Culture Change: Models in Prehistory*, C. Renfrew (ed.). University of Pittsburgh Press, Pittsburgh, pp. 227–254.
Birdsell, J. (1973) A basic demographic unit. *Curr. Anthropol.* **14**:337–356.
Bunge, M. (1977) Levels and reduction. *Am. J. Physiol.* **2**: R75-R82.
Cavalli-Sforza, L. (1974) The genetics of human populations. *Sci. Am.* **231**:81–89.
Dewey, E., and E. Dakin (1947) *Cycles: The Science of Prediction.* Holt, New York.
Eisenberg, J., N. Muckenhirn, and R. Rudran (1972) The relation between ecology and social structure in primates. *Science* **176**:863–874.
Eliade, M. (1978) *A History of Religious Ideas.* University of Chicago Press, Chicago.
Garland, T. (1983) Scaling the ecological cost of transport in terrestrial mammals. *Am. Nat.* **121**:571–587.
Gurel, O., and O. Rössler (eds.) (1979) *Bifurcation theory and Applications in Scientific Disciplines. Ann. N.Y. Acad. Sci.* **316**.
Hamblin, D. (1973) *The First Cities.* Time, New York.
Helleman, R. (ed.) (1980) *Nonlinear Dynamics. Ann. N.Y. Acad. Sci.* **357**.
Hirschfelder, J., C. Curtiss, and B. Bird (1964) *Molecular Theory of Gases and Liquids.* Wiley, New York.
Hoffman, M. (1979) *Egypt Before the Pharoahs.* Knopf, New York.
Iberall, A. (1973) On the neurophysiological basis of war. *Gen. Syst.* **18**:161–166.
Iberall, A. (1984) Contributions to a physical science for the study of civilization. *J. Social Biol. Struct.* **7**:259–283.
Iberall, A., and S. Cardon (1980) Urban and regional systems modeling. Final report of Gen. Tech.

Serv., Inc., to Transportation Systems Center, DOT, Cambridge, Mass. (October). Contract DOT-TSC-1734.

Iberall, A., and W. McCulloch (1969) The organizing principle of complex living systems. *J. Basic Engr.* **91**:290–294.

Iberall, A., and H. Soodak (1978) Physical basis for complex systems—some propositions relating levels of organizations. *Collect. Phenom.* **3**:9–24.

Iberall, A., and D. Wilkinson (1984) Human sociogeophysics: Explaining the macroscopic patterns of man on earth—Phase I;—Phase II. *GeoJournal* **8.2**:171–179; **8.4**:387–391.

Iberall, A., H. Soodak, and C. Arensberg (1980) Homeokinetic physics of societies—A new discipline: Autonomous groups, cultures, polities. In: *Perspectives in Biomechanics*, Vol. 1, Part A, H. Reul, D. Ghista, and G. Rau (eds.). Harwood, New York, pp. 433–528.

Lasswell, H. (1958) *Politics: Who Gets What, When, How.* World, New York.

Lee, R., and I. DeVore (eds.) (1968) *Man the Hunter.* Aldine, Chicago.

Mellaart, J. (1965) *Earliest Civilizations of the Near East.* McGraw-Hill, New York.

Montroll, E., and W. Badger (1974) *Introduction to Quantitative Aspects of Social Phenomena.* Gordon & Breach, New York.

Murdock, G. (1967) *Ethnographic Atlas.* University of Pittsburgh Press, Pittsburgh.

Pfeiffer, J. (1969) *The Emergence of Man.* Harper & Row, New York.

Quigley, C. (1961) *The Evolution of Civilizations: An Introduction to Historical Analysis.* Macmillan Co., New York.

Sahlins, M. (1972) *Stone Age Economics.* Aldine, Chicago.

Schmandt-Besserat, D. (1978) The earliest precursor of writing. *Sci. Am.* **238**:50–59.

Scott, J. (1962) Introduction to animal behaviour. In: *The Behavior of Domestic Animals*, E. Havez (ed.). Williams & Wilkins, Baltimore, pp. 3–20.

Soodak, H., and A. Iberall (1978) Homeokinetics: A physical science for complex systems. *Science* **201**:579–582.

Sorokin, P. (1937) *Social and Cultural Dynamics*, 4 vols. American Book Company, New York.

Steward, J. (1936) The economic and social basis of primitive bands. In: *Essays in Anthropology Presented to A. L. Kroeber*, R. Lowie (ed.). University of California Press, Berkeley, 331–350.

van Doren Stern, P. (1969) *Prehistoric Europe.* Norton, New York.

Wendorf, F., and R. Schild (1976) *Prehistory of the Nile Valley.* Academic Press, New York.

Wendorf, F., and R. Schild (1980) *Prehistory of the Eastern Sahara.* Academic Press, New York.

IX

Topological Representation of Self-Organization

Gregory B. Yates

The language of mathematical topology is used increasingly to describe the phenomena of self-organization. The emerging topological theory of dynamical bifurcation, like the related but distinct Elementary Catastrophe and Classical Bifurcation theories, promises rigorous qualitative solutions to the intractable equations of complex systems. It promises, for example, to allow proof that a particular self-organizing system will exhibit some kind of oscillatory behavior even when the precise oscillation cannot be derived. Qualitative dynamics of this kind are clearly important in the study of self-organization.

Because Dynamical Bifurcation Theory is topological, many of its concepts may be introduced visually, using examples in low dimensions. This unusual approach has the advantage of making the concepts accessible to the nonmathematician. In the first chapter of this section, Abraham and Shaw present such a visual, intuitive introduction to dynamics.

The great attraction of Dynamical Bifurcation Theory lies in its promise of a mechanistic explanation of self-organization. In other words, it may provide a mathematical rationale for the restriction of a system to a specific evolutionary path. In the second chapter of this section, Abraham explores the relationship between dynamics and self-organization. After giving a brief history of dynamics he outlines a program by which dynamical models perhaps may be extended to explain a variety of self-organization phenomena, potentially including irreversibility, fluctuation, coherence, symmetry-breaking, and complementarity. He then surveys the outstanding problems of Dynamical Bifurcation Theory and proposes that their resolution lies in a strong parallel development of theoretical and empirical branches of the mathematics.

GREGORY B. YATES • Crump Institute for Medical Engineering, University of California, Los Angeles, California 90024.

Abraham concludes that, although certain deep problems complicate the application of Dynamical Bifurcation Theory to the study of self-organization, this application nevertheless holds great promise of both temporary and permanent benefits to science. Among the temporary benefits he includes the practice of new patterns of thought that may prepare the philosophical climate for a future morphodynamics. As permanent benefits Abraham cites the exposure of attractors and stable bifurcations as essential features of morphodynamics.

29

Dynamics
A Visual Introduction

Ralph H. Abraham and Christopher D. Shaw

ABSTRACT

A dynamical system is one whose state may be represented as a point in a space, where each point is assigned a vector specifying the evolution. The basic ideas of the mathematical theory of dynamical systems are presented here visually, with a minimum of discussion, using examples in low dimensions. The "AB portrait" is introduced as a record of attractors and basins. The basic dynamical bifurcations also are given, including examples of bifurcations with two controls. Extensions of dynamical concepts are proposed in order to allow modeling of hierarchical and complex systems. These extensions include serial and parallel coupling of dynamical systems in networks.

The references for the ideas in this chapter can be found in Chapter 30. —THE EDITOR

While working together on the illustrations for a book, we discovered that we could explain mathematical ideas visually, within an easy and pleasant working partnership. Our efforts to illustrate "dynamics and self-organization" expanded inevitably into the work presented here. We use an animation technique familiar from *Scientific American* to develop the main ideas of dynamical systems theory, while relying as little as possible upon verbal descriptions. This style of presentation is at least ten times more costly than the usual verbal one. But then, a picture is worth a thousand words.

The ideas included are a mixture of ones familiar from the recent literature of dynamics and new ones based upon personal reflection. The reader should keep in mind that this is a personal view, and that the field of dynamics now is undergoing rapid evolution.

RALPH H. ABRAHAM • Division of Natural Sciences, University of California, Santa Cruz, California 95064. **CHRISTOPHER D. SHAW** • Department of Mathematics, University of California, Santa Cruz, California 95064.

Whitehead's (1925) *Science and the Modern World* describes the early history of dynamics in two periods—Galileo to Newton and Newton to Einstein—of the origins of modern science. Two further periods extend into the 20th century—Poincaré to Thom and Thom to the present. The ideas of Poincaré, originator of geometric dynamics, depart radically from the earlier concepts of Galileo and Newton. It is ironic that he is not mentioned by Whitehead, because Poincaré's great follower, George D. Birkhoff, was Whitehead's colleague at Harvard and was carrying on the new approach at the very moment Whitehead was writing his history. After Birkhoff, dynamics was dormant in the West, while the followers of Lyapunov, a Russian contemporary of Poincaré, continued the development of geometric techniques and concepts. This line of study was revived in the United States by the topologist Solomon Lefshetz (1950). Since then the field of dynamics has experienced tremendous expansion.

We present the basic concepts of dynamics in four historical groups: Galilean, Newtonian, Poincaréan, and Thomian. From antiquity to Galileo, general physical concepts of kinematics were developed, especially *space, time, curve of motion in space, instantaneous velocity at a point on the curve*, and *final motion or asymptotic destination of the curve*—probably thought to be a *limit point*. From Newton to Poincaré, the mathematical expression of "local" concepts flowered: *Euclidean space-time domain, integral curve, vectorfield*, and *attractor* (taken to be a limit point, or a *limit cycle*).

From Poincaré to Thom, the global geometric perspective emerged; the state space (or mathematical domain) of the dynamical system expanded from an open region in a flat Euclidean space to a *manifold*, or smooth space of arbitrary geometric and topological type. The dynamical system came to be viewed globally also: analytically, as a "flow" (or group of motions of the space of states upon itself); and geometrically, through its *phase portrait*. More complicated limit sets, such as the ergodic two-dimensional *torus of irrational rotation*, became known and the revolutionary concepts of *structural stability, generic property, and bifurcation* emerged. These concepts will be described in more detail below.

Thom developed the idea of *stable bifurcation*, as well as an even more global "big picture" of infinite dimension in which the stable bifurcations became geometric objects, amenable to classification. In the same period a new class of attractors, the *chaotic attractors*, was discovered experimentally, and a veritable industry of applications began. A parallel development, based on the analysis of invariant measures, has taken place in Russia. (The ideas of the Russian school are not included in this survey, although they are quite important.)

Dynamics

Mathematical dynamical systems (hereafter simply "dynamical systems") consist of deterministic equations, including ordinary differential equations, partial differential equations of the evolution type, or finite difference equations. The equations may occur singly or in sets.

Since Poincaré, dynamical systems have been studied using topological and

29. Dynamics

geometric methods, and dynamical systems theory has diverged from the classical analysis of differential equations and operators. A main goal of the geometric, or "qualitative," theory has been to understand the "final motions," or asymptotic limit sets, of a dynamical system. Here we present the most important ideas of dynamics through examples in low dimensions. The main idea is the "AB portrait" of a dynamical system, which records those aspects of the dynamics that are most evident in the qualitative point of view, the attractors and basins.

We proceed by presenting and describing 84 figures. In Chapter 30 the concepts will be applied to the problem of describing self-organizing behaviors.

Basic Concepts

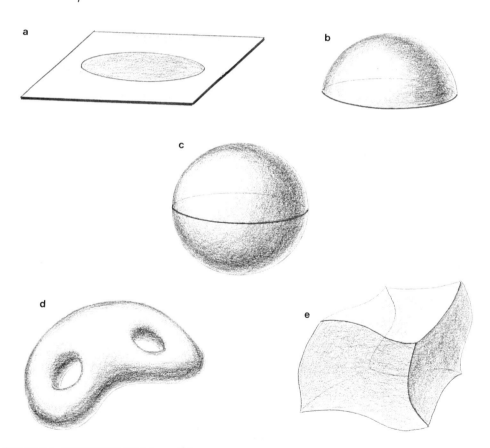

FIGURE 1. STATE SPACE. Many different spaces may be considered to be the domain of a dynamical system. Examples include: an open region in the plane (a); the upper hemisphere (b); the entire two-sphere (c); a closed surface with two holes (d); an open region of three-dimensional space (e); or a higher-dimensional surface. Each point of the state space corresponds to a "virtual state" of a system being modeled.

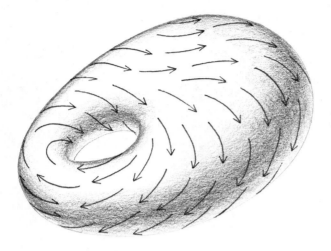

FIGURE 2. A DYNAMICAL SYSTEM on a state space consists of a vector assigned to each point. Each point in the state space is interpreted as a virtual state of the system modeled, as described above. Each of these based vectors is interpreted as a dynamical rule: the state must evolve with the speed and direction of the vector based there.

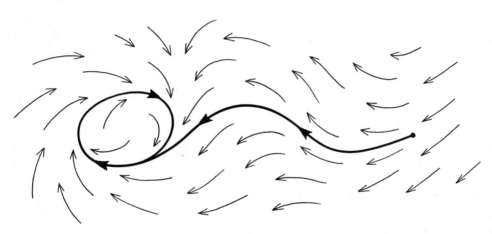

FIGURE 3. Starting at any point (a) in the state space, there is a uniquely determined curve (b) following the dynamical rules at each point it passes. The starting point is called the *initial state* (a); the curve (b) is its *trajectory*, and the asymptotic limit set of the curve (c) is the *limit set*. In these illustrations the curved arrows represent the *flow* of the dynamic, i.e., the simultaneous movement of all initial states along their trajectories.

29. Dynamics

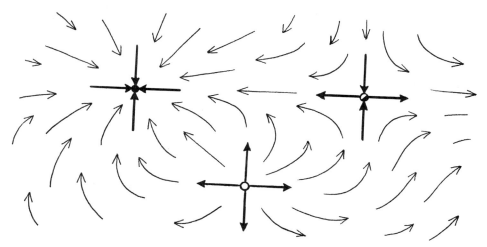

FIGURE 4. LIMIT POINTS IN TWO DIMENSIONS. The simplest limit set is a *limit point*. In two dimensions there are only three types of generic limit point: the point attractor (a); the point repellor (b); and the saddle point (c). In this context, the *inset* of a limit point refers to the set of all initial states which asymptotically approach the point in the future. The *outset* of the limit point comprises all initial states approaching the limit point as time (the parameter along the trajectory) goes backward. The saddle (c) has both inset and outset of one dimension. The attractive point (a) has a two-dimensional inset, called its *basin*. The repellor (b) has a two-dimensional outset.

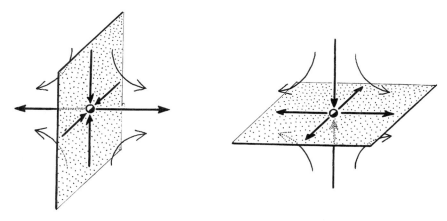

FIGURE 5. LIMIT POINTS IN THREE DIMENSIONS. Here attractors have three-dimensional insets, or basins, while repellors have three-dimensional outsets. There are two types of generic saddle: one has a two-dimensional inset (a); the other (b) has a one-dimensional inset.

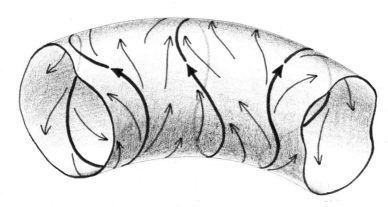

FIGURE 6. LIMIT CYCLES IN TWO DIMENSIONS. A circular limit set is called a *periodic limit set*, or a *limit cycle*. In two dimensions there are only two generic types, attracting (the *periodic attractor*) and repelling (the *periodic repellor*).

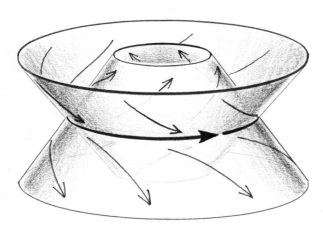

FIGURE 7. LIMIT CYCLES IN THREE DIMENSIONS. Here we have again the periodic attractor (three-dimensional inset) and the periodic repellor (three-dimensional outset), as well as a new type, the *limit cycle of saddle type*. Its inset is a two-dimensional cylinder, as is its outset. The two cylinders intersect in a circle, which is the limit cycle itself.

29. Dynamics

Basins and Separatrices

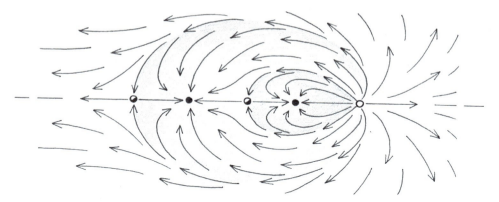

FIGURE 8. A *basin* is the inset of an attractor. A dynamical system usually has basins with one attractor in each. The state space is decomposed into a set of basins. The *probability of an attractor* is the relative area (or volume) of its basin. Here a dynamical system in two dimensions is shown, with three basins; two are shaded, and the third surrounds them.

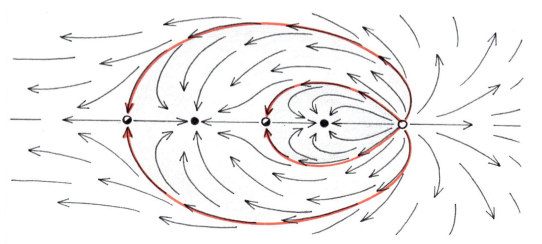

FIGURE 9. SEPARATRICES IN TWO DIMENSIONS. The *separatrices* are the boundaries of the basins. Usually we shall show these in red. By definition, separatrix points are not in basins. Thus, they belong to the insets of limit sets which are not attractors, the *exceptional limit sets*. Ideally, the probability of an initial point belonging to a separatrix should be zero. In this case, we may say that the separatrix is *thin*. The red separatrices here are the insets of saddle points. This picture of a state space, divided into basins by separatrices, with an attractor in each basin, is the *AB portrait* of the dynamical system.

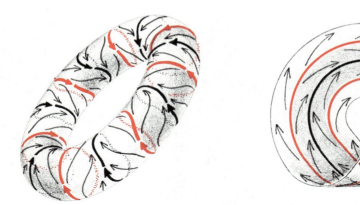

FIGURE 10. REPELLORS AS ACTUAL SEPARATRICES. In addition to the insets of exceptional (nonattracting) limit sets, the separatrix contains all the repellors of the dynamical system. These are the limit sets which would be the attractors if the arrow of time were reversed. Separatrices may be *actual* or *virtual*. Periodic repellors are shown here in dimension two. Because they actually divide the state space into distinct basins, they are actual separatrices. This occurs only in the two-dimensional case.

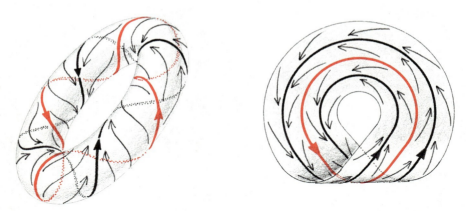

FIGURE 11. REPELLORS AS VIRTUAL SEPARATRICES. The periodic repellors in this figure do not divide the state space into distinct basins, because the state space with the red cycle removed has only one piece, the basin of the unique (black) periodic attractor. These periodic repellors therefore are virtual separatrices. Repelling limit points are also virtual separatrices.

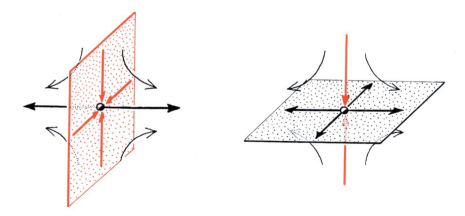

FIGURE 12. SEPARATRICES IN THREE DIMENSIONS. In two dimensions the separatrices are one-dimensional. In three dimensions they must be two-dimensional surfaces in order actually to separate the state space into disjoint regions. More generally, in "n-dimensions," they should have dimension "$n - 1$." Geometers call these "codimension one hypersurfaces."

This figure (similar to Figure 5) illustrates typical separatrices in three dimensions. The red surface, the inset of a saddle point, is called an actual separatrix because it actually separates the three-dimensional state space. The red curve, also the inset of a saddle point, is likewise a separatrix, but because it is not a codimension one hypersurface (it has codimension two) and cannot actually separate the space, it is called a virtual separatrix.

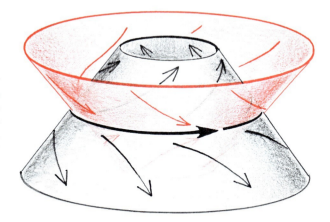

FIGURE 13. PERIODIC INSET SEPARATRICES. In three dimensions the periodic saddle (see Figure 7) is characterized by two cylindrical surfaces which pass through it, crossing through each other: the inset and the outset. The inset is a separatrix (as always) and therefore is shown in red. Furthermore, this inset has codimension one and is an actual separatrix.

Chaotic Attractors

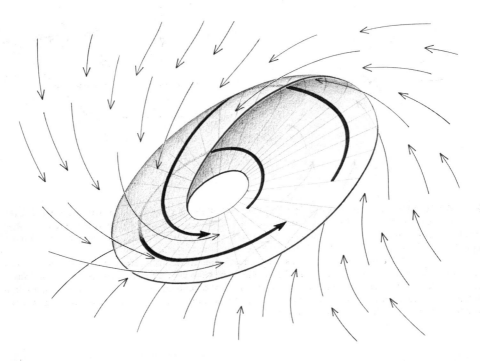

FIGURE 14. THE RÖSSLER. Beyond the point attractor (dimension zero) and the periodic attractor (dimension one) lives a little-known world of more complicated generic attractors. Most of these have been discovered by experimentalists; the one shown here was found by Rössler (an experimental dynamicist) with an analog computer. Like all of the attractors of dimensions greater than one, it is *chaotic* in the sense of power spectrum analysis; it emits broadband noise (Rössler, 1979). It is a *thick attractor*; although it looks two-dimensional, microscopic analysis reveals a *fractal* thickness. This attractor has "fractal dimension" 2+, i.e., a fraction more than two! The next six illustrations describe this "microscopic analysis."

29. Dynamics

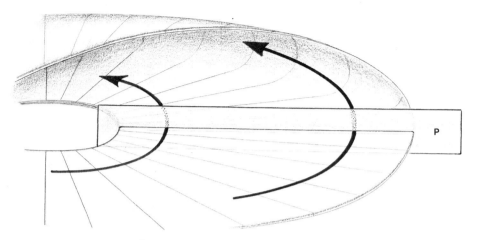

FIGURE 15. THE POINCARÉ SECTION. We cut through the attractor, in order to study its cross section, called the Poincaré section.

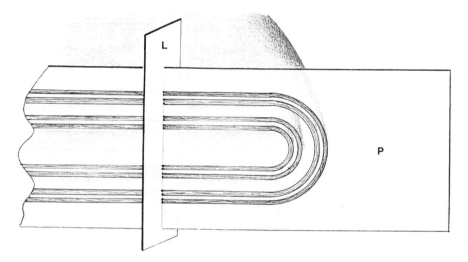

FIGURE 16. THE LORENZ SECTION. Enlarging the Poincaré section, we see that it consists of many curves compressed together. We cut through these curves with another cross section, as in Lorenz (1963).

FIGURE 17. THE LAYER SET. The second cut results in a line of dots, one for each curve of the Poincaré section, or, equivalently, one for each sheet of the Rössler attractor (Rössler, 1979). This result is a *Cantor set*.

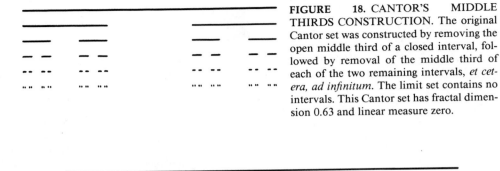

FIGURE 18. CANTOR'S MIDDLE THIRDS CONSTRUCTION. The original Cantor set was constructed by removing the open middle third of a closed interval, followed by removal of the middle third of each of the two remaining intervals, *et cetera, ad infinitum*. The limit set contains no intervals. This Cantor set has fractal dimension 0.63 and linear measure zero.

FIGURE 19. THE MIDDLE FIFTHS CONSTRUCTION. Another Cantor set can be constructed by removing two fifths at each stage. This construction produces a Cantor set of fractal dimension 0.68 and zero linear measure. By removing smaller sectors at each stage, however, a limit set can by constructed which has positive linear measure, yet which contains no intervals. These are *thick sets*.

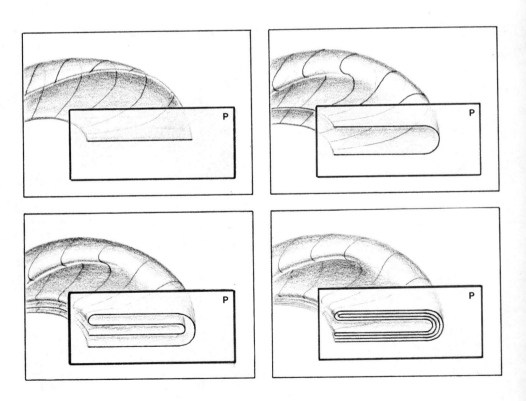

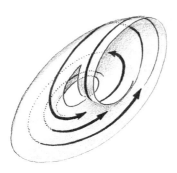 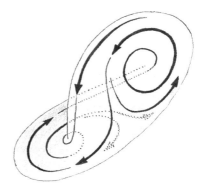

FIGURE 21. OTHER CHAOTIC ATTRACTORS. In addition to the Rössler attractor, many other attractors have been found by experiment. On the left is shown the *funnel*, also discovered by Rössler. Other attractors found by Rössler look amazingly like the seashells called tops, turbans, and sundials. On the right is the *Lorenz attractor*, the first to be found. This attractor was discovered in a digital simulation during Lorenz's study of atmospheric turbulence in 1961. These all seem to be *ergodic*, i.e., the averaging procedures of statistical physics apply to them.

←

FIGURE 20. THICKNESS OF THE ATTRACTOR. A Cantor set of two-dimensional sheets is a *Cantor 2-manifold* and has fractal dimension $2 + T$, where T is the fractal dimension of the Cantor set, measured across the layers. In this illustration we see why the Rössler attractor is a Cantor manifold. To the original Poincaré section across the Rössler attractor, we apply the *Poincaré section map*. Each point is carried once around the attractor, following the unique trajectory of the dynamical system, until it crosses through the Poincaré section again. Three iterations of this map are shown here. With each iteration, the section of the Rössler attractor is pulled out double-width, folded over, flattened, and reinserted into the section. The limit of this process is a Cantor manifold, and it covers the entire Poincaré section. We say the attractor is *fractal* if the fractal dimension is not an integer, and *thick* if the measure is nonzero (implying the same, integer dimension as the state space).

Separatrices also can have a Cantor structure; these are called *fractal* or *thick separatrices* by the same criteria of dimension and measure. In a thick separatrix, the probability of an arbitrary initial point belonging to the separatrix will be nonzero, but, one hopes, small. The limit sets of such initial points, which are not attractors, yet which have insets of positive volume, may be called *vague attractors*. In fact, some authors (e.g., Pugh and Shub, 1980) call even these "attractors."

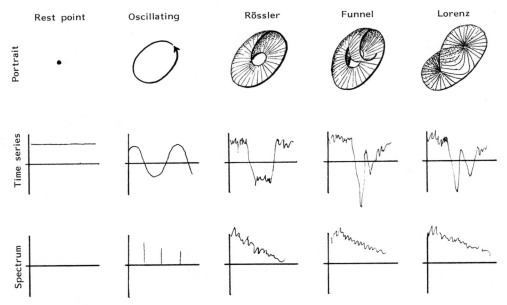

FIGURE 22. A LIST OF ATTRACTORS. Dynamicists would like to classify all generic attractors in some taxonomic scheme. While the end is not in sight, we have the beginning of a list. Here the list starts on the left, extending to the right beyond the five steps shown. Beneath each *attractor portrait* is its corresponding *time series* (or time record of a single coordinate of a trajectory along the attractor). Its corresponding *power spectrum* is shown beneath the time series. (The power spectrum is a plot of power versus frequency in the harmonic analysis of the time series.) Beneath the spectrum we could imagine a list of further attributes of the attractors across the top, adequate to distinguish them from one another.

Structural Stability

FIGURE 23. UNSTABLE ATTRACTORS. The system on the spheroid, at left, has a point attractor and a point repellor. It is *structurally stable*, in that any dynamical system obtained from it by a small perturbation will have essentially the same AB portrait. On the other hand, the system on the toroid, shown on the right, has no attractor other than the entire toroid. Its trajectories wind forever around it, like a solenoid. It is *structurally unstable*, because a small perturbation can change the AB portrait into a finite number of basins, each containing a closed orbit which winds several times around it (details are shown later).

29. Dynamics

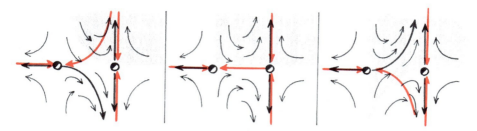

FIGURE 24. UNSTABLE BASINS. The dynamical system shown in the center has a *saddle connection*. A perturbation can produce the AB portrait on the left or the one on the right; these are essentially different (i.e., topologically nonconjugate) portraits. The attractors may be the same, but the basins are not. This sequence, read from left to right, is an example of *basin bifurcation*.

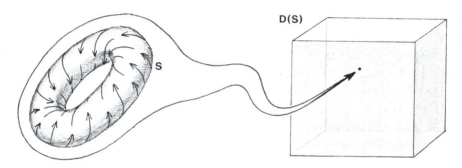

FIGURE 25. THE BIG PICTURE. An important overview of all dynamical systems on a given state space has been developed by Thom. Let S denote the given state space, e.g., a toroid, and $D(S)$ the set of all smooth dynamical systems on it. In this "big picture," a specific dynamical system corresponds to a single point.

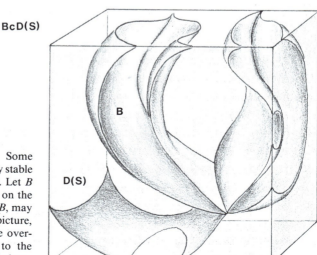

FIGURE 26. THE BAD SET. Some dynamical systems are structurally stable (good) while others are not (bad). Let B denote the set of all bad systems, on the given state space, S. The bad set, B, may be visualized within the big picture, $D(S)$. This account completes the overview of Thom, which is basic to the description of bifurcations given below.

Coupled Oscillators

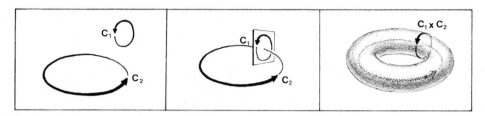

FIGURE 27. THE STATE SPACE FOR TWO OSCILLATORS. The state space for a single oscillator may be assumed to be a circle, C, of unit circumference. The dynamical system can be described as a constant velocity in the direction of the arrow, so that one full cycle has a period of V seconds and a frequency of $F = 1/V$ cycles per second. Now consider two such oscillators with state spaces denoted by C^1 and C^2 and frequencies F^1 and f^2. The state space for the combined system of the two oscillators is the torus, $C^1 \times C^2$.

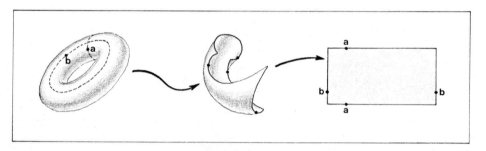

FIGURE 28. THE FLAT TORUS. For easier visualization we cut the torus along two circles, corresponding to the C^1 and C^2 "axes," and flatten it. To wrap it up again, identify (paste together) the two vertical edges, then the two horizontal circles. Thus, the two points labeled "a" represent the same point on the torus, as do the two points "b." Other than the boundary points, every point of the flat torus specifies a unique instantaneous state for each of the oscillators.

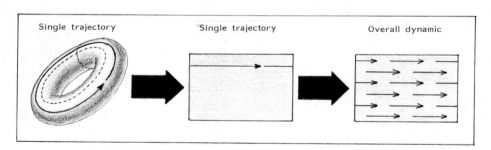

FIGURE 29. THE DYNAMIC FOR TWO OSCILLATORS. Suppose, for example, that the *second* oscillator is jammed or stuck ($F^2 = 0$). A trajectory on the real torus is then a horizontal circle, corresponding to a horizontal line on the flat torus, proceeding to the right with speed $V^1 = 1/F^1$.

29. Dynamics

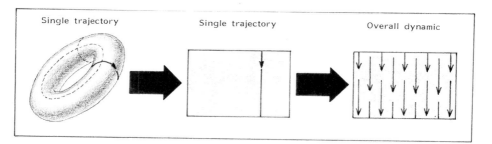

FIGURE 30. ANOTHER EXAMPLE. If the *first* oscillator is stuck, the dynamic is vertical, as shown here.

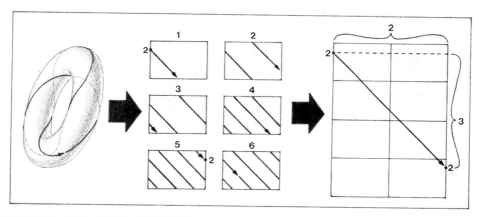

FIGURE 31. THE GENERAL UNCOUPLED CASE. When both oscillators are running, F^1 and $F^2 > 0$; the trajectories on the flat torus are still straight lines, with slope $V^2/V^1 = F^1/F^2$. For example, suppose this ratio of frequencies is $-3/2$. Thus, the first oscillator completes two cycles clockwise, while the second completes three cycles counterclockwise. The trajectories on the real torus are all closed cycles, or periodic trajectories, which wrap three times around the small waist (corresponding to C^2) and twice around the large waist. A full cycle of this compound oscillation is visualized best upon six or more copies of the flat torus, as shown here. This picture is modified easily to display the dynamic for coupled oscillators having any rational ratio of frequencies, or *rotation number*. The case of an irrational rotation number requires drawing a line of irrational slope across a doubly infinite array of flat tori, because the trajectories on the real torus never close.

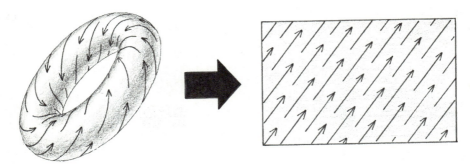

FIGURE 32. STRUCTURAL INSTABILITY. The uncoupled oscillators are structurally unstable, because a slight perturbation of the frequency of either (or both) will cause the rotation number to change through infinitely many rational and irrational values, each corresponding to an essentially different (topologically nonconjugate) portrait.

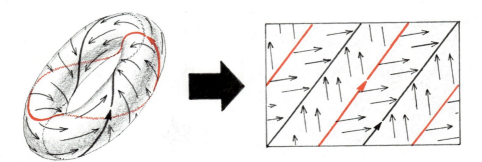

FIGURE 33. COUPLED OSCILLATORS. Coupling refers to a small (and unspecified) generic perturbation of the uncoupled dynamic previously described. The trajectories become wavy. According to the historically important *Peixoto theorem* (special to the two-dimensional case), the generic coupled dynamic must be structurally stable, having a finite number of basins, each containing a periodic attractor, and bounded by a periodic repellor. All of these limit cycles have the same period, determined by the rational rotation number of the dynamic. Because only these attractors can be found (as stable equilibria of the coupled oscillators), the oscillators are said to be *entrained*. The two oscillators, observed separately, are found to have frequencies in a rational ratio; therefore, an integral number of cycles of one oscillator takes the same time as another integral number of the other. Here we show a simple case, with only one basin and rotation number 1/2.

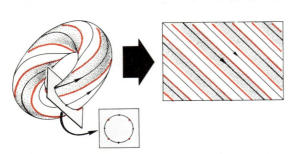

FIGURE 34. BRAIDS. The typical portrait for entrained (structurally stable or generically coupled) oscillators has many basins, braided together around the real torus. Their visualization is facilitated by cutting through the torus with a plane, obtaining a circular cross section, the Poincaré section. This circle is punctuated by alternating red points (where separatrices cross) and black points (corresponding to attractors). Here we show a 2-braid. There are two distinct basins, one of which is shaded. Each has rotation number -3/2, as in Figure 31.

29. Dynamics

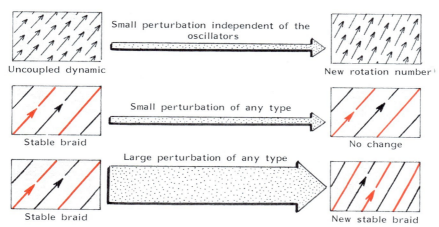

FIGURE 35. ENTRAINMENT IS STABLE. While uncoupled oscillators are structurally unstable (Figure 32), two generically coupled oscillators are stable, according to Peixoto's theorem. Even so, a large perturbation can change one stable braid into another.

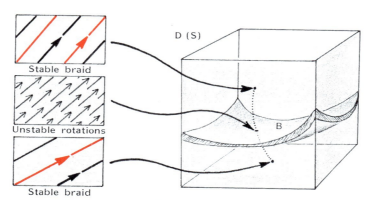

FIGURE 36. THE BIG PICTURE OF ENTRAINMENT. Because each dynamic on the torus is represented by a single point in the big picture, the transition of one stable braid into another by perturbation can be represented by a dotted line. An intermediate dynamic with an irrational rotation number must belong to the bad set.

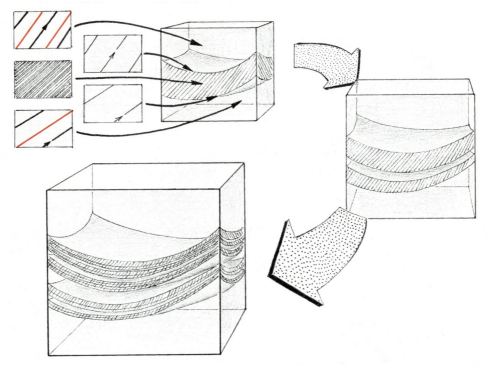

FIGURE 37. THE MICROSTRUCTURE OF THE BAD SET. Suppose the transition of one stable braid into another is made continuously. The rotation number of the initial braid then changes continuously into the rotation number of the final braid. Between these two rational numbers are infinitely many irrationals. The bad set, therefore, must have an infinite set of layers, pressed very closely together. It has been shown that this bad set is a thick Cantor manifold (see Figure 20) in some general sense.

Bifurcations

Dynamical bifurcation refers to the continuous deformation of one dynamical system into another inequivalent one, through structural instability. One must be careful not to confuse Dynamical Bifurcation theory (created by Poincaré in 1885), which is the subject of this survey, with the somewhat similar subjects of Elementary Catastrophe theory or Classical Bifurcation theory. These are compared in Chapter 30.

The most important applications of dynamical systems theory demand the inclusion of controls in the model. The full accommodation of this demand into dynamical systems theory first occurs in Thom's (1972) *Structural Stability and Morphogenesis*. Here we present a visual account of the basic notions of dynamical bifurcation, along with a selection of the simplest bifurcations.

29. Dynamics

Catastrophic Bifurcations

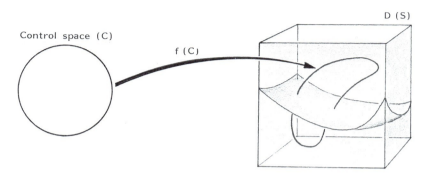

FIGURE 38. DYNAMICS WITH ONE CONTROL. The states of the controller are modeled by an auxiliary space, the control space, C. In this section, we consider only the case of a single control; the dimension of C is one. Here, C is shown as a circle. An auxiliary function, f, maps the control space into the big picture. For each "setting of the control knob," represented by a point c of C, there is, therefore, a corresponding dynamical system $f(c)$ on the given state space, S. As c is moved along C, $f(c)$ moves smoothly within the big picture, and the dynamical system on S changes smoothly as well.

As long as $f(c)$ does not cross the bad set, B, the AB portrait of the dynamic on S does not change in any essential way. After traversing B, however, the portrait is essentially different (i.e., it is topologically nonconjugate). These transformations are known as *bifurcations*. In the best cases, the image of the control space within the big picture, $f(C)$, crosses cleanly through the bad set, B. These cases, called *generic bifurcations*, are the main target of Thom's program. The generic bifurcations with one control are also called *generic arcs*.

In the applications of dynamics with controls, the modeling function—or *morphogenetic field*, as Thom called it—cannot be specified exactly. Thus, *stable bifurcations*, which are insensitive to perturbations of the modeling function, f, are the events most important to know.

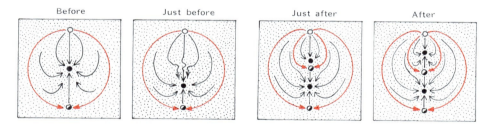

FIGURE 39. STATIC CREATION. Here are four frames from a "movie" of the simplest stable arc, from left to right. At first there is a single point attractor in the center of its basin. Next, the trajectories approaching it from above are pinched together, like a pigtail. This change is caused by a turn of the "knob" and, as yet, there has been no bifurcation. In the third frame there are two point attractors, each in its own basin. The new one is just above a new saddle point, which was created at the same moment, or control setting, called the *bifurcation point* in the control space. In the last frame, the new attractor has receded further from its companion saddle, and, thus, also from the separatrix bounding its basin. This event, sometimes called the *saddle-node bifurcation*, is similar to the *fold catastrophe* of Elementary Catastrophe theory.

When this "movie" is shown in reverse sequence, the upper attractor drifts toward the edge of its basin. At the bifurcation point, it collides with a saddle in the separatrix. The attractor and separatrix simultaneously vaporize, or disappear "into the blue." This event is called *static annihilation*.

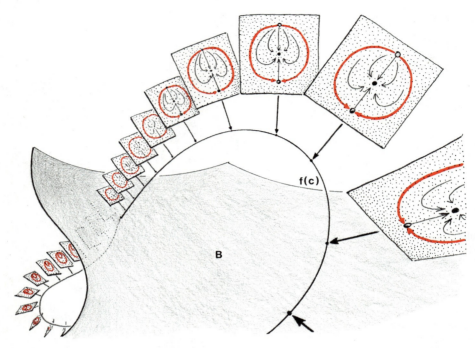

FIGURE 40. STATIC CREATION IN THE BIG PICTURE. Here we embed the four frames into a complete movie, with each frame attached to its associated point on the generic arc in the big picture.

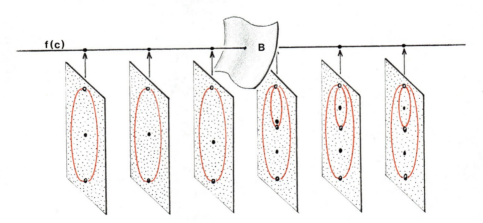

FIGURE 41. THE BIFURCATION DIAGRAM. Here is an alternative visualization of a bifurcation movie, sometimes called the "little big picture." The frames are stacked side by side, like the pages of a book. Interpolating between these, the attractors fill in attractor surfaces or curves (black) while the separatrices trace separating surfaces or curves (red).

29. Dynamics

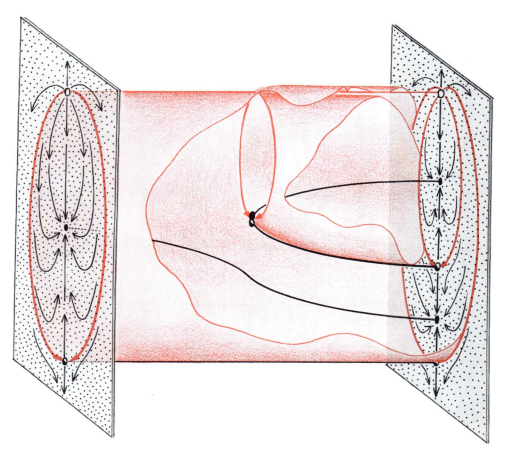

FIGURE 42. BIFURCATION DIAGRAM OF STATIC CREATION. Here is the filled-in portrait for the static creation bifurcation of Figure 39. This is a *catastrophic bifurcation* in the sense that the track of the upper attractor appears "out of the blue," i.e., far from any other attractor.

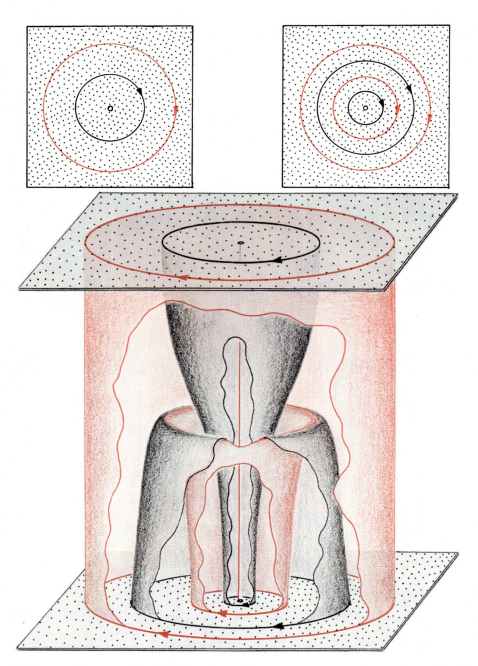

FIGURE 43. CREATION OF AN OSCILLATION. This is the second of the simple bifurcations. In this event a periodic attractor appears catastrophically. Before the creation event, a periodic attractor (black) is shown in a basin bounded by a periodic repellor (actual separatrix, red) and a point repellor (virtual separatrix, red). Without change in these three elements, the AB portrait after the event has two new elements, a periodic attractor and a periodic repellor. The bifurcation diagram shows how these two elements were created simultaneously, as the control moved past the bifurcation point in the control space. The new periodic repellor is the separatrix of the new basin of attraction. Were this movie to be viewed in reverse, we would see the inner attractor growing, until it approached the boundary of its basin and touched it, causing both elements to vaporize at once. This is a typical annihilation catastrophe and is also known as *hard self-excitation*.

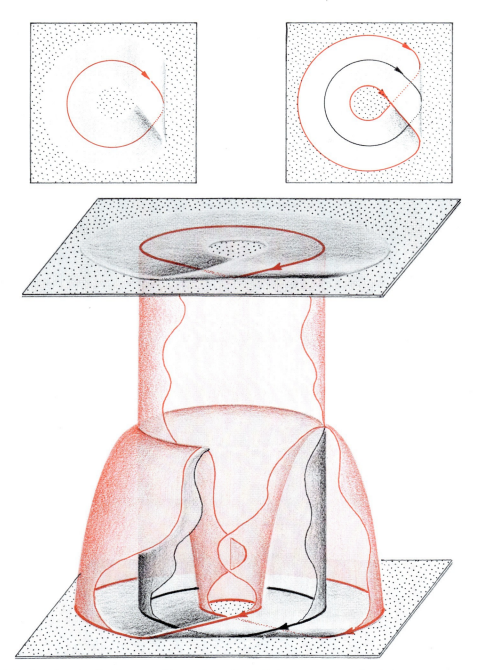

FIGURE 44. CREATION OF AN OSCILLATION. This is another type of catastrophic creation of an oscillation. The state space, S, is a Möbius band. Before the creation event, a periodic repellor circles the Möbius band. Empirically invisible, it is a virtual separatrix; hence, it is shown red here. After the event it has become an attractor, and thus is shown in black. The new separatrix bounding the basin of this attractor circuits the Möbius band twice before closing. The bifurcation diagram reveals the details of the transformation process. At the instant the red cycle turns black, the new separatrix branches off, rather like a paraboloid of revolution. We could view this bifurcation as an explosion, in which a virtual separatrix thickens, creating a new basin. A new attractor has appeared suddenly.

Hysteresis

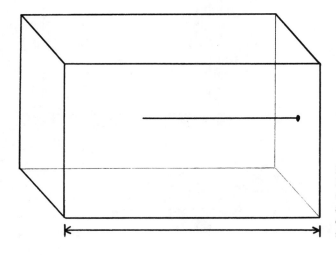

FIGURE 45. CATASTROPHIC BIFURCATION involves the sudden creation of a new basin and attractor, as in the three examples of the preceding section.

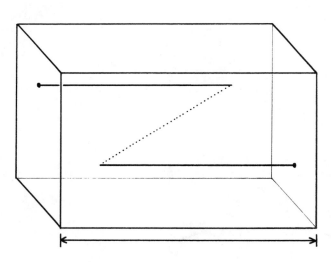

FIGURE 46. HYSTERESIS occurs in a bifurcation diagram containing at least two catastrophic bifurcations, back to back. In this example a static creation is followed by a static annihilation, as in the *fold catastrophe* of Elementary Catastrophe theory. We shall refer to this configuration as a *hysterical kink*.

29. Dynamics

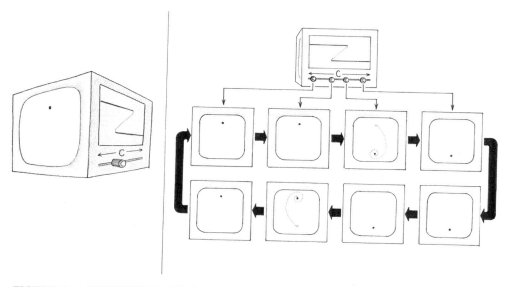

FIGURE 47. A HYSTERESIS LOOP is shown here, in a mechanical representation, realizing the preceding bifurcation diagram. Beginning with the control full left, the knob is moved to the right. The sequence of observations in the top row results. Initially the system is caught in the upper attractor. After the knob passes the first bifurcation point in the control interval, a second attractor exists. The machine, however, does not reveal it; there is no change of behavior. After the knob passes the second bifurcation point, the upper attractor vanishes (after colliding with its separatrix, which slides along the dotted track), and the machine finds itself in the far reaches of the basin of the lower attractor. The dynamic asserts itself, the current state rushes toward its new attractor far below, the transient dies away, and the machine settles down to its new equilibrium. The knob is full right.

Next the knob is pushed slowly back to the left. After a similar sequence, the screen shows that the machine again has settled down in the original attractor, near the top. The transition upward occurs near the left bifurcation point, however, after passing the control point of the downward transition. This is the classical hysteresis loop behavior. In the context of Dynamical Bifurcation theory it has many more complicated forms.

Subtle Bifurcations

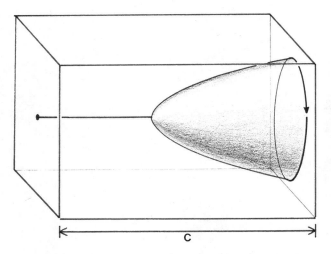

FIGURE 48. SUBTLE BIFURCATIONS differ from catastrophic ones in that nothing new appears suddenly out of the blue. Instead, an existing attractor changes gently into another type of attractor. In the bifurcation diagram shown here (and discussed below) a point attractor changes into a periodic attractor. The change is noticed only after the amplitude of the oscillation has grown large enough to be observed.

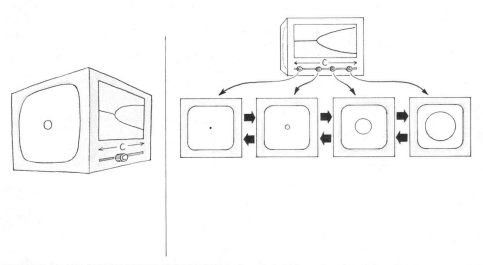

FIGURE 49. MACHINE REPRESENTATION of a subtle bifurcation shows that the attractor under observation, a point attractor, changes into an oscillation as the control is moved to the right. The amplitude increases as the control is moved more to the right. As the control is returned to the left, the same sequence is observed in reverse. The amplitude of the oscillation decreases, and the circle shrinks to a point. The death of the oscillator occurs at the same bifurcation point at which it first appeared. There is *no hysteresis* with subtle bifurcations.

29. Dynamics

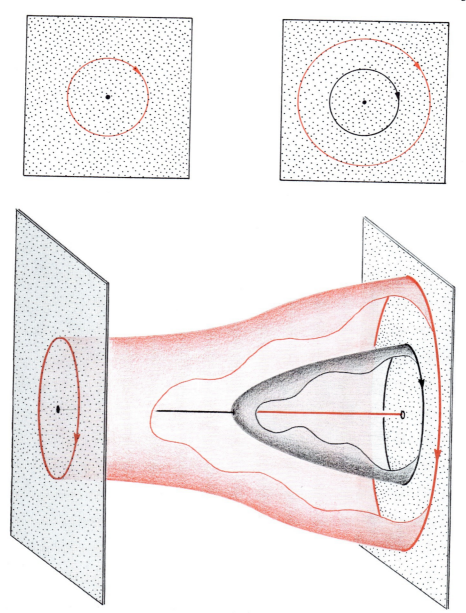

FIGURE 50. EXCITATION. This event, also known as the *Poincaré-Andronov-Hopf bifurcation*, is an outstanding example of subtle bifurcation. Without actual change of basin, the point attractor turns into a periodic attractor. An oscillation has been born. Again, the bifurcation diagram shows how the amplitude of the new oscillation grows parabolically as the control moves to the right. This event may provide a dynamical model for the morphogenesis of the flat rings of Saturn.

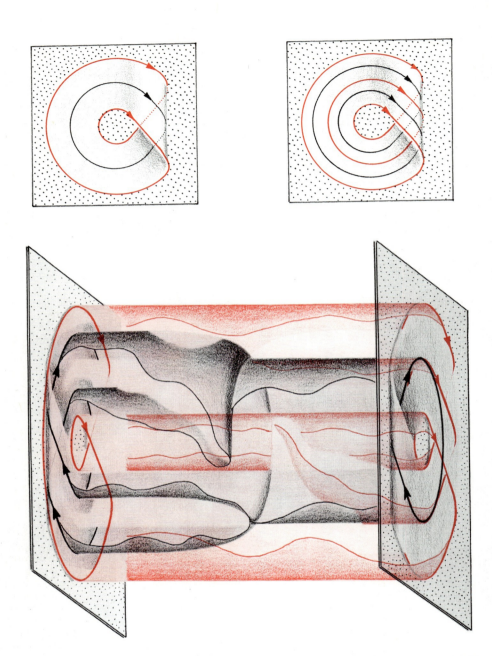

FIGURE 51. THE OCTAVE JUMP. Before bifurcation a periodic attractor (black) circles the Möbius band in a basin bounded by an actual separatrix (periodic repellor, red) which circuits twice before closing. After the event the original attractor becomes a virtual separatrix (periodic repellor, red) in the basin of a new periodic attractor (black). The outer boundary of this basin is the same separatrix as previously. The new attractor has twice the period and half the frequency of the original one, and is appropriately named. The bifurcation diagram shows the whole sequence of this event, which is a subtle bifurcation in the sense that the jump down an octave becomes noticeable gradually, as in a slow yodel. No attractor track has appeared "out of the blue."

Thick Bifurcations

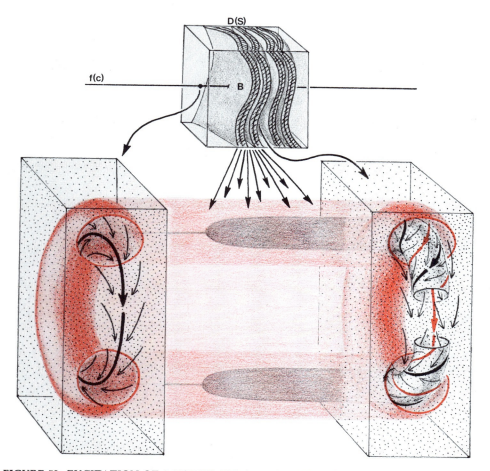

FIGURE 52. EXCITATION OF A TORUS. This is a subtle bifurcation, closely related to the excitation of an oscillation (Poincaré–Andronov–Hopf bifurcation, Figure 50), shown previously. Coarsely observed, it resembles a single event. Before the event there is a single periodic attractor in a three-dimensional basin. For simplicity the boundary of its basin is shown as a torus, although this is not frequent. After the subtle bifurcation, the periodic orbit is changed to a repellor (red circle) within a thin torus (black) on which a braid (see Figure 34) is shown. The torus is *attracting, yet not an attractor*. The new attractors, after the event, are the periodic attractors within the braid on the torus. These fluctuate, however, among many different (and topologically inequivalent) types of braids immediately after the creation of the torus, as described in the discussion of the coupled oscillator (see Figure 37). This event thus includes a Cantor set of bifurcation points in the control space, illustrated here by a small piece of the big picture. We call this a *thick bifurcation* because the set B^f of bifurcation points in the control space, C, is a thick Cantor set. The probability of observing a nonstructurally stable attractor (in this case it would be an attracting torus—a "miracle" from the point of view of structural stability dogma) is the measure of the bifurcation set, B^f. Because this is a positive number, we have here a model for the expectation of "miracles." This multiple bifurcation event was discovered, bit by bit, in the work of Neimark (1959), Peixoto (1962), Sotomayor (1974), and Herman (1979), and may be called the *fluctuating braid bifurcation*. This event may provide a dynamical model for the morphogenesis of the braided rings of Saturn.

There are further excitations in which an attractive, two-dimensional torus becomes an attractive, three-dimensional torus, and so on, through higher modes of oscillation. The relaxation of the dynamics on these tori into structurally stable systems comprises the extension of the entrainment idea to coupled systems of several oscillators. The details are as yet unknown.

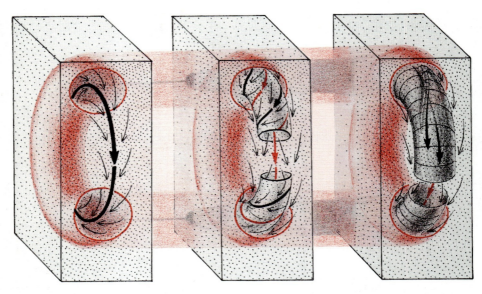

FIGURE 53. THE ONSET OF CHAOS. This phrase refers to a class of bifurcation events in which a simple (point or periodic) attractor turns into a chaotic attractor. Only a few of these have been studied, some by experimentalists, others by theoreticians. Here we show a typical example of the subclass known as *omega explosions*. The sequence begins with a fluctuating braid, as described in Figure 52. The onset of chaos involves changes in the underlying invariant torus (shaded black) of the braid. As the control is moved to the right, this torus develops a wave, which progresses into a fold and then a crease. As the pleated fold of the torus is pressed back down onto the torus by its attraction within the dynamic, the torus suddenly thickens, becoming a Cantor manifold (dimension $2+$), which is the new attractor. Before the event there is a braid of periodic attractors. Afterward these are lost amid a thick toral attractor of the chaotic type. This example was discovered by R. Shaw in a study of forced oscillation on an analog computer.

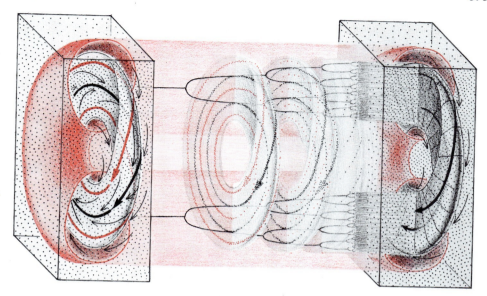

FIGURE 54. THE OCTAVE CASCADE. In another closely related context, an infinite sequence of octave jumps (see Figure 51) has been proven to be generic and to converge to an onset of chaos (Feigenbaum, 1978). After this limit point in the bifurcation set in control space, another bifurcation sequence converges from the right (Lorenz, 1980). This is a similar cascade of octave jumps, but all the attractors involved are chaotic. There is experimental evidence for the existence of this phenomenon in the omega explosion as well (Shaw, 1981), but it has not been proven to be actually an infinite sequence. Here we show the cascade of octave jumps for a periodic attractor traced on a substrate shaped like the Rössler attractor (see Figure 14), as suggested by the experimental work. The limit of the sequence is an onset of chaos bifurcation, in which the Rössler-looking substrate emerges as the attractor.

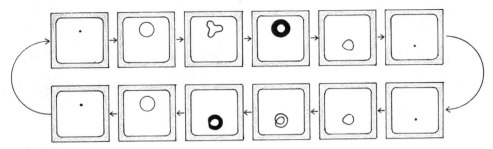

FIGURE 55. TYPICAL BIFURCATION SEQUENCES. In many experimental situations described by dynamical models with one control, moving the control in one direction results in a sequence of the bifurcations described singly in the preceding illustrations. Moving the control back again to the original value results in a similar, but different sequence. Here we illustrate some commonly observed sequences. Above, presented as the control moves to the right, a point attractor becomes periodic by excitation, then braided and chaotic—this is the *Ruelle–Takens sequence* (Ruelle and Takens, 1971). The chaotic attractor then disappears (by collision with its separatrix—a bifurcation anticipated by theory but not yet studied, or illustrated here). Below, a similar sequence is seen as the control is returned to the left, but different attractors are involved. Hysteresis may be observed. The dynamical model behind this phenomenon is shown in Figure 56.

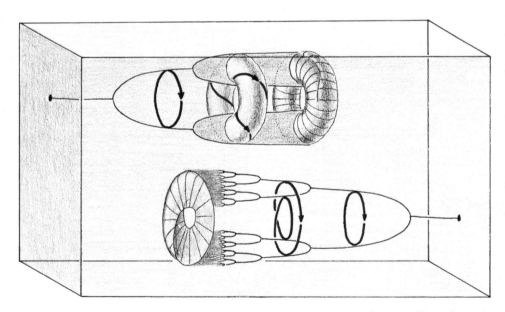

FIGURE 56. CHAOTIC HYSTERESIS. The bifurcation diagrams shown above are all atomic events. In real systems, compound models are encountered in which several of the atomic events are connected in a single bifurcation diagram. In these maps it is easy to understand hysteresis—the sure sign of catastrophic (as opposed to subtle) bifurcations. Here we show a fictitious bifurcation diagram in which two attractors undergo bifurcation separately, each in its own basin (except at the ends, where one or the other basin has disappeared). The observed behavior of the system is described in the preceding figure.

29. Dynamics

An Exemplary Application: Fluid Dynamics

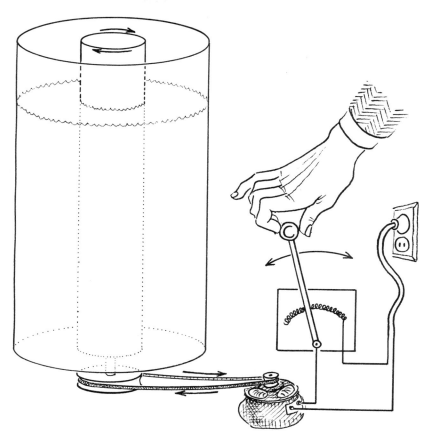

FIGURE 57. A STIRRING MACHINE. One way (not necessarily the best) to stir a cup of fluid is to rotate a swizzle stick in it. Great progress has been made in the study of turbulent fluid dynamics using this image. Since its earliest description a century ago, this experiment of Mallock and Couette has been repeated over and over, with ever-improved rods and cups. The onset of turbulence is carefully reproduced and observed with this apparatus. As the speed of stirring is increased beyond a critical (bifurcation) value, the expected flow of the fluid (in concentric, cylindrical lamellae) is replaced by *annular vortices*, patterns discovered by Taylor in 1923. Additional increase in the rate of stirring produces more bifurcations and, finally, turbulence.

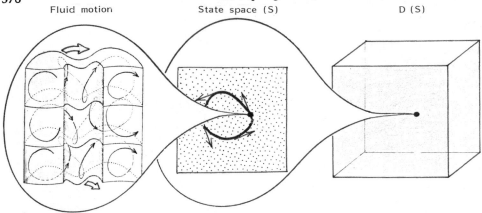

Fluid motion   State space (S)   D (S)

FIGURE 58. THE DYNAMICAL MODEL FOR TURBULENCE. To model this experimental situation with a dynamical system, the first step is the definition of the state space. One way to do this (due to Lagrange) is to record the fluid velocity vector at each point within the cup. This velocity vectorfield is considered to be the (instantaneous) state of the fluid motion and corresponds to a single point in the state space, S. This space, unfortunately, is infinite-dimensional. The change in the state of fluid motion as time passes is modeled by the partial-differential equations of Navier–Stokes. These may be considered (roughly) as a dynamical system on the state space, S. Thus, the Navier–Stokes equations are considered to be a system of ordinary differential equations on an infinite-dimensional state space and, therefore, a single point in the big picture, $D(S)$, instead of a system of partial differential equations on a three-dimensional fluid domain.

The behaviors of the fluid which are actually observed—Taylor vortices, for example—are attractors of this dynamical system on S. Furthermore, the rotation of the stirring rod is part of the Navier–Stokes system (the "boundary conditions"); thus, changes in the speed of rotation, c, move the dynamic $f(c)$ around in the big picture, $D(S)$. There are many technical problems involved in this model, which is still an active research area. Nevertheless, this model, studied by Ehrenfest and extensively used by Ruelle, makes the results of Dynamical Bifurcation theory applicable to the Couette–Taylor experiments and to fluid dynamics in general. The chaotic attractor in this dynamical model (from Ruelle and Takens, 1971) corresponds to turbulence in the fluid.

FIGURE 59. LAMELLAR FLOW in concentric cylinders occurs at slow stirring rates. Here we begin a sequence illustrating the successive events observed as the speed of stirring is gradually increased from zero. Initially the fluid is still. This condition is illustrated by the column on the left. The photo at top left shows the cylinder of fluid at rest, seen from the front. The drawing at left center shows the state as a velocity vectorfield; all velocity vectors are zero. The drawing at lower left shows the dynamical model: a point attractor at the origin (all velocity vectors zero) of the state space, S.

Next, in the column on the right, the stirring rate has increased slightly, and the flow is slow and lamellar. The photo of the fluid (top right) seen from the front is unchanged. But the velocity vectorfield (right center) is a pattern of horizontal, concentric circles, slowly rotating clockwise (as seen from above). The dynamical model (lower right) is again a point attractor in the infinite-dimensional state space, S, but it is no longer at the origin. The attractive point has drifted from the origin to a nearby point, corresponding to the lamellar flow, as the control (stirring speed) has been changed. This drift is indicated by the curve between the two lower drawings. There has been no bifurcation yet. (We are grateful to Rob Shaw for providing these photos of his work with Russ Donnelly's Couette machine at the University of Oregon.)

29. Dynamics

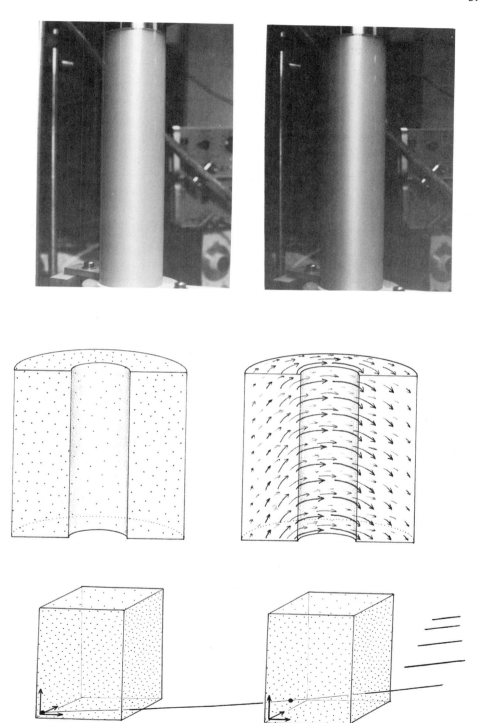

IX. Topological Representation of Self-Organization

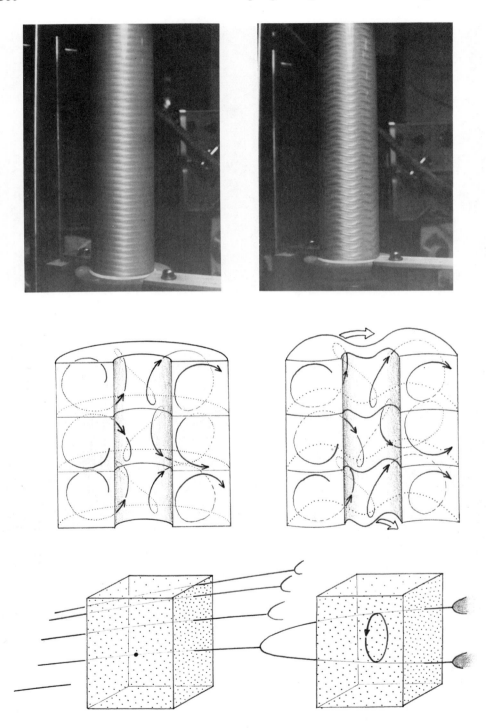

FIGURE 60. TAYLOR CELLS appear after additional increase in the rate of stirring. The column on the left illustrates the state of the fluid after the appearance of the annular vortices: photo of the fluid from the front; velocity vectorfield; and dynamical model (from top to bottom.) Even now the state of the fluid is still a rest point, because the velocity vectorfield is constant with respect to time, although it varies from point to point. It is possible that there has been no bifurcation yet in the dynamical model, because the attractor is still a point. Experimentalists, however, report that hysteresis is observed in the formation of the cells; the critical value of the speed of stirring for formation of the Taylor cells is higher than the critical value for their disappearance. Because hysteresis is the certain sign of catastrophic bifurcation, we suppose a kink (a static annihilation linked to a static creation bifurcation) in the track of the point attractor, from the lower right drawing of the preceding state (Figure 59) to the lower left here. Thus, we are creating a bifurcation portrait from the lower drawings of each column.

The bifurcation indicated by the kink between the preceding figure and this one is an outstanding example of the emergence of form through the *breaking of symmetry*. The lamellar flow has the symmetry of the vertical axis of the stirring rod, a one-dimensional symmetry. The stack of Taylor vortices has the symmetry of a discrete set of points within that axis, a zero-dimensional symmetry. One dimension of symmetry is lost when the attractor leaps from the lower branch of the kink to the upper branch.

The column on the right shows a new phenomenon, wavy vortices. Following increase in the stirring rate, the Taylor cells show waves which seem to rotate within the fluid (in the photo at top right). The velocity vectorfield (right center) now shows a periodic variation at each point within the fluid. The dynamical model (lower right) is, therefore, a periodic attractor. Between the two columns a Hopf bifurcation (Figure 50) has subtly taken place.

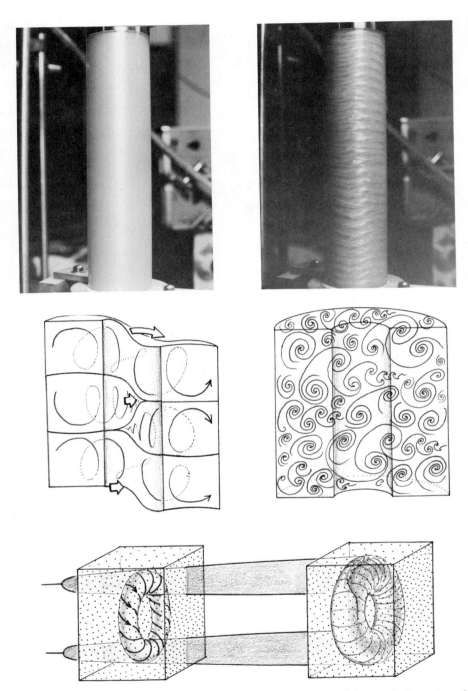

FIGURE 61. THE ONSET OF CHAOS follows additional increase of the speed of rotation of the central cylinder. After the Hopf bifurcation to wavy vortices, another bifurcation (observed by R. Shaw) produces pairs of dislocations of the annular cells. On the left this effect is shown as a braid bifurcation, which is pure speculation on our part, although experiments indicate the excitation of a new oscillatory mode. On the right, weakly turbulent fluid motion is shown where short bits of dislocated cells move chaotically about.

Bifurcations with Two Controls

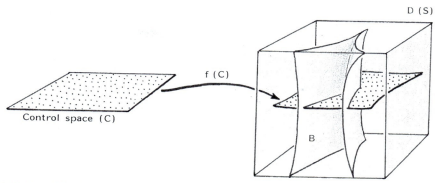

FIGURE 62. A DYNAMICAL SYSTEM WITH TWO CONTROLS. The knowledge we have gained of "generic arcs" in the big picture is small, but in some applications can be very powerful. Most applications, however, have more than one control, and our understanding of "generic disks" in the big picture—dynamical systems having multiple controls—has hardly begun. Here is an illustration of a generic two-dimensional disk in the big picture, corresponding to the *cusp catastrophe* of Elementary Catastrophe theory.

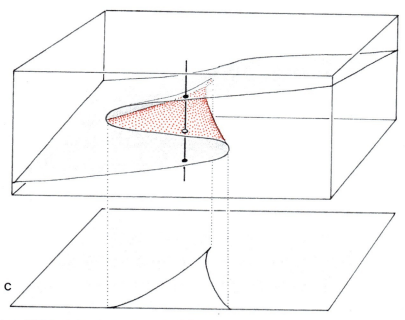

FIGURE 63. BIFURCATION DIAGRAM OF THE STATIC CUSP CATASTROPHE. In this example the state space is one-dimensional (vertical), and the control plane has two dimensions (horizontal); thus, the bifurcation diagram is actually three-dimensional, as shown here. The bifurcation set, B^f in C, is a cusp. For control values within this cusp, the one-dimensional AB portrait has two point attractors (black), separated by a point repellor (red). For control values outside the cusped bifurcation set there is only one attractor. There is another version of this event, the *periodic cusp catastrophe*, in which a periodic attractor bifurcates into two periodic attractors and a saddle cycle (see Figure 7). The two basins are separated by the periodic inset of this saddle cycle (see Figure 13). This periodic bifurcation has been found by Zeeman (1977) in a study of the forced Duffing equation.

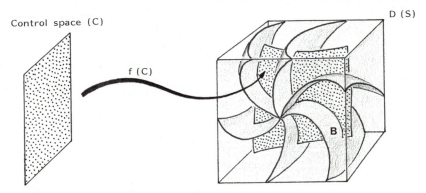

FIGURE 64. STARS IN THE BIG PICTURE also have been discovered. Here is a generic two-dimensional disk, stably intersecting a star of hypersurfaces of codimension one. This portrait corresponds to one of the few studied to date, the Andronov–Takens $(2,-)$ bifurcation. It requires a state space of dimension two or more (Takens, 1973).

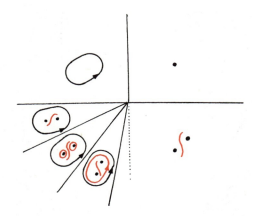

FIGURE 65. THE ANDRONOV–TAKENS BIFURCATION TABLEAU. With two controls and two state variables, the bifurcation diagram must be four-dimensional. Because direct visualization of this diagram is not possible, a simplified scheme has evolved, called the *method of tableaux*. Here we show the tableau for the Andronov–Takens bifurcation. The bifurcation set, B^f in C, consists of six rays radiating from a point. A generic arc bifurcation takes place across each ray. In each wedge the AB portrait is drawn, as if it lived in the control plane (which it does not). If one could steer the two control knobs very well, one could force the control point to pass through the vertex of the bifurcation set. For example, the point attractor (top) could be exploded into a periodic attractor and two attractive points (bottom). This event would never occur in a generic arc bifurcation or in an empirical device.

Complex Systems

Dynamics and studies of self-organization are two fields lying between mathematics and the sciences. They have been growing simultaneously, interacting and enriching each other. The concept of morphogenetic field and the big picture of Thom are products of this interaction, but the models provided so far by dynamics may be too simple to support the ideas now emerging in self-organization theory. We now propose some extensions of the concepts of dynamics that may prove useful in modeling hierarchical and complex systems.

29. Dynamics

Self-Regulation and Guidance Systems

FIGURE 66. SELF-REGULATION OF ONE CONTROL. Given that we have a dynamical system depending on a control parameter, who turns the control knob? If it turns itself, through action of a dynamical subsystem on the control space itself, the model is called a self-regulating system. Here a typical dynamical system with one control, producing a Hopf bifurcation, is shown with a generic self-regulation vectorfield on the one-dimensional control space. Because this AB portrait on the line has only two point attractors, their basins separated by a point repellor, the control knob will seek one of these two attractive settings.

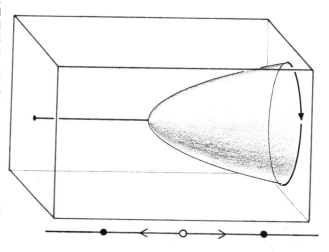

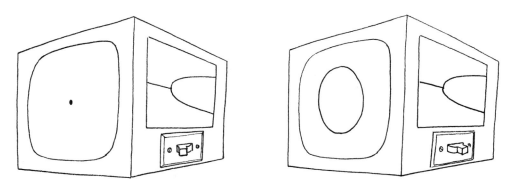

FIGURE 67. A MACHINE REPRESENTATION of the self-regulation system described above is this "electric stopwatch."

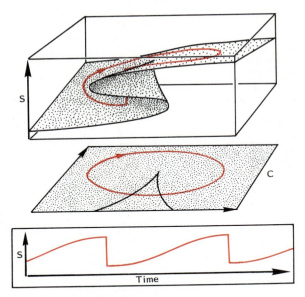

FIGURE 68. SELF-REGULATION OF TWO CONTROLS. Here a simple dynamical system with two-dimensional control space, the cusp of Figures 62 and 63, is endowed with a self-regulation vectorfield. This particular system has a single periodic attractor in the control space, C, with a virtual separatrix (a point repellor) within it. We always assume in these contexts that *the separatrices of the self-regulation dynamic are in "general position" with respect to the bifurcation set of the controlled dynamic*, in the sense that their intersection is as small as possible. The behavior of this system, introduced by Zeeman (1977) in a study of the electrical activity of neurons, is shown in the sawtooth-like time series recording the motion in the one-dimensional state space, S.

29. Dynamics

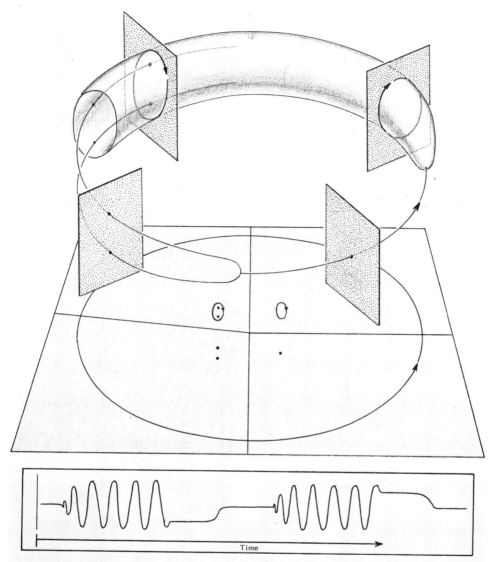

FIGURE 69. SELF-REGULATION WITH PERIODIC BEHAVIOR. The preceding example had a periodic attractor for its controls, but only a point attractor for its state dynamic. We show here a richer example, with the same control space and self-regulation dynamic, but with a state space of two dimensions. The bifurcation diagram is that of Andronov and Takens, shown in tableau form in Figures 64 and 65. Here we show the actual bifurcation diagram, obtained when the controls are restricted to the values along the periodic attractor on the control space. The time series corresponding to one of the two state variables also is shown. It resembles that of a stringed instrument rhythmically plucked. The direction in which it is plucked varies from cycle to cycle.

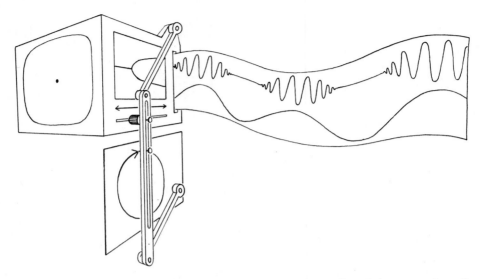

FIGURE 70. GUIDANCE SYSTEMS FOR ONE CONTROL. The self-regulation system is too limited for many applications, as this example shows. In this figure a dynamical system with one knob is controlled by a regulating vectorfield on an auxiliary space of two dimensions. A smoothly periodic variation of the control knob therefore is possible. In this example the controlled dynamic is again a Hopf bifurcation diagram, and the output time series of one of the two state variables is an intermittent oscillation. A similar system, described by Crutchfield, emits intermittent noise.

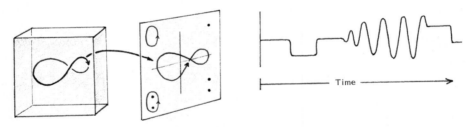

FIGURE 71. GUIDANCE SYSTEMS FOR MORE CONTROLS. By guidance system we mean a regulation vectorfield on an auxiliary space, B, and a *linkage map* from B to C, the control space of a dynamical system with controls, on a state space, S. Here is an example in which B, C, and S all have dimension two. The regulation dynamic on B has a periodic attractor, but the projection of this cycle from B into C by the linkage map is a loop that crosses itself. We shall call the projection of an attractor, in general, a *macron*. The controlled dynamic illustrated here is again the Andronov–Takens system of Figures 64, 65, and 69; the output is the time series shown. This is a periodic sequence of square waves and intermittent oscillation. Many other possibilities can be constructed with a guidance system as simple as this one.

29. Dynamics

Serial Coupling and Hierarchies

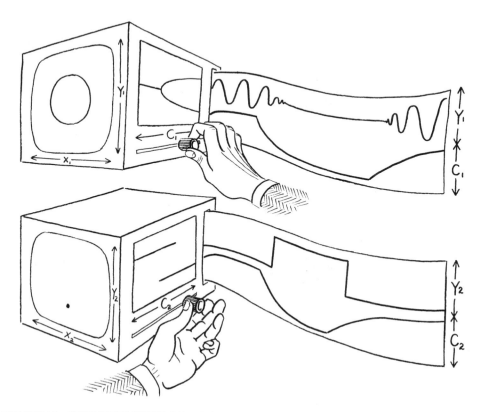

FIGURE 72. GUIDANCE WITH CONTROLS. In our sequence of portraits of gradually increasing complexity, we now consider two dynamical systems with controls. For example, let one be a subtle Hopf bifurcation (Figures 48–50) and the other a catastrophic kink exhibiting hysteresis (Figure 46, 47).

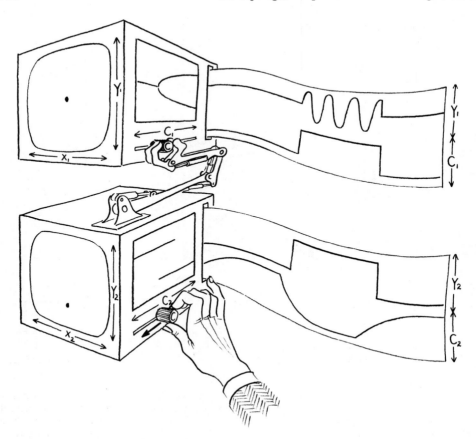

FIGURE 73. SERIAL COUPLING of these two systems means that a linking map is given from the state space of one, S^1, to the control space of the other, C^2. Thus, for a fixed value of the first control knob, the fixed dynamic on S^1 is a guidance system for the second system. But moving the first control knob changes the guidance dynamic. In the example illustrated here, moving the lower control knob full right, then full left, produces the time series shown for the guided system above. The lower control turns the oscillation abruptly on and off, with hysteresis.

29. Dynamics

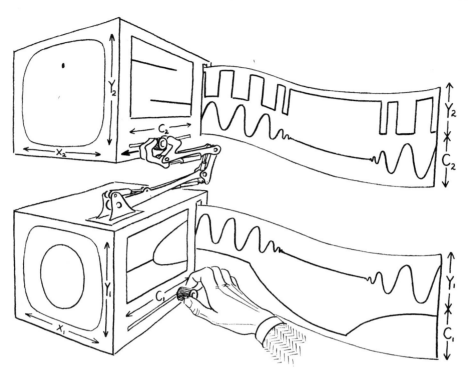

FIGURE 74. ANOTHER EXAMPLE is obtained by interchanging the roles of the two systems previously illustrated. Here the Hopf bifurcation controls the hysterical kink system. The lower control turns the square wave oscillator abruptly on and off, without hysteresis. (This type of system could be used, for example, to model the cAMP pulse relay activity of cellular slime mold—see Chapter 10.)

FIGURE 75. SYMBOLS for controlled dynamical systems may be made schematically by representing the control space as a horizontal line segment and the state space as a vertical line segment, regardless of the actual dimensions of these spaces. A single dynamical system with controls then is represented by a box with a line beneath it.

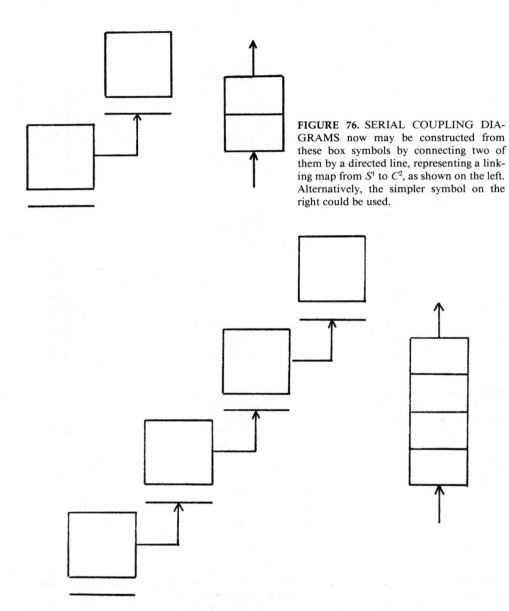

FIGURE 76. SERIAL COUPLING DIAGRAMS now may be constructed from these box symbols by connecting two of them by a directed line, representing a linking map from S^1 to C^2, as shown on the left. Alternatively, the simpler symbol on the right could be used.

FIGURE 77. HIERARCHICAL SYSTEMS may be modeled by a sequence of serially coupled dynamical systems with controls, as shown here in two equivalent symbols.

29. Dynamics

Parallel Coupling, Complex Systems, and Networks

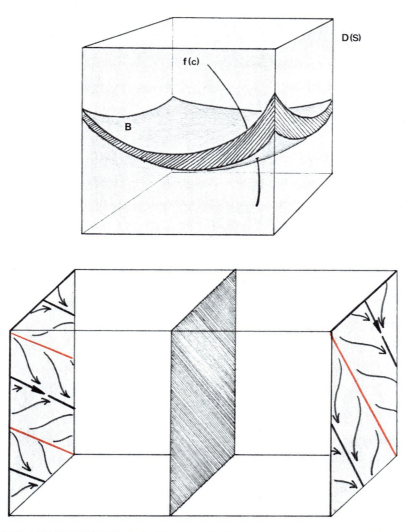

FIGURE 78. AN UNFOLDING of a bad (not structurally stable) dynamical system, X, in the bad set, B, denotes an embedding of X into a generic, controlled dynamical system, f, with control space, C; the result is that $f(C)$ in the big picture, $D(S)$, passes through B at X. This behavior requires that the dimension of C be not less than the codimension of B at X. The diagrams of catastrophic and subtle bifurcations (Figures 38–44 and 48–51) all represent unfoldings of a bad dynamical system. Figure 36 represents the fluctuating braid bifurcation, obtained by generically coupling two oscillators, as an unfolding of the bad dynamic for the uncoupled oscillators. We may consider the unfolding of the uncoupled oscillators as a sort of "dynamical multiplication" of them.

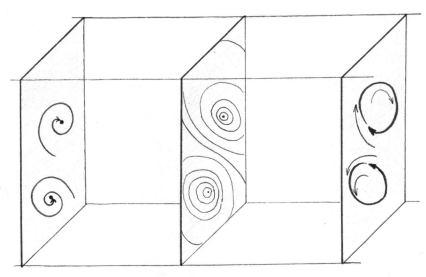

FIGURE 79. THE PARALLEL COUPLING OF DYNAMICAL SYSTEMS generalizes this dynamical multiplication to any dynamical system. Let X^1 be a dynamical system on a state space S^1, and let X^2 be another dynamic, on another state space, S^2. The product system, $X^1 \times X^2$, on the product space, $S^1 \times S^2$, probably will be bad. An unfolding of this product dynamic is called a parallel (or flexible) coupling of X^1 and X^2.

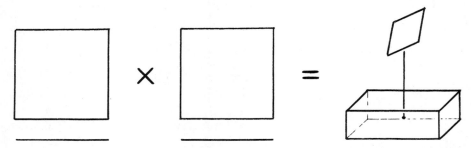

FIGURE 80. THE PARALLEL COUPLING OF CONTROLLED DYNAMICAL SYSTEMS adds a new control, C^0, for unfolding, to the controls C^1, and C^2, of the two controlled systems. The addition permits complete unfolding of the product of the interactions, as shown here.

29. Dynamics

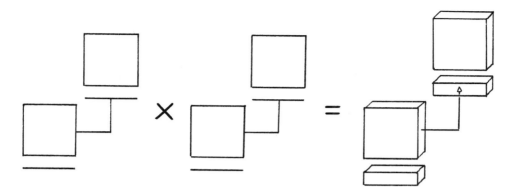

FIGURE 81. THE PARALLEL COUPLING OF GUIDANCE SYSTEMS repeats the coupling scheme for controlled systems, illustrated in the preceding figure, on both levels. To simplify the picture, we suppose that one new control, C^0, simultaneously unfolds both the guidance and the guided levels.

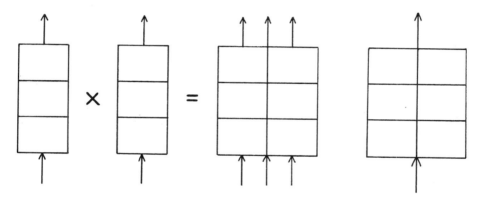

FIGURE 82. THE PARALLEL COUPLING OF HIERARCHICAL SYSTEMS (of the same number of levels) repeats the coupling scheme for guidance systems, illustrated in the preceding figure, on all levels of the hierarchy. (See Figure 75 for an explanation of the box symbol.) Again, one new set of controls, C^0, is assumed to control the flexible couplings on each level simultaneously and to unfold all of the couplings completely.

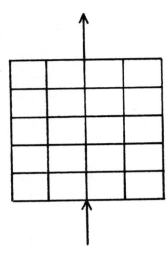

FIGURE 83. A COMPLEX SYSTEM may be made up of a set of hierarchical systems of the same length by assuming a parallel coupling between each pair of hierarchical systems in the set. The unfolding controls again are lumped into a single new control space, C^0. These controls are assumed to unfold all the parallel couplings at once.

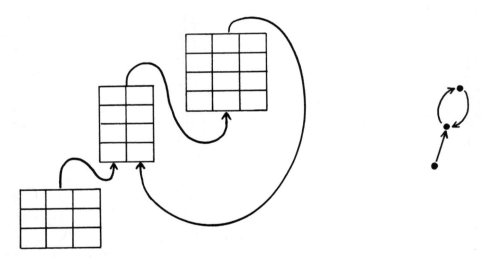

FIGURE 84. NETWORKS of complex systems are formed by directing the outputs of some to the controls of others. Here are two symbolic representations of a (simple) dynamical network. Chapter 30 relates the images and ideas of this chapter to the problem of self-organization.

ACKNOWLEDGMENTS. Some of the ideas presented here have been sharpened through discussion with colleagues. In particular, it is a pleasure to thank Alan Garfinkel for his emphasis on entrainment, Christopher Zeeman for his ideas on Cantor sets, and Rob Shaw for his description and photos of annular vortices in fluids.

References

Feigenbaum, M. (1978) Quantitative universality for a class of nonlinear transformations. *J. Stat. Phys.* **19**:25–52.

Herman, M. R. (1979) Sur la Conjugaison Différentiable des Difféomorphismes du Cerole a des Rotations. *Publ. Math. IHES* **49**.

Lefshetz, S. (1950) *Contribution to the Theory of Nonlinear Oscillations.* Princeton University Press, Princeton, N.J.

Lorenz, E. (1963) Deterministic nonperiodic flow. *J. Atmos. Sci.* **20**:130–141.

Lorenz, E. (1980) Noisy periodicity and reverse bifurcation. *Ann. N.Y. Acad. Sci.* **357**:282–291.

Neimark, Y. I. (1959) On some cases of dependence of periodic motions on parameters. *Dokl. Akad. Nauk SSSR* **129**:736–739.

Peixoto, M. (1962) Structural stability on two-dimensional manifolds. *Topology* **2**:101–121.

Poincaré, H. (1885) Sur l'équilibre d'une masse fluide animée d'un mouvement de rotation. *Acta Math.* **7**:259–380.

Pugh, C. C., and M. Shub (1980) Differentiability and continuity of invariant manifolds. Preprint, Queens College, Flushing, N.Y.

Rössler, O. E. (1979) Chaos. In: *Structural Stability in Physics*, W. Güttinger and E. Eikemeyer (eds.). Springer-Verlag, Berlin, pp. 290–309.

Ruelle, D., and F. Takens (1971) On the nature of turbulence. *Commun. Math. Phys.* **20**:167–192.

Shaw, R. (1981) Strange attractors, chaotic behavior, and information flow. *Z. Naturforsch.* 36a:80–112.

Sotomayor, J. (1974) Generic one-parameter families of vector fields in two-dimensional manifolds. *Publ. Math. I. H. E. S.* **43**:5–46.

Takens, F. (1973) Introduction to global analysis. *Commun. Math. Inst., Rijksuniversiteit Utrecht* **2**:1–111.

Thom, R. (1972) *Stabilité structurelle et morphogénèse: Essai d'une théorie générale des modeles.* Benjamin, New York.

Whitehead, A. N. (1925) *Science and the Modern World.* Macmillan Co., New York.

Zeeman, C. (1976) The classification of elementary catastrophes of codimension < 5. In: *Structural Stability, the Theory of Catastrophes, and Applications in the Sciences*, P. Hilton (ed.). Springer-Verlag, Berlin, pp. 263–327.

Zeeman, E. C. (1977) *Catastrophe Theory: Selected Papers.* Addison-Wesley, Reading, Mass.

30

Dynamics and Self-Organization

Ralph H. Abraham

ABSTRACT

Modern views of qualitative dynamics seem to promise simple geometric models of complex behavior and a mathematical rationale for constraint of evolutionary processes to particular paths. Dynamical bifurcation theory should not be confused with elementary catastrophe theory or with classical bifurcation theory. A short history of dynamics is given. A new category of dynamical models for complex systems is proposed, based on networks of serially coupled dynamical systems. These models may potentially be extended to account for irreversibility, fluctuation, coherence, symmetry-breaking, complementarity, and other phenomena of self-organization. Finally, ten outstanding problems for dynamics that are central to the development of self-organization theory are described. —THE EDITOR

There are various, distinct mathematical viewpoints applicable to the description of self-organizing systems. Here my concern will be the assessment of (and speculation on) just one of these: the viewpoint of *dynamics*. This subject lies between mathematics and the sciences, and has been central to increasing hopes for a rigorous mathematical theory of morphogenesis. Optimism was generated by the visionaries—Turing and Thom—and justified by successful and spectacular applications to physics, especially to hydrodynamics. Extrapolating from these applications, dynamics seems to promise:

- Simple geometric models for complex and chaotic behavior
- A complete taxonomy of dynamic states or their attractors, and of developmental events (bifurcations) for morphogenetic sequences
- A mathematical rationale for the constraint of complex self-organizing sys-

RALPH H. ABRAHAM • Division of Natural Sciences, University of California, Santa Cruz, California 95064.

tems to a simple "homeorhesis" (in Waddington's terminology), or evolutionary process

The overpowering allure of dynamics for self-organization theory is the promise of a mechanical explanation for morphogenesis, a kind of rebirth of Greek rationalism, or scientific materialism. In fact, simultaneous with the emergence of dynamics as a field, dynamicism is emerging as a prospective scientific cosmology, a philosophical new wave.

We must be careful not to confuse *dynamical bifurcation theory*, which is the basis of this essay, with the rather similar subjects of *elementary catastrophe theory* (see, e.g., Zeeman, 1976, or Poston and Stewart, 1978), or *classical bifurcation theory* (surveyed in Marsden, 1978). The bifurcations of gradient dynamical systems, including elementary catastrophes, are less general than the dynamical bifurcations of dynamical systems, and these are in turn less general than the classical bifurcations of nonlinear operators and partial differential equations. All three fields use the same vocabulary in different ways—a circumstance that can lead to confusion. Furthermore, all three viewpoints may be applied to the same partial differential equations!

Dynamics has three aspects: mathematical; experimental; and applied. The history of experimental dynamics divides naturally into the subcategories of

- Real machines (since Galileo, 1600)
- Analog machines (since Bush, 1931)
- Digital machines (since Lorenz, 1963)

These are described in Abraham and Marsden (1978) and shown in Table 1. Before the widespread use of analog and digital computing machines in dynamics (i.e., before 1960), it was difficult to distinguish between experimental and applied dynamics, because real machines usually are not sufficiently tractable to permit serious exploration of their detailed dynamics. In fact, the capability of a machine to admit preparation in an arbitrary initial state effectively *defines* it as an analog computer. The evolution of this capability for nonlinear electrical oscillation ushered in the purely experimental period (Hayashi, 1953, 1964).

By a *real machine* we mean any system of the phenomenal universe that behaves sufficiently like a dynamical system, with states that may be related to an idealized state space, evolution from an initial state along a trajectory to a final motion, dynamics that can be changed by control "knobs," and so on. The real machines that have engaged dynamicists most seriously involve fluids, gases, elastic solids, chemical reactions, neurons, and so on.

In this chapter, I discuss the potential of applied dynamics for modeling self-organizing systems. First, I propose some dynamical models that in the future might be useful to describe self-organizing systems. Then I present some unsolved problems of dynamical systems theory that are important for the program suggested in the first part. [All of the concepts of dynamics essential for this discussion are presented in an elementary, intuitive way in Chapter 29, and at greater length in Abraham and Shaw (1983, 1984, 1985), in a similar style.]

TABLE 1. The History of Dynamics

Date	Mathematical dynamics	Digital	Analog	Real	Applied dynamics
1600					Galileo
					Kepler
	Newton				
	Leibnitz				
1700					
	Lagrange			Chladni	
1800					
				Faraday	
			Thompson		
	Lie			Rayleigh	
	Poincaré				
1900	Lyapunov				Ehrenfest
	Julia				Duffing
				Van der Pol	
	Birkhoff		Bush		
	Hopf		Philbrick		
	Peixoto		Hayashi		Turing
	Smale	Lorenz			Thom
			Rössler	Gollub/Swinney	Ruelle
2000					

Models for Self-Organization Suggested by Dynamics

I propose here a concordance between the concepts of dynamics and those of self-organization. I do not expect this to be a convincing case, because the worked examples necessary for proof would require an effort such as the 5 years Zeeman (1977) devoted to documentation of the Thom (1972) case for Elementary Catastrophe Theory models for morphogenesis.

Experimental Dynamics: A Canonical Example

Although hydrodynamics is a classical subject, its firm connection to dynamics began only recently, with Arnol'd (1966). Ruelle has traced the roots of the idea back to Ehrenfest's thesis, a century ago.

Three hydrodynamical machines have been very important in experimental dynamics: the Couette–Taylor stirring machine; the Rayleigh–Bénard simmering machine; and the Chladni–Faraday vibrating machine. Detailed descriptions and bibliographies may be found in the books cited in the References and in Abraham (1976b) and in Fenstermacher *et al.* (1979). The behavior of all three machines show great similarities, so here I shall consider explicitly only the Couette device, in which two concentric cylinders are mounted on a common vertical axis. Through the transparent outer cylinder, we observe a fluid contained between the two cylinders. The inner cylinder may be rotated by a motor, with the speed parameter controlled by the experimentalist. If we naively assume that there is an abstract dynamical system with a finite-dimensional state space that we can visualize as a plane, then exploration of the machine and direct observation of its attractors yields a partial map of the bifurcation diagram.

Initially the system has an equilibrium (a point attractor) corresponding to the rotation of concentric cylindrical lamellae that show a one-dimensional symmetry group of vertical translations, and another of cylindrical symmetry.

As the speed parameter is increased (visualized as moving to the right), the equilibrium ends, and the state leaps "catastrophically" to a new point attractor, corresponding to spiraling motion of discrete annular rings of fluid, the "Taylor cells." Taylor cells have a zero-dimensional, or discrete, symmetry group of vertical translations. This catastrophe is a prototypical symmetry-breaking bifurcation. Further extension of the control parameter (speed of rotation of the inner boundary cylinder) produces a new excitation in a sequence leading to turbulence.

Several specific sequences leading to turbulence have been suggested, and there are several proposals for a deterministic scheme for describing turbulence, based on finite-dimensional, dynamical bifurcation theory. This development, originating with Lorenz (1963) and Ruelle and Takens (1971), is of great interest to physics. The link between bifurcation diagrams and the classical partial differential equations of hydrodynamics (the Navier–Stokes equations) has been elaborated in considerable detail, especially in Marsden and McCracken (1976), Marsden (1977), Ratiu (1977), Smale (1977), Bowen (1977), Abraham and Marsden (1978), Ruelle (1980), Pugh and Shub (1980), and Rand (1980).

I shall describe briefly three problems of this treatment of the Couette machine, not only because of its role as an important example of morphogenesis itself, but because similar treatment of more general partial differential equations may be expected in the near future, especially equations of the "reaction–diffusion" type.

The first problem is technically very severe: many kinds of partial differential equations on a finite-dimensional domain may be viewed easily as dynamical systems on an infinite-dimensional space. But they turn out to be *rough dynamical systems* (described later) instead of the smooth sort to which dynamical bifurcation theory easily applies. Specialists have taken two approaches to this obstacle: the easy way around consists of reducing the domain to finite dimensions by some special tricks (thereby losing credibility); the hard way consists of re-proving the necessary results of dynamical bifurcation theory in the rough context (losing readability). Both approaches have been pursued successfully (Bernard and Ratiu, 1977).

The second problem is equally severe: because the experimental equipment (e.g., a Couette–Taylor machine) is a real machine, *it cannot be prepared in an arbitrary initial state* (i.e., arbitrary instantaneous fluid motion). In fact, the fluid is always at rest between the cylinders (the zero point of the infinite-dimensional state space) when the experiment begins. Therefore, only a small section of the global bifurcation diagram may be discovered by experimental exploration. But the model, the global bifurcation diagram of the Navier–Stokes equations of fluid dynamics, could be mapped completely by digital simulation, if a large-enough computer were built. The techniques for this project comprise an entire field of numerical analysis.

The third problem seems potentially more tractable, although not yet solved. This is the *problem of observation*. Indeed, we may distinguish two problems of observation: *ignorance* and *error*. By "ignorance" I mean that we look at an infinite-dimensional state, but can record only a small number of parameters. Even if we measure without error, the data describe only a point in a finite-dimensional space. Thus, the observation procedure, at best, defines a projection map from the infinite-dimensional state space, S, to a record space, R, of finite dimension. I shall refer to this as the *output projection map*.

For example, in the recent work with the Couette machine described by Fenstermacher et al. (1979), observation is restricted to a single direction of fluid velocity at a single point in the fluid region, measured by laser–Doppler velocimetry. In other words, the dimension of R is one, an example of extreme ignorance. This difficulty is inescapable in any application of dynamical bifurcation theory to partial differential equations. Therefore, I have coined the word *macron* for the image of an attractor in S, projected into R (Abraham, 1976a). The question arises: can we recognize the attractor, having observed only its projected macron? We have not learned to recognize attractors yet, so this question is open.

By the problem of error I mean the problem of determining a point exactly, in a finite dimensional space. Errors arise either in the measurement of the coordinates, or in the storage of the data. This problem usually is handled by techniques of statistical physics or by information theory; Shaw (1981) presents a particularly interesting discussion of it.

In the future we hope for mathematical classifications of attractors and macrons that are: 1) experimentally identifiable, in spite of ignorance and error; 2) well-founded in the rough context of partial differential equations of evolution, viewed as dynamical systems on infinite-dimensional spaces; and 3) tractable in digital simulation. Substantial progress is being made, and I am assuming a satisfactory resolution of the above three problems in making the following prediction: *dynamical bifurcation theory will be useful for self-organization theory.*

Complex Dynamical Systems and Self-Organization

The Couette machine discussed above provides a canonical example of the role of dynamical bifurcation theory in modeling a simple self-organizing system. But to move to a more complicated example, such as slime mold aggregation (Chapter 10), would reveal quickly the inadequacy of our models based on a single dynamical system with controls. Rather, a continuum of dynamical systems with controls might be required, coupled together in a space-dependent way, with the continuum of controls manipulated by another such system! Therefore, I have proposed a richer class of models, called *complex dynamical systems* (Chapter 29). These new models combine dynamical systems that have controls into networks, by means of serial and parallel coupling. The full elaboration of the generic behavior and bifurcations of these complex systems—through theory and experiment—is still ahead. Yet I conjecture that their behavior will provide adequate metaphors and useful models for the homeorhesis of some self-organizing systems found in nature. For the present, I shall have to be satisfied with brief indications for the possible unification into this framework of some of the concepts described in the literature of self-organization theory.

The main ideas underlying this accommodation are that:

- Mathematical models for morphogenesis involve partial differential equations of the "reaction–diffusion" type (see the original pioneering papers of Rashevsky, 1940a,b,c, and Turing, 1952)
- The model equations may be viewed as "rough" dynamical systems on an infinite-dimensional state space (see Guckenheimer, 1980)
- These dynamical systems have controls, and may be combined into complex systems to model a given self-organizing system

For specific examples, see the articles by Carpenter (1976), Conley and Smoller (1976), Guckenheimer (1976), and Rinzel (1976); also Part III of Gurel and Rössler (1979); the master equation (Gardiner *et al.*, 1979) and the Fokker-Planck equation (Haken, 1977). The equations of elastodynamics have been treated in this way by Marsden and Hughes (1978) and Holmes and Marsden (1979). Applications of Elementary Catastrophe Theory in this area by Thompson and Hunt (1973), Zeeman (1976), and Poston and Stewart (1978) are relevant.

I now complete my case for the complex dynamical system scheme, by relating it to specific models emerging in self-organization theory.

My scheme is an extension of Thom's, and so his morphogenetic field is automatically accommodated. The proposal of Turing (1952), the dissipative structures of Prigogine (1978), and the synergetics of Haken (1977) all are based upon partial differential equations of the reaction–diffusion type; they all fit directly into this scheme. I suspect that the homeokinetics of Iberall and Soodak (Chapter 27) and the models of Winfree (1980) also fit.

The ideas of *irreversibility* and *fluctuation* emphasized by Prigogine are included in my scheme: irreversibility in the error problem (Shaw, 1981); and fluctuation in the fluctuating braid bifurcation, either in serial or parallel coupling (Chapter 29).

The ideas of *coherence* and *order parameter* emphasized by Haken (Chapter 21) are also natural in my complex scheme; coherence is an aspect of entrainment, and a special case of order parameter is known in dynamics under the name *slow manifold of an attracting point*. In fact, slow manifolds exist for a large class of attractors—those satisfying Axiom A (see Smale, 1967, 1971, or Irwin, 1980), and not just for rest points. Complementary *fast foliations* define a projection of the state space onto the slow manifold, so order parameters are included in this scheme as an output projection map. An outstanding pedagogic example of the slow manifold and fast foliation concepts is found in the heart and neuron model of Zeeman (1977, p. 81).

The concept of *symmetry-breaking bifurcation* fits the complex dynamical scheme well. The *hypercycle* idea of Eigen (Schuster and Sigmund, Chapter 5) has a dynamic interpretation as a limit cycle, and this has been generalized by Zeeman. But I would suggest a more extensive model for it, as a closed cycle of a complex dynamical system (Chapter 29, Figure 84).

Microcosmic/macrocosmic complementarity (see Prigogine, 1980, 1981) fits into my scheme as part of the problem of observation, as the output projection map of a model translates the microcosmic dynamics into macrocosmic observables. *But here I see the need for the further development of an extensive statistical and information theory of dynamical systems, along with near-Hamiltonian dynamics, to place thermodynamical laws on a firmer foundation* (see Shaw, 1981; Smale, 1980; Ruelle, 1980).

The *symbol/material complementarity* (Pattee, 1981; also Chapter 17) also may be discussed in the context of complex dynamical systems. An attractor functions as a symbol when it is viewed through an output projection map by a *slow observer*. If the dynamic along the attractor is too fast to be recorded by the slow-reading observer, he then may recognize the attractor only by its averaged attributes, fractal dimension, power spectrum, and so on, but fail to recognize the trajectory along the attractor as a deterministic system. See Chapter 29, Figure 69, for an example of such a material model for a cyclic symbol sequence with random replacements in a self-regulating dynamical system.

Tomović's idea (Chapter 20) of nonparametric control is accommodated naturally in my scheme. I rest my case here, for the present.

Critique of Dynamical Models

As explained above, dynamicism is encroaching on our scientific cosmology. Let us suppose that the problems mentioned were solved, the properties of complex dynamical systems worked out, and specific models for self-organizing systems at hand, satisfactory from the point of view of explanation and prediction. Even then we would know precious little about the mechanisms of self-organization, without a parallel development of the qualitative, geometric theory of partial differential equations. But this effort has hardly begun, although classical bifurcation theory is a beginning (see, e.g., Haken, 1977). I shall refer to this emerging branch of mathematics as *morphodynamics*.

Compare my model for the Couette machine (which claims that patterns will change only in certain ways) with the morphodynamic analysis of Haken (1977, and Chapter 21) that actually discovers the patterns! We may regard dynamics (i.e., dynamic bifurcation theory, together with all its extensions described above) as an intermediary step. The development of this theory has its own importance in mathematics. Numerous applications in the physical, biological, social, and information sciences await its maturity. Yet for evolution, self-organization theory, morphogenesis, and related scientific subjects, it seems to me that dynamics will play a preparatory role, paving the way for a fuller morphodynamics in the future.

The benefits of using dynamical concepts at the present stage of development of self-organization theory fall in two classes: permanent ones—the acquisition of concepts to be embedded in morphodynamics, guiding its development; and temporary ones—the practice of new patterns of thought. In the first category, I would place the attractors, the stable bifurcations, and their global bifurcation diagrams, as essential features of morphodynamics. These may be regarded as guidelines, exclusion rules, and topological restrictions on the full complexity of morphodynamic sequences. The temporary category would include the specific models in which dynamics is applied to rough systems on infinite-dimensional state spaces, in order to accommodate partial differential equations. This application is valuable, because the basic concepts of our scientific cosmology—noise, fluctuation, coherence, symmetry, explanation, mathematical models, determinism, causality—are challenged by dynamical models and essentially altered. Thus, the philosophical climate for the emergence of morphodynamics is created.

Elementary Catastrophe Theory has similarly provided some excellent examples of scientific explanation, permanent members in the applied mathematics Hall of Fame. Yet its most important function in the history of mathematics may turn out to be the *practice* it provides scientists in geometric and visual representation of dynamical concepts, paving the way for understanding of dynamics, bifurcations, and the rest.

In summary, dynamicism is without doubt an important intellectual trend, challenging the fundamental concepts of mathematics and the sciences. I see its importance for self-organizing system theory as temporary and preparatory for a more complete morphodynamics of the future. And yet dynamicism even now promises a permanent legacy of restrictions, a taxonomy of legal, universal restraints on morphogenetic processes—a Platonic idealism.

We must be careful not to cast aside dynamics, as Newton cast out wave theory, for not explaining forms. Whitehead (1925) wrote at the end of *Science and The Modern World*:

> ... a general danger [is] inherent in modern science. Its methodological procedure is exclusive and intolerant, and rightly so. It fixes attention on a definite group of abstractions, neglects everything else, and elicits every scrap of information and theory which is relevant to what it has retained. This method is triumphant, provided that the abstractions are judicious. But, however triumphant, the triumph is within limits. The neglect of these limits leads to disastrous oversights ... true rationalism must always transcend itself by recurrence to the concrete in search of inspiration.

Problems for Dynamics Suggested by Self-Organization

I now shall describe ten open problems for mathematical dynamics that are especially significant to the program proposed here for self-organization theory.

1. Taxonomy of Attractors

The observed "states" for a dynamical system are its attractors (Chapter 29). The need for a complete taxonomy of equivalence classes of generic chaotic attractors is well appreciated, and several attacks on this problem are under way. The meanings of "equivalent" and "generic" are still evolving. I think that a scheme based on the geometric features of experimentally discovered attractors will triumph eventually. These features include: the distribution of critical points and other particulars of shape (especially "ears"); the gross reinsertion maps (as described by Rössler, 1979); the fractal dimension; and the shuffling map of the Cantor section. Analytic features, such as the power spectra, entropy, and Lyapunov characteristic exponents determined by these attractors, may be important (but probably insufficient) for a complete classification.

2. Separatrices and AB Portraits

A dynamical system determines a *separatrix* in its domain that separates the domain into distinct open basins, each containing a unique attractor. The separatrix may contain vague attractors (Chapter 29, Figure 20) in vague basins (not open, yet of positive volume). These attractors and basins are of primary importance in applied dynamics. The decomposition of the domain, by the separatrix, into basins, and the location and identity of an attractor in each, comprises the *AB portrait* of the dynamical system. The *yin-yang problem* for AB portraits consists of finding a set of dynamical systems, the *good set*, which is large (yin) enough to be "generic" and small (yang) enough to be classifiable into "equivalence" classes of AB portraits.

As shown by the pioneering program of Smale (1967, 1970; see also Abraham and Marsden, 1978), an enormous simplification of this problem may be expected by taking into account topological restrictions on generic separatrices. Still, this problem is so difficult that it has been solved only for orientable surfaces (Peixoto, 1962). Even on the Möbius band it is still an open problem.

3. Local Bifurcations

In $D(S)$, the set of all smooth dynamical systems on a given state space S, the bad set B consists of the dynamical systems not structurally stable. The bad set is very bad (Chapter 29, Figure 37), and the *problem of local bifurcations* consists of discovering the full structure of B by the method of constructing local cross sections, called "generic bifurcations" or "full unfoldings." To Poincaré or Hopf, before the discovery of chaotic attractors (and exceptional limit sets), this problem looked mathematically tractable. Opinion changed after the discovery of the braid bifurcation by Sotomayor (1974). Now our hopes are based on experimental exploration. We can expect the discovery of most important bifurcations of equilibria and limit cycles, elucidation of new universal bifurcation sequences, and a growing insight into the pathways for the onset of chaos (Chapter 29).

4. Global Bifurcations

Dynamical bifurcation theory, up to the present, has been concerned mainly with the study of local bifurcations. When the control parameters are extended over large ranges of values, as they must be in most applications, we get *global bifurcation diagrams* (Chapter 29, Figure 56). The outstanding problem here is to find generic properties and topological restrictions for these diagrams. A historic first step was taken by Mallet-Paret and Yorke (1982). The invariants they discovered for "snakes" of limit cycles may be extended to families of tori and chaotic attractors. Even a partial resolution of this problem may have tremendous implications for future applications, such as sociodynamics and brain theory.

5. Symmetry-Breaking Bifurcations

The discovery of generic properties of dynamical systems with symmetry, and the classification of the symmetry-breaking bifurcations, has begun only recently. For the early results, see Schechter (1976), Golubitsky and Schaeffer (1978, 1979a,b,c, 1980), Buzano et al. (1982), and Fields (1980).

6. Near-Conservative Systems

We identify a *conservative* subset $C(S)$ in $D(S)$, the set of all smooth dynamical systems in state space S. This consists of the canonical Hamiltonian equations of classical mechanics, each determined by an energy function. Many famous equations of physics can be described as Hamiltonian systems on an infinite-dimensional state space (Abraham and Marsden, 1978): for example, the Schrödinger equation; Korteweg–de Vries equation; Euler equations of a perfect fluid; Lagrangian field theory; and Einstein equations of general relativity. (Precise details are given in texts on mechanics, e.g., Abraham and Marsden, 1978.) However, many important applications of dynamics address dynamical systems that are not in the set $C(S)$ of conservative Hamiltonian systems, but only near it. We call these the *near-Hamiltonian systems*. For a simple example, see Holmes (1980). The applications under consideration here, in fact, will involve near-Hamiltonian systems on an infinite-dimensional state space, S.

Dissipations (that appear at any small perturbation of the dynamic away from the energy-conserving form) validate the ergodic hypothesis. This postulate seems to be correct for near-Hamiltonian systems, but wrong for Hamiltonian systems, as explained by Smale (1980). Peixoto (1962) first publicized the problems of near-Hamiltonian systems. How can it be that near-Hamiltonian systems appear (approximately) to conserve energy? In other words, can the ergodic hypothesis and the law of conservation of energy both be derived from dynamics, or not? I am not going to hazard a prediction for the resolution of this problem, but I can imagine that it may be related to another hard one, called geometric quantization.

7. Infinite-Dimensional State Spaces

This problem consists of the extension of the preceding ideas to the context of partial differential equations of evolution, viewed as "rough" dynamical systems on infinite-dimensional state spaces. Two techniques have been described above: reduction to finite dimensions (to recover the topological exclusion principles of "snakes"); and direct mathematical assault on the properties of the rough system. It is at this point that we could imagine a renaissance of the art of applied mathematics.

8. Macroscopic Observation

The identification of a chaotic attractor in a real machine, in a simulation device, or anywhere else in the phenomenal universe—in the sense of matching an observation of "noise" to the taxonomy of attractors—is very restricted by the limitations of observation: ignorance (i.e., projection down to a few dimensions); and error, as described above. Thus, even if we someday recognize that an extraterrestrial radio source has generated a signal and not noise, will we be able to decode it as a sequence of symbols, i.e., as a bifurcation sequence of chaotic attractors? It seems likely that the classification of macrons (lower-dimensional projections of attractors) by dynamical topology will follow closely the classification of the attractors themselves. Some progress has been made (Froehling *et al.*, 1981; Frederickson *et al.*, 1983; Russell *et al.*, 1980; Takens, 1980).

The problem of errors seems unsolvable, limited as we are by the uncertainty principle. Thus, the theory behind the preceding problems needs to be recast in a *decidable scheme*, as described in Abraham (1979). An important step in this direction is found in Shaw (1981). What is needed here is a precise *theory of observation*: the relationship between the ideal mathematical model—the reversible, microscopic, dynamical system with its numerous dimensions and its allowable states, attractors—and the low-dimensional, irreversible, *macroscopic* world, with its observed macrons, finite-state memory media, and so on.

9. Parallel Coupling

The coupling of two dynamical systems is a perturbation of the Cartesian product of the two systems. Elsewhere (Abraham, 1976b) I have introduced a version of this, called *flexible coupling*, in which the perturbation depends upon control parameters. This version is obviously appropriate for many applications, and also for Thom's "big picture" concept of *unfolding* an unstable system into a stable family. We call the full unfolding of a coupled system by flexible coupling a *parallel coupling*, in a generalization of the entrainment of coupled oscillators (Chapter 29). To understand fully the possible results of such a coupling process, it seems sufficient to couple the attractors of one to the attractors of the other, in pairs. The bifurcation portraits obtained by coupling any two attractors may be viewed as a sort of *multiplication table* for attractors. The full determination of this table is the *parallel coupling problem*. For example, a point "times" a point is a point (no bifurcations); a point times a cycle is a cycle; but a cycle times a cycle is a *braid bifurcation*, as the unfolding of the coupled oscillators is complicated by entrainment (Chapter 29). This result is the first nontrivial example of the parallel coupling problem. Circle "times" chaos and chaos "times" chaos are open problems, perhaps important in forced oscillations, in the behavior of Langevin's equation (Haken, 1977), and in future applications in which noise becomes recognizable as signal. These instances may provide an understanding of the massive coupling and entrainment (coherence) phenomena of biological and social interaction.

10. Serial Coupling

Charting the behavior of complex dynamical systems is the newest problem of this list. Probably the first step will be experimental work, primarily to discover the basic properties of serially coupled hierarchical systems. The simplest case, two dynamical systems with controls, with a linkage map from the output projection of one to the controls of the other (Chapter 29, Figure 68), involves a "generic coupling" hypothesis for the linkage map, problems of observation on both levels, and new questions of entrainment between levels.

There are miles to go before we sleep....

ACKNOWLEDGMENTS. It is a pleasure to thank Eugene Yates, organizer, and the sponsors of the 1979 Dubrovnik Conference for their invitation; the National Science Foundation for support; everyone in the bibliography for sharing manuscripts and information; and Michael Arbib, Alan Garfinkel, Norman Packard, Stephen Smale, Joel Smoller, and Christopher Zeeman for extensively criticizing the first draft of this manuscript.

References

Abraham, R. H. (1976a) Vibrations and the realization of form. In: *Evolution and Consciousness: Human Systems in Transition.* E. Jantsch (ed.). Addison–Wesley, Reading, Mass., pp. 131–149.

Abraham, R. H. (1976b) Macroscopy of resonance. In: *Structural Stability, the Theory of Catastrophes, and Applications in the Sciences,* P. Hilton (ed.). Springer-Verlag, Berlin, pp. 1–9.

Abraham, R. H. (1979) Dynasim: Exploratory research in bifurcations using interactive computersim graphics. *Ann. N.Y. Acad. Sci.* **316**:676–684.

Abraham, R. H., and C. D. Shaw (1983) *Dynamics, the Geometry of Behavior,* Part 1. Aerial Press, Santa Cruz, CA.

Abraham, R. H., and C. D. Shaw (1984) *Dynamics, the Geometry of Behavior,* Part 2. Aerial Press, Santa Cruz, CA.

Abraham, R. H., and C. D. Shaw (1985) *Dynamics, the Geometry of Behavior,* Part 3. Aerial Press, Santa Cruz, CA.

Abraham, R. H., and J. E. Marsden (1978) *Foundations of Mechanics.* 2nd ed., Cummings, Menlo Park, Calif.

Arnol'd, V. I. (1966) Sur la géometrie différentielle des groupes de Lie de dimension infinie et ses applications a l'hydrodynamique des fluides parfaits. *Ann. Inst. Fourier Grenoble* **16(1)**:319–361.
Bernard, P., and T. Ratiu (eds.) (1977) *Turbulence Seminar*. Springer-Verlag, Berlin.
Bowen, R. (1977) A model for Couette flow data. In: *Turbulence Seminar*. P. Bernard and T. Ratiu (eds.). Springer-Verlag, Berlin, pp. 117–135.
Buzano, E., G. Geymonat, and T. Poston (1985) Post-buckling behavior of a non-linearly hyperelastic thin rod with invariant under the dihedral group D_n cross-section. *Arch. Rat. Mech.* **89**:307–388.
Carpenter, G. A. (1976) Nerve impulse equations. In: *Structural Stability, the Theory of Catastrophes, and Applications in the Sciences*. P. Hilton (ed.). Springer-Verlag, Berlin, pp. 58–76.
Conley, C. C., and J. Smoller (1976) Remarks on traveling wave solutions of non-linear diffusion equations. In: *Structural Stability, the Theory of Catastrophes, and Applications in the Sciences*. P. Hilton (ed.). Springer-Verlag, Berlin, pp. 77–89.
Fenstermacher, P. R., H. L. Swinney, S. V. Benson, and J. P. Gollub (1979) Bifurcations to periodic, quasi-periodic, and chaotic regimes in rotating and convecting fluids. *Ann. N.Y. Acad. Sci.* **316**:652–666.
Fields, M. J. (1980) Equivariant dynamical systems. *Am. Math. Soc. Trans.* **259**:185–205.
Frederickson, P., J. L. Kaplan, E. D. Yorke, and J. A. Yorke (1983) The Liapunov dimension of strange attractors. *J. Diff. Eq.* **49**:185–207.
Froehling, H., J. P. Crutchfield, J. D. Farmer, N. H. Packard, and R. Shaw (1981) On determining the dimension of chaotic flows. *Phys. D* **3**:605–617.
Gardiner, W., S. Chaturvedi, and D. F. Walls (1979) Master equations, stochastic differential equations, and bifurcations. *Ann. N.Y. Acad. Sci.* **316**:453–462.
Golubitsky, M., and D. Schaeffer (1978) Bifurcation near a double eigenvalue for a model chemical reactor. Math. Res. Center Report 1844.
Golubitsky, M., and D. Schaeffer (1979a) A theory for imperfect bifurcation via singularity theory. *Commun. Pure Appl. Math.* **32**:21–98.
Golubitsky, M., and D. Schaeffer (1979b) Imperfect bifurcation in the presence of symmetry. *Commun. Math. Phys.* **67**.
Golubitsky, M., and D. Schaeffer (1979c) Mode jumping in the buckling of a rectangular plate. *Commun. Math. Phys.* **69**.
Golubitsky, M., and D. Schaeffer (1982) Bifurcations with $O_{(3)}$ symmetry, including applications to the Bénard problem. *Comm. Pure Appl. Math.* **35**:81–111.
Guckenheimer, J. (1976) Constant velocity waves in oscillating chemical reactions. In: *Structural Stability, the Theory of Catastrophes, and Applications in the Sciences*. P. Hilton (ed.). Springer-Verlag, Berlin, pp. 99–103.
Guckenheimer, J. (1980) On quasiperiodic flow with three independent frequencies. Preprint, University of California, Santa Cruz.
Gurel, O., and O. E. Rössler (eds.) (1979) *Bifurcation Theory and Applications in Scientific Diciplines*. *Ann. N.Y. Acad. Sci.* **316**.
Haken, H. (1977) *Synergetics—An Introduction*. Springer-Verlag, Berlin.
Hayashi, C. (1953) *Forced Oscillation in Nonlinear Systems*. Nippon, Osaka.
Hayashi, C. (1964) *Nonlinear Oscillations in Physical Systems*. McGraw-Hill, New York.
Helleman, R. H. G. (ed.) (1980) *Nonlinear Dynamics*. *Ann. N.Y. Acad. Sci.* **357**.
Hilton, P. (ed.) (1976) *Structural Stability, the Theory of Catastrophes, and Applications in the Sciences*. Springer-Verlag, Berlin.
Holmes, P. J. (1980) Averaging and chaotic motion in forced oscillations. *SIAM J. Appl. Math.* **38**:65–80.
Holmes, P., and J. E. Marsden (1979) Qualitative techniques for bifurcation analysis of complex systems. *Ann. N.Y. Acad. Sci.* **316**:608–622.
Irwin, M. C. (1980) *Smooth Dynamical Systems*. Academic Press, New York.
Jantsch, E. (ed.) (1981) *The Evolutionary Vision*. Selected Symposia Series, American Association for the Advancement of Science, Washington, D.C.
Jantsch, E., and C. H. Waddington (eds,) (1976) *Evolution and Consciousness: Human Systems in Transition*. Addison–Wesley, Reading, Mass.
Lorenz, E. (1963) Deterministic nonperiodic flow. *J. Atmos. Sci.* **20**:130–141.

Mallet-Paret, J., and J. A. Yorke (1982) Snakes: Oriented families of periodic orbits, their sources, sinks, and continuation. *J. Diff. Eq.* **43**:419–450.

Marsden, J. E. (1977) Attempts to relate the Navier–Stokes equations to turbulence. In: *Turbulence Seminar.* P. Bernard and T. Ratiu (eds.). Springer-Verlag, Berlin, pp. 1–22.

Marsden, J. E. (1978) Qualitative methods in bifurcation theory. *Bull. Am. Math. Soc.* **64**:1125–1148.

Marsden, J. E., and T. J. R. Hughes (1978) Topics in the mathematical foundations of elasticity. In: *Nonlinear Analysis and Mechanics: Heriot-Watt Symposium,* Vol. II, R. J. Knops (ed.). Pitman, New York, pp. 30–285.

Marsden, J. E., and M. McCracken (1976) *The Hopf Bifurcation and Its Applications.* Springer-Verlag, Berlin.

Pattee, H. H. (1981) Symbol-structure complementarity in biological evolution. In: *The Evolutionary Vision.* E. Jantsch (ed.). American Association for the Advancement of Science, Washington, D.C., pp. 117–128.

Peixoto, M. (1962) Structural stability on two-dimensional manifolds. *Topology* **2**:101–121.

Poston, T., and I. N. Stewart (1978) *Catastrophe Theory and Its Applications.* Pitman, New York.

Prigogine, I. (1978) Time, structure, and fluctuations. *Science,* **201**:777–785.

Prigogine, I. (1980) *On Being and Becoming.* Freeman, San Francisco.

Prigogine, I. (1981) Time, irreversibility, and randomness. In: *The Evolutionary Vision,* E. Jantsch (ed.). American Association for the Advancement of Science, Washington, D.C., pp. 73–82.

Pugh, C. C., and M. Shub (1980) Differentiability and continuity of invariant manifolds. Preprint, Queens College, Flushing, N.Y.

Rand, D. (1980) The pre-turbulent transitions and flows of a viscous fluid between concentric rotating cylinders. Preprint, University of Warwick.

Rashevsky, S. (1940a) An approach to the mathematical biophysics of biological self-regulation and of cell polarity. *Bull. Math. Biophys.* **2**:15–25.

Rashevsky, S. (1940b) Further contribution to the theory of cell polarity and self-regulation. *Bull. Math. Biophys.* **2**:65–67.

Rashevsky, S. (1940c) Physicomathematical aspects of some problems of organic form. *Bull. Math. Biophys.* **2**:109–121.

Ratiu, T. S. (1977) Bifurcations, semiflows, and Navier–Stokes equations. In: *Turbulence Seminar.* P. Bernard and T. Ratiu (eds.). Springer-Verlag, Berlin, pp. 23–35.

Rinzel, J. (1976) Nerve signalling and spatial stability of wave trains. In: *Structural Stability, the Theory of Catastrophes, and Applications in the Sciences.* P. Hilton (ed.). Springer-Verlag, Berlin, pp. 127–142.

Rössler, O. E. (1979) Chaos. In: *Structural Stability in Physics.* W. Guttinger and H. Eikemeyer (eds.). Springer-Verlag, Berlin, pp. 290–309.

Ruelle, D. (1980) Measures describing a turbulent flow. *Ann. N.Y. Acad. Sci.* **357**:1–9.

Ruelle, D., and F. Takens (1971) On the nature of turbulence. *Commun. Math. Phys.* **20**:167–192; **23**:343–344.

Russell, D. A., J. D. Hanson, and E. Ott (1980) The dimension of strange attractors. Univ. Phys. Publ. 81–014.

Schechter, S. (1976) Bifurcations with symmetry. In: *The Hopf Bifurcation and Its Applications.* J. E. Marsden and M. McCracken (eds.). Springer-Verlag, Berlin, pp. 224–249.

Shaw, R. (1981) Strange attractors, chaotic behavior, and information flow. *Z. Naturforsch.* **36a**:80–112.

Smale, S. (1967) Differentiable dynamical systems. *Bull. Am. Math. Soc.* **73**:747–817.

Smale, S. (1970) Stability and genericity in dynamical systems. In: *Séminaire Bourbaki 1969–1970.* Springer-Verlag, Berlin, pp. 177–185.

Smale, S. (1977) Dynamical systems and turbulence. In: *Turbulence Seminar.* P. Bernard and T. Ratiu (eds.). Springer-Verlag, Berlin, pp. 48–70.

Smale, S. (1980) On the problem of reviving the ergodic hypothesis of Boltzmann and Birkhoff. *Ann. N.Y. Acad. Sci.* **357**:260–266.

Sotomayor, J. (1974) Generic one-parameter families of vector fields in two-dimensional manifolds. *Publ. Math. I.H.E.S.* **43**:5–46.

Takens, F. (1980) The dimension of chaotic attractors. Preprint.

Thom, R. (1972) *Stabilité structurelle et morphogénèse: Essai d'une théorie générale des modèles.* Benjamin, Menlo Park, Calif.

Thompson, J. M. T., and G. W. Hunt (1973) *A General Theory of Elastic Stability.* Wiley, New York.

Turing, A. (1952) The chemical basis of morphogenesis. *Philos. Trans. R. Soc. London. Ser. B.* **237**:37.

Whitehead, A. N. (1925) *Science and the Modern World.* Macmillan Co., New York.

Winfree, A. T. (1980) *The Geometry of Biological Time.* Springer-Verlag, Berlin.

Zeeman, C. (1976) The classification of elementary catastrophes of codimension ≤ 5. In: *Structural Stability, the Theory of Catastrophes, and Applications in the Sciences.* P. Hilton (ed.). Springer-Verlag, Berlin, pp. 263–327.

Zeeman, E. C. (1977) *Catastrophe Theory: Selected Papers.* Addison–Wesley, Reading, Mass.

Epilogue

Quantumstuff and Biostuff
A View of Patterns of Convergence in Contemporary Science [1]

F. Eugene Yates

ABSTRACT

The achievements of science in the 17th, 18th, and 19th centuries depended on a shift in emphasis from: (1) the general to the particular with the emergence of specialties; (2) pure rationality to empirical tests of causality; (3) man the actor within nature to the scientist as objective spectator. In the 20th century these trends have reversed. Specialties are giving way to convergence of methods and concepts. But the gulf between physics and biology remains wide: While physics continues to draw great strength from mathematics, biology works within the framework of natural language and remains intractable to useful mathematization. In this summary I examine the effectiveness of mathematics in natural science and consider what might be done to reduce biology to physics, in the light of earlier efforts by Bohr and Delbrück. Physical DNA, cell receptors, the development of an embryo, and evolution of life and the Cosmos are examined as four arenas for the possible convergence of biology and physics. I suggest that the gulf between these two sciences is actually that between an informational view and a dynamical view of form and process. Criteria for the characterization of any artificial, man-made system as "living" are examined, and Monod's behavioral (dynamic) criteria are contrasted with von Neumann's (informational) requirements for a self-reproducing automaton. A new thought-collective is now self-organizing and attracting physicists, topologists and biologists to the problem of the self-organization of complex systems. I find in that new collective the hope for ultimate convergence of the sciences.

In 1611 John Donne's poem "An Anatomie of the World" complained of the collapse of the established astrocosmological images handed down from Aristotle, Dante . . .

---

[1]Subtitle based on M. Gell-Mann's keynote address at the 1985 meeting of the American Association for the Advancement of Science, Los Angeles.

F. EUGENE YATES • Crump Institute for Medical Engineering, University of California, Los Angeles, California 90024.

> And new Philosophy calls all in doubt,
> The Element of fire is quite put out;
> The Sun is lost, and th' earth, and no mans wit
> Can well direct him where to looke for it.
> And freely men confesse that this world's spent,
> When in the Planets, and the Firmament
> They seeke so many new; they see that this
> Is crumbled out againe to his Atomies.
> 'Tis all in peeces, all cohaerance gone . . .

The dismantlers of "cohaerance" were Kepler, Copernicus, and Galileo. Newton's *Principia Mathematica Philosophiae Naturalis*, which completed the job, had not yet appeared. Stephen Toulmin, beginning with Donne's lament, has explored the trends of modern physical science during the 17th, 18th and 19th centuries and identified surprising reversals in the emergence of 20th century postmodern science (relativity and quantum mechanics). I have based this introduction on his chapter "All Cohaerance Gone" (1982, pp. 217–274).

In the 17th century the unity of the astronomical order (cosmos) and the human social order (polis) as envisioned by Plato in The Republic and later subsumed by the Stoics as "cosmopolis", gave way to the dualisms separating Mind from Matter, Rationality from Causality, and Humanity from Nature. Under these new conditions science flourished, chiefly because of two shifts: that from interest in the general to the particular; and that from a stance within nature to the posture of the objective spectator, outside the phenomenon being studied. Aristotle had proposed a general theory of motion and change "in all things", but it proved sterile when compared to results of empirical studies of the particular motions of planets, or vibrating strings, or of changes among selected chemical reactants or reproducing organisms. The experimental method and scientific specialties were born and the machine became the reductionistic, root metaphor underlying scientific progress (Pepper, 1942). Deterministic causal mechanism, as championed by Newton, Descartes and Laplace, was the ideal. These shifts proved so explosively productive that it is with some amazement that we regard the condition of postmodern physics today:

> The objective spectator is dead (Toulmin, 1982).
> Strict causal determinism has been abandoned (Pagels, 1982; Herbert, 1985).
> Empirical data cannot, even in principle, test all predictions of the latest theories (Lowenstein, 1985).
> Specialties are disappearing into each other (Gell-Mann, 1985).

Patterns of Convergence in Contemporary Science

The overlapping and merging of the specialties is a striking trend of contemporary science illuminated by Gell-Mann in his keynote address at the 1985 meeting of the American Association for the Advancement of Science (Gell-Mann, 1985). As cosmology merges with elementary particle physics and they both draw upon pure mathematics for their discoveries, we see also growing similarities in studies of biochemistry, biological ultrastructure, genetics, immunology, endocrinology, neurotransmission, growth, development, reproduction, repair, adaptation, aging, and cell–cell interactions and motions of all kinds. The older biological specialties have become quaint; scientists who would have formerly worked in Departments of Physiology, Bacteriology, or Anatomy have joined Institutes of Molecular Biology.

Nevertheless, there is more than one style of scientific inquiry and the different specialties have developed different ways of arranging and acquiring knowledge (Collins, 1975; Whitley, 1985). Surely no one can dispute that the sciences are organized differently with respect to both the relative importance of theory versus empirical fact and the manner in which they obtain financial resources for their work. Furthermore, the different branches of science study different things:

> All real features of the world can be studied directly on their own terms. They do not have to be approached indirectly, by finding their mirror images in a pattern studied by physicists... The scientific temper is one that looks for the appropriate method in each field, that carefully distinguishes different sorts of questions for differing treatment. To become obsessed with a method for its own sake and to try to use it where it is unsuitable is thoroughly unscientific (Midgley, 1978).

But in both physics and biology the Assumption of Simplicity continues to hold strong. Their mutual slogan is: Surface complexity arises from deep simplicity.

> ... There is a scientific problem even more fundamental than the origin of the universe. We want to know the origin of the rules that have governed the universe and everything in it. Physicists, or at least some of us, believe that there is a simple set of laws of nature, of which all our complicated present physical and chemical laws are just mathematical consequences. We do not know these underlying laws, but, as an act of faith if you like, we expect that eventually we will... (Weinberg, 1985)

(I hope there is also always room for the opposite view—that surface simplicity arises from deep complexity through the action of constraints.)

Trends in Physics

One modern view of a truth in physics has been aptly described by Misner (1977):

> I think physics has a model for immutable truth. A truth that we do not expect is ever going to change. A truth that represents a permanent and final grasp of some limited aspects of nature. Most people would say that is incompatible with the expectation that our theories will be falsified. I adhere to the expectation that our theories will be falsified, and look for the immutable truth only in those theories that have already been falsified, that is, in what I would call mythical theories. Newtonian mechanics, special relativity, Maxwell's equations, the Schrödinger-equation theory of atomic structure, and the like. Every one of these theories remains in constant everyday use in spite of the fact that it is known to be false in certain applications. In some cases, its concepts have been utterly discarded or eliminated in more refined theories which we believe to be more correct. Nevertheless, I think these theories adequately grasp certain aspects of nature. For instance, I expect geometrical optics will continue to be used by people who use hand lenses, align binoculars, and so forth. It is much easier to think of a light ray as moving in a straight line for certain specific applications than to try to operate in that context

with the wave theory of light. For these applications the quantum electrodynamical, field operator theory of light would simply not clarify one's grasp of the relevant physics but only becloud it with so many complications that one would be unable to focus on those aspects of reality that could be helpful guides to action . . .

I believe that once a theory has been falsified and its limits are known, it is a better theory. It is a theory whose truth is more clearly defined and in that sense more true. . . . In some well-known theories of physics, such as Schrödinger-equation atomic structure, we have an important insight that will continue in use even after better theories (quantum electrodynamics) are in their turn falsified. Physicists expect to continue teaching the old, immutable, infallible myths of Newtonian mechanics, special relativity, and Maxwell's equations while searching for failings in the theories that have supplanted these. I want no suggestion here that the myth of Newtonian mechanics is false or inadequate. It is an example of the most certain and permanent truth man has ever achieved. Its only failing is its scope; it does not cover everything. But within its now well-recognized domain it conceptualizes its portion of external physical reality better than its successors, and that is why it continues in use.

Today the physics of the very large has merged with the physics of the very small, together contemplating the joining of the four forces of nature into one, giving us a story of the evolution of the universe starting earlier than 10^{-10} seconds after the Big Bang (Weinberg, 1985) and proceeding past the death of protons at 10^{31} seconds to the end of black holes at 10^{107} seconds (Table 1). Physics has come even closer to mathematics, as quantumstuff becomes more fundamental than matter and energy. (Feynman's 1965 declaration that "physics is not mathematics" has a fainter sound today.)

Quantum mechanical fluctuations are more deeply probabilistic than classical, statistical mechanical fluctuations. Many attempts have been made to base quantum mechanics on a classical deterministic theory (that is not a hidden variable theory). In an example developed by Nelson (1985), quantum phenomena are described in terms of a stochastic process resembling statistical mechanical diffusion as the primary construct, rather than on the more deeply probabilistic wave functions. In his model quantum fluctuations arise from direct physical interactions rather than from an uncertainty principle. But all deterministic interpretations of quantum mechanics carry a severe penalty—the notion of locality (within the setting of Bell's theorem for quantum mechanics) is shown not to hold (Challifour, 1985; Herbert, 1985). Locality is fundamental to a physicist's view of reality.

At the moment no interpretation of quantum reality prevails over its rivals, nor is it even agreed that any interpretation is a correct description of nature, in spite of the thorough validation of quantum mechanics itself by experiment. Even so, quantum mechanics may be only an approximation at the atomic scale of a more fundamental theory, yet to be developed, that would be valid at subnuclear scales. Quantum theory peculiarly describes a measured atom in a very different manner than an unmeasured atom. The famous Copenhagen interpretation holds that the unmeasured atom is not real: its dynamic attributes are created or realized in the act of measurement. The unmeasured world is merely semireal.

> One of the inevitable facts of life is that all of our choices are real choices. Taking one path means forsaking all others. Ordinary human experience does not encompass simultaneous contradictory events or multiple histories. For us, the world possesses a

TABLE 1. Cosmic Evolution[a]

	Age of the universe (seconds)	Events
	10^{107}	Last black holes vanish
	10^{100}	
	10^{90}	
	10^{80}	
	10^{70}	Black holes start to vanish
	10^{60}	
	10^{50}	
	10^{40}	Protons decaying; solid matter vanishing
	10^{30}	
		Stars escape from galaxies or are lost in central black holes.
		Stars go out.
	10^{20}	
		NOW
		Stars and planets form from atoms.
		Atoms form from nuclei and electrons.
	10^{10}	Nuclei form from protons and neutrons.
	10^{0}	
	10^{-10}	Protons form from quarks.
		em force separates from weak.
	10^{-20}	
	10^{-30}	
		GUT era, Higgs field fluctuations; GUT freezing—strong force separates from electroweak.
		Gravity + 1 subatomic grand force; bosons and fermions
Expansive cooling ↑	10^{-40}	
	10^{-43}	(Planck time)
		SGUT era—physical laws precipitate out as rules of the game.
	0	Big Bang (or multiple bangs if there are bubbles)
		?

[a] Modified from Poundstone, 1985, page 149. Proton decay has not yet been observed experimentally. If the proposed fifth force of nature—supercharge—is confirmed it may be possible, and necessary, to revise the account of cosmic evolution, eliminating the supposed decay of protons. The last word on cosmology has not been heard.

singularity and concreteness apparently absent in the atomic realm. Only one event at a time happens here; but that event really happens. The quantum world, on the other hand, is not a world of actual events like our own but a world full of numerous unrealized tendencies for action. (Herbert, 1985)

These two worlds are bridged by a special interaction—that of measurement. During the measurement act one quantum possibility is singled out, "abandons its shadowy sisters, and surfaces in our own ordinary world as an actual event"

(Herbert, 1985). We are entitled to wonder at this point whether or not biological entities such as T-cell receptors, surprising DNA introns bearing codes for the enzyme reverse transcriptase (most introns are non-coding), antibodies or human consciousness, are, or arise from, classical, ordinary objects in physics that do not require dipping into quantum weirdness for their description or explanation. Or will we fail to account for the apparent intentionality, the teleonomic behavior, of biological systems, their goal-directedness, unless we commit to some interpretation of quantum mechanics? As I shall discuss later, quantum chemistry (free of any interpretation about reality) helps us understand DNA and receptors, and we biologists seem to be doing well by regarding most of our objects as physically ordinary even though we bathe them with intentionality and anthropomorphize them as controllers, regulators, executors, proofreaders, and so forth.

Because measurement is a privileged act in the "Copenhagen" interpretation of quantum mechanics, and creates all microscopic reality observed by experiments on a prepared system, in physics as well as biology we often take out loans of intelligence that we seem unable to repay. I believe Dennett's warning applies to physics, as to biology:

> Any time a theory builder proposes to call any event, state, structure, etc., in any system (say the brain of an organism) a *signal* or *message* or *command* or otherwise endows it with content, he takes out a *loan of intelligence* ... This loan must be repaid eventually by finding and analyzing away these readers or comprehenders; for, failing this, the theory will have among its elements unanalyzed manalogs endowed with enough intelligence to read the signals, etc., and thus the theory will *postpone* answering the major question: What makes for intelligence? (Dennett, 1978)

In his account of quantum theory, *Mathematische Grundlagen der Quantenmechanik* (1932), von Neumann reluctantly concluded that he had to locate the collapse of the electron's wave function—the jump that creates the reality of its dynamic attributes, inside human consciousness itself (see Herbert, 1985, pages 145–148). At that point naive realism in physics was dead.

Reducing Biology to Physics

Fortunately, modern biology claims reduction mainly to the physics of ordinary objects and generally ignores confusions surrounding quantum reality. Even molecular biology is primarily a science of macroscopic phenomena as biologists usually talk about it. But is it possible that consciousness and memory may involve an act of "classicalization" of quantum effects in the brain that were delocalized prior to "formulation" (measurement) of a thought or memory—that we should look for macroscopic manifestations of quantum coherence, as in lasing? Is it possible that some human cortical neurons specialize in capturing subatomic quantum fluctuations and propagating them as amplified, statistical mechanical fluctuations through neuronal networks? I can't support these speculations; the scales seem wrong. Neurotransmission and ion fluxes, the smallest scale machinery of the brain, operate at the level of atoms, molecules, cellular cooperatives, and other supra-atomic structures, at which quantum effects are presumably already classicalized. Quantum coherence requires a tight coupling between the energy flow and the effect; even the new, relatively low-energy chemical laser hosts require this tight coupling, and coherence collapses quickly if the external source of energy fails. Life does not run that closely tied to the external environmental potentials. (However, it must be admitted that consciousness fails rather fast when the supply of glucose or oxygen to the brain is acutely impaired. Slow impairments though can be tolerated, in many cases.)

Just after the "completion" of quantum mechanics, Niels Bohr immediately saw philosophical implications for biology arising out of the fundamental changes in the conceptions of natural law (1933). The relaxation of complete causal determinism in physics opened up new views of the relation of life sciences to the natural sciences. This opportunity was explored by Schrödinger (1956), and repeatedly by Delbrück (1949, 1970, 1986) who developed the implications of Bohr's epistemology into a philosophical stance for the new science of molecular biology. But life scientists in general have not responded; they have continued to pursue mechanism in the style of the physics of the pre-quantum era of the 19th century. After the loss of the machine root metaphor, physics came up with no other. It offered just mathematics as a way of viewing reality. Meanwhile, biology cloaked itself in the raiment of a special kind of encounter not covered in physical collision theory—the communicative interaction. Its metaphor is the information processing and computing machine, but without mathematics.

The gulf between biological explanation and physical explanation is that between descriptions in natural language and operations in formal language. Some physical concepts and theories can be expressed in natural language but it takes a mathematical model (formal language) that satisfies a theory to specify it so that it can be used by physicists to probe nature. In contrast, little of biological knowledge requires specification by formal language in order to be usable by biologists; natural language almost always suffices (see, for example, "The Molecules of Life"—a collection of eleven articles in *Scientific American*, Vol. 253, October, 1985, pp. 48–173). Because of that difference in style, I fear that biology and physics are diverging instead of converging. Yet, both sciences rely on logic as a way of probing nature. For example, the development of gene-splicing as a research and technological tool did not require mathematics, but it did require logic expressed in natural language.

The logic behind experiments in modern biology sounds very much like the logic of earlier physical experimentation (17th to 19th century). For example, consider the following five-step chain of reasoning, worked out over the last 30 years. (1) Because specific immune responses in mammals can be mounted against a very wide variety of antigens, it followed either that the antibody molecule must have its conformation determined by the antigen (an idea later proved false by experiment) or else the antigen must in some way select the antibody (actually the cells that produce them). (2) It then followed logically that cells involved in arranging antibody production to match the challenge of an antigen must have some kind of receptors that recognize the antigens, just as do the antibodies themselves that result eventually from that recognition. (3) It could then be reasoned that the genes specifying the receptors (in this case T-cell receptors—

T-cells regulate the activities of other immune cells including the antibody producing B-cells) may have adopted some of the same strategies that belong to the large "supergene" family that codes for a variety of proteins that are required for the normal immune response. Indeed, recent studies of T-cell receptor genes show that they have much in common with the antibody genes. Both sets of genes accomplish the assembly of complete genes from smaller coding segments, but whereas antibody genes are likely to undergo a mutation in the dividing B-cells that produce them, the sequence data for (the beta-chain part of) the T-cell receptor indicate that analogous mutations there are rare. (4) That finding logically forces the question of how the required diversity is then generated in T-cell receptor proteins, if not by mutation or by using a very large number of gene segments (also ruled out empirically). Another mechanism must be used. (5) Guesses can be made as to what it might be and appropriate experiments designed. Such experiments have in fact led to the discovery that the manner in which the various segments of the T-cell receptor gene are joined together is very flexible and that the joining of the segments may result in the use of any of three possible reading frames for some of the codons, each of which would give a different amino acid sequence in the resulting protein (Bier *et al.*, 1985; Marx, 1985a). Thus a new basis for generation of diversity is found. The chains of logic behind many such experiments in molecular biology have sharpened to the point that natural language decision trees can be set up in advance.

While physics searches for even deeper hidden symmetries via mathematics, biology continues to accumulate empirically inspired tales that presumably reflect some underlying physical reality (ordinary or quantum?) not worked out. Can physics and biology be brought closer together? Blumenfeld commented on reduction of biology to physics as follows (1981):

> The word "biophysics" is quite popular nowadays ... This word, however, may lead ... to a misunderstanding. Is there indeed "the biological physics", i.e., the special physics of living matter with laws differing from those of conventional physics? The answer to this question is: ... *The known principal laws of physics are quite sufficient for a complete description and understanding of the structure and functioning of all existing biological systems.*
>
> When a scientist states that he understands the chemical properties of benzene it means that it is possible for him, if only in principle, to cover logically all the distance from the postulates of quantum mechanics, electrodynamics, and statistical physics to the explanation of the reactivity and other chemical properties of the complex system containing 6 carbon nuclei, 6 protons and 42 electrons. Practically, nobody actually tries to cover the whole distance (and, due to immense mathematical difficulties, it cannot be covered without crude approximations).... It would mean a conscious refusal to use many rules formulated by chemists, which in the end, constitute the science of chemistry. The boring and, to a considerable degree, meaningless question about the possibility of reducing the laws of chemistry to those of physics can be, probably, answered in this way: it can be done but it is rather unnecessary. (My) conviction is that in biophysics, as in chemistry, it will not be necessary to introduce new postulates and world constants in order to understand the structure and functioning of biological systems. However, the future of biophysical theory of biological evolution must take into account the essential difference between phenotype and genotype....
>
> Some people, full of enthusiasm as regards the recent achievements of molecular biology, think that the development of the biochemistry of nucleic acids, of protein synthesis and its regulation, are leading us steadfastly to the true understanding of the mech-

anisms of ontogenesis and elementary hereditary phenotype changes. Here, however, one should bear in mind that in this branch of biology all we know at the molecular level concerns only the mechanisms of hereditary changes in protein molecules. It is a long way from proteins to characters...

And, so far, that way is not paved with mathematics. Mathematics is greater than physics in generality. Yet biology seems *too special* for physics. Biological organisms are not the sort of thing you would look at if you were seeking overarching principles. Organisms are made special by their initial conditions, boundary conditions and constraints that must be superimposed upon more general physical laws, and which must be independently stipulated before those laws bear directly on the organic realm. Traditionally, the determination of such supplementary constraints has been thought to be the job of biologists and not that of physicists, though neither has ever doubted that, "since physics is the essence of material nature in all of its manifestations, the relationship of physics to biology is that of general to particular" (Rosen, 1986).

According to this view, the next step in the reduction of biology to physics would require that the constraints that shape biological dynamics be given a physical or mathematical account. Polanyi argued that this reduction was impossible (Polanyi, 1968), but Causey (1969) rejected that impossibility argument by pointing out that the reduction might be accomplished through appeal to an evolutionary theory as intermediary. Thus, as we near the end of the 20th century, the convergence of biology and physics seems to depend on the following points:

1. Physics had to relax its hopes for strict determinism which had seemed to rule out the possibility of biological novelty, contrary to the facts of biological evolution and diversity.
2. Both physics and biology had to strengthen their understanding of evolutionary, historical, irreversible processes in which the macroscopic (including macromolecular) dynamics are not time symmetric, and small factors or fluctuations may have amplified effects, especially at bifurcation points where symmetry can easily be broken.
3. The uses of terms and concepts cast in the mold of information-communication metaphors that now dominate biological explanations have to be justified by accounting for information as a byproduct or side effect of dynamics. In this account information will provide the constraints that particularize the physically lawful biological processes. Such informational constraints usually are structures. The dichotomies of structure and function, form and process, genotype and phenotype, biology and physics, are seen to be variations on that of information and dynamics (Yates, 1982b).

Information versus Dynamics

At present there is a conspicuous seam in the pattern of convergence of the sciences, a large wrinkle that separates physics and biology, even though in both sciences a central problem is explication of evolutionary, historical, self-organizing systems. The two styles of approach to problems are very different, as can be illustrated by comparing physical organic chemistry with biological chemistry. Both are examples of carbon chemistry; both address the subject of catalysis.

Organic chemistry describes catalytic and other reaction mechanisms according to rules of chemistry reducible in principle to quantum mechanics (Goddard, 1985), but in biochemistry the emphasis is peculiar: Reactions are treated as though they had "intentions" other than going to equilibrium. This different flavor of biochemistry can be sampled in some of the phrases and topic headings to be found in a textbook of biochemistry (Stryer, 1981), viz.: "Communication within a protein molecule"; "In the step from myoglobin to hemoglobin, a macromolecule capable of perceiving information from its environment has emerged"; "Molecular pathology of hemoglobin"; "The complexity of the replication apparatus may be necessary to assure very high fidelity"; "Lesions in DNA are continually being repaired"; "RNA polymerase takes instructions from a DNA template"; "It is evident that the high fidelity of protein synthesis is critically dependent on the hydrolytic proofreading action of many aminoacyl-tRNA synthetases"; "Cyclic AMP is an ancient hunger signal"; "Information flows through methylated proteins in bacterial chemotaxis".

My choice of the above examples is not meant to detract from the splendid account in Stryer's book of biochemical processes from a mechanistic basis. Rather, I wish to illustrate the teleonomic mind-set of the biologist as he addresses the phenomena of interest to him. The apparent intentionality of the biochemistry drives him away from simple machine metaphors to information metaphors, and especially to the computer as information processing machine that can join the two metaphors. But how does carbon chemistry support the concept of error? Is that property of biochemical networks any more than the inevitable consequence of random variation and selection (Kimura, 1961), viewed anthromorphically? It is still hard for many to grasp the idea that a blind but cumulative selection process acting on random variations can have such creative power. The residues of such processes are easy for us to imagine as being a result of design, purpose or plan, because only by such means can we, in a shorter time, make workable devices or systems. But in assuming an *a priori* plan for an after-the-fact result of chance and selection, we make a serious category error. Pattee (1982) has forthrightly tackled the problem of the use of concepts such as "symbol", "referent" and "meaning" in the descriptions of biochemical and evolutionary phenomena. He sees cells as symbol-matter systems transforming strings from a small set of elements, by rules, into other strings, in turn transformed by laws into functional machines. In his account information and dynamics are complementary.

Because we take satisfaction in anthropomorphizing DNA as a selfish commander or director (Dawkins, 1976) (as well as in objectifying man in economics and the social sciences), the information metaphor will continue to follow the camps of scientific advance, never reputable enough to gather close around the fire at the center, but always beckoning from the periphery toward easy pleasures. Thus we confront the linguistic properties of the immune system and construe its chemical and cellular kinetics as "generative grammar".

> Immunologists sometimes use words they have borrowed from linguistics, such as 'immune response'. Looking at languages, we find that all of them make do with a vocabulary of roughly one hundred thousand words or less. These vocabulary sizes are a hundred times smaller than the estimates of the size of the antibody repertoire available to our immune system. But if we consider that the variable region that characterizes an antibody molecule is made up of two polypeptides, each about 100 amino acid residues long, and that its 3-dimensional structure displays a set of several combining sites, we may find a more reasonable analogy between language and the immune system by regarding the variable region of a given antibody molecule not as a word but as a sentence or a phrase . . . (Jerne, 1985)

If we seek a physical–mathematical account of life, can we achieve that account within *one* explanatory framework only, or must we use at least *two*—an informational or linguistic theoretic, and a dynamic theoretic? Whether the monists who would invoke only dynamics (Yates, 1982a), or the pluralists, who urge information-dynamic complementarities (Pattee, 1977; Chapter 17), prove to give the more valuable account in the long run, the scientific themes presented in this book should succeed ultimately in placing life firmly within physical nature, leaving perhaps only the one miracle of the Big Bang outside the reach of science (and some physicists even see smoothness all the way to the beginning, without a singularity as originator). If we try to step outside of nature as science explains it each time we are stuck for an explanation, then we are free to invoke as many miracles as we need: one for Existence itself, one for the origination of life on earth and one for the appearance of Man's consciousness and his gods. But to those with scientific temperaments such behavior is shockingly loose. In science miracles are not to be multiplied unnecessarily; we must make no appeals to gods outside the machine or to ghosts within it. Vitalism has no place. The Big Bang gave us the pieces and the board, and we are left to formulate the game that the pieces and their interactions express. To eliminate residual vitalism from molecular biology, and the intentionality from DNA, we might start with the physical aspects of DNA.

Physical DNA

At the beginnings of modern molecular biology thirty-five years ago, the melting points of double strands of DNA were matters of great interest in biophysics; now it is such things as the quantum mechanical vibrational modes of these molecules and the transitions from one conformation, B, to another, Z (Irikura, Tidor, Brooks, and Karplus, 1985). Starting with x-ray structure of the left-handed Z conformation of DNA (the duplex $dCdG_3$), estimates of internal motions have been made. From the vibrational normal modes of the molecular dynamics, vibrational entropy at a specified temperature was calculated. The various calculated harmonic entropies, though approximate, seemed to demonstrate that these internal motions are significant for the thermodynamics of DNA conformational transitions. The Z conformer is more rigid than the B conformer and that rigidity diminishes the probability of intercalation of certain compounds. (It is not yet certain that the Z form exists in vivo, or functions there.)

It is essential to understand DNA conformation in this kind of detail to get at the function of those regulatory proteins that help to turn genes on or off. These proteins match, by binding, some short DNA sequences from an entire genome. The chemical affinities that lead to specific associations between these two classes of macromolecules have been elucidated by cocrystallization of the DNA-binding region of a bacteriophage regulatory protein and the DNA sequence to which it can bind. [An example of such studies, by John Anderson, Mark Ptashne and Stephen Harrison, has been clearly described for the nonspecialist by Marx (1985b).]

The 3-dimensional picture of DNA structure is hierarchical. Supercoils and knots appear from the linear arrays of the single strands and the twisted double strands. The mechanism of site-specific genetic recombination catalyzed by the enzyme Tn3 has been studied by extending a model of synapsis and strand exchange. Simulation predicted the formation of a DNA product with a specific knotted structure, and this structure was subsequently found by 2-dimensional gel electrophoresis (Wasserman, Dungan, and Cozzarelli, 1985). Confirmation of the mathematically predicted DNA knot has provided strong support for topological modeling of DNA and geometrical analysis of this molecule.

In these current studies of DNA as a molecule, this biological object becomes soundly anchored in physics and chemistry through chemical thermodynamics, quantum chemistry and topology. And at the same time that mathematicians have offered topological linkage numbers, twist numbers and writhing numbers (and a theorem connecting them) as descriptors of higher-level DNA structures, and pointed out the possibility of knots, they have also advanced their own methods to distinguish different types of knots by making bridges between operator algebras and knot theory.

Developmental Biology—Is It Physical?

Although structural, mathematical studies do not directly address the "informational" aspects of DNA, biologists have long understood that as far as the gene goes, structure *is* information. Thus, one bridge between physics-as-dynamics and biology-as-information is the structure of DNA at all of its levels of elaboration. Nevertheless, topological and quantum mechanical accounts of DNA stop far short of fully justifying phrases common to molecular biology such as: "Linear information contained in the DNA generates a specific 3-dimensional organism in the course of development from the fertilized egg", or, "Master genes in homeoboxes (segments of DNA controlling spatial organization) and chronogenes (genes controlling the timing of events in embryogenesis) together programmatically direct development."

The casual assumption that biological development is *program-driven* forces use of information metaphors, not rationalized physically, to account for the wonders of biology. (See Chapter 18 for Stent's version of this problem.) Perhaps development would be better connected to physics if we biologists took a more *execution-driven* view in which some genes serve as modifiable constraints on cooperative kinetic processes that generate new structures as new constraints at the next stage. It requires a "loan of intelligence" to assert that a DNA "program" called the homeobox helps to "orchestrate development" in a startling array of animals. (An account of the homeobox for the nonspecialist has been given by Gehring, 1985.) Of course, in a very narrow sense almost every computer program seems execution-driven because the state of some switches may be determined by the outcome of the operation of an OR gate or an IF–THEN element of computation, etc. However, I am using the term more broadly to mean that the execution of one stage of processing leads to a change in the master program itself, or in the hardware available for subsequent stages. (Program-driven computations can simulate many execution-driven processes through subroutines and intermittent calling-up of peripheral devices, etc. But in such cases everything must be on hand and ready in advance—contrary to the conditions of truly execution-driven processes, which are self-organizing.)

Biologists have to answer how, in multicellular organisms, cells become destined to develop along a certain pathway at certain phases. This restriction in trajectory is called "determination." During early cell divisions in most embryos there are gradual reductions in developmental potentiality for the individual cells. Sooner or later in all embryos most of the cells lose their plural potentialities. They then differentiate. Differentiation is an all encompassing term that designates the processes whereby the differences that were "determined" become manifest. Differentiation is the selective "expression" of genetic "information" to produce the characteristic form and functions of the complex, fully developed embryo.

An account has to be given for the fact that the cells of the embryo not only become "committed" to a certain fate, and realize these fates, but they do so in the "correct" place at the "correct" time in the formation of pattern. Determination, differentiation and pattern formation are the outstanding features of normal embryonic development of all plants and animals. (Davenport (1979) has summarized many of the experiments in animals that demonstrate the dependence of these processes on both epigenetic and genetic influences. The series edited by Browder (1985, 1986) offers an updated account covering a wide range of living systems.)

It is the nature of the problem of self-organization in embryology that constraints appear and disappear (some cells die and get out of the way; genes turn on and off). But DNA is no ordinary constraint on kinetics; all mammals have approximately the same genome size, from mice to men. How do you get a man instead of a mouse from the same amount of genetic "information"? The molecular biologist today would say that the answer lies in the homeoboxes (homeotic genes, see Harding *et al.*, 1985) and chronogenes, but that view fails to acknowledge the flow of reverse influences running from enzymatic activities and product formation, back through regulatory proteins or metabolites, to the masking or unmasking of DNA surfaces or grooves and then on to the proteins that replicate the DNA or map it into RNA and then to protein again. Development consists of flow processes with circular causalities shot full of cooperative and competitive chemical phenomena and spatial and biochemical canalizations. For example, it may require modulable enzymatic methylation of DNA as a secondary mechanism changing DNA structure and turning genes on and off (Rozin *et al.*, 1984;

also see Kolata, 1985, for a nonspecialist's account of the controversy surrounding the view that methylation is crucial for development). At every stage we find activation and inhibition, cooperation and competition, symmetry-breaking and chemical complementarities as six major types of processes.

The information metaphor that I claim is inimical to clear thinking in attempts to reduce biology to physics will probably never disappear; it is too convenient a shorthand. But there is little (or no) more substance in the statement that "homeoboxes and chronogenes direct development" than there is in the statement "balloons rise by levity". Many explanatory constructs in conversational molecular biology remain at the low level of Kiplingesque "Just So" stories (How the Camel Got His Hump; How the Elephant Got His Trunk . . .).

The unsatisfactory state of developmental biology and its distance from natural science is now being relieved by further applications of topological approaches to the problem of morphogenesis. René Thom (1986) has commented that "all modern biological thought has been trapped in the fallacious homonym associated with the phrase 'genetic code', and this abuse of language has resulted in a state of conceptual sterility, out from which there is little hope of escape." But there may be some escape through a new extension of his original account of 1972. Working with four morphologies of flow (birth, stopping, confluence or ramification), Thom shows that the ancient millwheel expresses many dynamic features needed to describe embryonic morphogenesis in metazoa: canalization of dynamics, coupling to a potential to obtain a direct flow, entrainment, nonlinear catastrophic escapements during a (fictive) retroflux, dissipation of free energy, and periodic behavior. He argues that life is itself a spatially and biochemically canalized phenomenon with dynamics shaped by membranes, cytoskeletons and macromolecules. Cell replication requires duplication of a singularity in a flow field undergoing a continuous deformation. Thom defines a "meromorphic" potential for the migrations and transformations of the genome during replication.

Metabolic processes involve both flows of small metabolites (ions, small molecules) and the assembly flows of polymerization of macromolecules. These two flows are seen by Thom as a direct flux and a retroflux in a hysteretic loop, expressible as a simple cusp singularity that is transformed to a butterfly bifurcation locus during cell replication. Genomic DNA appears here as a constraint on flows, serving to connect spatial and biochemical canalizations. These fresh abstractions bring models of fundamental aspects of living systems within the reach of mathematics and computer simulation and thus perhaps also of physics, and loosen the bondage arising from the overdependence of developmental biology on natural language narratives.

Evolutionary Scenarios as Arena for Convergence between Biology and Physics

The fluctuations and phase changes required by modern physical theory to account for the development of the expanding, cooling Cosmos invoke both deep symmetries and broken symmetries as sufficient, dynamical bases of construction. Although the laws of physics themselves may have in some sense precipitated as "this time-around" rules, that given a different fluctuation might have been different, the evolution of the Cosmos is not ordinarily regarded as being information- or program-driven. It is described in mathematical or dynamical terms that address physical laws, symmetries, initial conditions, boundary conditions and constraints—all explored, of course, by means of approximations simulated on information-processing machinery. Therein lies a paradox.

Landauer has noted that computation is a physical process inevitably utilizing physical degrees of freedom and therefore restricted by the laws of physics and by the construction materials and operating environments available in our actual universe. In practice we use algorithms for information processing to express physical law. Therefore we have a conundrum: The ultimate form of physical laws must be consistent with the restrictions on the physical executability of algorithms, and these restrictions are in turn dependent on physical law (Landauer, 1986). As physics becomes even more deeply mathematical, and analytic solutions are given up for approximations, and the approximations become computationally more demanding, one could be pardoned for feeling that we are leaving physics behind. (However, Manthey and Moret (1983) analogize computation as quantum mechanics!)

In place of physics we find treatments of the evolution of the Cosmos as if it were a variant of John Horton Conway's game called "Life" (devised in 1970 and introduced by Martin Gardner's columns in *Scientific American* in October 1970 and February 1971). For example, Poundstone (1985) has interpreted modern cosmology as a rule-defined game executed on a finite-state automaton. The dynamics are arbitrary or suppressed, and information is everything. At this point it is well to remind ourselves that however they arose, the rules of the game are the laws of physics and that the world of physical possibility and actuality, being limited by these laws, is a small subset of the range of mathematical possibility. Mathematical possibility is limited only by the need to avoid logical contradictions in the constructive use of tautologies that permit the transformation of one set of statements into another.

In contrast to cosmic evolution (Table 1) which is at least crudely mathematized (and also translated into a narrative), biological evolution resists that degree of illumination through mathematical modeling. Biology's Big Bang may have begun in a slurry of molecules in a pre-chicken broth, or perhaps instead from reactions on crystalline clay minerals—an old idea recently given updated and thoughtful treatment by Cairns-Smith (1985). He takes issue with the argument that because the central molecules of life are the same in all organisms on the earth today and because at least some of these molecules can be made under conditions that might well have existed on primitive earth—therefore these molecules were present at the creation. (But these are the assumptions implicit in Chapter 4 of this book.) Cairns-Smith develops the caveat that "we should doubt whether amino acids or any other of the now critical biochemicals would have been at all useful right at the start," and provides an alternative scenario.

Whatever the guesses about startup conditions, the position of modern biology with respect to subsequent evolution remains founded in the idea that random variations (and catastrophes?) occurring both within the prebiotic system and in its environment led by selection to branchings and extinctions, to continuity and diversity. To those classical Darwinian notions, three others, at least, have been added: (1) the punctuated equilibrium hypothesis of Niles Eldredge and Stephen Gould (see Chapter 6) that views evolution as a saltatory process ("Falconer-like evolution"), at least at certain epochs, rather than as a gradual process. (2) The hierarchical evolution hypothesis that natural selection acts not only on individuals (phenotypes) within species, as traditionally supposed, but at many levels, even between species. The species now becomes the new "individual" (Lewin, 1985). (3) The hypothesis that most mutations are biased and not random. These ideas, combined with evidence that microevolutionary changes may drive macroevolution, and vice versa, have extended conventional neo-Darwinism, but not without continuing controversy.

Biological evolution is rich in mechanisms and physicists are finding similar richness as they try to resolve the sequence of events that led to stars, globular clusters, galaxies and supra-galactic clusters. There, too, fluctuations and chance apparently were "determining." In both physics and biology we find randomness, uncertainty and noise turning into the necessity of self-organization and the emergence of order, at all levels, as an expression of exuberance of an energetic, probabilistic universe still in the spring of its years.

Receptors as Bridge between Information and Dynamics, Biology and Physics

If the informational and the dynamic aspects of living systems were complementary modes of explanation, as argued by Pattee in Chapter 17, we could accept Campbell's claim (1982) that:

> Evidently nature can no longer be seen as matter and energy alone. Nor can all her secrets be unlocked with the keys of chemistry and physics, brilliantly successful as these two branches of science have been in our century. A third component is needed for any explanation of the world that claims to be complete. To the powerful theories and physics must be added a late arrival: a theory of information. Nature must be interpreted as matter, energy and information.

In physics and mathematics, matter/energy and probabilities are all conserved and perhaps even adumbrated by quantum fields inside the classical vacuum (Boyer, 1985). That fluctuating field is prior to both matter and energy; it is probabilistic—the ultimate stuff of reality, according to at least one conjectural extension of quantum theory. In contrast, information and entropy are not conserved. Information can be created and destroyed, even though it can be related to probabilities through analogy with the Boltzmann statistical mechanical entropy. Physics is traditionally more at home with conservative fields (e.g., Hamiltonians) than with nonconservative processes, whereas biology seems dependent on gaining order out of dissipative, entropy-producing processes. This difference in regard to the seeming importance of the second law of thermodynamics provides additional grounds for the present and continuing gulf between trends in biology and in physics. As biology pursues the information metaphor and physics abandons all metaphors and settles for mathematical abstraction, the two sciences seem to diverge.

The linguistic or informational aspects of biology, that now seem so extra-physical, can perhaps be brought within the reach of physical dynamics through the notion of catalysis and amplification as a basis for "languages", as suggested by Iberall and Soodak in Chapter 27. Or we may find that seemingly deep problems like those of direct perception, or willful control of motor actions, may turn on the specialization of nervous systems for creating and parsing kinematic flow fields as abstractions of the kinetic world of Newtonian forces that play upon the ordinary objects of an organism and its surround (Kugler and Turvey, 1986). In that cooperation between kinetics and kinematics the rhythmic movements of animals and plants may become self-organized, with the semiotic significance of perceptions, their information, arising from singularities in these fields (see also Thom, 1986; Yates and Kugler, 1984).

The convergence of biology and physics, if it ever occurs, will result, I believe, from rationalization of information using terms and concepts of physics. At the moment it does not seem profitable to attempt that demonstration in the fields of neurosciences, immunology or genetics because these branches of biology are too thoroughly permeated with the information metaphor. A better opportunity lies in analysis of receptors (Yates, 1982b). Receptors bind chemicals in their environment (the "recognition" step) and modulate dynamic contingencies or actions within the cells bearing the receptors, the "activation" step. Activation usually involves chemical cascades as amplifiers, and mappings from variations of one chemical species into those of another, the "second messenger". Second messengers in the chain of events centered on the receptor are very often fluxes of calcium ions or cyclic nucleotides (Alberts *et al.*, 1983, pages 733–753; Carafoli and Penniston, 1985). But not all receptor–ligand induced activations require second messengers; some amplifications are accomplished by more direct chemical actions of receptors at the genome. Whatever the mechanistic details, receptors stand between environmental, chemical variations and those cellular dynamic changes that biologists informally regard as "interpretations" of the environment. In the field of receptor biology the techniques of quantum chemistry, chemical kinetics, cell physiology and gene-cloning-probes intersect. It is here, if anywhere, that information emerges from dynamics.

Of all the dynamic bifurcations in biology, the transition from life to death strikes us as being the most dramatic. (This transition is not always irreversible—the field of cryptobiology studies multicellular organisms such as the British tardigrade that can reversibly, on hydration and dehydration, gain and lose life.) Now John Young (see Marx, 1985c) believes he has found a protein made by one species of amoeba as well as by a variety of immune "killer" cells, that can punch holes in the membranes of victim cells. This protein belongs to a well-known family, the complement system, that has long been recognized as contributing to immune responses by helping to kill foreign bacteria. The theatrical "killing" process turns out to be membrane instabilities following protein–membrane interactions that open up aqueous channels that permit lethal ion fluxes to arrest cell processes. We have in this case a story in natural language that sounds more physical when physicochemical terms are substituted for theatrical terms. But, in contrast to the constructions of mathematics expressed in formal language (which is not only a language but also a guide to subsequent correct thinking after formulation of a statement), the natural language stories of biology are purely descriptive even when physical terms and concepts are invoked. That is the state of most reductions of biology to physics today. Mathematics doesn't fit modern biology.

Resolution of the perplexities about the relation of information to dynamics, of biology to physics, would require some understanding of why mathematics works in the physical sciences.

Why Does Mathematics Work in the Natural Sciences?

Eugene Wigner (1960) commented on "the unreasonable effectiveness of mathematics" in an essay that explored the intimate relation between some mathematics and physics. That association had already been remarked upon by Heinrich Hertz: "One cannot escape the feeling that these mathematical formulae have an independent existence and an intelligence of their own, that they are wiser than we, wiser even than their discoverers, that we get more out of them than was originally put into them" (from Pagels, 1982, page 301).

Feinberg (1985) argues that the effectiveness of mathematics in science reflects the fact that concepts of both disciplines grew out of a set of intuitions that originated in the same kinds of human experiences. Both mathematicians and physicists start from ordinary experience, but then build on concepts that have been introduced earlier within their own fields. As the concepts in both physics and mathematics become increasingly abstract, that is, increasingly removed from ordinary experience, they are no longer easily shared with workers in other fields. Because of the removal, Feinberg thinks it is even likely that mathematics and natural science will begin to diverge and find less common ground. Accordingly, biology could become even more widely separated from physics and mathematics than at present.

Kline (1980) reviews the history of varying opinions on the effectiveness of mathematics in the natural sciences, noting that mathematics deals only with the simplest concepts and phenomena of the physical world; it does not deal with men but with inanimate matter, introducing limited and even artificial concepts to institute some sense of order in nature. Man's mathematics may be no more than a workable scheme while nature itself may be far more complex or even lack inherent design. "Though it is purely a human creation, the access mathematics has given us to some domains of nature enables us to progress far beyond all expectations. Indeed it is paradoxical that abstractions so remote from reality should achieve so much. Artificial the mathematical account may be; a fairy tale perhaps, but one with a moral. Human reason has a power even if it is not readily explained."

Davis and Hersh (1981) see mathematical models as being imposed by fiat on the problem of interest because there is no systematic way of telling *a priori* what should or should not be taken into account. In support of their position they examine various aspects of addition, finding that there is, and there can be, no comprehensive systematization of all the situations in which it is appropriate to add. For example—Question: one can of tuna fish costs $1.05. How much do two cans of tuna fish cost? Answer: at my supermarket one can of tuna fish sells for $1.05 but two cans sell for $2.00—a discounted price.

I believe that human mathematical skills emerged from the same evolutionary process that created a central nervous system capable of running internal simulations of acts for which there is no immediate external stimulus. Most behavior of animal and plant species, except those of hominid lines, is strongly stimulus-bound. Squirreling away nuts during fall to guarantee survival through the coming winter is not a plan of action but a response to ongoing changes in light and temperature. In contrast, man can plan tomorrow's hunt and explore its possibilities through internal simulations not connected to any immediate hunger stimulus.

An organism's chief contact with reality lies in the getting of food (or money as token). It is not hard to imagine that natural selection could have favored the emergence of any capacity to make abstractions that would increase the chances of obtaining food or shelter necessary for the survival of the tribe. It *is* hard to imagine a selection pressure directly favoring the discovery of a Hilbert space. However, once the capability for abstract thinking and internal simulations emerged, it is perhaps not such a large step from the idea of tomorrow and tomorrow's events to quarks and photinos. Abstractions that first enhance the yield of food for the body can later serve as metaphorical food for the mind, starting with naming, counting, triangulating a relationship between it as tool, me as user, and the function between. A brain that can abstract one thing can to some extent abstract another, including self and tools. Erich Harth (1982) has brought together new perspectives on traditional mind-body problems that help to tie such aspects of cognition to physical science.

As for their mathematics, physicists find that some of the absurd results (infinities) arising in applications of quantum field theories to describe or define subatomic particles can be got rid of by the injection of an extra number that can take on different values. But that seemingly ad hoc procedure used to rescue the mathematics, as it were, looks less bizarre in the context of an interpretation of the extra number as a measure of the minimum distance for separation between two objects in space. A "finagle factor" sets up a justification for the lattice space of quantum chromodynamics. Just when physicists look as if they are committing the Fallacy of Misplaced Concreteness, they emerge with a possible new success (though the issue is still open).

Such activities of physicists and mathematicians remind us that all our endeavors smell of their human origins. We cannot scrub our sciences free of that odor nor is it worthwhile to try. But what we can do is eliminate redundancies and seek further unities, in the spirit of simplicity and parsimony that has motivated modern physics. Physicists seek the answers to two kinds of questions: (1) What is the world made of? and (2) What are the fundamental laws of nature? Can the answers to these two questions, in principle, provide explanations for all the beings and the becomings in the universe, including those in biology? Biologists don't care—they make more (successful) Just So stories while physicists look for missing mass, photinos, gravitons, superstrings in three macro dimensions and six super-microdimensions, grainy space–time lattices and superforce (Schwartz, 1985; Davies, 1984; Disney, 1984), and computations to explore them.

It may be that mathematics and computation have seized too much power over our minds. They show signs of aggressing beyond their traditional place as effective means to express thinking and explore implications (their handmaiden-of-physics role) and of becoming the *only way* to think or invent in physics. Their limitations or preferences (for example, for sequential processing, linear formulations, conservative fields, symmetries and simple systems) may impede our discovery of new means to comprehend complex systems. If surface complexity arises from deep simplicity we still have to show how that happens. The reductionistic path is easier than the constructionist (Anderson, 1972). Surface complexity has its own sovereignty, needing description at its own level.

Such limitations could damage scientific enterprise and hold up the reduction of biology to physics, but I don't think there is much cause for alarm. Those mysterious creative jumps of human minds that generate what Einstein called "absolute postulates" (see Pagels, p. 40) continually arise as new starting points for explanation, description and exploration. These inventions need not originate in verbal, mathematical or visual capabilities of the brain—they seem instead to well up out of pools of preconscious thought, not confined by the earthenworks consciousness uses to shape figures of thought. The operations of consciousness are perhaps just afterthoughts (Harnad, 1982). These afterthoughts can be constricted by convention, thereby becoming censors of invention—but even so, the wellspring of new ideas never dries up. Whatever the ultimate limitations on human thought or on conceptualization and invention in science, the autocatalytic character of discovery has opened up wonders of life and cosmos beyond anything imaginable before the 20th century, and there is no limit in sight. Some people wonder if physics is reaching "the end of the road" (Gardner, 1985), but we have heard that question before. Superstrings and superforce are not necessarily all that is new in what physics can say to us about man's nature and his place in the universe. Indeed, why not suppose that when physics has its unified field, and a quantumstuffed vacuum, it will be ready at last to leave the domain of simple systems and tackle the physics of life and mind? The unification of biology and physics could be the ultimate ornament in the crown of scientific achievement, which is motivated by a desire for parsimony in explanations. I expect that the understanding of self-organizing systems will be its setting and mathematics the cement that holds it in place.

Do We Need a Revolution to Accelerate the Convergence of the Sciences?

Since the publication in 1962 of "The Structure of Scientific Revolutions"—a carefully executed examination of the development of science by T. S. Kuhn, casual readers of the book have had too easy a job accounting for intellectual progress. It seems comfortable to attribute conceptual advances to dramatic changes in explanatory constructs brought about either by great leaders or by great movements. Cohen (1985) extends the revolutionary model of progress and specifies four stages of scientific revolutions: (1) creation of a new idea, (2) working out the idea, largely in private, (3) dissemination to professionals and (4) acceptance (by conversion rather than persuasion) and incorporation. Sterman (1985) has even run a computer simulation testing a theory of scientific revolutions.

Newton evidently did not regard himself as a revolutionary, having stood to see further, as he said, on the shoulders of giants. (This remark has been forever transformed and hilariously fixed by Merton (1985).) Discovery and advance in science have a smoother character than that of a revolution; more like a continuity with fluctuations, with surges and pauses, than like fractures or abrupt replacements of paradigms. Perhaps scientific advance is itself a punctuated evolution; but even so, in science generally as in biology, all that is, is necessarily the descendant of what was before. Revolutions are largely invented retrospectively when we have forgotten chancey details and are consolidating our gains. Revolutions make good stories but there is always a residue of the fabulous in our accounts of aleatory phenomena. (These remarks are not meant to deny the powerful intuitive leap that turned gravity from a force into the geometry of a space–time manifold, but only to respect the historical ripeness of the circumstances in which it occurred.)

Far from picturing a revolution, the patterns of convergence in modern science suggest the less grandiose image of a self-organizing system. Physics looks biological as it loses determinism and assigns chance a constructive role in the evolution of the Cosmos. Biology looks physical as it peers into the quantum chemistry of its chief actors, the nucleic acids and proteins, and even as biologists turn to quantum chemistry to understand enzymes, physicists are turning to the problem of the origin of life (Demetrius, 1984; Anderson, 1983). A new thought-collective is self-organizing, and I suspect it will be called revolutionary when looked at retrospectively 50 years from now.

A thought collective was defined by Fleck in 1935 as a community of persons mutually exchanging ideas or maintaining intellectual interaction that provides the special carrier for the historical development of any field of thought, as well as for the given stock of knowledge and level of culture. Thought collectives not only establish fads and fashions in science, and drive intellectual advances, but they also gradually reformulate and perhaps even redefine the very nature of science. As with every scientific concept and theory, so is every fact culturally conditioned—features of the scientific enterprise brilliantly set forth by Fleck's examination of the genesis and development of a particular scientific fact.

Summary and Predictions

To sum up the themes of this book I offer a short list of predictions and opinions:

1. At the structural level of enzymes, antibodies, ionophores, receptors and genes, biology belongs comfortably within the domain of quantum chemistry. Advances will be made by extending current simulations based at least remotely

on the Schrödinger equation, from which qualitative principles responsible for the results of particular biological experiments will follow (for example, see Weinstein *et al.*, 1981). The principles, starting with the quantum theory of unimolecular reactions (Pritchard, 1984), which is currently being strengthened to give new accounts of phenomenological rate constants, will lead to predictions of how biological chemicals will act under other circumstances. Chemically useful concepts from electronic wave functions have put theorists in the mainstream of chemistry. Simulations of chemical systems with competing reactions taking place simultaneously at various reaction sites have become possible because of advances in chemical dynamics and irreversible statistical mechanics and in computer technology. The promise of such simulation has been illustrated, for example, by a study of the enzyme thermolysin (Goddard, 1985). Biology will not require any particular interpretation of quantum reality for further advances of this kind; computational applications of quantum chemistry will suffice. Oscillatory phenomena, both as intramolecular vibrational relaxations and as properties of chemical networks (Field and Burger, 1985) will be seen to express the stability properties of the systems. In multicellular structures such reactions will account for organizing centers of waves in excitable media (Winfree and Strogatz, 1984). The detailed structure of such reactions will be proved by spectrophotometric techniques enhanced by digital image reconstruction (Müller, et al., 1985). The chemical kinetics of life will be represented by nonlinear, asymptotic orbital stability (homeodynamics) rather than through relaxational trajectories to point attractors (homeostasis).

2. Biological objects larger than macromolecules can safely be treated as classical objects and quantum effects can be ignored, though as an analogy, quantum coherence may remain provocative for descriptions of correlations at a distance (without local correlation) in neurobiology. Two physical approaches will be used to account for the qualitative behavior of macroscopic biological systems: (a) that of classical statistical mechanics and irreversible thermodynamics, and (b) mappings between kinetic force fields and kinematic flow fields. Either or both of these approaches will prove adequate to account for the emergence of self-organized structures from multiple cell–cell interactions, that even prokaryotes can accomplish (Dworkin and Kaiser, 1985). Mathematical models at these levels will emphasize qualitative dynamics; the fatuous quantitative mathematics that traveled in the wake of General Systems Theory and its sisters (see Berlinski, 1976, for the criticism) will play little if any role, whereas dynamical systems theory, bifurcation theory, catastrophe theory all will have increasing influence.

3. Detailed, rigorous and quantitative statistical mechanical treatments will not be feasible for elaborate biochemical systems, which are thermodynamically nonideal as a result of excluded volume effects and interactions that cause activities to differ significantly from concentrations, leading to strikingly nonlinear behavior.

4. Competitive and cooperative phenomena, activation and inhibition, and structural complementarities will be nearly universal elements of all mathematical models of the qualitative behavior of biological systems. The statistical mechanical approach to cooperative phenomena and biochemistry is already

well-developed for equilibrium and steady-state biochemical systems (Hill, 1985). It needs to be extended to include bifurcation phenomena. A case for the next round of attention might be the hypothesis that supposes that eukaryotic cells started out as primitive organisms without mitochondria or chloroplasts and then established a stable endosymbiotic relation (a cooperative with one previously independent system inside the other) with a bacterium, whose oxidative phosphorylation system they co-opted to drive their own machinery. We find evidence for that hypothesis in the separate genetic systems of mitochondria and chloroplasts, different from that in the nucleus of animal and plant cells (Alberts *et al.*, 1984).

5. At its largest scale, that of the terrestrial biosphere, biological processes will yield to statistical thermodynamic modeling of the overall biosphere–lithosphere–hydrosphere–atmosphere ecological system in which each component shapes the others and is in turn shaped by them. The global, structural stability of the whole system will again be seen to express six main processes: activation or inhibition, cooperation or competition, chemical complementarities and broken symmetries—all features that can be broadly formulated mathematically. The specifics are, as always, the hard part and progress will be, for a long time to come, case-by-case. Many of the chapters in this book illustrate working examples.

6. Computation of approximate models of the behavior of biological objects based on dynamical systems theory will benefit, as will computation of all complex models, by advances in computational hardware and software (Buzbee and Sharp, 1985). The models will be run on hypercubes, vector processors, massively parallel machines—or just on smaller personal computers. Each machine will have its place for a selected class of problems.

7. Dynamical systems theory will provide the link between models *on* theory that satisfy and specify and formalize the text of physical theory, and models *of* data that arise from empirical measurements on biological systems. It is standard scientific activity in physics, and will become so in biology, to find a logical connection between models on theory and models of data. (The decisions about what to look at, and the absolute assumptions that drive the creation of theory will remain, as always, inductive mysteries of the human mind. The so-called "inference engines" of artificial intelligence expert systems will not soon, if ever, be a fruitful source of inventiveness to generate theoretical models in biology.)

8. Biology will wean itself from the poisonous computer metaphor, particularly as it has been misapplied to neurosciences. Churchland has given a strong statement protesting those misapplications (Churchland, 1982). In contrast, however, artificial intelligence will prosper by borrowing some principles (not substances) from biology to arrive at new styles of computing. Computer hardware and software will be designed so they can self-organize as a problem is run. A potentially powerful step in that direction has been taken by Hillis (1985).

9. The laboratory creation of synthetic life, starting only with inanimate chemicals, is likely to be preceded by a mathematical model demonstrating by simulation the minimal set of carbon-bond reactions required. The model will specify amplification by replication, evolution toward complex forms and func-

tions and the emergence of autonomous morphogenesis in heterogeneous media, nearly invariant reproduction and teleonomic behavior—the three chief behavioral characteristics of living systems noted by Monod (1971). A physical perturbation, an environmental fluctuation—e.g., jabbing the reaction mixture with a hot wire, will be required to set off the self-organizing and sustaining reactions, just as is now the case with the Belousov–Zhabotinskii reaction (Müller et al., 1985). It is likely that a simulated inorganic clay will serve as a template to constrain the chemical kinetics at startup. The mathematical model will guide choices of reactants and boundary conditions for the actual synthesis of life in the laboratory. It will not be an artificial intelligence model with string processing of 0's and 1's representing arbitrary data and arbitrary rules. Instead the model will detail the emergence of constraints in successive stages, each constraint arising because of a new form which is a major product of the preceding set of reactions. The dynamic evolution of the system will be execution-driven; there will be nothing resembling a program, except the structure of the clay template as initial condition. That template will not necessarily be active at each stage in the dynamic evolution. New forms will constrain new processes; the new processes will maintain the new forms. The evolution will proceed along one or more of the lines now being investigated by Manfred Eigen, Peter Schuster, Hans Kuhn, Philip Anderson, Lloyd Demetrius, Sidney Fox, Michael Conrad, René Thom . . .

The nine predictions above are almost linear extrapolations from present scientific knowledge and technical art. Breakthroughs cannot be predicted, but I do not think that breakthroughs in either biology or physics will be required to accomplish the laboratory synthesis of a living system.

I am less sanguine about our ability to know we have accomplished the task when it is done. I foresee great disputes about whether or not an accomplished synthetic, minimal system is or is not "living". I would settle if the synthetic system met the three behavioral conditions asserted by Monod, listed above, as the minimum properties of a living system. Others might insist that the system meet the requirements of von Neumann's theory of self-reproducing automata (1966) which have been well paraphrased by Poundstone (1985, p. 191) as follows:

1. A living system encapsulates a complete description of itself.
2. It avoids the paradox seemingly inherent in (1) by not trying to include a description *of* the description *in* the description.
3. Instead, the description serves a dual role. It is a *coded* description of the rest of the system. At the same time, it is a sort of working model (which need not be decoded) of itself.
4. Part of the system, a supervisory unit, "knows" about the dual role of the description and makes sure that the description is interpreted both ways during reproduction.
5. Another part of the system, a universal constructor, can build any of a large class of objects—including the living system itself—provided that it is given the proper directions.
6. Reproduction occurs when the supervisory unit instructs the universal constructor to build a new copy of the system, including a description.

A purely behavioral definition of life is handicapped by the fact that many forms of life have little behavior. Some bacteria and spores do not exhibit irritability, or even metabolism for extended periods. Poundstone argues that the cogency of the information-theory definition of life is better seen when applied to problematic cases: "Whenever biologists try to formulate definitions of life, they are troubled by the following: a virus; a growing crystal; Penrose's tiles; a mule; a dead body of something that *was* indisputably alive; an extraterrestrial creature whose biochemistry is not based on carbon; an intelligent computer or robot." To his list I would add flames and the tardigrade I mentioned earlier, that is unmistakably alive when wet and dead when dry, and can cycle back and forth between the two states.

We already have discussions of "man-made life" (Cherfas, 1982), which are expositions on the technology and commerce of genetic engineering as we now do it. The level of that technology consists of insertions of nucleic acids into complex, already-organized, dynamical systems, to constrain them to synthesize products we want for scientific research or commerce. That's not nearly the synthesis of life that I am discussing here.

In designing his cellular automata to act as self-reproducing machines, von Neumann at first bogged down trying to specify transport processes that could bring the components needed to build the robot within reach of his universal constructor. His kinematic model even had the robot floating in a lake. To avoid the problem of locating a needed component out of a random soup, and transporting it to the site of assembly, von Neumann finally decided just to allow his cellular model to find components or call them into existence whenever and wherever they were needed. His recursive rules permitted "new matter" to be created in response to pulses sent out by the robot brain. Giving yourself an infinite supply of energy and materials located just where you need them may not be unreasonable for the initial conditions, but at some stage the "living" system must generate on-board energy stores and the means to tap environmental potentials, chemical and physical, episodically but over long times, to replenish them. It need not move about—individual bristlecone pines in the White Mountains of California have done well for millenia while staying in one spot.

The current maturity of physical, biological and computational sciences has produced, in my opinion, a ripeness for the formulation of the problem of synthesizing life in the laboratory. Admittedly physicists, biologists and computer scientists are so busy riding the surging waves of their flooding fields that such a project would be seen as professionally irrelevant or distracting. So, it is not likely to be undertaken systematically, but will probably happen en passant. Nevertheless, in the formulation of a quantum mechanical or statistical thermodynamical model of cooperative and competitive reactions and their constraints, using primarily activations and inhibitions to create behaviorally rich networks of cells and receptors, we are already finding home ground for the convergence of biology, computation and physics. As that happens some physicists may be at least briefly distracted from their headlong pursuit of the noumenal Zero Moment, and biologists from their murky information metaphors, being together dazzled at the possibilities latent in the new formulation.

The formulation of a problem is often more essential than its solution, which may be merely a matter of mathematical or experimental skill. To raise new questions, new possibilities, to regard old problems from a new angle, requires creative imagination and marks real advance in science. (Einstein and Infeld, 1938)

References

Alberts, B., D. Bray, J. Lewis, M. Raff, K. Roberts, and J. D. Watson (1983) *Molecular Biology of the Cell.* Garland Publishing, Inc., New York.
Anderson, P. W. (1972) More is different. *Science* **177**:393–396.
Anderson, P. W. (1983) Suggested model for prebiotic evolution: The use of chaos. *Proceedings National Academy of Sciences (U.S.A.)* **80**:3386–3390.
Berlinski, D. (1976) *On Systems Analysis: An Essay Concerning the Limitations of Some Mathematical Methods in the Social, Political, and Biological Sciences.* MIT Press, Cambridge, Massachusetts.
Bier, E., Y. Hashimoto, M. I. Greene, and A. M. Maxam (1985) Active T-cell receptor genes have intron deoxyribonuclease hypersensitive sites. *Science* **229**:528–534.
Blumenfeld, L. A. (1981) *Problems of Biological Physics.* Springer-Verlag, Berlin.
Bohr, N. (1933) Light and life. *Nature* **131**:421–423, 457–459.
Boyer, T. H. (1985) The classical vacuum. *Scientific American* **253**:70–78.
Browder, L. W. (Ed.) (1985) *Developmental Biology. A Comprehensive Synthesis.* Volume 1, Oogenesis. Plenum, New York (Vols. 2, 3 and 4 are in press).
Buzbee, B. L., and D. H. Sharp (1985) Perspectives on supercomputing. *Science* **227**:591–597.
Cairns-Smith, A. G. (1985) The first organisms. *Scientific American* **252**:90–100.
Campbell, J. (1982) *Grammatical Man: Information, Entropy, Language and Life.* Simon and Schuster, New York, p. 16.
Carafoli, E., and J. T. Penniston (1985) The calcium signal. *Scientific American* **253**:70–78.
Causey, R. (1969) Polanyi on structure and reduction. *Synthèse* **20**:230–237.
Challifour, J. L. (1985) Review of "Quantum Fluctuations" by Edward Nelson. *Science* **229**:645–646.
Cherfas, J. (1982) *Man-Made Life: An Overview of the Science, Technology and Commerce of Genetic Engineering.* Pantheon Books, New York.
Churchland, P. M. (1982) Is *thinker* a natural kind? *Dialogue* **21**:223–238.
Cohen, I. B. (1985) *Revolution in Science.* Harvard University Press, Cambridge, Massachusetts.
Collins, R. (1975) *Conflict Sociology.* Academic Press, New York.
Davenport, R. (1979) *An Outline of Animal Development.* Addison-Wesley Publishing Company, Reading, Massachusetts.
Davies, P. (1984) *Superforce: The Search for a Grand Unified Theory of Nature.* Simon and Schuster, New York.
Davis, P. J., and R. Hersh (1981) *The Mathematical Experience.* Birkhaüser, Boston.
Dawkins, R. (1976) *The Selfish Gene.* Oxford University Press, London.
Delbrück, M. (1949) A physicist looks at biology. *Trans. of the Connecticut Academy of Science* **38**:173–190.
Delbrück, M. (1970) A physicist's renewed look at biology: 20 years later. *Science* **168**:1312–1315.
Delbrück, M. (1986) *Mind from Matter? An Essay on Evolutionary Epistemology.* (Edited by G. S. Stent, E. P. Fischer, S. W. Golomb, D. Presti, and H. Seiler, completing material written in approximately 1975). Blackwell Scientific Publications, Inc., Palo Alto, California.
Demetrius, L. (1984) Self-organization of macromolecular systems: The notion of adaptive value. *Proc. Nat. Acad. Sci. USA* **81**:6068–6072.
Dennett, D. C. (1978) *Brainstorms: Philosophical Essays on Mind and Psychology.* Bradford Books Publishers, Inc. Montgomery, Vermont, p. 12.
Disney, M. (1984) *The Hidden Universe.* Macmillan Publishing Company, New York.
Dworkin, M., and D. Kaiser (1985) Cell interactions in Myxobacterial growth and development. *Science* **230**:18–24.

Einstein, A., and L. Infeld (1938) *The Evolution of Physics: The Growth of Ideas from Early Concepts to Relativity and Quanta.* Simon and Schuster, New York.
Feinberg, G. (1985) *Solid Clues: Quantum Physics, Molecular Biology, and the Future of Science.* Simon and Schuster, New York.
Feynman, R. (1965) *The Character of Physical Law.* MIT Press, Cambridge, Massachusetts.
Field, R. J., and M. Burger (Eds.) (1985) *Oscillations and Traveling Waves in Chemical Systems.* Wiley-Interscience, New York.
Fleck, L. (1979) (English Edition) *Genesis and Development of a Scientific Fact.* University of Chicago Press, Chicago. (Originally published in German in 1935: Entstehung und Entwicklung einer wissenschaftlichen Tatsache: Einführung in die Lehre vom Denkstil und Denkkollectiv. Benno Schwabe, Basel, Switzerland.)
Gardner, M. (1985) Physics: the end of the road? *New York Review* June 13: 31–34.
Gehring, W. J. (1985) The molecular basis of development. *Scientific American* **253**:153–162.
Gell-Mann, M. (1985) Patterns of convergence in contemporary science. Keynote address, 1985 meeting of the American Association for the Advancement of Science, Los Angeles, California. (Available on tape from AAAS Headquarters, 1776 Massachusetts Avenue, N.W., Washington, D.C., 10036.)
Goddard III, W. A. (1985) Theoretical chemistry comes alive: The full partner with experiment. *Science* **227**:917–923.
Harding, K., C. Wedeen, W. McGinnis, and M. Levine (1985) Spatially regulated expression of homeotic genes in Drosophila. *Science* **229**:1236–1242.
Harnad, S. (1982) Consciousness: An afterthought. *Cognition and Brain Theory* **5**:29–47.
Harth, E. (1983) *Windows on the Mind: Reflections on the Physical Basis of Consciousness.* Quill, New York.
Herbert, N. (1985) *Quantum Reality: Beyond the New Physics.* Anchor Press/Doubleday. Garden City, New York.
Hill, T. L. (1985) *Cooperativity Theory in Biochemistry.* Springer-Verlag, New York.
Hillis, D. (1985) *The Connection Machine.* MIT Press, Cambridge, Massachusetts.
Irikura, K. K., B. Tidor, B. R. Brooks, and M. Karplus (1985) Transition from B to Z DNA: Contribution of internal fluctuations to the configurational entropy difference. *Science* **229**:571–572.
Jerne, N. K. (1985) The generative grammar of the immune system. *Science* **229**:1057–1059.
Kimura, M. (1961) Natural selection as the process of accumulating genetic information in adaptive evolution. *Genet. Res. Camb.* **2**:127–140.
Kline, M. (1980) *Mathematics: The Loss of Certainty.* Oxford University Press, New York.
Kolata, G. (1985) Fitting methylation into development. *Science* **228**:1183–1184.
Kugler, P. N., and M. Turvey (1987) *Information, Natural Law and the Self-Assembly of Rhythmic Movement: Theoretical and Experimental Investigations.* L. Erlbaum Publishing Company, Hillsdale, New Jersey.
Kuhn, T. S. (1962) *The Structure of Scientific Revolutions.* University of Chicago Press, Chicago.
Landauer, R. (1986) Computation and physics. In: *Foundations of Physics* (in press).
Lewin, R. (1985) Pattern and process in life's history. *Science* **229**:151–153.
Lowenstein, J. H. (1985) Review of "Renormalization: An Introduction to Renormalization, the Renormalization Group, and Operator-Product Expansion" by John C. Collins. *Science* **229**:44.
Manthey, M. J., and B. M. E. Moret (1983) The computational metaphor and quantum physics. *Comm. ACM.* **26**:137–145.
Marx, J. L. (1985a) The T-cell receptor. *Science* **227**:733–735.
Marx, J. L. (1985b) A crystalline view of protein-DNA binding. *Science* **229**:846–848.
Marx, J. L. (1985c) A potpourri of membrane receptors. *Science* **230**:649–651.
Merton, R. K. (1985) *On the Shoulders of Giants: A Shandean Postscript.* Harcourt Brace, Jovanovich, New York. The Vicennial Edition. (The original edition was 1965 by The Free Press, New York.)
Midgley, M. (1978) *Beast and Man: The Roots of Human Nature.* Cornell University Press, Ithaca, New York.
Misner, C. W. (1977) Cosmology and theology. In: *Cosmology, History and Theology*, W. Yourgrau and A. D. Breck (eds.), Plenum, New York, pp. 98–99.
Monod, J. (1971) *Chance and Necessity.* A. A. Knopf, New York.

Müller, S. C., T. Plesser, and B. Hess (1985) The structure of the core of the spiral wave in the Belousov-Zhabotinskii reaction. *Science* **230**:661–663.
Nelson, E. (1985) *Quantum Fluctuations*. Princeton University Press, Princeton, New Jersey.
Pagels, H. R. (1982) *The Cosmic Code: Quantum Physics as the Language of Nature*. Simon and Schuster, New York.
Pattee, H. (1977) Dynamic and linguistic modes of complex systems. *Int. J. Gen. Sys.* **3**:259–266.
Pattee, H. H. (1982) Cell psychology: An evolutionary approach to the symbol-matter problem. *Cognition and Brain Theory* **5**:325–341.
Pepper, S. C. (1942) *World Hypotheses*. University of California Press, Berkeley.
Polanyi, M. (1968) Life's irreducible structure. *Science* **160**:1308–1312.
Poundstone, W. (1985) *The Recursive Universe: Cosmic Complexity and the Limits of Scientific Knowledge*. William Morrow and Company, Inc., New York.
Pritchard, H. O. (1984) *The Quantum Theory of Unimolecular Reactions*. Cambridge University Press, New York.
Rosen, R. (1986) Some epistemological issues in physics and biology. In: *Quantum Theory and Beyond*, B. J. Hiley and F. T. Peat (eds.), Routledge and Kegan Paul, Oxfordshire, England (in press).
Rozin, A., H. Cedar, and A. D. Riggs (Eds.) (1984) *DNA Methylation*. Springer-Verlag, New York.
Schrödinger, E. (1956) *What Is Life?* Doubleday, New York.
Schwartz, J. H. (1985) Completing Einstein. *Science* **85**:60–64.
Sterman, J. D. (1985) The growth of knowledge: Testing a theory of scientific revolutions with a formal model. *Technological Forecasting and Social Change* **28**:93–122.
Stryer, L. (1981) *Biochemistry (Second Edition)*. W. H. Freeman, San Francisco.
Thom, R. (1972) *Stabilité Structurelle et Morphogénèse: Essai d'une Théorie Genérale des Modèles*. Benjamin, Reading, Massachusetts, and Intereditions, Paris.
Thom, R. (1986) Organs and tools: A common theory of morphogenesis. In: *Complexity, Language and Life: Mathematical Approaches*. J. Casti (ed.), IASA, Luxembourg, Springer (in press).
Toulmin, S. (1982) *The Return to Cosmology: Postmodern Science and the Theology of Nature*. University of California Press, Berkeley.
von Neumann, J. (1932) *Mathematische Grundlagen der Quantenmechanik*. Springer, Berlin. (English translation by Robert Beyer, Princeton Univ. Press, New Jersey, 1955.)
von Neumann, J. (with A. W. Burks) (1966) *Theory of Self-Reproducing Automata*. University of Illinois Press, Urbana.
Wasserman, S. A., J. M. Dungan, and N. R. Cozzarelli (1985) Discovery of a predicted DNA knot substantiates a model for site-specific recombination. *Science* **229**:171–174.
Weinberg, S. (1985) Origins. *Science* **230**:15–18.
Weinstein, H., R. Osman, and J. P. Green (1981) Quantum chemical studies on molecular determinants for drug action. *Ann. N.Y. Acad. Sci.* **367**:434–451.
Whitley, R. (1985) *The Intellectual and Social Organization of the Sciences*. Oxford University Press, New York.
Wigner, E. P. (1960) The unreasonable effectiveness of mathematics. *Communications on Pure and Applied Mathematics* **13**:1–14.
Winfree, A., and S. Strogatz (1984) Organizing centers for waves in excitable media. *Nature* **311**:611–615.
Yates, F. E. (1982a) Outline of a physical theory of physiological systems. *Can. J. Physiol. Pharmacol* **60**:217–248.
Yates, F. E. (1982b) Systems analysis of hormone action: Principles and strategies. In: *Biological Regulation and Development*, Vol. 3A: Hormone Action, R. F. Goldberger and K. R. Yamamoto (eds.), Plenum Press, New York, pp. 25–97.
Yates, F. E., and P. N. Kugler (1984) Signs, singularities and significance: A physical model for semiotics. *Semiotica* **52**:49–77.

Associative Index

The Associative Index identifies and cross-references 30 concepts, ideas, or themes pertinent to the study of self-organizing systems, which appear in many different places in the volume, but are not explicitly identified by section or chapter headings. It contains more general items and headings than does the detailed index and it serves to cross-correlate in a way that the conventional index cannot.

The Associative Index consists of a field of 41 columns (headings of the General Introduction, section Introductions, chapters, and Epilogue) that identify elements of the volume and 30 rows representing the major themes of the volume. The rows are arranged in an order that seemed logical to the editors—it is not alphabetical.

Associative Index

Global topic	General Introduction (F. E. Yates)	Section I. Evolving Physical Systems			Section II. Evolution of Life					Section III. Morphogenesis of Organisms					
		Introduction (F. E. Yates)	Chapter 1 (Soodak)	Chapter 2 (Iberall)	Introduction (Walter)	Chapter 3 (Morowitz)	Chapter 4 (Orgel)	Chapter 5 (Schuster and Sigmund)	Chapter 6 (Gould)	Introduction (G. B. Yates)	Chapter 7 (Wood)	Chapter 8 (Glisin)	Chapter 9 (Goodwin)	Chapter 10 (Garfinkel)	Chapter 11 (Finch)
Origin of life		o			o	●	●	●							o
Evolution, competition, and selection	o						o	●	●						
Atomistic and discrete models										o	o		o	●	●
Field and continuous models		o	o										o	o	
Physical forces			o	o											
Symmetry and symmetry-breaking	●					o									
Quantum mechanics									o						
Macro- and microscopic coupling			o	o		o			●				o	o	o
Hierarchies										o					
Cooperativity and order parameters			o	o		o									
Thermodynamics		o	o												
Fluctuations and stochastic processes	o	o											o	o	
Dissipation, irreversibility, and entropy	●			o							o			o	
Catalysis							o	o					o	●	
Topological dynamics								o						o	
Nonlinearity								●						●	o
Cycles (material, modes) and oscillations						o		●							o
Regulation and control							o	●						o	o
Stability, reliability, and efficiency	o		●	●	o					o	●	●	●	●	
Morphogenesis					o							o			
Information and computation								●							o
Networks											o	o		●	
Neural biology															o
Programmatic phenomena													o		
Complexity										o		o		o	
Emergent properties	o													o	
Reductionism	o												o	o	
Philosophy of science				●										●	
Specific explanatory model			●	●		●		●	●					●	●
General explanatory model															

Key to entries: Solid circles (●) indicate that, in the opinion of the editors, the theme (row heading) makes up a substantial aspect of that book element (column heading). Open circles (o) indicate that the theme is represented, but not treated in detail, in the designated book element.

Associative Index 647

	Section IV. *Neural Biology and Networks*				Section V. *Epistemology*			Section VI. *Control Theory*			Section VII. *Physics of Self-Organization*					Section VIII. *Physics of Complexity*				Section IX. *Topology*			
	Introduction (Walter)	Chapter 12 (Bellman and Roosta)	Chapter 13 (Stent)	Chapter 14 (von der Malsburg)	Chapter 15 (Arbib)	Introduction (G. B. Yates)	Chapter 16 (Ayala)	Chapter 17 (Pattee)	Chapter 18 (Stent)	Introduction (F. E. Yates)	Chapter 19 (Stear)	Chapter 20 (Tomović)	Introduction (F. E. Yates)	Chapter 21 (Haken)	Chapter 22 (Landauer)	Chapter 23 (Anderson and Stein)	Chapter 24 (Soodak and Iberall)	Introduction (F. E. Yates)	Chapter 25 (Caianiello)	Chapter 26 (Musès)	Chapter 27 (Iberall and Soodak)	Chapter 28 (Iberall)	Introduction (G. B. Yates)	Chapter 29 (Abraham and Shaw)	Chapter 30 (Abraham)	Epilogue (F. E. Yates)

Subject Index

Absolute postulate, origins, 636
Acrasin, as slime mold organizer, 184
Action
  discretized, 509
  modes, 510
Adam Smith and model market, 207
Adaptation, 50, 220, 250
  a. behavior in control theory, 404
  in species evolution, 125
  two general strategies, 435
Adenosine triphosphate (ATP), 56–57
Adiabatic
  elimination, 421
  process, 62
Aggregation
  defined, 183
  in slime mold, 181, 193, 194
Aging, 213–231 passim
Allometry, 218–219
Allosteric enzymes, 387–394 passim
Altruism and Prisoner's Dilemma, 207
Amphiphilic molecules, 56
Analogy
  brain as computer, 265
  computer as poor a. to brain, 274
  dangers of biological a., 204
  ordered modes as good brain a., 275
Anticodons, 103
Aristotelian physics, 500
Artificial intelligence, 280–304 passim
  inference engines of, 639
Atomisms
  defined for societies, 522
  as general elements in s.o., 459, 460, 461
Atomistic models of slime mold criticized, 191, 201, 202

n, footnote; a, abstract.

Attractors, 429, 509, 549–596 passim
  as essential features of s.o., 606–609
  as symbols, 605
Autocatalysis, 78–79, 454, 507
Autonomy, as life-defining, 447

Bacteriophage, 98, 99, 136–150 passim, 628
Basins (of attraction), 549–596 passim
Behavioral modes, 330
Bellman (quotation from), 347
Bell's theorem, 620
Belousov–Zhabotinsky reaction, 78, 79, 640
Bénard cells, as s.o. example, 462
Bénard instability, 25, 452, 454
Bifurcations, 544–596 passim
  as alternative behaviors, 331, 423–424
  change in state-space description, 329
  criticality values and, 505–506
  denied by Social Darwinism, 208
  as essential features of s.o., 606
  life to death example, 633
  of limit cycles, 429–430
  mathematics of b. in s.o., 607
  in reaction-diffusion, 182
  as simplest catastrophe, 452
  in slime mold behavior, 189
  small role in species evolution, 436
  as symmetry-breaking, 182, 184, 272
  theory, 3, 9, 412
  in topological dynamics, 603, 605
Big Bang, 18, 620, 627
Biocontrol (*see also* Control), 399
Bistability, 422–423
Blumenfeld (quotation from), 624–625
Boltzmann entropy, 7
Boundary conditions, 430, 431, 447, 453, 518, 631
Boundary layer, 37–38, 46, 507
Boundary value problem, two-point, 368

Brain theory, 279–307 passim
Brain
    breakdown of determinism in, 265, 276
    as communication network, 274
    as computer, 265, 266
    cooperativity in, 275
    is not a computer, 266
    ordered modes in, 274, 275–276
    s.o. not programmatic in, 275
    thoughts are discontinuous, 276
Broken symmetry, spontaneous (definition of), 449
Browne (quotation from), 181
Brownian motion, 187, 503, 508, 518
    as society metaphor, 522, 535
Bulk viscosity, 25, 459, 515, 525

Campbell (quotation from), 632
Cancer, and control theory, 404
Cantor sets, 554
Cardinal cells, failure to explain unity, 275
Carson (quotation from), 120
Catalysis, 80–81
    heterogeneous, 26
    by mineral surfaces, 100
    Pb^{2+} as, 72
    in prebiotic chemistry, 70, 72
    role of language, 509
    in slime mold structures, 198
    Zn^{2+} as, 72
Catastrophes, 9, 631
    basic to self-organization, 210
    cusp c. in slime mold, 193
    theory, 329, 452, 563, 638
Causality
    circular, 427
    in river evolution, 44
    untraceable in ordered modes, 276
Cerebellum, Boylls's model, 298–299, 300
Cerebral hermeneutics (see Hermeneutics)
Certainty-equivalence principle, 377
Chance, 329–335 passim
    determining events in evolution, 632
    genetic code as "frozen accident," 415
    irreducibility of c. in ordered modes, 276
    as necessity, 12
    suppressed in some models, 10
Chaos, 555–596 passim
    arising from dynamic instability, 452–453
    attractors in s.o., 609
    in cosmological models, 20
    dynamics as deterministic but unpredictable, 10
    far-from-equilibrium conditions, 419
    following parameter extension, 425–426

Chaos (*cont.*)
    power law formulation, 390
    Prigogine error concerning c., 454
    as source of order, 3
Chemiosmotic hypothesis, 57
Chemistry, as source of time delays in nature, 523
Chemotaxis, in slime mold, 184
Chicken and egg problem, 456
Chladni-Faraday machine, s.o. in, 602
CHNOPS (carbon, hydrogen, nitrogen, oxygen, phosphorous, sulphur) system, 57, 62, 63, 517
Chronogenes, 629
Circadian rhythm, 221
Civilization (see also Societies)
    as nonreproductive organism, 440
Clay, as template for startup of life, 517, 631, 640
Clocks (see also Oscillations), 335
    molecular, 217
Close-to-equilibrium (see Near-equilibrim)
Clustering, 479
Coalitions, evolution of, 207
Codons, 103
Coherence (see also Entrainment), 427, 503, 618
    vs. noise in dynamical systems, 440
    in topological dynamics, 604
Collective processes, 433
Compartment formation, 100, 101, 105, 106
Competition (see also Natural selection)
    in self-replicating molecules, 103
Complementarity, 329–337 passim
    evidence of limitation of brain, 3
    information/dynamics as c., 627
    under topological dynamics, 605
Complementary replication, 81, 82
Complementation, chemical, 140, 639
Complex dynamical systems, defined, 604
Complex systems, conservations in, 523
Complexity, 473, 499, 508, 596, 636
    arising from simplicity, 446
    basic to organisms, 318
    bulk viscosity as measure of, 459, 461
    of different sciences, 344
    and distance from equilibrium, 436
    of environment and gene number, 441
    importance of environmental c. in self-repair, 439
    levels of c. in s.o., 273
    limit to c. achievable by evolution, 441
    physical measure proposed, 461
    and relaxation processes, 462
    thermodynamics and, 459

Computation, 631, 636
  parallel, 303
  reliable, 284
  theory, 287
Computers, 332, 334, 337, 631
  as brain analogy, 265–266
  brain as distributed, 286
  digital as metaphor, 238, 639
  flow diagrams, 287, 288–289
  networked, 239
  as poor brain analogy, 274–275
  simulation, 279, 281, 301, 349, 630
Condensation
  occurs at all scales, 459
  process, 23, 413, 438, 506, 507
  in societies, 521, 532
  symmetry-breaking in, 459
Consciousness, as afterthought, 636
Conservations
  in complex systems, 523
  of momentum in societies, 523
  in natural hierarchies, 528
  near-conservative systems, 608
  number limited in homeokinetics, 522
  of population, 525
  in societies, 521, 522
  underlie ensemble statistics, 524
  of value in exchange, 535
Constraints, 168, 453, 625, 629, 631, 640
  boundary conditions as potentials and fluxes, 33
  boundary values, 178–179
  on forms, genetic program, 168
  informational, 326, 336, 337, 414
  kinetic, 147–148, 149–150
  nonholonomic, nonintegrable, 334, 335, 414
  as organizing rules, 170
  like superselection rules, 174
  symbolic information as, 332
Context, spatial structure of the world, 305
Continuum
  description of societies, 525
  in macroscopic descriptions, 460
Control
  in enzymatic reactions, 380–393 passim
  feedback (closed-loop), 354–359 passim, 369
  genetic, 393
  multiloop and multivariable, 363–364
  muscle c. is nonnumerical, 402
  nonnumerical c. appropriate to biology, 399, 401
  numerical c. not too useful in biology, 403
  open-loop, 354–384 passim
  optimal, 365–369, 370
  optimal stochastic feedback, 377

Control (cont.)
  pattern recognition in, 403
  three aspects of biological c., 402
Control theory, 286, 297, 347–350 passim, 351–416 passim
  and cancer, 404
  goals vs. selection rules in, 400, 401
  hard to derive from physics, 404
  and natural selection, 400, 401
  optimal control and biochemical systems, 394–395
  and physical theory, 378
  relation to biology, 379–380
  and self-reproduction, 399, 400
  system concepts, 384
  transcends physical theory, 369
  weakness of c.t. in biology, 406
Convection
  in rivers, 37
  trade as social convection, 535
Cooperation, enhances generalized rigidity in s.o., 459
Cooperative computation, 280–306 passim
Cooperative signaling, in slime mold, 199
Cooperative systems, common features of, 265, 638
Cooperativity, 283, 419–433 passim
  among brain cells, 237
  in nerve organization, 265
  as relaxation process, 266
Coordination, visuomotor, 302, 305
Correspondence, between predictions and measurements, 328
Cosmic evolution (see Evolution of cosmos)
Cosmic radiation, background, 20, 21
Cosmic strings, 27
Cosmic turbulence, 20, 21, 25
Couette–Taylor machine, s.o. in, 602
Counting
  Boltzmann, 477
  Bose–Einstein, 477
Couplings
  of dynamical systems in s.o., 609
  nonlinear, 414
  soft, internal, 410
Criticality, 411, 449, 505–506
  in Reynolds number yields structure, 464
  in s.o., 462
Culture, defined, 467
Curvature of the universe, 29
Cybernetics, 475
Cycles (see also Oscillations), 455, 515, 518
  of action and perception, 287, 508
  among action modes, 511–512

Cycles (*cont.*)
  catalytic, 103
  chemical, 63
  coupling, 104
  ecological, 63
  entrainment of ovulation c., 201
  hydrological, 33–35, 46
  inhibition networks, 251, 253
  interaction of sciences, 280
  intracellular calcium ions, 247
  kinetics, 97
  mechanism for phage replication, 84
  mechanism of polynucleotide replication, 83
  metabolic, 515
  meteorological, 33, 35, 36
  modal, 518
  nonlinear thermodynamic, 411, 516
  ovulatory, 223–229 passim
  polarization, 247
  self-excitatory oscillations in networks, 248
  in societies, 524
  symmetry, 94
  temperature, 454
  in topological dynamics, 610
Cyclic AMP, guiding self-organization, 181
Cycling theorem, 54, 62

Darwin (quotation from), 124
Darwinism
  as dynamic regulation, 349
  ended Enlightenment ideal, 170
  as magical power, 454
  and modern synthesis, 50–51, 114, 115, 126
  neo-D., 169
  new additions to, 632
  not compatible with high order kinetics, 106
  as optimization, 102
  and polynucleotide replication, 98
  replicator-mutation equation aid, 89
  restored by compartmentalization, 105
  selection, and origin of life, 456
  selection at molecular level, 100
  simple equation for, 94
  social D. denies bifurcation, 208
  as survival of fittest, 425
  and symbol-referent relationship, 336
Death, 214, 455
  general causes of, 537
  of societies, 537
Degrees of freedom, 424, 631
  noise as, 437
Delays, time d. seem to distort conservations, 523
Demographics, thermodynamics of, 469
Dennett (quotation from), 622

Density, as order parameter, 209
Determination (embryological), 629
Determinism, 327–331 passim
  breakdown of d. in brain, 265, 276
  maximum size of mappable operations, 473
  microscopic, 330
  of Newton, Descartes, Laplace, 618
  no longer dominant in physics, 625
  not possible for brain, 474
  and Social Darwinism, 208
Deus ex machina, 5, 348
  implied by optimal control, 401
  in self-repair, 435
Development
  nongenomic constraints, 132
  two mechanisms, 131
Dictyostelium discoideum (*see also* Slime mold, as model of self-organization), 183
Differentiation
  atomistic vs. field views, 155
  cellular d. and cell polarity, 153, 154
  controlled by maternal mRNA, 160
  embryological, 629
  inducers in cellular d., 154
  piano model of, 156
  preformation in, 157
  and preformed histones, 158
  and preformed mRNA, 159
  as programmatic, 156
  in slime mold, 192
Diffusion
  in slime mold aggregation, 185, 191
  in societies, 527
Discontinuities
  in geological record, 122
  Leibniz on d. in nature, 210
  in species evolution, 50, 76, 113
  in thought processes, 276
Discreteness, as source of stability, 476
Dissipation
  in Brownian motion, 518
  and bulk viscosity, 25, 459
  creates and destroys life, 514–515
  as degradation, 513
  with fluctuations, 504, 508, 632
  of infall energies, galactic, 23
  in information storage, 436
  in living systems, 447
  as macroscopic field process, 411
  and maximum entropy principle, 454
  and system complexity, 461
  in topological dynamics, 608
Dissipative structures
  arising from reaction-diffusion equations, 604
  conditions for, 334, 413

Dissipative structures (*cont.*)
  defined, 330, 445–457 passim
  of the first kind, 415
  may not exist, 456
  of the second kind, 415
  in s.o., 462
  shortcomings of theory, 445
DNA, 66–103 passim
  condensation, 144
  as constraint, 630
  genetic material, 134, 352, 455
  homeobox, 629
  physical aspects, 627–628
  product as catalyst, 454
  repair, 215–216
  sequences in aging, 230
  selfish, 626
Dynamic programming
  allows network s.o., 241–242
  general power of, 244
Dynamical bifurcation theory (*see also* Dynamics, topological)
  distinguished from other theories, 599
Dynamical systems
  basic to control theory, 400
  for information storage, 435
  rivers as example, 35
  theory, 429, 543–596, 638, 639
Dynamics
  bifurcation in topological d., 605
  complementarity in topological d., 605
  dissipation in topological d., 608
  fluctuations in topological d., 604
  history of, 600
  irreversibility in topological d., 604
  micro/macro coupling in, 605
  qualitative d. and s.o., 599
  s.o. in topological d., 601, 607
  supplies valuable taxonomy, 606
  symmetry–breaking in topological d., 605
  ten problems in s.o., 607
  topological d.
    and hydrodynamics, 601
    may subsume thermodynamics, 605
    of slime mold, 603
    of turbulence, 602
    subsumes dissipative structures, 604
    subsumes synergetics, 604
  weakness of qualitative d. in evolution study,

Economics, 207
  thermodynamics of, 407, 535
Efficiency
  catalytic, 101
  of prebiotic chemistry, 72, 73

Élan vital, 5, 438
Elasticity, general property of s.o. systems, 459, 465
Embryogenesis
  historical, not programmatic process, 342
  as self-organization, 153
Emergence
  defined, 182
  epistemology of, 317
  of social structure, 205
Endocrine cells, entrainment of, 201
Endocrine mechanisms, in aging, 219
Ensemble physics, in societies, 525
Ensemble statistics, reflect conservations, 524
Entrainment, 561
  basic to self-organization, 210
  frequency vs. phase, 200
  in insects, 200
  of oscillators, 195
  in self-organization, 181
  in slime mold, 200
  as source of slime mold structure, 199
Entropy, 7
  accumulates in cyclic processes, 512–513
  and bulk viscosity, $515n$
  changes as basis for biological order, 9
  classical concept, 419
  cycles of action modes minimize production, 511
  different concepts of, 8
  increases with time, 327, 411
  not usual in living systems, 415, 447, 454, 518
  relationship with information, 632
  versus structure, 182
  of water, 8
Epigenesis, 285
Epistemology
  evolution implies limit to intuition, 442
  subjectivity in control theory, 400
Equations of change, 502, 514
Equations of state in s.o., 502, 524
Equilibrium
  being far from thermal, 418, 422, 424, 517
  dynamic e. in societies, 524
  evolution not possible near e., 436
  macroscopic thermodynamic, 20
  moving, 421
  population changes occur near-e., 523
  Reynolds number reflects disequilibrium, 464
  systems, 446
  thermodynamic, 41–42, 55
  top-down/bottom-up meaningless in e. structures, 276
Equivalence classes, 476, 483
Ergodicity, 34, 512

Error
  accuracy of base incorporation, 89, 91, 92, 98, 99, 103, 105, 131
  in chemistry, 626
  measurement, 375
  probability, 455
  threshold, 99, 102
Error catastrophe (Orgel), 216
Estradiol, and brain damage, 226
Euclidian space-time, 544
Evolution
  of coalitions, 207
  of cosmos, 18, 19, 621, 631
  of societies, 530
Evolution of species (see also Darwinism)
  discontinuities in, 113
  environment as information source in, 402, 404
  equations of, 609
  holism in, 204
  as phase transition, 436
  and preadaptation, 124
  requires far from equilibrium system, 436, 441
  as subject in dynamics, 605
  thermodynamic account of, 459
  trends in, 122
Excitability, in slime mold, 187, 189
Execution-driven processes, 149, 628, 640
Expanding universe, 18, 26, 28–30, 630

Fact, scientific, culturally conditioned, 637
Factory day
  in biology is usually earth day, 468
  defined, 524
  in s.o. systems, 461
  in societies, 467, 524
Fallacy of Misplaced Concreteness, 635
Far-from-equilibrium process, 410, 411, 412, 447, 454
Far-from-equilibrium thermodynamics, 330
Fast foliation, 421
Ferromagnetism, 282, 449
Field models
  of embryogenesis, 167–180 passim
  oscillations in, 194
  of slime mold, 184–203 passim
Fields
  natural systems as, 33
  ordering, 272
Filtering, optimal nonlinear, 375
Fitness function, 94
Flow processes, 25, 507
  constructive, 414
  driven by potentials in s.o., 529
  four morphologies, 630

Flow processes (cont.)
  in generalized condensation, 459
  general property of s.o. systems, 459, 461, 465
  generates form, 465
  groundwater, 43
  in natural hierarchies, 528
  turbulent, 37
Fluctuations (see also Noise), 421–431 passim, 504–508
  adiabatic, 21
  amplified at bifurcations, 625
  in birth and death rates linked, 514
  break symmetry, 449
  cooperative f. in s.o., 272
  in cosmic processes, 630
  density inhomogeneities, 19–20, 27
  determining, 632
  and dissipative structures, 334
  and dynamic stability, 395, 435
  flow-initiated, 25
  in geophysics, 466
  isothermal, 22
  in metastates guide evolution, 437
  quantum mechanical, 10, 473, 620
  in rainfall, 34
  role in cooperativity, 265
  role in nonlinear phenomena, 411
  some are creative, 517–518
  statistical mechanical, 10, 620
  thermal, 335
  thermodynamic, 19, 23
  in topological dynamics, 604
  of vacuum energy, 489
Fluid, as metaphor of society, 522
Fluids, topological dynamics of, 603
Fokker–Planck equation, 376
Force systems, 512
Forces, physical, 500
Fourier harmonic analysis, 513
Fourier transform, 357
Frames, theory of, 292
Free radicals, 59–60, 61, 62
Free will, overwhelmed by societal thermodynamics, 469
Frequency response, 357
Frozen accident, 97, 106, 415

Gain, 361, 362
  feedback loop, 355, 388, 390, 394
Galactic superclusters, 26–27
Galaxies, thermodynamics of their s.o., 460
Galton's polyhedron, as evolution metaphor, 127
Game theory, and self-organization, 95, 205, 206
Gauge symmetry, 449

Gaussian process, 375, 377, 378
Genes
  expression, and control theory, 405
  expression, and information theory, 406
  number and environmental complexity, 441
General relativity, 18
General systems theory, 638
Genetic code
  catalytic aspects, 509
  degeneracies, 100
  deviations in mitochondria, 77
  no physical basis, 333, 415
  problem of origin, 66, 103
  required for translation, 80
  term is fallacious homonym, 630
Genetics
  bacteriophage, 139
  information and nervous structure, 273
  mutations, largely neutral effect of, 117
  programs, 168, 169, 214, 229
Genome, intrinsic geometry of, 162
Genotype, 179
Geophysics
  s.o. in, 466
  susceptible to generalized thermodynamics, 465
Goals
  basic to control theory, 399, 400
  vs. selection in control theory, 401
Goldschmidt (quotation from), 121
Gompertz function, 214
Grand Unified Field Theory, 26

Hamilton–Jacobi theory, 370
Hamiltonian, 368, 450, 632
Harmonic functions, 171–172, 176, 177, 178
Hayflick limit, 215
Heat transfer
  as mass transfer, 40
  in rivers, 38, 39, 42
Hermeneutics
  cerebral, 342
  defined, 342
  of historical phenomena, 339
  large role of h. in soft sciences, 344
  preunderstanding in, 343
Heterochrony, 229
Heterocooperativity, 149
Hierarchical modular systems, 476–488 passim
Hierarchies, 476–483
  atomic vs. continuum levels, 501–502
  basic to s.o., 272
  clustering, 23
  in control theory, 401
  flows and potentials in living h., 528
  in generalized thermodynamics, 461

Hierarchies (cont.)
  intelligence and social h., 208
  new qualities at each level, 418
  in organisms, 318
  physical law in natural h., 522
  piecewise continuous, 518
  principle of, in complex systems, 508
  reflect discontinuities, 116
  in slime mold models, 202
  in species evolution, 113a, 116, 122, 126
  in thermodynamics, 459
High-energy phosphate bonds, 58
Historical models, in biological explanation, 324
Historical phenomena, 339
Hobbes, and emergence of social structure, 205, 207
Holism, in self-organization, 202, 204, 210
Homeobox, 629
Homeokinetics
  approach of h. to cosmic origins, 17–31
  approach of h. to river dynamics, 47
  as extension of homeostasis, 512, 518
  as extension of physics, 500, 509, 518
  and hierarchical control, 395–396
  limits number of conservations, 522
  and societies, 522
Homeostasis, 219, 220, 379, 395, 512
Homeotic genes, 629
Homologies, in deep structure, 177
Homomorphism, 326
Hopeful monsters, 124, 125
Hubble's constant, 29
Hydrodynamics
  as social metaphor, 537
  and topological dynamics, 601
Hylozoism, and origin of life, 489
Hypercycles, 94, 95–96, 97, 98, 104
Hysteresis, 252, 568–569

Immune system, 215, 221, 395
Information, 625–633 passim
  in bacteriophage, 99
  content of biological systems, 413
  in controller, 348
  does not explain stability, 435
  environment as i. source in evolution, 402, 404
  flow, 297, 306
  genetic, 103
  in hierarchical systems, 477–479
  molecular, 454
  nonthreshold transfer, 238
  of parts of a whole, 168
  in polymers, 455
  as poor metaphor in biology, 641
  positional, 177, 430

Information (*cont.*)
  in self-organization, 332
  semantic, 333
  and stability, 436
  storage, static vs. dynamic, 435, 436
  symbolic, 326, 331–332, 336
  theory, 406, 605
  transmission in societies, 521, 527
  versus dynamics, 625–627
Information-dependent systems, 326
Inhibition
  competitive, enzymatic, 384
  end product, 388
  noncompetitive, enzymatic, 385–387
  in rhythm control, 247
  uncompetitive, enzymatic, 385
Insect colonies, behave like slime molds, 183
Insects, entrainment of behavior in, 200
Instabilities, 325–338 passim, 431, 433
  of attractors, 556
  Bénard, 452
  and broken symmetries, 449
  in emergence, 419, 504, 506
  in feedback control, 362
  in lasers, 425–426
  life as dynamic i., 447
  of membranes, 663
Intelligence, and social stratification, 208
Intentionality, 349, 626
Invariance, 332, 473, 484
  characteristic of organisms, 171
  in transformations, 168, 170
Ionophores, 412, 413
Irreversibility, and topological dynamics, 599, 604

James, William (quotation from), 471
Jeans mass, 22, 24, 30–31
Jerne (quotation from), 627
Jump-like (*see also* Discontinuities)

Kinematics, 544, 638
Kinetics
  chemical and self-organization, 335
  complex, 390n
Kipling's *Just So Stories*, 630, 635
Knot theory, 628

Lagrangian, 367
Lamarckism, and self-repair, 440
Language, 504, 509, 626
  as catalyst, 633
  formal, 510, 623
  natural, 476, 509, 623
Laplace transform, 357

Laplace, and mechanism, 182
Laplacian universe, 327
Laser, 419–424 passim, 425, 453, 454
  phase entrainment in, 201
Laws, as explanations, 325
Leibniz, says nature doesn't jump, 210
Life (game), 631
Life span, 214, 215, 217, 218, 220
Life
  as automaton, 640–641
  autonomy as life-defining, 447
  as an emergent property, 446
  physical characteristics, 414
  properties of, behavioral, 640–641
  start-up, fluctuations and flow processes, 25
Limit cycle, 429, 430, 544–548 passim
Limit point, 544–547 passim
Limit sets, 545–596 passim
Linear models, embody laissez-faire ideology, 209
Linguistics, mathematical, 483
Local optima, trapping in, 268
Lotka-Volterra systems, 95
Luria, quotation from, 6

Macroevolution, 50, 51
Malthus, population law, 513–514
Manifolds, 544
  center, 421, 429
  invariant, 421
  slow, 421, 429
Markovian process, 378, 508
Master equations, 428, 518
Master theorem (of Duistermaat), 421
Materialism, 6
Mathematics
  character, 634–636
  compared to physics, 625
  unreasonable effectiveness of, 634
Matter, three basic properties of, 459, 465
Mayr (quotation from), 115
Meanders, 34, 43, 45, 46, 512
Measurement, 332–333, 335, 348
Mechanics, 352
  variational theory, 366, 367, 369, 375
Membranes, 100, 105, 633
  bilayer, 54, 56, 58
  potentials, 56, 57
Mendelism, 50
Menopause, 220
Mesogranularity, and aggregation, 183
Metaphors
  four root m., 2
  as prescience, 2
Meteorological theory, 36

Subject Index

Micelles (see compartments), membranes, 516
Michaelis constant, 380
Michaelis–Menton kinetics, 390
Michelangelo, unfinished statues "Prigione," 4
Micro/macro coupling
  in control theory, 405
  in generalized thermodynamics, 460
  often complex in s.o., 273
  in quantum mechanics of s.o., 494
  of Reynolds number, 463
  in societies, 467
  in topological dynamics, 605
Microevolution, 50
Mind
  limited by evolution, 442
  and origin of life, 490
  processes are discontinuous, 276
Mind-body problem, 6, 275
Misner (quotation from), 619–620
Mitochondria, 77
Models
  of data, 639
  on theory, 639
Modern synthesis (of Darwinism and Mendelism), 50
Modes, dynamic, 509
Momentum, in societies, 523
Monetary systems, 485–488
Morphogenesis, in topological dynamics, 604
Mutations, 91, 94, 100, 105, 134, 140, 169, 221, 436, 440
Myvart (quotation from), 127

Nash equilibria, defined, 205
Natural selection (see also Darwinism)
  basic to s.o., 272
  and control theory, 400–402
  in cooperativity, 265
  in prebiotic chemistry, 68
  in species evolution, 114
Navier–Stokes equations, topological dynamics of, 603
Near-equilibrium (see also Equilibrium), 33, 35, 45, 411, 413
  evaporation as measure, 41
  in rivers, 35
Negentropy, 9
Nervous system
  development historical, not programmatic, 342
  and evolution of societies, 529
  result of program?, 340
Networks
  brain as communication n., 274
  of complex systems, 596
Networks (cont.)
  dynamic programming in, 241
  inhibitory, 252
  neural, 427, 473
  oscillatory, 250
  self-excitatory, 249
  s.o. communication nets, 241
  s.o. in n. of coupled dynamical systems, 599
Neurobiology, as hard and soft science, 344
Neuromimes, 258, 260
Neutrinos, 27
Newtonian physics, 12, 28, 170, 326, 500, 501, 503, 513, 618, 633
Noise (see also Chance; Fluctuations)
  in biosynthetic pathway, 382
  in control systems, 353, 360–364, 371–372
  as degrees of freedom, 437
  has independent source, 11
  internal vs. external, 437
  in measurements, 376, 377
  $1/f$, 414, 510, 511
  and reliability, 284, 334
  role of variable n. in evolution, 438
  in sensors, 374, 375
  in species evolution, 114
  and stability, 208
Nonequilibrium systems, 63
Nonlinear coupling, drives self-organization, 208, 210
Nothing-but fallacy, 202, 317, 323
Nucleation phenomena, in biology, 436, 440
Nucleic acids (see DNA, RNA), 65–66
  sugars, 72
Nucleotide triphosphate, 84–85
Nucleotides, origin of, 66, 67

Observability, 374
Onsager relations, 411
Operon, 394
Optical pumping, 173
Optic flow, 286, 293, 294–296, 306
Optimality, in control theory, 399–402 passim
Order parameter, 424–433 passim, 448–453 passim, 504–507 passim
  density as, 209
  phase coherence as, 202
  in s.o., 272
  in topological dynamics, 604
Order, danger of too much, 4
Ordered modes
  basic to brain organization, 274
  as good brain analogy, 275
  thoughts as, 276
Oregonator, 78

Organism
    basic complexity of, 318
    as field, 176
    as fundamental biological entity, 176
    and Galton's polyhedron, 127
    hierarchy in, 318
    restoring concept of, 126
    whole-part relation in, 316
Origin of life
    and hylozoism, 489
    and nucleic acids, 66
    philosophy of, 489
Oscillations (*see also* Clocks; Cycles; Periodicity; Rhythms; Temporal patterns; Waves, of slime mold aggregation), 453, 508, 515n, 566–567
    basic to self-organization, 210
    cellular, 253
    in chemical concentrations, 79
    in dipole moments, 420
    as frustrated aggregation, 199
    importance to life, 188
    interneurons, 257
    network, 246, 248
    and relay devices, 193
    in slime molds, 187, 188, 193
    spontaneous in neurons, 247
    temporal, 507
    thermodynamic engine cycles, 395
Oscillators (*see also* Clocks), 558–560
    central nervous system, 246, 249
    coupled o. make patterns, 210
    neuronal, 260–261
    swimming, 257

Pacemakers
    of aging, 221–222
    in slime mold, 191, 193
    and spiral waves, 196, 198
Pattern formation
    in coupled oscillators, 210
    in thermodynamic systems, 459
Pattern recognition, in nonnumerical control, 403
Patterns, spatial, 79
Pauli exclusion principle, 54, 62
Periodicity (*see also* Oscillations), 429, 453, 551
    of boundary conditions, 432
Perturbation theory, second-order, 89, 91
Phase separation, 55, 63
Phase transitions, 423, 506
    equilibrium, 430
    nonequilibrium, 430
    second-order, 449
    in species evolution, 436, 440

Phenocopying, 169
Phenotype, 179
Philosophy of science, 3, 339–340
Photosynthesis, 57, 58, 77
Physical forces, 15
Physical laws, may not imply biocontrol, 404
Physics
    basis for comprehending life, 415–416
    compared to mathematics, 412, 620
    may meet psychology, 489
    parsimony of explanation in, 409
    as source of strategic principles, 474
Planck time, 18
Planck's constant, 473, 474
Poincaré section, 553
Politics, thermodynamics of, 467, 535
Pontryagin Maximum Principle, 367
Population
    changes occur near equilibrium, 523
    conserved in societies, 523
Positive feedback, 249
Possibility, mathematical, physical, 631
Potentials
    drive flow in s.o, 529
    in natural hierarchies, 528
    problem of origin in societies, 529, 532
    thermodynamic p. in s.o., 525
Preformation and epigenesis, new debate, 153, 155
Pressure, social, 503
Preunderstanding
    in hermeneutics, 343
    of nervous system, 345
Primal sketch (Marr), 290, 292
Primordial oil slick, Onsager, 56
Prisoner's Dilemma
    defined, 205
    as self-organization model, 206
Program, as confusing term, 339
Program-driven processes, 628, 631
Programmatic phenomena
    defined, 340
    in societies, 527
Progress, 513
Protein synthesis
    and control theory, 405
    as programmatic, 341
Protein-folding, spontaneous, 473
Protons, flow, 57, 58
Psychoanalysis, epistemological weakness of, 344
Psychology, may meet physics, 489
Psychrometric constant, 38
Psychrometric process, 25, 36, 507
Punctualist model (*see* Discontinuities)

Qualitative dynamics, defined, 189
Quantum mechanics, 620–628 passim, 632–641 passim
  of "aperiodic crystal," 335
  as computation, 631
  and discrete time in control theory, 400
  of gravity, 26
  ground state, 450
  interpretations of, 473, 620
  phase of wave functions, 449
  rules in chemistry, 53–54
  and s.o., 494

Racemization, of proteins, 217
Rayleigh–Bénard machine (see also entries under Bénard), 602
Reaction graphs, 59, 60, 61
Reaction network, 63
Reaction-diffusion equations, tractability of, 602
Reaction-diffusion processes, 182, 198, 282–283, 286, 334, 425, 428, 507
Receptors
  T-cell, 623
  bridge between information and dynamics, 632–634
Recombination epoch, 20, 21, 22, 25
Red shift, 18
Reductionism, 5, 176, 418, 636
  biological, 204, 315
  biology to physics, 622
  complete r. may be impossible, 324
  criticized in slime mold models, 202
  in Darwinian evolution, 115
  epistemological, 320, 321, 404, 522
  epistemological r. unlikely, 316
  methodological r. defined, 319, 320
  moderate r. defined, 522
  ontological r. defined, 317
  opposed by symmetry-breaking, 203
  three kinds, 315
  versus hierarchy, 116
Redundancy, importance in self-repair, 439
Reeves (quotation from), 1
Regeneration, 155, 177–178
Regulation, 347–350 passim, 378
Relaxation oscillators, in slime mold, 201
Relaxation processes, characterize complex systems, 266, 462–463, 467, 523
Reliability
  of communication, 241
  vs. cost in communication nets, 242, 243
Renormalization, 511, 518
Repellor, 429, 550
Replicator, 87, 89, 94
Reticular formation, and modal behavior, 281

Retinotectal projection, as model s.o. system, 265–266
Retinotopy, 283–284, 287, 294, 306
Revolutions, scientific, 636–637
Reynolds number, 38, 474, 505, 506, 507, 508
  critical in s.o., 462
  generalized 459, 464
  micro/macro aspects of, 463
Rhythms (see also Oscillations)
  endogenous, 247
  motor, 245–261 passim, 306
Ribose, 72
Richards (quotation from), 114
Rigidity, 447, 450–454 passim
  general property of s.o. systems, 459, 465
RNA, 66–104 passim, 352
  in aging brain, 215
  as catalyst, 456
  double helix, 68
  earliest true biomolecule, 456
  early proof of existence, 134
  part of circular process, 629
Robson (quotation from), 114
Romanes (quotation from), 115
Rössler attractor, 552
Rough dynamical systems
  Couette machine as, 602
  evolution in, 609
Rules
  restrictive, 63
  simple for ordering, 59
  structural, 476
  syntactical, 476

Saltation (see discontinuities)
Saltatory (punctuated) equilibrium, 632
Scaling
  of developmental time, 230
  of time in aging, 222
Schema, 287, 288, 289, 291, 297, 301
Schlössberg (quotation from), 12
Sciences
  complexity of various, 344
  hard vs. soft, 344
Second law of thermodynamics, 7, 12
Second messengers, 633
Second-best, problem of, 208
Selection rules, 176
Selection (see also Natural selection), 101, 331, 455, 635
  in brain model, 303
  Darwinian, 106, 217
  Darwinian at molecular level, 89, 100, 106
  neutral, 336
Selectionist hypothesis, molecular evolution, 50

Self-amplification
  basic to s.o., 272
  in cooperativity, 265
Self-assembly, in embryogenesis, 163
Self-organization models, general strategy, 208
Self-organization
  defined, 182
  driven by nonlinear coupling, 208, 210
  general requirements of, 272
  not programmatic in brain, 275
  in topological dynamics, 607
  weakness of term, 435, 438
Self-regulation, 585–587
Self-repair, and environmental complexity, 439
Self-replication
  and control theory, 399, 400
  of macromolecules, 65
  of oligonucleotides, 67
  origin of, 65, 66
  vs. technological reproduction, 403
Senescence (see Aging)
Sensitivity function, for control system, 360
Sensitivity, of feedback control, 361
Separatrices, 549–596 passim
  in s.o., 607
Signals
  in communication nets, 241, 242
  cooperativity in brain, 275
  in retinotectal organization, 267, 273
Similitude, theory of, 40
Simulations (see Computers)
Singularity, 27, 177, 326, 452
Slaving (synergetic), 396, 410, 421, 424, 425, 427
Slime mold, as model of self-organization, 183, 603
Slow manifold, 421
Societies
  Brownian motion in, 535
  bulk viscosity in, 525
  condensation in, 532
  conservations in, 521
  continuum description of, 525
  death of, 537
  diffusion in, 527
  dominated by trade and war, 521
  factory day in, 467
  five basic variables in, 468
  as fluidlike, 522
  micro/macro coupling in, 467
  as physical systems, 521
  programmatic phenomena in, 527
  relaxation processes in, 467
  temporal spectrum of, 526
  thermodynamics of, 459, 467, 521, 537

Software and hardware, in network s.o., 244
Solid harmonics, 177
Somatotopy, 284, 286, 299
Spatial averaging, 171
Spatial pattern, nervous system, 95
Speciation
  in Darwinian modern synthesis, 118
  discontinuities in, 119, 120, 123
  as phase transition, 436, 440
Spectral gap, 421
Spherical harmonics, 174
Spirals
  and pacemakers, 198
  from reaction-diffusion equations, 198
  in slime mold, 194, 197
Stability
  and Brownian motion, 187
  in evolution, 206
  far from equilibrium, 437
  requires more than structure, 462, 464
  and self-repair, 435
  societies intrinsically unstable, 537
  weakness of linear analysis, 187
State space, 328, 545–546
State vector, 374, 428
Statistical mechanics, 375, 378, 472, 478, 502, 620, 638, 641
Statistics, of dynamical systems needed, 605
Steady-state response, 357
Stereopsis, 286, 292, 293
Stiffness, 451
Stochastic processes, 373–378 passim
Strategic principals, physical, 500, 519
Streams, in slime mold, 199
String processes, 640
Strings
  cosmic, 27
  as slime mold structure, 196
Structural homologies, in vertebrate limbs, 170
Structural stability, 544, 639
Structuralist biology, 178
Structure
  formation requires critical Reynolds number, 464–465
  and function, 426, 427, 455, 625
  as information, 426
  surface, 177
  symmetry-breaking in s. formation, 462
  viscosity and s. formation, 462
Switching, 509, 510
Symbol systems, 332–338 passim
Symbols, as attractors, 605
Symmetries of space, 448
Symmetry-breaking
  as antireductionistic, 203

Symmetry-breaking *(cont.)*
  basic to self-organization, 181, 210, 272, 608
  from bifurcation, 272
  in condensation, 459
  in dissipative structures, 462
  in reaction-diffusion, 182
  in slime mold streaming, 199
  small role in species evolution, 436
  and topological dynamics, 599, 605
Synapses, behavior rules of, 267
Synergetics, 395, 396, 417–433, 604
Synthetic life, 639

Taylor cells, 581
Teleology, 324, 350
Teleonomy, 626, 640
Template, for structure, 484
Template-directed synthesis, 73
Templates, 72–89 passim, 92, 98
Temporal patterns *(see also* Oscillations), 168, 172, 286
Terrestrial processes, historical timetable, 516
Territories, in slime mold, 186, 192
Thermodynamics
  and complexity, 459
  domain of, 412–413
  hierarchies in, 459
  inadequate for biology, 418
  may be subsumed under topological dynamics, 605
  second law, 514
  and structure, 182
  third law, 7
  zeroth law, 483
Thompson (quotation from), 127
Thought collectives (Fleck), 637
Thoughts, as brain ordered modes, 276
Time scales, in complex systems, 468, 521, 525
Time, origin of, 515n
Topology *(see also* Dynamics)
  constrains embryogenesis, 162
  constrains morphogenesis, 274
Torus, 430, 558, 573
Tractability
  intractability of slime mold equations, 201
  of reaction-diffusion equations, 602
  trapping in local optima, 268

Trade, as convection in societies, 521, 535
Transitions
  first-order, 412, 450
  second-order, 412, 452
Translation, polynucleotide-polypeptide interactions, 104
Transport coefficients, 413
Transport processes, three kinds, 502–503
Truth, physical, 619
Turbulence, 37, 447, 462, 507, 578, 602
Turing, morphogenesis, 282, 283, 430

Urey–Miller experiments, 59, 66

Vacuum, 27, 472, 492, 496, 636
van der Waals forces, generalized in s.o., 460
Variational principle, 173
Viruses *(see* Bacteriophage)
Viscosity
  bulk v. as measure of complexity 459, 515, 525
  effective, 45, 46
  general role in s.o., 460–464 passim
  in rivers, 45
  shear, 515
Vision, theories, 290
Vitalism, 12, 315, 317, 350, 627
von Neumann, on self-replication, 282, 335
Vortices, as structures, 182

War, as physical process, 521
Water, structure, 54
Watson–Crick pairing, 68, 70
Waves, of slime mold aggregation, 190
Wear, in machines, 438
Whirlpools, as slime mold structure, 196
Whitehead (quotation from), 606
Wholes and parts, of organisms, 316
Wilson (quotation from), 116
Wright (quotation from), 123

Yates (quotation from), 11–12

Zipf's law, 476, 484
Zipf's distribution, 510, 511